INTEGRALS CONTAINING $\sqrt{a + bu}$

14. $\displaystyle\int u\sqrt{a + bu}\, du = \frac{2}{15b^2}(3bu - 2a)(a + bu)^{3/2} + C$

15. $\displaystyle\int u^2\sqrt{a + bu}\, du = \frac{2}{105b^3}(15b^2u^2 - 12abu + 8a^2)(a + bu)^{3/2} + C$

16. $\displaystyle\int u^n\sqrt{a + bu}\, du = \frac{2u^n(a + bu)^{3/2}}{b(2n + 3)} - \frac{2an}{b(2n + 3)}\int u^{n-1}\sqrt{a + bu}\, du$

17. $\displaystyle\int \frac{u\, du}{\sqrt{a + bu}} = \frac{2}{3b^2}(bu - 2a)\sqrt{a + bu} + C$

18. $\displaystyle\int \frac{u^2\, du}{\sqrt{a + bu}} = \frac{2}{15b^3}(3b^2u^2 - 4abu + 8a^2)\sqrt{a + bu} + C$

19. $\displaystyle\int \frac{u^n\, du}{\sqrt{a + bu}} = \frac{2u^n\sqrt{a + bu}}{b(2n + 1)} - \frac{2an}{b(2n + 1)}\int \frac{u^{n-1}\, du}{\sqrt{a + bu}}$

20. $\displaystyle\int \frac{du}{u\sqrt{a + bu}} = \begin{cases} \dfrac{1}{\sqrt{a}} \ln\left|\dfrac{\sqrt{a + bu} - \sqrt{a}}{\sqrt{a + bu} + \sqrt{a}}\right| + C \quad (a > 0) \\[3mm] \dfrac{2}{\sqrt{-a}} \tan^{-1}\sqrt{\dfrac{a + bu}{-a}} + C \quad (a < 0) \end{cases}$

21. $\displaystyle\int \frac{du}{u^n\sqrt{a + bu}} = -\frac{\sqrt{a + bu}}{a(n - 1)u^{n-1}} - \frac{b(2n - 3)}{2a(n - 1)}\int \frac{du}{u^{n-1}\sqrt{a + bu}}$

22. $\displaystyle\int \frac{\sqrt{a + bu}\, du}{u} = 2\sqrt{a + bu} + a\int \frac{du}{u\sqrt{a + bu}}$

23. $\displaystyle\int \frac{\sqrt{a + bu}\, du}{u^n} = -\frac{(a + bu)^{3/2}}{a(n - 1)u^{n-1}} - \frac{b(2n - 5)}{2a(n - 1)}\int \frac{\sqrt{a + bu}\, du}{u^{n-1}}$

INTEGRALS CONTAINING $a^2 \pm u^2$ $\quad(a > 0)$

24. $\displaystyle\int \frac{du}{a^2 + u^2} = \frac{1}{a}\tan^{-1}\frac{u}{a} + C$

25. $\displaystyle\int \frac{du}{a^2 - u^2} = \frac{1}{2a}\ln\left|\frac{u + a}{u - a}\right| + C$

26. $\displaystyle\int \frac{du}{u^2 - a^2} = \frac{1}{2a}\ln\left|\frac{u - a}{u + a}\right| + C$

INTEGRALS CONTAINING $\sqrt{u^2 \pm a^2}$ $\quad(a > 0)$

27. $\displaystyle\int \frac{du}{\sqrt{u^2 \pm a^2}} = \ln|u + \sqrt{u^2 \pm a^2}| + C$

28. $\displaystyle\int \sqrt{u^2 \pm a^2}\, du = \frac{u}{2}\sqrt{u^2 \pm a^2} \pm \frac{a^2}{2}\ln|u + \sqrt{u^2 \pm a^2}| + C$

29. $\displaystyle\int u\sqrt{u^2 \pm a^2}\, du = \frac{1}{3}(u^2 \pm a^2)^{3/2} + C$

30. $\displaystyle\int u^2\sqrt{u^2 \pm a^2}\, du = \frac{u}{8}(2u^2 \pm a^2)\sqrt{u^2 \pm a^2} - \frac{a^4}{8}\ln|u + \sqrt{u^2 \pm a^2}| + C$

31. $\displaystyle\int \frac{\sqrt{u^2 + a^2}\, du}{u} = \sqrt{u^2 + a^2} - a\ln\left|\frac{a + \sqrt{u^2 + a^2}}{u}\right| + C$

32. $\displaystyle\int \frac{\sqrt{u^2 - a^2}\, du}{u} = \sqrt{u^2 - a^2} - a\sec^{-1}\frac{u}{a} + C$

33. $\displaystyle\int \frac{\sqrt{u^2 \pm a^2}\, du}{u^2} = -\frac{\sqrt{u^2 \pm a^2}}{u} + \ln|u + \sqrt{u^2 \pm a^2}| + C$

34. $\displaystyle\int \frac{u^2\, du}{\sqrt{u^2 \pm a^2}} = \frac{u}{2}\sqrt{u^2 \pm a^2} \mp \frac{a^2}{2}\ln|u + \sqrt{u^2 \pm a^2}| + C$

35. $\displaystyle\int \frac{du}{u\sqrt{u^2 + a^2}} = -\frac{1}{a}\ln\left|\frac{a + \sqrt{u^2 + a^2}}{u}\right| + C$

36. $\displaystyle\int \frac{du}{u\sqrt{u^2 - a^2}} = \frac{1}{a}\sec^{-1}\frac{u}{a} + C$

37. $\displaystyle\int \frac{du}{u^2\sqrt{u^2 \pm a^2}} = \mp\frac{\sqrt{u^2 \pm a^2}}{a^2u} + C$

38. $\displaystyle\int (u^2 \pm a^2)^{3/2}\, du = \frac{u}{8}(2u^2 \pm 5a^2)\sqrt{u^2 \pm a^2} + \frac{3a^4}{8}\ln|u + \sqrt{u^2 \pm a^2}| + C$

39. $\displaystyle\int \frac{du}{(u^2 \pm a^2)^{3/2}} = \pm\frac{u}{a^2\sqrt{u^2 \pm a^2}} + C$

INTEGRALS CONTAINING $\sqrt{a^2 - u^2}$ $\quad(a > 0)$

40. $\displaystyle\int \frac{du}{\sqrt{a^2 - u^2}} = \sin^{-1}\frac{u}{a} + C$

41. $\displaystyle\int \sqrt{a^2 - u^2}\, du = \frac{u}{2}\sqrt{a^2 - u^2} + \frac{a^2}{2}\sin^{-1}\frac{u}{a} + C$

42. $\displaystyle\int u^2\sqrt{a^2 - u^2}\, du = \frac{u}{8}(2u^2 - a^2)\sqrt{a^2 - u^2} + \frac{a^4}{8}\sin^{-1}\frac{u}{a} + C$

43. $\displaystyle\int \frac{\sqrt{a^2 - u^2}\, du}{u} = \sqrt{a^2 - u^2} - a\ln\left|\frac{a + \sqrt{a^2 - u^2}}{u}\right| + C$

44. $\displaystyle\int \frac{\sqrt{a^2 - u^2}\, du}{u^2} = -\frac{\sqrt{a^2 - u^2}}{u} - \sin^{-1}\frac{u}{a} + C$

45. $\displaystyle\int \frac{u^2\, du}{\sqrt{a^2 - u^2}} = -\frac{u}{2}\sqrt{a^2 - u^2} + \frac{a^2}{2}\sin^{-1}\frac{u}{a} + C$

46. $\displaystyle\int \frac{du}{u\sqrt{a^2 - u^2}} = -\frac{1}{a}\ln\left|\frac{a + \sqrt{a^2 - u^2}}{u}\right| + C$

47. $\displaystyle\int \frac{du}{u^2\sqrt{a^2 - u^2}} = -\frac{\sqrt{a^2 - u^2}}{a^2u} + C$

48. $\displaystyle\int (a^2 - u^2)^{3/2}\, du = -\frac{u}{8}(2u^2 - 5a^2)\sqrt{a^2 - u^2} + \frac{3a^4}{8}\sin^{-1}\frac{u}{a} + C$

49. $\displaystyle\int \frac{du}{(a^2 - u^2)^{3/2}} = \frac{u}{a^2\sqrt{a^2 - u^2}} + C$

Preface

This book consists of the multivariable portion of my text, *Calculus with Analytic Geometry*, 4th Edition (Chapters 14–18), as well as Chapter 19 on second-order differential equations. Also included as appendices are reprints of the chapter on infinite series (Chapter 11) and the section on first-order differential equations (Section 7.7). In addition, there is an appendix that covers the basic concepts of complex numbers.

This book, in combination with the *Brief Edition of Calculus with Analytic Geometry*, 4th Edition, covers all of the material in *Calculus with Analytic Geometry*, 4th edition, plus the added material on complex numbers.

It is assumed that the reader has completed a course on single variable calculus, but for reference a brief review of key concepts from single-variable calculus is given at the front of the text.

Reviewing and Class Testing

This edition is the outgrowth of twelve years of classroom use and input from hundreds of students and instructors who have written to me with constructive comments over the years. In addition, the exposition and exercise sets have been refined and polished in response to input from a team of professors who carefully critiqued every single line of the third edition as they used it in the classroom. It is my hope that this painstaking attention to pedagogical detail has resulted in a book that surpasses the earlier editions for clarity and accuracy.

Rigor

The challenge of writing a good calculus textbook is to strike the right balance of rigor and clarity. It is my goal in this text to present precise mathematics to the fullest extent possible for the freshman audience. The theory is presented in a style tailored for beginners, but in those places where precision conflicts with clarity, the exposition is designed for clarity. However, I believe it to be of fundamental importance for the student to understand the distinction between a careful proof and an informal argument; thus, when the arguments are informal, I make this clear to the reader.

Illustrations

This text is more heavily illustrated than most calculus books because illustrations play a special role in my philosophy of pedagogical exposition. Freshmen have great difficulty in reading mathematics and extracting concepts from mathematical formulas, yet they can often grasp a concept immediately when the right picture is presented. This is not surprising since mathematics is a language that must be learned. Until the language is mastered one cannot hope to understand the ideas that the language conveys. I use illustrations to help teach the language of mathematics by supporting the theorems and formulas with illustrations that help the reader understand the concepts embodied in the mathematical symbolism. In keeping with current trends, I have used modern four-color typography with a consistency of style and color selection throughout the text.

Exercises

The exercises in this new edition have been extensively modified and expanded. Each exercise set begins with routine drill problems and progresses gradually toward problems of greater difficulty. I have tried to construct well-balanced exercise sets with more variety than is available in most calculus texts. Each chapter ends with a set of supplementary exercises to help the student consolidate his or her mastery of the chapter. In addition, many of the exercise sets now contain problems requiring a calculator as well as so-called "spiral" problems (problems that use concepts from earlier chapters). Answers are given to odd-numbered problems, and at the beginning of each exercise set there is a list of those exercises that require a calculator; these are labeled with the icon $\boxed{\text{C}}$.

☐ A NOTE FROM THE AUTHOR

I am gratified that the third edition of this text continued the tradition of its predecessors as the most widely used textbook in calculus and that after twelve years in print users and reviewers have continued to praise the clarity of the exposition.

We are now on the threshold of major changes in the way calculus and, indeed, all of mathematics will be taught. Fueled by rapid advances in technology and a reevaluation of traditional course content, we are entering a period of exploration in the teaching of calculus. This new edition reflects a clear, but cautious, commitment to the newer visions of calculus. My goal in this revision is to provide instructors with all of the *supplemental* resources required to begin experimenting with new ideas and new technology within the framework of the traditional course structure.

Users of the earlier editions will be able to ease comfortably into the new edition, but will sense a more contemporary philosophy in the exposition and exercises, with increased attention to numerical computations and estimation, as well as flexibility for varying the order and emphasis of topics. For instructors on the cutting edge of calculus reform, there is an extensive array of supplements that use new technology: symbolic algebra software, calculators, Hypercard stacks, and even CD-ROM.

Although there are many changes in this new edition, I remain committed to the philosophy that the heart and soul of a fine textbook is the clarity of its exposition; there are very few sections in this new edition that have not been polished and refined. My goal, as in earlier editions, is to *teach* the material in the clearest possible way with a level of rigor that is suitable for the mainstream calculus audience.

Supplements

Graphing Calculator Supplement
The following supplement contains a collection of problems that are intended to be solved on a graphing calculator. The problems are not specific to a particular brand of calculator. The manual also provides an overview of the types of calculators available, general instructions for calculator use, and a discussion of the numerical pitfalls of roundoff and truncation error.

- *Discovering Calculus with Graphing Calculators*
  Joan McCarter, *Arizona State University*
  ISBN: 0-471-55609-2

Symbolic Algebra Supplements
The following supplements are collections of problems for the student to solve. Each contains a brief set of instructions for using the software as well as an extensive set of problems utilizing the capabilities of the software. The problems range from very basic to those involving real-world applications.

- *Discovering Calculus with DERIVE*
  Jerry Johnson and Benny Evans, *Oklahoma State University*
  ISBN: 0-471-55155-4
- *Discovering Calculus with MAPLE*
  Kent Harris, *Western Illinois University*
  ISBN: 0-471-55156-2
- *Discovering Calculus with MATHEMATICA*
  Bert Braden, Don Krug, Steve Wilkinson, *Northern Kentucky University*
  ISBN: 0-471-53969-4

Macintosh Hypercard Stacks
This supplement is a set of Hypercard 2.0 stacks that are designed primarily for lecture demonstrations. Each stack is self-contained, but all have a common interface. There is an initial set of six stacks (with more to be developed). The six initial stacks are concerned with limits, Newton's Method, the definition of the definite integral, convergence of Taylor polynomials, polar coordinates, and mathematics history.

- *HYPERCALCULUS*
  Chris Rorres and Loren Argabright, *Drexel University*
  ISBN: 0-471-57052-4

CD-ROM Version of Calculus for IBM Compatible Computers
This supplement is an electronic version of the entire text and the *Student's Solutions Manual* on compact disk for use with IBM compatible computers equipped with a CD-ROM drive. All text material and illustrations are stored on disk with an interconnecting network of hyperlinks that allows the student to access related items that do not appear in proximity in the text. A complete keyword glossary and step-by-step discussions of key concepts are also included.

- *CD-ROM Version of Anton Calculus/4E: An Electronic Study Environment*
  Developed by Electric Book Company
  ISBN: 0-471-55803-6

Linear Algebra Supplement
The following supplement gives a brief introduction to those aspects of linear algebra that are of immediate concern to the calculus student. The emphasis is on methods rather than proof.

- *Linear Algebra Supplement to Accompany Anton Calculus/4E*
  ISBN: 0-471-56893-7

Student Study Resources
The following supplement is a tutorial, review, and study aid for the student.

- *The Calculus Companion to Accompany Anton Calculus/4E, Vols. 1 and 2*
  William H. Barker and James E. Ward, *Bowdoin College*
  ISBN: 0-471-55139-2 (Volume 1); ISBN: 0-471-55138-4 (Volume 2)

The following supplement contains detailed solutions to all odd-numbered exercises.

- *Student's Solutions Manual to Accompany Anton Calculus/4E*
  Albert Herr, *Drexel University*
  ISBN: 0-471-55140-6

Resources for the Instructor
There is a resource package for the instructor that includes hard copy and electronic test banks and other materials. These can be obtained by writing on your institutional letterhead to Susan Elbe, Senior Marketing Manager, John Wiley & Sons, Inc., 605 Third Avenue, New York, N.Y., 10158-0012.

Acknowledgments

It has been my good fortune to have the advice and guidance of many talented people, whose knowledge and skills have enhanced this book in many ways. For their valuable help I thank:

Reviewers and Contributors to Earlier Editions
Edith Ainsworth, *University of Alabama*
David Armacost, *Amherst College*
Larry Bates, *University of Calgary*
Marilyn Blockus, *San Jose State University*
David Bolen, *Virginia Military Institute*
George W. Booth, *Brooklyn College*
Mark Bridger, *Northeastern University*
John Brothers, *Indiana University*
Robert C. Bueker, *Western Kentucky University*
Robert Bumcrot, *Hofstra University*
Chris Christensen, *Northern Kentucky University*
David Cohen, *University of California, Los Angeles*
Michael Cohen, *Hofstra University*
Robert Conley, *Precision Visuals*
A. L. Deal, *Virginia Military Institute*
Charles Denlinger, *Millersville State College*
Dennis DeTurck, *University of Pennsylvania*
Jacqueline Dewar, *Loyola Marymount University*
Irving Drooyan, *Los Angeles Pierce College*
Hugh B. Easler, *College of William and Mary*
Joseph M. Egar, *Cleveland State University*
Garret J. Etgen, *University of Houston*
James H. Fife, *University of Richmond*
Barbara Flajnik, *Virginia Military Institute*
Nicholas E. Frangos, *Hofstra University*
Katherine Franklin, *Los Angeles Pierce College*
Michael Frantz, *University of La Verne*
William R. Fuller, *Purdue University*
Raymond Greenwell, *Hofstra University*
Gary Grimes, *Mt. Hood Community College*
Jane Grossman, *University of Lowell*
Michael Grossman, *University of Lowell*

Douglas W. Hall, *Michigan State University*
Nancy A. Harrington, *University of Lowell*
Albert Herr, *Drexel University*
Peter Herron, *Suffolk County Community College*
Robert Higgins, *Quantics Corporation*
Louis F. Hoelzle, *Bucks County Community College*
Harvey B. Keynes, *University of Minnesota*
Paul Kumpel, *SUNY, Stony Brook*
Leo Lampone, *Quantics Corporation*
Bruce Landman, *Hofstra University*
Benjamin Levy, *Lexington H.S., Lexington, Mass.*
Phil Locke, *University of Maine, Orono*
Stanley M. Lukawecki, *Clemson University*
Nicholas Macri, *Temple University*
Melvin J. Maron, *University of Louisville*
Thomas McElligott, *University of Lowell*
Judith McKinney, *California State Polytechnic University, Pomona*
Joseph Meier, *Millersville State College*
David Nash, *VP Research, Autofacts, Inc.*
Mark A. Pinsky, *Northeastern University*
William H. Richardson, *Wichita State University*
David Sandell, *U.S. Coast Guard Academy*
Donald R. Sherbert, *University of Illinois*
Wolfe Snow, *Brooklyn College*
Norton Starr, *Amherst College*
Richard B. Thompson, *The University of Arizona*
William F. Trench, *Trinity University*
Walter W. Turner, *Western Michigan University*
Richard C. Vile, *Eastern Michigan University*
James Warner, *Precision Visuals*
Candice A. Weston, *University of Lowell*
Yihren Wu, *Hofstra University*
Richard Yuskaitis, *Precision Visuals*

Content Reviewers
The following people critiqued the previous edition and recommended many
of the changes that found their way into the new edition:
Ray Boersma, *Front Range Community College*
Terrance Cremeans, *Oakland Community College*
Tom Drouet, *East Los Angeles College*
Ken Dunn, *Dalhousie University*
Kent Harris, *Western Illinois University*
Maureen Kelly, *Northern Essex Community College*
Richard Nowakowski, *Dalhousie University*
Robert Phillips, *University of South Carolina at Aiken*
David Randall, *Oakland Community College*
George Shapiro, *Brooklyn College*
Ian Spatz, *Brooklyn College*

Accuracy Reviewers of the Fourth Edition
The following people worked with me in reading the manuscript, galley proofs, and page proofs for mathematical accuracy. Their perceptive comments have improved the text immeasurably:
Irl C. Bivens, *Davidson College*
Hannah Clavner, *Drexel University*
Garret J. Etgen, *University of Houston*
Daniel Flath, *University of South Alabama*
Susan L. Friedman, *Bernard M. Baruch College, CUNY*
Evelyn Weinstock, *Glassboro State College*

Problem Contributors
The following people contributed numerous new and imaginative problems to the text:
Irl C. Bivens, *Davidson College*
Daniel Bonar, *Denison University*
James Caristi, *Valparaiso University*
G. S. Gill, *Brigham Young University*
Albert Herr, *Drexel University*
Herbert Kasube, *Bradley University*
Phil Kavanaugh, *Illinois Wesleyan University*
John Lucas, *University of Wisconsin–Oshkosh*
Ron Moore, *Ryerson Polytechnical Institute*
Barbara Moses, *Bowling Green State University*
David Randall, *Oakland Community College*
Jean Springer, *Mount Royal College*
Peter Waterman, *Northern Illinois University*

Answers, Solutions, and Index
The following people assisted with the critically important job of preparing the index and obtaining answers for the text and solutions for the *Student's Solutions Manual*:
Harry N. Bixler, *Bernard M. Baruch College, CUNY*
Hannah Clavner, *Drexel University*
Michael Dagg
Stephen L. Davis, *Davidson College*
Susan L. Friedman, *Bernard M. Baruch College, CUNY*
Shirley Wakin, *University of New Haven*

Computer Illustrations
Many of the illustrations involving two-dimensional mathematical graphs were generated electronically using software developed by Techsetters, Inc. A number of people contributed to the development of the software:
Edward A. Burke, *Hudson River Studio*—color separation
John R. DiStefano (student), *Drexel University*—programmer
Theo Gray, *Wolfram Research*—technical assistance
Julie Varbalow (student), *Skidmore College*—testing
   Some of the three-dimensional surfaces were generated by Mark Bridger of Northeastern University using the SURFS software package that he developed.

The Wiley Staff
There are so many people at Wiley who contributed in special ways to this project that it is impossible to mention them all. However, a word of appreciation is due to the people I worked with very closely: Joan Carrafiello, Lucille Buonocore, Ann Berlin, Steve Kraham, Susan Elbe, and my editor Barbara Holland.

Other Supplementary Materials
The following people provided additional material for tests and other supplements:

Pasquale Condo, *University of Lowell*
Maureen Kelley, *Northern Essex Community College*
Catherine H. Pirri, *Northern Essex Community College*

Special Contributions
I owe an enormous debt of gratitude to Albert Herr of Drexel University, who worked so closely with me on this edition that his name appears on the title page. Professor Herr is a gifted, award-winning teacher of mathematics with years of experience in the calculus classroom. His careful attention to technical detail and his dogged challenges to the clarity of virtually every line of exposition fostered hours of debate over pedagogy and helped bring this edition to a level of quality I could not have achieved alone. Professor Herr is also the source for many of the imaginative new exercises in this edition. I feel fortunate that Al was by my side in preparing this revision.

If I had the power to grant an honorary degree in mathematics, I would give it to my assistant, Mary Parker, who worked on virtually every aspect of this text—from the tedious task of xeroxing tens of thousands of pages to the technical tasks of accuracy checking and coordination. Her warm sense of humor and dedication to the quality of this book were a constant source of inspiration.

H. A.

Contents

Introduction

The creation of calculus is attributed to two of the greatest geniuses in the history of mathematics and science—Gottfried Wilhelm Leibniz and Isaac Newton.

☐ **GOTTFRIED WILHELM**
LEIBNIZ (1646–1716)

This gifted genius was one of the last people to have mastered most major fields of knowledge—an impossible accomplishment in our own era of specialization. He was an expert in law, religion, philosophy, literature, politics, geology, metaphysics, alchemy, history, and mathematics.

Leibniz was born in Leipzig, Germany. His father, a professor of moral philosophy at the University of Leipzig, died when Leibniz was six years old. The precocious boy then gained access to his father's library and began reading voraciously on a wide range of subjects, a habit that he maintained throughout his life. At age 15 he entered the University of Leipzig as a law student and by the age of 20 received a doctorate from the University of Altdorf. Subsequently, Leibniz followed a career in law and international politics, serving as counsel to kings and princes.

During his numerous foreign missions, Leibniz came in contact with outstanding mathematicians and scientists who stimulated his interest in mathematics—most notably, the physicist Christian Huygens. In mathematics Leibniz was self-taught, learning the subject by reading papers and journals. As a result of this fragmented mathematical education, Leibniz often duplicated the results of others, and this ultimately led to a raging conflict as to whether he or Isaac Newton should be regarded as the inventor of calculus. The argument over this question engulfed the scientific circles of England and Europe, with most scientists on the continent supporting Leibniz and those in England supporting Newton. The conflict was unfortunate, and both sides suffered in the end. The continent lost the benefit of Newton's discoveries in astronomy and physics for more than 50 years, and for a long period England became a second-rate country mathematically because its mathematicians were hampered by Newton's inferior calculus notation. It is of interest to note that Newton and Leibniz never went to the lengths of vituperation of their advocates—both were sincere admirers of each other's work. The fact is that both men invented calculus independently; Leibniz invented it 10 years after Newton, in 1685, but he published his results 20 years before Newton published his own work on the subject.

Leibniz never married. He was moderate in his habits, quick-tempered, but easily appeased, and charitable in his judgment of other people's work. In spite of his great achievements, Leibniz never received the honors showered on Newton, and he spent his final years as a lonely embittered man. At his funeral there was one mourner, his secretary. An eyewitness stated, "He was buried more like a robber than what he really was—an ornament of his country."

□ ISAAC NEWTON
(1642–1727)

Newton was born in the village of Woolsthorpe, England. His father died before he was born and his mother raised him on the family farm. As a youth he showed little evidence of his later brilliance, except for an unusual talent with mechanical devices—he apparently built a working water clock and a toy flour mill powered by a mouse. In 1661 he entered Trinity College in Cambridge with a deficiency in geometry. Fortunately, Newton caught the eye of Isaac Barrow, a gifted mathematician and teacher. Under Barrow's guidance Newton immersed himself in mathematics and science, but he graduated without any special distinction. Because the Plague was spreading rapidly through London, Newton returned to his home in Woolsthorpe and stayed there during the years of 1665 and 1666. In those two momentous years the entire framework of modern science was miraculously created in Newton's mind—he discovered calculus, recognized the underlying principles of planetary motion and gravity, and determined that "white" sunlight was composed of all colors, red to violet. For some reason he kept his discoveries to himself. In 1667 he returned to Cambridge to obtain his Master's degree and upon graduation became a teacher at Trinity. Then in 1669 Newton succeeded his teacher, Isaac Barrow, to the Lucasian chair of mathematics at Trinity, one of the most honored chairs of mathematics in the world. Thereafter, brilliant discoveries flowed from Newton steadily. He formulated the law of gravitation and used it to explain the motion of the moon, the planets, and the tides; he formulated basic theories of light, thermodynamics, and hydrodynamics; and he devised and constructed the first modern reflecting telescope.

Gottfried Leibniz
(Culver Pictures)

Isaac Newton
(Culver Pictures)

Throughout his life Newton was hesitant to publish his major discoveries, revealing them only to a select circle of friends, perhaps because of a fear of criticism or controversy. In 1687, only after intense coaxing by the astronomer, Edmond Halley (Halley's comet), did Newton publish his masterpiece, *Philosophiae Naturalis Principia Mathematica* (The Mathematical Principles of Natural Philosophy). This work is generally considered to be the most important and influential scientific book ever written. In it Newton explained the workings of the solar system and formulated the basic laws of motion which to this day are fundamental in engineering and physics. However, not even the pleas of his friends could convince Newton to publish his discovery of calculus. Only after Leibniz published his results did Newton relent and publish his own work on calculus.

After 35 years as a professor, Newton suffered depression and a nervous breakdown. He gave up research in 1695 to accept a position as warden and later master of the London mint. During the 25 years that he worked at the mint, he did virtually no scientific or mathematical work. He was knighted in 1705 and on his death was buried in Westminster Abbey with all the honors his country could bestow. It is interesting to note that Newton was a learned theologian who viewed the primary value of his work to be its support of the existence of God. Throughout his life he worked passionately to date biblical events by relating them to astronomical phenomena. He was so consumed with this passion that he spent years searching the Book of Daniel for clues to the end of the world and the geography of hell.

Newton described his brilliant accomplishments as follows: "I seem to have been only like a boy playing on the seashore and diverting myself in now and then finding a smoother pebble or prettier shell than ordinary, whilst the great ocean of truth lay all undiscovered before me."

The Discoverers
of Calculus

Gottfried Wilhelm Leibniz   (1646-1716)

Sir Isaac Newton   (1642-1727)

Preliminaries

REVIEW OF SELECTED CONCEPTS FROM SINGLE-VARIABLE CALCULUS

We assume that the reader has previously studied single-variable calculus, but for convenience we have provided lists of the key single-variable concepts required in each chapter, followed by appropriate extracts from the Brief Edition of this text. The numbering of theorems and figures in this review is from the Brief Edition. Most references in the text to material in the Brief Edition can be found in this review. This is not intended to provide a complete summary of required material, and should you need a more in-depth review, refer to the text from which you studied single-variable calculus.

CHAPTER 14 REVIEW TOPICS

- Distance between points in the plane
- Midpoint of a line segment
- Equations of circles
- Graph of $Ax^2 + Ay^2 + Dx + Ey + F = 0$
- Parametric equations
- Graphs of conic sections
- Translation of coordinate axes
- Relationship between polar and rectangular coordinates
- Work

1.6.1   THEOREM.   *The distance d between two points (x_1, y_1) and (x_2, y_2) in a coordinate plane is given by*

$$d = \sqrt{(x_2 - x_1)^2 + (y_2 - y_1)^2}$$

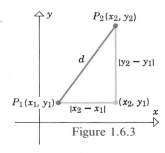

Figure 1.6.3

1.6.2   THEOREM (*The Midpoint Formula*).   *The midpoint of the line segment joining two points (x_1, y_1) and (x_2, y_2) in a coordinate plane is*

$$\left(\tfrac{1}{2}(x_1 + x_2), \tfrac{1}{2}(y_1 + y_2)\right)$$

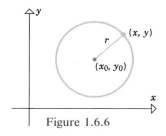

Figure 1.6.6

The *standard form of the equation of a circle.*

$$(x - x_0)^2 + (y - y_0)^2 = r^2$$

1.6.3   THEOREM.   *An equation of the form*

$$Ax^2 + Ay^2 + Dx + Ey + F = 0$$

where $A \neq 0$, represents a circle, or a point, or else has no graph.

Conic Sections

ORIENTATION	DESCRIPTION	STANDARD EQUATION
y axis with curve opening right, point $(p, 0)$, line $x = -p$	• Vertex at the origin. • Parabola opens in the positive *x*-direction. • Symmetric about the *x*-axis.	$y^2 = 4px$
y axis with curve opening left, point $(-p, 0)$, line $x = p$	• Vertex at the origin. • Parabola opens in the negative *x*-direction. • Symmetric about the *x*-axis.	$y^2 = -4px$
y axis with curve opening up, point $(0, p)$, line $y = -p$	• Vertex at the origin. • Parabola opens in the positive *y*-direction. • Symmetric about the *y*-axis.	$x^2 = 4py$
y axis with curve opening down, line $y = p$, point $(0, -p)$	• Vertex at the origin. • Parabola opens in the negative *y*-direction. • Symmetric about the *y*-axis.	$x^2 = -4py$

ORIENTATION	DESCRIPTION	STANDARD EQUATION
	• Foci and major axis on the x-axis. • Minor axis on the y-axis. • Center at the origin. • x-intercepts: $\pm a$. • y-intercepts: $\pm b$. • $a \geq b$	$\dfrac{x^2}{a^2} + \dfrac{y^2}{b^2} = 1$
	• Foci and major axis on the y-axis. • Minor axis on the x-axis. • Center at the origin. • x-intercepts: $\pm b$. • y-intercepts: $\pm a$. • $a \geq b$	$\dfrac{x^2}{b^2} + \dfrac{y^2}{a^2} = 1$

ORIENTATION	DESCRIPTION	STANDARD EQUATION	ASYMPTOTE EQUATIONS
	• Foci on the x-axis. • Conjugate axis on the y-axis. • Center at the origin.	$\dfrac{x^2}{a^2} - \dfrac{y^2}{b^2} = 1$	$y = \dfrac{b}{a}x$ $y = -\dfrac{b}{a}x$
	• Foci on the y-axis. • Conjugate axis on the x-axis. • Center at the origin.	$\dfrac{y^2}{a^2} - \dfrac{x^2}{b^2} = 1$	$y = \dfrac{a}{b}x$ $y = -\dfrac{a}{b}x$

Translation Equations

$$x' = x - h, \quad y' = y - k$$

$$x = x' + h, \quad y = y' + k$$

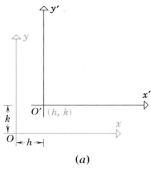

(a)

Figure 12.2.7

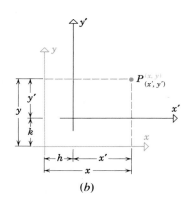

(b)

Parabola with Vertex (h, k) and Axis Parallel to y-axis

$$(x - h)^2 = \pm 4p(y - k)$$

where the + sign occurs if the parabola opens in the positive y-direction and the − sign if it opens in the negative y-direction.

Parabola with Vertex (h, k) and Axis Parallel to x-axis

$$(y - k)^2 = \pm 4p(x - h)$$

where the + sign occurs if the parabola opens in the positive x-direction and the − sign if it opens in the negative x-direction

Ellipse with Center (h, k) and Major Axis Parallel to x-axis

$$\frac{(x - h)^2}{a^2} + \frac{(y - k)^2}{b^2} = 1 \quad (a \geq b)$$

Ellipse with Center (h, k) and Major Axis Parallel to y-axis

$$\frac{(x - h)^2}{b^2} + \frac{(y - k)^2}{a^2} = 1 \quad (a \geq b)$$

Hyperbola with Center (h, k) and Focal Axis Parallel to x-axis

$$\frac{(x - h)^2}{a^2} - \frac{(y - k)^2}{b^2} = 1$$

Hyperbola with Center (h, k) and Focal Axis Parallel to y-axis

$$\frac{(y - k)^2}{a^2} - \frac{(x - h)^2}{b^2} = 1$$

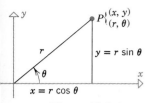

Figure 13.1.4

Relationship between Polar and Rectangular Coordinates

$$x = r \cos \theta$$
$$y = r \sin \theta$$

and

$$r^2 = x^2 + y^2$$
$$\tan \theta = \frac{y}{x}$$

Also review the concept of work.

CHAPTER 15 REVIEW TOPICS

- Functions, domain, and range
- Concept of a limit
- Properties of limits
- Continuity
- Definition of a derivative
- Tangent lines
- Properties of derivatives
- Chain rule
- Indefinite integrals and properties
- Definite integrals and properties
- Arc length
- Velocity, speed, and acceleration

2.1.1 DEFINITION.   A *function* is a rule that assigns to each element in a nonempty set A one and only one element in a set B. The set A is called the *domain* of the function, and the set of all values of $f(x)$ is called the *range*.

*If a function is defined by a formula and there is no domain explicitly stated, then it is understood that the domain consists of all real numbers for which the formula makes sense, and the function has a real value. This is called the **natural domain** of the function.*

2.4.1 NOTATION. If the value of $f(x)$ approaches the number L_1 as x approaches x_0 from the right side, we write

$$\lim_{x \to x_0^+} f(x) = L_1$$

which is read, "the limit of $f(x)$ as x approaches x_0 from the right is equal to L_1."

2.4.2 NOTATION. If the value of $f(x)$ approaches the number L_2 as x approaches x_0 from the left side, we write

$$\lim_{x \to x_0^-} f(x) = L_2$$

which is read, "the limit of $f(x)$ as x approaches x_0 from the left is equal to L_2."

2.4.3 NOTATION. If the limit from the left side is the same as the limit from the right side, that is, if

$$\lim_{x \to x_0^-} f(x) = \lim_{x \to x_0^+} f(x) = L$$

then we write

$$\lim_{x \to x_0} f(x) = L$$

which is read, "the limit of $f(x)$ as x approaches x_0 is equal to L."

Also, review limits involving

$$\lim_{x \to +\infty} f(x), \qquad \lim_{x \to -\infty} f(x), \qquad \lim_{x \to x_0} f(x) = \pm\infty$$

2.5.1 THEOREM. *Let* $\lim$ *stand for one of the limits* $\lim\limits_{x \to a}$, $\lim\limits_{x \to a^-}$, $\lim\limits_{x \to a^+}$, $\lim\limits_{x \to +\infty}$, *or* $\lim\limits_{x \to -\infty}$. *If* $L_1 = \lim f(x)$ *and* $L_2 = \lim g(x)$ *both exist, then*

(a) $\lim [f(x) + g(x)] = \lim f(x) + \lim g(x) = L_1 + L_2$

(b) $\lim [f(x) - g(x)] = \lim f(x) - \lim g(x) = L_1 - L_2$

(c) $\lim [f(x)g(x)] = \lim f(x) \lim g(x) = L_1 L_2$

(d) $\lim \dfrac{f(x)}{g(x)} = \dfrac{\lim f(x)}{\lim g(x)} = \dfrac{L_1}{L_2} \quad$ *if* $L_2 \neq 0$

(e) $\lim \sqrt[n]{f(x)} = \sqrt[n]{\lim f(x)} = \sqrt[n]{L_1} \quad$ *provided* $L_1 \geq 0$ *if n is even.*

2.7.1   DEFINITION.   A function f is said to be ***continuous at a point c*** if the following conditions are satisfied:

1. $f(c)$ is defined.

2. $\lim\limits_{x \to c} f(x)$ exists.

3. $\lim\limits_{x \to c} f(x) = f(c)$.

If one or more of the conditions in this definition fails to hold, then f is called ***discontinuous at c*** and c is called a ***point of discontinuity*** of f. If f is continuous at all points of an open interval (a, b), then f is said to be ***continuous on (a, b)***. A function that is continuous on $(-\infty, +\infty)$ is said to be ***continuous everywhere*** or simply ***continuous***.

Also review continuity on a closed interval.

2.7.3   THEOREM.   *If the functions f and g are continuous at c, then*

(a) *$f + g$ is continuous at c;*
(b) *$f - g$ is continuous at c;*
(c) *$f \cdot g$ is continuous at c;*
(d) *f/g is continuous at c if $g(c) \neq 0$ and is discontinuous at c if $g(c) = 0$.*

2.7.5   THEOREM.   *Let* $\lim$ *stand for one of the limits* $\lim\limits_{x \to c}$, $\lim\limits_{x \to c^-}$, $\lim\limits_{x \to c^+}$, $\lim\limits_{x \to +\infty}$, *or* $\lim\limits_{x \to -\infty}$. *If* $\lim g(x) = L$ *and if the function f is continuous at L, then* $\lim f(g(x)) = f(L)$. *That is,* $\lim f(g(x)) = f(\lim g(x))$.

2.7.6   THEOREM.   *If the function g is continuous at the point c and the function f is continuous at the point g(c), then the composition $f \circ g$ is continuous at c.*

3.2.1   DEFINITION.   The function f' defined by the formula

$$f'(x) = \lim_{h \to 0} \frac{f(x + h) - f(x)}{h}$$

is called the ***derivative with respect to x*** of the function f. The domain of f' consists of all x for which the limit exists.

Geometric Interpretation of the Derivative
f' is the function whose value at x is the slope of the tangent line to the graph of f at x.

Rate of Change Interpretation of the Derivative
If $y = f(x)$, then f' is the function whose value at x is the instantaneous rate of change of y with respect to x at the point x.

Derivative Notations for $y = f(x)$

$$f'(x) = \lim_{\Delta x \to 0} \frac{\Delta f}{\Delta x}$$

$$\frac{dy}{dx} = \lim_{\Delta x \to 0} \frac{f(x + \Delta x) - f(x)}{\Delta x}$$

$$\frac{dy}{dx} = \lim_{\Delta x \to 0} \frac{\Delta y}{\Delta x}$$

$$\frac{dy}{dx} = f'(x)$$

$$\frac{d}{dx}[f(x)] = f'(x)$$

$$\left.\frac{dy}{dx}\right|_{x=x_0} = f'(x_0)$$

$$\left.\frac{d}{dx}[f(x)]\right|_{x=x_0} = f'(x_0)$$

Tangent Line Equation to $y = f(x)$ at x_0

$$y - y_0 = f'(x_0)(x - x_0)$$

Derivative Formulas

$$\frac{d}{dx}[c] = 0$$

$$\frac{d}{dx}[f(x) + g(x)] = \frac{d}{dx}[f(x)] + \frac{d}{dx}[g(x)]$$

$$\frac{d}{dx}[cf(x)] = c\frac{d}{dx}[f(x)]$$

$$\frac{d}{dx}[f(x) - g(x)] = \frac{d}{dx}[f(x)] - \frac{d}{dx}[g(x)]$$

$$\frac{d}{dx}[f(x)g(x)] = f(x)\frac{d}{dx}[g(x)] + g(x)\frac{d}{dx}[f(x)]$$

$$\frac{d}{dx}\left[\frac{f(x)}{g(x)}\right] = \frac{g(x)\dfrac{d}{dx}[f(x)] - f(x)\dfrac{d}{dx}[g(x)]}{[g(x)]^2}$$

3.5.2 THEOREM (*The Chain Rule*). *If g is differentiable at the point x and f is differentiable at the point g(x), then the composition f ∘ g is differentiable at the point x. Moreover, if*

$$y = f(g(x)) \quad and \quad u = g(x)$$

then y = f(u) and

$$\frac{dy}{dx} = \frac{dy}{du} \cdot \frac{du}{dx}$$

4.2.2 THEOREM. *Let f be a function that is continuous on a closed interval [a, b] and differentiable on the open interval (a, b).*

(a) *If $f'(x) > 0$ for every value of x in (a, b), then f is increasing on [a, b].*
(b) *If $f'(x) < 0$ for every value of x in (a, b), then f is decreasing on [a, b].*
(c) *If $f'(x) = 0$ for every value of x in (a, b), then f is constant on [a, b].*

5.2.1 DEFINITION. A function F is called an *antiderivative* of a function f if $F'(x) = f(x)$ on some interval.

5.2.2 THEOREM. *If F(x) is any antiderivative of f(x), then for any value of C, the function F(x) + C is also an antiderivative of f(x); moreover, on any interval, every antiderivative of f(x) is expressible in the form F(x) plus a constant.*

The relationship in Theorem 5.2.2 can be expressed using an indefinite integral:

$$\int f(x)\,dx = F(x) + C$$

5.6.3 DEFINITION. Let the interval $[a, b]$ be divided into n subintervals. Let the k-th subinterval have width Δx_k, let x_k^* be any point in the k-th subinterval, and let max Δx_k be the largest of the widths of the subintervals. If the function f is defined on $[a, b]$, then f is called *Riemann integrable* on $[a, b]$ or more simply *integrable* on $[a, b]$ if the limit

$$\lim_{\max \Delta x_k \to 0} \sum_{k=1}^{n} f(x_k^*)\,\Delta x_k$$

exists. If f is integrable on $[a, b]$, then we define the **definite integral** of f from a to b by

$$\int_a^b f(x)\, dx = \lim_{\max \Delta x_k \to 0} \sum_{k=1}^{n} f(x_k^*)\, \Delta x_k$$

5.6.6   THEOREM.   *Let f be a function that is defined at all points in the interval $[a, b]$.*

(a)   *If f is continuous on $[a, b]$, then f is integrable on $[a, b]$.*
(b)   *If f is bounded on $[a, b]$ and has only finitely many points of discontinuity on $[a, b]$, then f is integrable on $[a, b]$.*
(c)   *If f is not bounded on $[a, b]$, then f is not integrable on $[a, b]$.*

Area Interpretation of the Definite Integral

If the function f is continuous on $[a, b]$ and $f(x) \geq 0$ for all x in $[a, b]$, then

$$\int_a^b f(x)\, dx$$

is the area under the curve $y = f(x)$ over the interval $[a, b]$. If $f(x)$ has both positive and negative values on $[a, b]$, then this integral represents the difference of two areas: the area under $y = f(x)$ above $[a, b]$ minus the area over $y = f(x)$ below $[a, b]$.

$$\int_a^b f(x)\, dx = (A_I + A_{III}) - A_{II} = \begin{bmatrix} \text{area above} \\ [a, b] \end{bmatrix} - \begin{bmatrix} \text{area below} \\ [a, b] \end{bmatrix}$$

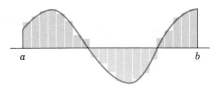 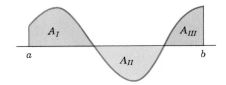

Figure 5.6.11

5.7.1   THEOREM (***The First Fundamental Theorem of Calculus***).   *If f is continuous on $[a, b]$ and if F is an antiderivative of f on $[a, b]$, then*

$$\int_a^b f(x)\, dx = F(b) - F(a)$$

5.9.3   THEOREM (*The Second Fundamental Theorem of Calculus*).   *Let f be a continuous function on an interval I, and let a be any point in I. If F is defined by*

$$F(x) = \int_a^x f(t)\, dt$$

then $F'(x) = f(x)$ *at each point x in the interval I.*

It follows from Theorems 5.7.1 and 5.9.3 that

$$\int_a^b f(x)\, dx = \left[\int f(x)\, dx \right]_a^b \qquad \frac{d}{dx}\left[\int_a^x f(t)\, dt \right] = f(x)$$

5.7.2   THEOREM.   *If f and g are integrable on [a, b] and if c is a constant, then cf, f + g, and f − g are integrable on [a, b] and*

(a) $\quad\displaystyle\int_a^b cf(x)\, dx = c \int_a^b f(x)\, dx$

(b) $\quad\displaystyle\int_a^b [f(x) + g(x)]\, dx = \int_a^b f(x)\, dx + \int_a^b g(x)\, dx$

(c) $\quad\displaystyle\int_a^b [f(x) - g(x)]\, dx = \int_a^b f(x)\, dx - \int_a^b g(x)\, dx$

The same equalities that hold in Theorem 5.7.2 also for indefinite integrals.

5.7.3   THEOREM.   *If f is integrable on a closed interval containing the three points a, b, and c, then*

$$\int_a^b f(x)\, dx = \int_a^c f(x)\, dx + \int_c^b f(x)\, dx$$

no matter how the points are ordered.

Integration by Parts for Indefinite Integrals

$$\int u\, dv = uv - \int v\, du$$

Integration by Parts for Definite Integrals

$$\int_a^b u\, dv = uv \Big]_a^b - \int_a^b v\, du$$

13.4.1   THEOREM (*Arc Length of Parametric Curves*).   *If $x'(t)$ and $y'(t)$ are continuous functions for $a \le t \le b$, then the parametric curve*

$$x = x(t), \quad y = y(t), \quad a \le t \le b$$

has arc length L given by

$$L = \int_a^b \sqrt{[x'(t)]^2 + [y'(t)]^2}\, dt = \int_a^b \sqrt{\left(\frac{dx}{dt}\right)^2 + \left(\frac{dy}{dt}\right)^2}\, dt$$

4.11.1   DEFINITION.   If $s(t)$ is the position function of a particle moving on a coordinate line, then the **instantaneous velocity** at time t is defined by

$$v(t) = s'(t) = \frac{ds}{dt}$$

and the **instantaneous acceleration** at time t is defined by

$$a(t) = v'(t) = \frac{dv}{dt} = s''(t)$$

The **instantaneous speed** of a particle moving on a coordinate line is defined to be the absolute value of the velocity. Thus,

$$\begin{bmatrix} \text{instantaneous} \\ \text{speed} \end{bmatrix} = |v(t)| = |s'(t)| = \left|\frac{ds}{dt}\right|$$

CHAPTER 16

- Triangle inequality
- $\delta - \epsilon$ definition of a limit
- Relationship between continuity and differentiability
- Mean-value theorem
- Differentials
- Relative and absolute extrema
- First and second derivative tests

1.2.6   THEOREM (*Triangle Inequality*).   *If a and b are any real numbers, then*

$$|a + b| \le |a| + |b|$$

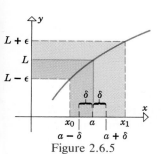

Figure 2.6.5

2.6.1 DEFINITION. Let $f(x)$ be defined for all x in some open interval containing the number a, with the possible exception that $f(x)$ may or may not be defined at a. We shall write

$$\lim_{x \to a} f(x) = L$$

if, given any number $\epsilon > 0$, we can find a number $\delta > 0$ such that $f(x)$ satisfies

$$|f(x) - L| < \epsilon$$

whenever x satisfies

$$0 < |x - a| < \delta$$

3.2.2 THEOREM. *If f is differentiable at a point x_0, then f is also continuous at x_0.*

Theorem 3.2.2 shows that differentiability at a point implies continuity at that point. The converse, however, is false—*a function may be continuous at a point but not differentiable there.* Whenever the graph of a function has a corner at a point, but no break or gap there, we have a point where the function is continuous, but not differentiable.

4.10.2 THEOREM (*Mean-Value Theorem*). *Let f be differentiable on (a, b) and continuous on $[a, b]$. Then there is at least one point c in (a, b) where*

$$f'(c) = \frac{f(b) - f(a)}{b - a}$$

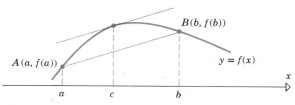

Figure 4.10.4

Increments

If $y = f(x)$, and x changes from an initial value x_0 to a final value x_1, then there is a corresponding change in the value of y from $y_0 = f(x_0)$ to $y_1 = f(x_1)$. The change in x, denoted by Δx, is called the ***increment in x*** and the change in y, denoted by Δy, is called the ***increment in y***:

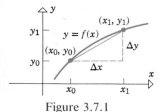

Figure 3.7.1

$$\Delta x = x_1 - x_0 \qquad \Delta y = y_1 - y_0 = f(x_1) - f(x_0)$$

Differentials

Regard x as fixed and *define dx* to be an independent variable that can be assigned an arbitrary value. If f is differentiable at x, then we *define dy* by the formula

$$dy = f'(x)\,dx$$

It is important to understand the distinction between the increment Δy and the differential dy. To see the difference, let us assign the independent variables dx and Δx the same value, so $dx = \Delta x$. Then Δy represents the change in y that occurs when we start at x and travel *along the curve* $y = f(x)$ until we have moved $\Delta x\,(= dx)$ units in the x-direction, while dy represents the change in y that occurs if we start at x and travel *along the tangent* line until we have moved $dx\,(= \Delta x)$ units in the x-direction (Figure 3.7.4).

For $\Delta x = dx$ near zero, dy is commonly used as an approximation to Δy:

$$\Delta y \approx dy$$

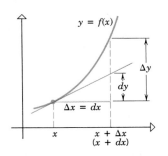

$y = f(x)$

Δy

dy

$\Delta x = dx$

x

$x + \Delta x$
$(x + dx)$

Figure 3.7.4

4.3.1   DEFINITION.   A function f is said to have a **relative maximum** at x_0 if $f(x_0) \geq f(x)$ for all x in some open interval containing x_0.

4.3.2   DEFINITION.   A function f is said to have a **relative minimum** at x_0 if $f(x_0) \leq f(x)$ for all x in some open interval containing x_0.

4.3.3   DEFINITION.   A function f is said to have a **relative extremum** at x_0 if it has either a relative maximum or a relative minimum at x_0.

4.3.4   THEOREM.   *If f has a relative extremum at x_0, then either $f'(x_0) = 0$ or f is not differentiable at x_0.*

4.3.5   DEFINITION.   A **critical point** for a function f is any value of x in the domain of f at which $f'(x) = 0$ or at which f is not differentiable; the critical points where $f'(x) = 0$ are called **stationary points** of f.

4.3.6   THEOREM (*First Derivative Test*).   *Suppose f is continuous at a critical point x_0.*

(a)   *If $f'(x) > 0$ on an open interval extending left from x_0 and $f'(x) < 0$ on an open interval extending right from x_0, then f has a relative maximum at x_0.*

(b)   *If $f'(x) < 0$ on an open interval extending left from x_0 and $f'(x) > 0$ on an open interval extending right from x_0, then f has a relative minimum at x_0.*

(c) *If $f'(x)$ has the same sign [either $f'(x) > 0$ or $f'(x) < 0$] on an open interval extending left from x_0 and on an open interval extending right from x_0, then f does not have a relative extremum at x_0.*

4.3.7   THEOREM (*Second Derivative Test*).   *Suppose f is twice differentiable at a stationary point x_0.*

(a) *If $f''(x_0) > 0$, then f has a relative minimum at x_0.*
(b) *If $f''(x_0) < 0$, then f has a relative maximum at x_0.*

4.6.1   DEFINITION.   If $f(x_0) \geq f(x)$ for all x in the domain of f, then $f(x_0)$ is called the ***maximum value*** or ***absolute maximum value*** of f.

4.6.2   DEFINITION.   If $f(x_0) \leq f(x)$ for all x in the domain of f, then $f(x_0)$ is called the ***minimum value*** or ***absolute minimum value*** of f.

4.6.3   DEFINITION.   A number that is either the maximum or the minimum value of a function f is called an ***extreme value*** or ***absolute extreme value*** of f. Sometimes the terms ***extremum*** or ***absolute extremum*** are also used.

4.6.4   THEOREM (*Extreme-Value Theorem*).   *If a function f is continuous on a closed interval $[a, b]$, then f has both a maximum value and a minimum value on $[a, b]$.*

4.6.5   THEOREM.   *If a function f has an extreme value (either a maximum or a minimum) on an open interval (a, b), then the extreme value occurs at a critical point of f.*

4.6.6   THEOREM.   *Let f be continuous on an interval I and assume that f has exactly one relative extremum on I, say at x_0.*

(a) *If f has a relative minimum at x_0, then $f(x_0)$ is the minimum value of f on the interval I.*
(b) *If f has a relative maximum at x_0, then $f(x_0)$ is the maximum value of f on the interval I.*

CHAPTER 17

- Graphs in polar coordinates
- Volumes by slicing (cross-sections)
- Surface area

Lines in Polar Coordinates

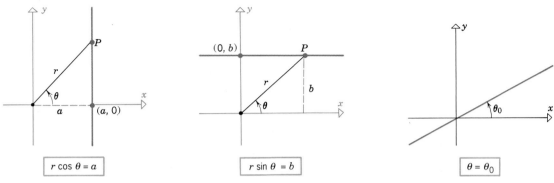

Figure 13.2.1   Figure 13.2.2   Figure 13.2.3

Circles in Polar Coordinates

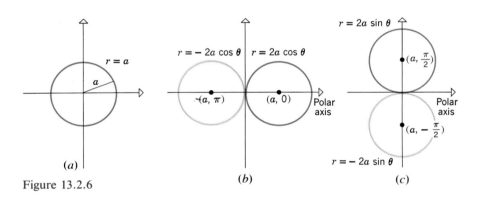

Figure 13.2.6   (b)   (c)

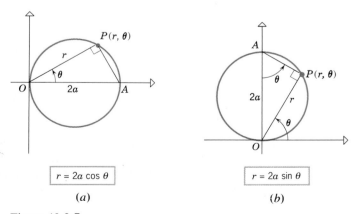

Figure 13.2.7

Limaçons and Cardioids

$$r = a + b \sin\theta, \qquad r = a - b \sin\theta$$
$$r = a + b \cos\theta, \qquad r = a - b \cos\theta$$

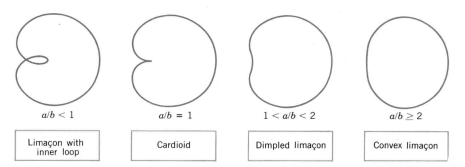

$a/b < 1$	$a/b = 1$	$1 < a/b < 2$	$a/b \geq 2$
Limaçon with inner loop	Cardioid	Dimpled limaçon	Convex limaçon

Figure 13.2.10

Lemniscates

$$r^2 = a^2 \cos 2\theta, \qquad r^2 = -a^2 \cos 2\theta$$
$$r^2 = a^2 \sin 2\theta, \qquad r^2 = -a^2 \sin 2\theta$$

A lemniscate

Figure 13.2.12

Rose Curves

$$r = a \sin n\theta \quad,$$

$$r = a \cos n\theta$$

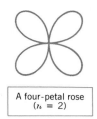

A three-petal rose $(n = 3)$

A four-petal rose $(n = 2)$

Figure 13.2.16

6.2.1 VOLUME FORMULA. Let S be a solid bounded by two parallel planes perpendicular to the x-axis at $x = a$ and $x = b$. If, for each x in $[a, b]$, the cross-sectional area of S perpendicular to the x-axis is $A(x)$, then the volume of the solid is

$$V = \int_a^b A(x)\,dx$$

provided $A(x)$ is integrable.

There is a similar result for cross sections perpendicular to the y-axis.

6.2.2   VOLUME FORMULA.   Let S be a solid bounded by two parallel planes perpendicular to the y-axis at $y = c$ and $y = d$. If, for each y in $[c, d]$, the cross-sectional area of S perpendicular to the y-axis is $A(y)$, then the volume of the solid is

$$V = \int_c^d A(y)\,dy$$

provided $A(y)$ is integrable.

6.5.2   SURFACE AREA FORMULAS.   Let f be nonnegative function on $[a, b]$, and let f' be continuous on $[a, b]$. Then the **surface area S** generated the portion of the curve $y = f(x)$ between $x = a$ and $x = b$ about the x-axis is

$$S = \int_a^b 2\pi f(x)\sqrt{1 + [f'(x)]^2}\,dx$$

For a curve expressed in the form $x = g(y)$, where g' is continuous on $[c, d]$, and $g(y) \geq 0$ for $c \leq y \leq d$, the surface area S generated by revolving the portion of the curve from $y = c$ to $y = d$ about the y-axis is given by

$$S = \int_c^d 2\pi g(y)\sqrt{1 + [g'(y)]^2}\,dy$$

CHAPTER 18

- See review material for Chapters 14–17

CHAPTER 19

- First order linear and separable differential equations (see Section 7.7 in Appendix C).

14

Three-Dimensional Space; Vectors

Jakob Bernoulli   (1654-1705)

■ **14.1** RECTANGULAR COORDINATES IN 3-SPACE; SPHERES; CYLINDRICAL SURFACES

In this section we shall discuss coordinate systems in three-dimensional space and some basic facts about surfaces in three dimensions.

□ **RECTANGULAR COORDINATE SYSTEMS**

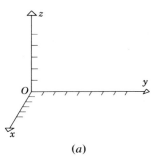

(a)

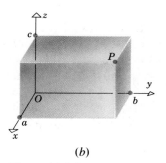

(b)

Figure 14.1.1

Just as points in a plane can be placed in one-to-one correspondence with pairs of real numbers by using two perpendicular coordinate lines, so points in three-dimensional space can be placed in one-to-one correspondence with triples of real numbers by using three mutually perpendicular coordinate lines. To obtain this correspondence, we choose the coordinate lines so that they intersect at their origins, and we call these lines the *x-axis,* the *y-axis,* and the *z-axis* (Figure 14.1.1*a*).

The three coordinate axes form a three-dimensional *rectangular* or *Cartesian coordinate system,* and the point of intersection of the coordinate axes is called the *origin* of the coordinate system.

Each pair of coordinate axes determines a plane called a *coordinate plane.* These are referred to as the *xy-plane,* the *xz-plane,* and the *yz-plane.* To each point P in 3-space we assign a triple of numbers (a, b, c), called the *coordinates of P,* by passing three planes through P parallel to the coordinate planes, and letting a, b, and c be the coordinates of the intersections of these planes with the x, y, and z axes, respectively (Figure 14.1.1*b*). The notation $P(a, b, c)$ will sometimes be used to denote a point P with coordinates (a, b, c).

In Figure 14.1.2 we have constructed the points whose coordinates are $(4, 5, 6)$ and $(-3, 2, -4)$.

In this text we shall call three-dimensional space *3-space,* two-dimensional space (a plane) *2-space,* and one-dimensional space (a line) *1-space.*

Rectangular coordinate systems in 3-space fall into two categories: *left-handed* and *right-handed.* A right-handed system has the property that when the fingers

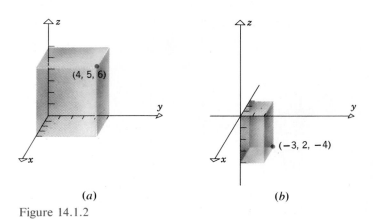

(a)   (b)

Figure 14.1.2

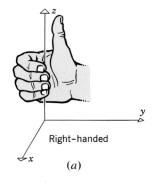

Right-handed

(a)

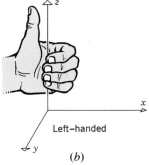

Left-handed

(b)

Figure 14.1.3

of the right hand are cupped so that they curve from the positive x-axis toward the positive y-axis, the thumb points (roughly) in the direction of the positive z-axis (Figure 14.1.3a). A system that is not right-handed is called left-handed (Figure 14.1.3b). We shall use only right-handed coordinate systems.

Just as the coordinate axes in a two-dimensional coordinate system divide 2-space into four quadrants, so the coordinate planes of a three-dimensional coordinate system divide 3-space into eight parts, called **octants** (count them). Those points having three positive coordinates form the **first octant;** the remaining octants have no standard numbering.

The reader should be able to visualize the following results about three-dimensional rectangular coordinate systems:

REGION	DESCRIPTION
xy-plane	Consists of all points of the form $(x, y, 0)$
xz-plane	Consists of all points of the form $(x, 0, z)$
yz-plane	Consists of all points of the form $(0, y, z)$
x-axis	Consists of all points of the form $(x, 0, 0)$
y-axis	Consists of all points of the form $(0, y, 0)$
z-axis	Consists of all points of the form $(0, 0, z)$

☐ **DISTANCE FORMULA IN 3-SPACE**

To obtain a formula for the distance between two points in 3-space, we first consider a box (rectangular parallelepiped) with dimensions a, b, and c (Figure 14.1.4a). By the Theorem of Pythagoras, the length of the diagonal of the base is $\sqrt{a^2 + b^2}$. Moreover, the diagonal of the box is the hypotenuse of a right triangle having a diagonal of the base for one side and a vertical edge for the other side (Figure 14.1.4b). Thus, by the Theorem of Pythagoras, the length of the diagonal of the box is the square root of $[\sqrt{a^2 + b^2}]^2 + c^2$ or more simply

$$\begin{bmatrix} \text{the length of a} \\ \text{diagonal of a} \\ \text{box} \end{bmatrix} = \sqrt{a^2 + b^2 + c^2} \tag{1}$$

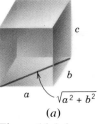

(a)

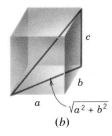

(b)

Figure 14.1.4

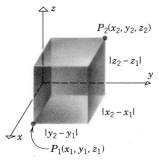

Figure 14.1.5

Suppose now that $P_1(x_1, y_1, z_1)$ and $P_2(x_2, y_2, z_2)$ are points in 3-space. As shown in Figure 14.1.5, these points are on diagonally opposite corners of a box whose dimensions are

$$|x_2 - x_1|, \quad |y_2 - y_1|, \quad \text{and} \quad |z_2 - z_1|$$

Thus, from (1) the distance d between P_1 and P_2 is

$$d = \sqrt{|x_2 - x_1|^2 + |y_2 - y_1|^2 + |z_2 - z_1|^2}$$

or equivalently

$$d = \sqrt{(x_2 - x_1)^2 + (y_2 - y_1)^2 + (z_2 - z_1)^2} \tag{2}$$

Example 1   The distance between the points $(4, -1, 3)$ and $(2, 3, -1)$ is

$$d = \sqrt{(4 - 2)^2 + (-1 - 3)^2 + (3 + 1)^2} = \sqrt{36} = 6 \quad \blacktriangleleft$$

In 2-space the midpoint of the line segment joining $P_1(x_1, y_1)$ and $P_2(x_2, y_2)$ is

$$\left(\tfrac{1}{2}(x_1 + x_2), \tfrac{1}{2}(y_1 + y_2)\right)$$

In 3-space the formula is similar:

$$\begin{bmatrix} \text{The midpoint of the} \\ \text{line segment joining} \\ P_1(x_1, y_1, z_1) \text{ and } P_2(x_2, y_2, z_2) \end{bmatrix} = \left(\tfrac{1}{2}(x_1 + x_2), \tfrac{1}{2}(y_1 + y_2), \tfrac{1}{2}(z_1 + z_2)\right) \tag{3}$$

(We shall prove this result later.)

Example 2   The midpoint of the line segment joining the points $(-1, 3, -8)$ and $(3, 1, 0)$ is

$$\left(\tfrac{1}{2}(-1 + 3), \tfrac{1}{2}(3 + 1), \tfrac{1}{2}(-8 + 0)\right) = (1, 2, -4) \quad \blacktriangleleft$$

□ **SPHERES**

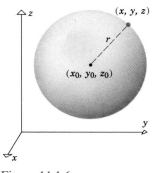

Figure 14.1.6

In 2-space, the graph of an equation relating the variables x and y is the set of all points (x, y) whose coordinates satisfy the equation. Usually, such graphs are curves. Similarly, in 3-space the graph of an equation relating x, y, and z is the set of all points (x, y, z) whose coordinates satisfy the equation. Usually, such graphs are surfaces in 3-space. For example, in 3-space, a sphere with center (x_0, y_0, z_0) and radius r consists of all points (x, y, z) at a distance of r units from (x_0, y_0, z_0) (Figure 14.1.6). Thus, from (2)

$$\sqrt{(x - x_0)^2 + (y - y_0)^2 + (z - z_0)^2} = r$$

or equivalently

$$(x - x_0)^2 + (y - y_0)^2 + (z - z_0)^2 = r^2 \tag{4}$$

This equation represents the sphere with center (x_0, y_0, z_0) and radius r; it is called the ***standard form of the equation*** of a sphere. Some examples are given in the following table:

EQUATION	GRAPH
$(x - 3)^2 + (y - 2)^2 + (z - 1)^2 = 9$	Sphere with center $(3, 2, 1)$ and radius 3
$(x + 1)^2 + y^2 + (z + 4)^2 = 5$	Sphere with center $(-1, 0, -4)$ and radius $\sqrt{5}$
$x^2 + y^2 + z^2 = 1$	Sphere with center $(0, 0, 0)$ and radius 1

Example 3 Describe the graph of

$$x^2 + y^2 + z^2 - 2x - 4y + 8z + 17 = 0$$

Solution. We can put the equation in the form of (4) by completing the squares:

$$(x^2 - 2x) + (y^2 - 4y) + (z^2 + 8z) = -17$$
$$(x^2 - 2x + 1) + (y^2 - 4y + 4) + (z^2 + 8z + 16) = -17 + 21$$
$$(x - 1)^2 + (y - 2)^2 + (z + 4)^2 = 4$$

Thus, the graph is a sphere of radius 2 centered at $(1, 2, -4)$. ◀

By squaring out and collecting terms in (4), we can rewrite the equation of a sphere in the form

$$x^2 + y^2 + z^2 + Gx + Hy + Iz + J = 0 \tag{5}$$

However, it is not true that every equation of form (5) has a sphere as its graph, for if we begin with an equation of form (5) and complete the squares as in Example 2, we shall obtain an equation of the form

$$(x - x_0)^2 + (y - y_0)^2 + (z - z_0)^2 = k$$

where k is a constant. If $k > 0$, then the equation represents a sphere of radius $\sqrt{k}$. However, if $k = 0$, the graph is the single point (x_0, y_0, z_0), and if $k < 0$, the equation is not satisfied by any real values of x, y, and z; thus, there is no graph. To summarize:

14.1.1 THEOREM. *An equation of the form*

$$x^2 + y^2 + z^2 + Gx + Hy + Iz + J = 0$$

represents either a sphere or a point, or else has no graph.

□ **CYLINDRICAL SURFACES**   Now that we have introduced graphs in 3-space, an equation in x and y such as

$$x + y = 1 \qquad (6)$$

becomes ambiguous. Should we graph this equation in 2-space (Figure 14.1.7a) or should we take the position that (6) is really the equation

$$x + y + 0z = 1 \qquad (7)$$

in which case it should be graphed in 3-space? If we take the latter position, then the graph is the plane shown in Figure 14.1.7b since (x, y, z) will satisfy (7) for *every* z as long as x and y satisfy (6).

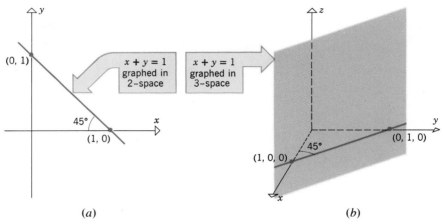

Figure 14.1.7

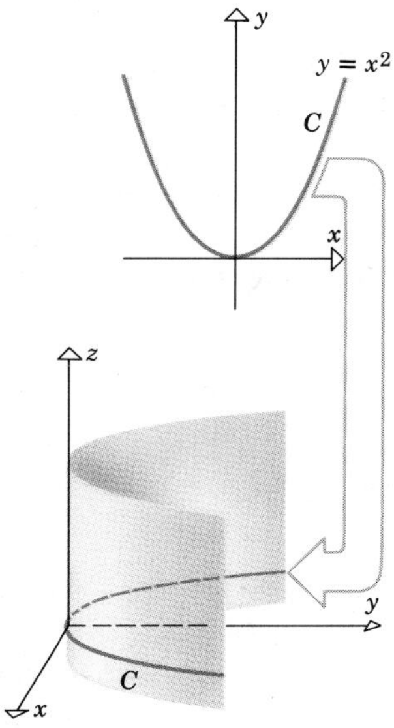

Actually, it is correct to graph (6) either in 2-space or 3-space—the appropriate choice will usually be clear from the context in which the equation arises.

A three-dimensional surface generated by a line traversing a plane curve and moving parallel to a fixed line is sometimes called a ***cylindrical surface;*** the moving line is called the ***generator*** of the surface. In general, if an equation in x and y has a curve C for its graph in the xy-plane, then the same equation graphed in 3-space produces the cylindrical surface that is traced out as a generator parallel to the z-axis traverses the curve C in the xy-plane. Similarly, an equation in x and z only represents a cylindrical surface in 3-space with generator parallel to the y-axis; and an equation in y and z alone represents a cylindrical surface in 3-space with generator parallel to the x-axis. In brief:

> *An equation containing only two of the three variables x, y, and z represents a cylindrical surface in 3-space. The generator of the surface is parallel to the axis corresponding to the missing variable.*

Example 4   Sketch the graph of $y = x^2$ in 3-space.

Solution.   (See Figure 14.1.8.)   ◄

Figure 14.1.8

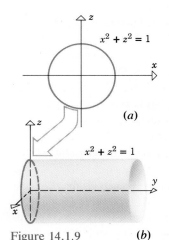

Figure 14.1.9   **(b)**

Example 5 Sketch the graph of $x^2 + z^2 = 1$ in 3-space.

Solution. Since y does not appear in this equation, the graph is a cylindrical surface with generator parallel to the y-axis. It is helpful to begin with a sketch of the equation in 2-space. In the xz-plane the curve $x^2 + z^2 = 1$ is a circle (Figure 14.1.9a). Thus, in 3-space this equation represents a right-circular cylinder parallel to the y-axis (Figure 14.1.9b). ◀

Example 6 Sketch the graph of $z = \sin y$ in 3-space.

Solution. (See Figure 14.1.10.) ◀

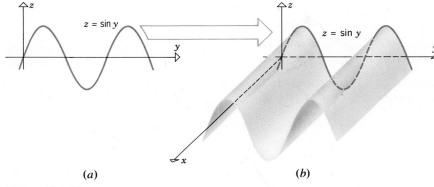

(a)   **(b)**

Figure 14.1.10

▶ Exercise Set 14.1

1. Plot the points P and Q in a right-handed coordinate system. Then, find the distance between them and the midpoint of the line segment joining them.
   (a) $P(0, 0, 0)$; $Q(2, 1, 3)$
   (b) $P(5, 2, 3)$; $Q(4, 1, 6)$
   (c) $P(-2, -1, 3)$; $Q(3, 0, 5)$
   (d) $P(-1, -1, -3)$; $Q(4, 3, -2)$.

2. A cube of side 4 has its geometric center at the origin and its faces parallel to the coordinate planes. Sketch the cube and give the coordinates of the vertices.

3. A rectangular parallelepiped has its faces parallel to the coordinate planes and has $(4, 2, -2)$ and $(-6, 1, 1)$ as endpoints of a diagonal. Sketch the parallelepiped and give the coordinates of the vertices.

4. Show that $(4, 5, 2)$, $(1, 7, 3)$, and $(2, 4, 5)$ are vertices of an equilateral triangle.

5. (a) Show that $(2, 1, 6)$, $(4, 7, 9)$, and $(8, 5, -6)$ are the vertices of a right triangle.

   (b) Which vertex is at the 90° angle?
   (c) Find the area of the triangle.

6. Find the distance from the point $(-5, 2, -3)$ to the
   (a) xy-plane   (b) xz-plane   (c) yz-plane
   (d) x-axis   (e) y-axis   (f) z-axis.

7. Show that the distance from a point (x_0, y_0, z_0) to the z-axis is $\sqrt{x_0^2 + y_0^2}$, and find the distances from the point to the x and y axes.

In Exercises 8–11, find an equation for the sphere with center C and radius r.

8. $C(0, 0, 0)$; $r = 8$.   9. $C(-2, 4, -1)$; $r = 6$.

10. $C(5, -2, 4)$; $r = \sqrt{7}$.   11. $C(0, 1, 0)$; $r = 3$.

12. In each part find an equation for the sphere with center $(-3, 5, -4)$ and satisfying the given condition.
    (a) Tangent to the xy-plane
    (b) Tangent to the xz-plane
    (c) Tangent to the yz-plane.

13. In each part, find an equation for the sphere with center $(2, -1, -3)$ and satisfying the given condition.
   (a) Tangent to the xy-plane
   (b) Tangent to the xz-plane
   (c) Tangent to the yz-plane.

In Exercises 14–19, find the standard equation of the sphere satisfying the given conditions.

14. Center $(1, 0, -1)$; diameter $= 8$.

15. A diameter has endpoints $(-1, 2, 1)$ and $(0, 2, 3)$.

16. Center $(-1, 3, 2)$ and passing through the origin.

17. Center $(3, -2, 4)$ and passing through $(7, 2, 1)$.

18. Center $(-3, 5, -4)$; tangent to the sphere of radius 1 centered at the origin (two answers).

19. Center $(0, 0, 0)$; tangent to the sphere of radius 1 centered at $(3, -2, 4)$ (two answers).

In Exercises 20–25, describe the surface whose equation is given.

20. $x^2 + y^2 + z^2 - 2x - 6y - 8z + 1 = 0$.

21. $x^2 + y^2 + z^2 + 10x + 4y + 2z - 19 = 0$.

22. $x^2 + y^2 + z^2 - y = 0$.

23. $2x^2 + 2y^2 + 2z^2 - 2x - 3y + 5z - 2 = 0$.

24. $x^2 + y^2 + z^2 + 2x - 2y + 2z + 3 = 0$.

25. $x^2 + y^2 + z^2 - 3x + 4y - 8z + 25 = 0$.

26. Find the largest and smallest distances between the point $P(1, 1, 1)$ and the sphere $x^2 + y^2 + z^2 - 2y + 6z - 6 = 0$.

27. Find the largest and smallest distances between the origin and the sphere $x^2 + y^2 + z^2 + 2x - 2y - 4z - 3 = 0$.

28. Describe the set of all points in 3-space whose coordinates satisfy the inequality $x^2 + y^2 + z^2 - 2x + 8z \leq 8$.

29. Describe the set of all points in 3-space whose coordinates satisfy the inequality $y^2 + z^2 + 6y - 4z > 3$.

30. The distance between a point $P(x, y, z)$ and the point $A(1, -2, 0)$ is twice the distance between P and the point $B(0, 1, 1)$. Show that the set of all such points is a sphere, and find the center and radius of the sphere.

31. A bowling ball of radius R is placed inside a box just large enough to hold it, and it is secured for shipping by packing a Styrofoam sphere into each corner of the box. Find the radius of the largest Styrofoam sphere that can be used. [*Hint:* Take the origin of a Cartesian coordinate system at a corner of the box with the coordinate axes along the edges.] (See Figure 14.1.11.)

In Exercises 32–39, sketch the surface whose equation is given.

32. (a) $y = x$ (b) $y = z$ (c) $x = z$.

33. (a) $x^2 + y^2 = 25$ (b) $y^2 + z^2 = 25$
   (c) $x^2 + z^2 = 25$.

34. (a) $y = x^2$ (b) $z = x^2$ (c) $y = z^2$.

35. (a) $y = e^x$ (b) $x = \ln z$ (c) $yz = 1$.

36. (a) $2x + 3y = 6$ (b) $2x + z = 3$.

37. (a) $y = \sin x$ (b) $z = \cos x$.

38. (a) $z = 1 - y^2$ (b) $z = \sqrt{3 - x}$.

39. (a) $4x^2 + 9z^2 = 36$ (b) $y^2 - 4z^2 = 4$.

40. In each part of Figure 14.1.12 find an equation for the right-circular cylinder of radius a shown.

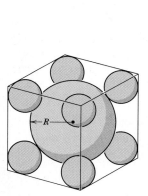

Figure 14.1.11

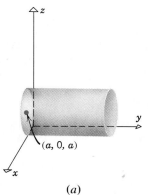

(a)

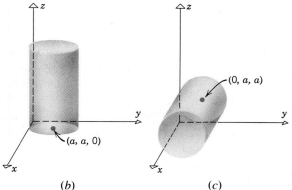

(b) (c)

Figure 14.1.12

41. Show that for all values of θ and ϕ, the point $(a \sin \phi \cos \theta, \; a \sin \phi \sin \theta, \; a \cos \phi)$ lies on the sphere $x^2 + y^2 + z^2 = a^2$.

42. Consider the equation

$$x^2 + y^2 + z^2 + Gx + Hy + Iz + J = 0$$

and let $K = G^2 + H^2 + I^2 - 4J$.

(a) Prove that the equation represents a sphere, a point, or no graph, according to whether $K > 0$, $K = 0$, or $K < 0$.

(b) In the case where $K > 0$, find the center and radius of the sphere.

■ **14.2** VECTORS

*Many physical quantities such as area, length, mass, and temperature are completely described once the magnitude of the quantity is given. Such quantities are called scalars. Other physical quantities, called **vectors**, are not completely determined until both a magnitude and a direction are specified. For example, wind movement is usually described by giving the speed and the direction, say 20 mi/hr northeast. The wind speed and wind direction together form a vector quantity called the wind velocity. Other examples of vectors are force and displacement. In this section we shall develop the basic mathematical properties of vectors.*

□ **VECTORS VIEWED GEOMETRICALLY**

Vectors can be represented geometrically as directed line segments or arrows in two- or three-dimensional space; the direction of the arrow specifies the direction of the vector and the length of the arrow describes its magnitude. The tail of the arrow is called the *initial point* of the vector, and the tip of the arrow the *terminal point*. We shall denote vectors by lowercase boldface type such as **a**, **k**, **v**, **w**, and **x**. When discussing vectors, we shall refer to real numbers as *scalars*. Scalars will be denoted by lowercase italic type such as a, k, v, w, and x.

If, as in Figure 14.2.1a, the initial point of a vector **v** is A and the terminal point is B, we write

$$\mathbf{v} = \overrightarrow{AB}$$

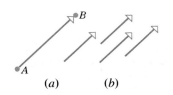

(a) $\qquad$ (b)

Figure 14.2.1

Vectors having the same length and same direction, such as those in Figure 14.2.1b, are called *equivalent*. Since we want a vector to be determined solely by its length and direction, equivalent vectors are regarded as *equal* even though they may be located in different positions. If **v** and **w** are equivalent, we write

$$\mathbf{v} = \mathbf{w}$$

14.2.1 DEFINITION. If **v** and **w** are any two vectors, then the sum **v** + **w** is the vector determined as follows. Position the vector **w** so that its initial point coincides with the terminal point of **v**. The vector **v** + **w** is represented by the arrow from the initial point of **v** to the terminal point of **w** (Figure 14.2.2a).

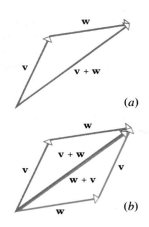

Figure 14.2.2

In Figure 14.2.2*b* we have constructed two sums, $\mathbf{v} + \mathbf{w}$ (purple arrows) and $\mathbf{w} + \mathbf{v}$ (green arrows). It is evident that

$$\mathbf{v} + \mathbf{w} = \mathbf{w} + \mathbf{v}$$

and that the sum coincides with the diagonal of the parallelogram determined by $\mathbf{v}$ and $\mathbf{w}$ when these vectors are located so that they have the same initial point.

The vector of length zero is called the **zero vector** and is denoted by $\mathbf{0}$. We define

$$\mathbf{0} + \mathbf{v} = \mathbf{v} + \mathbf{0} = \mathbf{v}$$

Since there is no natural direction for the zero vector, we shall agree that it can be assigned any direction that is convenient for the problem at hand.

If $\mathbf{v}$ is any nonzero vector, then $-\mathbf{v}$, the **negative** of $\mathbf{v}$, is defined to be the vector having the same magnitude as $\mathbf{v}$, but oppositely directed (Figure 14.2.3). This vector has the property

$$\mathbf{v} + (-\mathbf{v}) = \mathbf{0}$$

(Why?) In addition, we define $-\mathbf{0} = \mathbf{0}$.

Figure 14.2.3

14.2.2 DEFINITION. If $\mathbf{v}$ and $\mathbf{w}$ are any two vectors, then **subtraction** of $\mathbf{w}$ from $\mathbf{v}$ is defined by

$$\mathbf{v} - \mathbf{w} = \mathbf{v} + (-\mathbf{w})$$

(See Figure 14.2.4*a*.)

To obtain $\mathbf{v} - \mathbf{w}$ without constructing $-\mathbf{w}$, position $\mathbf{v}$ and $\mathbf{w}$ so that their initial points coincide; the vector from the terminal point of $\mathbf{w}$ to the terminal point of $\mathbf{v}$ is then the vector $\mathbf{v} - \mathbf{w}$ (Figure 14.2.4*b*).

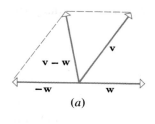

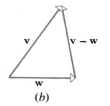

Figure 14.2.4

14.2.3 DEFINITION. If $\mathbf{v}$ is a nonzero vector and k is a nonzero real number (scalar), then the **product** $k\mathbf{v}$ is defined to be the vector whose length is $|k|$ times the length of $\mathbf{v}$ and whose direction is the same as that of $\mathbf{v}$ if $k > 0$ and opposite to that of $\mathbf{v}$ if $k < 0$. We define $k\mathbf{v} = \mathbf{0}$ if $k = 0$ or $\mathbf{v} = \mathbf{0}$.

Figure 14.2.5 illustrates the relationship between a vector $\mathbf{v}$ and the vectors $\frac{1}{2}\mathbf{v}$, $(-1)\mathbf{v}$, $2\mathbf{v}$, and $(-3)\mathbf{v}$. Note that the vector $(-1)\mathbf{v}$ has the same length as $\mathbf{v}$ but is oppositely directed. Thus, $(-1)\mathbf{v}$ is just the negative of $\mathbf{v}$; that is,

$$(-1)\mathbf{v} = -\mathbf{v}$$

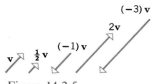

Figure 14.2.5

☐ **VECTORS IN COORDINATE SYSTEMS**

Up to this point everything we have said about vectors applies equally well to vectors in 2-space or 3-space. However, problems involving vectors can often be simplified by introducing a rectangular coordinate system. For this purpose it will be necessary for us to distinguish between vectors in 2-space (the plane) and vectors in 3-space.

If $\mathbf{v}$ is a vector in 2-space or 3-space with its initial point at the origin of a rectangular coordinate system (Figure 14.2.6), then the coordinates (v_1, v_2) or (v_1, v_2, v_3) of the terminal point are called the **components** of $\mathbf{v}$ and we write

$$\mathbf{v} = \langle v_1, v_2 \rangle \quad \text{or} \quad \mathbf{v} = \langle v_1, v_2, v_3 \rangle$$

depending on whether the vector is in 2-space or 3-space.

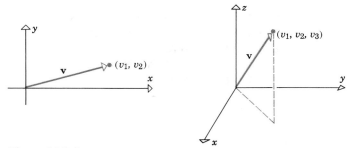

Figure 14.2.6

Because the zero vector $\mathbf{0}$ has length zero, its terminal point and initial point coincide. Thus,

$$\mathbf{0} = \langle 0, 0 \rangle \quad \text{(in 2-space)}$$

$$\mathbf{0} = \langle 0, 0, 0 \rangle \quad \text{(in 3-space)}$$

☐ **EQUALITY OF VECTORS**

If equivalent vectors, $\mathbf{v}$ and $\mathbf{w}$, are located so that their initial points fall at the origin, then it is obvious that their terminal points must coincide (since the vectors have the same length and direction). The vectors thus have the same components. Conversely, vectors with the same components are equivalent, since they have the same length and same direction. In summary, the vectors

$$\mathbf{v} = \langle v_1, v_2 \rangle \quad \text{and} \quad \mathbf{w} = \langle w_1, w_2 \rangle$$

in 2-space are equivalent (i.e., $\mathbf{v} = \mathbf{w}$) if and only if

$$v_1 = w_1 \quad \text{and} \quad v_2 = w_2$$

and the vectors

$$\mathbf{v} = \langle v_1, v_2, v_3 \rangle \quad \text{and} \quad \mathbf{w} = \langle w_1, w_2, w_3 \rangle$$

in 3-space are equivalent if and only if

$$v_1 = w_1, \quad v_2 = w_2, \quad \text{and} \quad v_3 = w_3$$

☐ ARITHMETIC
OPERATIONS ON
VECTORS

The following theorem shows how to perform arithmetic operations on vectors using components.

14.2.4 THEOREM. *If* $\mathbf{v} = \langle v_1, v_2 \rangle$ *and* $\mathbf{w} = \langle w_1, w_2 \rangle$ *are vectors in 2-space and k is any scalar, then*

$$\mathbf{v} + \mathbf{w} = \langle v_1 + w_1, v_2 + w_2 \rangle \tag{1a}$$
$$\mathbf{v} - \mathbf{w} = \langle v_1 - w_1, v_2 - w_2 \rangle \tag{1b}$$
$$k\mathbf{v} = \langle kv_1, kv_2 \rangle \tag{1c}$$

Similarly, if $\mathbf{v} = \langle v_1, v_2, v_3 \rangle$ *and* $\mathbf{w} = \langle w_1, w_2, w_3 \rangle$ *are vectors in 3-space and k is any scalar, then*

$$\mathbf{v} + \mathbf{w} = \langle v_1 + w_1, v_2 + w_2, v_3 + w_3 \rangle \tag{2a}$$
$$\mathbf{v} - \mathbf{w} = \langle v_1 - w_1, v_2 - w_2, v_3 - w_3 \rangle \tag{2b}$$
$$k\mathbf{v} = \langle kv_1, kv_2, kv_3 \rangle \tag{2c}$$

We shall not prove this theorem. However, results (1a) and (1c) should be evident from Figure 14.2.7. Similar figures in 3-space can be used to motivate (2a) and (2c). Formulas (1b) and (2b) can be obtained from parts (a) and (c) by writing

$$\mathbf{v} - \mathbf{w} = \mathbf{v} + (-1)\mathbf{w}$$

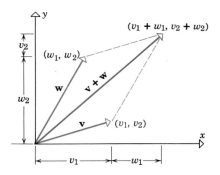

Figure 14.2.7

Example 1   If $\mathbf{v} = \langle 1, -2 \rangle$ and $\mathbf{w} = \langle 7, 6 \rangle$, then

$$\mathbf{v} + \mathbf{w} = \langle 1, -2 \rangle + \langle 7, 6 \rangle = \langle 1 + 7, -2 + 6 \rangle = \langle 8, 4 \rangle$$
$$4\mathbf{v} = 4\langle 1, -2 \rangle = \langle 4(1), 4(-2) \rangle = \langle 4, -8 \rangle$$
$$-\mathbf{v} = (-1)\mathbf{v} = (-1)\langle 1, -2 \rangle = \langle -1, 2 \rangle$$
$$\mathbf{v} - \mathbf{w} = \langle 1, -2 \rangle - \langle 7, 6 \rangle = \langle 1 - 7, -2 - 6 \rangle = \langle -6, -8 \rangle \qquad \blacktriangleleft$$

Example 2 If $\mathbf{v} = \langle -2, 0, 1 \rangle$ and $\mathbf{w} = \langle 3, 5, -4 \rangle$, then

$$\mathbf{v} + \mathbf{w} = \langle -2, 0, 1 \rangle + \langle 3, 5, -4 \rangle = \langle 1, 5, -3 \rangle$$
$$-3\mathbf{v} = \langle 6, 0, -3 \rangle$$
$$-\mathbf{w} = \langle -3, -5, 4 \rangle$$
$$\mathbf{w} - 2\mathbf{v} = \langle 3, 5, -4 \rangle - \langle -4, 0, 2 \rangle = \langle 7, 5, -6 \rangle \quad \blacktriangleleft$$

REMARK. It should be evident from Theorem 14.2.4 and the foregoing examples that, except for the number of components, there is no difference between arithmetic computations on vectors in 2-space and 3-space.

□ **VECTORS WITH INITIAL POINT NOT AT THE ORIGIN**

Sometimes a vector is positioned so that its initial point is not at the origin (Figure 14.2.8).

If the coordinates of the initial and terminal points of a vector are known, then the components of the vector can be obtained using the following theorem.

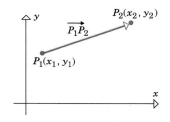

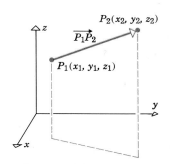

Figure 14.2.8

14.2.5 THEOREM. *If $\overrightarrow{P_1P_2}$ is a vector in 2-space with initial point $P_1(x_1, y_1)$ and terminal point $P_2(x_2, y_2)$, then*

$$\overrightarrow{P_1P_2} = \langle x_2 - x_1, y_2 - y_1 \rangle \tag{3a}$$

Similarly, if $\overrightarrow{P_1P_2}$ is a vector in 3-space with initial point $P_1(x_1, y_1, z_1)$ and terminal point $P_2(x_2, y_2, z_2)$, then

$$\overrightarrow{P_1P_2} = \langle x_2 - x_1, y_2 - y_1, z_2 - z_1 \rangle \tag{3b}$$

We shall give the proof in 2-space. The proof in 3-space is similar.

Proof. The vector $\overrightarrow{P_1P_2}$ is the difference of vectors $\overrightarrow{OP_2}$ and $\overrightarrow{OP_1}$ (Figure 14.2.9). Thus,

$$\overrightarrow{P_1P_2} = \overrightarrow{OP_2} - \overrightarrow{OP_1} = \langle x_2, y_2 \rangle - \langle x_1, y_1 \rangle = \langle x_2 - x_1, y_2 - y_1 \rangle \quad \blacksquare$$

To paraphrase the foregoing theorem loosely, *the components of any vector are the coordinates of its terminal point minus the coordinates of its initial point.*

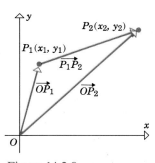

Figure 14.2.9

Example 3 In 2-space the vector with initial point $P_1(1, 3)$ and terminal point $P_2(4, -2)$ is

$$\overrightarrow{P_1P_2} = \langle 4 - 1, -2 - 3 \rangle = \langle 3, -5 \rangle$$

and in 3-space the vector with initial point $A(0, -2, 5)$ and terminal point $B(3, 4, -1)$ is

$$\overrightarrow{AB} = \langle 3 - 0, 4 - (-2), -1 - 5 \rangle = \langle 3, 6, -6 \rangle \quad \blacktriangleleft$$

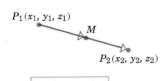

$P_1(x_1, y_1, z_1)$

M

$P_2(x_2, y_2, z_2)$

$$\overrightarrow{P_1M} = \tfrac{1}{2}\overrightarrow{P_1P_2}$$

Figure 14.2.10

In the previous section we stated without proof that the midpoint of the line segment joining the points $P_1(x_1, y_1, z_1)$ and $P_2(x_2, y_2, z_2)$ has coordinates $M(\bar{x}, \bar{y}, \bar{z})$ given by

$$\bar{x} = \tfrac{1}{2}(x_1 + x_2), \quad \bar{y} = \tfrac{1}{2}(y_1 + y_2), \quad \bar{z} = \tfrac{1}{2}(z_1 + z_2) \tag{4}$$

To see that this is so we need only observe that

$$\overrightarrow{P_1M} = \tfrac{1}{2}\overrightarrow{P_1P_2}$$

(Figure 14.2.10), so that

$$\langle \bar{x} - x_1, \bar{y} - y_1, \bar{z} - z_1 \rangle = \tfrac{1}{2}\langle x_2 - x_1, y_2 - y_1, z_2 - z_1 \rangle$$

or, on equating components,

$$\bar{x} - x_1 = \tfrac{1}{2}(x_2 - x_1), \quad \bar{y} - y_1 = \tfrac{1}{2}(y_2 - y_1), \quad \bar{z} - z_1 = \tfrac{1}{2}(z_2 - z_1)$$

from which (4) follows.

☐ **RULES OF VECTOR ARITHMETIC**

The following theorem shows that many of the familiar rules of ordinary arithmetic also hold for vector arithmetic.

14.2.6 THEOREM.   *For any vectors* **u**, **v**, *and* **w** *and any scalars* k *and* l, *the following relationships hold*:

(*a*) $\mathbf{u} + \mathbf{v} = \mathbf{v} + \mathbf{u}$

(*b*) $(\mathbf{u} + \mathbf{v}) + \mathbf{w} = \mathbf{u} + (\mathbf{v} + \mathbf{w})$

(*c*) $\mathbf{u} + \mathbf{0} = \mathbf{0} + \mathbf{u} = \mathbf{u}$

(*d*) $\mathbf{u} + (-\mathbf{u}) = \mathbf{0}$

(*e*) $k(l\mathbf{u}) = (kl)\mathbf{u}$

(*f*) $k(\mathbf{u} + \mathbf{v}) = k\mathbf{u} + k\mathbf{v}$

(*g*) $(k + l)\mathbf{u} = k\mathbf{u} + l\mathbf{u}$

(*h*) $1\mathbf{u} = \mathbf{u}$

Before discussing the proof, we note that we have developed two approaches to vectors: *geometric*, in which vectors are represented by arrows or directed line segments, and *analytic*, in which vectors are represented by pairs or triples of numbers called components. As a consequence, the results in this theorem can be established either geometrically or analytically. As an illustration, we shall prove part (*b*) both ways. The remaining proofs are left as exercises.

Proof (b) (Analytic in 2-space).   Let $\mathbf{u} = \langle u_1, u_2 \rangle$, $\mathbf{v} = \langle v_1, v_2 \rangle$, and $\mathbf{w} = \langle w_1, w_2 \rangle$. Then

$$
\begin{aligned}
(\mathbf{u} + \mathbf{v}) + \mathbf{w} &= (\langle u_1, u_2 \rangle + \langle v_1, v_2 \rangle) + \langle w_1, w_2 \rangle \\
&= \langle u_1 + v_1, u_2 + v_2 \rangle + \langle w_1, w_2 \rangle \\
&= \langle (u_1 + v_1) + w_1, (u_2 + v_2) + w_2 \rangle \\
&= \langle u_1 + (v_1 + w_1), u_2 + (v_2 + w_2) \rangle \\
&= \langle u_1, u_2 \rangle + \langle v_1 + w_1, v_2 + w_2 \rangle \\
&= \mathbf{u} + (\mathbf{v} + \mathbf{w})
\end{aligned}
$$

Proof (b) (Geometric). Let **u**, **v**, and **w** be represented by $\overrightarrow{PQ}$, $\overrightarrow{QR}$, and $\overrightarrow{RS}$ as shown in Figure 14.2.11. Then

$$\mathbf{v} + \mathbf{w} = \overrightarrow{QS} \quad \text{and} \quad \mathbf{u} + (\mathbf{v} + \mathbf{w}) = \overrightarrow{PS}$$
$$\mathbf{u} + \mathbf{v} = \overrightarrow{PR} \quad \text{and} \quad (\mathbf{u} + \mathbf{v}) + \mathbf{w} = \overrightarrow{PS}$$

Therefore,

$$(\mathbf{u} + \mathbf{v}) + \mathbf{w} = \mathbf{u} + (\mathbf{v} + \mathbf{w}) \qquad \blacksquare$$

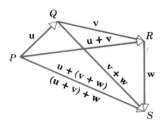

Figure 14.2.11

REMARK.   In light of part (*b*) of this theorem, the symbol **u** + **v** + **w** is unambiguous since the same result is obtained no matter where parentheses are inserted. Moreover, if the vectors **u**, **v**, and **w** are placed "tip to tail," then the sum **u** + **v** + **w** is the vector from the initial point of **u** to the terminal point of **w** (Figure 14.2.11).

☐ **LENGTH OF A VECTOR**

Geometrically, the *length* of a vector **v**, also called the *norm* of **v**, is the distance between its initial and terminal points. The length (or norm) of **v** is denoted by $\|\mathbf{v}\|$. It follows from the distance formulas in 2-space and 3-space that the norm of a vector $\mathbf{v} = \langle v_1, v_2 \rangle$ in 2-space is given by

$$\|\mathbf{v}\| = \sqrt{v_1^2 + v_2^2} \tag{5a}$$

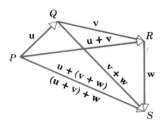

(*a*)

and the norm of a vector $\mathbf{v} = \langle v_1, v_2, v_3 \rangle$ in 3-space is given by

$$\|\mathbf{v}\| = \sqrt{v_1^2 + v_2^2 + v_3^2} \tag{5b}$$

(Figures 14.2.12*a* and 14.2.12*b*).

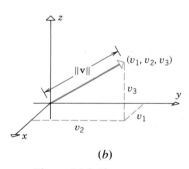

(*b*)

Figure 14.2.12

Example 4   Find the norm of

    (a) $\mathbf{v} = \langle -2, 3 \rangle$ (b) $\mathbf{w} = \langle 2, 3, 6 \rangle$

Solution.   From (5a) and (5b)

$$\|\mathbf{v}\| = \sqrt{(-2)^2 + 3^2} = \sqrt{13}$$
$$\|\mathbf{w}\| = \sqrt{2^2 + 3^2 + 6^2} = \sqrt{49} = 7 \qquad \blacktriangleleft$$

Recall from Definition 14.2.3 that the length of $k\mathbf{v}$ is $|k|$ times the length of **v**. Expressed as an equation, this statement says that

$$\|k\mathbf{v}\| = |k|\,\|\mathbf{v}\| \tag{6}$$

This formula applies to both vectors in 2-space and 3-space.

☐ **UNIT VECTORS**

It follows from (6) that if a nonzero vector **v** is multiplied by $1/\|\mathbf{v}\|$ (the reciprocal of its length), then the result is a vector of length 1 in the same direction as **v**. (Why?) This process of multiplying **v** by $1/\|\mathbf{v}\|$ to obtain a vector of length 1 is called *normalizing* **v**.

Example 5   A vector of length 1 in the same direction as $\mathbf{v} = \langle 3, 4\rangle$ is

$$\frac{1}{\|\mathbf{v}\|}\mathbf{v} = \frac{1}{\sqrt{3^2 + 4^2}}\langle 3, 4\rangle = \frac{1}{5}\langle 3, 4\rangle = \left\langle \frac{3}{5}, \frac{4}{5}\right\rangle \quad \blacktriangleleft$$

A vector of length 1 is called a **unit vector**. Of special importance are the unit vectors that run along the positive coordinate axes of a rectangular coordinate system. In both 2-space and 3-space the unit vectors along the x and y–axes are denoted by $\mathbf{i}$ and $\mathbf{j}$, respectively, and in 3-space the unit vector along the z-axis is denoted by $\mathbf{k}$. Thus,

$$\mathbf{i} = \langle 1, 0\rangle, \qquad \mathbf{j} = \langle 0, 1\rangle \qquad \boxed{\text{In 2-space}}$$

$$\mathbf{i} = \langle 1, 0, 0\rangle, \quad \mathbf{j} = \langle 0, 1, 0\rangle, \quad \mathbf{k} = \langle 0, 0, 1\rangle \quad \boxed{\text{In 3-space}}$$

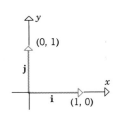

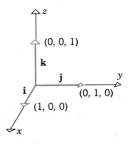

Figure 14.2.13

(See Figure 14.2.13.)

Every vector $\mathbf{v} = \langle v_1, v_2\rangle$ in 2-space is expressible uniquely in terms of $\mathbf{i}$ and $\mathbf{j}$ since we can write

$$\mathbf{v} = \langle v_1, v_2\rangle = \langle v_1, 0\rangle + \langle 0, v_2\rangle = v_1\langle 1, 0\rangle + v_2\langle 0, 1\rangle = v_1\mathbf{i} + v_2\mathbf{j}$$

and similarly every vector $\mathbf{v} = \langle v_1, v_2, v_3\rangle$ in 3-space is expressible uniquely in terms of $\mathbf{i}$, $\mathbf{j}$, and $\mathbf{k}$ since we can write

$$\mathbf{v} = \langle v_1, v_2, v_3\rangle = v_1\langle 1, 0, 0\rangle + v_2\langle 0, 1, 0\rangle + v_3\langle 0, 0, 1\rangle = v_1\mathbf{i} + v_2\mathbf{j} + v_3\mathbf{k}$$

REMARK.   The notations $\langle v_1, v_2, v_3\rangle$ and $v_1\mathbf{i} + v_2\mathbf{j} + v_3\mathbf{k}$ are interchangeable, and we shall use both of them. For example, (5b) can be written as

$$\|v_1\mathbf{i} + v_2\mathbf{j} + v_3\mathbf{k}\| = \sqrt{v_1^2 + v_2^2 + v_3^2}$$

Similarly, the notations $\langle v_1, v_2\rangle$ and $v_1\mathbf{i} + v_2\mathbf{j}$ are interchangeable.

Example 6

$$\langle 2, 3\rangle = 2\mathbf{i} + 3\mathbf{j}$$
$$\langle 5, -4\rangle = 5\mathbf{i} + (-4)\mathbf{j} = 5\mathbf{i} - 4\mathbf{j}$$
$$\langle 0, 2\rangle = 0\mathbf{i} + 2\mathbf{j} = 2\mathbf{j}$$
$$\langle -4, 0\rangle = -4\mathbf{i} + 0\mathbf{j} = -4\mathbf{i}$$

$$\langle 0, 0\rangle = 0\mathbf{i} + 0\mathbf{j} = \mathbf{0}$$
$$(3\mathbf{i} + 2\mathbf{j}) + (4\mathbf{i} + \mathbf{j}) = 7\mathbf{i} + 3\mathbf{j}$$
$$5(6\mathbf{i} - 2\mathbf{j}) = 30\mathbf{i} - 10\mathbf{j}$$
$$\|2\mathbf{i} - 3\mathbf{j}\| = \sqrt{2^2 + (-3)^2} = \sqrt{13} \quad \blacktriangleleft$$

Example 7

$$\langle 2, -3, 4\rangle = 2\mathbf{i} - 3\mathbf{j} + 4\mathbf{k}$$
$$\langle 0, 3, 0\rangle = 3\mathbf{j}$$
$$(3\mathbf{i} + 2\mathbf{j} - \mathbf{k}) - (4\mathbf{i} - \mathbf{j} + 2\mathbf{k}) = -\mathbf{i} + 3\mathbf{j} - 3\mathbf{k}$$
$$2(\mathbf{i} + \mathbf{j} - \mathbf{k}) + 4(\mathbf{i} - \mathbf{j}) = 6\mathbf{i} - 2\mathbf{j} - 2\mathbf{k}$$
$$\|\mathbf{i} + 2\mathbf{j} - 3\mathbf{k}\| = \sqrt{1^2 + 2^2 + (-3)^2} = \sqrt{14} \quad \blacktriangleleft$$

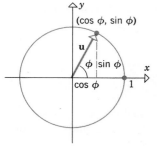

Figure 14.2.14

Example 8 If **u** is a unit vector in 2-space and if ϕ is the angle from the positive x-axis to **u**, then

$$\mathbf{u} = \langle \cos\phi, \sin\phi \rangle = (\cos\phi)\mathbf{i} + (\sin\phi)\mathbf{j}$$

(Figure 14.2.14); and if **v** is an arbitrary vector making an angle ϕ with the positive x-axis, then **v** has the same direction as **u**, but is $\|\mathbf{v}\|$ times as long. Thus,

$$\mathbf{v} = \|\mathbf{v}\| \langle \cos\phi, \sin\phi \rangle \quad \text{or equivalently} \quad \mathbf{v} = \|\mathbf{v}\| \cos\phi\,\mathbf{i} + \|\mathbf{v}\| \sin\phi\,\mathbf{j}$$

For example, the vector of length 2 making an angle of $\pi/4$ with the positive x-axis is

$$\mathbf{v} = 2\cos\frac{\pi}{4}\mathbf{i} + 2\sin\frac{\pi}{4}\mathbf{j} = \sqrt{2}\mathbf{i} + \sqrt{2}\mathbf{j} \quad \blacktriangleleft$$

▶ Exercise Set 14.2

In Exercises 1–4, sketch the vectors with the initial point at the origin.

1. (a) $\langle 2, 5 \rangle$
(b) $\langle -5, -4 \rangle$
(c) $\langle 2, 0 \rangle$.

2. (a) $\langle -3, 7 \rangle$
(b) $\langle 6, -2 \rangle$
(c) $\langle 0, -8 \rangle$.

3. (a) $\langle 1, -2, 2 \rangle$
(b) $\langle 2, 2, -1 \rangle$.

4. (a) $\langle -1, 3, 2 \rangle$
(b) $\langle 3, 4, 2 \rangle$.

In Exercises 5–8, sketch the vectors with the initial point at the origin.

5. (a) $-5\mathbf{i} + 3\mathbf{j}$
(b) $3\mathbf{i} - 2\mathbf{j}$
(c) $-6\mathbf{j}$.

6. (a) $4\mathbf{i} + 2\mathbf{j}$
(b) $-2\mathbf{i} - \mathbf{j}$
(c) $4\mathbf{i}$.

7. (a) $-\mathbf{i} + 2\mathbf{j} + 3\mathbf{k}$
(b) $2\mathbf{i} + 3\mathbf{j} - \mathbf{k}$.

8. (a) $2\mathbf{j} - \mathbf{k}$
(b) $\mathbf{i} - \mathbf{j} + 2\mathbf{k}$.

In Exercises 9–12, find the components of the vector $\overrightarrow{P_1P_2}$.

9. (a) $P_1(3, 5)$, $P_2(2, 8)$
(b) $P_1(7, -2)$, $P_2(0, 0)$
(c) $P_1(-6, -2)$, $P_2(-4, -1)$
(d) $P_1(0, 0)$, $P_2(-8, 7)$.

10. (a) $P_1(1, 3)$, $P_2(4, 1)$
(b) $P_1(6, -4)$, $P_2(0, 0)$
(c) $P_1(-8, -1)$, $P_2(-3, -2)$
(d) $P_1(0, 0)$, $P_2(-3, -5)$.

11. (a) $P_1(5, -2, 1)$, $P_2(2, 4, 2)$
(b) $P_1(-1, 3, 5)$, $P_2(0, 0, 0)$.

12. (a) $P_1(0, 0, 0)$, $P_2(-1, 6, 1)$
(b) $P_1(4, 1, -3)$, $P_2(9, 1, -3)$.

13. Find the terminal point of $\mathbf{v} = 3\mathbf{i} - 2\mathbf{j}$ if the initial point is $(1, -2)$.

14. Find the terminal point of $\mathbf{v} = \langle 7, 6 \rangle$ if the initial point is $(2, -1)$.

15. Find the initial point of $\mathbf{v} = \langle -2, 4 \rangle$ if the terminal point is $(2, 0)$.

16. Find the terminal point of $\mathbf{v} = \mathbf{i} + 2\mathbf{j} - 3\mathbf{k}$ if the initial point is $(-2, 1, 4)$.

17. Find the initial point of $\mathbf{v} = \langle -3, 1, 2 \rangle$ if the terminal point is $(5, 0, -1)$.

18. Let $\mathbf{u} = \langle 1, 3 \rangle$, $\mathbf{v} = \langle 2, 1 \rangle$, and $\mathbf{w} = \langle 4, -1 \rangle$. Find
(a) $\mathbf{u} - \mathbf{w}$
(b) $7\mathbf{v} + 3\mathbf{w}$
(c) $-\mathbf{w} + \mathbf{v}$
(d) $3(\mathbf{u} - 7\mathbf{v})$
(e) $-3\mathbf{v} - 8\mathbf{w}$
(f) $2\mathbf{v} - (\mathbf{u} + \mathbf{w})$.

19. Let $\mathbf{u} = 2\mathbf{i} + 3\mathbf{j}$, $\mathbf{v} = 4\mathbf{i}$, $\mathbf{w} = -\mathbf{i} - 2\mathbf{j}$. Find
(a) $\mathbf{w} - \mathbf{v}$
(b) $6\mathbf{u} + 4\mathbf{w}$
(c) $-\mathbf{v} - 2\mathbf{w}$
(d) $4(3\mathbf{u} + \mathbf{v})$
(e) $-8(\mathbf{v} + \mathbf{w}) + 2\mathbf{u}$
(f) $3\mathbf{w} - (\mathbf{v} - \mathbf{w})$.

20. Let $\mathbf{u} = \langle 2, -1, 3 \rangle$, $\mathbf{v} = \langle 4, 0, -2 \rangle$, $\mathbf{w} = \langle 1, 1, 3 \rangle$. Find
(a) $\mathbf{u} - \mathbf{w}$
(b) $7\mathbf{v} + 3\mathbf{w}$
(c) $-\mathbf{w} + \mathbf{v}$
(d) $3(\mathbf{u} - 7\mathbf{v})$
(e) $-3\mathbf{v} - 8\mathbf{w}$
(f) $2\mathbf{v} - (\mathbf{u} + \mathbf{w})$.

21. Let $\mathbf{u} = 3\mathbf{i} - \mathbf{k}$, $\mathbf{v} = \mathbf{i} - \mathbf{j} + 2\mathbf{k}$, $\mathbf{w} = 3\mathbf{j}$. Find
(a) $\mathbf{w} - \mathbf{v}$
(b) $6\mathbf{u} + 4\mathbf{w}$
(c) $-\mathbf{v} - 2\mathbf{w}$
(d) $4(3\mathbf{u} + \mathbf{v})$
(e) $-8(\mathbf{v} + \mathbf{w}) + 2\mathbf{u}$
(f) $3\mathbf{w} - (\mathbf{v} - \mathbf{w})$.

In Exercises 22–25, compute the norm of **v**.

22. (a) $\mathbf{v} = \langle 3, 4 \rangle$ (b) $\mathbf{v} = -\mathbf{i} + 7\mathbf{j}$
(c) $\mathbf{v} = -3\mathbf{j}$.

23. (a) $\mathbf{v} = \langle 1, -1 \rangle$ (b) $\mathbf{v} = \langle 2, 0 \rangle$
(c) $\mathbf{v} = \sqrt{2}\mathbf{i} - \sqrt{7}\mathbf{j}$.

24. (a) $\mathbf{v} = \mathbf{i} + \mathbf{j} + \mathbf{k}$ (b) $\mathbf{v} = \langle -1, 2, 4 \rangle$.

25. (a) $\mathbf{v} = -3\mathbf{i} + 2\mathbf{j} + \mathbf{k}$ (b) $\mathbf{v} = \langle 0, -3, 0 \rangle$.

26. Let $\mathbf{u} = \langle 1, -3 \rangle$, $\mathbf{v} = \langle 1, 1 \rangle$, and $\mathbf{w} = \langle 2, -4 \rangle$. Find
(a) $\|\mathbf{u} + \mathbf{v}\|$ (b) $\|\mathbf{u}\| + \|\mathbf{v}\|$
(c) $\|-2\mathbf{u}\| + 2\|\mathbf{v}\|$ (d) $\|3\mathbf{u} - 5\mathbf{v} + \mathbf{w}\|$.

27. Let $\mathbf{u} = 2\mathbf{i} - 5\mathbf{j}$, $\mathbf{v} = 2\mathbf{i}$, and $\mathbf{w} = 3\mathbf{i} + 4\mathbf{j}$. Find
(a) $\|\mathbf{v} + \mathbf{w}\|$ (b) $\|\mathbf{v}\| + \|\mathbf{w}\|$
(c) $\|-3\mathbf{u}\| + 4\|\mathbf{v}\|$ (d) $\|\mathbf{u} - \mathbf{v} - \mathbf{w}\|$
(e) $\dfrac{1}{\|\mathbf{w}\|}\mathbf{w}$ (f) $\left\| \dfrac{1}{\|\mathbf{w}\|}\mathbf{w} \right\|$.

28. Let $\mathbf{u} = \langle 2, -1, 0 \rangle$ and $\mathbf{v} = \langle 0, 1, -1 \rangle$. Find
(a) $\|\mathbf{u} + \mathbf{v}\|$ (b) $\|\mathbf{u}\| + \|\mathbf{v}\|$
(c) $\|3\mathbf{u}\|$ (d) $\|2\mathbf{u} - 3\mathbf{v}\|$.

29. Let $\mathbf{u} = \mathbf{i} - 3\mathbf{j} + 2\mathbf{k}$, $\mathbf{v} = \mathbf{i} + \mathbf{j}$, and $\mathbf{w} = 2\mathbf{i} + 2\mathbf{j} - 4\mathbf{k}$. Find
(a) $\|\mathbf{u} + \mathbf{v}\|$ (b) $\|\mathbf{u}\| + \|\mathbf{v}\|$
(c) $\|-2\mathbf{u}\| + 2\|\mathbf{v}\|$ (d) $\|3\mathbf{u} - 5\mathbf{v} + \mathbf{w}\|$
(e) $\dfrac{1}{\|\mathbf{w}\|}\mathbf{w}$ (f) $\left\| \dfrac{1}{\|\mathbf{w}\|}\mathbf{w} \right\|$.

30. Let $\mathbf{u} = \langle -1, 1 \rangle$, $\mathbf{v} = \langle 0, 1 \rangle$, and $\mathbf{w} = \langle 3, 4 \rangle$. Find the vector $\mathbf{x}$ that satisfies $\mathbf{u} - 2\mathbf{x} = \mathbf{x} - \mathbf{w} + 3\mathbf{v}$.

31. Let $\mathbf{u} = \langle 1, 3 \rangle$, $\mathbf{v} = \langle 2, 1 \rangle$, $\mathbf{w} = \langle 4, -1 \rangle$. Find the vector $\mathbf{x}$ that satisfies $2\mathbf{u} - \mathbf{v} + \mathbf{x} = 7\mathbf{x} + \mathbf{w}$.

32. Find $\mathbf{u}$ and $\mathbf{v}$ if $\mathbf{u} + \mathbf{v} = \langle 2, -3 \rangle$ and $3\mathbf{u} + 2\mathbf{v} = \langle -1, 2 \rangle$.

33. Find $\mathbf{u}$ and $\mathbf{v}$ if $\mathbf{u} + 2\mathbf{v} = 3\mathbf{i} - \mathbf{k}$ and $3\mathbf{u} - \mathbf{v} = \mathbf{i} + \mathbf{j} + \mathbf{k}$.

34. Give a geometric argument to show that if $\mathbf{u}$ and $\mathbf{v}$ are nonzero and $\mathbf{u}$ is not parallel to $\mathbf{v}$, then any vector $\mathbf{w}$ in the plane of $\mathbf{u}$ and $\mathbf{v}$ can be written as $\mathbf{w} = c_1\mathbf{u} + c_2\mathbf{v}$ for a suitable choice of c_1 and c_2.

35. Give a geometric argument to show that if $\mathbf{u}$, $\mathbf{v}$, and $\mathbf{w}$ are not coplanar, then any vector $\mathbf{z}$ can be written as $\mathbf{z} = c_1\mathbf{u} + c_2\mathbf{v} + c_3\mathbf{w}$ for a suitable choice of scalars c_1, c_2, and c_3.

36. Let $\mathbf{u} = 2\mathbf{i} - \mathbf{j}$ and $\mathbf{v} = 4\mathbf{i} + 2\mathbf{j}$. Find scalars c_1 and c_2 such that $c_1\mathbf{u} + c_2\mathbf{v} = -4\mathbf{j}$.

37. Let $\mathbf{u} = \langle 1, -3 \rangle$ and $\mathbf{v} = \langle -2, 6 \rangle$. Show that there do not exist scalars c_1 and c_2 such that $c_1\mathbf{u} + c_2\mathbf{v} = \langle 3, 5 \rangle$.

38. Let $\mathbf{u} = \langle 1, 0, 1 \rangle$, $\mathbf{v} = \langle 3, 2, 0 \rangle$, and $\mathbf{w} = \langle 0, 1, 1 \rangle$. Find scalars c_1, c_2, and c_3 such that $c_1\mathbf{u} + c_2\mathbf{v} + c_3\mathbf{w} = \langle -1, 1, 5 \rangle$.

39. Let $\mathbf{u} = \mathbf{i} - \mathbf{j}$, $\mathbf{v} = 3\mathbf{i} + \mathbf{k}$, and $\mathbf{w} = 4\mathbf{i} - \mathbf{j} + \mathbf{k}$. Show that there do not exist scalars c_1, c_2, and c_3 such that $c_1\mathbf{u} + c_2\mathbf{v} + c_3\mathbf{w} = 2\mathbf{i} + \mathbf{j} - \mathbf{k}$.

40. Let $\mathbf{v} = 4\mathbf{i} - 3\mathbf{j}$. Find all scalars k such that $\|k\mathbf{v}\| = 3$.

41. Verify parts (b), (e), (f), and (g) of Theorem 14.2.6 for $\mathbf{u} = \langle 1, -3 \rangle$, $\mathbf{v} = \langle 6, 6 \rangle$, $\mathbf{w} = \langle -8, 1 \rangle$, $k = 3$, and $l = 6$.

42. Find a unit vector having the same direction as $-\mathbf{i} + 4\mathbf{j}$.

43. Find a unit vector oppositely directed to $3\mathbf{i} - 4\mathbf{j}$.

44. Find a unit vector having the same direction as $2\mathbf{i} - \mathbf{j} - 2\mathbf{k}$.

45. Find a unit vector oppositely directed to $6\mathbf{i} - 4\mathbf{j} + 2\mathbf{k}$.

46. Find a unit vector having the same direction as the vector from the point $A(-3, 2)$ to the point $B(1, -1)$.

47. Find a unit vector having the same direction as the vector from the point $A(-1, 0, 2)$ to the point $B(3, 1, 1)$.

48. Find a vector having the same direction as the vector $\mathbf{v} = -2\mathbf{i} + 3\mathbf{j}$ but with three times the length of $\mathbf{v}$.

49. Find a vector oppositely directed to $\mathbf{v} = \langle 3, -4 \rangle$ but with half the length of $\mathbf{v}$.

50. Find a vector with the same direction as $\mathbf{v} = \langle 7, 0, -6 \rangle$ but with twice the length of $\mathbf{v}$.

51. Find a vector oppositely directed to $\mathbf{v} = -3\mathbf{i} + 4\mathbf{j} + \mathbf{k}$ but with twice the length of $\mathbf{v}$.

52. Let $\mathbf{r} = \langle x, y \rangle$. Describe the set of points (x, y) for which $\|\mathbf{r}\| = 1$.

53. Let $\mathbf{r}_0 = \langle x_0, y_0 \rangle$ and $\mathbf{r} = \langle x, y \rangle$. Describe the set of all points (x, y) for which $\|\mathbf{r} - \mathbf{r}_0\| = 1$.

54. Let $\mathbf{r}_1 = \langle x_1, y_1 \rangle$, $\mathbf{r}_2 = \langle x_2, y_2 \rangle$, and $\mathbf{r} = \langle x, y \rangle$. Describe the set of all points (x, y) for which $\|\mathbf{r} - \mathbf{r}_1\| + \|\mathbf{r} - \mathbf{r}_2\| = k$, where $k > \|\mathbf{r}_2 - \mathbf{r}_1\|$.

55. Let $\mathbf{r}_0 = \langle x_0, y_0, z_0 \rangle$ and $\mathbf{r} = \langle x, y, z \rangle$. Describe the set of all points (x, y, z) for which
(a) $\|\mathbf{r}\| = 2$ (b) $\|\mathbf{r} - \mathbf{r}_0\| = 3$
(c) $\|\mathbf{r} - \mathbf{r}_0\| \leq 1$.

56. Find two unit vectors in 2-space parallel to the line $y = 3x + 2$.

57. (a) Find two unit vectors in 2-space parallel to the line $x + y = 4$.
(b) Find two unit vectors in 2-space perpendicular to the line in part (a).

58. Let P be the point $(2, 3)$ and Q the point $(7, -4)$. Use vectors to find the point on the line segment joining P and Q that is $\frac{3}{4}$ of the way from P to Q.

59. For the points P and Q in Exercise 58, use vectors to find the point on the line segment joining P and Q that is $\frac{3}{4}$ of the way from Q to P.

60. Find a unit vector in 2-space making an angle of $135°$ with the x-axis.

61. (a) Find a unit vector in 2-space making an angle of $\pi/3$ with the positive x-axis.
(b) Find a vector of length 4 in 2-space making an angle of $3\pi/4$ with the positive x-axis.

62. Use vectors to find the length of the diagonal of the parallelogram determined by $\mathbf{i} + \mathbf{j}$ and $\mathbf{i} - 2\mathbf{j}$.

63. Use vectors to find the fourth vertex of a parallelogram, three of whose vertices are $(0, 0)$, $(1, 3)$, and $(2, 4)$. [*Note:* There is more than one answer.]

64. Prove: $\|\mathbf{u} + \mathbf{v}\| \leq \|\mathbf{u}\| + \|\mathbf{v}\|$ geometrically.

65. Prove parts (*a*), (*c*), and (*e*) of Theorem 14.2.6 analytically in 2-space.

66. Prove parts (*d*), (*g*), and (*h*) of Theorem 14.2.6 analytically in 2-space.

67. Prove part (*f*) of Theorem 14.2.6 geometrically.

68. Use vectors to prove that the midpoints of the sides of a quadrilateral are the vertices of a parallelogram.

69. Use vectors to prove that the line segment joining the midpoints of two sides of a triangle is parallel to the third side and half as long.

70. Find $\overrightarrow{AB} + \overrightarrow{BC} + \overrightarrow{CA}$, where A, B, and C are the vertices of a triangle.

In Exercises 71–73, let A, B, C, and D be any four points in 3-space.

71. If M is the midpoint of BC, show that $\overrightarrow{AB} + \overrightarrow{AC} = 2\overrightarrow{AM}$.

72. If M and N are the midpoints of AC and BD, show that $\overrightarrow{AB} + \overrightarrow{CD} = 2\overrightarrow{MN}$.

73. If M and N are the midpoints of AC and BD, show that $\overrightarrow{AB} + \overrightarrow{AD} + \overrightarrow{CB} + \overrightarrow{CD} = 4\overrightarrow{MN}$.

74. Let A and B be distinct points on a straight line L. If a point P different from B is on L, then $\overrightarrow{AP} = t\overrightarrow{PB}$ for some value of the scalar t. Let $\mathbf{a}$, $\mathbf{b}$, and $\mathbf{r}$ be vectors from the origin to the points A, B, and P, respectively. Show that $\mathbf{r} = (\mathbf{a} + t\mathbf{b})/(1 + t)$.

75. Let $\mathbf{r}_1, \mathbf{r}_2, \ldots, \mathbf{r}_n$ be vectors from the origin to points $P_1, P_2, \ldots, P_n$, respectively. The *centroid* of points $P_1, P_2, \ldots, P_n$ is defined as the point P for which $\sum_{k=1}^{n} \overrightarrow{PP_k} = \mathbf{0}$. Let $\mathbf{r}$ be the vector from the origin to P. Show that $\mathbf{r} = \dfrac{1}{n} \sum_{k=1}^{n} \mathbf{r}_k$. [*Hint:* Write $\overrightarrow{PP_k}$ as a difference of vectors $\mathbf{r}$ and $\mathbf{r}_k$.]

76. Let $P_1, P_2, \ldots, P_n$ be consecutive vertices of a polygon in 2-space, all of whose interior angles are less than π. Let $\mathbf{r}_k$ be the vector from the origin to the point P_k for $k = 1, 2, \ldots, n$. From Exercise 75, the endpoint of the vector

$$\mathbf{r} = \frac{1}{n} \sum_{k=1}^{n} \mathbf{r}_k$$

drawn from the origin is the centroid of the points $P_1, P_2, \ldots, P_n$. It can be shown that the centroid is the balance point of the polygon.
(a) Find the centroid of the triangle with vertices $P_1(1, 1)$, $P_2(3, 3)$, and $P_3(5, 0)$.
(b) Cut the polygon described in part (a) out of cardboard and show that it balances when the tip of a pencil is placed at the centroid.

77. Follow the directions of Exercise 76 for the quadrilateral with vertices $P_1(-1, 2)$, $P_2(2, 3)$, $P_3(5, -2)$, and $P_4(0, -1)$.

■ **14.3** DOT PRODUCT; PROJECTIONS

> *In this section we shall introduce a type of multiplication of vectors in 2-space and 3-space. We shall also discuss the arithmetic properties of this multiplication and give some of its applications.*

□ **ANGLE BETWEEN VECTORS**

Let **u** and **v** be two nonzero vectors in 2-space or 3-space, and assume these vectors have been positioned so that their initial points coincide. By the ***angle between* u *and* v,** we shall mean the angle θ determined by **u** and **v** that satisfies $0 \leq \theta \leq \pi$ (Figure 14.3.1).

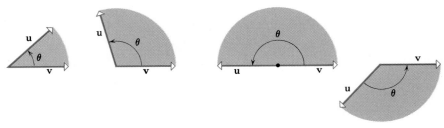

Figure 14.3.1

> **14.3.1** DEFINITION. If **u** and **v** are vectors in 2-space or 3-space and θ is the angle between **u** and **v**, then the ***dot product*** or ***Euclidean inner product*** **u · v** is defined by
>
> $$\mathbf{u} \cdot \mathbf{v} = \begin{cases} \|\mathbf{u}\| \, \|\mathbf{v}\| \cos \theta, & \text{if } \mathbf{u} \neq \mathbf{0} \text{ and } \mathbf{v} \neq \mathbf{0} \\ 0, & \text{if } \mathbf{u} = \mathbf{0} \text{ or } \mathbf{v} = \mathbf{0} \end{cases}$$

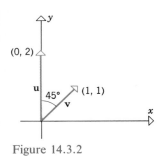

Figure 14.3.2

Example 1 As shown in Figure 14.3.2, the angle between vectors $\mathbf{u} = \langle 0, 2 \rangle$ and $\mathbf{v} = \langle 1, 1 \rangle$ is 45°. Thus,

$$\mathbf{u} \cdot \mathbf{v} = \|\mathbf{u}\| \, \|\mathbf{v}\| \cos \theta = \sqrt{0^2 + 2^2} \sqrt{1^2 + 1^2} \cos 45° = (2)(\sqrt{2}) \frac{1}{\sqrt{2}} = 2 \quad \blacktriangleleft$$

□ **FORMULA FOR THE DOT PRODUCT**

For purposes of computation, it is desirable to have a formula that expresses the dot product of two vectors in terms of the components of the vectors. We shall derive such a formula for vectors in 3-space and just state the corresponding formula for vectors in 2-space.

Let $\mathbf{u} = \langle u_1, u_2, u_3 \rangle$ and $\mathbf{v} = \langle v_1, v_2, v_3 \rangle$ be two nonzero vectors. If, as in Figure 14.3.3, θ is the angle between **u** and **v**, then the law of cosines yields

$$\|\overrightarrow{PQ}\|^2 = \|\mathbf{u}\|^2 + \|\mathbf{v}\|^2 - 2\|\mathbf{u}\| \, \|\mathbf{v}\| \cos \theta \tag{1}$$

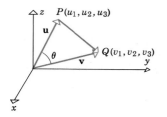

Figure 14.3.3

Since $\overrightarrow{PQ} = \mathbf{v} - \mathbf{u}$, we can rewrite (1) as

$$\|\mathbf{u}\| \, \|\mathbf{v}\| \cos \theta = \tfrac{1}{2}(\|\mathbf{u}\|^2 + \|\mathbf{v}\|^2 - \|\mathbf{v} - \mathbf{u}\|^2)$$

or

$$\mathbf{u} \cdot \mathbf{v} = \tfrac{1}{2}(\|\mathbf{u}\|^2 + \|\mathbf{v}\|^2 - \|\mathbf{v} - \mathbf{u}\|^2)$$

Substituting

$$\|\mathbf{u}\|^2 = u_1^2 + u_2^2 + u_3^2, \quad \|\mathbf{v}\|^2 = v_1^2 + v_2^2 + v_3^2$$

and

$$\|\mathbf{v} - \mathbf{u}\|^2 = (v_1 - u_1)^2 + (v_2 - u_2)^2 + (v_3 - u_3)^2$$

we obtain, after simplifying,

$$\mathbf{u} \cdot \mathbf{v} = u_1 v_1 + u_2 v_2 + u_3 v_3 \tag{2a}$$

If $\mathbf{u} = \langle u_1, u_2 \rangle$ and $\mathbf{v} = \langle v_1, v_2 \rangle$ are two vectors in 2-space, then the formula corresponding to (2a) is

$$\mathbf{u} \cdot \mathbf{v} = u_1 v_1 + u_2 v_2 \tag{2b}$$

If $\mathbf{u}$ and $\mathbf{v}$ are nonzero vectors, the formula in Definition 14.3.1 can be written as

$$\cos \theta = \frac{\mathbf{u} \cdot \mathbf{v}}{\|\mathbf{u}\| \, \|\mathbf{v}\|} \tag{3}$$

Example 2   Consider the vectors

$$\mathbf{u} = 2\mathbf{i} - \mathbf{j} + \mathbf{k} \quad \text{and} \quad \mathbf{v} = \mathbf{i} + \mathbf{j} + 2\mathbf{k}$$

Find $\mathbf{u} \cdot \mathbf{v}$ and determine the angle θ between $\mathbf{u}$ and $\mathbf{v}$.

Solution.

$$\mathbf{u} \cdot \mathbf{v} = u_1 v_1 + u_2 v_2 + u_3 v_3 = (2)(1) + (-1)(1) + (1)(2) = 3$$

For the given vectors, $\|\mathbf{u}\| = \|\mathbf{v}\| = \sqrt{6}$, so that

$$\cos \theta = \frac{3}{\sqrt{6}\sqrt{6}} = \frac{1}{2}$$

Thus, $\theta = 60°$ ◀

Example 3   Find the angle between a diagonal of a cube and one of its edges.

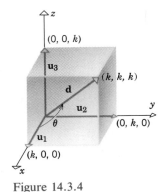

Figure 14.3.4

Solution.   Let k be the length of an edge and let us introduce a coordinate system as shown in Figure 14.3.4.

If we let $\mathbf{u}_1 = \langle k, 0, 0\rangle$, $\mathbf{u}_2 = \langle 0, k, 0\rangle$, and $\mathbf{u}_3 = \langle 0, 0, k\rangle$, then the vector

$$\mathbf{d} = \langle k, k, k\rangle = \mathbf{u}_1 + \mathbf{u}_2 + \mathbf{u}_3$$

is a diagonal of the cube. The angle θ between $\mathbf{d}$ and the edge $\mathbf{u}_1$ satisfies

$$\cos\theta = \frac{\mathbf{u}_1 \cdot \mathbf{d}}{\|\mathbf{u}_1\|\,\|\mathbf{d}\|} = \frac{k^2}{(k)(\sqrt{3k^2})} = \frac{1}{\sqrt{3}}$$

Thus, with the help of a calculator

$$\theta = \cos^{-1}\frac{1}{\sqrt{3}} \approx 54°44' \quad \blacktriangleleft$$

The sign of the dot product provides useful information about the angle between two vectors.

14.3.2   THEOREM.   *If* $\mathbf{u}$ *and* $\mathbf{v}$ *are nonzero vectors in 2-space or 3-space, and if* θ *is the angle between them, then*

θ *is acute*	*if and only if* $\mathbf{u} \cdot \mathbf{v} > 0$
θ *is obtuse*	*if and only if* $\mathbf{u} \cdot \mathbf{v} < 0$
$\theta = \pi/2$	*if and only if* $\mathbf{u} \cdot \mathbf{v} = 0$

Proof.   Since $\mathbf{u}$ and $\mathbf{v}$ are nonzero vectors, $\|\mathbf{u}\| > 0$ and $\|\mathbf{v}\| > 0$. Thus,

$$\mathbf{u} \cdot \mathbf{v} = \|\mathbf{u}\|\,\|\mathbf{v}\|\cos\theta$$

is positive, negative, or zero according to whether $\cos\theta$ is positive, negative, or zero. Since $0 \le \theta \le \pi$, it follows that θ is acute if and only if $\cos\theta > 0$; θ is obtuse if and only if $\cos\theta < 0$; and $\theta = \pi/2$ if and only if $\cos\theta = 0$.   ∎

Example 4   If $\mathbf{u} = \mathbf{i} - 2\mathbf{j} + 3\mathbf{k}$, $\mathbf{v} = -3\mathbf{i} + 4\mathbf{j} + 2\mathbf{k}$, and $\mathbf{w} = 3\mathbf{i} + 6\mathbf{j} + 3\mathbf{k}$, then

$$\mathbf{u} \cdot \mathbf{v} = (1)(-3) + (-2)(4) + (3)(2) = -5$$
$$\mathbf{v} \cdot \mathbf{w} = (-3)(3) + (4)(6) + (2)(3) = 21$$
$$\mathbf{u} \cdot \mathbf{w} = (1)(3) + (-2)(6) + (3)(3) = 0$$

Therefore $\mathbf{u}$ and $\mathbf{v}$ make an obtuse angle, $\mathbf{v}$ and $\mathbf{w}$ make an acute angle, and $\mathbf{u}$ and $\mathbf{w}$ are perpendicular.   ◀

□ ORTHOGONAL VECTORS   Perpendicular vectors are also called *orthogonal* vectors. In light of Theorem 14.3.2, two nonzero vectors are orthogonal if and only if their dot product is zero. If we agree to consider $\mathbf{u}$ and $\mathbf{v}$ to be perpendicular when either or both of these vectors is $\mathbf{0}$, then we can state without exception that two vectors $\mathbf{u}$ and $\mathbf{v}$ are orthogonal (perpendicular) if and only if $\mathbf{u} \cdot \mathbf{v} = 0$.

Example 5   Show that in 2-space the vector $a\mathbf{i} + b\mathbf{j}$ is perpendicular to the line $ax + by + c = 0$.

Solution.   Let $P_1(x_1, y_1)$ and $P_2(x_2, y_2)$ be distinct points on the line so that

$$ax_1 + by_1 + c = 0$$
$$ax_2 + by_2 + c = 0 \tag{4}$$

Since the vector $\overrightarrow{P_1P_2} = (x_2 - x_1)\mathbf{i} + (y_2 - y_1)\mathbf{j}$ runs along the line, we need only show that $a\mathbf{i} + b\mathbf{j}$ and $\overrightarrow{P_1P_2}$ are perpendicular. But on subtracting the equations in (4), we obtain

$$a(x_2 - x_1) + b(y_2 - y_1) = 0$$

which can be expressed in the form

$$(a\mathbf{i} + b\mathbf{j}) \cdot [(x_2 - x_1)\mathbf{i} + (y_2 - y_1)\mathbf{j}] = 0$$

or

$$(a\mathbf{i} + b\mathbf{j}) \cdot \overrightarrow{P_1P_2} = 0$$

so that $a\mathbf{i} + b\mathbf{j}$ and $\overrightarrow{P_1P_2}$ are perpendicular.   ◀

□ **DIRECTION COSINES**

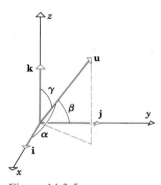

Figure 14.3.5

Of special interest are the angles α, β, and γ that a vector $\mathbf{u}$ in 3-space makes with the vectors $\mathbf{i}$, $\mathbf{j}$, and $\mathbf{k}$ (Figure 14.3.5). These are called the *direction angles* of $\mathbf{u}$. The numbers $\cos \alpha$, $\cos \beta$, and $\cos \gamma$ are called the *direction cosines* of $\mathbf{u}$. Formulas for the direction cosines follow easily from (3).

14.3.3   THEOREM.   *The three direction cosines of a nonzero vector* $\mathbf{u} = u_1\mathbf{i} + u_2\mathbf{j} + u_3\mathbf{k}$ *in 3-space are*

$$\cos \alpha = \frac{u_1}{\|\mathbf{u}\|}, \quad \cos \beta = \frac{u_2}{\|\mathbf{u}\|}, \quad \cos \gamma = \frac{u_3}{\|\mathbf{u}\|}$$

Proof.   Since $\mathbf{u} \cdot \mathbf{i} = (u_1)(1) + (u_2)(0) + (u_3)(0) = u_1$, it follows that

$$\cos \alpha = \frac{\mathbf{u} \cdot \mathbf{i}}{\|\mathbf{u}\| \|\mathbf{i}\|} = \frac{u_1}{\|\mathbf{u}\|}$$

Similarly for $\cos \beta$ and $\cos \gamma$.   ∎

The direction cosines of a vector $\mathbf{u} = u_1\mathbf{i} + u_2\mathbf{j} + u_3\mathbf{k}$ can be obtained by simply reading off the components of the unit vector $\mathbf{u}/\|\mathbf{u}\|$ since

$$\frac{\mathbf{u}}{\|\mathbf{u}\|} = \frac{u_1}{\|\mathbf{u}\|}\mathbf{i} + \frac{u_2}{\|\mathbf{u}\|}\mathbf{j} + \frac{u_3}{\|\mathbf{u}\|}\mathbf{k} = (\cos \alpha)\mathbf{i} + (\cos \beta)\mathbf{j} + (\cos \gamma)\mathbf{k}$$

Example 6   Find the direction cosines of the vector $\mathbf{u} = 2\mathbf{i} - 4\mathbf{j} + 4\mathbf{k}$, and estimate the direction angles to the nearest degree.

Solution. $\|\mathbf{u}\| = \sqrt{4 + 16 + 16} = 6$, so that $\mathbf{u}/\|\mathbf{u}\| = \frac{1}{3}\mathbf{i} - \frac{2}{3}\mathbf{j} + \frac{2}{3}\mathbf{k}$. Thus,

$$\cos\alpha = \frac{1}{3}, \quad \cos\beta = -\frac{2}{3}, \quad \cos\gamma = \frac{2}{3}$$

With the help of a calculator that can compute inverse trigonometric functions, one obtains

$$\alpha = \cos^{-1}\left(\frac{1}{3}\right) \approx 71°, \quad \beta = \cos^{-1}\left(-\frac{2}{3}\right) \approx 132°, \quad \gamma = \cos^{-1}\left(\frac{2}{3}\right) \approx 48°$$

◄

☐ **PROPERTIES OF THE DOT PRODUCT**

The following arithmetic properties of the dot product are useful in calculations involving vectors.

14.3.4 THEOREM. *If* $\mathbf{u}$, $\mathbf{v}$, *and* $\mathbf{w}$ *are vectors in 2- or 3-space and k is a scalar, then*

(a) $\mathbf{u} \cdot \mathbf{v} = \mathbf{v} \cdot \mathbf{u}$

(b) $\mathbf{u} \cdot (\mathbf{v} + \mathbf{w}) = \mathbf{u} \cdot \mathbf{v} + \mathbf{u} \cdot \mathbf{w}$

(c) $k(\mathbf{u} \cdot \mathbf{v}) = (k\mathbf{u}) \cdot \mathbf{v} = \mathbf{u} \cdot (k\mathbf{v})$

(d) $\mathbf{v} \cdot \mathbf{v} = \|\mathbf{v}\|^2$

We shall prove parts (c) and (d) for vectors in 3-space and omit the remaining proofs.

Proof (c). Let $\mathbf{u} = \langle u_1, u_2, u_3 \rangle$ and $\mathbf{v} = \langle v_1, v_2, v_3 \rangle$; then

$$k(\mathbf{u} \cdot \mathbf{v}) = k(u_1 v_1 + u_2 v_2 + u_3 v_3) = (ku_1)v_1 + (ku_2)v_2 + (ku_3)v_3 = (k\mathbf{u}) \cdot \mathbf{v}$$

Similarly, $k(\mathbf{u} \cdot \mathbf{v}) = \mathbf{u} \cdot (k\mathbf{v})$.

Proof (d). $\mathbf{v} \cdot \mathbf{v} = v_1 v_1 + v_2 v_2 + v_3 v_3 = v_1{}^2 + v_2{}^2 + v_3{}^2 = \|\mathbf{v}\|^2$. ∎

Part (d) of the last theorem is sometimes expressed in the following alternative form:

$$\|\mathbf{v}\| = \sqrt{\mathbf{v} \cdot \mathbf{v}} \tag{5}$$

☐ **ORTHOGONAL PROJECTIONS OF VECTORS**

In many applications it is of interest to "decompose" a vector $\mathbf{u}$ into a sum of two terms, one parallel to a specified nonzero vector $\mathbf{b}$ and the other perpendicular to $\mathbf{b}$. If $\mathbf{u}$ and $\mathbf{b}$ are positioned so that their initial points coincide at a point Q, we may decompose the vector $\mathbf{u}$ as follows (Figure 14.3.6): Drop a perpendicular from the tip of $\mathbf{u}$ to the line through $\mathbf{b}$ and construct the vector $\mathbf{w}_1$ from Q to the foot of this perpendicular; next, form the difference

$$\mathbf{w}_2 = \mathbf{u} - \mathbf{w}_1$$

As indicated in Figure 14.3.6, the vector $\mathbf{w}_1$ is parallel to $\mathbf{b}$, the vector $\mathbf{w}_2$ is

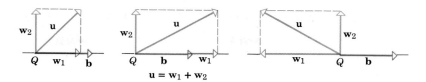

$$u = w_1 + w_2$$

Figure 14.3.6

perpendicular to **b**, and

$$w_1 + w_2 = w_1 + (u - w_1) = u$$

The vector w_1 is called the ***orthogonal projection of u on b*** or sometimes the ***vector component of u along b.*** It is denoted by

$$\text{proj}_b\, u \tag{6}$$

The vector w_2 is called the ***vector component of u orthogonal to b.*** Since $w_2 = u - w_1$, this vector may be written in notation (6) as

$$w_2 = u - \text{proj}_b\, u$$

The following theorem gives formulas for calculating the vectors $\text{proj}_b\, u$ and $u - \text{proj}_b\, u$.

14.3.5 THEOREM. *If **u** and **b** are vectors in 2-space or 3-space and if* $b \neq 0$, *then*

$$\text{proj}_b\, u = \frac{u \cdot b}{\|b\|^2}\, b \qquad (\textit{vector component of } \mathbf{u} \textit{ along } \mathbf{b})$$

$$u - \text{proj}_b\, u = u - \frac{u \cdot b}{\|b\|^2}\, b \qquad \begin{array}{l}(\textit{vector component of } \mathbf{u} \\ \textit{orthogonal to } \mathbf{b})\end{array}$$

Proof. Let $w_1 = \text{proj}_b\, u$ and $w_2 = u - \text{proj}_b\, u$. Since w_1 is parallel to **b**, it must be a scalar multiple of **b**, so it can be written in the form $w_1 = k\mathbf{b}$. Thus,

$$u = w_1 + w_2 = k\mathbf{b} + w_2 \tag{7}$$

Taking the dot product of both sides of (7) with **b** and using Theorem 14.3.4 yields

$$u \cdot b = (k\mathbf{b} + w_2) \cdot b = k\|b\|^2 + w_2 \cdot b \tag{8}$$

But $w_2 \cdot b = 0$ since w_2 is perpendicular to **b**; thus, (8) yields

$$k = \frac{u \cdot b}{\|b\|^2}$$

Since $\text{proj}_b\, u = w_1 = k\mathbf{b}$, we obtain

$$\text{proj}_b\, u = \frac{u \cdot b}{\|b\|^2}\, b \qquad \blacksquare$$

Example 7 Let $\mathbf{u} = 2\mathbf{i} - \mathbf{j} + 3\mathbf{k}$ and $\mathbf{b} = 4\mathbf{i} - \mathbf{j} + 2\mathbf{k}$. Find the vector component of $\mathbf{u}$ along $\mathbf{b}$ and the vector component of $\mathbf{u}$ orthogonal to $\mathbf{b}$.

Solution.

$$\mathbf{u} \cdot \mathbf{b} = (2)(4) + (-1)(-1) + (3)(2) = 15$$
$$\|\mathbf{b}\|^2 = 4^2 + (-1)^2 + 2^2 = 21$$

Thus, the vector component of $\mathbf{u}$ along $\mathbf{b}$ is

$$\text{proj}_\mathbf{b}\, \mathbf{u} = \frac{\mathbf{u} \cdot \mathbf{b}}{\|\mathbf{b}\|^2}\mathbf{b} = \frac{15}{21}(4\mathbf{i} - \mathbf{j} + 2\mathbf{k}) = \frac{20}{7}\mathbf{i} - \frac{5}{7}\mathbf{j} + \frac{10}{7}\mathbf{k}$$

and the vector component of $\mathbf{u}$ orthogonal to $\mathbf{b}$ is

$$\mathbf{u} - \text{proj}_\mathbf{b}\, \mathbf{u} = (2\mathbf{i} - \mathbf{j} + 3\mathbf{k}) - \left(\frac{20}{7}\mathbf{i} - \frac{5}{7}\mathbf{j} + \frac{10}{7}\mathbf{k}\right) = -\frac{6}{7}\mathbf{i} - \frac{2}{7}\mathbf{j} + \frac{11}{7}\mathbf{k}$$

As a check, the reader may wish to verify that the vectors $\mathbf{u} - \text{proj}_\mathbf{b}\, \mathbf{u}$ and $\mathbf{b}$ are perpendicular by showing that their dot product is zero. ◀

A formula for the length of the vector component of $\mathbf{u}$ along $\mathbf{b}$ may be obtained by writing

$$\|\text{proj}_\mathbf{b}\, \mathbf{u}\| = \left\|\frac{\mathbf{u} \cdot \mathbf{b}}{\|\mathbf{b}\|^2}\mathbf{b}\right\| = \left|\frac{\mathbf{u} \cdot \mathbf{b}}{\|\mathbf{b}\|^2}\right|\|\mathbf{b}\| \quad \boxed{\text{Since } \frac{\mathbf{u} \cdot \mathbf{b}}{\|\mathbf{b}\|^2} \text{ is a scalar}}$$

$$= \frac{|\mathbf{u} \cdot \mathbf{b}|}{\|\mathbf{b}\|^2}\|\mathbf{b}\| \quad \boxed{\text{Since } \|\mathbf{b}\|^2 > 0}$$

which yields

$$\|\text{proj}_\mathbf{b}\, \mathbf{u}\| = \frac{|\mathbf{u} \cdot \mathbf{b}|}{\|\mathbf{b}\|} \tag{9}$$

If θ denotes the angle between $\mathbf{u}$ and $\mathbf{b}$, then $\mathbf{u} \cdot \mathbf{b} = \|\mathbf{u}\|\,\|\mathbf{b}\| \cos\theta$, so that (9) can also be written as

$$\|\text{proj}_\mathbf{b}\, \mathbf{u}\| = \|\mathbf{u}\|\,|\cos\theta| \tag{10}$$

(Verify.) A geometric interpretation of this result is given in Figure 14.3.7.

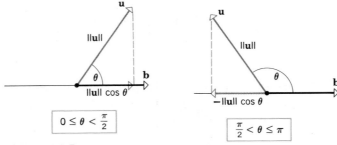

Figure 14.3.7

As an application of these formulas we shall use vector methods to derive a formula for the distance from a point in the *xy*-plane to a line in the plane.

Example 8 Find a formula for the distance D between the point $P_0(x_0, y_0)$ and the line $ax + by + c = 0$.

Solution. Let $Q(x_1, y_1)$ be any point on the line and position the vector

$$\mathbf{n} = a\mathbf{i} + b\mathbf{j}$$

so that its initial point is at Q.

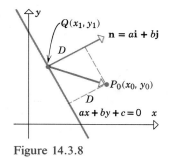

Figure 14.3.8

By virtue of Example 5, the vector $\mathbf{n}$ is perpendicular to the line (Figure 14.3.8). As indicated in the figure, the distance D is equal to the length of the orthogonal projection of the vector $\overrightarrow{QP_0}$ on $\mathbf{n}$; thus, from (9),

$$D = \|\text{proj}_\mathbf{n}\, \overrightarrow{QP_0}\| = \frac{|\overrightarrow{QP_0} \cdot \mathbf{n}|}{\|\mathbf{n}\|}$$

But

$$\overrightarrow{QP_0} = \langle x_0 - x_1, y_0 - y_1 \rangle$$
$$\overrightarrow{QP_0} \cdot \mathbf{n} = a(x_0 - x_1) + b(y_0 - y_1)$$
$$\|\mathbf{n}\| = \sqrt{a^2 + b^2}$$

so that

$$D = \frac{|a(x_0 - x_1) + b(y_0 - y_1)|}{\sqrt{a^2 + b^2}} \tag{11}$$

Since the point $Q(x_1, y_1)$ lies on the line, its coordinates satisfy the equation of the line, so

$$ax_1 + by_1 + c = 0 \qquad \text{or} \qquad c = -ax_1 - by_1$$

Substituting this expression in (11) yields the formula

$$D = \frac{|ax_0 + by_0 + c|}{\sqrt{a^2 + b^2}} \qquad \blacktriangleleft \tag{12}$$

As an illustration of Formula (12), the distance D from the point $(1, -2)$ to the line $3x + 4y - 6 = 0$ is

$$D = \frac{|(3)(1) + 4(-2) - 6|}{\sqrt{3^2 + 4^2}} = \frac{|-11|}{\sqrt{25}} = \frac{11}{5}$$

☐ **WORK**

Recall from Section 6.7 that the work W done by a constant force $\mathbf{F}$ of magnitude $\|\mathbf{F}\|$ acting in the direction of motion on a particle moving from P to Q on a line is

$$W = (\text{force}) \cdot (\text{distance}) = \|\mathbf{F}\| \, \|\overrightarrow{PQ}\|$$

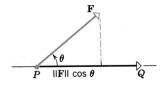

Figure 14.3.9

If the force **F** is constant, but makes an angle θ with the direction of motion (Figure 14.3.9), then we define the work done by **F** to be

$$W = \mathbf{F} \cdot \overrightarrow{PQ} = (\|\mathbf{F}\| \cos \theta) \, \|\overrightarrow{PQ}\| \qquad (13)$$

The quantity $\|\mathbf{F}\| \cos \theta$ is the "component" of force in the direction of motion and $\|\overrightarrow{PQ}\|$ is the distance traveled by the particle.

Example 9 A wagon is pulled horizontally by exerting a continual force of 10 lb on the handle at an angle of 60° with the horizontal. How much work is done in moving the wagon 50 ft?

Solution. Introduce an *xy*-coordinate system so that the wagon moves from $P(0, 0)$ to $Q(50, 0)$ along the *x*-axis (Figure 14.3.10). In this coordinate system

$$\overrightarrow{PQ} = 50\mathbf{i}$$

and

$$\mathbf{F} = (10 \cos 60°)\mathbf{i} + (10 \sin 60°)\mathbf{j} = 5\mathbf{i} + 5\sqrt{3}\mathbf{j}$$

so that the work done is

$$W = \mathbf{F} \cdot \overrightarrow{PQ} = (5\mathbf{i} + 5\sqrt{3}\mathbf{j}) \cdot (50\mathbf{i}) = 250 \text{ (foot-pounds)} \quad \blacktriangleleft$$

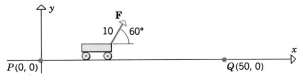

Figure 14.3.10

▶ Exercise Set 14.3

1. Find $\mathbf{u} \cdot \mathbf{v}$.
   (a) $\mathbf{u} = \mathbf{i} + 2\mathbf{j}$, $\mathbf{v} = 6\mathbf{i} - 8\mathbf{j}$
   (b) $\mathbf{u} = \langle -7, -3 \rangle$, $\mathbf{v} = \langle 0, 1 \rangle$
   (c) $\mathbf{u} = \mathbf{i} - 3\mathbf{j} + 7\mathbf{k}$, $\mathbf{v} = 8\mathbf{i} - 2\mathbf{j} - 2\mathbf{k}$
   (d) $\mathbf{u} = \langle -3, 1, 2 \rangle$, $\mathbf{v} = \langle 4, 2, -5 \rangle$.

2. In each part of Exercise 1, find the cosine of the angle θ between $\mathbf{u}$ and $\mathbf{v}$.

3. Determine whether $\mathbf{u}$ and $\mathbf{v}$ make an acute angle, an obtuse angle, or are orthogonal.
   (a) $\mathbf{u} = 7\mathbf{i} + 3\mathbf{j} + 5\mathbf{k}$, $\mathbf{v} = -8\mathbf{i} + 4\mathbf{j} + 2\mathbf{k}$
   (b) $\mathbf{u} = 6\mathbf{i} + \mathbf{j} + 3\mathbf{k}$, $\mathbf{v} = 4\mathbf{i} - 6\mathbf{k}$
   (c) $\mathbf{u} = \langle 1, 1, 1 \rangle$, $\mathbf{v} = \langle -1, 0, 0 \rangle$
   (d) $\mathbf{u} = \langle 4, 1, 6 \rangle$, $\mathbf{v} = \langle -3, 0, 2 \rangle$.

4. Find the orthogonal projection of $\mathbf{u}$ on $\mathbf{a}$.
   (a) $\mathbf{u} = 2\mathbf{i} + \mathbf{j}$, $\mathbf{a} = -3\mathbf{i} + 2\mathbf{j}$

   (b) $\mathbf{u} = \langle 2, 6 \rangle$, $\mathbf{a} = \langle -9, 3 \rangle$
   (c) $\mathbf{u} = -7\mathbf{i} + \mathbf{j} + 3\mathbf{k}$, $\mathbf{a} = 5\mathbf{i} + \mathbf{k}$
   (d) $\mathbf{u} = \langle 0, 0, 1 \rangle$, $\mathbf{a} = \langle 8, 3, 4 \rangle$.

5. In each part of Exercise 4, find the vector component of $\mathbf{u}$ orthogonal to $\mathbf{a}$.

6. Find $\|\text{proj}_{\mathbf{a}} \mathbf{u}\|$.
   (a) $\mathbf{u} = 2\mathbf{i} - \mathbf{j}$, $\mathbf{a} = 3\mathbf{i} + 4\mathbf{j}$
   (b) $\mathbf{u} = \langle 4, 5 \rangle$, $\mathbf{a} = \langle 1, -2 \rangle$
   (c) $\mathbf{u} = 2\mathbf{i} - \mathbf{j} + 3\mathbf{k}$, $\mathbf{a} = \mathbf{i} + 2\mathbf{j} + 2\mathbf{k}$
   (d) $\mathbf{u} = \langle 4, -1, 7 \rangle$, $\mathbf{a} = \langle 2, 3, -6 \rangle$.

7. Verify Theorem 14.3.4(*c*) for $\mathbf{u} = 6\mathbf{i} - \mathbf{j} + 2\mathbf{k}$, $\mathbf{v} = 2\mathbf{i} + 7\mathbf{j} + 4\mathbf{k}$, and $k = -5$.

8. Find two vectors in 2-space of norm 1 that are orthogonal to $3\mathbf{i} - 2\mathbf{j}$.

9. Let $\mathbf{u} = \langle 1, 2 \rangle$, $\mathbf{v} = \langle 4, -2 \rangle$, and $\mathbf{w} = \langle 6, 0 \rangle$. Find
   (a) $\mathbf{u} \cdot (7\mathbf{v} + \mathbf{w})$ (b) $\|(\mathbf{u} \cdot \mathbf{w})\mathbf{w}\|$
   (c) $\|\mathbf{u}\|(\mathbf{v} \cdot \mathbf{w})$ (d) $(\|\mathbf{u}\|\mathbf{v}) \cdot \mathbf{w}$.

10. Explain why each of the following expressions makes no sense.
    (a) $\mathbf{u} \cdot (\mathbf{v} \cdot \mathbf{w})$ (b) $(\mathbf{u} \cdot \mathbf{v}) + \mathbf{w}$
    (c) $\|\mathbf{u} \cdot \mathbf{v}\|$ (d) $k \cdot (\mathbf{u} + \mathbf{v})$.

11. Use vectors to find the cosines of the interior angles of the triangle with vertices $(-1, 0)$, $(2, -1)$, and $(1, 4)$.

12. Find two unit vectors in 2-space that make an angle of $45°$ with $4\mathbf{i} + 3\mathbf{j}$.

13. Show that $A(2, -1, 1)$, $B(3, 2, -1)$, and $C(7, 0, -2)$ are vertices of a right triangle. At which vertex is the right angle?

14. Find k so that the vector from the point $A(1, -1, 3)$ to the point $B(3, 0, 5)$ is perpendicular to the vector from A to the point $P(k, k, k)$.

15. Let $\mathbf{r}_0 = \langle x_0, y_0 \rangle$ and $\mathbf{r} = \langle x, y \rangle$. Describe the set of all points (x, y) for which
    (a) $\mathbf{r} \cdot \mathbf{r}_0 = 0$ (b) $(\mathbf{r} - \mathbf{r}_0) \cdot \mathbf{r}_0 = 0$
    (c) $\mathbf{r} \cdot (\mathbf{r} - \mathbf{r}_0) = 0$.

16. Suppose that $\mathbf{a} \cdot \mathbf{b} = \mathbf{a} \cdot \mathbf{c}$ and $\mathbf{a} \neq \mathbf{0}$. Does it follow that $\mathbf{b} = \mathbf{c}$? Explain.

17. Let $\mathbf{a} = k\mathbf{i} + \mathbf{j}$ and $\mathbf{b} = 4\mathbf{i} + 3\mathbf{j}$. Find k so that
    (a) $\mathbf{a}$ and $\mathbf{b}$ are orthogonal
    (b) the angle between $\mathbf{a}$ and $\mathbf{b}$ is $\pi/4$
    (c) the angle between $\mathbf{a}$ and $\mathbf{b}$ is $\pi/6$
    (d) $\mathbf{a}$ and $\mathbf{b}$ are parallel.

18. Find the direction cosines of $\mathbf{u}$ and estimate the direction angles to the nearest degree.
    (a) $\mathbf{u} = \mathbf{i} + \mathbf{j} - \mathbf{k}$ (b) $\mathbf{u} = 2\mathbf{i} - 2\mathbf{j} + \mathbf{k}$
    (c) $\mathbf{u} = 3\mathbf{i} - 2\mathbf{j} - 6\mathbf{k}$ (d) $\mathbf{u} = 3\mathbf{i} - 4\mathbf{k}$.

19. Prove: The direction cosines of a vector satisfy the equation $\cos^2 \alpha + \cos^2 \beta + \cos^2 \gamma = 1$.

20. Prove: Two nonzero vectors $\mathbf{u}_1$ and $\mathbf{u}_2$ are perpendicular if and only if their direction cosines satisfy
    $$\cos \alpha_1 \cos \alpha_2 + \cos \beta_1 \cos \beta_2 + \cos \gamma_1 \cos \gamma_2 = 0$$

21. Use Formula (12) to calculate the distance between the point and the line.
    (a) $3x + 4y + 7 = 0$; $(1, -2)$
    (b) $y = -2x + 1$; $(-3, 5)$
    (c) $2x + y = 8$; $(2, 6)$.

22. Given the points $A(1, 1, 0)$, $B(-2, 3, -4)$, and $P(-3, 1, 2)$,

(a) find $\|\text{proj}_{\overrightarrow{AB}} \overrightarrow{AP}\|$
(b) use the Pythagorean Theorem and the result of part (a) to find the distance from P to the line through A and B.

23. Follow the directions of Exercise 22 for the points $A(2, 1, -3)$, $B(0, 2, -1)$, and $P(4, 3, 0)$.

24. A boat travels 100 meters due north while the wind exerts a force of 50 newtons toward the northeast. How much work does the wind do?

25. Find the work done by a force $\mathbf{F} = -3\mathbf{j}$ (pounds) applied to a point that moves on a line from $(1, 3)$ to $(4, 7)$. Assume that distance is measured in feet.

26. Let $\mathbf{u}$ and $\mathbf{v}$ determine a parallelogram. Use vectors to prove that the diagonals of the parallelogram are perpendicular if and only if the sides are equal in length.

27. Let $\mathbf{u}$ and $\mathbf{v}$ determine a parallelogram. Use vectors to prove that the parallelogram is a rectangle if and only if the diagonals are equal in length.

28. Prove: $\|\mathbf{u} + \mathbf{v}\|^2 + \|\mathbf{u} - \mathbf{v}\|^2 = 2\|\mathbf{u}\|^2 + 2\|\mathbf{v}\|^2$.

29. Prove: $\mathbf{u} \cdot \mathbf{v} = \frac{1}{4}\|\mathbf{u} + \mathbf{v}\|^2 - \frac{1}{4}\|\mathbf{u} - \mathbf{v}\|^2$.

30. Find, to the nearest degree, the angle between the diagonal of a cube and a diagonal of one of its faces.

31. Find, to the nearest degree, the acute angle formed by two diagonals of a cube.

32. Find, to the nearest degree, the angles that a diagonal of a box with dimensions 10 in. by 15 in. by 25 in. makes with the edges of the box.

33. Prove: If vectors $\mathbf{v}_1$, $\mathbf{v}_2$, and $\mathbf{v}_3$ are nonzero and mutually perpendicular, then any vector $\mathbf{v}$ can be written as
    $$\mathbf{v} = c_1\mathbf{v}_1 + c_2\mathbf{v}_2 + c_3\mathbf{v}_3$$
    where $c_i = (\mathbf{v} \cdot \mathbf{v}_i)/\|\mathbf{v}_i\|^2$, $i = 1, 2, 3$.

34. Show that the three vectors $\mathbf{v}_1 = 3\mathbf{i} - \mathbf{j} + 2\mathbf{k}$, $\mathbf{v}_2 = \mathbf{i} + \mathbf{j} - \mathbf{k}$, and $\mathbf{v}_3 = \mathbf{i} - 5\mathbf{j} - 4\mathbf{k}$ are mutually perpendicular. Use the result of Exercise 33 to find scalars c_1, c_2, and c_3 so that
    $c_1\mathbf{v}_1 + c_2\mathbf{v}_2 + c_3\mathbf{v}_3 = \mathbf{i} - \mathbf{j} + \mathbf{k}$.

35. Prove: If $\mathbf{v}$ is orthogonal to $\mathbf{w}_1$ and $\mathbf{w}_2$, then $\mathbf{v}$ is orthogonal to $k_1\mathbf{w}_1 + k_2\mathbf{w}_2$ for all scalars k_1 and k_2.

36. Let $\mathbf{u}$ and $\mathbf{v}$ be nonzero vectors, and let $k = \|\mathbf{u}\|$ and $l = \|\mathbf{v}\|$. Prove that
    $$\mathbf{w} = l\mathbf{u} + k\mathbf{v}$$
    bisects the angle between $\mathbf{u}$ and $\mathbf{v}$.

■ 14.4 CROSS PRODUCT

> *In many applications of vectors to problems in geometry, physics, and engineering, it is of interest to construct a vector in 3-space that is perpendicular to two given vectors. In this section we shall introduce a type of vector multiplication that facilitates this construction.*

☐ DETERMINANTS

We begin with some notation. If a, b, c, and d are real numbers, then the symbol

$$\begin{vmatrix} a & b \\ c & d \end{vmatrix}$$

called a **2 × 2 (*two-by-two*) *determinant***, denotes the number

$$\begin{vmatrix} a & b \\ c & d \end{vmatrix} = ad - bc$$

For example,

$$\begin{vmatrix} 3 & -2 \\ 4 & 5 \end{vmatrix} = (3)(5) - (-2)(4) = 15 + 8 = 23$$

A **3 × 3 *determinant***, which is denoted by

$$\begin{vmatrix} a_1 & a_2 & a_3 \\ b_1 & b_2 & b_3 \\ c_1 & c_2 & c_3 \end{vmatrix}$$

is defined in terms of 2 × 2 determinants by the formula

$$\begin{vmatrix} a_1 & a_2 & a_3 \\ b_1 & b_2 & b_3 \\ c_1 & c_2 & c_3 \end{vmatrix} = a_1 \begin{vmatrix} b_2 & b_3 \\ c_2 & c_3 \end{vmatrix} - a_2 \begin{vmatrix} b_1 & b_3 \\ c_1 & c_3 \end{vmatrix} + a_3 \begin{vmatrix} b_1 & b_2 \\ c_1 & c_2 \end{vmatrix}$$

The right side of this formula is easily remembered by noting that a_1, a_2, and a_3 are the entries in the first "row" of the left side, and the 2 × 2 determinants on the right side arise by deleting the first row and an appropriate column from the left side. The pattern is as follows:

$$\begin{vmatrix} a_1 & a_2 & a_3 \\ b_1 & b_2 & b_3 \\ c_1 & c_2 & c_3 \end{vmatrix} = a_1 \begin{vmatrix} a_1 & a_2 & a_3 \\ b_1 & b_2 & b_3 \\ c_1 & c_2 & c_3 \end{vmatrix} - a_2 \begin{vmatrix} a_1 & a_2 & a_3 \\ b_1 & b_2 & b_3 \\ c_1 & c_2 & c_3 \end{vmatrix} + a_3 \begin{vmatrix} a_1 & a_2 & a_3 \\ b_1 & b_2 & b_3 \\ c_1 & c_2 & c_3 \end{vmatrix}$$

For example,

$$\begin{vmatrix} 3 & -2 & -5 \\ 1 & 4 & -4 \\ 0 & 3 & 2 \end{vmatrix} = 3 \begin{vmatrix} 4 & -4 \\ 3 & 2 \end{vmatrix} - (-2) \begin{vmatrix} 1 & -4 \\ 0 & 2 \end{vmatrix} + (-5) \begin{vmatrix} 1 & 4 \\ 0 & 3 \end{vmatrix}$$

$$= 3(20) + 2(2) - 5(3) = 49$$

□ **CROSS PRODUCT**

> **14.4.1** DEFINITION. If $\mathbf{u} = \langle u_1, u_2, u_3 \rangle$ and $\mathbf{v} = \langle v_1, v_2, v_3 \rangle$ are vectors in 3-space, then the ***cross product*** $\mathbf{u} \times \mathbf{v}$ is the vector defined by
>
> $$\mathbf{u} \times \mathbf{v} = \begin{vmatrix} u_2 & u_3 \\ v_2 & v_3 \end{vmatrix} \mathbf{i} - \begin{vmatrix} u_1 & u_3 \\ v_1 & v_3 \end{vmatrix} \mathbf{j} + \begin{vmatrix} u_1 & u_2 \\ v_1 & v_2 \end{vmatrix} \mathbf{k} \qquad (1)$$

Formula (1) can be remembered by writing it in the form

$$\mathbf{u} \times \mathbf{v} = \begin{vmatrix} \mathbf{i} & \mathbf{j} & \mathbf{k} \\ u_1 & u_2 & u_3 \\ v_1 & v_2 & v_3 \end{vmatrix} \qquad (2)$$

However, this is just a mnemonic device since the entries in a determinant must be numbers, not vectors.

Example 1 Find $\mathbf{u} \times \mathbf{v}$, where $\mathbf{u} = \langle 1, 2, -2 \rangle$ and $\mathbf{v} = \langle 3, 0, 1 \rangle$.

Solution.

$$\mathbf{u} \times \mathbf{v} = \begin{vmatrix} \mathbf{i} & \mathbf{j} & \mathbf{k} \\ 1 & 2 & -2 \\ 3 & 0 & 1 \end{vmatrix}$$

$$= \begin{vmatrix} 2 & -2 \\ 0 & 1 \end{vmatrix} \mathbf{i} - \begin{vmatrix} 1 & -2 \\ 3 & 1 \end{vmatrix} \mathbf{j} + \begin{vmatrix} 1 & 2 \\ 3 & 0 \end{vmatrix} \mathbf{k} = 2\mathbf{i} - 7\mathbf{j} - 6\mathbf{k} \qquad \blacktriangleleft$$

Observe that the cross product of two vectors is another vector, whereas the dot product of two vectors is a scalar. Moreover, the cross product is defined only for vectors in 3-space, whereas the dot product is defined for vectors in 2-space and 3-space.

The following theorem gives an important relationship between dot product and cross product and also shows that $\mathbf{u} \times \mathbf{v}$ is orthogonal to both $\mathbf{u}$ and $\mathbf{v}$.

> **14.4.2** THEOREM. *If* $\mathbf{u}$ *and* $\mathbf{v}$ *are vectors in 3-space, then*
>
> (*a*) $\mathbf{u} \cdot (\mathbf{u} \times \mathbf{v}) = 0$ $\qquad\qquad\qquad$ ($\mathbf{u} \times \mathbf{v}$ *is orthogonal to* $\mathbf{u}$)
>
> (*b*) $\mathbf{v} \cdot (\mathbf{u} \times \mathbf{v}) = 0$ $\qquad\qquad\qquad$ ($\mathbf{u} \times \mathbf{v}$ *is orthogonal to* $\mathbf{v}$)
>
> (*c*) $\|\mathbf{u} \times \mathbf{v}\|^2 = \|\mathbf{u}\|^2 \|\mathbf{v}\|^2 - (\mathbf{u} \cdot \mathbf{v})^2$ $\qquad$ (*Lagrange's identity*)

Proof. Let $\mathbf{u} = \langle u_1, u_2, u_3 \rangle$ and $\mathbf{v} = \langle v_1, v_2, v_3 \rangle$.

(*a*) By definition

$$\mathbf{u} \times \mathbf{v} = \begin{vmatrix} u_2 & u_3 \\ v_2 & v_3 \end{vmatrix} \mathbf{i} - \begin{vmatrix} u_1 & u_3 \\ v_1 & v_3 \end{vmatrix} \mathbf{j} + \begin{vmatrix} u_1 & u_2 \\ v_1 & v_2 \end{vmatrix} \mathbf{k}$$

which may be rewritten as

$$\mathbf{u} \times \mathbf{v} = \langle u_2 v_3 - u_3 v_2, \ u_3 v_1 - u_1 v_3, \ u_1 v_2 - u_2 v_1 \rangle \qquad (3)$$

so that

$$\mathbf{u} \cdot (\mathbf{u} \times \mathbf{v}) = u_1(u_2v_3 - u_3v_2) + u_2(u_3v_1 - u_1v_3) + u_3(u_1v_2 - u_2v_1) = 0$$

(b) Similar to (a).

(c) From (3) we obtain

$$\|\mathbf{u} \times \mathbf{v}\|^2 = (u_2v_3 - u_3v_2)^2 + (u_3v_1 - u_1v_3)^2 + (u_1v_2 - u_2v_1)^2 \qquad (4)$$

Moreover,

$$\|\mathbf{u}\|^2 \|\mathbf{v}\|^2 - (\mathbf{u} \cdot \mathbf{v})^2 = (u_1{}^2 + u_2{}^2 + u_3{}^2)(v_1{}^2 + v_2{}^2 + v_3{}^2)$$
$$- (u_1v_1 + u_2v_2 + u_3v_3)^2 \qquad (5)$$

Lagrange's identity can be established by "multiplying out" the right sides of (4) and (5) and verifying their equality. ∎

Example 2 Let $\mathbf{u} = \langle 1, 2, -2 \rangle$ and $\mathbf{v} = \langle 3, 0, 1 \rangle$. In Example 1 we showed that

$$\mathbf{u} \times \mathbf{v} = \langle 2, -7, -6 \rangle$$

Thus,

$$\mathbf{u} \cdot (\mathbf{u} \times \mathbf{v}) = (1)(2) + (2)(-7) + (-2)(-6) = 0$$

and

$$\mathbf{v} \cdot (\mathbf{u} \times \mathbf{v}) = (3)(2) + (0)(-7) + (1)(-6) = 0$$

so that the vector $\mathbf{u} \times \mathbf{v}$ is orthogonal to both $\mathbf{u}$ and $\mathbf{v}$ as guaranteed by Theorem 14.4.2. ◀

If $\mathbf{u}$ and $\mathbf{v}$ are nonzero vectors in 3-space, then the length of $\mathbf{u} \times \mathbf{v}$ has a useful geometric interpretation. Lagrange's identity, given in Theorem 14.4.2, states that

$$\|\mathbf{u} \times \mathbf{v}\|^2 = \|\mathbf{u}\|^2 \|\mathbf{v}\|^2 - (\mathbf{u} \cdot \mathbf{v})^2 \qquad (6)$$

If θ denotes the angle between $\mathbf{u}$ and $\mathbf{v}$, then $\mathbf{u} \cdot \mathbf{v} = \|\mathbf{u}\| \|\mathbf{v}\| \cos \theta$, so (6) can be rewritten as

$$\|\mathbf{u} \times \mathbf{v}\|^2 = \|\mathbf{u}\|^2 \|\mathbf{v}\|^2 - \|\mathbf{u}\|^2 \|\mathbf{v}\|^2 \cos^2 \theta$$
$$= \|\mathbf{u}\|^2 \|\mathbf{v}\|^2 (1 - \cos^2 \theta)$$
$$= \|\mathbf{u}\|^2 \|\mathbf{v}\|^2 \sin^2 \theta$$

Since $0 \leq \theta \leq \pi$, it follows that $\sin \theta \geq 0$, so

$$\|\mathbf{u} \times \mathbf{v}\| = \|\mathbf{u}\| \|\mathbf{v}\| \sin \theta \qquad (7)$$

But $\|\mathbf{v}\| \sin \theta$ is the altitude of the parallelogram determined by $\mathbf{u}$ and $\mathbf{v}$ (Figure 14.4.1). Thus, from (7), the area A of this parallelogram is given by

$$A = (\text{base})(\text{altitude}) = \|\mathbf{u}\| \|\mathbf{v}\| \sin \theta = \|\mathbf{u} \times \mathbf{v}\|$$

In other words, the length of $\mathbf{u} \times \mathbf{v}$ is numerically equal to the area of the parallelogram determined by $\mathbf{u}$ and $\mathbf{v}$.

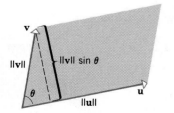

Figure 14.4.1

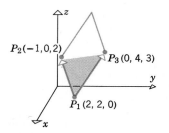

Figure 14.4.2

Example 3 Find the area of the triangle that is determined by the points $P_1(2, 2, 0)$, $P_2(-1, 0, 2)$, and $P_3(0, 4, 3)$.

Solution. The area A of the triangle is half the area of the parallelogram determined by the vectors $\overrightarrow{P_1P_2}$ and $\overrightarrow{P_1P_3}$ (Figure 14.4.2). But $\overrightarrow{P_1P_2} = \langle -3, -2, 2 \rangle$ and $\overrightarrow{P_1P_3} = \langle -2, 2, 3 \rangle$, so

$$\overrightarrow{P_1P_2} \times \overrightarrow{P_1P_3} = \langle -10, 5, -10 \rangle$$

(verify), and consequently

$$A = \frac{1}{2} \|\overrightarrow{P_1P_2} \times \overrightarrow{P_1P_3}\| = \frac{15}{2} \quad \blacktriangleleft$$

It follows from (7) that $\mathbf{u} \times \mathbf{v} = \mathbf{0}$ if and only if

$$\mathbf{u} = \mathbf{0}, \quad \text{or} \quad \mathbf{v} = \mathbf{0}, \quad \text{or} \quad \sin \theta = 0$$

In all three cases the vectors $\mathbf{u}$ and $\mathbf{v}$ are parallel. For the first two cases this is true because $\mathbf{0}$ is parallel to every vector, and in the third case, $\sin \theta = 0$ implies that the angle θ between $\mathbf{u}$ and $\mathbf{v}$ is $\theta = 0$ or $\theta = \pi$. In summary, we have the following result.

14.4.3 THEOREM. *If $\mathbf{u}$ and $\mathbf{v}$ are vectors in 3-space, then $\mathbf{u} \times \mathbf{v} = \mathbf{0}$ if and only if $\mathbf{u}$ and $\mathbf{v}$ are parallel vectors.*

The main arithmetic properties of the cross product are listed in the next theorem.

14.4.4 THEOREM. *If $\mathbf{u}$, $\mathbf{v}$, and $\mathbf{w}$ are any vectors in 3-space and k is any scalar, then*

(a) $\mathbf{u} \times \mathbf{v} = -(\mathbf{v} \times \mathbf{u})$

(b) $\mathbf{u} \times (\mathbf{v} + \mathbf{w}) = (\mathbf{u} \times \mathbf{v}) + (\mathbf{u} \times \mathbf{w})$

(c) $(\mathbf{u} + \mathbf{v}) \times \mathbf{w} = (\mathbf{u} \times \mathbf{w}) + (\mathbf{v} \times \mathbf{w})$

(d) $k(\mathbf{u} \times \mathbf{v}) = (k\mathbf{u}) \times \mathbf{v} = \mathbf{u} \times (k\mathbf{v})$

(e) $\mathbf{u} \times \mathbf{0} = \mathbf{0} \times \mathbf{u} = \mathbf{0}$

(f) $\mathbf{u} \times \mathbf{u} = \mathbf{0}$

We shall prove (a) and leave the remaining proofs as exercises.

Proof (a). First note that interchanging the rows of a 2×2 determinant changes the sign of the determinant, since

$$\begin{vmatrix} c & d \\ a & b \end{vmatrix} = bc - ad = -(ad - bc) = - \begin{vmatrix} a & b \\ c & d \end{vmatrix}$$

It follows that interchanging $\mathbf{u}$ and $\mathbf{v}$ in (1) interchanges the rows of the three determinants on the right side of (1) and thereby changes the sign of each component in the cross product. Thus, $\mathbf{u} \times \mathbf{v} = -(\mathbf{v} \times \mathbf{u})$. ∎

Cross products of the unit vectors **i**, **j**, and **k** are of special interest. We obtain, for example,

$$\mathbf{i} \times \mathbf{j} = \begin{vmatrix} \mathbf{i} & \mathbf{j} & \mathbf{k} \\ 1 & 0 & 0 \\ 0 & 1 & 0 \end{vmatrix} = \begin{vmatrix} 0 & 0 \\ 1 & 0 \end{vmatrix} \mathbf{i} - \begin{vmatrix} 1 & 0 \\ 0 & 0 \end{vmatrix} \mathbf{j} + \begin{vmatrix} 1 & 0 \\ 0 & 1 \end{vmatrix} \mathbf{k} = \mathbf{k}$$

The reader should have no trouble obtaining the following list of cross products:

$$\mathbf{i} \times \mathbf{j} = \mathbf{k} \qquad \mathbf{j} \times \mathbf{k} = \mathbf{i} \qquad \mathbf{k} \times \mathbf{i} = \mathbf{j}$$
$$\mathbf{j} \times \mathbf{i} = -\mathbf{k} \qquad \mathbf{k} \times \mathbf{j} = -\mathbf{i} \qquad \mathbf{i} \times \mathbf{k} = -\mathbf{j}$$
$$\mathbf{i} \times \mathbf{i} = \mathbf{0} \qquad \mathbf{j} \times \mathbf{j} = \mathbf{0} \qquad \mathbf{k} \times \mathbf{k} = \mathbf{0}$$

The diagram in Figure 14.4.3 is helpful for remembering these results. In this diagram, the cross product of two consecutive vectors going clockwise is the next vector around, and the cross product of two consecutive vectors going counterclockwise is the negative of the next vector around.

WARNING.   It is *not* true in general that $\mathbf{u} \times (\mathbf{v} \times \mathbf{w}) = (\mathbf{u} \times \mathbf{v}) \times \mathbf{w}$. For example, we have

$$\mathbf{i} \times (\mathbf{j} \times \mathbf{j}) = \mathbf{i} \times \mathbf{0} = \mathbf{0} \quad \text{and} \quad (\mathbf{i} \times \mathbf{j}) \times \mathbf{j} = \mathbf{k} \times \mathbf{j} = -\mathbf{i}$$

so that

$$\mathbf{i} \times (\mathbf{j} \times \mathbf{j}) \neq (\mathbf{i} \times \mathbf{j}) \times \mathbf{j}$$

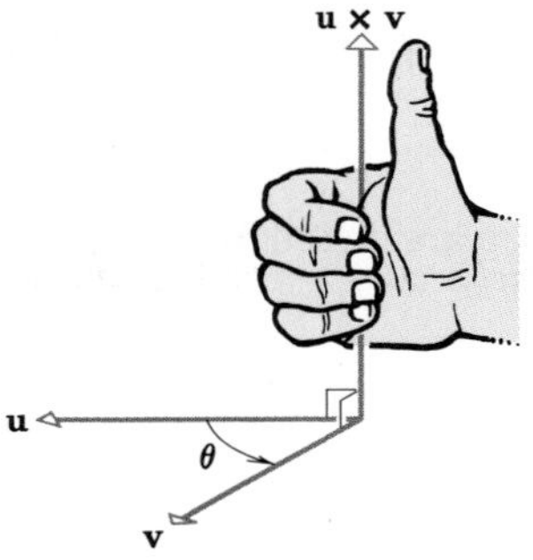

We know from Theorem 14.4.2 that $\mathbf{u} \times \mathbf{v}$ is orthogonal to both $\mathbf{u}$ and $\mathbf{v}$. It can be shown that if $\mathbf{u}$ and $\mathbf{v}$ are nonzero vectors, then the direction of $\mathbf{u} \times \mathbf{v}$ can be determined using the following "right-hand rule"* (Figure 14.4.4). Let θ be the angle between $\mathbf{u}$ and $\mathbf{v}$, and suppose $\mathbf{u}$ is rotated through the angle θ until it coincides with $\mathbf{v}$. If the fingers of the right hand are cupped so that they point in the direction of rotation, then the thumb indicates (roughly) the direction of $\mathbf{u} \times \mathbf{v}$. The reader may find it instructive to practice this rule with the products

$$\mathbf{i} \times \mathbf{j} = \mathbf{k} \qquad \mathbf{j} \times \mathbf{k} = \mathbf{i} \qquad \mathbf{k} \times \mathbf{i} = \mathbf{j}$$

Initially, we defined a vector to be a directed line segment or arrow in 2-space or 3-space; coordinate systems and components were introduced later in order to simplify computations with vectors. Thus, a vector has a "mathematical existence" regardless of whether a coordinate system has been introduced. Furthermore, the components of a vector are not determined by the vector alone; they depend as well on the coordinate system chosen. For example, in Figure 14.4.5 we have indicated a fixed-plane vector $\mathbf{v}$ and two different coordinate systems. In the xy-coordinate system the components of $\mathbf{v}$ are $\langle 1, 1 \rangle$, and in the $x'y'$-system they are $\langle \sqrt{2}, 0 \rangle$.

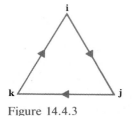

Figure 14.4.3

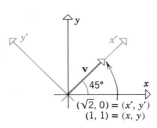

Figure 14.4.4

Figure 14.4.5

---

*Recall that we agreed to consider only right-handed coordinate systems in this text. Had we used left-handed systems instead, a "left-hand rule" would apply here.

This raises an important question about our definition of cross product. Since we defined the cross product $\mathbf{u} \times \mathbf{v}$ in terms of the components of $\mathbf{u}$ and $\mathbf{v}$, and since these components depend on the coordinate system chosen, it seems possible that two *fixed* vectors $\mathbf{u}$ and $\mathbf{v}$ might yield different vectors for $\mathbf{u} \times \mathbf{v}$ in different coordinate systems. Fortunately, this is not the case. To see this, we need only recall that

(i) $\mathbf{u} \times \mathbf{v}$ is perpendicular to both $\mathbf{u}$ and $\mathbf{v}$.

(ii)   The orientation of $\mathbf{u} \times \mathbf{v}$ is determined by the right-hand rule.

(iii) $\|\mathbf{u} \times \mathbf{v}\| = \|\mathbf{u}\| \, \|\mathbf{v}\| \sin \theta$.

These three properties completely determine the vector $\mathbf{u} \times \mathbf{v}$. Properties (i) and (ii) determine the direction, and property (iii) determines the length. Since these properties depend only on the lengths and relative positions of $\mathbf{u}$ and $\mathbf{v}$ and not on the particular right-handed coordinate system being used, the vector $\mathbf{u} \times \mathbf{v}$ will remain unchanged if a different right-handed coordinate system is introduced. Thus, we say that the definition of $\mathbf{u} \times \mathbf{v}$ is *coordinate free*. This result is of importance to physicists and engineers who often work with many coordinate systems in the same problem.

☐ TRIPLE SCALAR
PRODUCTS

If $\mathbf{a} = \langle a_1, a_2, a_3 \rangle$, $\mathbf{b} = \langle b_1, b_2, b_3 \rangle$, and $\mathbf{c} = \langle c_1, c_2, c_3 \rangle$ are vectors in 3-space, then the number

$$\mathbf{a} \cdot (\mathbf{b} \times \mathbf{c})$$

is called the *triple scalar product* of $\mathbf{a}$, $\mathbf{b}$, and $\mathbf{c}$. The triple scalar product may be conveniently calculated from the formula

$$\mathbf{a} \cdot (\mathbf{b} \times \mathbf{c}) = \begin{vmatrix} a_1 & a_2 & a_3 \\ b_1 & b_2 & b_3 \\ c_1 & c_2 & c_3 \end{vmatrix} \tag{8}$$

The validity of this formula may be seen by writing

$$\mathbf{a} \cdot (\mathbf{b} \times \mathbf{c}) = \mathbf{a} \cdot \left(\begin{vmatrix} b_2 & b_3 \\ c_2 & c_3 \end{vmatrix} \mathbf{i} - \begin{vmatrix} b_1 & b_3 \\ c_1 & c_3 \end{vmatrix} \mathbf{j} + \begin{vmatrix} b_1 & b_2 \\ c_1 & c_2 \end{vmatrix} \mathbf{k} \right)$$

$$= \begin{vmatrix} b_2 & b_3 \\ c_2 & c_3 \end{vmatrix} a_1 - \begin{vmatrix} b_1 & b_3 \\ c_1 & c_3 \end{vmatrix} a_2 + \begin{vmatrix} b_1 & b_2 \\ c_1 & c_2 \end{vmatrix} a_3$$

$$= \begin{vmatrix} a_1 & a_2 & a_3 \\ b_1 & b_2 & b_3 \\ c_1 & c_2 & c_3 \end{vmatrix}$$

Example 4   Calculate the triple scalar product $\mathbf{a} \cdot (\mathbf{b} \times \mathbf{c})$ of the vectors

$$\mathbf{a} = 3\mathbf{i} - 2\mathbf{j} - 5\mathbf{k}, \quad \mathbf{b} = \mathbf{i} + 4\mathbf{j} - 4\mathbf{k}, \quad \mathbf{c} = 3\mathbf{j} + 2\mathbf{k}$$

Solution.

$$\mathbf{a} \cdot (\mathbf{b} \times \mathbf{c}) = \begin{vmatrix} 3 & -2 & -5 \\ 1 & 4 & -4 \\ 0 & 3 & 2 \end{vmatrix}$$

$$= 3 \begin{vmatrix} 4 & -4 \\ 3 & 2 \end{vmatrix} - (-2) \begin{vmatrix} 1 & -4 \\ 0 & 2 \end{vmatrix} + (-5) \begin{vmatrix} 1 & 4 \\ 0 & 3 \end{vmatrix}$$

$$= 60 + 4 - 15 = 49 \quad \blacktriangleleft$$

REMARK.   The symbol $(\mathbf{a} \cdot \mathbf{b}) \times \mathbf{c}$ makes no sense since we cannot form the cross product of a scalar and a vector. Thus, no ambiguity arises if we write $\mathbf{a} \cdot \mathbf{b} \times \mathbf{c}$ rather than $\mathbf{a} \cdot (\mathbf{b} \times \mathbf{c})$. However, for clarity we shall usually keep the parentheses.

From the following calculations, we see that interchanging the rows of a 2×2 determinant changes the sign of its value and that converting the rows into columns has no effect on the value.

$$\begin{vmatrix} c & d \\ a & b \end{vmatrix} = bc - ad = -(ad - bc) = - \begin{vmatrix} a & b \\ c & d \end{vmatrix}$$

$$\begin{vmatrix} a & c \\ b & d \end{vmatrix} = ad - bc = \begin{vmatrix} a & b \\ c & d \end{vmatrix}$$

The following analogous result holds for 3×3 determinants. We omit the proof.

14.4.5   THEOREM.

(a)   *Interchanging any two rows of a 3×3 determinant changes the sign of its value.*

(b) $\begin{vmatrix} a_1 & a_2 & a_3 \\ b_1 & b_2 & b_3 \\ c_1 & c_2 & c_3 \end{vmatrix} = \begin{vmatrix} a_1 & b_1 & c_1 \\ a_2 & b_2 & c_2 \\ a_3 & b_3 & c_3 \end{vmatrix}$

It follows from part (a) of this theorem that

$$\mathbf{a} \cdot (\mathbf{b} \times \mathbf{c}) = \mathbf{c} \cdot (\mathbf{a} \times \mathbf{b}) = \mathbf{b} \cdot (\mathbf{c} \times \mathbf{a}) \tag{9}$$

since the 3×3 determinants that represent these products can be obtained from one another by *two* row interchanges. (Verify.) These relationships may be remembered by moving the vectors $\mathbf{a}$, $\mathbf{b}$, and $\mathbf{c}$ clockwise around the vertices of the triangle in Figure 14.4.6.

The triple scalar product $\mathbf{a} \cdot (\mathbf{b} \times \mathbf{c})$ has a useful geometric interpretation. If we assume, for the moment, that the vectors $\mathbf{a}$, $\mathbf{b}$, $\mathbf{c}$ do not all lie in the same plane when they are positioned with a common initial point, then the three vectors form adjacent sides of a parallelepiped (Figure 14.4.7). If the parallelogram determined by $\mathbf{b}$ and $\mathbf{c}$ is regarded as the base of the parallelepiped, then

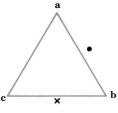

Figure 14.4.6

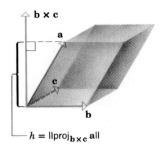

Figure 14.4.7

the area of the base is $\|\mathbf{b} \times \mathbf{c}\|$, and the height h is the length of the orthogonal projection of $\mathbf{a}$ on $\mathbf{b} \times \mathbf{c}$ (Figure 14.4.7). Therefore, by Formula (9) of Section 14.3 we have

$$h = \|\text{proj}_{\mathbf{b} \times \mathbf{c}} \, \mathbf{a}\| = \frac{|\mathbf{a} \cdot (\mathbf{b} \times \mathbf{c})|}{\|\mathbf{b} \times \mathbf{c}\|}$$

It follows that the volume V of the parallelepiped is

$$V = (\text{area of base}) \cdot \text{height} = \|\mathbf{b} \times \mathbf{c}\| \frac{|\mathbf{a} \cdot (\mathbf{b} \times \mathbf{c})|}{\|\mathbf{b} \times \mathbf{c}\|}$$

or, on simplifying,

$$V = \begin{bmatrix} \text{volume of parallelepiped with} \\ \text{adjacent sides } \mathbf{a}, \, \mathbf{b}, \text{ and } \mathbf{c} \end{bmatrix} = |\mathbf{a} \cdot (\mathbf{b} \times \mathbf{c})| \tag{10}$$

REMARK.   It follows from this formula that

$$\mathbf{a} \cdot (\mathbf{b} \times \mathbf{c}) = \pm V$$

where the $+$ or $-$ results depending on whether $\mathbf{a}$ makes an acute or obtuse angle with $\mathbf{b} \times \mathbf{c}$ (Theorem 14.3.2).

If the vectors $\mathbf{a}$, $\mathbf{b}$, and $\mathbf{c}$ do not lie in a plane when they are positioned with a common initial point, then they determine a parallelepiped of positive volume. Thus, from (10), $\mathbf{a} \cdot (\mathbf{b} \times \mathbf{c}) \neq 0$ if $\mathbf{a}$, $\mathbf{b}$, and $\mathbf{c}$ do not lie in a plane. It follows, therefore, that if $\mathbf{a} \cdot (\mathbf{b} \times \mathbf{c}) = 0$, then $\mathbf{a}$, $\mathbf{b}$, and $\mathbf{c}$ lie in a plane. Conversely, it can be shown that if $\mathbf{a}$, $\mathbf{b}$, and $\mathbf{c}$ lie in a plane, then $\mathbf{a} \cdot (\mathbf{b} \times \mathbf{c}) = 0$. In summary, we have the following result.

14.4.6   THEOREM.   *If the vectors* $\mathbf{a} = \langle a_1, a_2, a_3 \rangle$, $\mathbf{b} = \langle b_1, b_2, b_3 \rangle$, *and* $\mathbf{c} = \langle c_1, c_2, c_3 \rangle$ *have the same initial point, then they lie in a plane if and only if*

$$\mathbf{a} \cdot (\mathbf{b} \times \mathbf{c}) = \begin{vmatrix} a_1 & a_2 & a_3 \\ b_1 & b_2 & b_3 \\ c_1 & c_2 & c_3 \end{vmatrix} = 0$$

▶ Exercise Set 14.4

In Exercises 1–4, find $\mathbf{a} \times \mathbf{b}$.

1. $\mathbf{a} = \langle 1, 2, -3 \rangle$, $\mathbf{b} = \langle -4, 1, 2 \rangle$.

2. $\mathbf{a} = 3\mathbf{i} + 2\mathbf{j} - \mathbf{k}$, $\mathbf{b} = -\mathbf{i} - 3\mathbf{j} + \mathbf{k}$.

3. $\mathbf{a} = \langle 0, 1, -2 \rangle$, $\mathbf{b} = \langle 3, 0, -4 \rangle$.

4. $\mathbf{a} = 4\mathbf{i} + \mathbf{k}$, $\mathbf{b} = 2\mathbf{i} - \mathbf{j}$.

5. Let $\mathbf{u} = \langle 2, -1, 3 \rangle$, $\mathbf{v} = \langle 0, 1, 7 \rangle$, and $\mathbf{w} = \langle 1, 4, 5 \rangle$. Find

(a) $\mathbf{u} \times (\mathbf{v} \times \mathbf{w})$

(b) $(\mathbf{u} \times \mathbf{v}) \times \mathbf{w}$

(c) $\mathbf{u} \times (\mathbf{v} - 2\mathbf{w})$

(d) $(\mathbf{u} \times \mathbf{v}) - 2\mathbf{w}$

(e) $(\mathbf{u} \times \mathbf{v}) \times (\mathbf{v} \times \mathbf{w})$

(f) $(\mathbf{v} \times \mathbf{w}) \times (\mathbf{u} \times \mathbf{v})$.

6. Find a vector orthogonal to both **u** and **v**.
  (a) $\mathbf{u} = -7\mathbf{i} + 3\mathbf{j} + \mathbf{k}$, $\mathbf{v} = 2\mathbf{i} + 4\mathbf{k}$
  (b) $\mathbf{u} = \langle -1, -1, -1 \rangle$, $\mathbf{v} = \langle 2, 0, 2 \rangle$.

7. Verify Theorem 14.4.2 for the vectors $\mathbf{u} = \mathbf{i} - 5\mathbf{j} + 6\mathbf{k}$ and $\mathbf{v} = 2\mathbf{i} + \mathbf{j} + 2\mathbf{k}$.

8. Verify Theorem 14.4.4 for $\mathbf{u} = \langle 2, 0, -1 \rangle$, $\mathbf{v} = \langle 6, 7, 4 \rangle$, $\mathbf{w} = \langle 1, 1, 1 \rangle$, and $k = -3$.

9. Find all unit vectors parallel to the yz-plane that are perpendicular to the vector $3\mathbf{i} - \mathbf{j} + 2\mathbf{k}$.

10. Prove: If θ is the angle between **u** and **v** and $\mathbf{u} \cdot \mathbf{v} \neq 0$, then $\tan\theta = \|\mathbf{u} \times \mathbf{v}\| / (\mathbf{u} \cdot \mathbf{v})$.

11. Simplify $(\mathbf{u} + \mathbf{v}) \times (\mathbf{u} - \mathbf{v})$.

12. Find the area of the parallelogram determined by the vectors **u** and **v**.
  (a) $\mathbf{u} = \mathbf{i} - \mathbf{j} + 2\mathbf{k}$, $\mathbf{v} = 3\mathbf{j} + \mathbf{k}$
  (b) $\mathbf{u} = 2\mathbf{i} + 3\mathbf{j}$, $\mathbf{v} = -\mathbf{i} + 2\mathbf{j} - 2\mathbf{k}$.

13. Find the area of the triangle having vertices P, Q, and R.
  (a) $P(1, 5, -2)$, $Q(0, 0, 0)$, $R(3, 5, 1)$
  (b) $P(2, 0, -3)$, $Q(1, 4, 5)$, $R(7, 2, 9)$.

14. Use the cross product to find the sine of the angle between the vectors $\mathbf{a} = 2\mathbf{i} + 3\mathbf{j} - 6\mathbf{k}$ and $\mathbf{b} = 2\mathbf{i} + 3\mathbf{j} + 6\mathbf{k}$.

15. (a) Find the area of the triangle having vertices $A(1, 0, 1)$, $B(0, 2, 3)$, and $C(2, 1, 0)$.
  (b) Use the result of part (a) to find the length of the altitude from vertex C to side AB.

16. Show that if **u** is a vector from any point on a line to a point P not on the line, and **v** is a vector parallel to the line, then the distance between P and the line is given by $\|\mathbf{u} \times \mathbf{v}\| / \|\mathbf{v}\|$.

17. Use the result of Exercise 16 to find the distance between the point P and the line through the points A and B.
  (a) $P(-3, 1, 2)$, $A(1, 1, 0)$, $B(-2, 3, -4)$
  (b) $P(4, 3, 0)$, $A(2, 1, -3)$, $B(0, 2, -1)$.

18. Find all unit vectors in the plane of $\mathbf{u} = 3\mathbf{i} + \mathbf{k}$ and $\mathbf{v} = \mathbf{i} - \mathbf{j} - \mathbf{k}$ that are perpendicular to the vector $\mathbf{w} = \mathbf{i} + 2\mathbf{j}$.

19. What is wrong with the expression $\mathbf{u} \times \mathbf{v} \times \mathbf{w}$?

In Exercises 20–23, find $\mathbf{a} \cdot (\mathbf{b} \times \mathbf{c})$.

20. $\mathbf{a} = \langle 1, -2, 2 \rangle$, $\mathbf{b} = \langle 0, 3, 2 \rangle$, $\mathbf{c} = \langle -4, 1, -3 \rangle$.

21. $\mathbf{a} = 2\mathbf{i} - 3\mathbf{j} + \mathbf{k}$, $\mathbf{b} = 4\mathbf{i} + \mathbf{j} - 3\mathbf{k}$, $\mathbf{c} = \mathbf{j} + 5\mathbf{k}$.

22. $\mathbf{a} = \langle 2, 1, 0 \rangle$, $\mathbf{b} = \langle 1, -3, 1 \rangle$, $\mathbf{c} = \langle 4, 0, 1 \rangle$.

23. $\mathbf{a} = \mathbf{i}$, $\mathbf{b} = \mathbf{i} + \mathbf{j}$, $\mathbf{c} = \mathbf{i} + \mathbf{j} + \mathbf{k}$.

24. Suppose that $\mathbf{u} \cdot (\mathbf{v} \times \mathbf{w}) = 3$. Find
  (a) $\mathbf{u} \cdot (\mathbf{w} \times \mathbf{v})$ (b) $(\mathbf{v} \times \mathbf{w}) \cdot \mathbf{u}$
  (c) $\mathbf{w} \cdot (\mathbf{u} \times \mathbf{v})$ (d) $\mathbf{v} \cdot (\mathbf{u} \times \mathbf{w})$
  (e) $(\mathbf{u} \times \mathbf{w}) \cdot \mathbf{v}$ (f) $\mathbf{v} \cdot (\mathbf{w} \times \mathbf{w})$.

25. Find the volume of the parallelepiped with sides **a**, **b**, and **c**.
  (a) $\mathbf{a} = \langle 2, -6, 2 \rangle$, $\mathbf{b} = \langle 0, 4, -2 \rangle$, $\mathbf{c} = \langle 2, 2, -4 \rangle$
  (b) $\mathbf{a} = 3\mathbf{i} + \mathbf{j} + 2\mathbf{k}$, $\mathbf{b} = 4\mathbf{i} + 5\mathbf{j} + \mathbf{k}$, $\mathbf{c} = \mathbf{i} + 2\mathbf{j} + 4\mathbf{k}$.

26. Determine whether **u**, **v**, and **w** lie in the same plane.
  (a) $\mathbf{u} = \langle 1, -2, 1 \rangle$, $\mathbf{v} = \langle 3, 0, -2 \rangle$, $\mathbf{w} = \langle 5, -4, 0 \rangle$
  (b) $\mathbf{u} = 5\mathbf{i} - 2\mathbf{j} + \mathbf{k}$, $\mathbf{v} = 4\mathbf{i} - \mathbf{j} + \mathbf{k}$, $\mathbf{w} = \mathbf{i} - \mathbf{j}$
  (c) $\mathbf{u} = \langle 4, -8, 1 \rangle$, $\mathbf{v} = \langle 2, 1, -2 \rangle$, $\mathbf{w} = \langle 3, -4, 12 \rangle$.

27. Consider the parallelepiped with sides
$$\mathbf{a} = 3\mathbf{i} + 2\mathbf{j} + \mathbf{k}$$
$$\mathbf{b} = \mathbf{i} + \mathbf{j} + 2\mathbf{k}$$
$$\mathbf{c} = \mathbf{i} + 3\mathbf{j} + 3\mathbf{k}$$
  (a) Find the volume.
  (b) Find the area of the face determined by **a** and **c**.
  (c) Find the angle between **a** and the plane containing the face determined by **b** and **c**.

28. Find a vector **n** perpendicular to the plane determined by the points $A(0, -2, 1)$, $B(1, -1, -2)$, and $C(-1, 1, 0)$.

29. Prove:
  (a) $(\mathbf{u} + k\mathbf{v}) \times \mathbf{v} = \mathbf{u} \times \mathbf{v}$
  (b) $\mathbf{u} \cdot \mathbf{v} \times \mathbf{z} = \mathbf{u} \times \mathbf{v} \cdot \mathbf{z}$.

30. Let **u**, **v**, and **w** be vectors in 3-space with the same initial point and such that no two are collinear. Prove:
  (a) $\mathbf{u} \times (\mathbf{v} \times \mathbf{w})$ lies in the plane determined by **v** and **w**
  (b) $(\mathbf{u} \times \mathbf{v}) \times \mathbf{w}$ lies in the plane determined by **u** and **v**.

31. Prove parts (b) and (c) of Theorem 14.4.4.

32. Prove parts (d), (e), and (f) of Theorem 14.4.4.

33. Prove that $\mathbf{x} \times (\mathbf{y} \times \mathbf{z}) = (\mathbf{x} \cdot \mathbf{z})\mathbf{y} - (\mathbf{x} \cdot \mathbf{y})\mathbf{z}$.
  [*Hint:* First prove the result for the case $\mathbf{z} = \mathbf{i}$, then for $\mathbf{z} = \mathbf{j}$, and then for $\mathbf{z} = \mathbf{k}$. Finally, prove it for an arbitrary vector $\mathbf{z} = z_1\mathbf{i} + z_2\mathbf{j} + z_3\mathbf{k}$.]

34. For the vectors $\mathbf{a} = \mathbf{i} + 3\mathbf{j} - \mathbf{k}$, $\mathbf{b} = \mathbf{i} + \mathbf{j} + 2\mathbf{k}$, and

$c = 3i - j + 2k$, calculate $a \times (b \times c)$ using Exercise 33, then check your result by calculating directly.

35. Prove: If a, b, c, and d lie in the same plane when positioned with a common initial point, then

$$(a \times b) \times (c \times d) = 0$$

36. It is a theorem of solid geometry that the volume of a tetrahedron is $\frac{1}{3}$(area of base) · (height). Use this result to prove that the volume of a tetrahedron whose sides are the vectors a, b, and c is $\frac{1}{6}|a \cdot (b \times c)|$.

37. Use the result of Exercise 36 to find the volume of the tetrahedron with vertices P, Q, R, and S.

(a) $P(-1, 2, 0)$, $Q(2, 1, -3)$, $R(1, 0, 1)$, $S(3, -2, 3)$

(b) $P(0, 0, 0)$, $Q(1, 2, -1)$, $R(3, 4, 0)$, $S(-1, -3, 4)$.

■ 14.5 PARAMETRIC EQUATIONS OF LINES

In this section we shall discuss parametric equations of lines in 2-space and 3-space. In 3-space, parametric equations of lines are especially important because they are generally the most convenient form for representing such lines algebraically.

□ LINES DETERMINED BY
A POINT AND A VECTOR

A line in 2-space or 3-space can be determined uniquely by specifying a point on the line and a nonzero vector parallel to the line (Figure 14.5.1). The following theorem gives parametric equations of the line through a point P_0 and parallel to a nonzero vector v.

14.5.1 THEOREM.

(a) *The line in 2-space that passes through the point $P_0(x_0, y_0)$ and is parallel to the nonzero vector $v = \langle a, b \rangle = ai + bj$ has parametric equations*

$$x = x_0 + at, \quad y = y_0 + bt \tag{1a}$$

(b) *The line in 3-space that passes through the point $P_0(x_0, y_0, z_0)$ and is parallel to the nonzero vector $v = \langle a, b, c \rangle = ai + bj + ck$ has parametric equations*

$$x = x_0 + at, \quad y = y_0 + bt, \quad z = z_0 + ct \tag{1b}$$

We shall prove part (b). The proof of (a) is similar.

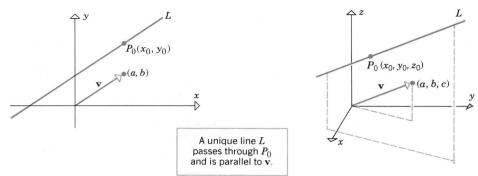

A unique line L passes through P_0 and is parallel to $\mathbf{v}$.

Figure 14.5.1

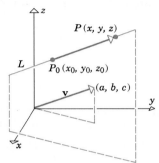

Figure 14.5.2

Proof (b). If L is the line in 3-space passing through the point $P_0(x_0, y_0, z_0)$ and parallel to the nonzero vector $\mathbf{v} = \langle a, b, c \rangle$, it is clear (Figure 14.5.2) that L consists precisely of those points $P(x, y, z)$ for which the vector $\overrightarrow{P_0 P}$ is parallel to $\mathbf{v}$. In other words, the point $P(x, y, z)$ is on L if and only if $\overrightarrow{P_0 P}$ is a scalar multiple of $\mathbf{v}$, say

$$\overrightarrow{P_0 P} = t\mathbf{v}$$

This equation can be written as

$$\langle x - x_0, y - y_0, z - z_0 \rangle = \langle ta, tb, tc \rangle$$

Equating components yields

$$x - x_0 = ta, \quad y - y_0 = tb, \quad z - z_0 = tc$$

from which (1b) follows. ■

REMARK.   Although it is not stated explicitly, it is understood in (1a) and (1b) that $-\infty < t < +\infty$, which reflects the fact that lines extend indefinitely.

Example 1   Find parametric equations of the line

(a)   passing through $(4, 2)$ and parallel to $\mathbf{v} = \langle -1, 5 \rangle$;

(b)   passing through $(1, 2, -3)$ and parallel to $\mathbf{v} = 4\mathbf{i} + 5\mathbf{j} - 7\mathbf{k}$;

(c)   passing through the origin in 3-space and parallel to $\mathbf{v} = \langle 1, 1, 1 \rangle$.

Solution (a).   From (1a) with $x_0 = 4$, $y_0 = 2$, $a = -1$, and $b = 5$ we obtain

$$x = 4 - t, \quad y = 2 + 5t$$

Solution (b).   From (1b) we obtain

$$x = 1 + 4t, \quad y = 2 + 5t, \quad z = -3 - 7t$$

Solution (c).   From (1b) with $x_0 = 0$, $y_0 = 0$, $z_0 = 0$, $a = 1$, $b = 1$, and $c = 1$ we obtain

$$x = t, \quad y = t, \quad z = t \quad \blacktriangleleft$$

Example 2

(a)   Find parametric equations of the line L passing through the points $P_1(2, 4, -1)$ and $P_2(5, 0, 7)$.

(b)   Where does the line intersect the xy-plane?

Solution (*a*).   Since the vector $\overrightarrow{P_1P_2} = \langle 3, -4, 8 \rangle$ is parallel to L and $P_1(2, 4, -1)$ lies on L, the line L is given by

$$x = 2 + 3t, \quad y = 4 - 4t, \quad z = -1 + 8t$$

Solution (*b*).   The line intersects the xy-plane at the point where $z = -1 + 8t = 0$; that is, when $t = \frac{1}{8}$. Substituting this value of t in the parametric equations for L yields the point of intersection

$$(x, y, z) = (\tfrac{19}{8}, \tfrac{7}{2}, 0) \quad \blacktriangleleft$$

Example 3   Let L_1 and L_2 be the lines

$$L_1: x = 1 + 4t, \quad y = 5 - 4t, \quad z = -1 + 5t$$
$$L_2: x = 2 + 8t, \quad y = 4 - 3t, \quad z = 5 + t$$

(a)   Are the lines parallel?

(b)   Do the lines intersect?

Solution (*a*).   The line L_1 is parallel to the vector $4\mathbf{i} - 4\mathbf{j} + 5\mathbf{k}$, and the line L_2 is parallel to the vector $8\mathbf{i} - 3\mathbf{j} + \mathbf{k}$. These vectors are not parallel since neither is a scalar multiple of the other. Thus, the lines are not parallel.

Solution (*b*).   In order for the lines to intersect at some point (x_0, y_0, z_0) these coordinates would have to satisfy the equations of both L_1 and L_2. In other words, there would have to exist values t_1 and t_2 for the parameters such that

$$x_0 = 1 + 4t_1, \quad y_0 = 5 - 4t_1, \quad z_0 = -1 + 5t_1$$

and

$$x_0 = 2 + 8t_2, \quad y_0 = 4 - 3t_2, \quad z_0 = 5 + t_2$$

This leads to three conditions on t_1 and t_2,

$$1 + 4t_1 = 2 + 8t_2$$
$$5 - 4t_1 = 4 - 3t_2 \tag{2}$$
$$-1 + 5t_1 = 5 + t_2$$

We shall try to solve these equations for t_1 and t_2. If we obtain a solution, then the lines intersect; if we find that there is no solution, then the lines do not intersect, since the three conditions cannot be satisfied.

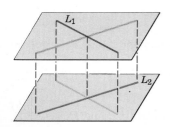

Parallel planes containing skew lines L_1 and L_2 may be determined by translating each line until it intersects the other.

Figure 14.5.3

The first two equations in (2) may be solved by adding them together to obtain

$$6 = 6 + 5t_2$$

or $t_2 = 0$. Substituting $t_2 = 0$ in the first equation yields

$$1 + 4t_1 = 2$$

or $t_1 = \frac{1}{4}$. However, the values $t_1 = \frac{1}{4}$, $t_2 = 0$ do not satisfy the third equation in (2), so there is no simultaneous solution to the three equations. Thus, the lines do not intersect. ◄

Two lines in 3-space that are not parallel and do not intersect (such as those in Example 3) are called *skew* lines. As illustrated in Figure 14.5.3, any two skew lines lie in parallel planes.

☐ **LINE SEGMENTS**

Sometimes one is not interested in an entire line, but rather some *segment* of a line. Parametric equations of a line segment can be obtained by finding parametric equations for the entire line, then restricting the parameter appropriately so that only the desired segment is generated as the parameter varies.

Example 4   Find parametric equations for the line segment joining the points $P_1(2, 4, -1)$ and $P_2(5, 0, 7)$.

Solution.   From Example 2, the line through P_1 and P_2 has parametric equations $x = 2 + 3t$, $y = 4 - 4t$, $z = -1 + 8t$. With these equations, the point P_1 corresponds to $t = 0$ and P_2 to $t = 1$. Thus, the line segment from P_1 to P_2 is given by

$$x = 2 + 3t, \quad y = 4 - 4t, \quad z = -1 + 8t \qquad (0 \leq t \leq 1) \qquad ◄$$

☐ **VECTOR EQUATIONS OF LINES**

We shall now show how vector notation can be used to express the parametric equations of a line in a more compact form. Because two vectors are equal if and only if their components are equal, (1a) and (1b) can be written as

$$\langle x, y \rangle = \langle x_0 + at, y_0 + bt \rangle$$

and

$$\langle x, y, z \rangle = \langle x_0 + at, y_0 + bt, z_0 + ct \rangle$$

or, equivalently, as

$$\langle x, y \rangle = \langle x_0, y_0 \rangle + t \langle a, b \rangle \qquad (3a)$$

and

$$\langle x, y, z \rangle = \langle x_0, y_0, z_0 \rangle + t \langle a, b, c \rangle \qquad (3b)$$

In 2-space let us introduce the vectors $\mathbf{r}$, $\mathbf{r}_0$, and $\mathbf{v}$ given by

$$\mathbf{r} = \langle x, y \rangle, \quad \mathbf{r}_0 = \langle x_0, y_0 \rangle, \quad \mathbf{v} = \langle a, b \rangle \qquad (4a)$$

and in 3-space

$$\mathbf{r} = \langle x, y, z \rangle, \quad \mathbf{r_0} = \langle x_0, y_0, z_0 \rangle, \quad \mathbf{v} = \langle a, b, c \rangle \qquad (4b)$$

Substituting (4a) and (4b) in (3a) and (3b), respectively, yields the equation

$$\mathbf{r} = \mathbf{r_0} + t\mathbf{v} \qquad (5)$$

in both cases. We call this the *vector equation of a line* in 2-space or 3-space. In this equation $\mathbf{v}$ is a nonzero vector parallel to the line and $\mathbf{r_0}$ is a vector whose components are the coordinates of a point on the line.

Example 5   The equation

$$\langle x, y, z \rangle = \langle -1, 0, 2 \rangle + t\langle 1, 5, -4 \rangle$$

is of form (5) with

$$\mathbf{r_0} = \langle -1, 0, 2 \rangle \quad \text{and} \quad \mathbf{v} = \langle 1, 5, -4 \rangle$$

Thus, the equation represents the line in 3-space that passes through the point $(-1, 0, 2)$ and is parallel to the vector $\langle 1, 5, -4 \rangle$.   ◀

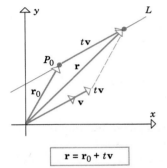

$$\boxed{\mathbf{r} = \mathbf{r_0} + t\mathbf{v}}$$

Figure 14.5.4

Figure 14.5.4 illustrates how Equation (5) can be interpreted geometrically: $\mathbf{r_0}$ can be viewed as a vector from the origin to a point P_0 on L, $t\mathbf{v}$ is a scalar multiple of a nonzero vector $\mathbf{v}$ that is parallel to L, and $\mathbf{r} = \mathbf{r_0} + t\mathbf{v}$ can be interpreted as a vector from the origin to a point on L. As the parameter t varies from $-\infty$ to $+\infty$, the terminal point of $\mathbf{r}$ traces out the line L.

Example 6   Find a vector equation of the line in 3-space that passes through the points $P_1(2, 4, -1)$ and $P_2(5, 0, 7)$.

Solution.   The vector

$$\overrightarrow{P_1P_2} = \langle 3, -4, 8 \rangle$$

is parallel to the line, so it can be used as $\mathbf{v}$ in (5). For $\mathbf{r_0}$ we can use either the vector from the origin to P_1 or the vector from the origin to P_2. Using the former yields

$$\mathbf{r_0} = \langle 2, 4, -1 \rangle$$

Thus, a vector equation of the line through P_1 and P_2 is

$$\langle x, y, z \rangle = \langle 2, 4, -1 \rangle + t\langle 3, -4, 8 \rangle$$

(Note that equating corresponding components on the two sides of this vector equation yields the parametric equations for the line that was obtained in Example 2.)   ◀

Line segments in 2-space or 3-space can be represented by vector equations with restrictions on the parameter.

Example 7   The vector equation

$$\langle x, y \rangle = \langle -3, 2 \rangle + t \langle 1, 1 \rangle \quad (0 \le t \le 3)$$

represents a segment of the line in 2-space that passes through the point $(-3, 2)$ and is parallel to the vector $\langle 1, 1 \rangle$. Since

$$\langle x, y \rangle = \langle -3, 2 \rangle \quad \text{if } t = 0$$

and

$$\langle x, y \rangle = \langle -3, 2 \rangle + \langle 3, 3 \rangle = \langle 0, 5 \rangle \quad \text{if } t = 3$$

the equation represents the portion of the line between the points $(-3, 2)$ and $(0, 5)$ (Figure 14.5.5).   ◄

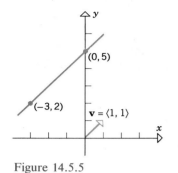

Figure 14.5.5

▶ Exercise Set 14.5   Ⓒ *53, 54, 57*

In Exercises 1–8, find parametric equations for the line through P_1 and P_2.

1. $P_1(3, -2)$, $P_2(5, 1)$.   **2.** $P_1(0, 1)$, $P_2(-3, -4)$.

3. $P_1(4, 1)$, $P_2(4, 3)$.   **4.** $P_1(5, 2)$, $P_2(3, 7)$.

5. $P_1(5, -2, 1)$, $P_2(2, 4, 2)$.

6. $P_1(-1, 3, 5)$, $P_2(-1, 3, 2)$.

7. $P_1(0, 0, 0)$, $P_2(-1, 6, 1)$.

8. $P_1(4, 0, 7)$, $P_2(-1, -1, 2)$.

In Exercises 9–16, find parametric equations for the line segment joining P_1 and P_2.

9. $P_1(3, -2)$, $P_2(5, 1)$.   **10.** $P_1(0, 1)$, $P_2(-3, -4)$.

11. $P_1(4, 1)$, $P_2(4, 3)$.   **12.** $P_1(5, 2)$, $P_2(3, 7)$.

13. $P_1(5, -2, 1)$, $P_2(2, 4, 2)$.

14. $P_1(-1, 3, 5)$, $P_2(-1, 3, 2)$.

15. $P_1(0, 0, 0)$, $P_2(-1, 6, 1)$.

16. $P_1(4, 0, 7)$, $P_2(-1, -1, 2)$.

In Exercises 17–25, find parametric equations for the line.

17. The line through $(-5, 2)$ and parallel to $2\mathbf{i} - 3\mathbf{j}$.

18. The line through $(0, 3)$ and parallel to the line $x = -5 + t$, $y = 1 - 2t$.

19. The line tangent to the circle $x^2 + y^2 = 25$ at the point $(3, -4)$.

20. The line tangent to the parabola $y = x^2$ at the point $(-2, 4)$.

21. The line through $(-1, 2, 4)$ and parallel to $3\mathbf{i} - 4\mathbf{j} + \mathbf{k}$.

22. The line through $(2, -1, 5)$ and parallel to $\langle -1, 2, 7 \rangle$.

23. The line through $(-2, 0, 5)$ and parallel to the line $x = 1 + 2t$, $y = 4 - t$, $z = 6 + 2t$.

24. The line through the origin and parallel to the line $x = t$, $y = -1 + t$, $z = 2$.

25. The line through $(3, 7, 0)$ and parallel to the x-axis.

26. Where does the line $x = 1 + 3t$, $y = 2 - t$ intersect
(a)   the x-axis   (b)   the y-axis?

27. Where does the line $x = 2t$, $y = 3 + 4t$ intersect the parabola $y = x^2$?

28. Where does the line $x = -1 + 2t$, $y = 3 + t$, $z = 4 - t$ intersect
(a)   the xy-plane   (b)   the xz-plane
(c)   the yz-plane?

29. Where does the line $x = -2$, $y = 4 + 2t$, $z = -3 + t$ intersect
(a)   the xy-plane   (b)   the xz-plane
(c)   the yz-plane?

30. Where does the line $x = 2 - t$, $y = 3t$, $z = -1 + 2t$ intersect the plane $2y + 3z = 6$?

31. Where does the line $x = 1 + t$, $y = 3 - t$, $z = 2t$ intersect the cylinder $x^2 + y^2 = 16$?

32. Find parametric equations for the line through (x_0, y_0, z_0) and (x_1, y_1, z_1).

33. Find parametric equations for the line through (x_1, y_1, z_1) and parallel to the line $x = x_0 + at$, $y = y_0 + bt$, $z = z_0 + ct$.

34. Prove: If a, b, and c are nonzero, then each point on the line $x = x_0 + at$, $y = y_0 + bt$, $z = z_0 + ct$ satisfies

$$\frac{x - x_0}{a} = \frac{y - y_0}{b} = \frac{z - z_0}{c}$$

and conversely, each point (x, y, z) satisfying these equations lies on the line. (These are called the *symmetric equations* of the line.)

35. Show that the lines

$$x = 2 + t, \quad y = 2 + 3t, \quad z = 3 + t$$
and
$$x = 2 + t, \quad y = 3 + 4t, \quad z = 4 + 2t$$

intersect and find the point of intersection.

36. Show that the lines

$$x + 1 = 4t, \quad y - 3 = t, \quad z - 1 = 0$$
and
$$x + 13 = 12t, \quad y - 1 = 6t, \quad z - 2 = 3t$$

intersect and find the point of intersection.

37. Show that the lines

$$x = 1 + 7t, \quad y = 3 + t, \quad z = 5 - 3t$$
and
$$x = 4 - t, \quad y = 6, \quad z = 7 + 2t$$

are skew.

38. Show that the lines

$$x = 2 + 8t, \quad y = 6 - 8t, \quad z = 10t$$
and
$$x = 3 + 8t, \quad y = 5 - 3t, \quad z = 6 + t$$

are skew.

39. Determine whether P_1, P_2, and P_3 lie on the same line.
  (a) $P_1(6, 9, 7)$, $P_2(9, 2, 0)$, $P_3(0, -5, -3)$
  (b) $P_1(1, 0, 1)$, $P_2(3, -4, -3)$, $P_3(4, -6, -5)$.

40. Find k_1 and k_2 so that the point $(k_1, 1, k_2)$ lies on the line passing through $(0, 2, 3)$ and $(2, 7, 5)$.

41. Find the point on the line segment joining $P_1(1, 4, -3)$ and $P_2(1, 5, -1)$ that is $\frac{2}{3}$ of the way from P_1 to P_2.

42. In each part, determine whether the lines are parallel.

  (a) $x = 3 - 2t, \quad y = 4 + t, \quad z = 6 - t$
    and
    $x = 5 - 4t, \quad y = -2 + 2t, \quad z = 7 - 2t$
  (b) $x = 5 + 3t, \quad y = 4 - 2t, \quad z = -2 + 3t$
    and
    $x = -1 + 9t, \quad y = 5 - 6t, \quad z = 3 + 8t$.

43. Show that the equations

$$x = 3 - t, \quad y = 1 + 2t$$
and
$$x = -1 + 3t, \quad y = 9 - 6t$$

represent the same line.

44. Show that the equations

$$x = 1 + 3t, \quad y = -2 + t, \quad z = 2t$$
and
$$x = 4 - 6t, \quad y = -1 - 2t, \quad z = 2 - 4t$$

represent the same line.

45. Find a vector equation of the line in 2-space that passes through the points P_1 and P_2.
  (a) $P_1(2, -1)$, $P_2(-5, 3)$
  (b) $P_1(0, 3)$, $P_2(4, 3)$.

46. Find a vector equation of the line in 3-space that passes through the points P_1 and P_2.
  (a) $P_1(3, -1, 2)$, $P_2(0, 1, 1)$
  (b) $P_1(2, 4, 1)$, $P_2(2, -1, 1)$.

47. Describe the line segment represented by the vector equation $\langle x, y \rangle = \langle 1, 0 \rangle + t\langle -2, 3 \rangle$, $0 \le t \le 2$.

48. Describe the line segment represented by the vector equation $\langle x, y, z \rangle = \langle -2, 1, 4 \rangle + t\langle 3, 0, -1 \rangle$, $0 \le t \le 3$.

In Exercises 49–52, use the result in Exercise 16, Section 14.4.

49. Find the distance between the point $(-2, 1, 1)$ and the line $x = 3 - t$, $y = t$, $z = 1 + 2t$.

50. Find the distance between the point $(1, 4, -3)$ and the line $x = 2 + t$, $y = -1 - t$, $z = 3t$.

51. Verify that the lines $x = 2 - t$, $y = 2t$, $z = 1 + t$ and $x = 1 + 2t$, $y = 3 - 4t$, $z = 5 - 2t$ are parallel, and find the distance between them.

52. Verify that the lines $x = 2t$, $y = 3 + 4t$, $z = 2 - 6t$ and $x = 1 + 3t$, $y = 6t$, $z = -9t$ are parallel, and find the distance between them.

53. Let L_1 and L_2 be the lines whose parametric equations are

$$x = 1 + 2t, \quad y = 2 - t, \quad z = 4 - 2t$$

and

$$x = 9 + t, \quad y = 5 + 3t, \quad z = -4 - t$$

(a) Show that L_1 and L_2 intersect at the point $(7, -1, -2)$.

(b) Find, to the nearest degree, the acute angle between L_1 and L_2 at their intersection.

(c) Find parametric equations for the line that is perpendicular to L_1 and L_2 and passes through their point of intersection.

54. Let L_1 and L_2 be the lines whose parametric equations are

$$x = 4t, \quad y = 1 - 2t, \quad z = 2 + 2t$$

and

$$x = 1 + t, \quad y = 1 - t, \quad z = -1 + 4t$$

(a) Show that L_1 and L_2 intersect at the point $(2, 0, 3)$.

(b) Find, to the nearest degree, the acute angle between L_1 and L_2 at their intersection.

(c) Find parametric equations for the line that is perpendicular to L_1 and L_2 and passes through their point of intersection.

55. Find parametric equations for the line that contains the point $(0, 2, 1)$ and intersects the line $x = 2t$, $y = 1 - t$, $z = 2 + t$ at a right angle.

56. Find parametric equations for the line that contains the point $(3, 1, -2)$ and intersects the line $x = -2 + 2t$, $y = 4 + 2t$, $z = 2 + t$ at a right angle.

57. Let $x = 4 - t$, $y = 1 + 2t$, $z = 2 + t$ and $x = t$, $y = 1 + t$, $z = 1 + 2t$ be the straight-line paths of motion of two particles. Suppose that t is the time in seconds and that x, y, and z are measured in centimeters.

(a) How far apart are the particles when $t = 0$?

(b) How close can the particles get?

■ 14.6 PLANES IN 3-SPACE

> *In this section we shall use vectors to derive equations of planes in 3-space, and we shall use these equations to solve some basic geometric problems.*

☐ **PLANES PARALLEL TO THE COORDINATE PLANES**

The plane parallel to the *xy*-plane and passing through the point $(0, 0, k)$ on the *z*-axis consists of all points (x, y, z) for which $z = k$ (Figure 14.6.1); thus, the equation of this plane is

$$z = k$$

Similarly, $x = k$ represents a plane parallel to the *yz*-plane and passing through $(k, 0, 0)$, while $y = k$ represents a plane parallel to the *xz*-plane and passing through $(0, k, 0)$.

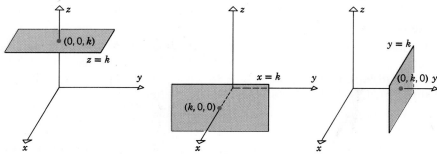

Figure 14.6.1

☐ PLANES DETERMINED BY A POINT AND A NORMAL VECTOR

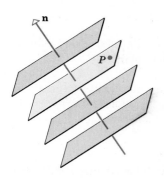

The colored plane is uniquely determined by the point P and the vector $\mathbf{n}$ perpendicular to the plane

Figure 14.6.2

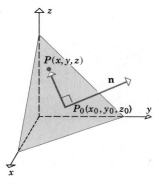

Figure 14.6.3

A plane in 3-space is uniquely determined by specifying a point in the plane and a vector perpendicular to the plane (Figure 14.6.2). A vector perpendicular to a plane is called a *normal* to the plane.

Suppose that we want the equation of the plane passing through the point $P_0(x_0, y_0, z_0)$ and perpendicular to the nonzero vector $\mathbf{n} = \langle a, b, c \rangle$. It is evident from Figure 14.6.3 that the plane consists precisely of those points $P(x, y, z)$ for which the vector $\overrightarrow{P_0P}$ is perpendicular to $\mathbf{n}$; or, phrased as an equation,

$$\mathbf{n} \cdot \overrightarrow{P_0P} = 0 \tag{1}$$

Since $\overrightarrow{P_0P} = \langle x - x_0, y - y_0, z - z_0 \rangle$, (1) can be rewritten as

$$a(x - x_0) + b(y - y_0) + c(z - z_0) = 0 \tag{2}$$

We shall call this the *point-normal form* of the equation of a plane.

Example 1 Find an equation of the plane passing through the point $(3, -1, 7)$ and perpendicular to the vector $\mathbf{n} = \langle 4, 2, -5 \rangle$.

Solution. From (2), a point-normal form of the equation is

$$4(x - 3) + 2(y + 1) - 5(z - 7) = 0 \quad \blacktriangleleft$$

By multiplying out and collecting terms, (2) can be rewritten in the form

$$ax + by + cz + d = 0 \tag{3}$$

where $a, b, c,$ and d are constants, and $a, b,$ and c are not all zero. To illustrate, the equation in Example 1 can be rewritten as

$$4x + 2y - 5z + 25 = 0$$

As our next theorem shows, every equation of form (3) represents a plane in 3-space.

> **14.6.1** THEOREM. *If $a, b, c,$ and d are constants, and $a, b,$ and c are not all zero, then the graph of the equation*
>
> $$ax + by + cz + d = 0$$
>
> *is a plane having the vector $\mathbf{n} = \langle a, b, c \rangle$ as a normal.*

Proof. By hypothesis, $a, b,$ and c are not all zero. Assume, for the moment, that $a \neq 0$. The equation $ax + by + cz + d = 0$ can be rewritten in the form $a[x + (d/a)] + by + cz = 0$. But this is a point-normal form of the plane passing through the point $(-d/a, 0, 0)$ and having $\mathbf{n} = \langle a, b, c \rangle$ as a normal.

If $a = 0$, then either $b \neq 0$ or $c \neq 0$. A straightforward modification of the above argument will handle these other cases. ∎

The equation in Theorem 14.6.1 is called the *general form* of the equation of a plane.

Example 2 Determine whether the planes

$$3x - 4y + 5z = 0 \quad \text{and} \quad -6x + 8y - 10z - 4 = 0$$

are parallel.

Solution. It is clear geometrically that two planes are parallel if and only if their normals are parallel vectors. A normal to the first plane is

$$\mathbf{n_1} = \langle 3, -4, 5 \rangle$$

and a normal to the second plane is

$$\mathbf{n_2} = \langle -6, 8, -10 \rangle$$

Since $\mathbf{n_2}$ is a scalar multiple of $\mathbf{n_1}$, the normals are parallel. Thus, the planes are also parallel. ◀

A plane cannot be specified by giving a point on it and *one* vector parallel to it, since there are infinitely many such planes (Figure 14.6.4*a*). However, as shown in Figure 14.6.4*b*, a plane is uniquely determined by giving a point in the plane and *two* nonparallel vectors that are parallel to the plane. A plane is also uniquely determined by specifying three noncollinear points in the plane (Figure 14.6.4*c*).

Example 3 Find an equation of the plane through the points $P_1(1, 2, -1)$, $P_2(2, 3, 1)$, and $P_3(3, -1, 2)$.

Solution. Since the points P_1, P_2, and P_3 lie in the plane, the vectors $\overrightarrow{P_1P_2} = \langle 1, 1, 2 \rangle$ and $\overrightarrow{P_1P_3} = \langle 2, -3, 3 \rangle$ are parallel to the plane. Therefore,

$$\overrightarrow{P_1P_2} \times \overrightarrow{P_1P_3} = \begin{vmatrix} \mathbf{i} & \mathbf{j} & \mathbf{k} \\ 1 & 1 & 2 \\ 2 & -3 & 3 \end{vmatrix} = 9\mathbf{i} + \mathbf{j} - 5\mathbf{k}$$

is normal to the plane, since it is perpendicular to both $\overrightarrow{P_1P_2}$ and $\overrightarrow{P_1P_3}$. By using this normal and the point $P_1(1, 2, -1)$ in the plane, we obtain the point-normal form

$$9(x - 1) + (y - 2) - 5(z + 1) = 0$$

which may be rewritten as

$$9x + y - 5z - 16 = 0 \quad ◀$$

Example 4 Determine whether the line

$$x = 3 + 8t, \quad y = 4 + 5t, \quad z = -3 - t$$

is parallel to the plane $x - 3y + 5z = 12$.

Solution. The vector $\mathbf{v} = \langle 8, 5, -1 \rangle$ is parallel to the line and the vector

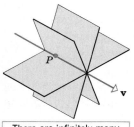

There are infinitely many planes containing P and parallel to $\mathbf{v}$.

(a)

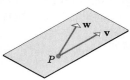

There is a unique plane through P that is parallel to both $\mathbf{v}$ and $\mathbf{w}$.

(b)

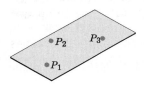

There is a unique plane through three noncollinear points.

(c)

Figure 14.6.4

$\mathbf{n} = \langle 1, -3, 5 \rangle$ is normal to the plane. In order for the line and plane to be parallel, the vectors $\mathbf{v}$ and $\mathbf{n}$ must be perpendicular. But this is not so, since the dot product

$$\mathbf{v} \cdot \mathbf{n} = (8)(1) + (5)(-3) + (-1)(5) = -12$$

is nonzero. Thus, the line and plane are not parallel. ◀

Example 5   Find the intersection of the line and plane in Example 4.

Solution.   If we let (x_0, y_0, z_0) be the point of intersection, then the coordinates of this point satisfy both the equation of the plane and the parametric equations of the line. Thus,

$$x_0 - 3y_0 + 5z_0 = 12 \tag{4}$$

and for some value of t, say $t = t_0$,

$$x_0 = 3 + 8t_0, \quad y_0 = 4 + 5t_0, \quad z_0 = -3 - t_0 \tag{5}$$

Substituting (5) in (4) yields

$$(3 + 8t_0) - 3(4 + 5t_0) + 5(-3 - t_0) = 12$$

Solving for t_0 yields $t_0 = -3$ and on substituting this value in (5), we obtain

$$(x_0, y_0, z_0) = (-21, -11, 0) \quad ◀$$

Two intersecting planes determine two angles of intersection, an acute angle θ ($0 \le \theta \le 90°$) and its supplement $180° - \theta$ (Figure 14.6.5a). If $\mathbf{n}_1$ and $\mathbf{n}_2$ are normals to the planes, then the angle between $\mathbf{n}_1$ and $\mathbf{n}_2$ is θ or $180° - \theta$, depending on the directions of the normals (Figure 14.6.5b). Thus, the angles of intersection of two planes may be determined from the normals.

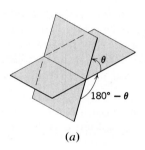

(a)

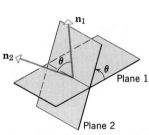

In this figure the angle between $\mathbf{n}_1$ and $\mathbf{n}_2$ is θ. If the direction of one of the normals is reversed, then the angle between the normals would be $180° - \theta$

(b)

Figure 14.6.5

Example 6   Find the acute angle of intersection between the two planes $2x - 4y + 4z = 7$ and $6x + 2y - 3z = 2$.

Solution.   From the given equations, we obtain the normals $\mathbf{n}_1 = \langle 2, -4, 4 \rangle$ and $\mathbf{n}_2 = \langle 6, 2, -3 \rangle$. Since the dot product $\mathbf{n}_1 \cdot \mathbf{n}_2 = -8$ is negative, the normals make an obtuse angle. To obtain normals making an acute angle, we can reverse the direction of either $\mathbf{n}_1$ or $\mathbf{n}_2$. To be specific, let us reverse the direction of $\mathbf{n}_1$. Thus, the acute angle θ between the planes will be the angle between $-\mathbf{n}_1 = \langle -2, 4, -4 \rangle$ and $\mathbf{n}_2 = \langle 6, 2, -3 \rangle$. From (3) of Section 14.3 we obtain

$$\cos \theta = \frac{(-\mathbf{n}_1) \cdot \mathbf{n}_2}{\|-\mathbf{n}_1\| \, \|\mathbf{n}_2\|} = \frac{8}{\sqrt{36}\sqrt{49}} = \frac{4}{21}$$

With the aid of a calculator one obtains

$$\theta = \cos^{-1}\left(\frac{4}{21}\right) \approx 79° \quad ◀$$

DISTANCE PROBLEMS INVOLVING PLANES

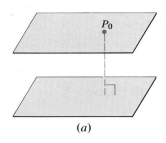

(a)

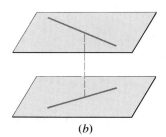

(b)

Figure 14.6.6

We conclude this section by discussing three basic "distance problems" in 3-space:

- Find the distance between a point and a plane.
- Find the distance between two parallel planes.
- Find the distance between two skew lines.

The three problems are related. If we can find the distance between a point and a plane, then we can find the distance between parallel planes by computing the distance between one of the planes and an arbitrary point P_0 in the other plane (Figure 14.6.6a). Moreover, we can find the distance between two skew lines by computing the distance between parallel planes containing them (Figure 14.6.6b).

14.6.2 THEOREM. *The distance D between a point $P_0(x_0, y_0, z_0)$ and the plane $ax + by + cz + d = 0$ is*

$$D = \frac{|ax_0 + by_0 + cz_0 + d|}{\sqrt{a^2 + b^2 + c^2}} \tag{6}$$

Proof. Let $Q(x_1, y_1, z_1)$ be any point in the plane, and position the normal $\mathbf{n} = \langle a, b, c \rangle$ so that its initial point is at Q. As illustrated in Figure 14.6.7, the distance D is equal to the length of the orthogonal projection of $\vec{QP_0}$ on $\mathbf{n}$. Thus, from (9) of Section 14.3,

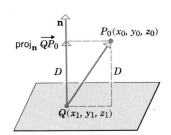

Figure 14.6.7

$$D = \|\text{proj}_{\mathbf{n}} \vec{QP_0}\| = \frac{|\vec{QP_0} \cdot \mathbf{n}|}{\|\mathbf{n}\|}$$

But

$$\vec{QP_0} = \langle x_0 - x_1, y_0 - y_1, z_0 - z_1 \rangle$$
$$\vec{QP_0} \cdot \mathbf{n} = a(x_0 - x_1) + b(y_0 - y_1) + c(z_0 - z_1)$$
$$\|\mathbf{n}\| = \sqrt{a^2 + b^2 + c^2}$$

Thus,

$$D = \frac{|a(x_0 - x_1) + b(y_0 - y_1) + c(z_0 - z_1)|}{\sqrt{a^2 + b^2 + c^2}} \tag{7}$$

Since the point $Q(x_1, y_1, z_1)$ lies in the plane, its coordinates satisfy the equation of the plane, so that

$$ax_1 + by_1 + cz_1 + d = 0$$

or

$$d = -ax_1 - by_1 - cz_1$$

Substituting this expression in (7) yields (6). ∎

REMARK. Note the similarity between (6) and the formula for the distance between a point and a line in 2-space [(12) of Section 14.3].

Example 7 Find the distance D between the point $(1, -4, -3)$ and the plane $2x - 3y + 6z = -1$.

Solution. To apply (6), we first rewrite the equation of the plane in the form

$$2x - 3y + 6z + 1 = 0$$

Then

$$D = \frac{|(2)(1) + (-3)(-4) + 6(-3) + 1|}{\sqrt{2^2 + (-3)^2 + 6^2}} = \frac{|-3|}{7} = \frac{3}{7} \quad \blacktriangleleft$$

Example 8 The planes

$$x + 2y - 2z = 3 \quad \text{and} \quad 2x + 4y - 4z = 7$$

are parallel since their normals, $\langle 1, 2, -2 \rangle$ and $\langle 2, 4, -4 \rangle$, are parallel vectors. Find the distance between these planes.

Solution. To find the distance D between the planes, we may select an arbitrary point in one of the planes and compute its distance to the other plane. By setting $y = z = 0$ in the equation $x + 2y - 2z = 3$, we obtain the point $P_0(3, 0, 0)$ in this plane. From (6), the distance from P_0 to the plane $2x + 4y - 4z = 7$ is

$$D = \frac{|(2)(3) + 4(0) + (-4)(0) - 7|}{\sqrt{2^2 + 4^2 + (-4)^2}} = \frac{1}{6} \quad \blacktriangleleft$$

Example 9 It was shown in Example 3 of Section 14.5 that the lines

$$L_1: x = 1 + 4t, \quad y = 5 - 4t, \quad z = -1 + 5t$$

and

$$L_2: x = 2 + 8t, \quad y = 4 - 3t, \quad z = 5 + t$$

are skew. Find the distance between them.

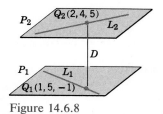

Figure 14.6.8

Solution. Let P_1 and P_2 denote parallel planes containing L_1 and L_2, respectively (Figure 14.6.8). To find the distance D between L_1 and L_2, we shall calculate the distance from a point in P_1 to the plane P_2. Setting $t = 0$ in the equations of L_1 yields the point $Q_1(1, 5, -1)$ in plane P_1. (The value $t = 0$ was chosen for simplicity. Any value of t will suffice.) Next we shall obtain an equation for plane P_2.

The vector $\mathbf{u_1} = \langle 4, -4, 5 \rangle$ is parallel to line L_1, and therefore also parallel to planes P_1 and P_2. Similarly, $\mathbf{u_2} = \langle 8, -3, 1 \rangle$ is parallel to L_2 and hence parallel to P_1 and P_2. Therefore, the cross product

$$\mathbf{n} = \mathbf{u}_1 \times \mathbf{u}_2 = \begin{vmatrix} \mathbf{i} & \mathbf{j} & \mathbf{k} \\ 4 & -4 & 5 \\ 8 & -3 & 1 \end{vmatrix} = 11\mathbf{i} + 36\mathbf{j} + 20\mathbf{k}$$

is normal to both P_1 and P_2. Using this normal and the point $Q_2(2, 4, 5)$ found by setting $t = 0$ in the equations of L_2, we obtain an equation for P_2:

$$11(x - 2) + 36(y - 4) + 20(z - 5) = 0$$

or

$$11x + 36y + 20z - 266 = 0$$

The distance between $Q_1(1, 5, -1)$ and this plane is

$$D = \frac{|(11)(1) + (36)(5) + (20)(-1) - 266|}{\sqrt{11^2 + 36^2 + 20^2}} = \frac{95}{\sqrt{1817}}$$

This is also the distance between L_1 and L_2. ◀

▶ Exercise Set 14.6   Ⓒ 11

In Exercises 1–4, find an equation of the plane passing through P and having $\mathbf{n}$ as a normal.

1. $P(2, 6, 1)$; $\mathbf{n} = \langle 1, 4, 2 \rangle$.
2. $P(-1, -1, 2)$; $\mathbf{n} = \langle -1, 7, 6 \rangle$.
3. $P(1, 0, 0)$; $\mathbf{n} = \langle 0, 0, 1 \rangle$.
4. $P(0, 0, 0)$; $\mathbf{n} = \langle 2, -3, -4 \rangle$.
5. Find an equation of the plane passing through the given points.
   (a) $(-2, 1, 1)$, $(0, 2, 3)$, and $(1, 0, -1)$
   (b) $(3, 2, 1)$, $(2, 1, -1)$, and $(-1, 3, 2)$.
6. Determine whether the planes are parallel.
   (a) $3x - 2y + z = 4$ and $6x - 4y + 3z = 7$
   (b) $2x - 8y - 6z - 2 = 0$ and
       $-x + 4y + 3z - 5 = 0$
   (c) $y = 4x - 2z + 3$ and $x = \frac{1}{4}y + \frac{1}{2}z$.
7. Determine whether the line and plane are parallel.
   (a) $x = 4 + 2t$, $y = -t$, $z = -1 - 4t$;
       $3x + 2y + z - 7 = 0$
   (b) $x = t$, $y = 2t$, $z = 3t$; $x - y + 2z = 5$.
8. Determine whether the planes are perpendicular.
   (a) $x - y + 3z - 2 = 0$, $2x + z = 1$
   (b) $3x - 2y + z = 1$, $4x + 5y - 2z = 4$.
9. Determine whether the line and plane are perpendicular.
   (a) $x = -1 + 2t$, $y = 4 + t$, $z = 1 - t$;
       $4x + 2y - 2z = 7$

(b) $x = 3 - t$, $y = 2 + t$, $z = 1 - 3t$;
    $2x + 2y - 5 = 0$.
10. Find the point of intersection of the line and plane.
    (a) $x = t$, $y = t$, $z = t$; $3x - 2y + z - 5 = 0$
    (b) $x = 1 + t$, $y = -1 + 3t$, $z = 2 + 4t$;
        $x - y + 4z = 7$
    (c) $x = 2 - t$, $y = 3 + t$, $z = t$;
        $2x + y + z = 1$.
11. Find the acute angle of intersection of the planes (to the nearest degree).
    (a) $x = 0$ and $2x - y + z - 4 = 0$
    (b) $x + 2y - 2z = 5$ and $6x - 3y + 2z = 8$.
12. Find an equation of the plane through $(-1, 4, -3)$ and perpendicular to the line $x - 2 = t$,
    $y + 3 = 2t$, $z = -t$.
13. Find an equation of
    (a) the xy-plane   (b) the xz-plane
    (c) the yz-plane.
14. Find an equation of the plane that contains the point (x_0, y_0, z_0) and is
    (a) parallel to the xy-plane
    (b) parallel to the yz-plane
    (c) parallel to the xz-plane.
15. Find an equation of the plane through the origin that is parallel to the plane $4x - 2y + 7z + 12 = 0$.

16. Find an equation of the plane containing the line $x = -2 + 3t$, $y = 4 + 2t$, $z = 3 - t$ and perpendicular to the plane $x - 2y + z = 5$.

17. Find an equation of the plane through $(-1, 4, 2)$ and containing the line of intersection of the planes $4x - y + z - 2 = 0$ and $2x + y - 2z - 3 = 0$.

18. Show that the points $(1, 0, -1)$, $(0, 2, 3)$, $(-2, 1, 1)$, and $(4, 2, 3)$ lie in the same plane.

19. Find parametric equations of the line through $(5, 0, -2)$ that is parallel to the planes $x - 4y + 2z = 0$ and $2x + 3y - z + 1 = 0$.

20. Find an equation of the plane through $(-1, 2, -5)$ and perpendicular to the planes $2x - y + z = 1$ and $x + y - 2z = 3$.

21. Find an equation of the plane through $(1, 2, -1)$ and perpendicular to the line of intersection of the planes $2x + y + z = 2$ and $x + 2y + z = 3$.

22. Find a plane through the points $P_1(-2, 1, 4)$, $P_2(1, 0, 3)$ and perpendicular to the plane $4x - y + 3z = 2$.

23. Show that the lines
$$x = -2 + t, \quad y = 3 + 2t, \quad z = 4 - t$$
and
$$x = 3 - t, \quad y = 4 - 2t, \quad z = t$$
are parallel and find an equation of the plane they determine.

24. Find an equation of the plane that contains the point $(2, 0, 3)$ and the line $x = -1 + t$, $y = t$, $z = -4 + 2t$.

25. Find an equation of the plane, each of whose points is equidistant from $(2, -1, 1)$ and $(3, 1, 5)$.

26. Find an equation of the plane containing the line $x = 3t$, $y = 1 + t$, $z = 2t$ and parallel to the intersection of the planes $2x - y + z = 0$ and $y + z + 1 = 0$.

27. Show that the line $x = 0$, $y = t$, $z = t$
(a) lies in the plane $6x + 4y - 4z = 0$
(b) is parallel to and below the plane $5x - 3y + 3z = 1$
(c) is parallel to and above the plane $6x + 2y - 2z = 3$.

28. Show that the lines
$$x + 1 = 4t, \quad y - 3 = t, \quad z - 1 = 0$$
and
$$x + 13 = 12t, \quad y - 1 = 6t, \quad z - 2 = 3t$$
intersect and find an equation of the plane they determine.

29. Find parametric equations of the line of intersection of the planes
(a) $-2x + 3y + 7z + 2 = 0$ and $x + 2y - 3z + 5 = 0$
(b) $3x - 5y + 2z = 0$ and $z = 0$.

30. Show that the plane whose intercepts with the coordinate axes are $x = a$, $y = b$, and $z = c$ has equation
$$\frac{x}{a} + \frac{y}{b} + \frac{z}{c} = 1$$
provided a, b, and c are nonzero.

In Exercises 31–33, find the distance between the point and the plane.

31. $(1, -2, 3)$; $2x - 2y + z = 4$.

32. $(0, 1, 5)$; $3x + 6y - 2z - 5 = 0$.

33. $(7, 2, -1)$; $20x - 4y - 5z = 0$.

In Exercises 34–36, find the distance between the given parallel planes.

34. $2x - 3y + 4z = 7$, $4x - 6y + 8z = 3$.

35. $-2x + y + z = 0$, $6x - 3y - 3z - 5 = 0$.

36. $x + y + z = 1$, $x + y + z = -1$.

In Exercises 37–39, find the distance between the given skew lines.

37. $x = 1 + 7t$, $y = 3 + t$, $z = 5 - 3t$; $x = 4 - t$, $y = 6$, $z = 7 + 2t$.

38. $x = 3 - t$, $y = 4 + 4t$, $z = 1 + 2t$; $x = t$, $y = 3$, $z = 2t$.

39. $x = 2 + 4t$, $y = 6 - 4t$, $z = 5t$; $x = 3 + 8t$, $y = 5 - 3t$, $z = 6 + t$.

40. Show that the line $x = -1 + t$, $y = 3 + 2t$, $z = -t$ and the plane $2x - 2y - 2z + 3 = 0$ are parallel, and find the distance between them.

41. Prove: The planes $a_1x + b_1y + c_1z = d_1$ and $a_2x + b_2y + c_2z = d_2$ are perpendicular if and only if $a_1a_2 + b_1b_2 + c_1c_2 = 0$.

42. Let $\mathbf{r}_0 = \langle x_0, y_0, z_0 \rangle$ and $\mathbf{r} = \langle x, y, z \rangle$. Describe the set of all points (x, y, z) for which
(a) $\mathbf{r} \cdot \mathbf{r}_0 = 0$ (b) $(\mathbf{r} - \mathbf{r}_0) \cdot \mathbf{r}_0 = 0$.

43. Find an equation of the sphere with its center at $(2, 1, -3)$ and tangent to the plane $x - 3y + 2z = 4$.

44. Find where the line that passes through the point $(3, 1, 0)$ and is normal to the plane $2x + y - z = 0$ intersects the plane.

■ 14.7   QUADRIC SURFACES

> *In this section we shall study an important class of surfaces that are the three-dimensional analogs of the conic sections.*

☐ **MESH PLOTS OF SURFACES**

In 2-space the general shape of a curve can be obtained by plotting points. However, for surfaces in 3-space, point-plotting is not generally helpful since too many points are needed to obtain even a crude picture of the surface. It is better to build up the shape of the surface using curves obtained by cutting the surface with planes parallel to the coordinate planes. This is called a **mesh plot** of the surface. For example, Figure 14.7.1 shows a mesh plot of the surface $z = x^3 - 3xy^2$. (This surface is called a "monkey saddle" because a monkey sitting astride the x-axis has a place for its two legs and tail.)

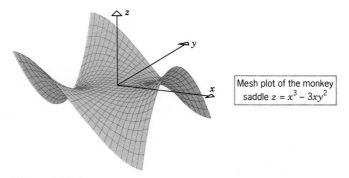

Mesh plot of the monkey saddle $z = x^3 - 3xy^2$

Figure 14.7.1

☐ **TRACES OF SURFACES**

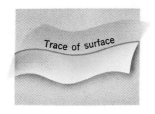

Trace of surface

Figure 14.7.2

The curve of intersection of a surface with a plane is called the **trace** of the surface in the plane (Figure 14.7.2). Mesh plots are usually built up from traces in planes parallel to the coordinate planes. Equations for such traces can be obtained by substituting the equation of the plane into the equation of the surface. For example, the trace of the monkey saddle of Figure 14.7.1 in the plane $x = 3$ is obtained by substituting $x = 3$ into

$$z = x^3 - 3xy^2$$

which yields

$$z - 27 = -9y^2 \quad (x = 3) \tag{1}$$

This is a parabola with vertex at the point $(x, y, z) = (3, 0, 27)$ opening in the negative z-direction (why?). Observe that this result is consistent with Figure 14.7.1.

REMARK.   In (1) we explicitly noted the restriction $x = 3$ in parentheses. This is necessary because the equation by itself does not convey the information that the trace lies in the plane $x = 3$.

□ **THE QUADRIC SURFACES**

Earlier in the text, we saw that the graph in two dimensions of a second-degree equation in x and y,

$$Ax^2 + Bxy + Cy^2 + Dx + Ey + F = 0$$

is a conic section (possibly degenerate). In three dimensions, the graph of a second-degree equation in x, y, and z,

$$Ax^2 + By^2 + Cz^2 + Dxy + Exz + Fyz + Gx + Hy + Iz + J = 0$$

is called a *quadric surface* or a *quadric*. The most important quadric surfaces, shown in Figure 14.7.3, are the *ellipsoids, hyperboloids of one and two sheets, elliptic cones, elliptic paraboloids,* and *hyperbolic paraboloids.*

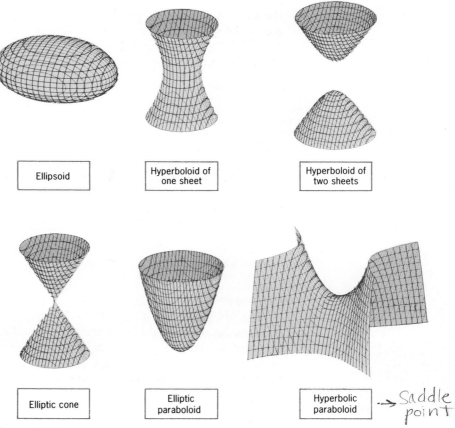

Ellipsoid

Hyperboloid of one sheet

Hyperboloid of two sheets

Elliptic cone

Elliptic paraboloid

Hyperbolic paraboloid → Saddle point

Figure 14.7.3

The simplest equations for the quadric surfaces result when the surfaces are positioned in certain "standard positions" relative to the coordinate axes. Table 14.7.1 illustrates some typical standard positions and the equations that result. The table also describes the traces of the quadric surfaces in planes parallel to the coordinate planes. As we shall see, such traces are important in studying properties of quadric surfaces. The constants a, b, and c that appear in the equations in the table are all assumed to be positive.

Table 14.7.1

SURFACE	EQUATION	SURFACE	EQUATION
ELLIPSOID	$$\frac{x^2}{a^2} + \frac{y^2}{b^2} + \frac{z^2}{c^2} = 1$$ The traces in the coordinate planes are ellipses, as are the traces in planes parallel to the coordinate planes.	ELLIPTIC CONE	$$z^2 = \frac{x^2}{a^2} + \frac{y^2}{b^2}$$ The trace in the xy-plane is a point (the origin), and the traces in planes parallel to the xy-plane are ellipses. The traces in the yz- and xz-planes are pairs of lines intersecting at the origin. The traces in planes parallel to these are hyperbolas.
HYPERBOLOID OF ONE SHEET	$$\frac{x^2}{a^2} + \frac{y^2}{b^2} - \frac{z^2}{c^2} = 1$$ The trace in the xy-plane is an ellipse, as are the traces in planes parallel to the xy-plane. The traces in the yz-plane and xz-plane are hyperbolas, as are the traces in planes parallel to these.	ELLIPTIC PARABOLOID	$$z = \frac{x^2}{a^2} + \frac{y^2}{b^2}$$ The trace in the xy-plane is a point (the origin), and the traces in planes parallel to and above the xy-plane are ellipses. The traces in the yz- and xz-planes are parabolas, as are the traces in planes parallel to these.
HYPERBOLOID OF TWO SHEETS	$$\frac{x^2}{a^2} + \frac{y^2}{b^2} - \frac{z^2}{c^2} = -1$$ There is no trace in the xy-plane. In planes parallel to the xy-plane, which intersect the surface, the traces are ellipses. In the yz- and xz-planes, the traces are hyperbolas, as are the traces in planes parallel to these.	HYPERBOLIC PARABOLOID	$$z = \frac{y^2}{b^2} - \frac{x^2}{a^2}$$ The trace in the xy-plane is a pair of lines intersecting at the origin. The traces in planes parallel to the xy-plane are hyperbolas. The hyperbolas above the xy-plane open in the y-direction, and those below in the x-direction. The traces in the yz- and xz-planes are parabolas, as are the traces in planes parallel to these.

To illustrate how the traces described in Table 14.7.1 were obtained, we shall consider the case of the elliptic cone

$$z^2 = \frac{x^2}{a^2} + \frac{y^2}{b^2} \tag{2}$$

The analysis of the other cases is similar. The trace of (2) in the plane $z = 0$ (the xy-plane) is

$$\frac{x^2}{a^2} + \frac{y^2}{b^2} = 0 \qquad (z = 0)$$

which implies that $x = 0$, $y = 0$, $z = 0$. Thus, the trace is the single point $(0, 0, 0)$.

For $k \neq 0$, the trace of (2) in the plane $z = k$ is (after simplification)

$$\frac{x^2}{(ak)^2} + \frac{y^2}{(bk)^2} = 1 \qquad (z = k)$$

which is an ellipse. As $|k|$ increases, so do $(ak)^2$ and $(bk)^2$. Thus, the dimensions of these ellipses increase as the planes containing them recede from the xy-plane. This is consistent with the picture of the elliptic cone in Table 14.7.1.

The trace of (2) in the xz-plane is

$$z^2 = \frac{x^2}{a^2} \qquad (y = 0)$$

which is equivalent to the pair of equations

$$z = \frac{x}{a} \quad \text{and} \quad z = -\frac{x}{a} \qquad (y = 0)$$

These are the equations of two lines in the xz-plane that intersect at the origin. Similarly, the trace of (2) in the yz-plane is the pair of intersecting lines

$$z = \frac{y}{b} \quad \text{and} \quad z = -\frac{y}{b} \qquad (x = 0)$$

For $k \neq 0$, the trace of (2) in the plane $y = k$, which is parallel to the xz-plane, is (after simplification)

$$\frac{z^2}{k^2/b^2} - \frac{x^2}{a^2k^2/b^2} = 1 \qquad (y = k)$$

This is a hyperbola in the plane $y = k$ opening along a line parallel to the z-axis. Similarly, for $k \neq 0$ the trace in the plane $x = k$, which is parallel to the yz-plane, is (after simplification)

$$\frac{z^2}{k^2/a^2} - \frac{y^2}{b^2k^2/a^2} = 1 \qquad (x = k)$$

This is a hyperbola in the plane $x = k$ opening along a line parallel to the z-axis.

REMARK. An elliptic cone in which the elliptical cross sections are circles is called a *circular cone;* similarly, elliptic paraboloids with circular cross sections are called *circular paraboloids.*

□ **TECHNIQUES FOR GRAPHING QUADRIC SURFACES**

Accurate graphs of quadric surfaces are best left to computers. However, the techniques that follow can be used to obtain rough sketches of these surfaces that are useful for many purposes.

ELLIPSOIDS   A sketch of the ellipsoid

$$\frac{x^2}{a^2} + \frac{y^2}{b^2} + \frac{z^2}{c^2} = 1 \quad (a > 0, \, b > 0, \, c > 0) \tag{3}$$

can be obtained by first plotting the intersections with the coordinate axes, then sketching the elliptical traces in the coordinate planes, and then sketching the surface itself using the traces as a guide.

Example 1   Sketch the graph of the ellipsoid

$$\frac{x^2}{4} + \frac{y^2}{16} + \frac{z^2}{9} = 1 \tag{4}$$

Solution.   The x-intercepts, obtained by setting $y = 0$ and $z = 0$ in (4), are $x = \pm 2$. Similarly, the y-intercepts are $y = \pm 4$, and the z-intercepts are $z = \pm 3$. From these intercepts we obtain the elliptical traces and the ellipsoid sketched in Figure 14.7.4.   ◄

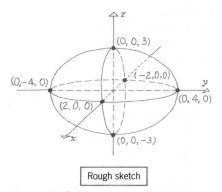

Figure 14.7.4

HYPERBOLOIDS OF ONE SHEET   A sketch of the hyperboloid of one sheet

$$\frac{x^2}{a^2} + \frac{y^2}{b^2} - \frac{z^2}{c^2} = 1 \quad (a > 0, \, b > 0, \, c > 0) \tag{5}$$

can be obtained by first sketching the elliptical trace in the xy-plane, then the elliptical traces in the planes $z = c$ and $z = -c$, and then the hyperbolic curves that join the endpoints of the axes of these ellipses.

Example 2 Sketch the graph of the hyperboloid of one sheet

$$x^2 + y^2 - \frac{z^2}{4} = 1 \tag{6}$$

Solution. The trace in the xy-plane, obtained by setting $z = 0$ in (6), is

$$x^2 + y^2 = 1 \qquad (z = 0)$$

which is a circle of radius 1 centered on the z-axis. The traces in the planes $z = 2$ and $z = -2$, obtained by setting $z = \pm 2$ in (6), are given by

$$x^2 + y^2 = 2 \qquad (z = \pm 2)$$

which are circles of radius $\sqrt{2}$ centered on the z-axis. Joining these circles by the hyperbolic traces in the vertical coordinate planes yields the sketch in Figure 14.7.5. ◄

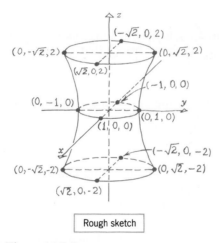

Rough sketch

Figure 14.7.5

HYPERBOLOIDS OF TWO SHEETS

A sketch of the hyperboloid of two sheets

$$\frac{x^2}{a^2} + \frac{y^2}{b^2} - \frac{z^2}{c^2} = -1 \quad (a > 0, \, b > 0, \, c > 0) \tag{7}$$

can be obtained by first plotting the intersections with the z-axis, then sketching the elliptical traces in the planes $z = 2c$ and $z = -2c$, and then sketching the hyperbolic traces that connect the z-axis intersections and the endpoints of the axes of the ellipses. (It is not essential to use the planes $z = \pm 2c$; any pair of horizontal planes above $z = c$ and below $z = -c$ would do just as well.)

Example 3   Sketch the graph of the hyperboloid of two sheets

$$x^2 + \frac{y^2}{4} - z^2 = -1 \tag{8}$$

Solution.   The z-intercepts, obtained by setting $x = 0$ and $y = 0$ in (8), are $z = \pm 1$. The traces in the planes $z = 2$ and $z = -2$, obtained by setting $z = \pm 2$ in (8), are given by

$$\frac{x^2}{3} + \frac{y^2}{12} = 1 \qquad (z = \pm 2)$$

Sketching these ellipses and the hyperbolic traces in the vertical coordinate planes yields Figure 14.7.6.   ◀

ELLIPTIC CONES   A sketch of the elliptic cone

$$z^2 = \frac{x^2}{a^2} + \frac{y^2}{b^2} \quad (a > 0, b > 0) \tag{9}$$

can be obtained by first sketching the elliptical traces in the planes $z = \pm 1$, then sketching the linear traces that connect the endpoints of the axes of the ellipses.

Example 4   Sketch the graph of the elliptic cone

$$z^2 = x^2 + \frac{y^2}{4} \tag{10}$$

Solution.   The traces of (10) in the planes $z = \pm 1$ are given by

$$x^2 + \frac{y^2}{4} = 1 \qquad (z = \pm 1)$$

Sketching these ellipses and the linear traces in the vertical coordinate planes yields the graph in Figure 14.7.7.   ◀

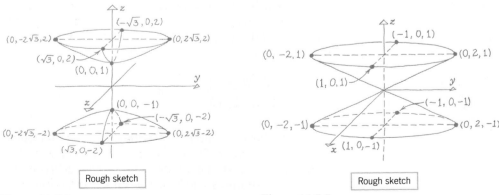

Figure 14.7.6   Figure 14.7.7

ELLIPTIC PARABOLOIDS   A sketch of the elliptic paraboloid

$$z = \frac{x^2}{a^2} + \frac{y^2}{b^2} \quad (a > 0, b > 0) \tag{11}$$

can be obtained by first sketching the elliptical trace in the plane $z = 1$, then sketching the parabolic traces (in the vertical coordinate planes) whose vertices are at the origin and pass through the endpoints of the axes of the ellipse.

Example 5   Sketch the graph of the elliptic paraboloid

$$z = \frac{x^2}{4} + \frac{y^2}{9} \tag{12}$$

Solution.   The trace of (12) in the plane $z = 1$ is

$$\frac{x^2}{4} + \frac{y^2}{9} = 1 \qquad (z = 1)$$

Sketching this ellipse and the parabolic traces in the vertical coordinate planes yields the graph in Figure 14.7.8.   ◀

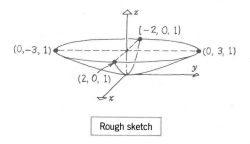

Rough sketch

Figure 14.7.8

HYPERBOLIC PARABOLOIDS   The graph of a hyperbolic paraboloid is difficult to draw, but fortunately a rough sketch showing the orientation of the surface relative to the coordinate axes suffices for most purposes. The orientation of the hyperbolic paraboloid

$$z = \frac{y^2}{b^2} - \frac{x^2}{a^2} \quad (a > 0, b > 0) \tag{13}$$

can be obtained by first sketching the two parabolic traces that pass through the origin; one of these results from setting $x = 0$ in (13) and the other from setting $y = 0$. After the parabolic traces are drawn, sketch the hyperbolic traces in the planes $z = \pm 1$ with their proper orientation. Finally, sketch in any missing edges.

Example 6   Sketch the graph of the hyperbolic paraboloid

$$z = \frac{y^2}{4} - \frac{x^2}{9} \qquad (14)$$

Solution.   Setting $x = 0$ in (14) yields

$$z = \frac{y^2}{4} \qquad (x = 0)$$

which is a parabola in the yz-plane with vertex at the origin and opening in the positive z-direction (why?), and setting $y = 0$ yields

$$z = -\frac{x^2}{9} \qquad (y = 0)$$

which is a parabola in the xz-plane with vertex at the origin and opening in the negative z-direction.

The trace in the plane $z = 1$ is

$$\frac{y^2}{4} - \frac{x^2}{9} = 1 \qquad (z = 1)$$

which is a hyperbola that opens along a line parallel to the y-axis (verify), and the trace in the plane $z = -1$ is

$$\frac{x^2}{9} - \frac{y^2}{4} = 1 \qquad (z = -1)$$

which is a hyperbola that opens along a line parallel to the x-axis. Combining all of the above information leads to the sketch in Figure 14.7.9.   ◄

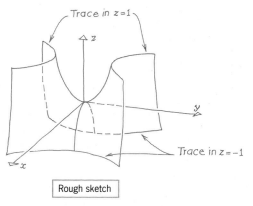

Trace in $z = 1$

Trace in $z = -1$

Rough sketch

Figure 14.7.9

REMARK.   The hyperbolic paraboloid (13) has an interesting behavior near the origin. The trace in the xz-plane has a relative maximum at the origin, and the trace in the yz-plane has a relative minimum there (see Figure 14.7.9, for example). On this surface the origin is commonly described as a *saddle point* or a *minimax point*.

REFLECTIONS OF QUADRIC SURFACES

In our earlier study of curves in the xy-plane, we observed that interchanging the variables x and y in an equation has the effect of reflecting the graph of the equation about the line $y = x$. Analogous results occur in xyz-coordinate systems. For example, interchanging the variables x and z in the equation of a surface has the geometric effect of reflecting that surface symmetrically about the plane $x = z$. Thus, the equation

$$x^2 = \frac{z^2}{a^2} + \frac{y^2}{b^2}$$

represents an elliptic cone opening along the x-axis rather than the z-axis as shown in Table 14.7.1.

We also note that replacing z by $-z$ in the equation of the elliptic paraboloid in Table 14.7.1 has the geometric effect of reflecting the surface symmetrically about the xy-plane, thereby producing an elliptic paraboloid that opens down. The equation for this surface can be written as

$$z = -\left(\frac{x^2}{a^2} + \frac{y^2}{b^2}\right) \tag{15}$$

TRANSLATION OF AXES IN 3-SPACE

Let an $x'y'z'$-coordinate system be obtained by translating an xyz-coordinate system so that the $x'y'z'$-origin is at the point whose xyz-coordinates are $(x, y, z) = (h, k, l)$ (Figure 14.7.10). It can be shown that the $x'y'z'$-coordinates and xyz-coordinates of a point P are related by

$$x' = x - h, \quad y' = y - k, \quad z' = z - l \tag{16}$$

[Compare this to (6) in Section 12.2.]

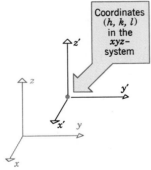

Coordinates (h, k, l) in the xyz-system

Figure 14.7.10

Example 7 Sketch the surface $z = 1 - x^2 - y^2$.

Solution. Rewrite the equation in the form

$$z - 1 = -(x^2 + y^2) \tag{17}$$

If we translate the coordinate axes so that the new origin is at the point

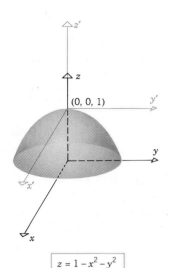

$$z = 1 - x^2 - y^2$$

Figure 14.7.11

$(h, k, l) = (0, 0, 1)$, then the translation equations (16) are

$$x' = x, \quad y' = y, \quad z' = z - 1$$

so that in $x'y'z'$-coordinates (17) becomes

$$z' = -(x'^2 + y'^2)$$

which is of form (15) with $a = b = 1$. Thus, the surface is a circular paraboloid, opening down, with vertex $(0, 0, 1)$ in xyz-coordinates (Figure 14.7.11). ◄

Example 8   Sketch the surface

$$4x^2 + 4y^2 + z^2 + 8y - 4z = -4$$

Solution.   Completing the squares yields

$$4x^2 + 4(y + 1)^2 + (z - 2)^2 = -4 + 4 + 4$$

or

$$x^2 + (y + 1)^2 + \frac{(z - 2)^2}{4} = 1 \tag{18}$$

If we translate the coordinate axes so that the new origin is at the point $(h, k, l) = (0, -1, 2)$, then the translation equations (16) are

$$x' = x, \quad y' = y + 1, \quad z' = z - 2$$

so that in $x'y'z'$-coordinates (18) becomes

$$x'^2 + y'^2 + \frac{z'^2}{4} = 1$$

which is of form (3) with $a = 1$, $b = 1$, and $c = 2$. Thus, the surface is the ellipsoid shown in Figure 14.7.12. ◄

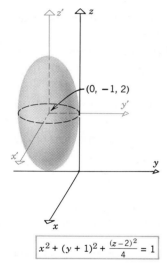

$$x^2 + (y + 1)^2 + \frac{(z - 2)^2}{4} = 1$$

Figure 14.7.12

► Exercise Set 14.7

In Exercises 1–6, find an equation for the trace of the surface in the given plane and identify the trace.

1. $4x^2 + y^2 + z^2 = 4$:
   (a) $z = 0$ (b) $x = 1/2$ (c) $y = 1$.

2. $9x^2 - y^2 + 4z^2 = 9$:
   (a) $z = 0$ (b) $x = 2$ (c) $y = 4$.

3. $9x^2 - y^2 - z^2 = 16$:
   (a) $y = 0$ (b) $x = 2$ (c) $z = 2$.

4. $x^2 + 4y^2 - 9z^2 = 0$:
   (a) $x = 0$ (b) $y = 1$ (c) $z = 1$.

5. $z = 9x^2 + 4y^2$:
   (a) $x = 0$ (b) $y = 2$ (c) $z = 4$.

6. $z = x^2 - 4y^2$:
    (a) $y = 0$ (b) $x = 1$ (c) $z = 4$.

In Exercises 7–28, name and sketch the quadric surface.

7. $x^2 + y^2/4 + z^2/9 = 1$. **8.** $x^2 + 4y^2 + 9z^2 = 36$.

9. $4x^2 + y^2 + 4z^2 = 16$. **10.** $4x^2 + 4y^2 + z^2 = 9$.

11. $x^2/4 + y^2/9 - z^2/16 = 1$.

12. $4x^2 - y^2 + 4z^2 = 16$.

13. $2y^2 - x^2 + 2z^2 = 8$. **14.** $x^2 + y^2 - z^2 = 9$.

15. $y^2 - 2x^2 - 2z^2 = 1$. **16.** $x^2 - 3y^2 - 3z^2 = 9$.

17. $9z^2 - 4y^2 - 9x^2 = 36$.

18. $y^2 - 4x^2 - z^2 = 4$.

19. $4z^2 = x^2 + 4y^2$.   **20.** $y^2 = x^2 + z^2$.

21. $x^2 - 3y^2 - 3z^2 = 0$. **22.** $9x^2 + 4y^2 - 36z^2 = 0$.

23. $y = x^2 + z^2$.   **24.** $z - 3x^2 - 3y^2 = 0$.

25. $4z = x^2 + 2y^2$.   **26.** $x - y^2 - 4z^2 = 0$.

27. $z = x^2/4 - y^2/9$.   **28.** $z = y^2 - x^2$.

29. The following equations represent quadric surfaces with orientations different from those in Table 14.7.1. Name and sketch the surface.

  (a) $\dfrac{z^2}{c^2} - \dfrac{y^2}{b^2} + \dfrac{x^2}{a^2} = 1$ (b) $\dfrac{x^2}{a^2} - \dfrac{y^2}{b^2} - \dfrac{z^2}{c^2} = 1$

  (c) $x = \dfrac{y^2}{b^2} + \dfrac{z^2}{c^2}$ (d) $x^2 = \dfrac{y^2}{b^2} + \dfrac{z^2}{c^2}$

  (e) $y = \dfrac{z^2}{c^2} - \dfrac{x^2}{a^2}$ (f) $y = -\left(\dfrac{x^2}{a^2} + \dfrac{z^2}{c^2}\right)$.

In Exercises 30–33, sketch the graph of the equation.

30. $z = \sqrt{1 - x^2 - y^2}$. **31.** $z = \sqrt{x^2 + y^2}$.

32. $z = \sqrt{1 + x^2 + y^2}$. **33.** $z = \sqrt{x^2 + y^2 - 1}$.

In Exercises 34–39, name and sketch the surface.

34. $\dfrac{(x-1)^2}{4} + \dfrac{(y-2)^2}{9} + \dfrac{(z-4)^2}{16} = 1$.

35. $z = (x+2)^2 + (y-3)^2 - 9$.

36. $4x^2 - y^2 + 16(z-2)^2 = 100$.

37. $9x^2 + y^2 + 4z^2 - 18x + 2y + 16z = 10$.

38. $z^2 = 4x^2 + y^2 + 8x - 2y + 4z$.

39. $z = 4 - x^2 - y^2 - 2y$.

40. Obtain the results in Table 14.7.1 for the ellipsoid $x^2/a^2 + y^2/b^2 + z^2/c^2 = 1$.

41. Obtain the results in Table 14.7.1 for the hyperboloid of one sheet $x^2/a^2 + y^2/b^2 - z^2/c^2 = 1$.

42. Obtain the results in Table 14.7.1 for the hyperboloid of two sheets $x^2/a^2 + y^2/b^2 - z^2/c^2 = -1$.

43. Obtain the results in Table 14.7.1 for the elliptic paraboloid $z = x^2/a^2 + y^2/b^2$.

44. Obtain the results in Table 14.7.1 for the hyperbolic paraboloid $z = y^2/b^2 - x^2/a^2$.

In Exercises 45–52, find an equation of the orthogonal projection onto the xy-plane of the curve of intersection of the surfaces. Identify the curve. [*Hint:* The values of z on both surfaces are the same along the curve of intersection.]

45. The paraboloids $z = x^2 + y^2$ and $z = 4 - x^2 - y^2$.

46. The paraboloids $z = x^2 + y^2$ and $z = 1 - 4x^2 - y^2$.

47. The paraboloid $z = x^2 + y^2$ and the plane $z = 2x$.

48. The paraboloid $z = 4 - x^2 - y^2$ and the parabolic cylinder $z = y^2$.

49. The cone $z^2 = x^2 + y^2$ and the plane $z = y + 1$.

50. The cone $z^2 = x^2 + y^2$ and the parabolic cylinder $z = 2\sqrt{y}$.

51. The ellipsoid $x^2 + y^2 + 4z^2 = 5$ and the parabolic cylinder $z = \sqrt{x}$.

52. The ellipsoid $x^2 + 2y^2 + z^2 = 2$ and the plane $z = x$.

53. For the elliptic paraboloid

$$z = \frac{x^2}{9} + \frac{y^2}{4}$$

  (a) find the focus and vertex of the (parabolic) trace in the plane $x = k$

  (b) find the foci and the endpoints of the major and minor axes of the (elliptic) trace in the plane $z = k$.

54. Use the method of slicing to find the volume of the ellipsoid

$$\frac{x^2}{a^2} + \frac{y^2}{b^2} + \frac{z^2}{c^2} = 1$$

[*Hint:* The area of the ellipse $x^2/a^2 + y^2/b^2 = 1$ is πab.]

55. Find an equation of the surface consisting of the points $P(x, y, z)$ for which the distance between P and the plane $z = -1$ is equal to the distance between P and the point $(0, 0, 1)$. Identify the surface.

56. Find an equation of the surface consisting of the points $P(x, y, z)$ for which the distance between P and the plane $z = -1$ is twice the distance between P and the point $(0, 0, 1)$. Identify the surface.

57. (a) Show that the lines
$$x = 3 + t, \quad y = 2 + t, \quad z = 5 + 2t$$
and
$$x = 3 + t, \quad y = 2 - t, \quad z = 5 + 10t$$
both lie completely on the hyperbolic paraboloid $z = x^2 - y^2$.

(b) Let $P_0(x_0, y_0, z_0)$ be any point on the surface $z = x^2 - y^2$. Show that it is always possible to

find two lines with equations of the form $x = x_0 + t, y = y_0 + at, z = z_0 + bt$ that pass through P_0 and lie completely on the surface.

58. (a) Show that the lines $x = 2 + \frac{3}{2}t, \; y = 1 + \frac{4}{5}t,$ $z = 2 + t$ and $x = 2 + t, \; y = 1, \; z = 2 + t$ both lie completely on the hyperboloid $x^2 + y^2 - z^2 = 1$.

(b) Let $P_0(x_0, y_0, z_0)$ be any point on the surface $x^2 + y^2 - z^2 = 1$. Show that it is always possible to find two lines with equations of the form $x = x_0 + at, y = y_0 + bt, z = z_0 + t$ that pass through P_0 and lie completely on the surface.

■ 14.8 CYLINDRICAL AND SPHERICAL COORDINATES

> *In this section we shall discuss two new types of coordinate systems in three dimensions that are extremely important in applied problems. These coordinate systems produce simpler equations than rectangular coordinate systems for surfaces with various kinds of symmetries.*

☐ **CYLINDRICAL AND SPHERICAL COORDINATES**

To have a useful coordinate system in three dimensions, each point in space must be associated with a unique triple of real numbers (the coordinates of the point), and each triple of real numbers must determine a unique point. Figure 14.8.1 illustrates three possible ways for doing this. The rectangular coordinates (x, y, z) of a point P are shown in part (a) of the figure, the **cylindrical coordinates** (r, θ, z) of P are shown in part (b), and the **spherical coordinates** (ρ, θ, ϕ) of P are shown in part (c).

The restrictions on the values of the cylindrical and spherical coordinates noted in Figure 14.8.1 are fairly standard; they ensure that each point, other than the origin, has a unique set of coordinates.

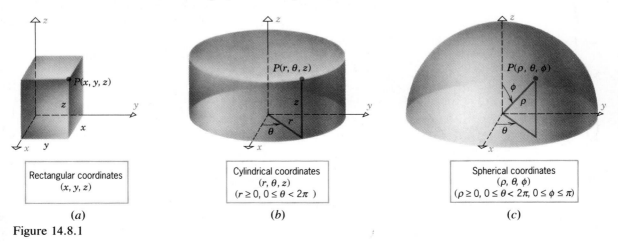

Rectangular coordinates (x, y, z)	Cylindrical coordinates (r, θ, z) $(r \geq 0, 0 \leq \theta < 2\pi)$	Spherical coordinates (ρ, θ, ϕ) $(\rho \geq 0, 0 \leq \theta < 2\pi, 0 \leq \phi \leq \pi)$
(a)	(b)	(c)

Figure 14.8.1

☐ **CONSTANT SURFACES**

In rectangular coordinates the surfaces represented by equations of the form

$$x = x_0, \quad y = y_0, \quad \text{and} \quad z = z_0$$

where x_0, y_0, and z_0 are constants, are planes parallel to the yz-plane, xz-plane, and xy-plane, respectively (Figure 14.8.2a). In cylindrical coordinates the surfaces represented by equations of the form

$$r = r_0, \quad \theta = \theta_0, \quad \text{and} \quad z = z_0$$

where r_0, θ_0, and z_0 are constants, are shown in Figure 14.8.2b.

- The surface $r = r_0$ is a right-circular cylinder of radius r_0 centered on the z-axis. At each point (r, θ, z) on this cylinder, r has the value r_0, but θ and z are unrestricted except for our blanket assumption that $0 \le \theta < 2\pi$.
- The surface $\theta = \theta_0$ is a half-plane attached along the z-axis and making an angle θ_0 with the positive x-axis. At each point (r, θ, z) on this surface, θ has the value θ_0, but r and z are unrestricted except for our blanket assumption that $r \ge 0$.
- The surface $z = z_0$ is a horizontal plane. At each point (r, θ, z) on this plane, z has the value z_0, but r and θ are unrestricted except for our blanket assumptions.

In spherical coordinates the surfaces represented by equations of the form

$$\rho = \rho_0, \quad \theta = \theta_0, \quad \text{and} \quad \phi = \phi_0$$

where ρ_0, θ_0, and ϕ_0 are constants, are shown in Figure 14.8.2c.

- The surface $\rho = \rho_0$ consists of all points whose distance ρ from the origin is ρ_0. Assuming ρ_0 to be nonnegative, this is a sphere of radius ρ_0 centered at the origin.
- As in cylindrical coordinates, the surface $\theta = \theta_0$ is a half-plane attached along the z-axis, making an angle of θ_0 with the positive x-axis.
- The surface $\phi = \phi_0$ consists of all points from which a line segment to the origin makes an angle of ϕ_0 with the positive z-axis. Depending on whether $0 < \phi_0 < \pi/2$ or $\pi/2 < \phi_0 < \pi$, this will be a cone opening up or opening down. (If $\phi_0 = \pi/2$, then the cone is flat and the surface is the xy-plane.)

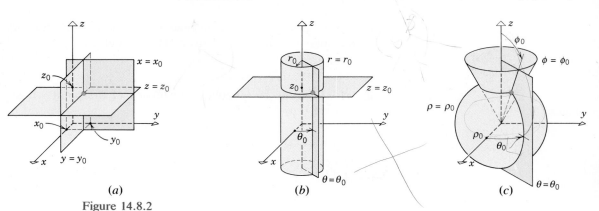

(a)   (b)   (c)

Figure 14.8.2

□ **CONVERTING
COORDINATES**

Frequently, the coordinates of a point are known in one type of coordinate system and it is of interest to find the coordinates in one of the other types. Table 14.8.1 lists the formulas for making such coordinate conversions.

Table 14.8.1

CONVERSION	FORMULAS
$(r, \theta, z) \rightarrow (x, y, z)$	$x = r \cos \theta, \quad y = r \sin \theta, \quad z = z$
$(x, y, z) \rightarrow (r, \theta, z)$	$r = \sqrt{x^2 + y^2}, \quad \tan \theta = y/x, \quad z = z$
$(\rho, \theta, \phi) \rightarrow (r, \theta, z)$	$r = \rho \sin \phi, \quad \theta = \theta, \quad z = \rho \cos \phi$
$(r, \theta, z) \rightarrow (\rho, \theta, \phi)$	$\rho = \sqrt{r^2 + z^2}, \quad \theta = \theta, \quad \tan \phi = r/z$
$(\rho, \theta, \phi) \rightarrow (x, y, z)$	$x = \rho \sin \phi \cos \theta, \quad y = \rho \sin \phi \sin \theta, \quad z = \rho \cos \phi$
$(x, y, z) \rightarrow (\rho, \theta, \phi)$	$\rho = \sqrt{x^2 + y^2 + z^2}, \quad \tan \theta = y/x, \quad \cos \phi = z/\sqrt{x^2 + y^2 + z^2}$

The formulas in this table can be derived by considering the diagrams in Figure 14.8.3. Part (*a*) of the figure illustrates the relationship between the rectangular coordinates (x, y, z) and the cylindrical coordinates (r, θ, z) of a point P; it shows that the value of z is the same in both coordinate systems and that (r, θ) is a pair of polar coordinates for the point (x, y) in the xy-plane. Thus, it follows from the relationship between polar and rectangular coordinates [Formulas (1a–b) of Section 13.1] that

$$x = r \cos \theta, \quad y = r \sin \theta, \quad z = z \tag{1}$$

Part (*b*) of Figure 14.8.3 illustrates the relationship between the spherical coordinates (ρ, θ, ϕ) and the cylindrical coordinates (r, θ, z) of a point P:

$$r = \rho \sin \phi, \quad \theta = \theta, \quad z = \rho \cos \phi \tag{2}$$

Substituting (2) into (1) yields the following relationships between the rectangular coordinates (x, y, z) and the spherical coordinates (ρ, θ, ϕ) of a point P:

$$x = \rho \sin \phi \cos \theta, \quad y = \rho \sin \phi \sin \theta, \quad z = \rho \cos \phi \tag{3}$$

We leave it as an exercise for the reader to deduce the remaining three conversion formulas in Table 14.8.1 from (1), (2), and (3).

Example 1   Find the rectangular coordinates of the point whose cylindrical coordinates are $(r, \theta, z) = (4, \pi/3, -3)$.

Solution.   From (1)

$$x = 4 \cos \frac{\pi}{3} = 2, \quad y = 4 \sin \frac{\pi}{3} = 2\sqrt{3}, \quad z = -3 \quad ◀$$

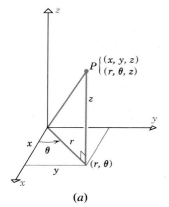

(a)

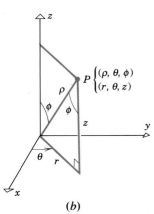

(b)

Figure 14.8.3

Example 2 Find an equation in cylindrical coordinates of the surface whose equation in rectangular coordinates is $z = x^2 + y^2 - 2x + y$.

Solution. From (1)

$$z = r^2 - 2r \cos \theta + r \sin \theta \quad \blacktriangleleft$$

Example 3 Find an equation in rectangular coordinates of the surface whose equation in cylindrical coordinates is $r = 4 \cos \theta$.

Solution. Multiplying both sides of the given equation by r yields the equation $r^2 = 4r \cos \theta$, then using the relationships $x^2 + y^2 = r^2$ and $x = r \cos \theta$, which follow from (1), yields

$$x^2 + y^2 = 4x \quad \text{or equivalently} \quad (x - 2)^2 + y^2 = 4$$

This is a right-circular cylinder parallel to the z-axis. $\quad \blacktriangleleft$

Example 4 Find the rectangular coordinates of the point whose spherical coordinates (ρ, θ, ϕ) are $(4, \pi/3, \pi/4)$.

Solution. From (3)

$$x = \rho \sin \phi \cos \theta = 4 \sin \frac{\pi}{4} \cos \frac{\pi}{3} = \sqrt{2}$$

$$y = \rho \sin \phi \sin \theta = 4 \sin \frac{\pi}{4} \sin \frac{\pi}{3} = \sqrt{6}$$

$$z = \rho \cos \phi = 4 \cos \frac{\pi}{4} = 2\sqrt{2} \quad \blacktriangleleft$$

Example 5 Find an equation of the paraboloid $z = x^2 + y^2$ in spherical coordinates.

Solution. Substituting (3) in this equation yields

$$\rho \cos \phi = \rho^2 \sin^2 \phi \cos^2 \theta + \rho^2 \sin^2 \phi \sin^2 \theta$$
$$\rho \cos \phi = \rho^2 \sin^2 \phi (\cos^2 \theta + \sin^2 \theta)$$
$$\rho \cos \phi = \rho^2 \sin^2 \phi$$

which simplifies to $\rho \sin^2 \phi = \cos \phi$. $\quad \blacktriangleleft$

☐ **SPHERICAL COORDINATES IN NAVIGATION**

Spherical coordinates are related to longitude and latitude coordinates used in navigation. Let us construct a right-hand rectangular coordinate system with origin at the earth's center, positive z-axis passing through the north pole, and positive x-axis passing through the prime meridian (Figure 14.8.4). If we assume the earth to be a perfect sphere of radius $\rho = 4000$ miles, then each point on the earth has spherical coordinates of the form $(4000, \theta, \phi)$, where ϕ and θ

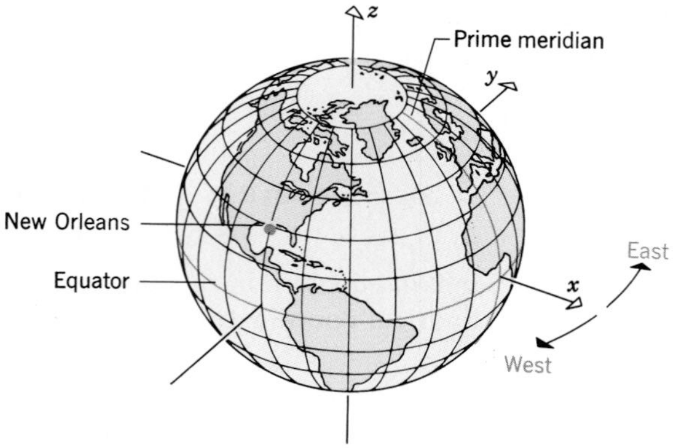

Figure 14.8.4

determine the latitude and longitude of the point. It is useful to specify longitudes in degrees east or west of the prime meridian and latitudes in degrees north or south of the equator. However, it is a simple matter to determine ϕ and θ from such data.

Example 6   The city of New Orleans is located at 90° West longitude and 30° North latitude. Find its spherical and rectangular coordinates relative to the coordinate axes of Figure 14.8.4. (Take miles as the unit of distance.)

Solution.   A longitude of 90° West corresponds to $\theta = 360° - 90° = 270°$ or $\theta = 3\pi/2$ radians; and a latitude of 30° North corresponds to $\phi = 90° - 30° = 60°$ or $\phi = \pi/3$ radians. Thus, the spherical coordinates (ρ, θ, ϕ) of New Orleans are $(4000, 3\pi/2, \pi/3)$.

From (3), the rectangular coordinates of New Orleans are

$$x = 4000 \sin \frac{\pi}{3} \cos \frac{3\pi}{2} = 4000 \frac{\sqrt{3}}{2} (0) = 0 \text{ miles}$$

$$y = 4000 \sin \frac{\pi}{3} \sin \frac{3\pi}{2} = 4000 \frac{\sqrt{3}}{2} (-1) = -2000\sqrt{3} \text{ miles}$$

$$z = 4000 \cos \frac{\pi}{3} = 4000 \left(\frac{1}{2}\right) = 2000 \text{ miles} \qquad \blacktriangleleft$$

▶ Exercise Set 14.8

1.   Convert from rectangular to cylindrical coordinates.
   (a) $(4\sqrt{3}, 4, -4)$ (b) $(-5, 5, 6)$
   (c) $(0, 2, 0)$ (d) $(4, -4\sqrt{3}, 6)$
   (e) $(\sqrt{2}, -\sqrt{2}, 1)$ (f) $(0, 0, 1)$.

2.   Convert from cylindrical to rectangular coordinates.
   (a) $(4, \pi/6, 3)$ (b) $(8, 3\pi/4, -2)$
   (c) $(5, 0, 4)$ (d) $(7, \pi, -9)$
   (e) $(6, 5\pi/3, 7)$ (f) $(1, \pi/2, 0)$.

3.   Convert from rectangular to spherical coordinates.
   (a) $(1, \sqrt{3}, -2)$ (b) $(1, -1, \sqrt{2})$
   (c) $(0, 3\sqrt{3}, 3)$ (d) $(-5\sqrt{3}, 5, 0)$
   (e) $(4, 4, 4\sqrt{6})$ (f) $(1, -\sqrt{3}, -2)$.

4.   Convert from spherical to rectangular coordinates.
   (a) $(5, \pi/6, \pi/4)$ (b) $(7, 0, \pi/2)$
   (c) $(1, \pi, 0)$ (d) $(2, 3\pi/2, \pi/2)$
   (e) $(1, 2\pi/3, 3\pi/4)$ (f) $(3, 7\pi/4, 5\pi/6)$.

5. Convert from cylindrical to spherical coordinates.
   (a) $(\sqrt{3}, \pi/6, 3)$ (b) $(1, \pi/4, -1)$
   (c) $(2, 3\pi/4, 0)$ (d) $(6, 1, -2\sqrt{3})$
   (e) $(4, 5\pi/6, 4)$ (f) $(2, 0, -2)$.

6. Convert from spherical to cylindrical coordinates.
   (a) $(5, \pi/4, 2\pi/3)$ (b) $(1, 7\pi/6, \pi)$
   (c) $(3, 0, 0)$ (d) $(4, \pi/6, \pi/2)$
   (e) $(5, \pi/2, 0)$ (f) $(6, 0, 3\pi/4)$.

In Exercises 7–14, an equation is given in cylindrical coordinates. Express the equation in rectangular coordinates and sketch the graph.

7. $r = 3$.   8. $\theta = \pi/4$.

9. $z = r^2$.   10. $z = r\cos\theta$.

11. $r = 4\sin\theta$.   12. $r = 2\sec\theta$.

13. $r^2 + z^2 = 1$.   14. $r^2\cos 2\theta = z$.

In Exercises 15–22, an equation is given in spherical coordinates. Express the equation in rectangular coordinates and sketch the graph.

15. $\rho = 3$.   16. $\theta = \pi/3$.

17. $\phi = \pi/4$.   18. $\rho = 2\sec\phi$.

19. $\rho = 4\cos\phi$.   20. $\rho\sin\phi = 1$.

21. $\rho\sin\phi = 2\cos\theta$.   22. $\rho - 2\sin\phi\cos\theta = 0$.

In Exercises 23–34, an equation of a surface is given in rectangular coordinates. Find an equation of the surface in (a) cylindrical coordinates and (b) spherical coordinates.

23. $z = 3$.   24. $y = 2$.
25. $z = 3x^2 + 3y^2$.   26. $z = \sqrt{3x^2 + 3y^2}$.
27. $x^2 + y^2 = 4$.   28. $x^2 + y^2 - 6y = 0$.
29. $x^2 + y^2 + z^2 = 9$.   30. $z^2 = x^2 - y^2$.
31. $2x + 3y + 4z = 1$.   32. $x^2 + y^2 - z^2 = 1$.
33. $x^2 = 16 - z^2$.   34. $x^2 + y^2 + z^2 = 2z$.

In Exercises 35–38, describe the three-dimensional region that satisfies the given inequalities.

35. $r^2 \le z \le 4$.

36. $0 \le r \le 2\sin\theta$, $\quad 0 \le z \le 3$.

37. $1 \le \rho \le 3$.

38. $0 \le \phi \le \pi/6$, $\quad 0 \le \rho \le 2$.

39. Leningrad, Russia, is located at 30° East longitude and 60° North latitude. Find its spherical and rectangular coordinates relative to the coordinate axes of Figure 14.8.4. Take miles as the unit of distance and assume the earth to be a sphere of radius 4000 miles.

40. (a) Show that the curve of intersection of the surfaces $z = \sin\theta$ and $r = a$ (cylindrical coordinates) is an ellipse.
    (b) Sketch a portion of the surface $z = \sin\theta$ for $0 \le \theta \le \pi/2$.

41. Sketch the surface whose equation in spherical coordinates is $\rho = a(1 - \cos\phi)$. [*Hint:* The surface is shaped like a familiar fruit.]

▶ SUPPLEMENTARY EXERCISES

1. Find $\overrightarrow{P_1P_2}$ and $\|\overrightarrow{P_1P_2}\|$.
   (a) $P_1(2, 3)$, $P_2(5, -1)$ (b) $P_1(2, -1)$, $P_2(1, 3)$.

In Exercises 2–7, find all vectors satisfying the given conditions.

2. A vector of length 1 in 2-space that is perpendicular to the line $x + y = -1$.

3. The vector oppositely directed to $3\mathbf{i} - 4\mathbf{j}$, and having the same length.

4. The vector obtained by rotating $\mathbf{i}$ counterclockwise through an angle θ in 2-space.

5. The vector with initial point $(1, 2)$ and a terminal point that is 3/5 of the way from $(1, 2)$ to $(5, 5)$.

6. A vector of length 2 that is parallel to the tangent to the curve $y = x^2$ at $(-1, 1)$.

7. The vector of length 12 in 2-space that makes an angle of 120° with the x-axis.

8. Solve for c_1 and c_2 given that $c_1\langle -2, 5\rangle + 3c_2\langle 1, 3\rangle = \langle -6, -51\rangle$.

9. Solve for $\mathbf{u}$ if $3\mathbf{u} - (\mathbf{i} + \mathbf{j}) = \mathbf{i} + \mathbf{u}$.

10. Solve for $\mathbf{u}$ and $\mathbf{v}$ if $3\mathbf{u} - 4\mathbf{v} = 3\mathbf{v} - 2\mathbf{u} = \langle 1, 2\rangle$.

11. Two forces $\mathbf{F}_1 = 2\mathbf{i} - \mathbf{j}$ and $\mathbf{F}_2 = -3\mathbf{i} - 4\mathbf{j}$ are applied at a point. What force $\mathbf{F}_3$ must be applied at the point to cancel the effect of $\mathbf{F}_1$ and $\mathbf{F}_2$?

12. Given the points $P(3, 4)$, $Q(1, 1)$, and $R(5, 2)$, use vector methods to find the coordinates of the fourth vertex of the parallelogram whose adjacent sides are $\overrightarrow{PQ}$ and $\overrightarrow{QR}$.

In Exercises 13 and 14, find
(a) $\|\mathbf{a}\|$ (b) $\mathbf{a} \cdot \mathbf{b}$
(c) $\mathbf{a} \times \mathbf{b}$ (d) $\mathbf{b} \times \mathbf{a}$
(e) the area of the triangle with sides $\mathbf{a}$ and $\mathbf{b}$
(f) $3\mathbf{a} - 2\mathbf{b}$.

13. $\mathbf{a} = \langle 1, 2, -1 \rangle$, $\mathbf{b} = \langle 2, -1, 3 \rangle$.

14. $\mathbf{a} = \langle 1, -2, 2 \rangle$, $\mathbf{b} = \langle 3, 4, -5 \rangle$.

In Exercises 15 and 16, find
(a) $\|\text{proj}_\mathbf{b}\, \mathbf{a}\|$ (b) $\|\text{proj}_\mathbf{a}\, \mathbf{b}\|$
(c) the angle between $\mathbf{a}$ and $\mathbf{b}$
(d) the direction cosines of $\mathbf{a}$.

15. $\mathbf{a} = 3\mathbf{i} - 4\mathbf{j}$, $\mathbf{b} = 2\mathbf{i} + 2\mathbf{j} - \mathbf{k}$.

16. $\mathbf{a} = -\mathbf{j}$, $\mathbf{b} = \mathbf{i} + \mathbf{j}$.

17. Verify the identity $\mathbf{a} \times (\mathbf{b} \times \mathbf{c}) = (\mathbf{a} \cdot \mathbf{c})\mathbf{b} - (\mathbf{a} \cdot \mathbf{b})\mathbf{c}$ for $\mathbf{a} = \mathbf{i} + \mathbf{j}$, $\mathbf{b} = 2\mathbf{i} - \mathbf{k}$, $\mathbf{c} = \mathbf{j} - \mathbf{k}$.

18. Find the vector with length 5 and direction angles $\alpha = 60°$, $\beta = 120°$, $\gamma = 135°$.

19. Find the vector with length 3 and direction cosines $-1/\sqrt{2}$, 0, and $1/\sqrt{2}$.

20. For the points $P(6, 5, 7)$ and $Q(7, 3, 9)$, find
(a) the midpoint of the line segment PQ
(b) the length and direction cosines of $\overrightarrow{PQ}$.

21. If $\mathbf{u} = \mathbf{i} + 2\mathbf{j} - 3\mathbf{k}$ and $\mathbf{v} = \mathbf{i} + \mathbf{j} + 2\mathbf{k}$, find
(a) the vector component of $\mathbf{u}$ along $\mathbf{v}$
(b) the vector component of $\mathbf{u}$ orthogonal to $\mathbf{v}$.

22. Find the vector component of $\mathbf{i}$ along $3\mathbf{i} - 2\mathbf{j} + \mathbf{k}$.

23. A diagonal of a box makes angles of $50°$ and $70°$ with two of its edges. Find, to the nearest degree, the angle that it makes with the third edge.

24. Consider the points $O(0, 0, 0)$, $A(0, a, a)$, and $B(-3, 4, 2)$. Find all nonzero values of a that make $\overrightarrow{OA}$ orthogonal to $\overrightarrow{AB}$.

25. Under what conditions are $\mathbf{u} + \mathbf{v}$ and $\mathbf{u} - \mathbf{v}$ orthogonal?

26. If $M(3, -1, 5)$ is the midpoint of the line segment PQ and if the coordinates of P are $(1, 2, 3)$, find the coordinates of Q.

27. Find all possible vectors of length 1 orthogonal to both $\mathbf{a} = \langle 3, -2, 1 \rangle$ and $\mathbf{b} = \langle -2, 1, -3 \rangle$.

28. Find the distance from the point $P(2, 3, 4)$ to the plane containing the points $A(0, 0, 1)$, $B(1, 0, 0)$, and $C(0, 2, 0)$.

In Exercises 29–32, find an equation for the plane that satisfies the given conditions.

29. The plane through $A(1, 2, 3)$ and $B(2, 4, 2)$ that is parallel to $\mathbf{v} = \langle -3, -1, -2 \rangle$.

30. The plane through $P_0(-1, 2, 3)$ that is perpendicular to the planes $2x - 3y + 5 = 0$ and $3x - y - 4z + 6 = 0$.

31. The plane that passes through $P(1, 1, 1)$, $Q(2, 3, 0)$, and $R(2, 1, 2)$.

32. The plane with intercepts $x = 2$, $y = -3$, $z = 10$.

33. Let L be the line through $P(1, 2, 8)$ that is parallel to $\mathbf{v} = \langle 3, -1, -4 \rangle$.
(a) For what values of k and l will the point $Q(k, 3, l)$ be on L?
(b) If L' has parametric equations $x = -8 - 3t$, $y = 5 + t$, $z = 0$, show that L' intersects L and find the point of intersection.
(c) Find the point at which L intersects the plane through $R(-4, 0, 3)$ having a normal vector $\langle 3, -2, 6 \rangle$.

34. Consider the lines L_1 and L_2 with symmetric equations

$$L_1: \frac{x - 1}{2} = \frac{y + \frac{3}{2}}{1} = \frac{z + 1}{2}$$

$$L_2: \frac{4 - x}{1} = \frac{3 - y}{2} = \frac{4 + z}{2}$$

(see Exercise 34, Section 14.5).
(a) Are L_1 and L_2 parallel? Perpendicular?
(b) Find parametric equations for L_1 and L_2.
(c) Do L_1 and L_2 intersect? If so, where?

35. Find parametric equations for the line through P_1 and P_2.
(a) $P_1(1, -1, 2)$, $P_2(3, 2, -1)$
(b) $P_1(1, -3, 4)$, $P_2(1, 2, -3)$.

36. For points $A(1, -1, 2)$, $B(2, -3, 0)$, $C(-1, -2, 0)$, and $D(2, 1, -1)$, find
   (a) $\overrightarrow{AB} \times \overrightarrow{AC}$ (b) the area of triangle ABC
   (c) the volume of the parallelepiped determined by the vectors $\overrightarrow{AB}$, $\overrightarrow{AC}$, $\overrightarrow{AD}$
   (d) the distance from D to the plane containing A, B, and C.

37. (a) Find parametric equations for the intersection of the planes $2x + y - z = 3$ and $x + 2y + z = 3$.
   (b) Find the acute angle between the two planes.

In Exercises 38–40, describe the region satisfying the given conditions.

38. (a) $x^2 + 9y^2 + 4z^2 > 36$
   (b) $x^2 + y^2 + z^2 - 6x + 2y - 6 < 0$.

39. (a) $z > 4x^2 + 9y^2$
   (b) $x^2 + 4y^2 + z^2 = 0$.

40. (a) $y^2 + 4z^2 = 4$, $0 \le x \le 2$
   (b) $9x^2 + 4y^2 + 36x - 8y = -60$.

In Exercises 41–45, identify the quadric surface whose equation is given.

41. $100x^2 + 225y^2 - 36z^2 = 0$.

42. $x^2 - z^2 + y = 0$.

43. $400x^2 + 25y^2 + 16z^2 = 400$.

44. $4x^2 - y^2 + 4z^2 = 4$.

45. $-16x^2 - 100y^2 + 25z^2 = 400$.

46. Identify the surface by completing the squares.
   (a) $x^2 + 4y^2 - z^2 - 6x + 8y + 4z = 0$
   (b) $x^2 + y^2 + z^2 + 6x - 4y + 12z = 0$.

47. Find the work done by a constant force $\mathbf{F} = 3\mathbf{i} - 4\mathbf{j} + \mathbf{k}$ (pounds) acting on a particle that moves along the line segment from $P(5, 7, 0)$ to $Q(6, 6, 6)$ (units in feet).

48. Two forces $\mathbf{F}_1 = \mathbf{i} - 3\mathbf{j} + \mathbf{k}$ and $\mathbf{F}_2 = \mathbf{i} + 2\mathbf{j} + 2\mathbf{k}$ (pounds) act on a particle as it moves in a straight line from $P(-1, -2, 3)$ to $Q(0, 2, 0)$ (units in feet). How much work is done?

49. Convert $(\sqrt{2}, \pi/4, 1)$ from cylindrical coordinates to
   (a) rectangular coordinates
   (b) spherical coordinates.

50. Convert from rectangular coordinates to (i) cylindrical coordinates, (ii) spherical coordinates.
   (a) $(2, 2, 2\sqrt{6})$ (b) $(1, \sqrt{3}, 0)$.

51. Express the equation in terms of rectangular coordinates.
   (a) $z = r^2 \cos 2\theta$ (b) $\rho^2 \sin \phi \cos \phi \cos \theta = 1$.

52. Sketch the set of points defined by the given conditions.
   (a) $0 \le \theta \le \pi/2$, $0 \le r \le \cos \theta$, $0 \le z \le 2$ (cylindrical coordinates)
   (b) $0 \le \theta \le \pi/2$, $0 \le \phi \le \pi/4$, $0 \le \rho \le 2 \sec \phi$ (spherical coordinates)
   (c) $r = 2 \sin \theta$, $0 \le z \le 2$ (cylindrical coordinates)
   (d) $\rho = 2 \cos \phi$ (spherical coordinates).

15
Vector-Valued Functions

Karl Weierstrass   (1815 -1897)

■ **15.1   INTRODUCTION TO VECTOR-VALUED FUNCTIONS**

> *In this section we shall show how vectors can be used to express parametric equations in a more compact form. As part of our work we shall discuss functions that associate vectors with real numbers. This new category of functions has important applications in science and engineering.*

☐ **VECTOR-VALUED FUNCTIONS**

Recall that a function is a rule that assigns to each element in its domain one and only one element in its range. Thus far, we have considered only functions for which the domain and range are sets of real numbers; such functions are called *real-valued functions of a real variable* or sometimes simply *real-valued functions*. In this section we shall consider functions for which the domain consists of real numbers and the range consists of vectors in 2-space or 3-space; such functions are called *vector-valued functions of a real variable* or more simply *vector-valued functions*. In 2-space such functions can be expressed in the form

$$\mathbf{r}(t) = \langle x(t), y(t) \rangle = x(t)\mathbf{i} + y(t)\mathbf{j}$$

and in 3-space in the form

$$\mathbf{r}(t) = \langle x(t), y(t), z(t) \rangle = x(t)\mathbf{i} + y(t)\mathbf{j} + z(t)\mathbf{k}$$

where $x(t)$, $y(t)$, and $z(t)$ are real-valued functions of the real variable t. These real-valued functions are called the *component functions* or *components* of **r**. As a matter of notation, we shall denote vector-valued functions with boldface type [$\mathbf{f}(t)$, $\mathbf{g}(t)$, and $\mathbf{r}(t)$] and real-valued functions, as usual, with lightface italic type [$f(t)$, $g(t)$, and $r(t)$].

Example 1   If

$$\mathbf{r}(t) = (\ln t)\mathbf{i} + \sqrt{t^2 + 2}\,\mathbf{j} + (\cos t\pi)\mathbf{k}$$

then the component functions are

$$x(t) = \ln t, \quad y(t) = \sqrt{t^2 + 2}, \quad \text{and} \quad z(t) = \cos t\pi$$

The vector that $\mathbf{r}(t)$ associates with $t = 1$ is

$$\mathbf{r}(1) = (\ln 1)\mathbf{i} + \sqrt{3}\,\mathbf{j} + (\cos \pi)\mathbf{k} = \sqrt{3}\,\mathbf{j} - \mathbf{k}$$

The function **r** is undefined if $t \le 0$ because $\ln t$ is undefined for such t.   ◄

If the domain of a vector-valued function is not stated explicitly, then it is understood to consist of all real numbers for which every component is defined and yields a real value. This is called the *natural domain* of the function. Thus, the natural domain of a vector-valued function is the intersection of the natural domains of its components.

□ **PARAMETRIC
EQUATIONS IN
VECTOR FORM**

Vector-valued functions can be used to express parametric equations in 2-space or 3-space in a compact form. For example, consider the parametric equations

$$x = x(t), \quad y = y(t) \tag{1}$$

Because two vectors are equivalent if and only if their corresponding components are equal, this pair of equations can be replaced by the single vector equation

$$x\mathbf{i} + y\mathbf{j} = x(t)\mathbf{i} + y(t)\mathbf{j} \tag{2}$$

Similarly, in 3-space the three parametric equations

$$x = x(t), \quad y = y(t), \quad z = z(t) \tag{3}$$

can be replaced by the single vector equation

$$x\mathbf{i} + y\mathbf{j} + z\mathbf{k} = x(t)\mathbf{i} + y(t)\mathbf{j} + z(t)\mathbf{k} \tag{4}$$

If we let

$$\mathbf{r} = x\mathbf{i} + y\mathbf{j} \qquad \text{and} \qquad \mathbf{r}(t) = x(t)\mathbf{i} + y(t)\mathbf{j}$$

in 2-space and let

$$\mathbf{r} = x\mathbf{i} + y\mathbf{j} + z\mathbf{k} \qquad \text{and} \qquad \mathbf{r}(t) = x(t)\mathbf{i} + y(t)\mathbf{j} + z(t)\mathbf{k}$$

in 3-space, then both (2) and (4) can be written as

$$\mathbf{r} = \mathbf{r}(t) \tag{5}$$

which is the vector form of the parametric equations in (1) and (3). Conversely, every vector equation of form (5) can be rewritten as parametric equations by equating components on the two sides.

REMARK.   In (5) we have used the letter $\mathbf{r}$ for both the dependent variable and the function name to reduce the number of different symbols being used. This dual use of $\mathbf{r}$ is common and rarely causes problems. We did the same thing with the parametric equations $x = x(t)$, $y = y(t)$, and $z = z(t)$.

Example 2   Express the given parametric equations as a single vector equation.

(a) $x = t^2, \quad y = 3t$ (b) $x = \cos t, \quad y = \sin t, \quad z = t$

Solution (a).   Using the two sides of the equations as components of a vector yields

$$x\mathbf{i} + y\mathbf{j} = t^2\mathbf{i} + 3t\mathbf{j}$$

Solution (b).   Proceeding as in part (a) yields

$$x\mathbf{i} + y\mathbf{j} + z\mathbf{k} = (\cos t)\mathbf{i} + (\sin t)\mathbf{j} + t\mathbf{k} \qquad \blacktriangleleft$$

Example 3   Find parametric equations that correspond to the vector equation

$$x\mathbf{i} + y\mathbf{j} + z\mathbf{k} = (t^3 + 1)\mathbf{i} + 3\mathbf{j} + e^t\mathbf{k}$$

Solution.   Equating corresponding components yields

$$x = t^3 + 1, \quad y = 3, \quad z = e^t \quad \blacktriangleleft$$

☐ **GRAPHS OF VECTOR-VALUED FUNCTIONS**

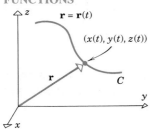

One method for interpreting a vector-valued function $\mathbf{r}(t)$ in 2-space or 3-space geometrically is to position the vector $\mathbf{r} = \mathbf{r}(t)$ with its initial point at the origin, and let C be the curve generated by the tip of the vector $\mathbf{r}$ as the parameter t varies (Figure 15.1.1). The vector $\mathbf{r}$, when positioned in this way, is called the *radius vector* or *position vector* of C, and C is called the **graph of the function $\mathbf{r}(t)$** or, equivalently, the **graph of the equation $\mathbf{r} = \mathbf{r}(t)$**. The vector equation $\mathbf{r} = \mathbf{r}(t)$ is equivalent to a set of parametric equations, so C is also called the **graph of these parametric equations.**

As t varies, the tip of the radius vector $\mathbf{r}$ traces out the curve C.

Figure 15.1.1

Example 4   Sketch the graph of the vector-valued function

$$\mathbf{r}(t) = (\cos t)\mathbf{i} + (\sin t)\mathbf{j}, \quad 0 \le t \le 2\pi$$

Solution.   The graph of $\mathbf{r}(t)$ is the graph of the vector equation

$$x\mathbf{i} + y\mathbf{j} = (\cos t)\mathbf{i} + (\sin t)\mathbf{j}, \quad 0 \le t \le 2\pi$$

or equivalently, it is the graph of the parametric equations

$$x = \cos t, \quad y = \sin t \quad (0 \le t \le 2\pi)$$

This is a circle of radius 1 that is centered at the origin with the direction of increasing t counterclockwise. The graph and a radius vector are shown in Figure 15.1.2.   ◄

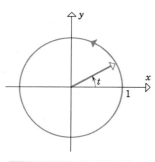

$$\mathbf{r} = (\cos t)\,\mathbf{i} + (\sin t)\,\mathbf{j}$$

Figure 15.1.2

Example 5   Sketch the graph of the vector-valued function

$$\mathbf{r}(t) = (\cos t)\mathbf{i} + (\sin t)\mathbf{j} + 2\mathbf{k}, \quad 0 \le t \le 2\pi$$

Solution.   The graph of $\mathbf{r}(t)$ is the graph of the vector equation

$$x\mathbf{i} + y\mathbf{j} + z\mathbf{k} = (\cos t)\mathbf{i} + (\sin t)\mathbf{j} + 2\mathbf{k}, \quad 0 \le t \le 2\pi$$

or, equivalently, it is the graph of the parametric equations

$$x = \cos t, \quad y = \sin t, \quad z = 2 \quad (0 \le t \le 2\pi)$$

From the last equation, the tip of the radius vector traces a curve in the plane $z = 2$, and from the first two equations and the preceding example, the curve is a circle of radius 1 centered on the z-axis and traced counterclockwise looking down the z-axis. The graph and a radius vector are shown in Figure 15.1.3.   ◄

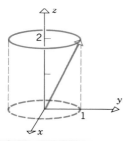

$$\mathbf{r} = (\cos t)\,\mathbf{i} + (\sin t)\,\mathbf{j} + 2\mathbf{k}$$

Figure 15.1.3

Example 6   Sketch the graph of the vector-valued function

$$\mathbf{r}(t) = (a \cos t)\mathbf{i} + (a \sin t)\mathbf{j} + (ct)\mathbf{k}$$

where a and c are positive constants.

Solution.   The graph of $\mathbf{r}(t)$ is the graph of the parametric equations

$$x = a \cos t, \quad y = a \sin t, \quad z = ct$$

As the parameter t increases, the value of $z = ct$ also increases, so the point (x, y, z) moves upward. However, as t increases, the point (x, y, z) also moves in a path directly over the circle

$$x = a \cos t, \quad y = a \sin t$$

in the xy-plane. The combination of these upward and circular motions produces a corkscrew-shaped curve that wraps around a right-circular cylinder of radius a centered on the z-axis (Figure 15.1.4). This curve is called a ***circular helix***.   ◀

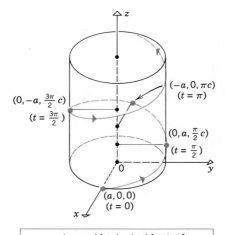

$$\mathbf{r} = (a \cos t)\mathbf{i} + (a \sin t)\mathbf{j} + (ct)\mathbf{k}$$

Figure 15.1.4

Example 7   Describe the graph of the vector equation

$$\mathbf{r} = (-2 + t)\mathbf{i} + 3t\mathbf{j} + (5 - 4t)\mathbf{k}$$

Solution.   The corresponding parametric equations are

$$x = -2 + t, \quad y = 3t, \quad z = 5 - 4t$$

From Theorem 14.5.1, the graph is the line in 3-space that passes through the point $(-2, 0, 5)$ and is parallel to the vector $\mathbf{i} + 3\mathbf{j} - 4\mathbf{k}$.   ◀

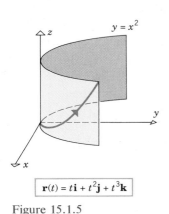

$$r(t) = ti + t^2j + t^3k$$

Figure 15.1.5

Example 8   The graph of the vector-valued function

$$\mathbf{r}(t) = t\mathbf{i} + t^2\mathbf{j} + t^3\mathbf{k}$$

is called a **twisted cubic**. Show that this curve lies on the parabolic cylinder $y = x^2$, and sketch the graph for $t \geq 0$.

Solution.   The corresponding parametric equations are

$$x = t, \quad y = t^2, \quad z = t^3$$

Eliminating the parameter t in the equations for x and y yields $y = x^2$, so the curve lies on the parabolic cylinder with this equation. The curve starts at the origin for $t = 0$; as t increases, so do x, y, and z, so the curve is traced in the upward direction, moving away from the origin along the cylinder (Figure 15.1.5). ◄

☐ **GRAPHS OF CONSTANT VECTOR-VALUED FUNCTIONS**

If $\mathbf{c}$ is a constant vector in the sense that it does not depend on a parameter, then the graph of $\mathbf{r} = \mathbf{c}$ is a single point since the radius vector remains fixed with its tip at $\mathbf{c}$. If $\mathbf{c} = x_0\mathbf{i} + y_0\mathbf{j}$ (in 2-space), then the graph is the point (x_0, y_0), and if $\mathbf{c} = x_0\mathbf{i} + y_0\mathbf{j} + z_0\mathbf{k}$ (in 3-space), then the graph is the point (x_0, y_0, z_0).

Example 9   The graph of the equation

$$\mathbf{r} = 2\mathbf{i} + 3\mathbf{j} - \mathbf{k}$$

is the point $(2, 3, -1)$ in 3-space. ◄

☐ **NORM OF A VECTOR-VALUED FUNCTION**

If $\mathbf{r}(t)$ is a vector-valued function, then for each value of the parameter t, the expression $\|\mathbf{r}(t)\|$ is a real-valued function of t because the norm (or length) of $\mathbf{r}(t)$ is a real number. For example, if

$$\mathbf{r}(t) = t\mathbf{i} + (t - 1)\mathbf{j}$$

then

$$\|\mathbf{r}(t)\| = \sqrt{t^2 + (t - 1)^2} = \sqrt{2t^2 - 2t + 1}$$

which is a real-valued function of t.

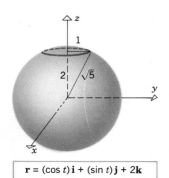

$$r = (\cos t)i + (\sin t)j + 2k$$

Figure 15.1.6

Example 10   In Example 5 we saw that the graph of

$$\mathbf{r}(t) = (\cos t)\mathbf{i} + (\sin t)\mathbf{j} + 2\mathbf{k}, \quad 0 \leq t \leq 2\pi$$

is a circle of radius 1 centered on the z-axis and lying in the plane $z = 2$. This circle lies on the surface of a sphere of radius $\sqrt{5}$ because for each value of t

$$\|\mathbf{r}(t)\| = \sqrt{\cos^2 t + \sin^2 t + 4} = \sqrt{1 + 4} = \sqrt{5}$$

which shows that each point on the circle is at a distance of $\sqrt{5}$ units from the origin (Figure 15.1.6). ◄

☐ SOME COMPLICATIONS

When it is desirable to emphasize that a curve C in 2-space or 3-space is the graph of a vector-valued function (or equivalently of a set of parametric equations), then we call C a ***parametric curve***. One of the complications in working with vector-valued functions is that a given parametric curve C is generally the graph of many different vector-valued functions. For example, the circle $x^2 + y^2 = 1$ is the graph of both

$$x\mathbf{i} + y\mathbf{j} = \cos t\,\mathbf{i} + \sin t\,\mathbf{j} \quad \text{and} \quad x\mathbf{i} + y\mathbf{j} = \cos(t^2)\mathbf{i} + \sin(t^2)\mathbf{j}$$

as may be seen by expressing the equations in parametric form and eliminating the parameter:

$$x^2 + y^2 = \cos^2 t + \sin^2 t = 1 \quad \text{and} \quad x^2 + y^2 = \cos^2(t^2) + \sin^2(t^2) = 1$$

The power of analytic geometry is derived from the fact that it allows us to deduce geometric properties of a curve from an equation of that curve and conversely to deduce properties of an equation from the geometric properties of its graph. However, there are complications that occur for vector-valued functions stemming from the fact that a given parametric curve C is generally the graph of many different vector-valued functions. As a result, properties of C inferred from any one of these functions may not be correct, since those same properties may not result if a different function is used. In later sections, we shall impose various restrictions on vector-valued functions to avoid this problem.

► Exercise Set 15.1

In Exercises 1–4, find the domain of **r** and the value of $\mathbf{r}(t_0)$.

1. $\mathbf{r}(t) = (\cos t)\mathbf{i} - 3t\mathbf{j}$; $t_0 = \pi$.
2. $\mathbf{r}(t) = \langle \sqrt{3t + 1}, t^2 \rangle$; $t_0 = 1$.
3. $\mathbf{r}(t) = (\cos \pi t)\mathbf{i} - (\ln t)\mathbf{j} + \sqrt{t - 2}\,\mathbf{k}$; $t_0 = 3$.
4. $\mathbf{r}(t) = \langle 2e^{-t}, \sin^{-1} t, \ln(1 - t) \rangle$; $t_0 = 0$.

In Exercises 5–8, express the parametric equations as a single vector equation of the form $\mathbf{r} = x(t)\mathbf{i} + y(t)\mathbf{j}$ or $\mathbf{r} = x(t)\mathbf{i} + y(t)\mathbf{j} + z(t)\mathbf{k}$.

5. $x = 3 \cos t$, $y = t + \sin t$.
6. $x = t^2 + 1$, $y = e^{-2t}$.
7. $x = 2t$, $y = 2 \sin 3t$, $z = 5 \cos 3t$.
8. $x = t \sin t$, $y = \ln t$, $z = \cos^2 t$.

In Exercises 9–12, find the parametric equations that correspond to the given vector equation.

9. $\mathbf{r} = 3t^2\mathbf{i} - 2\mathbf{j}$.
10. $\mathbf{r} = (\sin^2 t)\mathbf{i} + (1 - \cos 2t)\mathbf{j}$.
11. $\mathbf{r} = (2t - 1)\mathbf{i} - 3\sqrt{t}\,\mathbf{j} + (\sin 3t)\mathbf{k}$.
12. $\mathbf{r} = te^{-t}\mathbf{i} - 5t^2\mathbf{k}$.

In Exercises 13–18, describe the graph of the vector equation.

13. $\mathbf{r} = (2 - 3t)\mathbf{i} - 4t\mathbf{j}$.
14. $\mathbf{r} = (3 \sin 2t)\mathbf{i} + (3 \cos 2t)\mathbf{j}$.
15. $\mathbf{r} = 2t\mathbf{i} - 3\mathbf{j} + (1 + 3t)\mathbf{k}$.
16. $\mathbf{r} = 3\mathbf{i} + (2 \cos t)\mathbf{j} + (2 \sin t)\mathbf{k}$.

17. $\mathbf{r} = (3 \cos t)\mathbf{i} + (2 \sin t)\mathbf{j} - \mathbf{k}$.

18. $\mathbf{r} = -2\mathbf{i} + t\mathbf{j} + (t^2 - 1)\mathbf{k}$.

19. Find the slope of the line in 2-space that is represented by the vector equation

$$\mathbf{r} = (1 - 2t)\mathbf{i} - (2 - 3t)\mathbf{j}$$

20. Find the y-intercept of the line in 2-space that is represented by the vector equation

$$\mathbf{r} = (3 + 2t)\mathbf{i} + 5t\mathbf{j}$$

21. Find the coordinates of the point where the line $\mathbf{r} = (2 + t)\mathbf{i} + (1 - 2t)\mathbf{j} + 3t\mathbf{k}$ intersects the xz-plane.

22. Find the coordinates of the point where the line $\mathbf{r} = t\mathbf{i} + (1 + 2t)\mathbf{j} - 3t\mathbf{k}$ intersects the plane $3x - y - z = 2$.

In Exercises 23–34, sketch the graph of $\mathbf{r}(t)$ and show the direction of increasing t.

23. $\mathbf{r}(t) = 2\mathbf{i} + t\mathbf{j}$.

24. $\mathbf{r}(t) = \langle 3t - 4, 6t + 2 \rangle$.

25. $\mathbf{r}(t) = (1 + \cos t)\mathbf{i} + (3 - \sin t)\mathbf{j}, \ 0 \le t \le 2\pi$.

26. $\mathbf{r}(t) = \langle 2 \cos t, 5 \sin t \rangle, \ 0 \le t \le 2\pi$.

27. $\mathbf{r}(t) = (\cosh t)\mathbf{i} + (\sinh t)\mathbf{j}$.

28. $\mathbf{r}(t) = \sqrt{t}\,\mathbf{i} + (2t + 4)\mathbf{j}$.

29. $\mathbf{r}(t) = t\mathbf{i} + t\mathbf{j} + t\mathbf{k}$.

30. $\mathbf{r}(t) = (1 + 3t)\mathbf{i} + (-1 + t)\mathbf{j} + 2t\mathbf{k}$.

31. $\mathbf{r}(t) = (2 \cos t)\mathbf{i} + (2 \sin t)\mathbf{j} + t\mathbf{k}$.

32. $\mathbf{r}(t) = (9 \cos t)\mathbf{i} + (4 \sin t)\mathbf{j} + t\mathbf{k}$.

33. $\mathbf{r}(t) = t\mathbf{i} + t^2\mathbf{j} + 2\mathbf{k}$.

34. $\mathbf{r}(t) = t\mathbf{i} + t\mathbf{j} + (\sin t)\mathbf{k}, \ 0 \le t \le 2\pi$.

35. Show that the graph of

$$\mathbf{r} = (t \sin t)\mathbf{i} + (t \cos t)\mathbf{j} + t^2\mathbf{k}$$

lies on the paraboloid $z = x^2 + y^2$.

36. Show that the graph of

$$\mathbf{r} = t\mathbf{i} + \frac{1 + t}{t}\mathbf{j} + \frac{1 - t^2}{t}\mathbf{k}, \quad t > 0$$

lies in the plane $x - y + z + 1 = 0$.

37. Show that the graph of

$$\mathbf{r} = (\sin t)\mathbf{i} + (2 \cos t)\mathbf{j} + (\sqrt{3} \sin t)\mathbf{k}$$

is a circle, and find its center and radius. [*Hint:* Show that the curve lies on both a sphere and a plane.]

38. Show that the graph of

$$\mathbf{r} = (3 \cos t)\mathbf{i} + (3 \sin t)\mathbf{j} + (3 \sin t)\mathbf{k}$$

is an ellipse, and find the lengths of the major and minor axes. [*Hint:* Show that the graph lies on both a circular cylinder and a plane and use the result in Exercise 44 of Section 12.3.]

39. For the helix $\mathbf{r} = (a \cos t)\mathbf{i} + (a \sin t)\mathbf{j} + ct\mathbf{k}$, find c ($c > 0$) so that the helix will make one complete turn in a distance of 3 units measured along a line parallel to the z-axis.

40. How many revolutions will the circular helix $\mathbf{r} = (a \cos t)\mathbf{i} + (a \sin t)\mathbf{j} + 0.2t\mathbf{k}$ make in a distance of 10 units measured along a line parallel to the z-axis?

41. Show that the curve $\mathbf{r} = (t \cos t)\mathbf{i} + (t \sin t)\mathbf{j} + t\mathbf{k}$, $t \ge 0$, lies on the cone $z = \sqrt{x^2 + y^2}$. Describe the curve.

42. Describe the curve $\mathbf{r} = (a \cos t)\mathbf{i} + (b \sin t)\mathbf{j} + ct\mathbf{k}$, where a, b, and c are positive constants such that $a \ne b$.

In Exercises 43–48, find parametric equations of the curve of intersection of the surfaces. [*Note:* The answer is not unique.]

43. The cone $z = \sqrt{x^2 + y^2}$ and the plane $z = y + 2$. Identify the curve.

44. The paraboloid $z = x^2 + y^2$ and the plane $x = -2$. Identify the curve.

45. The circular cylinder $x^2 + y^2 = 9$ and the parabolic cylinder $z = x^2$.

46. The paraboloid $z = 4 - x^2 - y^2$ and the circular cylinder $x^2 + y^2 = 1$. Identify the curve.

47. The elliptic paraboloid $z = x^2 + 4y^2$ and the plane $z = 2x$. [*Hint:* Find the orthogonal projection of the curve onto the xy-plane.]

48. The cone $z = \sqrt{x^2 + y^2}$ and the parabolic cylinder $z = 2\sqrt{y}$. [*Hint:* See the hint in Exercise 47.]

■ **15.2** LIMITS AND DERIVATIVES OF VECTOR-VALUED FUNCTIONS

In this section we shall define limits and derivatives of vector-valued functions and give some basic interpretations and applications of these concepts.

□ **LIMITS OF VECTOR-VALUED FUNCTIONS**

The limit of a vector-valued function is defined to be the vector that results by taking the limit of each component. Thus, for a function $\mathbf{r}(t) = x(t)\mathbf{i} + y(t)\mathbf{j}$ in 2-space we define

$$\lim_{t \to a} \mathbf{r}(t) = \left(\lim_{t \to a} x(t)\right)\mathbf{i} + \left(\lim_{t \to a} y(t)\right)\mathbf{j} \tag{1}$$

and for a function $\mathbf{r}(t) = x(t)\mathbf{i} + y(t)\mathbf{j} + z(t)\mathbf{k}$ in 3-space we define

$$\lim_{t \to a} \mathbf{r}(t) = \left(\lim_{t \to a} x(t)\right)\mathbf{i} + \left(\lim_{t \to a} y(t)\right)\mathbf{j} + \left(\lim_{t \to a} z(t)\right)\mathbf{k} \tag{2}$$

If the limit of any component does not exist, then we shall agree that the limit of $\mathbf{r}(t)$ does not exist. These definitions are also applicable to the one-sided and infinite limits $\lim_{t \to a^+}$, $\lim_{t \to a^-}$, $\lim_{t \to +\infty}$, and $\lim_{t \to -\infty}$. It follows from (1) and (2) that

$$\lim_{t \to a} \mathbf{r}(t) = \mathbf{L}$$

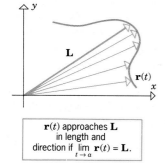

r(t) approaches **L** in length and direction if $\lim_{t \to a} \mathbf{r}(t) = \mathbf{L}$.

Figure 15.2.1

if and only if the components of $\mathbf{r}(t)$ approach the components of $\mathbf{L}$ as $t \to a$. Geometrically, this is equivalent to stating that the length and direction of $\mathbf{r}(t)$ approach the length and direction of $\mathbf{L}$ as $t \to a$ (Figure 15.2.1).

□ **CONTINUITY OF VECTOR-VALUED FUNCTIONS**

The definition of continuity for vector-valued functions is similar to that for real-valued functions. We shall say that $\mathbf{r}$ is *continuous at t_0* if

1. $\mathbf{r}(t_0)$ is defined;

2. $\lim_{t \to t_0} \mathbf{r}(t)$ exists;

3. $\lim_{t \to t_0} \mathbf{r}(t) = \mathbf{r}(t_0)$.

It can be shown that $\mathbf{r}$ is continuous at t_0 if and only if each component of $\mathbf{r}$ is continuous at t_0 (Exercise 59). As with real-valued functions, we shall call $\mathbf{r}$ *continuous everywhere* or simply *continuous* if $\mathbf{r}$ is continuous at all real values of t. Geometrically, the graph of a continuous vector-valued function is an unbroken curve.

☐ **DERIVATIVES OF VECTOR-VALUED FUNCTIONS**

The definition of a derivative for vector-valued functions is analogous to the definition for real-valued functions.

15.2.1   DEFINITION.   The derivative $\mathbf{r}'(t)$ of a vector-valued function $\mathbf{r}(t)$ is defined by

$$\mathbf{r}'(t) = \lim_{h \to 0} \frac{\mathbf{r}(t + h) - \mathbf{r}(t)}{h}$$

provided this limit exists.

For computational purposes the following theorem is extremely useful; it states that the derivative of a vector-valued function can be computed by differentiating each component.

15.2.2   THEOREM.

(a)   *If $\mathbf{r}(t) = x(t)\mathbf{i} + y(t)\mathbf{j}$ is a vector-valued function in 2-space, and if $x(t)$ and $y(t)$ are differentiable, then*

$$\mathbf{r}'(t) = x'(t)\mathbf{i} + y'(t)\mathbf{j}$$

(b)   *If $\mathbf{r}(t) = x(t)\mathbf{i} + y(t)\mathbf{j} + z(t)\mathbf{k}$ is a vector-valued function in 3-space, and if $x(t)$, $y(t)$, and $z(t)$ are differentiable, then*

$$\mathbf{r}'(t) = x'(t)\mathbf{i} + y'(t)\mathbf{j} + z'(t)\mathbf{k}$$

We shall prove part (a). The proof of (b) is similar.

Proof (a).   From Definition 15.2.1,

$$\mathbf{r}'(t) = \lim_{h \to 0} \frac{\mathbf{r}(t + h) - \mathbf{r}(t)}{h}$$

$$= \lim_{h \to 0} \frac{[x(t + h)\mathbf{i} + y(t + h)\mathbf{j}] - [x(t)\mathbf{i} + y(t)\mathbf{j}]}{h}$$

$$= \lim_{h \to 0} \frac{[x(t + h) - x(t)]}{h}\mathbf{i} + \lim_{h \to 0} \frac{[y(t + h) - y(t)]}{h}\mathbf{j}$$

$$= x'(t)\mathbf{i} + y'(t)\mathbf{j} \qquad \blacksquare$$

As with real-valued functions, there are various notations for the derivative of a vector-valued function. If $\mathbf{r} = \mathbf{r}(t)$, then some possibilities are

$$\frac{d}{dt}[\mathbf{r}(t)], \quad \frac{d\mathbf{r}}{dt}, \quad \mathbf{r}'(t), \quad \text{and} \quad \mathbf{r}'$$

Example 1 Let $\mathbf{r}(t) = t^2\mathbf{i} + t^3\mathbf{j}$. Find $\mathbf{r}'(t)$ and $\mathbf{r}'(1)$.

Solution.

$$\mathbf{r}'(t) = \frac{d}{dt}[t^2]\mathbf{i} + \frac{d}{dt}[t^3]\mathbf{j} = 2t\mathbf{i} + 3t^2\mathbf{j}$$

Substituting $t = 1$ yields $\mathbf{r}'(1) = 2\mathbf{i} + 3\mathbf{j}$. ◀

□ **TANGENT VECTORS AND TANGENT LINES**

The derivative of a vector-valued function has an important geometric interpretation which can be stated *informally* as follows:

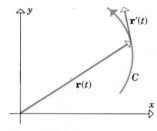

Figure 15.2.2

> **15.2.3** GEOMETRIC INTERPRETATION OF THE DERIVATIVE. *Suppose that C is the graph of a vector-valued function $\mathbf{r}(t)$ and that $\mathbf{r}'(t)$ exists and is nonzero for a given value of t. If the vector $\mathbf{r}'(t)$ is positioned with its initial point at the terminal point of the radius vector $\mathbf{r}(t)$ (Figure 15.2.2), then $\mathbf{r}'(t)$ is tangent to C and points in the direction of increasing parameter.*

To make this result plausible, let C be the graph of $\mathbf{r}(t)$, and for a fixed value of the parameter t construct the position vectors $\mathbf{r}(t)$ and $\mathbf{r}(t + h)$. If $h > 0$, then the tip of the vector $\mathbf{r}(t + h)$ is in the direction of increasing parameter from the tip of $\mathbf{r}(t)$, and if $h < 0$, it is the other way around (Figure 15.2.3). In either case, the difference $\mathbf{r}(t + h) - \mathbf{r}(t)$ coincides with the secant line through the tips of the vectors $\mathbf{r}(t)$ and $\mathbf{r}(t + h)$. Moreover, since h is a scalar, the vector

$$\frac{1}{h}[\mathbf{r}(t + h) - \mathbf{r}(t)] = \frac{\mathbf{r}(t + h) - \mathbf{r}(t)}{h} \tag{3}$$

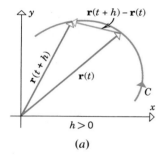

$h > 0$

(a)

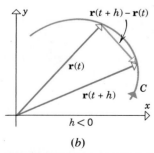

$h < 0$

(b)

Figure 15.2.3

also coincides with this secant line. If $h < 0$, then vector (3) is oppositely directed to $\mathbf{r}(t + h) - \mathbf{r}(t)$, and if $h > 0$, these vectors have the same direction. In both cases, vector (3) points in the direction of increasing parameter. As $h \to 0$, the secant lines through the tips of $\mathbf{r}(t + h)$ and $\mathbf{r}(t)$ tend toward the tangent line to C at the tip of $\mathbf{r}(t)$. Thus, the vector

$$\mathbf{r}'(t) = \lim_{h \to 0} \frac{\mathbf{r}(t + h) - \mathbf{r}(t)}{h} \tag{4}$$

is tangent to the curve C at the tip of $\mathbf{r}(t)$ and points in the direction of increasing parameter.

Motivated by the foregoing discussion, we make the following definition.

> **15.2.4** DEFINITION. Let P be a point on the graph of a vector-valued function $\mathbf{r}(t)$, and let $\mathbf{r}(t_0)$ be the radius vector from the origin to P (Figure 15.2.4). If $\mathbf{r}'(t_0)$ exists and $\mathbf{r}'(t_0) \neq \mathbf{0}$, then we call $\mathbf{r}'(t_0)$ the *tangent vector* to the graph of $\mathbf{r}$ at $\mathbf{r}(t_0)$.

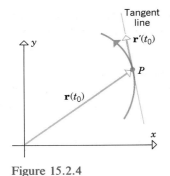

Figure 15.2.4

REMARK. Observe that the graph of a vector-valued function can fail to have a tangent vector at a point either because the derivative in (4) does not exist or because the derivative is zero at the point.

If a vector-valued function $\mathbf{r}(t)$ has a tangent vector $\mathbf{r}'(t_0)$ at a point on its graph, then the line that is parallel to $\mathbf{r}'(t_0)$ and passes through the tip of the radius vector $\mathbf{r}(t_0)$ is called the *tangent line* to the graph of $\mathbf{r}(t)$ at $\mathbf{r}(t_0)$ (Figure 15.2.4). It follows from Formula (5) of Section 14.5 that a vector equation of the tangent line is

$$\mathbf{r} = \mathbf{r}(t_0) + t\mathbf{r}'(t_0) \tag{5}$$

Example 2 Find parametric equations of the tangent line to the circular helix

$$x = \cos t, \quad y = \sin t, \quad z = t$$

at the point where $t = \pi/6$.

Solution. We shall first use Formula (5) to find a vector equation of the tangent line, then we shall equate components to obtain the parametric equations. A vector equation $\mathbf{r} = \mathbf{r}(t)$ of the helix is

$$x\mathbf{i} + y\mathbf{j} + z\mathbf{k} = (\cos t)\mathbf{i} + (\sin t)\mathbf{j} + t\mathbf{k}$$

Thus,

$$\mathbf{r}(t) = (\cos t)\mathbf{i} + (\sin t)\mathbf{j} + t\mathbf{k}$$
$$\mathbf{r}'(t) = (-\sin t)\mathbf{i} + (\cos t)\mathbf{j} + \mathbf{k}$$

At the point where $t = \pi/6$, these vectors are

$$\mathbf{r}\left(\frac{\pi}{6}\right) = \frac{\sqrt{3}}{2}\mathbf{i} + \frac{1}{2}\mathbf{j} + \frac{\pi}{6}\mathbf{k} \quad \text{and} \quad \mathbf{r}'\left(\frac{\pi}{6}\right) = -\frac{1}{2}\mathbf{i} + \frac{\sqrt{3}}{2}\mathbf{j} + \mathbf{k}$$

so from (5) with $t_0 = \pi/6$ a vector equation of the tangent line is

$$\mathbf{r} = \mathbf{r}\left(\frac{\pi}{6}\right) + t\mathbf{r}'\left(\frac{\pi}{6}\right) = \left(\frac{\sqrt{3}}{2}\mathbf{i} + \frac{1}{2}\mathbf{j} + \frac{\pi}{6}\mathbf{k}\right) + t\left(-\frac{1}{2}\mathbf{i} + \frac{\sqrt{3}}{2}\mathbf{j} + \mathbf{k}\right)$$

Simplifying, then equating the resulting components with the corresponding components of $\mathbf{r} = x\mathbf{i} + y\mathbf{j} + z\mathbf{k}$ yields the parametric equations

$$x = \frac{\sqrt{3}}{2} - \frac{1}{2}t, \quad y = \frac{1}{2} + \frac{\sqrt{3}}{2}t, \quad z = \frac{\pi}{6} + t \quad \blacktriangleleft$$

Example 3 The graph of $\mathbf{r}(t) = t^2\mathbf{i} + t^3\mathbf{j}$ is called a *semicubical parabola* (Figure 15.2.5). Find a vector equation of the tangent line to the graph of $\mathbf{r}(t)$ at

(a) the point $(0, 0)$ (b) the point $(1, 1)$

Solution (a). The derivative of $\mathbf{r}(t)$ is

$$\mathbf{r}'(t) = 2t\mathbf{i} + 3t^2\mathbf{j}$$

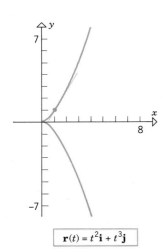

$$\mathbf{r}(t) = t^2\mathbf{i} + t^3\mathbf{j}$$

Figure 15.2.5

The point $(0, 0)$ on the graph of $\mathbf{r}$ corresponds to $t = 0$. At this point we have $\mathbf{r}'(0) = 0$, so there is no tangent vector at the point and consequently a tangent line does not exist at this point.

Solution (*b*).   The point $(1, 1)$ on the graph of $\mathbf{r}$ corresponds to $t = 1$, so from (5) a vector equation of the tangent line at this point is

$$\mathbf{r} = \mathbf{r}(1) + t\mathbf{r}'(1)$$

From the formulas for $\mathbf{r}(t)$ and $\mathbf{r}'(t)$ with $t = 1$, this equation becomes

$$\mathbf{r} = (\mathbf{i} + \mathbf{j}) + t(2\mathbf{i} + 3\mathbf{j}) \quad \blacktriangleleft$$

If $\mathbf{r}$ is a vector-valued function in 2-space or 3-space, then we say that $\mathbf{r}(t)$ is *smoothly parametrized* or that $\mathbf{r}$ is a *smooth function* of t if the components of $\mathbf{r}$ have continuous derivatives with respect to t and $\mathbf{r}'(t) \neq \mathbf{0}$ for any value of t. Thus, in 3-space

$$\mathbf{r}(t) = x(t)\mathbf{i} + y(t)\mathbf{j} + z(t)\mathbf{k}$$

is a smooth function of t if $x'(t)$, $y'(t)$, and $z'(t)$ are continuous and there is no value of t at which all three derivatives are zero. A parametric curve C in 2-space or 3-space will be called *smooth* if it is the graph of some smooth vector-valued function.

It can be shown that a smooth vector-valued function has a tangent line at every point on its graph. The circular helix in Example 2 is smooth, but the semicubical parabola in Example 3 is not.

☐ **PROPERTIES OF DERIVATIVES**

Because derivatives of vector-valued functions are computed by components, most basic theorems on differentiating real-valued functions carry over to vector-valued functions.

15.2.5   THEOREM (*Rules of Differentiation*).   *In either 2-space or 3-space let* $\mathbf{r}(t)$, $\mathbf{r}_1(t)$, *and* $\mathbf{r}_2(t)$ *be vector-valued functions,* $f(t)$ *a real-valued function,* k *a scalar, and* $\mathbf{c}$ *a fixed (constant) vector. Then the following rules of differentiation hold*:

$$\frac{d}{dt}[\mathbf{c}] = \mathbf{0}$$

$$\frac{d}{dt}[k\mathbf{r}(t)] = k\frac{d}{dt}[\mathbf{r}(t)]$$

$$\frac{d}{dt}[\mathbf{r}_1(t) + \mathbf{r}_2(t)] = \frac{d}{dt}[\mathbf{r}_1(t)] + \frac{d}{dt}[\mathbf{r}_2(t)]$$

$$\frac{d}{dt}[\mathbf{r}_1(t) - \mathbf{r}_2(t)] = \frac{d}{dt}[\mathbf{r}_1(t)] - \frac{d}{dt}[\mathbf{r}_2(t)]$$

$$\frac{d}{dt}[f(t)\mathbf{r}(t)] = f(t)\frac{d}{dt}[\mathbf{r}(t)] + \mathbf{r}(t)\frac{d}{dt}[f(t)]$$

The proofs are left as exercises.

In addition to the rules listed in the foregoing theorem, we have the following rules for differentiating dot products in 2-space or 3-space and cross products in 3-space:

$$\frac{d}{dt}[\mathbf{r}_1(t) \cdot \mathbf{r}_2(t)] = \mathbf{r}_1(t) \cdot \frac{d\mathbf{r}_2}{dt} + \frac{d\mathbf{r}_1}{dt} \cdot \mathbf{r}_2(t) \tag{6}$$

$$\frac{d}{dt}[\mathbf{r}_1(t) \times \mathbf{r}_2(t)] = \mathbf{r}_1(t) \times \frac{d\mathbf{r}_2}{dt} + \frac{d\mathbf{r}_1}{dt} \times \mathbf{r}_2(t) \tag{7}$$

The proofs of (6) and (7) are left as exercises.

REMARK.   In (6), the order of the factors in each term on the right does not matter, but in (7) it does.

In plane geometry one learns that a tangent line to a circle is perpendicular to the radius at the point of tangency. Consequently, if a point moves along a circular arc in 2-space, one would expect the radius vector and the tangent vector at any point on the arc to be perpendicular. This is the motivation for the following useful theorem, which is applicable in both 2-space and 3-space.

15.2.6   THEOREM.   *If* $\mathbf{r}(t)$ *is a vector-valued function in 2-space or 3-space and* $\|\mathbf{r}(t)\|$ *is constant for all t, then*

$$\mathbf{r}(t) \cdot \mathbf{r}'(t) = 0 \tag{8}$$

that is, $\mathbf{r}(t)$ *and* $\mathbf{r}'(t)$ *are orthogonal vectors for all t.*

Proof.   It follows from (6) with $\mathbf{r}_1(t) = \mathbf{r}_2(t) = \mathbf{r}(t)$ that

$$\frac{d}{dt}[\mathbf{r}(t) \cdot \mathbf{r}(t)] = \mathbf{r}(t) \cdot \frac{d\mathbf{r}}{dt} + \frac{d\mathbf{r}}{dt} \cdot \mathbf{r}(t)$$

or, equivalently,

$$\frac{d}{dt}[\|\mathbf{r}(t)\|^2] = 2\mathbf{r}(t) \cdot \frac{d\mathbf{r}}{dt} \tag{9}$$

But $\|\mathbf{r}(t)\|^2$ is constant, so its derivative is zero. Thus

$$2\mathbf{r}(t) \cdot \frac{d\mathbf{r}}{dt} = 0$$

from which (8) follows.   ∎

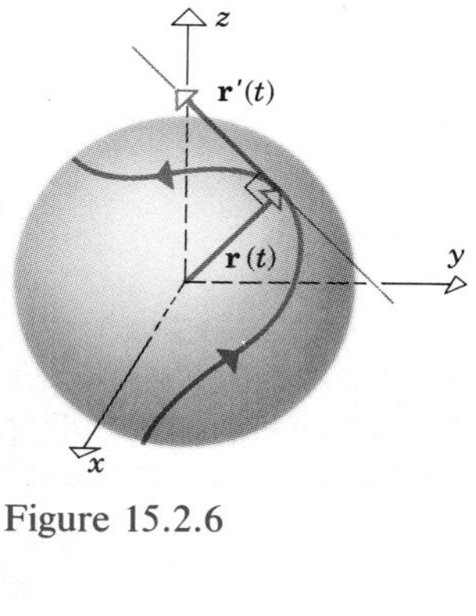

Figure 15.2.6

Example 4   Just as a tangent line to a circular arc in 2-space is perpendicular to the radius at the point of tangency, so a tangent line to a curve on the surface of a sphere in 3-space is perpendicular to the radius at the point of tangency (Figure 15.2.6). To see that this is so, suppose that the graph of $\mathbf{r}(t)$ lies on the surface of the sphere of radius $k > 0$ centered at the origin. For each value of t we have $\|\mathbf{r}(t)\| = k$, so by Theorem 15.2.6

$$\mathbf{r}(t) \cdot \mathbf{r}'(t) = 0$$

which implies that the radius vector $\mathbf{r}(t)$ and the tangent vector $\mathbf{r}'(t)$ are perpendicular. This completes the argument because the tangent line, where it exists, is parallel to the tangent vector. ◀

☐ **CHANGE OF PARAMETER**

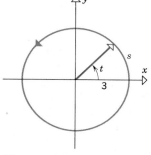

Figure 15.2.7

As noted in Section 15.1, it is possible for different vector-valued functions to have the same graph. For example, the graph of the function

$$\mathbf{r} = (3 \cos t)\mathbf{i} + (3 \sin t)\mathbf{j}, \quad 0 \le t \le 2\pi \tag{10}$$

is the circle of radius 3 centered at the origin with counterclockwise orientation. The parameter t can be interpreted geometrically as the positive angle in radians from the x-axis to the radius vector. For each value of t, let s be the length of the arc subtended by this angle on the circle (Figure 15.2.7). The parameters s and t are related by

$$t = s/3, \quad 0 \le s \le 6\pi \tag{11}$$

If we substitute this in (10), we obtain a vector-valued function of the parameter s, namely

$$\mathbf{r} = 3 \cos (s/3)\mathbf{i} + 3 \sin (s/3)\mathbf{j}, \quad 0 \le s \le 6\pi$$

whose graph is also the circle of radius 3 centered at the origin with counterclockwise orientation (verify).

In various problems it is helpful to change the parameter in a vector-valued function by making an appropriate substitution. For example, we changed the parameter above from t to s by substituting $t = s/3$ in (10). In general, if g is a real-valued function, then substituting

$$t = g(u)$$

in $\mathbf{r}(t)$ changes the parameter from t to u. When making such a change of parameter, it is important to ensure that the new vector-valued function of u is smooth if the original vector-valued function of t is smooth. It can be proved that this will be so if g satisfies the following conditions:

1. g is differentiable.
2. g' is continuous.
3. $g'(u) \neq 0$ for any u in the domain of g.
4. The range of g is the domain of $\mathbf{r}$.

If g satisfies these conditions, then we call $t = g(u)$ a ***smooth change of parameter***. We leave it for the reader to check that (11) is a smooth change of parameter. Henceforth, we shall assume that all changes of parameter are smooth, even if it is not stated explicitly.

The following version of the chain rule, which will be used extensively in later sections, is useful for finding derivatives when a change of parameter is made (Exercise 57).

15.2.7   THEOREM (*Chain Rule*).   *If* **r** *is a differentiable vector-valued function of* t, *and if* $t = g(u)$ *is a change of parameter, then*

$$\frac{d\mathbf{r}}{du} = \frac{d\mathbf{r}}{dt}\frac{dt}{du} \tag{12}$$

▶ **Exercise Set 15.2** $\boxed{C}$ 33, 35, 36

In Exercises 1–8, find the limit.

1. $\lim\limits_{t \to 3} (t^2\mathbf{i} + 2t\mathbf{j})$.

2. $\lim\limits_{t \to \pi/4} \langle \cos t, \sin t \rangle$.

3. $\lim\limits_{t \to 0^+} \left(\sqrt{t}\,\mathbf{i} + \dfrac{\sin t}{t}\mathbf{j} \right)$.

4. $\lim\limits_{t \to +\infty} \left\langle \dfrac{t^2 + 1}{3t^2 + 2}, \dfrac{1}{t} \right\rangle$.

5. $\lim\limits_{t \to 2} (t\mathbf{i} - 3\mathbf{j} + t^2\mathbf{k})$.

6. $\lim\limits_{t \to \pi} \langle \cos 3t, e^{-t}, \sqrt{t} \rangle$.

7. $\lim\limits_{t \to +\infty} \left(\tan^{-1} t\,\mathbf{i} + \dfrac{t}{t^2 + 3}\mathbf{j} + \cos\dfrac{2}{t}\mathbf{k} \right)$.

8. $\lim\limits_{t \to 1} \left\langle \dfrac{3}{t^2}, \dfrac{\ln t}{t^2 - 1}, \sin 2t \right\rangle$.

In Exercises 9 and 10, prove that **r** is continuous at t_0.

9. $\mathbf{r}(t) = (3 \sin t)\mathbf{i} - 2t\mathbf{j}; \ t_0 = \pi/2$.

10. $\mathbf{r}(t) = 5\mathbf{i} - \sqrt{3t + 1}\,\mathbf{j} + e^{2t}\mathbf{k}; \ t_0 = 1$.

In Exercises 11–14, find $\mathbf{r}'(t)$.

11. $\mathbf{r}(t) = (4 + 5t)\mathbf{i} + (t - t^2)\mathbf{j}$.

12. $\mathbf{r}(t) = 4\mathbf{i} - (\cos t)\mathbf{j}$.

13. $\mathbf{r}(t) = \dfrac{1}{t}\mathbf{i} + (\tan t)\mathbf{j} + e^{2t}\mathbf{k}$.

14. $\mathbf{r}(t) = (\tan^{-1} t)\mathbf{i} + (t \cos t)\mathbf{j} - \sqrt{t}\,\mathbf{k}$.

In Exercises 15–22, find $\mathbf{r}'(t_0)$; then sketch the graph of $\mathbf{r}(t)$ and the tangent vector $\mathbf{r}'(t_0)$.

15. $\mathbf{r}(t) = \langle t, t^2 \rangle; \ t_0 = 2$.

16. $\mathbf{r}(t) = (\cos t)\mathbf{i} + (\sin t)\mathbf{j}; \ t_0 = 3\pi/4$.

17. $\mathbf{r}(t) = \langle e^{-t}, e^{2t} \rangle; \ t_0 = \ln 2$.

18. $\mathbf{r}(t) = (\cos 2t)\mathbf{i} - (4 \sin t)\mathbf{j}; \ t_0 = \pi$.

19. $\mathbf{r}(t) = (2 \sin t)\mathbf{i} + \mathbf{j} + (2 \cos t)\mathbf{k}; \ t_0 = \pi/2$.

20. $\mathbf{r}(t) = (\cos t)\mathbf{i} + (\sin t)\mathbf{j} + t\mathbf{k}; \ t_0 = \pi/4$.

21. $\mathbf{r}(t) = 3\mathbf{i} + t\mathbf{j} + (2 - t^2)\mathbf{k}; \ t_0 = 1$.

22. $\mathbf{r}(t) = t\mathbf{i} + 2t\mathbf{j} + t^2\mathbf{k}; \ t_0 = 2$.

In Exercises 23–26, find parametric equations of the line tangent to the graph of $\mathbf{r}(t)$ at the point where $t = t_0$.

23. $\mathbf{r}(t) = t^2\mathbf{i} + (2 - \ln t)\mathbf{j}; \ t_0 = 1$.

24. $\mathbf{r}(t) = e^{2t}\mathbf{i} - (2 \cos 3t)\mathbf{j}; \ t_0 = 0$.

25. $\mathbf{r}(t) = (2 \cos \pi t)\mathbf{i} + (2 \sin \pi t)\mathbf{j} + 3t\mathbf{k}; \ t_0 = \frac{1}{3}$.

26. $\mathbf{r}(t) = (\ln t)\mathbf{i} + e^{-t}\mathbf{j} + t^3\mathbf{k}; \ t_0 = 2$.

In Exercises 27–30, find a vector equation of the line tangent to the graph of $\mathbf{r}(t)$ at the point P_0 on the curve.

27. $\mathbf{r}(t) = (2t - 1)\mathbf{i} + \sqrt{3t + 4}\,\mathbf{j}; \ P_0(-1, 2)$.

28. $\mathbf{r}(t) = (4 \cos t)\mathbf{i} - 3t\mathbf{j}; \ P_0(2, -\pi)$.

29. $\mathbf{r}(t) = t^2\mathbf{i} - \dfrac{1}{t + 1}\mathbf{j} + (4 - t^2)\mathbf{k}; \ P_0(4, 1, 0)$.

30. $\mathbf{r}(t) = (\sin t)\mathbf{i} + (\cosh t)\mathbf{j} + (\tan^{-1} t)\mathbf{k}; \ P_0(0, 1, 0)$.

31. Find an equation of the plane that is perpendicular to the curve $\mathbf{r} = (3 \sin t)\mathbf{i} - (2 \cos t)\mathbf{j} + t\mathbf{k}$ at the point where $t = \pi/2$. [*Note:* A plane is considered to be perpendicular to a curve at a point if it is perpendicular to the tangent line at that point.]

32. Find an equation of the plane that is perpendicular to the curve $\mathbf{r} = 3t^2\mathbf{i} + \sqrt{t + 5}\,\mathbf{j} - 2t\mathbf{k}$ at the point $P(3, 2, 2)$ on the curve. [See note in Exercise 31.]

33. (a) Find the points where the curve

$$\mathbf{r} = t\mathbf{i} + t^2\mathbf{j} - 3t\mathbf{k}$$

intersects the plane $2x - y + z = -2$.

(b) For the curve and plane in part (a), find, to the nearest degree, the acute angle that the tangent line to the curve makes with a line normal to the plane at each point of intersection.

34. Find where the tangent line to the curve

$$\mathbf{r} = e^{-2t}\mathbf{i} + (\cos t)\mathbf{j} + (3 \sin t)\mathbf{k}$$

at the point $(1, 1, 0)$ intersects the yz-plane.

In Exercises 35 and 36, show that the graphs of $\mathbf{r}_1(t)$ and $\mathbf{r}_2(t)$ intersect at the point P. Find, to the nearest degree, the acute angle between the tangent lines to the graphs of $\mathbf{r}_1(t)$ and $\mathbf{r}_2(t)$ at the point P.

35. $\mathbf{r}_1(t) = t^2\mathbf{i} + t\mathbf{j} + 3t^3\mathbf{k}$,
$\mathbf{r}_2(t) = (t - 1)\mathbf{i} + \frac{1}{4}t^2\mathbf{j} + (5 - t)\mathbf{k}$; $P(1, 1, 3)$.

36. $\mathbf{r}_1(t) = 2e^{-t}\mathbf{i} + (\cos t)\mathbf{j} + (t^2 + 3)\mathbf{k}$,
$\mathbf{r}_2(t) = (1 - t)\mathbf{i} + t^2\mathbf{j} + (t^3 + 4)\mathbf{k}$; $P(2, 1, 3)$.

37. Show that the graphs of $\mathbf{r}_1(t) = t\mathbf{i} + t^2\mathbf{j}$ and $\mathbf{r}_2(t) = t^3\mathbf{i} + t^6\mathbf{j}$ are the same, but that $\mathbf{r}_1'(t)$ is never zero, whereas $\mathbf{r}_2'(t) = \mathbf{0}$ for some value of t.

38. Show that the graphs of

$$\mathbf{r}_1(t) = (\cos t)\mathbf{i} + (\sin t)\mathbf{j} + t\mathbf{k}$$

and

$$\mathbf{r}_2(t) = \cos(t^3)\mathbf{i} + \sin(t^3)\mathbf{j} + t^3\mathbf{k}$$

are the same, but that $\mathbf{r}_1'(t)$ is never zero, whereas $\mathbf{r}_2'(t) = \mathbf{0}$ for some value of t.

In Exercises 39–42, determine whether $\mathbf{r}$ is a smooth function of the parameter t.

39. $\mathbf{r} = t^3\mathbf{i} + (3t^2 - 2t)\mathbf{j} + t^2\mathbf{k}$.

40. $\mathbf{r} = \cos(t^2)\mathbf{i} + \sin(t^2)\mathbf{j} + e^{-t}\mathbf{k}$.

41. $\mathbf{r} = te^{-t}\mathbf{i} + (t^2 - 2t)\mathbf{j} + \cos(\pi t)\mathbf{k}$.

42. $\mathbf{r} = \sin(\pi t)\mathbf{i} + (2t - \ln t)\mathbf{j} + (t^2 - t)\mathbf{k}$.

43. Calculate $(d/dt) [\mathbf{r}_1(t) \cdot \mathbf{r}_2(t)]$ two ways: first using Formula (6), and then by differentiating $\mathbf{r}_1(t) \cdot \mathbf{r}_2(t)$ directly.

(a) $\mathbf{r}_1(t) = 2t\mathbf{i} + 3t^2\mathbf{j} + t^3\mathbf{k}$, $\mathbf{r}_2(t) = t^4\mathbf{k}$

(b) $\mathbf{r}_1(t) = 3 \sec t\mathbf{i} - t\mathbf{j} + \ln t\mathbf{k}$,
$\mathbf{r}_2(t) = 4t\mathbf{i} - \sin t\mathbf{k}$.

44. Calculate $(d/dt) [\mathbf{r}_1(t) \times \mathbf{r}_2(t)]$ two ways: first using

Formula (7), and then by differentiating $\mathbf{r}_1(t) \times \mathbf{r}_2(t)$ directly.

(a) $\mathbf{r}_1(t) = 2t\mathbf{i} + 3t^2\mathbf{j} + t^3\mathbf{k}$, $\mathbf{r}_2(t) = t^4\mathbf{k}$

(b) $\mathbf{r}_1(t) = 3 \sec t\mathbf{i} - t\mathbf{j} + \ln t\mathbf{k}$,
$\mathbf{r}_2(t) = 4t\mathbf{i} - \sin t\mathbf{k}$.

In Exercises 45–47, calculate $d\mathbf{r}/du$ by the chain rule, and then check your result by expressing $\mathbf{r}$ in terms of u and differentiating.

45. $\mathbf{r} = t\mathbf{i} + t^2\mathbf{j}$; $t = 4u + 1$.

46. $\mathbf{r} = \langle 3 \cos t, 3 \sin t \rangle$; $t = \pi u$.

47. $\mathbf{r} = e^t\mathbf{i} + 4e^{-t}\mathbf{j}$; $t = u^2$.

48. Let $\mathbf{r}(t) = (a \cos t)\mathbf{i} + (a \sin t)\mathbf{j}$ $(a > 0)$.

(a) Show that $\mathbf{r}(t) \cdot \mathbf{r}'(t) = 0$.

(b) Give a geometric explanation of the result in part (a).

49. Prove: $(d/dt)[\mathbf{r}(t) \times \mathbf{r}'(t)] = \mathbf{r}(t) \times \mathbf{r}''(t)$. [*Hint:* Use Formula (7).]

50. Let $\mathbf{u} = \mathbf{u}(t)$, $\mathbf{v} = \mathbf{v}(t)$, and $\mathbf{w} = \mathbf{w}(t)$ be differentiable vector-valued functions. Use Formulas (6) and (7) to show that

$$\frac{d}{dt} [\mathbf{u} \cdot (\mathbf{v} \times \mathbf{w})]$$

$$= \frac{d\mathbf{u}}{dt} \cdot [\mathbf{v} \times \mathbf{w}] + \mathbf{u} \cdot \left[\frac{d\mathbf{v}}{dt} \times \mathbf{w} \right] + \mathbf{u} \cdot \left[\mathbf{v} \times \frac{d\mathbf{w}}{dt} \right]$$

51. Let $u_1, u_2, u_3, v_1, v_2, v_3, w_1, w_2,$ and w_3 be differentiable functions of t. Use Exercise 50 to show that

$$\frac{d}{dt} \begin{vmatrix} u_1 & u_2 & u_3 \\ v_1 & v_2 & v_3 \\ w_1 & w_2 & w_3 \end{vmatrix}$$

$$= \begin{vmatrix} u_1' & u_2' & u_3' \\ v_1 & v_2 & v_3 \\ w_1 & w_2 & w_3 \end{vmatrix} + \begin{vmatrix} u_1 & u_2 & u_3 \\ v_1' & v_2' & v_3' \\ w_1 & w_2 & w_3 \end{vmatrix} + \begin{vmatrix} u_1 & u_2 & u_3 \\ v_1 & v_2 & v_3 \\ w_1' & w_2' & w_3' \end{vmatrix}$$

52. Let $\mathbf{r} = \mathbf{r}(t)$. Show that

$$\frac{d}{dt} [\|\mathbf{r}\|] = \frac{1}{\|\mathbf{r}\|} \mathbf{r} \cdot \mathbf{r}'$$

[*Hint:* Consider $\mathbf{r} \cdot \mathbf{r}$.]

53. Use Theorem 15.2.5 and Exercise 52 to derive the formula

$$\frac{d}{dt} \left[\frac{\mathbf{r}}{\|\mathbf{r}\|} \right] = \frac{1}{\|\mathbf{r}\|} \mathbf{r}' - \frac{\mathbf{r} \cdot \mathbf{r}'}{\|\mathbf{r}\|^3} \mathbf{r}$$

54. Prove Theorem 15.2.5 for 2-space.

55. Derive Formulas (6) and (7) for 3-space.

56. Let $\mathbf{r}(t) = x(t)\mathbf{i} + y(t)\mathbf{j}$. Prove: If $\mathbf{r}'(t) = \mathbf{0}$ for all t in an interval, then $\mathbf{r}$ is constant on the interval. [*Hint:* See Theorem 4.2.2(c).]

57. Prove Theorem 15.2.7 for 2-space.

58. Let lim stand for any one of the limit symbols $\lim\limits_{t \to a}$, $\lim\limits_{t \to a^+}$, $\lim\limits_{t \to a^-}$, $\lim\limits_{t \to +\infty}$, or $\lim\limits_{t \to -\infty}$. Prove for 2-space:

(a) If k is a scalar and $\lim \mathbf{r}(t)$ exists, then $\lim k\mathbf{r}(t) = k \lim \mathbf{r}(t)$.

(b) If $\lim \mathbf{r}_1(t)$ and $\lim \mathbf{r}_2(t)$ exist, then
$$\lim [\mathbf{r}_1(t) + \mathbf{r}_2(t)] = \lim \mathbf{r}_1(t) + \lim \mathbf{r}_2(t),$$
$$\lim [\mathbf{r}_1(t) - \mathbf{r}_2(t)] = \lim \mathbf{r}_1(t) - \lim \mathbf{r}_2(t).$$

59. Prove for 2-space: $\mathbf{r}$ is continuous at t_0 if and only if each component of $\mathbf{r}$ is continuous at t_0.

■ 15.3 INTEGRATION OF VECTOR-VALUED FUNCTIONS; ARC LENGTH

In this section we shall define integrals of vector-valued functions and give some of their basic properties. We shall also derive some important formulas for the arc length of the graph of a vector-valued function.

Because derivatives of vector-valued functions are calculated by differentiating components, it is natural to define integrals of vector-valued functions in terms of components.

□ INTEGRALS OF VECTOR-VALUED FUNCTIONS

15.3.1 DEFINITION.

(a) If $\mathbf{r}(t) = x(t)\mathbf{i} + y(t)\mathbf{j}$ is a vector-valued function in 2-space, then we define

$$\int \mathbf{r}(t) \, dt = \left(\int x(t) \, dt \right) \mathbf{i} + \left(\int y(t) \, dt \right) \mathbf{j} \tag{1a}$$

$$\int_a^b \mathbf{r}(t) \, dt = \left(\int_a^b x(t) \, dt \right) \mathbf{i} + \left(\int_a^b y(t) \, dt \right) \mathbf{j} \tag{1b}$$

(b) If $\mathbf{r}(t) = x(t)\mathbf{i} + y(t)\mathbf{j} + z(t)\mathbf{k}$ is a vector-valued function in 3-space, then we define

$$\int \mathbf{r}(t) \, dt = \left(\int x(t) \, dt \right) \mathbf{i} + \left(\int y(t) \, dt \right) \mathbf{j} + \left(\int z(t) \, dt \right) \mathbf{k} \tag{2a}$$

$$\int_a^b \mathbf{r}(t) \, dt = \left(\int_a^b x(t) \, dt \right) \mathbf{i} + \left(\int_a^b y(t) \, dt \right) \mathbf{j} + \left(\int_a^b z(t) \, dt \right) \mathbf{k} \tag{2b}$$

Example 1 Let $\mathbf{r}(t) = 2t\mathbf{i} + 3t^2\mathbf{j}$. Find

(a) $\displaystyle\int \mathbf{r}(t)\, dt$ (b) $\displaystyle\int_0^2 \mathbf{r}(t)\, dt$

Solution (a).

$$\int \mathbf{r}(t)\, dt = \int (2t\mathbf{i} + 3t^2\mathbf{j})\, dt = \left(\int 2t\, dt\right)\mathbf{i} + \left(\int 3t^2\, dt\right)\mathbf{j}$$

$$= (t^2 + C_1)\mathbf{i} + (t^3 + C_2)\mathbf{j} = (t^2\mathbf{i} + t^3\mathbf{j}) + C_1\mathbf{i} + C_2\mathbf{j} = t^2\mathbf{i} + t^3\mathbf{j} + \mathbf{C}$$

where $\mathbf{C} = C_1\mathbf{i} + C_2\mathbf{j}$ is an arbitrary vector constant of integration.

Solution (b). $\displaystyle\int_0^2 \mathbf{r}(t)\, dt = \int_0^2 (2t\mathbf{i} + 3t^2\mathbf{j})\, dt$

$$= \left(\int_0^2 2t\, dt\right)\mathbf{i} + \left(\int_0^2 3t^2\, dt\right)\mathbf{j} = \Big[t^2\Big]_0^2\mathbf{i} + \Big[t^3\Big]_0^2\mathbf{j} = 4\mathbf{i} + 8\mathbf{j} \quad\blacktriangleleft$$

☐ **PROPERTIES OF INTEGRALS**

In the exercises we ask the reader to show that the standard properties of integrals carry over to vector-valued functions in 2-space and 3-space:

$$\int c\mathbf{r}(t)\, dt = c\int \mathbf{r}(t)\, dt \tag{3}$$

$$\int [\mathbf{r}_1(t) + \mathbf{r}_2(t)]\, dt = \int \mathbf{r}_1(t)\, dt + \int \mathbf{r}_2(t)\, dt \tag{4}$$

$$\int [\mathbf{r}_1(t) - \mathbf{r}_2(t)]\, dt = \int \mathbf{r}_1(t)\, dt - \int \mathbf{r}_2(t)\, dt \tag{5}$$

These properties also hold for definite integrals of vector-valued functions. In addition, we leave it for the reader to show that if $\mathbf{r}$ is a vector-valued function in 2-space or 3-space, then

$$\frac{d}{dt}\left[\int \mathbf{r}(t)\, dt\right] = \mathbf{r}(t) \tag{6}$$

This shows that an indefinite integral of $\mathbf{r}(t)$ is, in fact, the set of antiderivatives of $\mathbf{r}(t)$, just as for real-valued functions. We also leave it as an exercise to show that if $\mathbf{R}(t)$ is any antiderivative of $\mathbf{r}(t)$ in the sense that $\mathbf{R}'(t) = \mathbf{r}(t)$, then

$$\int \mathbf{r}(t)\, dt = \mathbf{R}(t) + \mathbf{C} \tag{7}$$

where **C** is an arbitrary vector constant of integration. Moreover,

$$\int_a^b \mathbf{r}(t)\, dt = \mathbf{R}(t) \Big]_a^b = \mathbf{R}(b) - \mathbf{R}(a) \tag{8}$$

which is the extension of the First Fundamental Theorem of Calculus (Theorem 5.7.1) to vector-valued functions.

Example 2 In part (a) of Example 1 we showed that $\mathbf{R}(t) = t^2\mathbf{i} + t^3\mathbf{j}$ is an antiderivative of $\mathbf{r}(t) = 2t\mathbf{i} + 3t^2\mathbf{j}$. (It is the antiderivative that results when $\mathbf{C} = \mathbf{0}$.) Thus, from (8)

$$\int_0^2 \mathbf{r}(t)\, dt = \int_0^2 (2t\mathbf{i} + 3t^2\mathbf{j})\, dt = \left[\mathbf{R}(t) \right]_0^2$$

$$= \left[t^2\mathbf{i} + t^3\mathbf{j} \right]_0^2 = (4\mathbf{i} + 8\mathbf{j}) - (0\mathbf{i} + 0\mathbf{j}) = 4\mathbf{i} + 8\mathbf{j}$$

which agrees with the result obtained in Example 1 by integrating term by term. ◄

In Theorem 13.4.1 we showed that if $x'(t)$ and $y'(t)$ are continuous for $a \le t \le b$, then the curve given by the parametric equations

$$x = x(t), \quad y = y(t) \qquad (a \le t \le b) \tag{9}$$

has arc length

$$L = \int_a^b \sqrt{\left(\frac{dx}{dt}\right)^2 + \left(\frac{dy}{dt}\right)^2}\, dt \tag{10}$$

This result generalizes to curves in 3-space exactly as one would expect: If $x'(t)$, $y'(t)$, and $z'(t)$ are continuous for $a \le t \le b$, then the curve given by the parametric equations

$$x = x(t), \quad y = y(t), \quad z = z(t) \qquad (a \le t \le b) \tag{11}$$

has arc length

$$L = \int_a^b \sqrt{\left(\frac{dx}{dt}\right)^2 + \left(\frac{dy}{dt}\right)^2 + \left(\frac{dz}{dt}\right)^2}\, dt \tag{12}$$

Example 3 Find the arc length of that portion of the circular helix

$$x = \cos t, \quad y = \sin t, \quad z = t$$

from $t = 0$ to $t = \pi$.

Solution.   **From (12) the arc length is**

$$L = \int_0^\pi \sqrt{\left(\frac{dx}{dt}\right)^2 + \left(\frac{dy}{dt}\right)^2 + \left(\frac{dz}{dt}\right)^2}\, dt$$

$$= \int_0^\pi \sqrt{(-\sin t)^2 + (\cos t)^2 + 1}\, dt = \int_0^\pi \sqrt{2}\, dt = \sqrt{2}\,\pi \qquad \blacktriangleleft$$

☐ **ARC LENGTH AS A PARAMETER**

For many purposes the best parameter to use for representing a curve in 2-space or 3-space parametrically is the length of arc measured along the curve from some fixed reference point. This can be done as follows:

Step 1.   Select an arbitrary point on the curve C to serve as a *reference point.*

Step 2.   Starting from the reference point, choose one direction along the curve to be the *positive direction* and the other to be the *negative direction.*

Step 3.   If P is a point on the curve, let s be the "signed" arc length along C from the reference point to P, where s is positive if P is in the positive direction from the reference point, and s is negative if P is in the negative direction. Figure 15.3.1 illustrates this idea.

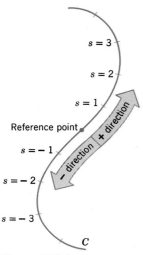

Figure 15.3.1

By this procedure, a unique point P on the curve is determined when a value for s is given. For example, $s = 2$ determines the point that is 2 units along the curve in the positive direction from the reference point, and $s = -\frac{3}{2}$ determines the point that is $\frac{3}{2}$ units along the curve in the negative direction from the reference point.

Let us now treat s as a variable. As the value of s changes, the corresponding point P moves along C and the coordinates of P become functions of s. Thus, in 2-space the coordinates of P are $(x(s), y(s))$, and in 3-space they are $(x(s), y(s), z(s))$. Therefore, in 2-space the curve C is given by the parametric equations

$$x = x(s), \quad y = y(s)$$

and in 3-space by

$$x = x(s), \quad y = y(s), \quad z = z(s)$$

REMARK.   When defining the parameter s, the choice of positive and negative directions is arbitrary. However, it may be that the curve C is already specified in terms of some other parameter t, in which case we shall agree always to take the direction of increasing t as the positive direction for the parameter s. By so doing, s will increase as t increases and vice versa.

The following theorem gives a formula for computing an arc-length parameter s when the curve C is expressed in terms of some other parameter t. This result will be used when we want to change the parametrization for C from t to s.

15.3.2 THEOREM.

(a) *Let C be a curve in 2-space given parametrically by*

$$x = x(t), \quad y = y(t)$$

where $x'(t)$ and $y'(t)$ are continuous functions. If an arc-length parameter s is introduced with its reference point at $(x(t_0), y(t_0))$, then the parameters s and t are related by

$$s = \int_{t_0}^{t} \sqrt{\left(\frac{dx}{du}\right)^2 + \left(\frac{dy}{du}\right)^2}\, du \tag{13a}$$

(b) *Let C be a curve in 3-space given parametrically by*

$$x = x(t), \quad y = y(t), \quad z = z(t)$$

where $x'(t)$, $y'(t)$, and $z'(t)$ are continuous functions. If an arc-length parameter s is introduced with its reference point at $(x(t_0), y(t_0), z(t_0))$, then the parameters s and t are related by

$$s = \int_{t_0}^{t} \sqrt{\left(\frac{dx}{du}\right)^2 + \left(\frac{dy}{du}\right)^2 + \left(\frac{dz}{du}\right)^2}\, du \tag{13b}$$

We shall prove (a) and leave the proof of (b) as an exercise.

Proof (a). If $t > t_0$, then from (10) (with u as the variable of integration rather than t) it follows that

$$\int_{t_0}^{t} \sqrt{\left(\frac{dx}{du}\right)^2 + \left(\frac{dy}{du}\right)^2}\, du \tag{14}$$

represents the arc length of that portion of the curve C that lies between $(x(t_0), y(t_0))$ and $(x(t), y(t))$. If $t < t_0$, then (14) is the negative of this arc length. In either case, integral (14) represents the "signed" arc length s between these points, which proves (13a). ∎

It follows from Formulas (13a) and (13b) and the Second Fundamental Theorem of Calculus (Theorem 5.9.3) that in 2-space

$$\frac{ds}{dt} = \frac{d}{dt}\left[\int_{t_0}^{t} \sqrt{\left(\frac{dx}{du}\right)^2 + \left(\frac{dy}{du}\right)^2}\, du\right] = \sqrt{\left(\frac{dx}{dt}\right)^2 + \left(\frac{dy}{dt}\right)^2}$$

and in 3-space

$$\frac{ds}{dt} = \frac{d}{dt}\left[\int_{t_0}^{t} \sqrt{\left(\frac{dx}{du}\right)^2 + \left(\frac{dy}{du}\right)^2 + \left(\frac{dz}{du}\right)^2}\, dt\right]$$

$$= \sqrt{\left(\frac{dx}{dt}\right)^2 + \left(\frac{dy}{dt}\right)^2 + \left(\frac{dz}{dt}\right)^2}$$

Thus, in 2-space and 3-space, respectively,

$$\frac{ds}{dt} = \sqrt{\left(\frac{dx}{dt}\right)^2 + \left(\frac{dy}{dt}\right)^2} \tag{15a}$$

$$\frac{ds}{dt} = \sqrt{\left(\frac{dx}{dt}\right)^2 + \left(\frac{dy}{dt}\right)^2 + \left(\frac{dz}{dt}\right)^2} \tag{15b}$$

REMARK.   Formulas (15a) and (15b) reveal two facts worth noting. First, ds/dt does not depend on t_0; that is, the value of ds/dt is independent of where the reference point for the parameter s is located. This is to be expected since changing the position of the reference point shifts each value of s by a constant (the arc length between the reference points), and this constant drops out when we differentiate. The second fact to be noted from (15a) and (15b) is that $ds/dt \geq 0$ for all t. This is also to be expected since s increases with t by the remark preceding Theorem 15.3.2. If the curve C is smooth, then it follows from (15a) and (15b) that $ds/dt > 0$ for all t (why?).

Example 4   Find parametric equations for the line

$$x = 2t + 1, \quad y = 3t - 2 \tag{16}$$

using arc length s as a parameter, where the reference point for s is the point $(1, -2)$.

Solution.   In Formula (13a) we used u as the variable of integration because t was needed as a limit of integration. To apply (13a), we first rewrite the given parametric equations with u in place of t; this gives

$$x = 2u + 1, \quad y = 3u - 2$$

from which we obtain

$$\frac{dx}{du} = 2, \quad \frac{dy}{du} = 3$$

From (16) we see that the reference point $(1, -2)$ corresponds to $t = t_0 = 0$, so (13a) yields

$$s = \int_{t_0}^{t} \sqrt{\left(\frac{dx}{du}\right)^2 + \left(\frac{dy}{du}\right)^2} \, du = \int_{0}^{t} \sqrt{13} \, du = \sqrt{13}u \Big]_{u=0}^{u=t} = \sqrt{13}t$$

Therefore,

$$t = \frac{1}{\sqrt{13}} s$$

Substituting this expression in the given parametric equations yields

$$x = 2\left(\frac{1}{\sqrt{13}} s\right) + 1 = \frac{2}{\sqrt{13}} s + 1$$

$$y = 3\left(\frac{1}{\sqrt{13}} s\right) - 2 = \frac{3}{\sqrt{13}} s - 2 \quad \blacktriangleleft$$

Example 5   Find parametric equations for the circle

$$x = a \cos t, \quad y = a \sin t \qquad (0 \le t \le 2\pi)$$

using arc length s as a parameter, with the reference point for s being $(a, 0)$, where $a > 0$.

Solution.   Proceeding as in Example 4, we first replace t by u in the given equations so that

$$x = a \cos u, \quad y = a \sin u$$

and

$$\frac{dx}{du} = -a \sin u, \quad \frac{dy}{du} = a \cos u$$

Since the reference point $(a, 0)$ corresponds to $t = 0$, we obtain from (13a),

$$s = \int_{t_0}^{t} \sqrt{\left(\frac{dx}{du}\right)^2 + \left(\frac{dy}{du}\right)^2} \, du = \int_{0}^{t} \sqrt{(-a \sin u)^2 + (a \cos u)^2} \, du$$

$$= \int_{0}^{t} a \, du = au \bigg]_{u=0}^{u=t} = at$$

Solving for t in terms of s yields

$$t = s/a$$

Substituting this in the given parametric equations and using the fact that $s = at$ ranges from 0 to $2\pi a$ as t ranges from 0 to 2π, we obtain

$$x = a \cos (s/a), \quad y = a \sin (s/a) \qquad (0 \le s \le 2\pi a) \quad \blacktriangleleft$$

▶ Exercise Set 15.3   Ⓒ 28

In Exercises 1–12, evaluate the integral.

1. $\int (3\mathbf{i} + 4t\mathbf{j})\, dt.$

2. $\int [(\cos t)\mathbf{i} + (\sin t)\mathbf{j}]\, dt.$

3. $\int_0^{\pi/3} \langle \cos 3t, -\sin 3t \rangle\, dt.$

4. $\int_0^1 (t^2\mathbf{i} + t^3\mathbf{j})\, dt.$ 5. $\int_1^9 (t^{1/2}\mathbf{i} + t^{-1/2}\mathbf{j})\, dt.$

6. $\int [(t \sin t)\mathbf{i} + \mathbf{j}]\, dt.$ 7. $\int \langle te^t, \ln t \rangle\, dt.$

8. $\int_0^2 \|t\mathbf{i} + t^2\mathbf{j}\|\, dt.$

9. $\int \left[t^2\mathbf{i} - 2t\mathbf{j} + \frac{1}{t}\mathbf{k} \right]\, dt.$

10. $\int \langle e^{-t}, e^t, 3t^2 \rangle\, dt.$

11. $\int_0^1 (e^{2t}\mathbf{i} + e^{-t}\mathbf{j} + t\mathbf{k})\, dt.$

12. $\int_{-3}^3 \langle (3 - t)^{3/2}, (3 + t)^{3/2}, 1 \rangle\, dt.$

13. Suppose that a particle moves through 3-space along the curve $\mathbf{r} = t\mathbf{i} - 3t^2\mathbf{j} + \mathbf{k}$ and that it is subjected to a force of $\mathbf{F} = 3x\mathbf{i} - 2\mathbf{j} + yz\mathbf{k}$ when it is at the point (x, y, z).
    (a) Find $\mathbf{F}$ in terms of t for points on the path.
    (b) Find $\int_0^2 \mathbf{F} \cdot \dfrac{d\mathbf{r}}{dt}\, dt.$ [*Note:* Later, we shall see that this is the work done by the force as the particle moves along the curve from the point where $t = 0$ to the point where $t = 2.$]

14. Find $\mathbf{r}(t)$ given that $\mathbf{r}'(t) = t^2\mathbf{i} + 2t\mathbf{j}$ and $\mathbf{r}(0) = \mathbf{i} + \mathbf{j}.$

15. Find $\mathbf{r}(t)$ given that $\mathbf{r}'(t) = (\cos t)\mathbf{i} + (\sin t)\mathbf{j}$ and $\mathbf{r}(0) = \mathbf{i} - \mathbf{j}.$

16. Find $\mathbf{r}(t)$ given that $\mathbf{r}''(t) = \mathbf{i} + e^t\mathbf{j}$, $\mathbf{r}(0) = 2\mathbf{i}$, and $\mathbf{r}'(0) = \mathbf{j}.$

17. Find $\mathbf{r}(t)$ given that $\mathbf{r}''(t) = 12t^2\mathbf{i} - 2\mathbf{j}$, $\mathbf{r}'(0) = \mathbf{0}$, and $\mathbf{r}(0) = 2\mathbf{i} - 4\mathbf{j}.$

18. Find $\mathbf{r}(t)$ given that $\mathbf{r}'(t) = e^{-2t}\mathbf{i} + (\cos t)\mathbf{j} - \mathbf{k}$ and $\mathbf{r}(0) = 3\mathbf{j} + 2\mathbf{k}.$

19. Find $\mathbf{r}(t)$ given that $\mathbf{r}'(t) = 2\mathbf{i} + \dfrac{t}{t^2 + 1}\mathbf{j} + t\mathbf{k}$ and $\mathbf{r}(1) = \mathbf{0}.$

20. Find $\mathbf{r}(t)$ given that $\mathbf{r}''(t) = (4 \sin 2t)\mathbf{i} + 6t\mathbf{j} + e^{-t}\mathbf{k}$, $\mathbf{r}(0) = 2\mathbf{i}$, and $\mathbf{r}'(0) = \mathbf{k}.$

In Exercises 21–26, find the arc length of the curve.

21. $\mathbf{r}(t) = (4 + 3t)\mathbf{i} + (2 - 2t)\mathbf{j} + (5 + t)\mathbf{k}; \; 3 \leq t \leq 4.$

22. $\mathbf{r}(t) = 3 \cos t\mathbf{i} + 3 \sin t\mathbf{j} + t\mathbf{k}; \; 0 \leq t \leq 2\pi.$

23. $\mathbf{r}(t) = t^3\mathbf{i} + t\mathbf{j} + \frac{1}{2}\sqrt{6}t^2\mathbf{k}; \; 1 \leq t \leq 3.$

24. $x = \cos^3 t, \; y = \sin^3 t, \; z = 2; \; 0 \leq t \leq \pi/2.$

25. $\mathbf{r}(t) = \langle e^t, e^{-t}, \sqrt{2}t \rangle; \; 0 \leq t \leq 1.$

26. $x = \frac{1}{2}t, \; y = \frac{1}{3}(1 - t)^{3/2}, \; z = \frac{1}{3}(1 + t)^{3/2}; \; -1 \leq t \leq 1.$

27. Find the arc length of the circular helix $x = a \cos t$, $y = a \sin t$, $z = ct$ for $0 \leq t \leq t_0.$

28. Copper tubing with an outside diameter of $\frac{1}{2}$ in. is to be wrapped in a circular helix around a cylindrical core that has a 12-in. diameter. What length of tubing will make one complete turn around the cylinder in a distance of 20 in. measured along the axis of the cylinder?

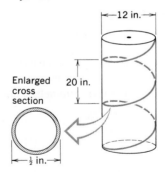

In Exercises 29–37, find parametric equations for the curve using arc length s as a parameter. Use the point on the curve where $t = 0$ as the reference point.

29. $\mathbf{r}(t) = (3t - 2)\mathbf{i} + (4t + 3)\mathbf{j}.$

30. $\mathbf{r}(t) = (3 \cos 2t)\mathbf{i} + (3 \sin 2t)\mathbf{j}; \; 0 \leq t \leq \pi.$

31. $\mathbf{r}(t) = (3 + \cos t)\mathbf{i} + (2 + \sin t)\mathbf{j}; \; 0 \leq t \leq 2\pi.$

32. $\mathbf{r}(t) = (\cos^3 t)\mathbf{i} + (\sin^3 t)\mathbf{j}; \; 0 \leq t \leq \pi/2.$

33. $\mathbf{r}(t) = \frac{1}{3}t^3\mathbf{i} + \frac{1}{2}t^2\mathbf{j}; \; t \geq 0.$

34. $\mathbf{r}(t) = (1 + t)^2\mathbf{i} + (1 + t)^3\mathbf{j}; \; 0 \leq t \leq 1.$

35. $\mathbf{r}(t) = (e^t \cos t)\mathbf{i} + (e^t \sin t)\mathbf{j}$; $0 \le t \le \pi/2$.

36. $\mathbf{r}(t) = \sin(e^t)\mathbf{i} + \cos(e^t)\mathbf{j} + \sqrt{3}e^t\mathbf{k}$; $t \ge 0$.

37. $\mathbf{r}(t) = (t \cos t)\mathbf{i} + (t \sin t)\mathbf{j} + \frac{2}{3}\sqrt{2}t^{3/2}\mathbf{k}$; $t \ge 0$.

38. Find parametric equations for the cycloid

$$x = at - a \sin t$$
$$y = a - a \cos t \qquad (0 \le t \le 2\pi)$$

using arc length as the parameter. Take $(0, 0)$ as the reference point.

39. Use the result in Exercise 27 to show that the circular helix

$$\mathbf{r} = (a \cos t)\mathbf{i} + (a \sin t)\mathbf{j} + ct\mathbf{k}$$

can be expressed as

$$\mathbf{r} = \left(a \cos \frac{s}{w}\right)\mathbf{i} + \left(a \sin \frac{s}{w}\right)\mathbf{j} + \frac{cs}{w}\mathbf{k}$$

where $w = \sqrt{a^2 + c^2}$ and s is an arc-length parameter with reference point at $(a, 0, 0)$.

40. Recall from Formula (5) of Section 14.5 that $\mathbf{r} = \mathbf{r}_0 + t\mathbf{v}$ is the vector equation of a line in 2-space or 3-space if $\mathbf{v} \ne \mathbf{0}$. Show that $\mathbf{r} = \mathbf{r}_0 + s\mathbf{v}/\|\mathbf{v}\|$ is a vector equation of the line in terms of an arc-length parameter s whose reference point is at the tip of $\mathbf{r}_0$.

41. A thread with negligible thickness is unwound from a spool of radius a, with the unwound piece always held straight.
   (a) Show that, in terms of the parameter θ ($\theta \ge 0$) as shown in the figure, the curve traced out by the free end of the thread has parametric equations

$$x = a(\cos \theta + \theta \sin \theta)$$
$$y = a(\sin \theta - \theta \cos \theta)$$

   (b) Find parametric equations for the curve of part (a) using arc length as the parameter, where $\theta = 0$ is the reference point.

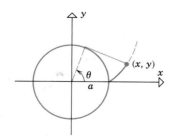

42. Show that in cylindrical coordinates a curve given by the parametric equations $r = r(t)$, $\theta = \theta(t)$, $z = z(t)$, for $a \le t \le b$, has arc length

$$L = \int_a^b \sqrt{\left(\frac{dr}{dt}\right)^2 + r^2\left(\frac{d\theta}{dt}\right)^2 + \left(\frac{dz}{dt}\right)^2}\, dt$$

[*Hint:* Use the relationships $x = r\cos\theta$, $y = r\sin\theta$.]

43. Use the formula in Exercise 42 to find the arc length of the following curves.
   (a) $r = e^{2t}$, $\theta = t$, $z = e^{2t}$; $0 \le t \le \ln 2$
   (b) $r = t^2$, $\theta = \ln t$, $z = \frac{1}{3}t^3$; $1 \le t \le 2$.

44. Show that in spherical coordinates a curve given by the parametric equations $\rho = \rho(t)$, $\theta = \theta(t)$, $\phi = \phi(t)$, for $a \le t \le b$, has arc length

$$L = \int_a^b \sqrt{\left(\frac{d\rho}{dt}\right)^2 + \rho^2 \sin^2\phi\left(\frac{d\theta}{dt}\right)^2 + \rho^2\left(\frac{d\phi}{dt}\right)^2}\, dt$$

[*Hint:* Use the relationships $x = \rho \sin\phi \cos\theta$, $y = \rho \sin\phi \sin\theta$, $z = \rho \cos\phi$.]

45. Use the formula in Exercise 44 to find the arc length of the following curves.
   (a) $\rho = e^{-t}$, $\theta = 2t$, $\phi = \pi/4$; $0 \le t \le 2$
   (b) $\rho = 2t$, $\theta = \ln t$, $\phi = \pi/6$; $1 \le t \le 5$.

46. Prove part (b) of Theorem 15.3.2.

47. Prove for 2-space: If $\mathbf{R}'(t) = \mathbf{r}(t)$ on an interval $[a, b]$, then
   (a) $\displaystyle\int \mathbf{r}(t)\, dt = \mathbf{R}(t) + \mathbf{C}$, where $\mathbf{C}$ is an arbitrary vector constant
   (b) $\displaystyle\int_a^b \mathbf{r}(t)\, dt = \mathbf{R}(b) - \mathbf{R}(a)$.

48. Prove for 2-space:
   (a) $\displaystyle\int c\mathbf{r}(t)\, dt = c\int \mathbf{r}(t)\, dt$, where c is a scalar constant
   (b) $\displaystyle\int [\mathbf{r}_1(t) + \mathbf{r}_2(t)]\, dt = \int \mathbf{r}_1(t)\, dt + \int \mathbf{r}_2(t)\, dt$.

15.4 UNIT TANGENT AND NORMAL VECTORS

> *In this section we shall discuss some geometric properties of vector-valued functions. Our work here will have important applications to the study of motion along a curved path in 2-space or 3-space.*

☐ **UNIT TANGENT VECTORS**

Recall from Section 15.2 that if $\mathbf{r}(t)$ is a vector-valued function in 2-space or 3-space with graph C, then the vector $\mathbf{r}'(t)$ is tangent to the graph of C if this vector is positioned so that its initial point is at the terminal point of the radius vector $\mathbf{r}(t)$. Moreover, $\mathbf{r}'(t)$ points in the direction of increasing t. Thus, if $\mathbf{r}'(t) \neq \mathbf{0}$, then the vector $\mathbf{T}(t)$ defined by

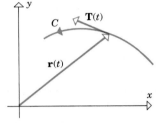

Figure 15.4.1

$$\mathbf{T}(t) = \frac{\mathbf{r}'(t)}{\|\mathbf{r}'(t)\|} \tag{1}$$

is tangent to C, points in the direction of increasing t, and has length 1 (Figure 15.4.1). We call $\mathbf{T}(t)$ the *unit tangent vector* to C at t.

Keep in mind that $\mathbf{T}(t)$ is defined only at those points where $\mathbf{r}'(t) \neq \mathbf{0}$. At each such point $\mathbf{T}(t)$ can be regarded as an indicator of the "direction" of the curve C.

Recall from Section 15.2 that if $\mathbf{r}(t)$ is a vector-valued function in 2-space or 3-space, then both $\mathbf{r}(t)$ and its graph C are called smooth if the components of $\mathbf{r}'(t)$ are continuous and $\mathbf{r}'(t) \neq \mathbf{0}$ for all values of t in the domain of $\mathbf{r}$. It follows from the second of these conditions that the graph of a smooth vector-valued function has a unit tangent vector at each point. Moreover, it follows from the first condition and (1) that the unit tangent vector $\mathbf{T}(t)$ is continuous (hence undergoes no "abrupt" changes of direction as t increases). For this reason, smooth curves are sometimes described as having *continuously turning tangents*. The function $\mathbf{r}(t) = t^2\mathbf{i} + t^3\mathbf{j}$ in Example 1 is not smooth since $\mathbf{r}'(t) = 2t\mathbf{i} + 3t^2\mathbf{j}$ is zero when $t = 0$. [Note the abrupt change in the direction of $\mathbf{T}(t)$ that occurs at the origin (Figure 15.4.2).]

REMARK. Unless stated otherwise, we shall always assume that the unit tangent vector $\mathbf{T}(t)$ is positioned with its initial point at the terminal point of $\mathbf{r}(t)$ as in Figure 15.4.1. This will ensure that $\mathbf{T}(t)$ is actually tangent to the graph of $\mathbf{r}(t)$ and not simply parallel to a tangent vector.

Example 1 Find the unit tangent vector to the graph of $\mathbf{r}(t) = t^2\mathbf{i} + t^3\mathbf{j}$ at the point where $t = 2$.

Solution. Since

$$\mathbf{r}'(t) = 2t\mathbf{i} + 3t^2\mathbf{j}$$

we obtain

$$\mathbf{T}(t) = \frac{\mathbf{r}'(t)}{\|\mathbf{r}'(t)\|} = \frac{2t\mathbf{i} + 3t^2\mathbf{j}}{\|2t\mathbf{i} + 3t^2\mathbf{j}\|} = \frac{2t\mathbf{i} + 3t^2\mathbf{j}}{\sqrt{4t^2 + 9t^4}}$$

Thus,

$$\mathbf{T}(2) = \frac{4\mathbf{i} + 12\mathbf{j}}{\sqrt{160}} = \frac{4\mathbf{i} + 12\mathbf{j}}{4\sqrt{10}} = \frac{1}{\sqrt{10}}\mathbf{i} + \frac{3}{\sqrt{10}}\mathbf{j}$$

The graph of $\mathbf{r}(t)$ and the vector $\mathbf{T}(2)$ are shown in Figure 15.4.2. ◀

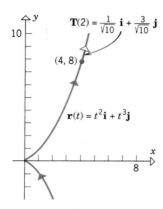

Figure 15.4.2

☐ **UNIT NORMAL VECTORS**   Let C be the graph of a smooth vector-valued function $\mathbf{r}(t)$ in 2-space or 3-space. In 2-space there are two unit vectors that are perpendicular to the unit tangent vector $\mathbf{T}(t)$, and in 3-space there are infinitely many such vectors (Figure 15.4.3). If $\mathbf{T}'(t) \neq \mathbf{0}$, then we define the *principal unit normal vector* to C at t to be the vector that is perpendicular to $\mathbf{T}(t)$ and has the same direction as $\mathbf{T}'(t)$. Thus,

$$\mathbf{N}(t) = \frac{\mathbf{T}'(t)}{\|\mathbf{T}'(t)\|} \qquad (2)$$

For simplicity, we shall often refer to $\mathbf{N}(t)$ as simply the *unit normal vector*. It

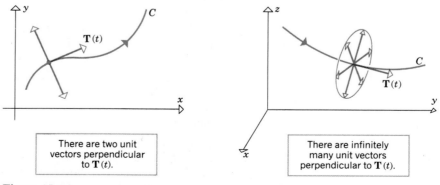

There are two unit vectors perpendicular to $\mathbf{T}(t)$.

There are infinitely many unit vectors perpendicular to $\mathbf{T}(t)$.

Figure 15.4.3

is evident that this vector has length 1 because it is obtained by normalizing $\mathbf{T}'(t)$. The following theorem shows that $\mathbf{N}$ and $\mathbf{T}$ are actually perpendicular.

15.4.1 THEOREM. *If $\mathbf{r}(t)$ is a vector-valued function in 2-space or 3-space, then for each value of t at which $\mathbf{T}(t)$ and $\mathbf{N}(t)$ exist, these vectors are perpendicular.*

Proof. The proof will be complete if we can show that $\mathbf{T}(t) \cdot \mathbf{N}(t) = 0$. But

$$\mathbf{T}(t) \cdot \mathbf{N}(t) = \mathbf{T}(t) \cdot \frac{\mathbf{T}'(t)}{\|\mathbf{T}'(t)\|} = \frac{1}{\|\mathbf{T}'(t)\|}\left(\mathbf{T}(t) \cdot \mathbf{T}'(t)\right)$$

so it suffices to show that $\mathbf{T}(t) \cdot \mathbf{T}'(t) = 0$. But this follows from Theorem 15.2.6 [with $\mathbf{T}(t)$ replacing $\mathbf{r}(t)$] and the fact that $\|\mathbf{T}(t)\| = 1$ for all t. ∎

REMARK. Unless stated otherwise, we shall always assume that $\mathbf{N}(t)$ is positioned with its initial point at the terminal point of $\mathbf{r}(t)$ (see Figure 15.4.4).

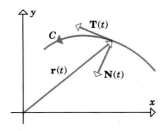

Figure 15.4.4

Example 2 Find $\mathbf{T}(t)$ and $\mathbf{N}(t)$ for the circular helix

$$x = a\cos t, \quad y = a\sin t, \quad z = ct$$

where $a > 0$.

Solution. The radius vector for the helix is

$$\mathbf{r}(t) = (a\cos t)\mathbf{i} + (a\sin t)\mathbf{j} + (ct)\mathbf{k}$$

Thus,

$$\mathbf{r}'(t) = (-a\sin t)\mathbf{i} + (a\cos t)\mathbf{j} + c\mathbf{k}$$

$$\|\mathbf{r}'(t)\| = \sqrt{(-a\sin t)^2 + (a\cos t)^2 + c^2} = \sqrt{a^2 + c^2}$$

$$\mathbf{T}(t) = \frac{\mathbf{r}'(t)}{\|\mathbf{r}'(t)\|} = -\frac{a\sin t}{\sqrt{a^2 + c^2}}\mathbf{i} + \frac{a\cos t}{\sqrt{a^2 + c^2}}\mathbf{j} + \frac{c}{\sqrt{a^2 + c^2}}\mathbf{k}$$

$$\mathbf{T}'(t) = -\frac{a\cos t}{\sqrt{a^2 + c^2}}\mathbf{i} - \frac{a\sin t}{\sqrt{a^2 + c^2}}\mathbf{j}$$

$$\|\mathbf{T}'(t)\| = \sqrt{\left(-\frac{a\cos t}{\sqrt{a^2 + c^2}}\right)^2 + \left(-\frac{a\sin t}{\sqrt{a^2 + c^2}}\right)^2} = \sqrt{\frac{a^2}{a^2 + c^2}} = \frac{a}{\sqrt{a^2 + c^2}}$$

$$\mathbf{N}(t) = \frac{\mathbf{T}'(t)}{\|\mathbf{T}'(t)\|} = (-\cos t)\mathbf{i} - (\sin t)\mathbf{j}$$

Because the $\mathbf{k}$ component of $\mathbf{N}(t)$ is zero, this vector lies in a horizontal plane for every value of t. Moreover, in the exercises we ask the reader to show that $\mathbf{N}(t)$ points directly toward the z-axis for all t (Figure 15.4.5). ◄

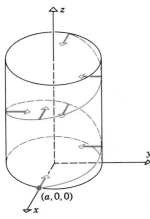

Figure 15.4.5

☐ **INWARD UNIT NORMAL
VECTORS IN 2-SPACE**

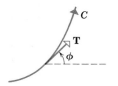

Figure 15.4.6

If **T** is a unit tangent vector to a curve C in 2-space, then, as noted earlier, there are two unit vectors perpendicular to **T** (Figure 15.4.3). To determine which of these is the principal unit normal **N**, we shall let $\phi = \phi(t)$ be the counterclockwise angle from the direction of the positive x-axis to $\mathbf{T} = \mathbf{T}(t)$ (Figure 15.4.6). Since **T** has length 1, we may write (see Example 8 of Section 14.2)

$$\mathbf{T} = (\cos \phi)\mathbf{i} + (\sin \phi)\mathbf{j} \tag{3}$$

It follows from this formula and the chain rule that

$$\mathbf{T}'(t) = \frac{d\mathbf{T}}{dt} = \frac{d\mathbf{T}}{d\phi}\frac{d\phi}{dt} = [(-\sin \phi)\mathbf{i} + (\cos \phi)\mathbf{j}]\frac{d\phi}{dt} = \mathbf{n}(t)\phi'(t) \tag{4}$$

where

$$\mathbf{n} = \mathbf{n}(t) = (-\sin \phi)\mathbf{i} + (\cos \phi)\mathbf{j} \tag{5}$$

Using the identities

$$\cos (\phi + \pi/2) = -\sin \phi \quad \text{and} \quad \sin (\phi + \pi/2) = \cos \phi$$

we can rewrite (5) as

$$\mathbf{n} = \mathbf{n}(t) = \cos \left(\phi + \frac{\pi}{2} \right)\mathbf{i} + \sin \left(\phi + \frac{\pi}{2} \right)\mathbf{j} \tag{6}$$

Comparing (3) and (6) we see that **n** is 90° counterclockwise from **T**. But from (4), $\mathbf{T}'(t)$ has the same direction as $\mathbf{n}(t)$ if $\phi'(t) > 0$ and the opposite direction if $\phi'(t) < 0$. The same is true of $\mathbf{N} = \mathbf{T}'(t)/\|\mathbf{T}'(t)\|$ (why?). (See Figure 15.4.7.)

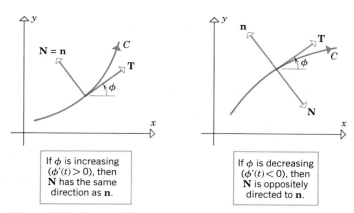

If ϕ is increasing
($\phi'(t) > 0$), then
N has the same
direction as **n**.

If ϕ is decreasing
($\phi'(t) < 0$), then
N is oppositely
directed to **n**.

Figure 15.4.7

It follows from the foregoing discussion that regardless of whether $\phi(t)$ is increasing or decreasing, the principal unit normal **N** points "inward" toward the concave side of the curve. For this reason **N** is called the **inward unit normal** in 2-space.

☐ **BINORMAL VECTORS IN 3-SPACE**

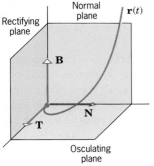

Figure 15.4.8

If $\mathbf{r}(t)$ is a vector-valued function in 3-space with graph C, then for each value of t at which $\mathbf{T}(t)$ and $\mathbf{N}(t)$ exist, we define

$$\mathbf{B}(t) = \mathbf{T}(t) \times \mathbf{N}(t) \tag{7}$$

to be the **binormal vector** to C at t. For each value of t the binormal vector $\mathbf{B}$ has length 1 and is perpendicular to both $\mathbf{T}$ and $\mathbf{N}$. At each point P on the curve C the vectors $\mathbf{T}$, $\mathbf{N}$, and $\mathbf{B}$ determine three mutually perpendicular planes through P whose names are shown in Figure 15.4.8. Moreover, as illustrated in Figure 14.4.4, the vectors are oriented by a "right-hand rule" in the sense that if the fingers of the right hand are cupped to point in the direction of rotation from $\mathbf{T}$ to $\mathbf{N}$, then the thumb indicates the direction of $\mathbf{B}$.

Three mutually perpendicular unit vectors in 3-space are sometimes called a **triad**. Whereas the **ijk**-triad is constant relative to the xyz-coordinate axes, the **TNB**-triad varies from point to point along the graph of $\mathbf{r}(t)$, since $\mathbf{T} = \mathbf{T}(t)$, $\mathbf{N} = \mathbf{N}(t)$, and $\mathbf{B} = \mathbf{B}(t)$ are functions of t (Figure 15.4.9). The **TNB**-triad is sometimes described as a *moving* triad.

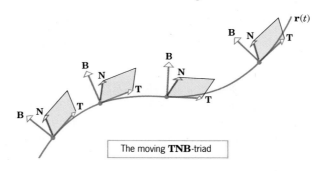

The moving **TNB**-triad

Figure 15.4.9

Exercise Set 15.4 ☐ 8

In Exercises 1–8, find the unit tangent vector $\mathbf{T}$ and the unit normal vector $\mathbf{N}$ for the given value of t. Then sketch the vectors and a portion of the curve containing the point of tangency.

1. $\mathbf{r}(t) = (5 \cos t)\mathbf{i} + (5 \sin t)\mathbf{j}$; $t = \pi/3$.

2. $\mathbf{r}(t) = 2t\mathbf{i} + 4t^2\mathbf{j}$; $t = 1$.

3. $\mathbf{r}(t) = (t^2 - 1)\mathbf{i} + t\mathbf{j}$; $t = 1$.

4. $\mathbf{r}(t) = e^t\mathbf{i} + e^{-t}\mathbf{j}$; $t = 0$.

5. $\mathbf{r}(t) = \frac{1}{2}t^2\mathbf{i} + \frac{1}{3}t^3\mathbf{j}$; $t = 1$.

6. $\mathbf{r}(t) = (\ln t)\mathbf{i} + t\mathbf{j}$; $t = e$.

7. $\mathbf{r}(t) = (4 \cos t)\mathbf{i} + (9 \sin t)\mathbf{j}$; $t = \pi/4$.

8. $\mathbf{r}(t) = (\ln \sin t)\mathbf{i} + (\ln \cos t)\mathbf{j}$; $t = \pi/6$.

In Exercises 9–16, find the unit tangent vector $\mathbf{T}$ and the unit normal vector $\mathbf{N}$ for the given value of t.

9. $\mathbf{r}(t) = 4 \cos t\mathbf{i} + 4 \sin t\mathbf{j} + t\mathbf{k}$; $t = \pi/2$.

10. $x = e^t$, $y = e^{-t}$, $z = t$; $t = 0$.

11. $\mathbf{r}(t) = t\mathbf{i} + \frac{1}{2}t^2\mathbf{j} + \frac{1}{3}t^3\mathbf{k}$; $t = 0$.

12. $x = \sin t$, $y = \cos t$, $z = \frac{1}{2}t^2$; $t = 0$.

13. $\mathbf{r}(t) = 3 \cos t\mathbf{i} + 4 \sin t\mathbf{j} + t\mathbf{k}$; $t = \pi/2$.

14. $x = e^t \cos t$, $y = e^t \sin t$, $z = e^t$; $t = 0$.

15. $\mathbf{r}(t) = \mathbf{i} + t\mathbf{j} + t^2\mathbf{k}$; $t = 1$.

16. $x = \cosh t$, $y = \sinh t$, $z = t$; $t = \ln 2$.

17. Show that the vector $\mathbf{N}(t) = (-\cos t)\mathbf{i} - (\sin t)\mathbf{j}$ found in Example 2 points directly toward the z-axis.

18. From Formula (2), the vector $\mathbf{N}$ points in the direction of

$$\mathbf{T}' = \frac{d}{dt}\left[\mathbf{r}'/\|\mathbf{r}'\|\right]$$

Replace $\mathbf{r}$ by $\mathbf{r}'$ in the formula of Exercise 53, Section 15.2, to show that $\mathbf{N}$ points in the direction of the vector $\mathbf{u} = \|\mathbf{r}'\|^2\mathbf{r}'' - (\mathbf{r}' \cdot \mathbf{r}'')\mathbf{r}'$ and hence that $\mathbf{N}$ can also be obtained from the formula $\mathbf{N} = \mathbf{u}/\|\mathbf{u}\|$.

In Exercises 19–26, use the formula in Exercise 18 to find **N** for the given value of t.

19. Exercise 1.   **20.** Exercise 2.

21. Exercise 3.   **22.** Exercise 4.

23. Exercise 9.   **24.** Exercise 10.

25. Exercise 11.   **26.** Exercise 12.

In Exercises 27 and 28, find $\mathbf{B} = \mathbf{T} \times \mathbf{N}$.

27. $\mathbf{r}(t) = (3 \sin t)\mathbf{i} + (3 \cos t)\mathbf{j} + 4t\mathbf{k}$.

28. $\mathbf{r}(t) = (a \cos t)\mathbf{i} + (a \sin t)\mathbf{j} + bt\mathbf{k}$ $(a \neq 0, b \neq 0)$.

29. Let S be the plane with the equation

$$a(x - x_0) + b(y - y_0) + c(z - z_0) = 0$$

and let $\mathbf{n} = a\mathbf{i} + b\mathbf{j} + c\mathbf{k}$ and $\mathbf{r}_0 = x_0\mathbf{i} + y_0\mathbf{j} + z_0\mathbf{k}$.

(a) Show that a curve with the vector equation $\mathbf{r} = \mathbf{r}(t)$ lies in the plane S if and only if $\mathbf{n} \cdot (\mathbf{r}(t) - \mathbf{r}_0) = 0$ for all t.

(b) Show that if the curve with the vector equation $\mathbf{r} = \mathbf{r}(t)$ lies in the plane S, then at each point where **T** and **N** are defined, these vectors also lie in S. [*Hint:* Use the result in part (a) to show that $\mathbf{r}'$ and $\mathbf{r}''$ are perpendicular to **n**, then use Formula (1) to show that $\mathbf{T} \cdot \mathbf{n} = 0$ and Exercise 18 to show that $\mathbf{N} \cdot \mathbf{n} = 0$.]

■ 15.5 CURVATURE

> *In this section we shall consider the problem of obtaining a numerical measure of how sharply a curve in 2-space or 3-space bends. Our results will have applications in geometry and in the study of motion along a curved path.*

□ **RELATIONSHIP BETWEEN DERIVATIVES OF RADIUS VECTORS AND ARC LENGTH**

Arc-length parameters play a special role in the study of curves in 2-space and 3-space because they are so closely related to the geometry of the curves they parametrize. We shall begin this section by establishing a relationship between the arc length of a curve and the derivative of the radius vector.

In the discussion to follow, we shall assume that $\mathbf{r}(t)$ is a smooth vector-valued function in 2-space or 3-space with graph C. As usual, we shall write

$$\mathbf{r}(t) = x(t)\mathbf{i} + y(t)\mathbf{j} \quad \text{or} \quad \mathbf{r}(t) = x(t)\mathbf{i} + y(t)\mathbf{j} + z(t)\mathbf{k}$$

depending on whether **r** is in 2-space or 3-space. Recall from Formulas (15a) and (15b) of Section 15.3 that if s is an arc-length parameter for C that increases with t, then

$$\frac{ds}{dt} = \sqrt{\left(\frac{dx}{dt}\right)^2 + \left(\frac{dy}{dt}\right)^2} \quad \text{or} \quad \frac{ds}{dt} = \sqrt{\left(\frac{dx}{dt}\right)^2 + \left(\frac{dy}{dt}\right)^2 + \left(\frac{dz}{dt}\right)^2} \quad \text{(1a–b)}$$

But,

$$\frac{d\mathbf{r}}{dt} = \left(\frac{dx}{dt}\right)\mathbf{i} + \left(\frac{dy}{dt}\right)\mathbf{j} \quad \text{or} \quad \frac{d\mathbf{r}}{dt} = \left(\frac{dx}{dt}\right)\mathbf{i} + \left(\frac{dy}{dt}\right)\mathbf{j} + \left(\frac{dz}{dt}\right)\mathbf{k} \quad \text{(2a–b)}$$

from which we obtain

$$\left\|\frac{d\mathbf{r}}{dt}\right\| = \sqrt{\left(\frac{dx}{dt}\right)^2 + \left(\frac{dy}{dt}\right)^2} \quad \text{or} \quad \left\|\frac{d\mathbf{r}}{dt}\right\| = \sqrt{\left(\frac{dx}{dt}\right)^2 + \left(\frac{dy}{dt}\right)^2 + \left(\frac{dz}{dt}\right)^2} \quad \text{(3a–b)}$$

Thus, it follows from (1a), (1b), (3a), and (3b) that in both 2-space and 3-space

$$\left\|\frac{d\mathbf{r}}{dt}\right\| = \frac{ds}{dt} \quad \text{or more compactly} \quad \|\mathbf{r}'(t)\| = \frac{ds}{dt} \quad \text{(4a–b)}$$

In Section 15.3 we obtained the arc-length formulas

$$L = \int_a^b \sqrt{\left(\frac{dx}{dt}\right)^2 + \left(\frac{dy}{dt}\right)^2}\, dt \quad \text{and} \quad L = \int_a^b \sqrt{\left(\frac{dx}{dt}\right)^2 + \left(\frac{dy}{dt}\right)^2 + \left(\frac{dz}{dt}\right)^2}\, dt$$

in 2-space and 3-space, respectively. It follows from (3a) and (3b) that these can be written as

$$L = \int_a^b \left\|\frac{d\mathbf{r}}{dt}\right\|\, dt \quad \text{or from (4)} \quad L = \int_a^b \left(\frac{ds}{dt}\right)\, dt \quad \text{(5a–b)}$$

These formulas are particularly nice because they are applicable in both 2-space and 3-space.

☐ **T EXPRESSED IN TERMS OF AN ARC-LENGTH PARAMETER**

Recall from Formula (1) in Section 15.4 that the unit tangent vector to the graph of $\mathbf{r}(t)$ is given by

$$\mathbf{T} = \frac{1}{\|\mathbf{r}'(t)\|}\frac{d\mathbf{r}}{dt} \tag{6}$$

In the case where the parameter is arc length, this formula simplifies to

$$\mathbf{T} = \frac{d\mathbf{r}}{ds} \tag{7}$$

This result can be derived from (6) using (4) and the chain rule (Theorem 15.2.7). We obtain

$$\mathbf{T} = \frac{1}{\|\mathbf{r}'(t)\|}\frac{d\mathbf{r}}{dt} = \frac{1}{ds/dt}\frac{d\mathbf{r}}{dt} = \frac{d\mathbf{r}/dt}{ds/dt} = \frac{d\mathbf{r}}{ds}$$

Example 1 In Example 5 of Section 15.3 we obtained the parametric equations

$$x = a\cos(s/a), \quad y = a\sin(s/a) \quad (0 \le s \le 2\pi a)$$

for a circle of radius a centered at the origin. In vector notation, this circle is the graph of

$$\mathbf{r} = a\cos(s/a)\mathbf{i} + a\sin(s/a)\mathbf{j}, \quad 0 \le s \le 2\pi a$$

so that for each value of s, the unit tangent vector to the circle is

$$\mathbf{T} = \frac{d\mathbf{r}}{ds} = -\sin(s/a)\mathbf{i} + \cos(s/a)\mathbf{j} \tag{8}$$

◀

☐ **CURVATURE**

Suppose that s is an arc-length parameter for a smooth curve C in 2-space or 3-space. Figure 15.5.1 suggests that for a curve in 2-space the "sharpness" of the bend in C is closely related to $d\mathbf{T}/ds$, which is the rate of change of the unit tangent vector $\mathbf{T}$ with respect to s. If C is a straight line (no bend), then the direction of $\mathbf{T}$ remains constant (Figure 15.5.1a); if C bends slightly, then $\mathbf{T}$ undergoes a gradual change of direction (Figure 15.5.1b); and if C bends sharply, then $\mathbf{T}$ undergoes a rapid change of direction (Figure 15.5.1c).

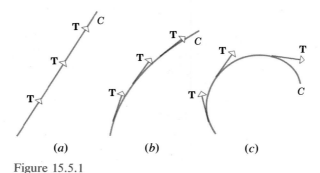

(a)　　　　(b)　　　　(c)

Figure 15.5.1

In 3-space, bends in a curve are not limited to a single plane—they can occur in all possible directions. In 3-space, $\mathbf{T}$ is perpendicular to the normal plane (Figure 15.4.8), so $d\mathbf{T}/ds$ relates to the rate at which the normal plane turns as s increases. Similarly, $d\mathbf{N}/ds$ and $d\mathbf{B}/ds$ relate to the rates at which the rectifying and osculating planes turn as s increases. In this text we shall focus primarily on $d\mathbf{T}/ds$, which is the most important of these derivatives in applications. We make the following definition.

15.5.1 DEFINITION. If C is a smooth curve in 2-space or 3-space, and if s is an arc-length parameter for C, then the **curvature** of C, denoted by $\kappa = \kappa(s)$ (κ = Greek "kappa"), is defined by

$$\kappa = \left\| \frac{d\mathbf{T}}{ds} \right\| \tag{9}$$

Observe that in (9) it is the *length* of the vector $d\mathbf{T}/ds$ that is the measure of curvature; thus, κ is a nonnegative real number. In general, the curvature varies from point to point along a curve; however, the following example shows that for circles in 2-space curvature is constant, as one might expect.

Example 2 In Example 1 we showed that the unit tangent vector to the circle of radius a centered at the origin is given by (8). Differentiating this with respect to s, then substituting in (9), yields

$$\frac{d\mathbf{T}}{ds} = -\frac{1}{a}\cos(s/a)\mathbf{i} - \frac{1}{a}\sin(s/a)\mathbf{j}$$

$$\kappa = \left\|\frac{d\mathbf{T}}{ds}\right\| = \sqrt{\left[-\frac{1}{a}\cos(s/a)\right]^2 + \left[-\frac{1}{a}\sin(s/a)\right]^2} = \frac{1}{a}$$

so the circle has constant curvature $1/a$. ◄

☐ FORMULAS FOR CURVATURE

Formula (9) is not useful for computational purposes unless the curve happens to be expressed in terms of an arc-length parameter. The following theorem provides two formulas for curvature in terms of a general parameter t.

15.5.2 THEOREM. *If $\mathbf{r}(t)$ is a smooth vector-valued function in 2-space or 3-space, then for each value of t at which $\mathbf{T}'(t)$ and $\mathbf{r}''(t)$ exist, the curvature κ can be expressed as*

(a) $$\kappa = \kappa(t) = \frac{\|\mathbf{T}'(t)\|}{\|\mathbf{r}'(t)\|} \qquad (10)$$

(b) $$\kappa = \kappa(t) = \frac{\|\mathbf{r}'(t) \times \mathbf{r}''(t)\|}{\|\mathbf{r}'(t)\|^3} \qquad (11)$$

Proof (a). It follows from (9), (4), and Theorem 15.2.7 (the chain rule) that

$$\kappa = \left\|\frac{d\mathbf{T}}{ds}\right\| = \left\|\frac{d\mathbf{T}/dt}{ds/dt}\right\| = \left\|\frac{d\mathbf{T}/dt}{\|d\mathbf{r}/dt\|}\right\| = \frac{\|\mathbf{T}'(t)\|}{\|\mathbf{r}'(t)\|}$$

Proof (b). It follows from Formula (1) of Section 15.4 that

$$\mathbf{r}'(t) = \|\mathbf{r}'(t)\| \, \mathbf{T}(t) \qquad (12)$$

so

$$\mathbf{r}''(t) = \|\mathbf{r}'(t)\|' \, \mathbf{T}(t) + \|\mathbf{r}'(t)\| \, \mathbf{T}'(t) \qquad (13)$$

But, from Formula (2) of Section 15.4 and part (a) of this theorem we have

$$\mathbf{T}'(t) = \|\mathbf{T}'(t)\| \, \mathbf{N}(t) \quad \text{and} \quad \|\mathbf{T}'(t)\| = \kappa \, \|\mathbf{r}'(t)\|$$

so

$$\mathbf{T}'(t) = \kappa \, \|\mathbf{r}'(t)\| \, \mathbf{N}(t)$$

Substituting this into (13) yields

$$\mathbf{r}''(t) = \|\mathbf{r}'(t)\|' \, \mathbf{T}(t) + \kappa \, \|\mathbf{r}'(t)\|^2 \, \mathbf{N}(t) \qquad (14)$$

Thus, from (12) and (14)

$$\mathbf{r}'(t) \times \mathbf{r}''(t) = \|\mathbf{r}'(t)\| \|\mathbf{r}'(t)\|' \, (\mathbf{T}(t) \times \mathbf{T}(t)) + \kappa \|\mathbf{r}'(t)\|^3 \, (\mathbf{T}(t) \times \mathbf{N}(t))$$

But the cross product of a vector with itself is zero [Theorem 14.4.4(f)], so this equation simplifies to

$$\mathbf{r}'(t) \times \mathbf{r}''(t) = \kappa \|\mathbf{r}'(t)\|^3 \, (\mathbf{T}(t) \times \mathbf{N}(t))$$

It follows from this equation and the fact that $\mathbf{T}(t) \times \mathbf{N}(t)$ is a unit vector (why?) that

$$\|\mathbf{r}'(t) \times \mathbf{r}''(t)\| = \kappa \|\mathbf{r}'(t)\|^3$$

Formula (11) now follows.   ▌

REMARKS.   Formula (10) is useful if $\mathbf{T}(t)$ happens to be known or is easy to obtain; however, Formula (11) will usually be easier to apply, since it involves only $\mathbf{r}(t)$ and its derivatives. We also note that cross products were defined only for vectors in 3-space, so to use Formula (12) in 2-space one must first write the 2-space function $\mathbf{r}(t) = x(t)\mathbf{i} + y(t)\mathbf{j}$ as the 3-space function $\mathbf{r}(t) = x(t)\mathbf{i} + y(t)\mathbf{j} + 0\mathbf{k}$ with a zero $\mathbf{k}$ component.

Example 3   Find $\kappa(t)$ for the circular helix

$$x = a \cos t, \quad y = a \sin t, \quad z = ct$$

where $a > 0$.

Solution.   The radius vector for the helix is

$$\mathbf{r} = (a \cos t)\mathbf{i} + (a \sin t)\mathbf{j} + (ct)\mathbf{k}$$

Thus,

$$\mathbf{r}'(t) = (-a \sin t)\mathbf{i} + (a \cos t)\mathbf{j} + c\mathbf{k}$$
$$\mathbf{r}''(t) = (-a \cos t)\mathbf{i} + (-a \sin t)\mathbf{j}$$

and

$$\mathbf{r}'(t) \times \mathbf{r}''(t) = \begin{vmatrix} \mathbf{i} & \mathbf{j} & \mathbf{k} \\ -a \sin t & a \cos t & c \\ -a \cos t & -a \sin t & 0 \end{vmatrix}$$

$$= (ac \sin t)\mathbf{i} - (ac \cos t)\mathbf{j} + a^2\mathbf{k}$$

Therefore,

$$\|\mathbf{r}'(t)\| = \sqrt{(-a \sin t)^2 + (a \cos t)^2 + c^2} = \sqrt{a^2 + c^2}$$

and

$$\|\mathbf{r}'(t) \times \mathbf{r}''(t)\| = \sqrt{(ac \sin t)^2 + (-ac \cos t)^2 + a^4}$$
$$= \sqrt{a^2c^2 + a^4} = a\sqrt{a^2 + c^2}$$

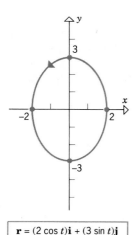

$$\mathbf{r} = (2 \cos t)\mathbf{i} + (3 \sin t)\mathbf{j}$$

Figure 15.5.2

so

$$\kappa(t) = \frac{\|\mathbf{r}'(t) \times \mathbf{r}''(t)\|}{\|\mathbf{r}'(t)\|^3} = \frac{a\sqrt{a^2 + c^2}}{(\sqrt{a^2 + c^2})^3} = \frac{a}{a^2 + c^2}$$

Note that κ does not depend on t, which means that the helix has constant curvature. ◄

Example 4 The graph of the vector equation

$$\mathbf{r} = (2 \cos t)\mathbf{i} + (3 \sin t)\mathbf{j}, \quad 0 \le t \le 2\pi$$

is the ellipse in Figure 15.5.2 (see Example 5 of Section 13.4). Find the curvature of the ellipse at the endpoints of the major and minor axes.

Solution. We have

$$\mathbf{r}'(t) = (-2 \sin t)\mathbf{i} + (3 \cos t)\mathbf{j} \quad \text{and} \quad \mathbf{r}''(t) = (-2 \cos t)\mathbf{i} + (-3 \sin t)\mathbf{j}$$

so

$$\mathbf{r}'(t) \times \mathbf{r}''(t) = \begin{vmatrix} \mathbf{i} & \mathbf{j} & \mathbf{k} \\ -2 \sin t & 3 \cos t & 0 \\ -2 \cos t & -3 \sin t & 0 \end{vmatrix} = [(6 \sin^2 t) + (6 \cos^2 t)]\mathbf{k} = 6\mathbf{k}$$

Therefore,

$$\|\mathbf{r}'(t)\| = \sqrt{(-2 \sin t)^2 + (3 \cos t)^2} = \sqrt{4 \sin^2 t + 9 \cos^2 t}$$

and

$$\|\mathbf{r}'(t) \times \mathbf{r}''(t)\| = 6$$

so

$$\kappa = \kappa(t) = \frac{\|\mathbf{r}'(t) \times \mathbf{r}''(t)\|}{\|\mathbf{r}'(t)\|^3} = \frac{6}{[4 \sin^2 t + 9 \cos^2 t]^{3/2}} \tag{15}$$

The endpoints of the minor axis are $(2, 0)$ and $(-2, 0)$, which correspond to $t = 0$ and $t = \pi$, respectively. Substituting these values in (15) yields the same curvature at both points, namely

$$\kappa = \kappa(0) = \kappa(\pi) = \frac{6}{9^{3/2}} = \frac{6}{27} = \frac{2}{9}$$

The endpoints of the major axis are $(0, 3)$ and $(0, -3)$, which correspond to $t = \pi/2$ and $t = 3\pi/2$, respectively; from (15) the curvature at these points is

$$\kappa = \kappa\left(\frac{\pi}{2}\right) = \kappa\left(\frac{3\pi}{2}\right) = \frac{6}{4^{3/2}} = \frac{3}{4}$$

Thus, the curvature is greater at the ends of the major axis than at the ends of the minor axis, as one would expect. ◄

☐ **RADIUS OF CURVATURE**

In the foregoing example, we found the curvature at the ends of the minor axis to be 2/9 and the curvature at the ends of the major axis to be 3/4. To obtain a better understanding of the meaning of these numbers, recall from Example 2 that a circle of radius a has a constant curvature of $1/a$; thus, the curvature of the ellipse at the ends of the minor axis is the same as that of a circle of radius 9/2, and the curvature at the ends of the major axis is the same as that of a circle of radius 4/3 (Figure 15.5.3).

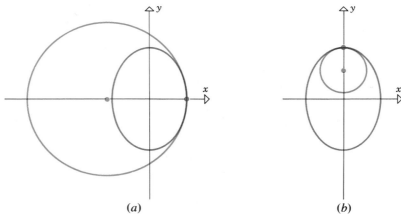

(a) (b)

Figure 15.5.3

In general, if a curve C in 2-space has nonzero curvature κ at a point P, then the circle of radius

$$\rho = \frac{1}{\kappa}$$

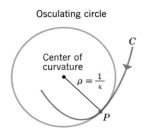

Osculating circle

Center of curvature
$\rho = \frac{1}{\kappa}$

C

P

Figure 15.5.4

sharing a common tangent with C at P, and centered on the concave side of the curve at P, is called the **circle of curvature** or **osculating circle** at P (Figure 15.5.4). The osculating circle and the curve C not only touch at P but they have equal curvatures at that point. In this sense, the osculating circle is the circle that best approximates the curve C near P. The radius ρ of the osculating circle at P is called the **radius of curvature** at P, and the center of the circle is called the **center of curvature** at P (Figure 15.5.4).

☐ **THE DERIVATIVE OF T WITH RESPECT TO ARC LENGTH**

The following formula, which will be used in the next section, follows from Formula (2) of Section 15.4 (with s as the parameter) and Formula (9) of this section (verify).

$$\frac{d\mathbf{T}}{ds} = \kappa\mathbf{N} \tag{16}$$

☐ **INTERPRETATION OF CURVATURE IN 2-SPACE**

A simple geometric interpretation of curvature in 2-space can be obtained by considering the angle ϕ measured counterclockwise from the direction of the positive x-axis to $\mathbf{T}$ (Figure 15.4.6). As noted in Formula (3) of Section 15.4,

we can write

$$\mathbf{T} = (\cos \phi)\mathbf{i} + (\sin \phi)\mathbf{j}$$

from which we obtain

$$\frac{d\mathbf{T}}{d\phi} = (-\sin \phi)\mathbf{i} + (\cos \phi)\mathbf{j}$$

By the chain rule

$$\frac{d\mathbf{T}}{ds} = \frac{d\mathbf{T}}{d\phi}\frac{d\phi}{ds}$$

so

$$\left\|\frac{d\mathbf{T}}{ds}\right\| = \left|\frac{d\phi}{ds}\right|\left\|\frac{d\mathbf{T}}{d\phi}\right\| = \left|\frac{d\phi}{ds}\right|\sqrt{(-\sin \phi)^2 + \cos^2 \phi} = \left|\frac{d\phi}{ds}\right|$$

It now follows from (9) that

$$\kappa = \left|\frac{d\phi}{ds}\right| \tag{17}$$

Thus, in 2-space curvature can be interpreted as the magnitude of the rate of change of ϕ with respect to s (Figure 15.5.5).

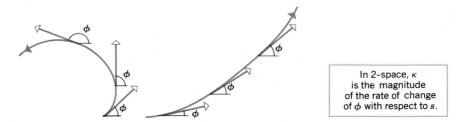

In 2-space, κ is the magnitude of the rate of change of ϕ with respect to s.

Figure 15.5.5

Formula (17) makes it evident geometrically that a line in 2-space has zero curvature at every point because the angle ϕ is constant (Figure 15.5.6) and consequently $\kappa = |d\phi/ds| = 0$.

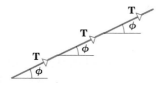

Figure 15.5.6

▶ Exercise Set 15.5   C 57, 58

In Exercises 1–14, use Formula (11) to find the curvature at the indicated point.

1. $r(t) = t^2\mathbf{i} + t^3\mathbf{j}; t = \frac{1}{2}.$

2. $r(t) = (4 \cos t)\mathbf{i} + (\sin t)\mathbf{j}; t = \pi/2.$

3. $r(t) = e^{3t}\mathbf{i} + e^{-t}\mathbf{j}; t = 0.$

4. $x = 1 - t^3, y = t - t^2; t = 1.$

5. $x = t \cos t, y = t \sin t; t = 1.$

6. $x = 2abt, y = b^2t^2; t = 1$ $(a > 0, b > 0).$

7. $r(t) = 4 \cos t\mathbf{i} + 4 \sin t\mathbf{j} + t\mathbf{k}; t = \pi/2.$

8. $x = e^t, y = e^{-t}, z = t; t = 0.$

9. $r(t) = t\mathbf{i} + \frac{1}{2}t^2\mathbf{j} + \frac{1}{3}t^3\mathbf{k}; t = 0.$

10. $x = \sin t, y = \cos t, z = \frac{1}{2}t^2; t = 0.$

11. $r(t) = 3 \cos t\mathbf{i} + 4 \sin t\mathbf{j} + t\mathbf{k}; t = \pi/2.$

12. $x = e^t \cos t, y = e^t \sin t, z = e^t; t = 0.$

13. $r(t) = \mathbf{i} + t\mathbf{j} + t^2\mathbf{k}; t = 1.$

14. $x = \cosh t, y = \sinh t, z = t; t = \ln 2.$

15. Use Formula (11) to show that for a plane curve described by $y = f(x)$ the curvature $\kappa(x)$ is

$$\kappa(x) = \frac{|d^2y/dx^2|}{[1 + (dy/dx)^2]^{3/2}}$$

[*Hint:* Let x be the parameter so that $r(x) = x\mathbf{i} + y\mathbf{j} = x\mathbf{i} + f(x)\mathbf{j}.$]

16. Show that the formula for κ in Exercise 15 can also be written as $\kappa = |y'' \cos^3 \phi|$, where ϕ is the angle of inclination of the tangent line to the graph of $y = f(x).$

In Exercises 17–22, use the formula of Exercise 15 to find the curvature at the indicated point.

17. $y = \sin x; x = \pi/2.$ 18. $y = x^3/3; x = 0.$

19. $y = 1/x; x = 1.$ 20. $y = e^{-x}; x = 1.$

21. $y = \tan x; x = \pi/4.$ 22. $y^2 - 4x^2 = 9; (2, 5).$

23. Use the formula of Exercise 15 to find the curvature of $y = \ln \cos x$ for $-\pi/2 < x < \pi/2.$ For what value of x is the curvature maximum?

24. Use Formula (11) to show that for a plane curve described by $r = x(t)\mathbf{i} + y(t)\mathbf{j}$ the curvature is

$$\kappa = \frac{|x'y'' - y'x''|}{(x'^2 + y'^2)^{3/2}}$$

where a prime denotes differentiation with respect to $t.$

In Exercises 25–30, use the formula of Exercise 24 to find the curvature in the indicated problem.

25. Exercise 1.   26. Exercise 2.

27. Exercise 3.   28. Exercise 4.

29. Exercise 5.   30. Exercise 6.

31. Use the formula of Exercise 24 to find the curvature at $t = 0$ and $t = \pi/2$ of the curve $x = a \cos t,$ $y = b \sin t$ $(a > 0, b > 0).$

32. Use Formula (11) to show that for a curve in polar coordinates described by $r = f(\theta)$ the curvature is

$$\kappa = \frac{\left| r^2 + 2\left(\dfrac{dr}{d\theta}\right)^2 - r\dfrac{d^2r}{d\theta^2} \right|}{\left[r^2 + \left(\dfrac{dr}{d\theta}\right)^2 \right]^{3/2}}$$

[*Hint:* Let θ be the parameter and use the relationships $x = r \cos \theta, y = r \sin \theta.$]

In Exercises 33–36, use the formula of Exercise 32 to find the curvature at the indicated point.

33. $r = 2 \sin \theta; \theta = \pi/6.$

34. $r = \theta; \theta = 1.$

35. $r = a(1 + \cos \theta); \theta = \pi/2.$

36. $r = e^{2\theta}; \theta = 1.$

37. Find $\kappa(t)$ for the cycloid

$$x = a(t - \sin t), y = a(1 - \cos t)$$

for $0 < t < \pi$ and sketch the graph of $\kappa(t).$

38. Find $\kappa(t)$ for the curve $x = e^{-t} \cos t, y = e^{-t} \sin t$ and sketch the graph of $\kappa(t).$

In Exercises 39–45, sketch the curve, calculate the radius of curvature at the indicated point, and sketch the osculating circle.

39. $y = \sin x; x = \pi/2.$ 40. $x = t, y = t^2; t = 1.$

41. $y = \frac{1}{2}x^2; x = -1.$ 42. $x = t, y = \sqrt{t}; t = 2.$

43. $y = \ln x; x = 1.$

44. $x = 2t, y = 4/t; t = 1.$

45. $x = t - \sin t$, $y = 1 - \cos t$; $t = \pi$.

46. Find the radius of curvature of $y = \cos x$ at $x = 0$ and $x = \pi$. Sketch the osculating circles at those points.

47. Find the radius of curvature of the ellipse $x = 2 \cos t$, $y = \sin t$, $0 \le t \le 2\pi$ at $t = 0$ and $t = \pi/2$. Sketch the osculating circles at those points.

48. Consider the curve $y = x^4 - 2x^2$.
(a) Find the radius of curvature at each relative extremum.
(b) Sketch the curve and show the osculating circles at the relative extrema.

49. Find the radius of curvature of the parabola $y^2 = 4px$ at $(0, 0)$.

50. At what point(s) does $y = e^x$ have maximum curvature?

51. At what point(s) does $4x^2 + 9y^2 = 36$ have minimum radius of curvature?

52. Find the value of x, $x > 0$, where $y = x^3$ has maximum curvature.

53. Find the maximum and minimum values of the radius of curvature for the curve $x = \cos t$, $y = \sin t$, $z = \cos t$.

54. Find the minimum value of the radius of curvature for the curve $x = e^t$, $y = e^{-t}$, $z = \sqrt{2}t$.

55. Show that the curvature of the polar curve $r = e^{a\theta}$ is inversely proportional to r.

56. Show that the curvature of the polar curve $r^2 = a^2 \cos 2\theta$ is directly proportional to r ($r > 0$).

57. Given that $d\mathbf{T}/ds = 0.03\mathbf{i} - 0.04\mathbf{j}$ at a certain point P on a curve, find the value of $|d\phi/ds|$, where ϕ is the angle of inclination of the tangent line to the curve at P. Assuming that s is measured in centimeters, express your answer in units of °/cm.

58. Assuming that arc length is measured in inches along the curve $y = x^3$, find the magnitude of the rate of change with respect to arc length of the angle of inclination of the tangent line to the curve in units of °/in. at the point where $x = 1$.

In Exercises 59–62, we shall say that a **smooth transition** occurs at a point P on a curve if the curvature κ is continuous at P. Use the formula for κ given in Exercise 15 to help solve the problem.

59. Show that the transition at $x = 0$ from the horizontal line $y = 0$ for $x \le 0$ to the parabola $y = x^2$ for $x > 0$ is not smooth, whereas the transition to $y = x^3$ for $x > 0$ is smooth.

60. (a) Sketch the graph of the curve defined piecewise by $y = x^2$ for $x < 0$, $y = x^4$ for $x \ge 0$.
(b) Show that for the curve in part (a) the transition at $x = 0$ is not smooth.

61. The figure below shows the arc of a circle of radius r with center at $(0, r)$. Find the value of a so that there is a smooth transition from the circle to the parabola $y = ax^2$ at the point where $x = 0$.

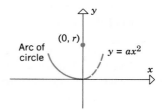

62. Find a, b, and c so that there is a smooth transition at $x = 0$ from the curve $y = e^x$ for $x \le 0$ to the parabola $y = ax^2 + bx + c$ for $x > 0$. [*Hint:* The curvature is continuous at those points where y'' is continuous.]

63. Let $\mathbf{r}(t) = (\ln t)\mathbf{i} + 2t\mathbf{j} + t^2\mathbf{k}$. Find
(a) $\|\mathbf{r}'(t)\|$ (b) $\dfrac{ds}{dt}$ (c) $\displaystyle\int_1^3 \|\mathbf{r}'(t)\|\, dt$.

64. Let $x = \cos t$, $y = \sin t$, $z = t^{3/2}$. Find
(a) $\|\mathbf{r}'(t)\|$ (b) $\dfrac{ds}{dt}$ (c) $\displaystyle\int_0^2 \|\mathbf{r}'(t)\|\, dt$.

65. Let C be the curve given parametrically in terms of arc length by
$$x = \tfrac{3}{5}s + 1, \quad y = \tfrac{4}{5}s - 2 \qquad (0 \le s \le 10)$$
(a) Find $\mathbf{T} = \mathbf{T}(s)$.
(b) Sketch the curve and the vector $\mathbf{T}(5)$.

66. Let C be the curve given parametrically in terms of arc length by
$$x = 2\cos\left(\frac{s}{2}\right), \quad y = 2\sin\left(\frac{s}{2}\right) \qquad (0 \le s \le 4\pi)$$
(a) Find $\mathbf{T} = \mathbf{T}(s)$ and $\mathbf{N} = \mathbf{N}(s)$.
(b) Sketch the curve and the vectors $\mathbf{T}(\pi/2)$ and $\mathbf{N}(\pi/2)$.

67. Let C be a smooth curve in the xy-plane that is tangent to the x-axis at the origin, and let s be an arc-length parameter for C with its reference point at the origin ($x = 0$ and $y = 0$ if $s = 0$). Suppose that the curvature κ is proportional to s and that s is chosen so that the constant of proportionality is positive ($\kappa = as$ where $a > 0$).

(a) Assuming that $d\phi/ds \geq 0$, use Formula (17) to show that $\phi = \frac{1}{2}as^2$.

(b) Use the result in part (a) and the fact that $\mathbf{T} = (\cos\phi)\mathbf{i} + (\sin\phi)\mathbf{j}$ to show that

$$x = \int_0^s \cos\left(\tfrac{1}{2}au^2\right) du, \quad y = \int_0^s \sin\left(\tfrac{1}{2}au^2\right) du$$

are parametric equations of the curve.

In Exercises 68–70, we assume that s is an arc-length parameter for a smooth curve C in 3-space and that $d\mathbf{T}/ds$ and $d\mathbf{N}/ds$ exist at each point on the curve. (This implies that $d\mathbf{B}/ds$ exists as well, since $\mathbf{B} = \mathbf{T} \times \mathbf{N}$.)

---

68. (a) Show that $d\mathbf{B}/ds$ is perpendicular to $\mathbf{B}$. [*Hint:* See Theorem 15.2.6.]

(b) Show that $d\mathbf{B}/ds$ is perpendicular to $\mathbf{T}$. [*Hint:* Use the fact that $\mathbf{B}$ is perpendicular to both $\mathbf{T}$ and $\mathbf{N}$, and differentiate $\mathbf{B} \cdot \mathbf{T}$ with respect to s.]

(c) Use the results in parts (a) and (b) to show that $d\mathbf{B}/ds$ is a scalar multiple of $\mathbf{N}$.

(d) It follows from part (c) that there is a real-valued function $\tau = \tau(s)$ (τ = Greek "tau"), called the *torsion,* such that $d\mathbf{B}/ds = -\tau\mathbf{N}$. Use the result in Exercise 29(b), Section 15.4, to show that if C lies entirely in a plane, then $\tau(s) = 0$ for all values of s. [*Note:* $d\mathbf{B}/ds$ is related to the rate at which the osculating plane turns (Figure 15.4.8), so the torsion is sometimes viewed as a measure of the tendency for C to twist out of the osculating plane.]

69. Let κ be the curvature of C and τ the torsion (defined in Exercise 68). By differentiating $\mathbf{N} = \mathbf{B} \times \mathbf{T}$ with respect to s, show that $d\mathbf{N}/ds = -\kappa\mathbf{T} + \tau\mathbf{B}$.

70. The following derivatives, known as the **Frenet-Serret formulas,** are fundamental in the theory of curves in 3-space:

$$d\mathbf{T}/ds = \kappa\mathbf{N} \qquad \text{[Formula (16)]}$$
$$d\mathbf{N}/ds = -\kappa\mathbf{T} + \tau\mathbf{B} \qquad \text{[Exercise 69]}$$
$$d\mathbf{B}/ds = -\tau\mathbf{N} \qquad \text{[Exercise 68(d)]}$$

Use the first two Frenet-Serret formulas and the fact that $\mathbf{r}'(s) = \mathbf{T}$ if $\mathbf{r} = \mathbf{r}(s)$ to show that

(a) $\kappa = \|\mathbf{r}''(s)\|$ and $\mathbf{N} = \mathbf{r}''(s)/\|\mathbf{r}''(s)\|$

(b) $\tau = \dfrac{[\mathbf{r}'(s) \times \mathbf{r}''(s)] \cdot \mathbf{r}'''(s)}{\|\mathbf{r}''(s)\|^2}$ and

$\mathbf{B} = \dfrac{\mathbf{r}'(s) \times \mathbf{r}''(s)}{\|\mathbf{r}''(s)\|}$.

71. Use the results in Exercise 70(b) and the results in Exercise 39 of Section 15.3 to show that for the circular helix

$$\mathbf{r} = (a\cos t)\mathbf{i} + (a\sin t)\mathbf{j} + ct\mathbf{k}$$

with $a > 0$ the torsion and the binormal vector are

$$\tau = \frac{c}{w^2} \quad \text{and}$$

$$\mathbf{B} = \left(\frac{c}{w}\sin\frac{s}{w}\right)\mathbf{i} - \left(\frac{c}{w}\cos\frac{s}{w}\right)\mathbf{j} + \left(\frac{a}{w}\right)\mathbf{k}$$

where $w = \sqrt{a^2 + c^2}$ and s has its reference point at $(a, 0, 0)$.

72. Suppose that the arc-length parameter in the Frenet-Serret formulas of Exercise 70 is a function $s = s(t)$ of a general parameter t and that primes are used to denote derivatives with respect to t.

(a) Use the first two Frenet-Serret formulas to show that $\mathbf{T}' = \kappa s'\mathbf{N}$ and $\mathbf{N}' = -\kappa s'\mathbf{T} + \tau s'\mathbf{B}$. [*Hint:* Use the chain rule.]

(b) Show that Formulas (12) and (14) can be written in the form

$$\mathbf{r}'(t) = s'\mathbf{T} \quad \text{and} \quad \mathbf{r}''(t) = s''\mathbf{T} + \kappa(s')^2\mathbf{N}$$

(c) Use the results in parts (a) and (b) to show that

$$\mathbf{r}'''(t) = [s''' - \kappa^2(s')^3]\mathbf{T} + [3\kappa s's'' + \kappa'(s')^2]\mathbf{N} + \kappa\tau(s')^3\mathbf{B}$$

(d) Use the results in parts (b) and (c) to show that

$$\tau = \frac{[\mathbf{r}'(t) \times \mathbf{r}''(t)] \cdot \mathbf{r}'''(t)}{\|\mathbf{r}'(t) \times \mathbf{r}''(t)\|^2}$$

In Exercises 73–76, use the formula in Exercise 72(d) to find the torsion $\tau = \tau(t)$.

---

73. The twisted cubic $\mathbf{r}(t) = 2t\mathbf{i} + t^2\mathbf{j} + \frac{1}{3}t^3\mathbf{k}$.

74. The circular helix $\mathbf{r}(t) = (a\cos t)\mathbf{i} + (a\sin t)\mathbf{j} + bt\mathbf{k}$.

75. $\mathbf{r}(t) = e^t\mathbf{i} + e^{-t}\mathbf{j} + \sqrt{2}t\mathbf{k}$.

76. $\mathbf{r}(t) = (t - \sin t)\mathbf{i} + (1 - \cos t)\mathbf{j} + t\mathbf{k}$.

■ 15.6 MOTION ALONG A CURVE

In Section 4.11 we considered the motion of a particle moving along a coordinate line. For such particles, there are only two possible directions of motion—the positive direction or the negative direction. For particles moving in 2-space or 3-space the situation is much more complicated, since there are infinitely many directions of motion possible. In this section we shall show how vectors can be used to study the motion of a particle moving along a curve in 2-space or 3-space.

□ VELOCITY, ACCELERATION, AND SPEED

Suppose that a particle is moving through 2-space or 3-space along a smooth curve C. At each instant of time we will regard the direction of motion to be the direction of the unit tangent vector, and we will regard the speed of the particle to be the instantaneous rate of change with respect to time of the arc length traveled by the particle, as measured from some arbitrary reference point (Figure 15.6.1a). The speed and the direction of motion together determine what is called the "velocity vector" of the particle. Mathematically, this vector can be written in the form

$$\mathbf{v}(t) = \frac{ds}{dt}\,\mathbf{T}(t) \tag{1}$$

which is a vector of length ds/dt with the same direction as $\mathbf{T}(t)$ (Figure 15.6.1b).

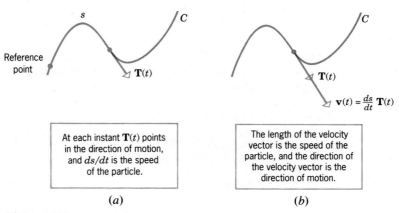

At each instant $\mathbf{T}(t)$ points in the direction of motion, and ds/dt is the speed of the particle.

(a)

The length of the velocity vector is the speed of the particle, and the direction of the velocity vector is the direction of motion.

(b)

Figure 15.6.1

Recall that for a particle moving on a coordinate line, the velocity function is the derivative of the position function. The same is true for a particle moving along a smooth curve in 2-space or 3-space. To see that this is so, suppose that the position function of the particle is $\mathbf{r}(t)$. From the definition of the unit tangent vector [Formula (1) of Section 15.4] and Formula (4) of Section 15.5 we have

$$\mathbf{T}(t) = \frac{\mathbf{r}'(t)}{\|\mathbf{r}'(t)\|} \quad \text{and} \quad \frac{ds}{dt} = \|\mathbf{r}'(t)\|$$

Substituting these expressions in (1) and simplifying yields

$$\mathbf{v}(t) = \mathbf{r}'(t)$$

which shows that the velocity function is the derivative of the position function. As was the case for motion along a coordinate line, the *acceleration function* of a particle moving along a curve in 2-space or 3-space is defined to be the derivative of the velocity function. Thus, we have the following definitions.

15.6.1   DEFINITION.   If a particle moves along a curve C in 2-space or 3-space so that its position vector at time t is $\mathbf{r}(t)$, then the *instantaneous velocity, instantaneous acceleration,* and *instantaneous speed* of the particle at time t are defined by

$$\text{velocity} = \mathbf{v}(t) = \frac{d\mathbf{r}}{dt} \tag{2}$$

$$\text{acceleration} = \mathbf{a}(t) = \frac{d\mathbf{v}}{dt} = \frac{d^2\mathbf{r}}{dt^2} \tag{3}$$

$$\text{speed} = \|\mathbf{v}(t)\| = \frac{ds}{dt} \tag{4}$$

---

where s is the total arc length traveled by the particle at time t, measured from some arbitrary reference point on C.

Example 1   A particle moves along a circular path in such a way that its x- and y-coordinates at time t are

$$x = 2 \cos t, \quad y = 2 \sin t$$

(a)   Find the instantaneous velocity and speed of the particle at time t.

(b)   Sketch the path of the particle, and show the position and velocity vectors at time $t = \pi/4$ with the velocity vector drawn so that its initial point is at the tip of the position vector.

(c)   Show that at each instant the acceleration vector is perpendicular to the velocity vector.

Solution (a).   At time t, the position vector is

$$\mathbf{r}(t) = 2\cos t\mathbf{i} + 2\sin t\mathbf{j}$$

so the instantaneous velocity and speed are

$$\mathbf{v}(t) = \frac{d\mathbf{r}}{dt} = -2\sin t\mathbf{i} + 2\cos t\mathbf{j}$$

$$\|\mathbf{v}(t)\| = \sqrt{(-2\sin t)^2 + (2\cos t)^2} = 2$$

Solution (b).   The graph of the parametric equations is a circle of radius 2 centered at the origin. At time $t = \pi/4$ the position and velocity vectors of the particles are

$$\mathbf{r}(\pi/4) = 2\cos(\pi/4)\mathbf{i} + 2\sin(\pi/4)\mathbf{j} = \sqrt{2}\mathbf{i} + \sqrt{2}\mathbf{j}$$
$$\mathbf{v}(\pi/4) = -2\sin(\pi/4)\mathbf{i} + 2\cos(\pi/4)\mathbf{j} = -\sqrt{2}\mathbf{i} + \sqrt{2}\mathbf{j}$$

These vectors and the circle are shown in Figure 15.6.2.

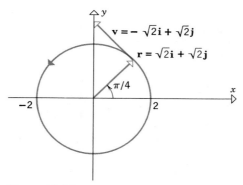

Figure 15.6.2

Solution (c).   At time t, the acceleration vector is

$$\mathbf{a}(t) = \frac{d\mathbf{v}}{dt} = -2\cos t\mathbf{i} - 2\sin t\mathbf{j}$$

One way of showing that $\mathbf{v}(t)$ and $\mathbf{a}(t)$ are perpendicular is to show that their dot product is zero (try it). However, it is easier to observe that $\mathbf{a}(t)$ is the negative of $\mathbf{r}(t)$, which implies that $\mathbf{v}(t)$ and $\mathbf{a}(t)$ are perpendicular, since at each point on a circle the radius and tangent line are perpendicular.   ◀

REMARKS.   In Example 1 we omitted all units for simplicity. We shall follow this practice throughout this section, except in some applied problems that appear later. Moreover, as in Example 1, we shall follow the standard convention of drawing velocity and acceleration vectors with their initial points at the tip of the position vector, rather than at the origin.

Since the velocity function $\mathbf{v}(t)$ of a particle moving in 2-space or 3-space can be obtained by differentiating the position function $\mathbf{r}(t)$, it follows that $\mathbf{r}(t)$ can be obtained by integrating $\mathbf{v}(t)$. Similarly, $\mathbf{v}(t)$ can be obtained by integrating $\mathbf{a}(t)$. However, integrating $\mathbf{v}(t)$ does not produce a unique position function, and integrating $\mathbf{a}(t)$ does not produce a unique velocity function, since constants of integration occur. To determine these constants, some additional information is needed. For example, if the position vector of the particle is known at some specific point in time, then the constant resulting from integrating $\mathbf{v}(t)$ can be determined from this information, and if the velocity vector of the particle is known at some point in time, then the constant resulting from integrating $\mathbf{a}(t)$ can be determined.

Example 2   A particle moves through 3-space in such a way that its velocity at time t is

$$\mathbf{v}(t) = \mathbf{i} + t\mathbf{j} + t^2\mathbf{k}$$

Find the coordinates of the particle at time $t = 1$ given that the particle is at the point $(-1, 2, 4)$ at time $t = 0$.

Solution.   Integrating the velocity function to obtain the position function yields

$$\mathbf{r}(t) = \int \mathbf{v}(t)\, dt = \int (\mathbf{i} + t\mathbf{j} + t^2\mathbf{k})\, dt = t\mathbf{i} + \frac{t^2}{2}\mathbf{j} + \frac{t^3}{3}\mathbf{k} + \mathbf{c} \qquad (5)$$

where $\mathbf{c}$ is a vector constant of integration. Since the coordinates of the particle at time $t = 0$ are $(-1, 2, 4)$, the position vector at time $t = 0$ is

$$\mathbf{r}(0) = -\mathbf{i} + 2\mathbf{j} + 4\mathbf{k} \qquad (6)$$

It follows on substituting $t = 0$ in (5) and equating the result with (6) that $\mathbf{c} = -\mathbf{i} + 2\mathbf{j} + 4\mathbf{k}$. Substituting this value of $\mathbf{c}$ in (5) and simplifying yields

$$\mathbf{r}(t) = (t - 1)\mathbf{i} + \left(\frac{t^2}{2} + 2\right)\mathbf{j} + \left(\frac{t^3}{3} + 4\right)\mathbf{k}$$

Thus, at time $t = 1$ the position vector of the particle is

$$\mathbf{r}(1) = 0\mathbf{i} + \frac{5}{2}\mathbf{j} + \frac{13}{3}\mathbf{k}$$

so its coordinates at that instant are $(0, \frac{5}{2}, \frac{13}{3})$.   ◀

◻ **DISPLACEMENT AND DISTANCE TRAVELED**

If a particle travels along a curve C in 2-space or 3-space, the **displacement** of the particle over the time interval $t_1 \leq t \leq t_2$ is commonly denoted by $\Delta\mathbf{r}$, and is defined as

$$\Delta\mathbf{r} = \mathbf{r}(t_2) - \mathbf{r}(t_1) \qquad (7)$$

(Figure 15.6.3). The displacement vector, which describes the change in position of the particle during the time interval, can be obtained by integrating

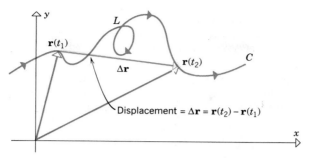

Figure 15.6.3

the velocity function from t_1 to t_2:

$$\Delta \mathbf{r} = \int_{t_1}^{t_2} \mathbf{v}(t) \, dt = \int_{t_1}^{t_2} \frac{d\mathbf{r}}{dt} \, dt = \mathbf{r}(t) \Big]_{t_1}^{t_2} = \mathbf{r}(t_2) - \mathbf{r}(t_1) \tag{8}$$

The distance L traveled by the particle along C during the time interval $t_1 \le t \le t_2$ can be obtained by integrating the magnitude of the velocity function from t_1 to t_2 [Formula (5a) of Section 15.5]:

$$L = \int_{t_1}^{t_2} \|\mathbf{v}(t)\| \, dt = \int_{t_1}^{t_2} \left\| \frac{d\mathbf{r}}{dt} \right\| \, dt \tag{9}$$

Example 3 Suppose that a particle travels along a circular helix in 3-space so that its position vector at time t is

$$\mathbf{r}(t) = (4 \cos \pi t)\mathbf{i} + (4 \sin \pi t)\mathbf{j} + t\mathbf{k}$$

Find the displacement and distance traveled by the particle during the time interval $1 \le t \le 5$.

Solution. We have

$$\mathbf{v}(t) = \frac{d\mathbf{r}}{dt} = (-4\pi \sin \pi t)\mathbf{i} + (4\pi \cos \pi t)\mathbf{j} + \mathbf{k}$$

$$\|\mathbf{v}(t)\| = \sqrt{(-4\pi \sin \pi t)^2 + (4\pi \cos \pi t)^2 + 1} = \sqrt{16\pi^2 + 1}$$

Since $\mathbf{r}(t)$ is known, the displacement can be found directly from (7); it is

$$\Delta \mathbf{r} = \mathbf{r}(5) - \mathbf{r}(1)$$
$$= [(4 \cos 5\pi)\mathbf{i} + (4 \sin 5\pi)\mathbf{j} + 5\mathbf{k}] - [(4 \cos \pi)\mathbf{i} + (4 \sin \pi)\mathbf{j} + \mathbf{k}]$$
$$= (-4\mathbf{i} + 5\mathbf{k}) - (-4\mathbf{i} + \mathbf{k}) = 4\mathbf{k}$$

which tells us that the change in the position of the particle over the time interval was 4 units straight up.

From (9), the distance L traveled by the particle during the time interval is

$$L = \int_{1}^{5} \sqrt{16\pi^2 + 1} \, dt = 4\sqrt{16\pi^2 + 1} \text{ units} \qquad \blacktriangleleft$$

In applications, it is often desirable to resolve the velocity and acceleration vectors into components that are parallel to the unit tangent and unit normal vectors. The following theorem explains how this can be done.

15.6.2   THEOREM.   *For a particle moving along a curve C in 2-space or 3-space, the velocity and acceleration vectors can be written as*

$$\mathbf{v} = \frac{ds}{dt}\,\mathbf{T} \tag{10}$$

$$\mathbf{a} = \frac{d^2s}{dt^2}\,\mathbf{T} + \kappa\left(\frac{ds}{dt}\right)^2\mathbf{N} \tag{11}$$

where s is an arc-length parameter for the curve, and $\mathbf{T} = \mathbf{T}(t)$, $\mathbf{N} = \mathbf{N}(t)$, *and* $\kappa = \kappa(t)$ *denote the unit tangent vector, unit normal vector, and curvature (Figure 15.6.4).*

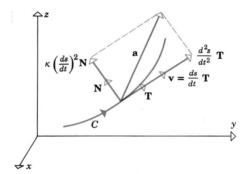

Figure 15.6.4

Proof.   Formula (10) is just a restatement of (1). To obtain (11), we differentiate both sides of (10) with respect to t; this yields

$$\mathbf{a} = \frac{d}{dt}\left(\frac{ds}{dt}\,\mathbf{T}\right) = \frac{d^2s}{dt^2}\,\mathbf{T} + \frac{ds}{dt}\frac{d\mathbf{T}}{dt}$$

$$= \frac{d^2s}{dt^2}\,\mathbf{T} + \frac{ds}{dt}\frac{d\mathbf{T}}{ds}\frac{ds}{dt}$$

$$= \frac{d^2s}{dt^2}\,\mathbf{T} + \left(\frac{ds}{dt}\right)^2\frac{d\mathbf{T}}{ds}$$

$$= \frac{d^2s}{dt^2}\,\mathbf{T} + \left(\frac{ds}{dt}\right)^2\kappa\mathbf{N} \quad \boxed{\begin{array}{l}\text{Formula (16)}\\ \text{of Section 15.5}\end{array}}$$

from which (11) follows.   ∎

Formula (11) should make sense from your experience as an automobile passenger. If a car increases its speed rapidly, the effect is to press the passenger back against the seat. This occurs because the rapid increase in speed results in a large value for

$$\frac{d}{dt}\left(\frac{ds}{dt}\right) = \frac{d^2s}{dt^2}$$

which produces a large tangential component of acceleration in Formula (11). On the other hand, if the car rounds a turn in the road, the passenger is thrown to the side: the greater the curvature in the road or the greater the speed of the car, the greater the force with which the passenger is thrown. This occurs because ds/dt (speed) and κ (curvature) both enter as factors in the normal component of acceleration in (11).

The coefficients of **T** and **N** in (11) are commonly denoted by

$$a_T = \frac{d^2s}{dt^2} \quad \text{and} \quad a_N = \kappa\left(\frac{ds}{dt}\right)^2 \tag{12a–b}$$

These numbers are called the ***scalar tangential component of acceleration*** and the ***scalar normal component of acceleration,*** respectively. Using this notation, (11) can be expressed as

$$\mathbf{a} = a_T\mathbf{T} + a_N\mathbf{N} \tag{13}$$

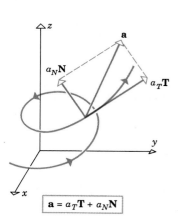

The vectors $a_T\mathbf{T}$ and $a_N\mathbf{N}$ are called the ***vector tangential component of acceleration*** and ***vector normal component of acceleration,*** respectively.

It should be noted that Formula (13) applies to motion in both 2-space and 3-space. What is interesting is that the 3-space formula does not involve the binormal vector **B**, so the acceleration vector always lies in the plane of **T** and **N** (the osculating plane), even for the most twisting paths of motion (Figure 15.6.5).

$$\mathbf{a} = a_T\mathbf{T} + a_N\mathbf{N}$$

Figure 15.6.5

Although Formulas (12a) and (12b) provide useful insight into the physical properties of the tangential and normal components of acceleration, they are not always the best formulas to use for computations. We shall now derive some more useful formulas that express the scalar normal and tangential components of acceleration directly in terms of the velocity and acceleration of the particle. If, as shown in Figure 15.6.6, we let θ be the angle between the vector **a** and the vector $a_T\mathbf{T}$, then we obtain

$$a_T = \|\mathbf{a}\| \cos\theta \quad \text{and} \quad a_N = \|\mathbf{a}\| \sin\theta$$

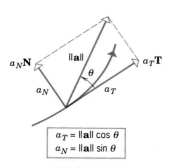

$$a_T = \|\mathbf{a}\| \cos\theta$$
$$a_N = \|\mathbf{a}\| \sin\theta$$

Figure 15.6.6

With the help of Definition 14.3.1 and Formula (7) of Section 14.4, these equations can be rewritten as

$$a_T = \frac{\|\mathbf{v}\|\,\|\mathbf{a}\| \cos\theta}{\|\mathbf{v}\|} = \frac{\mathbf{v}\cdot\mathbf{a}}{\|\mathbf{v}\|}$$

and

$$a_N = \frac{\|\mathbf{v}\|\,\|\mathbf{a}\| \sin\theta}{\|\mathbf{v}\|} = \frac{\|\mathbf{v}\times\mathbf{a}\|}{\|\mathbf{v}\|}$$

Summarizing, we have

$$a_T = \frac{\mathbf{v} \cdot \mathbf{a}}{\|\mathbf{v}\|} \quad \text{and} \quad a_N = \frac{\|\mathbf{v} \times \mathbf{a}\|}{\|\mathbf{v}\|} \tag{14a–b}$$

These formulas are applicable in both 2-space and 3-space, but to compute the cross product in 2-space, the vectors $\mathbf{v}$ and $\mathbf{a}$ must be treated as vectors in 3-space with a zero $\mathbf{k}$ component. Note also that a_N is nonnegative. Thus, for motion in 2-space the vector $a_N\mathbf{N}$ points "inward" toward the concave side of the curve because $\mathbf{N}$ does. (See the subsection of 15.4 on Inward Unit Normal Vectors in 2-Space.) For this reason, the vector $a_N\mathbf{N}$ is also called the **inward vector component of acceleration** in 2-space.

Before proceeding to a numerical example, it is worth noting that by rewriting Formula (12b) as

$$\kappa = \frac{a_N}{(ds/dt)^2} = \frac{a_N}{\|\mathbf{v}\|^2} \tag{15}$$

and combining it with (14b) we can write

$$\kappa = \frac{\|\mathbf{v} \times \mathbf{a}\|}{\|\mathbf{v}\|^3} \tag{16}$$

which expresses curvature in terms of velocity and acceleration. [Compare this result with Formula (11) of Section 15.5.]

Example 4 Suppose that a particle moves through 3-space so that its position vector at time t is

$$\mathbf{r}(t) = t\mathbf{i} + t^2\mathbf{j} + t^3\mathbf{k}$$

(The path is the twisted cubic shown in Figure 15.1.5.)

(a) Find the scalar tangential and normal components of acceleration at time t.

(b) Find the scalar tangential and normal components of acceleration at time $t = 1$.

(c) Find the vector tangential and normal components of acceleration at time $t = 1$.

(d) Find the curvature of the path at the point where the particle is located at time $t = 1$.

Solution (a). We have

$$\mathbf{v}(t) = \mathbf{r}'(t) = \mathbf{i} + 2t\mathbf{j} + 3t^2\mathbf{k}$$

$$\mathbf{a}(t) = \mathbf{v}'(t) = 2\mathbf{j} + 6t\mathbf{k}$$

$$\|\mathbf{v}(t)\| = \sqrt{1 + 4t^2 + 9t^4}$$

$$\mathbf{v}(t) \cdot \mathbf{a}(t) = 4t + 18t^3$$

$$v(t) \times a(t) = \begin{vmatrix} i & j & k \\ 1 & 2t & 3t^2 \\ 0 & 2 & 6t \end{vmatrix} = 6t^2i - 6tj + 2k$$

Thus, from (14a) and (14b)

$$a_T = \frac{v \cdot a}{\|v\|} = \frac{4t + 18t^3}{\sqrt{1 + 4t^2 + 9t^4}}$$

$$a_N = \frac{\|v \times a\|}{\|v\|} = \frac{\sqrt{36t^4 + 36t^2 + 4}}{\sqrt{1 + 4t^2 + 9t^4}} = 2\sqrt{\frac{9t^4 + 9t^2 + 1}{9t^4 + 4t^2 + 1}}$$

Solution (b). At time $t = 1$, the components a_T and a_N in part (a) are

$$a_T = \frac{22}{\sqrt{14}} \approx 5.88 \quad \text{and} \quad a_N = 2\sqrt{\frac{19}{14}} \approx 2.33$$

Solution (c). Since T and v have the same direction, T can be obtained by normalizing v, that is,

$$T(t) = \frac{v(t)}{\|v(t)\|}$$

At time $t = 1$ we have

$$T(1) = \frac{v(1)}{\|v(1)\|} = \frac{i + 2j + 3k}{\|i + 2j + 3k\|} = \frac{1}{\sqrt{14}}(i + 2j + 3k)$$

From this and part (b) we obtain the vector tangential component of acceleration:

$$a_T T(1) = \frac{22}{\sqrt{14}} T(1) = \frac{11}{7}(i + 2j + 3k) = \frac{11}{7}i + \frac{22}{7}j + \frac{33}{7}k$$

To find the normal vector component of acceleration, we rewrite $a = a_T T + a_N N$ as

$$a_N N = a - a_T T$$

Thus, at time $t = 1$ the normal vector component of acceleration is

$$a_N N(1) = a(1) - a_T T(1)$$

$$= (2j + 6k) - \left(\frac{11}{7}i + \frac{22}{7}j + \frac{33}{7}k\right)$$

$$= -\frac{11}{7}i - \frac{8}{7}j + \frac{9}{7}k$$

Solution (d). We shall apply Formula (16) with $t = 1$. From part (a)

$$\|v(1)\| = \sqrt{14} \quad \text{and} \quad v(1) \times a(1) = 6i - 6j + 2k$$

Thus, at time $t = 1$

$$\kappa = \frac{\|\mathbf{v} \times \mathbf{a}\|}{\|\mathbf{v}\|^3} = \frac{\sqrt{76}}{(\sqrt{14})^3} = \frac{1}{14}\sqrt{\frac{38}{7}} \approx 0.17 \quad \blacktriangleleft$$

☐ **MOTION OF A PROJECTILE**

Our work in this section provides many of the mathematical tools required to study such fundamental physical phenomena as planetary motion and the motion of objects in the earth's gravitational field. As an illustration, we shall investigate the following problem.

15.6.3 PROBLEM. *An object of mass m is fired, thrown, or released at some known point near the surface of the earth with a known initial velocity vector. Find the trajectory of the object.*

To solve this problem we need a result from physics known as **Newton's Second Law of Motion;** it states that when an object of mass m is subjected to a force $\mathbf{F}$, it undergoes an acceleration $\mathbf{a}$ satisfying

$$\mathbf{F} = m\mathbf{a} \tag{17}$$

We shall make three simplifying assumptions:

1. The mass of the object is constant.
2. The only force acting on the object is the force of the earth's gravity. (Thus, air resistance and the gravitational effect of other planets and celestial objects are ignored.)
3. The object remains sufficiently close to the earth that the earth's gravity can be assumed constant.

We shall assume that the mass of the object is m and that t (measured in seconds) is the time elapsed from the initial firing or release, so that this firing or release occurs at time $t = 0$. We shall denote the known initial velocity vector by $\mathbf{v}_0$ and the known initial position vector by $\mathbf{r}_0$.

As shown in Figure 15.6.7, we shall introduce an xy-coordinate system whose origin is on the surface of the earth and whose positive y-axis points up and passes through the initial position of the object. Thus, at time $t = 0$ the object has coordinates $(0, s_0)$, which are assumed known, and the initial position vector is $\mathbf{r}_0 = s_0\mathbf{j}$.

It is shown in physics that the downward force of the earth's gravity on an object of mass m is

$$\mathbf{F} = -mg\mathbf{j} \tag{18}$$

where g is a constant approximately equal to 32 ft/sec² if distance is measured

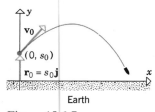

Earth

Figure 15.6.7

in feet and 9.8 m/sec^2 if distance is measured in meters. Substituting (18) into (17) yields

$$ma = -mg\mathbf{j}$$

or on canceling m from both sides

$$\mathbf{a} = -g\mathbf{j} \tag{19}$$

Observe that the acceleration vector $\mathbf{a}$ is constant because (19) does not involve t. Moreover, since the value of g is known, the acceleration vector $\mathbf{a}$ is also known; thus, we can find the position function of the object by integrating the known acceleration twice, once to obtain the velocity $\mathbf{v}(t)$, and again to obtain $\mathbf{r}(t)$. Thus, Problem 15.6.3 has been reduced to solving the vector differential equation

$$\frac{d^2\mathbf{r}}{dt^2} = -g\mathbf{j} \tag{20}$$

subject to the initial conditions

$$\mathbf{r}(0) = \mathbf{r}_0 = s_0\mathbf{j} \tag{21}$$
$$\mathbf{v}(0) = \mathbf{v}_0 \tag{22}$$

Integrating (20) with respect to t (keeping in mind that $-g\mathbf{j}$ is constant) yields

$$\mathbf{v}(t) = -gt\mathbf{j} + \mathbf{c}_1 \tag{23}$$

where $\mathbf{c}_1$ is a vector constant of integration. Substituting $t = 0$ in (23) and using initial condition (22) yields

$$\mathbf{v}_0 = \mathbf{c}_1$$

so that (23) may be written as

$$\mathbf{v}(t) = -gt\mathbf{j} + \mathbf{v}_0 \tag{24}$$

This formula specifies the velocity of the object at any time t. Integrating $\mathbf{v}(t)$ with respect to t (keeping in mind that $\mathbf{v}_0$ is constant) yields

$$\mathbf{r}(t) = -\frac{1}{2}gt^2\mathbf{j} + \mathbf{v}_0 t + \mathbf{c}_2 \tag{25}$$

where $\mathbf{c}_2$ is another vector constant of integration. Substituting $t = 0$ in (25) and using initial condition (21) yields

$$s_0\mathbf{j} = \mathbf{c}_2$$

so that (25) can be written as

$$\mathbf{r}(t) = \left(-\frac{1}{2}gt^2 + s_0\right)\mathbf{j} + \mathbf{v}_0 t \tag{26}$$

This is the result we were seeking—the position function of the object, expressed in terms of the initial velocity and position.

REMARK.   Observe that the mass of the object does not enter into the final formulas for velocity and position. Physically, this means that the mass has no influence on the trajectory or the velocity of the object—these are completely determined by the initial position and velocity. This explains the famous observation of Galileo that two objects of different mass, released from the same height, will reach the ground at the same time if air resistance is neglected.

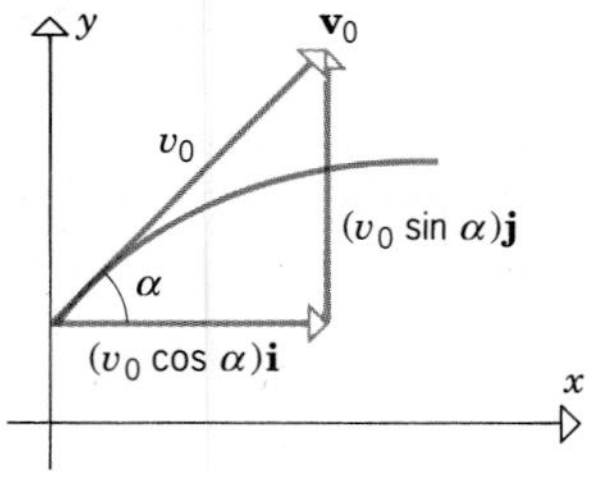

There is a useful alternative form of (26) that expresses the position function of the object in terms of its initial speed and the angle that the initial velocity vector makes with the x-axis. As shown in Figure 15.6.8, suppose that the initial speed $\|\mathbf{v_0}\|$ is denoted by v_0 and the angle that $\mathbf{v_0}$ makes with the x-axis is denoted by α. Thus, the vector $\mathbf{v_0}$ can be expressed as

$$\mathbf{v_0} = (v_0 \cos \alpha)\mathbf{i} + (v_0 \sin \alpha)\mathbf{j}$$

Substituting this expression in (26) and combining like components yields

$$\mathbf{r}(t) = (v_0 \cos \alpha)t\mathbf{i} + (s_0 + (v_0 \sin \alpha)t - \tfrac{1}{2}gt^2)\mathbf{j} \tag{27}$$

which is equivalent to the parametric equations

$$x = (v_0 \cos \alpha)t, \quad y = s_0 + (v_0 \sin \alpha)t - \tfrac{1}{2}gt^2 \tag{28}$$

The parametric equations reveal that the trajectory of the object is a parabolic arc. To see that this is so, we can solve the first equation for t, then substitute in the second to eliminate the parameter. This yields (verify)

$$y = s_0 + (\tan \alpha)x - \left(\frac{g}{2v_0^2 \cos^2 \alpha}\right)x^2$$

which is an equation of a parabola since the right side is a quadratic polynomial in x.

Example 5   A shell, fired from a cannon, has a muzzle speed (the speed as it leaves the barrel) of 800 ft/sec. The barrel makes an angle of 45° with the horizontal and, for simplicity, the barrel opening is assumed to be at ground level.

(a)   Find parametric equations for the shell's trajectory relative to the coordinate system in Figure 15.6.9.

(b)   How high does the shell rise?

(c)   How far does the shell travel horizontally?

(d)   What is the speed of the shell at its point of impact with the ground?

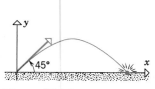

Figure 15.6.8

Figure 15.6.9

Solution (a).   From (28) with $v_0 = 800$, $\alpha = 45°$, $s_0 = 0$ (since the shell starts at ground level), and $g = 32$ (since distance is measured in feet), we obtain the parametric equations

$$x = (800 \cos 45°)t, \quad y = (800 \sin 45°)t - 16t^2 \qquad (t \geq 0)$$

which simplify to

$$x = 400\sqrt{2}\,t, \quad y = 400\sqrt{2}\,t - 16t^2 \qquad (t \geq 0) \tag{29}$$

Solution (b).   The maximum height of the shell is the maximum value of y in (29), which occurs when $dy/dt = 0$, that is, when

$$400\sqrt{2} - 32t = 0 \quad \text{or} \quad t = \frac{25\sqrt{2}}{2}$$

Substituting this value of t in (29) yields

$$y = 5000 \text{ ft}$$

as the maximum height of the shell.

Solution (c).   The shell will hit the ground when $y = 0$. From (29), this occurs when

$$400\sqrt{2}\,t - 16t^2 = 0 \quad \text{or} \quad t(400\sqrt{2} - 16t) = 0$$

The solution $t = 0$ corresponds to the initial position of the shell and the solution $t = 25\sqrt{2}$ to the time of impact. Substituting the latter value in the equation for x in (29) yields

$$x = 20{,}000 \text{ ft}$$

as the horizontal distance traveled by the shell.

Solution (d).   From (29), the position function of the shell is

$$\mathbf{r}(t) = (400\sqrt{2}\,t)\mathbf{i} + (400\sqrt{2}\,t - 16t^2)\mathbf{j}$$

so that the velocity function is

$$\mathbf{v}(t) = \mathbf{r}'(t) = 400\sqrt{2}\,\mathbf{i} + (400\sqrt{2} - 32t)\mathbf{j}$$

From part (c), impact occurs when $t = 25\sqrt{2}$, so that the velocity vector at this point is

$$\mathbf{v}(25\sqrt{2}) = 400\sqrt{2}\,\mathbf{i} + [400\sqrt{2} - 32(25\sqrt{2})]\mathbf{j} = 400\sqrt{2}\,\mathbf{i} - 400\sqrt{2}\,\mathbf{j}$$

Thus, the speed at impact is

$$\|\mathbf{v}(25\sqrt{2})\| = \sqrt{(400\sqrt{2})^2 + (-400\sqrt{2})^2} = 800 \text{ ft/sec} \qquad \blacktriangleleft$$

▶ Exercise Set 15.6   Ⓒ 17, 65–68, 72, 74, 75

In Exercises 1–6, $\mathbf{r}(t)$ is the position vector of a particle moving in the plane. Find the velocity, acceleration, and speed at an arbitrary time t; then sketch the path of the particle together with the velocity and acceleration vectors at the indicated time t.

1. $\mathbf{r}(t) = 3 \cos t\mathbf{i} + 3 \sin t\mathbf{j}$; $t = \pi/3$.

2. $\mathbf{r}(t) = t\mathbf{i} + t^2\mathbf{j}$; $t = 2$.

3. $\mathbf{r}(t) = e^t\mathbf{i} + e^{-t}\mathbf{j}$; $t = 0$.

4. $\mathbf{r}(t) = (2 + 4t)\mathbf{i} + (1 - t)\mathbf{j}$; $t = 1$.

5. $\mathbf{r}(t) = \cosh t\mathbf{i} + \sinh t\mathbf{j}$; $t = \ln 2$.

6. $\mathbf{r}(t) = 4 \cos t\mathbf{i} + 9 \sin t\mathbf{j}$; $t = \pi/2$.

In Exercises 7–12, find the velocity, speed, and acceleration at the given time t of a particle moving along the given curve.

7. $\mathbf{r}(t) = t\mathbf{i} + \frac{1}{2}t^2\mathbf{j} + \frac{1}{3}t^3\mathbf{k}$; $t = 1$.

8. $x = 1 + 3t$, $y = 2 - 4t$, $z = 7 + t$; $t = 2$.

9. $\mathbf{r}(t) = 2 \cos t\mathbf{i} + 2 \sin t\mathbf{j} + t\mathbf{k}$; $t = \pi/4$.

10. $\mathbf{r}(t) = 3t\mathbf{i} + 2t^2\mathbf{j} + \ln t\mathbf{k}$; $t = 1$.

11. $\mathbf{r}(t) = e^t \sin t\mathbf{i} + e^t \cos t\mathbf{j} + t\mathbf{k}$; $t = \pi/2$.

12. $\mathbf{r}(t) = 2t\mathbf{i} + t^2\mathbf{j} + \ln t\mathbf{k}$; $t = 2$.

13. Suppose that the position vector of a particle moving in the plane is $\mathbf{r} = 12\sqrt{t}\mathbf{i} + t^{3/2}\mathbf{j}$, $t > 0$. Find the minimum speed of the particle and its location when it has this speed.

14. Suppose that the motion of a particle is described by $\mathbf{r} = (t - t^2)\mathbf{i} - t^2\mathbf{j}$. Find the minimum speed of the particle and its location when it has this speed.

15. Find the maximum and minimum speeds of a particle whose motion is described by the position vector $\mathbf{r} = \sin 3t\mathbf{i} - 2 \cos 3t\mathbf{j}$.

16. Find the maximum and minimum speeds of a particle whose motion is described by the position vector $\mathbf{r} = 3 \cos 2t\mathbf{i} + \sin 2t\mathbf{j} + 4t\mathbf{k}$.

17. Find, to the nearest degree, the angle between $\mathbf{v}$ and $\mathbf{a}$ for $\mathbf{r} = t^3\mathbf{i} + t^2\mathbf{j}$ when $t = 1$.

18. Show that the angle between $\mathbf{v}$ and $\mathbf{a}$ is constant for $\mathbf{r} = e^t \cos t\mathbf{i} + e^t \sin t\mathbf{j}$. Find the angle.

19. Where on the path
$$\mathbf{r} = (t^2 - 5t)\mathbf{i} + (2t + 1)\mathbf{j} + 3t^2\mathbf{k}$$
are the velocity and acceleration vectors orthogonal?

20. Prove: If the speed of a particle is constant, then the acceleration and velocity vectors are orthogonal. [*Hint:* Consider $\mathbf{v} \cdot \mathbf{v}$.]

In Exercises 21 and 22, the position vectors of two particles are given. Show that the particles move along the same path but the speed of the first is constant and the speed of the second is not.

21. $\mathbf{r}_1 = 2 \cos 3t\mathbf{i} + 2 \sin 3t\mathbf{j}$,
    $\mathbf{r}_2 = 2 \cos (t^2)\mathbf{i} + 2 \sin (t^2)\mathbf{j}$ $(t \geq 0)$.

22. $\mathbf{r}_1 = (3 + 2t)\mathbf{i} + t\mathbf{j} + (1 - t)\mathbf{k}$,
    $\mathbf{r}_2 = (5 - 2t^3)\mathbf{i} + (1 - t^3)\mathbf{j} + t^3\mathbf{k}$.

In Exercises 23–30, use the given information to find the position and velocity vectors of the particle.

23. $\mathbf{a}(t) = -32\mathbf{j}$; $\mathbf{v}(0) = \mathbf{0}$; $\mathbf{r}(0) = \mathbf{0}$.

24. $\mathbf{a}(t) = t\mathbf{j}$; $\mathbf{v}(0) = \mathbf{i} + \mathbf{j}$; $\mathbf{r}(0) = \mathbf{0}$.

25. $\mathbf{a}(t) = -\cos t\mathbf{i} - \sin t\mathbf{j}$; $\mathbf{v}(0) = \mathbf{i}$; $\mathbf{r}(0) = \mathbf{j}$.

26. $\mathbf{a}(t) = \mathbf{i} + e^{-t}\mathbf{j}$; $\mathbf{v}(0) = 2\mathbf{i} + \mathbf{j}$; $\mathbf{r}(0) = \mathbf{i} - \mathbf{j}$.

27. $\mathbf{a}(t) = \mathbf{i} + t\mathbf{k}$; $\mathbf{v}(0) = \mathbf{0}$; $\mathbf{r}(0) = \mathbf{j}$.

28. $\mathbf{a}(t) = (\sin 2t)\mathbf{j}$; $\mathbf{v}(0) = \mathbf{i} - \mathbf{k}$; $\mathbf{r}(0) = \mathbf{0}$.

29. $\mathbf{a}(t) = \sin t\mathbf{i} + \cos t\mathbf{j} + e^t\mathbf{k}$; $\mathbf{v}(0) = \mathbf{k}$;
    $\mathbf{r}(0) = -\mathbf{i} + \mathbf{k}$.

30. $\mathbf{a}(t) = (t + 1)^{-2}\mathbf{j} - e^{-2t}\mathbf{k}$, $t > -1$; $\mathbf{v}(0) = 3\mathbf{i} - \mathbf{j}$;
    $\mathbf{r}(0) = 2\mathbf{k}$.

31. Prove: If the acceleration of a moving particle is zero for all t, then the particle moves on a straight line.

32. Show that for a particle in motion over the time interval $[t_1, t_2]$ the change in velocity $\mathbf{v}(t_2) - \mathbf{v}(t_1)$ is given by
$$\mathbf{v}(t_2) - \mathbf{v}(t_1) = \int_{t_1}^{t_2} \mathbf{a}(t)\, dt$$

In Exercises 33–38, find the displacement and the distance traveled over the indicated time interval.

33. $\mathbf{r}(t) = t^2\mathbf{i} + \frac{1}{3}t^3\mathbf{j}$; $1 \leq t \leq 3$.

34. $\mathbf{r}(t) = (1 - 3 \sin t)\mathbf{i} + 3 \cos t\mathbf{j}$; $0 \leq t \leq 3\pi/2$.

35. $\mathbf{r}(t) = 2 \sin 3t\mathbf{i} + 2 \cos 3t\mathbf{j}$; $0 \leq t \leq 2\pi$.

36. $\mathbf{r}(t) = 3 \sin 2t\mathbf{i} - 3 \cos 2t\mathbf{j} + 8t\mathbf{k}$; $0 \leq t \leq 3\pi/4$.

37. $\mathbf{r}(t) = e^t\mathbf{i} + e^{-t}\mathbf{j} + \sqrt{2}t\mathbf{k};\ 0 \le t \le \ln 3.$

38. $\mathbf{r}(t) = \cos 2t\mathbf{i} + (1 - \cos 2t)\mathbf{j} + (3 + \frac{1}{2}\cos 2t)\mathbf{k};$
     $0 \le t \le \pi.$

In Exercises 39–48, find the scalar tangential and normal components of acceleration at the indicated time t.

39. $\mathbf{r}(t) = 2\cos t\mathbf{i} + 2\sin t\mathbf{j};\ t = \pi/3.$

40. $\mathbf{r}(t) = t\mathbf{i} + t^2\mathbf{j};\ t = 1.$

41. $\mathbf{r}(t) = e^{-t}\mathbf{i} + e^t\mathbf{j};\ t = 0.$

42. $\mathbf{r}(t) = \cos(t^2)\mathbf{i} + \sin(t^2)\mathbf{j};\ t = \sqrt{\pi}/2.$

43. $\mathbf{r}(t) = (t^3 - 2t)\mathbf{i} + (t^2 - 4)\mathbf{j};\ t = 1.$

44. $\mathbf{r}(t) = e^t\cos t\mathbf{i} + e^t\sin t\mathbf{j};\ t = \pi/4.$

45. $\mathbf{r}(t) = t\mathbf{i} + t^2\mathbf{j} + t^3\mathbf{k};\ t = 1.$

46. $\mathbf{r}(t) = e^t\mathbf{i} + e^{-2t}\mathbf{j} + t\mathbf{k};\ t = 0.$

47. $\mathbf{r}(t) = 3\sin t\mathbf{i} + 2\cos t\mathbf{j} - \sin 2t\mathbf{k};\ t = \pi/2.$

48. $\mathbf{r}(t) = 2\mathbf{i} + t^3\mathbf{j} - 16\ln t\mathbf{k};\ t = 1.$

In Exercises 49–52, $\mathbf{v}$ and $\mathbf{a}$ are given at a certain instant of time. Find $a_T,\ a_N,\ \mathbf{T}$, and $\mathbf{N}$ at this instant.

49. $\mathbf{v} = -4\mathbf{j},\ \mathbf{a} = 2\mathbf{i} + 3\mathbf{j}.$

50. $\mathbf{v} = \mathbf{i} + 2\mathbf{j},\ \mathbf{a} = 3\mathbf{i}.$

51. $\mathbf{v} = 2\mathbf{i} + 2\mathbf{j} + \mathbf{k},\ \mathbf{a} = \mathbf{i} + 2\mathbf{k}.$

52. $\mathbf{v} = 3\mathbf{i} - 4\mathbf{k},\ \mathbf{a} = \mathbf{i} - \mathbf{j} + 2\mathbf{k}.$

In Exercises 53–56, the speed of a particle at an arbitrary time t is given. Find the scalar tangential component of acceleration at the indicated time.

53. $\sqrt{3t^2 + 4};\ t = 2.$

54. $\sqrt{t^2 + e^{-3t}};\ t = 0.$

55. $\sqrt{(4t - 1)^2 + \cos^2 \pi t};\ t = \frac{1}{4}.$

56. $\sqrt{t^4 + 5t^2 + 3};\ t = 1.$

In Exercises 57–60, find the curvature of the path of motion of the particle at the instant at which the velocity and acceleration vectors are given.

57. Exercise 49.   **58.** Exercise 50.

59. Exercise 51.   **60.** Exercise 52.

61. The nuclear accelerator at the Enrico Fermi Laboratory is circular with a radius of 1 km. Find the scalar normal component of acceleration of a proton moving around the accelerator with a constant speed of 3×10^5 km/sec.

In Exercises 62–64, use the formula for $\kappa(x)$ in Exercise 15 of Section 15.5 to help solve the problem.

62. Suppose that a particle moves with nonzero acceleration along the curve $y = f(x)$. Show that the acceleration vector is tangent to the curve at each point where $f''(x) = 0$.

63. A particle moves along the parabola $y = x^2$ with a constant speed of 3 units/sec. Find the normal component of acceleration as a function of x.

64. A particle moves along the curve $y = e^x$ with a constant speed of 2 units/sec. Find the normal component of acceleration as a function of x.

65. A shell is fired from ground level with a muzzle speed of 320 ft/sec and elevation angle of 60°. Find
     (a) parametric equations for the shell's trajectory
     (b) the maximum height reached by the shell
     (c) the horizontal distance traveled by the shell
     (d) the speed of the shell at impact.

66. Solve Exercise 65 assuming that the muzzle speed is 980 m/sec and the elevation angle is 45°.

67. A rock is thrown downward from the top of a building, 168 ft high, at an angle of 60° with the horizontal. How far from the base of the building will the rock land if its initial speed is 80 ft/sec?

68. Solve Exercise 67 assuming that the rock is thrown horizontally at a speed of 80 ft/sec.

69. A shell is to be fired from ground level at an elevation angle of 30°. What should the muzzle speed be in order for the maximum height of the shell to be 2500 ft?

70. A shell, fired from ground level at an elevation angle of 45°, hits the ground 24,500 m away. Calculate the muzzle speed of the shell.

71. Find two elevation angles that will enable a shell, fired from ground level with a muzzle speed of 800 ft/sec, to hit a ground-level target 10,000 ft away.

72. A ball rolls off a table 4 ft high while moving at a constant speed of 5 ft/sec.
     (a) How long does it take for the ball to hit the floor after it leaves the table?
     (b) At what speed does the ball hit the floor?
     (c) If a ball were dropped from table height (initial speed 0) at the same time the rolling ball leaves the table, which ball would hit the ground first?

73. A shell is fired from ground level at an elevation angle of α and a muzzle speed of v_0.

(a) Show that the maximum height reached by the shell is

$$\text{maximum height} = \frac{(v_0 \sin \alpha)^2}{2g}$$

(b) The **horizontal range** R of the shell is the horizontal distance traveled when the shell returns to ground level. Show that $R = (v_0^2 \sin 2\alpha)/g$. For what elevation angle will the range be maximum? What is the maximum range?

74. A shell is fired from ground level with an elevation angle α and a muzzle speed of v_0. Find the angle that should be used to hit a target at ground level that is at a distance of 3/4 the maximum range of the shell. Express your answer to the nearest tenth of a degree. [*Hint:* See Exercise 73(b).]

75. At time $t = 0$ a baseball that is 5 ft above the ground is hit with a bat. The ball leaves the bat with a speed of 80 ft/sec at an angle of $30°$ above the horizontal.

(a) How long will it take for the baseball to hit the ground? Express your answer to the nearest hundredth of a second.

(b) Use the result in part (a) to find the horizontal distance traveled by the ball. Express your answer to the nearest tenth of a foot.

76. At time $t = 0$ a projectile is fired from a height h above level ground at an elevation angle of α with a speed v. Let R be the horizontal distance to the point where the projectile hits the ground.

(a) Show that α and R must satisfy the equation

$$g(\sec^2 \alpha)R^2 - 2v^2(\tan \alpha)R - 2v^2 h = 0$$

(b) If g, h, and v are constant, then the equation in part (a) defines R implicitly as a function of α. Let R_0 be the maximum value of R and α_0 the value of α when $R = R_0$. Use implicit differentiation to find $dR/d\alpha$ and show that

$$\tan \alpha_0 = \frac{v^2}{gR_0}$$

[*Hint:* Assume that $dR/d\alpha = 0$ when R is maximum.]

(c) Use the results in parts (a) and (b) to show that

$$R_0 = \frac{v}{g}\sqrt{v^2 + 2gh} \quad \text{and}$$

$$\alpha_0 = \tan^{-1}\frac{v}{\sqrt{v^2 + 2gh}}$$

▶ **SUPPLEMENTARY EXERCISES**   © 39

In Exercises 1–3,
(a) find $\mathbf{v} = d\mathbf{r}/dt$ and $\mathbf{a} = d^2\mathbf{r}/dt^2$;
(b) sketch the graph of $\mathbf{r}(t)$, showing the direction of increasing t, and find the vectors $\mathbf{r}'(t)$ and $\mathbf{r}''(t)$ at the points corresponding to $t = t_0$ and $t = t_1$.

1. $\mathbf{r}(t) = \sqrt{t + 4}\,\mathbf{i} + 2t\mathbf{j};\ t_0 = -3,\ t_1 = 0.$

2. $\mathbf{r}(t) = \langle 2 + \cosh t, 1 - 2 \sinh t \rangle;\ t_0 = 0,\ t_1 = \ln 2.$

3. $\mathbf{r}(t) = \langle 2t^3 - 1, t^3 + 1 \rangle;\ t_0 = 0,\ t_1 = -\frac{1}{2}.$

4. Find the limits.

(a) $\lim_{t \to e} \langle t + \ln t^2, \ln t + t^2 \rangle$

(b) $\lim_{t \to \pi/6} (\cos 2t\mathbf{i} - 3t\mathbf{j}).$

5. Evaluate the integrals.

(a) $\displaystyle\int (k\mathbf{i} + m\mathbf{j})\, dt$

(b) $\displaystyle\int_0^{\ln 3} \langle e^{2t}, 2e^t \rangle\, dt$

(c) $\displaystyle\int_0^2 \|\cos t\mathbf{i} + \sin t\mathbf{j}\|\, dt$

(d) $\displaystyle\int \frac{d}{dt}[\sqrt{t^2 + 3}\,\mathbf{i} + \ln(\sin t)\mathbf{j}]\, dt.$

In Exercises 6–8, find (a) ds/dt and (b) parametric equations for the curve with arc length s as a parameter, assuming the point corresponding to t_0 is the reference point.

6. $\mathbf{r}(t) = (3e^t + 2)\mathbf{i} + (e^t - 1)\mathbf{j};\ t_0 = 0.$

7. $\mathbf{r}(t) = \left\langle \dfrac{t^2 + 1}{t}, \ln t^2 \right\rangle$, where $t > 0$; $t_0 = 1$.

8. $\mathbf{r}(t) = \langle t^3, t^2 \rangle$, where $t \geq 0$; $t_0 = 0$.

In Exercises 9 and 10, sketch the graph of the curve, showing the direction of increasing t.

9. $\mathbf{r}(t) = \langle t, t^2 + 1, 1 \rangle$.

10. $x = t$, $y = t$, $z = 2 \cos (\pi t/2)$, $0 \leq t \leq 2$.

11. Find the velocity, speed, acceleration, unit tangent vector, unit normal vector, and curvature when $t = 0$ for the motion given by $x = a \sin t$, $y = a \cos t$, $z = a \ln (\cos t)$ $(a > 0)$.

12. The position vector of a particle is
    $\mathbf{r}(t) = \sin (2t)\mathbf{i} + \cos (2t)\mathbf{j} + 2e^t\mathbf{k}$.
    (a) Find the velocity, acceleration, and speed as functions of t.
    (b) Find the scalar tangential and normal components of acceleration and the curvature when $t = 0$.

In Exercises 13 and 14, find the arc length of the curve.

13. $x = 2t$, $y = 4 \sin 3t$, $z = 4 \cos 3t$, $0 \leq t \leq 2\pi$.

14. $\mathbf{r}(t) = \langle e^{-t}, \sqrt{2}t, e^t \rangle$, $0 \leq t \leq \ln 2$.

15. Consider the curve whose position vector is

    $$\mathbf{r}(t) = \langle e^{-t}, e^{2t}, t^3 + 1 \rangle$$

    Find parametric equations for the tangent line to the curve at the point where $t = 0$.

16. (a) Show that the speed of a particle is constant if
    $\mathbf{r} = 3 \sin 2t\mathbf{i} - 3 \cos 2t\mathbf{j} - 8t\mathbf{k}$.
    (b) Show that $\mathbf{v}$ and $\mathbf{a}$ are orthogonal at each point on the path of part (a), and thus verify the result in Exercise 20, Section 15.6.

17. If $\mathbf{u} = \langle 2t, 3, -t^2 \rangle$ and $\mathbf{v} = \langle 0, t^2, t \rangle$, find

    (a) $\displaystyle\int_0^3 \mathbf{u}\, dt$ (b) $d(\mathbf{u} \times \mathbf{v})\big/dt$.

18. If $\mathbf{r}(t) = \langle \cos (\pi e^t), \sin (\pi e^t), \pi t \rangle$, find the angle between the acceleration $\mathbf{a}$ and the velocity $\mathbf{v}$ when $t = 0$.

For the curves given in Exercises 19–21, find (a) the unit tangent vector $\mathbf{T}$ at P_0 and (b) the curvature κ at P_0.

19. $\mathbf{r}(t) = (t^2 + 1)\mathbf{i} + (1/t)\mathbf{j}$; $P_0(2, 1)$.

20. $y = \ln x$; $P_0(1, 0)$.

21. $x = (y - 1)^2$; $P_0(0, 1)$.

In Exercises 22–25, find the curvature κ of the given curve at P_0.

22. $xy^2 = 1$; $P_0(1, 1)$.

23. $\mathbf{r}(t) = (t + t^3)\mathbf{i} + (t + t^2)\mathbf{j}$; $P_0(2, 2)$.

24. $y = a \cosh (x/a)$; $P_0(a, a \cosh 1)$.

25. $e^x = \sec y$; $P_0(0, 0)$.

26. Find the smallest radius of curvature and the point at which it occurs for $\mathbf{r}(t) = \langle e^{2t}, e^{-2t} \rangle$.

27. Find the equation of the osculating circle for the parabola $y = (x - 1)^2$ at the point $(1, 0)$. Verify that y' and y'' for the parabola are the same as y' and y'' for the osculating circle at $(1, 0)$.

In Exercises 28 and 29, calculate $d\mathbf{r}/du$ by the chain rule, and check the result by first expressing $\mathbf{r}$ in terms of u.

28. $\mathbf{r} = \langle \sin t, 2 \cos 2t \rangle$; $t = e^{u/2}$.

29. $\mathbf{r} = \langle e^t - 1, 2e^{2t} \rangle$; $t = \ln u$.

In Exercises 30 and 31, find the scalar tangential and normal components of acceleration.

30. $\mathbf{r}(t) = \langle \cosh 2t, \sinh 2t \rangle$, $t \geq 0$.

31. $\mathbf{r}(t) = \langle \sin t - t \cos t, \cos t + t \sin t \rangle$, $t \geq 0$.

For the motion described in Exercises 32 and 33,
(a) find $\mathbf{v}$, $\mathbf{a}$, and ds/dt at P_0;
(b) find κ at P_0;
(c) find a_T and a_N at P_0;
(d) describe the trajectory;
(e) find the center of the osculating circle at P_0.

32. $\mathbf{r}(t) = (1 - t^2)\mathbf{i} + 2t\mathbf{j}$; $P_0(0, 2)$.

33. $\mathbf{r}(t) = \langle e^{-t}, e^t \rangle$; $P_0(1, 1)$.

34. At $t = 0$, a particle of mass m is located at the point $(-2/m, 0)$ and has a velocity of $(2\mathbf{i} - 3\mathbf{j})/m$. Find the position function $\mathbf{r}(t)$ if the particle is acted upon by a force $\mathbf{F} = \langle 2 \cos t, 3 \sin t \rangle$, for $t \geq 0$.

35. The force acting on a particle of unit mass ($m = 1$) is $\mathbf{F} = (\sin t)\mathbf{i} + (4e^{2t})\mathbf{j}$. If the particle starts at the origin with an initial velocity $\mathbf{v}_0 = \mathbf{i} + 2\mathbf{j}$, find the position function $\mathbf{r}(t)$ at any $t \geq 0$.

36. A curve in a railroad track has the shape of the parabola $x = y^2/100$. If a train is loaded so that its scalar normal component of acceleration cannot exceed 25 units/sec², what is its maximum possible speed as it rounds the curve at $(0, 0)$?

37. A particle moves along the parabola $y = 2x - x^2$ with a constant x-component of velocity of 4 ft/sec. Find the scalar tangential and normal components of acceleration at the points (a) $(1, 1)$ and (b) $(0, 0)$.

38. If a particle moves along the curve $y = 2x^2$ with constant speed $ds/dt = 10$, what are a_T and a_N at $P(x, 2x^2)$?

39. A child twirls a weight at the end of a 2-meter string at a rate of 1 revolution/second. Find a_T and a_N for the motion of the weight.

For the motion described in Exercises 40 and 41, find (a) ds/dt and (b) the distance traveled over the interval described.

40. $\mathbf{r}(t) = \langle 2 \sinh t, \sinh^2 t \rangle$, $0 \le t \le 1$.

41. $\mathbf{r}(t) = e^t \langle \sin 2t, \cos 2t \rangle$, $0 \le t \le \ln 3$.

16
Partial Derivatives

Joseph Louis Lagrange   (1736 - 1813)

■ **16.1** FUNCTIONS OF TWO OR MORE VARIABLES

In previous sections we studied real-valued functions of a real variable and vector-valued functions of a real variable. In this section we shall consider real-valued functions of two or more real variables.

□ NOTATION AND
TERMINOLOGY

There are many familiar formulas in which a given variable depends on two or more other variables. For example, the area A of a triangle depends on the base length b and height h by the formula

$$A = \tfrac{1}{2}bh,$$

the volume V of a rectangular box depends on the length l, the width w, and the height h by the formula

$$V = lwh,$$

and the arithmetic average $\bar{x}$ of n real numbers, $x_1, x_2, \ldots, x_n$ depends on those numbers by the formula

$$\bar{x} = \frac{1}{n}(x_1 + x_2 + \cdots + x_n).$$

Thus, we say that

A is a function of the two variables b and h;
V is a function of the three variables l, w, and h;
$\bar{x}$ is a function of the n variables $x_1, x_2, \ldots, x_n$.

The terminology and notation for functions of two or more variables is similar to that used for functions of one variable. For example, the expression

$$z = f(x, y)$$

means that z is a function of x and y in the sense that a unique value of the *dependent variable* z is determined by specifying values for the *independent variables* x and y. Similarly,

$$w = f(x, y, z)$$

expresses w as a function of x, y, and z, and

$$u = f(x_1, x_2, \ldots, x_n)$$

expresses u as a function of $x_1, x_2, \ldots, x_n$.
The functional relationship

$$z = f(x, y) \tag{1}$$

has a useful geometric interpretation. When values of the independent variables x and y are specified, a point (x, y) in the xy-plane is determined. Thus, the dependent variable z in (1) may be viewed as a numerical value associated with the point (x, y). Similarly, the functional relationship

$$w = f(x, y, z)$$

associates the numerical value w with the point (x, y, z) in 3-space.

The following definitions summarize this discussion.

16.1.1 DEFINITION.   A ***function f of two real variables***, x and y, is a rule that assigns a unique real number $f(x, y)$ to each point (x, y) in some set D of the xy-plane.

16.1.2 DEFINITION.   A ***function f of three variables***, x, y, and z, is a rule that assigns a unique real number $f(x, y, z)$ to each point (x, y, z) in some set D of three-dimensional space.

The set D in these definitions is the ***domain*** of the function; it is the set of points at which the function is defined. If a function f is specified by a formula and the domain of f is not stated explicitly, then it is understood that the domain consists of all points at which the formula produces only real numbers and has no divisions by zero; this is called the ***natural domain*** of the function.

REMARK.   In more advanced courses the notion of "n-dimensional space" for $n > 3$ is defined, and a ***function f of n real variables***, $x_1, x_2, \ldots, x_n$, is regarded as a rule that assigns a unique real number $f(x_1, x_2, \ldots, x_n)$ to each "point" $(x_1, x_2, \ldots, x_n)$ in some set of n-dimensional space. However, we shall not pursue this idea in this text.

Example 1   Let

$$f(x, y) = 3x^2\sqrt{y} - 1$$

Find $f(1, 4)$, $f(0, 9)$, $f(t^2, t)$, $f(ab, 9b)$, and the natural domain of f.

Solution.   By substitution

$$f(1, 4) = 3(1)^2\sqrt{4} - 1 = 5$$
$$f(0, 9) = 3(0)^2\sqrt{9} - 1 = -1$$
$$f(t^2, t) = 3(t^2)^2\sqrt{t} - 1 = 3t^4\sqrt{t} - 1$$
$$f(ab, 9b) = 3(ab)^2\sqrt{9b} - 1 = 9a^2b^2\sqrt{b} - 1$$

Because of the $\sqrt{y}$, we must have $y \geq 0$ to avoid imaginary values for $f(x, y)$. Thus, the natural domain of f consists of all points in the xy-plane that are on or above the x-axis. (See Figure 16.1.1.)   ◄

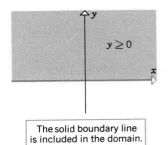

The solid boundary line is included in the domain.

Figure 16.1.1

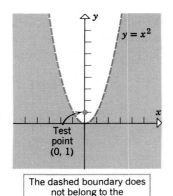

The dashed boundary does not belong to the domain.

Figure 16.1.2

Example 2 Sketch the natural domain of the function $f(x, y) = \ln(x^2 - y)$.

Solution. $\ln(x^2 - y)$ is defined only when $0 < x^2 - y$ or $y < x^2$. To sketch this region, we use the fact that the curve $y = x^2$ separates the region where $y < x^2$ from the region where $y > x^2$. To determine the region where $y < x^2$ holds, we can select an arbitrary "test point" off the boundary $y = x^2$ and determine whether $y < x^2$ or $y > x^2$ at the test point. For example, if we choose the test point $(x, y) = (0, 1)$, then $x^2 = 0$, $y = 1$, so that this point lies in the region where $y > x^2$. Thus, the region where $y < x^2$ is the one that does *not* contain the test point (Figure 16.1.2). ◄

Example 3 Let

$$f(x, y, z) = \sqrt{1 - x^2 - y^2 - z^2}$$

Find $f(0, \frac{1}{2}, -\frac{1}{2})$ and the natural domain of f.

Solution. By substitution,

$$f(0, \tfrac{1}{2}, -\tfrac{1}{2}) = \sqrt{1 - (0)^2 - (\tfrac{1}{2})^2 - (-\tfrac{1}{2})^2} = \sqrt{\tfrac{1}{2}}$$

Because of the square root sign, we must have $0 \le 1 - x^2 - y^2 - z^2$ in order to have a real value for $f(x, y, z)$. Rewriting this inequality in the form

$$x^2 + y^2 + z^2 \le 1$$

we see that the natural domain of f consists of all points on or within the sphere $x^2 + y^2 + z^2 = 1$. ◄

☐ **GRAPHS OF FUNCTIONS OF TWO VARIABLES**

Recall that for a function f of one variable, the graph of $f(x)$ in the xy-plane was defined to be the graph of the equation $y = f(x)$. Similarly, if f is a function of two variables, we define the *graph* of $f(x, y)$ in xyz-space to be the graph of the equation $z = f(x, y)$. In general, such a graph will be a surface in 3-space.

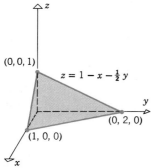

Figure 16.1.3

Example 4 Describe the graph of the function $f(x, y) = 1 - x - \frac{1}{2}y$ in xyz-space.

Solution. By definition, the graph of the given function is the graph of the equation

$$z = 1 - x - \tfrac{1}{2}y \quad \text{or equivalently} \quad x + \tfrac{1}{2}y + z = 1$$

which is a plane. A triangular portion of the plane can be sketched by plotting the intersections with the coordinate axes and joining them with line segments (Figure 16.1.3). ◄

Example 5 Sketch the graph of the function $f(x, y) = \sqrt{1 - x^2 - y^2}$ in xyz-space.

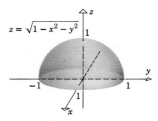

Figure 16.1.4

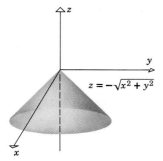

Figure 16.1.5

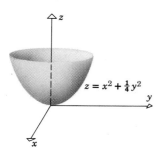

Figure 16.1.6

☐ **LEVEL CURVES**

Solution. By definition, the graph of the given function is the graph of the equation

$$z = \sqrt{1 - x^2 - y^2} \qquad (2)$$

After squaring both sides, this can be rewritten as

$$x^2 + y^2 + z^2 = 1$$

which represents a sphere of radius 1, centered at the origin. Since (2) imposes the added condition that $z \geq 0$, the graph is just the upper hemisphere (Figure 16.1.4). ◄

Example 6 Sketch the graph of the function $f(x, y) = -\sqrt{x^2 + y^2}$ in *xyz*-space.

Solution. The graph of the given function is the graph of the equation

$$z = -\sqrt{x^2 + y^2} \qquad (3)$$

After squaring, we obtain

$$z^2 = x^2 + y^2$$

which is the equation of a right-circular cone [(9) of Section 14.7]. Since (3) imposes the condition that $z \leq 0$, the graph is just the lower nappe of the cone (Figure 16.1.5). ◄

Example 7 Sketch the graph of the function $f(x, y) = x^2 + \frac{1}{4}y^2$ in *xyz*-space.

Solution. The graph of f is the graph of the equation

$$z = x^2 + \frac{1}{4}y^2$$

As discussed in Section 14.7 [Equation (11)], this is an elliptic paraboloid (Figure 16.1.6). ◄

When it is necessary to introduce a dependent variable for a function $f(x, y, z)$ of three variables, we shall usually use the letter w and write $w = f(x, y, z)$. Because this equation involves four variables, it cannot be graphed in three dimensions—"four dimensions" are needed. Thus, there is no *direct* way to represent a function of three or more variables geometrically. However, we shall now discuss some methods for representing functions geometrically that can be applied to functions of three variables.

We are all familiar with topographic (or contour) maps in which a three-dimensional landscape, such as a mountain range, is represented by two-dimensional contour lines or curves of constant elevation. Consider, for example, the model hill and its contour map shown in Figure 16.1.7. The contour map is constructed by passing planes of constant elevation through the hill, projecting the resulting contours onto a flat surface, and labeling the contours with their elevations. In Figure 16.1.7, note how the two gullies appear

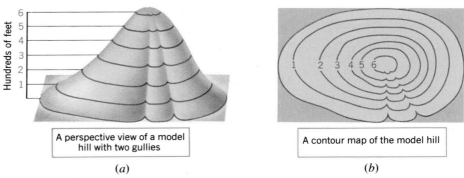

A perspective view of a model hill with two gullies

(a)

A contour map of the model hill

(b)

Figure 16.1.7

as indentations in the contour lines and how the curves are close together on the contour map where the hill has a steep slope and become more widely spaced where the slope is gradual.

Contour maps are useful for studying functions of two variables. If the surface

$$z = f(x, y) \tag{4}$$

is cut by the horizontal plane

$$z = k \tag{5}$$

and if the resulting curve is projected onto the xy-plane, then we obtain what is called the *level curve of height k* (or the *level curve with constant k*) for the function. It follows from (4) and (5) that this level curve has the equation $f(x, y) = k$ (see Figure 16.1.8).

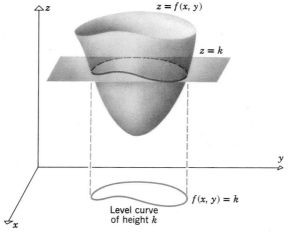

Figure 16.1.8

Example 8 The graph of the function $f(x, y) = x^2 + \frac{1}{4}y^2$ in xyz-space is the paraboloid shown in Figure 16.1.9a. The level curves have equations of the form

$$x^2 + \frac{1}{4}y^2 = k \tag{6}$$

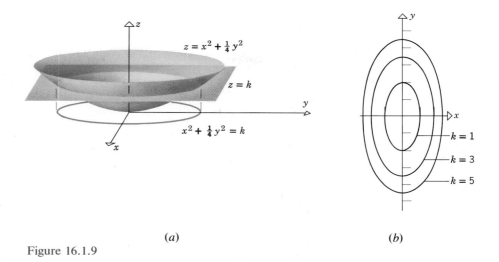

(a)   (b)

Figure 16.1.9

For $k > 0$ these are ellipses; for $k = 0$ it is the single point $(0, 0)$; and for $k < 0$ there are no level curves, since (6) is not satisfied by any real values of x and y. Some sample level curves are shown in Figure 16.1.9b. ◀

Example 9 The graph of the function $f(x, y) = 2 - x - y$ in xyz-space is the plane shown in Figure 16.1.10a. The level curves have equations of the form $2 - x - y = k$, or $y = -x + (2 - k)$. These form a family of parallel lines of slope -1 (Figure 16.1.10b). ◀

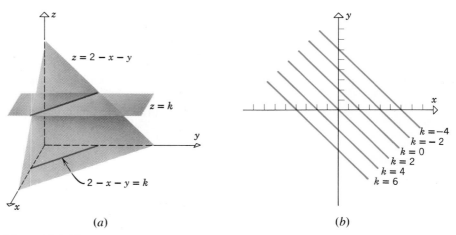

(a)   (b)

Figure 16.1.10

Example 10 The graph of the function $f(x, y) = y^2 - x^2$ in xyz-space is the hyperbolic paraboloid (saddle curve) shown in Figure 16.1.11a. The level curves have equations of the form $y^2 - x^2 = k$. For $k > 0$ these curves are hyperbolas opening along lines parallel to the y-axis; for $k < 0$ they are hyperbolas opening along lines parallel to the x-axis; and for $k = 0$ the level curve consists of the intersecting lines $y + x = 0$ and $y - x = 0$ (Figure 16.1.11b). ◀

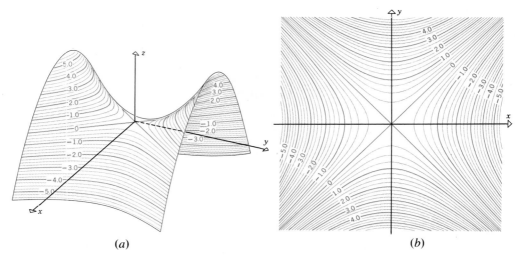

(a)   (b)

Figure 16.1.11

□ **LEVEL SURFACES**

The concept of a level curve for a function of two variables can be extended to functions of three variables. If k is a constant, then an equation of the form $f(x, y, z) = k$ will, in general, represent a surface in three-dimensional space (e.g., $x^2 + y^2 + z^2 = 1$ represents a sphere). The graph of this surface is called the *level surface with constant k* for the function f.

REMARK.   The term "level surface" can be confusing. A level surface need *not* be level in the sense of being horizontal. It is simply a surface on which all values of f are the same.

Example 11   Describe the level surfaces of $f(x, y, z) = x^2 + y^2 + z^2$.

Solution.   The level surfaces have equations of the form

$$x^2 + y^2 + z^2 = k$$

For $k > 0$ the graph of this equation is a sphere of radius $\sqrt{k}$, centered at the origin; for $k = 0$ the graph is the single point $(0, 0, 0)$; and for $k < 0$ there is no level surface (Figure 16.1.12).   ◄

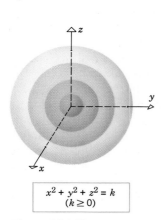

$$x^2 + y^2 + z^2 = k$$
$$(k \geq 0)$$

Figure 16.1.12

Example 12   Describe the level surfaces of $f(x, y, z) = z^2 - x^2 - y^2$.

Solution.   The level surfaces have equations of the form

$$z^2 - x^2 - y^2 = k$$

As discussed in Section 14.7, this equation represents a cone if $k = 0$, a hyperboloid of two sheets if $k > 0$, and a hyperboloid of one sheet if $k < 0$ (Figure 16.1.13).   ◄

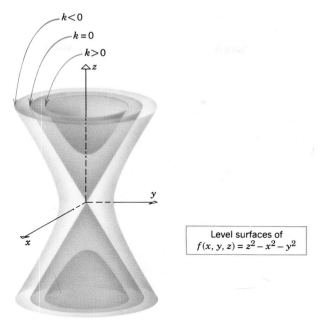

Figure 16.1.13

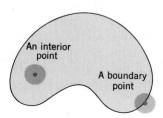

Figure 16.1.14

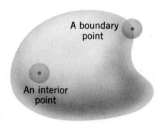

Figure 16.1.15

☐ **OPEN AND
CLOSED SETS**

In our study of functions of one variable, the domains of the functions we encountered were generally intervals. For functions of two or three variables the situation is more complicated, so we shall need to discuss some terminology about sets in 2-space and 3-space that will be helpful when we want to accurately describe the domain of a function of two or three variables.

If D is a set of points in 2-space, then a point (x_0, y_0) is called an *interior point* of D if there is *some* circular disk with positive radius, centered at (x_0, y_0), and containing only points in D (Figure 16.1.14). A point (x_0, y_0) is called a *boundary point* of D if *every* circular disk with positive radius and centered at (x_0, y_0) contains both points in D and points not in D (Figure 16.1.14). Similarly, if D is a set of points in 3-space, then a point (x_0, y_0, z_0) is called an *interior point* of D if there is some spherical ball with positive radius, centered at (x_0, y_0, z_0), and containing only points in D (Figure 16.1.15). A point (x_0, y_0, z_0) is called a *boundary point* of D if *every* spherical ball with positive radius and centered at (x_0, y_0, z_0) contains both points in D and points not in D (Figure 16.1.15).

For a set D in either 2-space or 3-space, the set of all boundary points of D is called the *boundary* of D and the set of all interior points of D is called the *interior* of D.

Recall that an open interval (a, b) on a coordinate line contains *neither* of its endpoints and a closed interval $[a, b]$ contains *both* of its endpoints. Analogously, a set D in 2-space or 3-space is called *open* if it contains *none* of its boundary points and *closed* if it contains *all* of its boundary points. The set D of all points in 2-space has no boundary; it is regarded as both open and closed. Similarly, the set D of all points in 3-space is both open and closed.

Example 13   Let D be the set of points in the xy-plane that are inside or on the circle of radius 1 centered at the origin (Figure 16.1.16a). In set notation

$$D = \{(x, y): x^2 + y^2 \leq 1\}$$

The interior of D is the set

$$I = \{(x, y): x^2 + y^2 < 1\}$$

and the boundary of D is the set

$$B = \{(x, y): x^2 + y^2 = 1\}$$

(Figures 16.1.16b and 16.1.16c). The set D is closed and the set I is open.   ◀

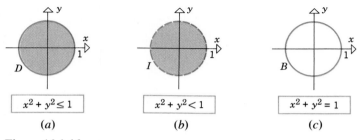

(a) $\quad$ (b) $\quad$ (c)

Figure 16.1.16

□ BOUNDED SETS

Just as we distinguished between finite intervals and infinite intervals on the real line, so we shall want to distinguish between regions of "finite extent" and regions of "infinite extent" in 2-space and 3-space. A set of points in 2-space is called **bounded** if the entire set can be contained within some rectangle, and is called **unbounded** if there is no rectangle that contains all the points of the set. Similarly, a set of points in 3-space is **bounded** if the entire set can be contained within some box, and is unbounded otherwise (Figure 16.1.17).

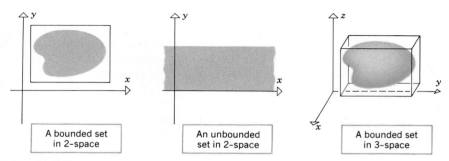

A bounded set in 2-space $\quad$ An unbounded set in 2-space $\quad$ A bounded set in 3-space

Figure 16.1.17

□ COMPUTER-GENERATED
SURFACES IN 3-SPACE

Computer programs are of enormous value for visualizing surfaces in 3-space. Figure 16.1.18 illustrates some of the complexity that such surfaces can have.

SOME COMPUTER-GENERATED SURFACES IN 3-SPACE

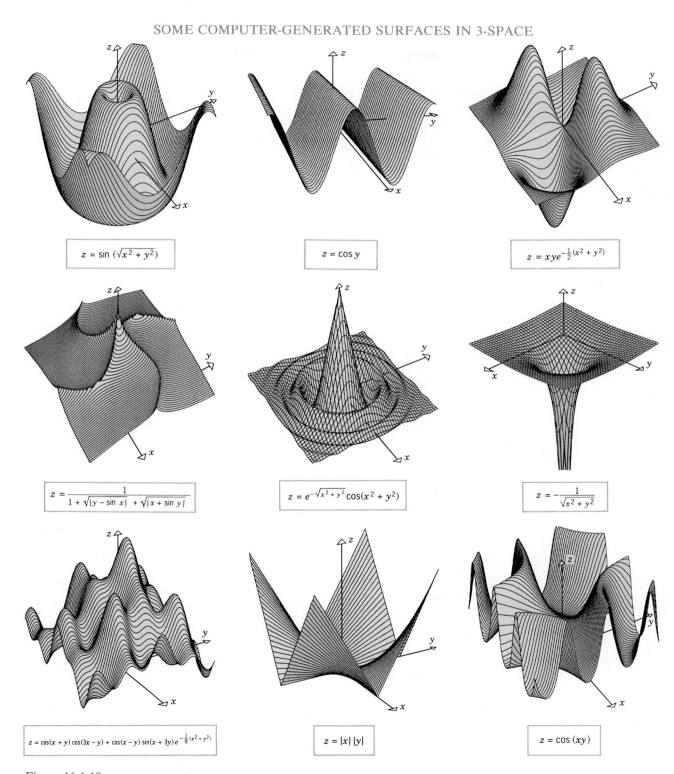

$z = \sin(\sqrt{x^2 + y^2})$

$z = \cos y$

$z = xye^{-\frac{1}{2}(x^2 + y^2)}$

$z = \dfrac{1}{1 + \sqrt{|y - \sin x|} + \sqrt{|x + \sin y|}}$

$z = e^{-\sqrt{x^2 + y^2}}\cos(x^2 + y^2)$

$z = -\dfrac{1}{\sqrt{x^2 + y^2}}$

$z = \cos(x + y)\cos(3x - y) + \cos(x - y)\sin(x + 3y)\,e^{-\frac{1}{8}(x^2 + y^2)}$

$z = |x|\,|y|$

$z = \cos(xy)$

Figure 16.1.18

☐ **BOUNDING BOXES FOR COMPUTER-GENERATED SURFACES**

Many computer programs draw surfaces within a box whose edges are parallel to the coordinate axes. As illustrated with the elliptic paraboloid in Figure 16.1.19, this sometimes produces artificial-looking cuts in the surface; however, this can be useful for visualizing the surface since the cuts are traces of the surface parallel to the coordinate planes.

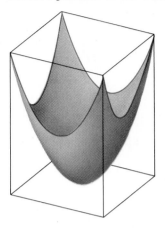

The box cuts the elliptic paraboloid, producing parabolic traces in the sides of the box.

Figure 16.1.19

▶ Exercise Set 16.1   Ⓒ *46, 47, 51, 61*

1. Let $f(x, y) = x^2y + 1$. Find
   (a) $f(2, 1)$ (b) $f(1, 2)$ (c) $f(0, 0)$
   (d) $f(1, -3)$ (e) $f(3a, a)$ (f) $f(ab, a - b)$.

2. Let $f(x, y) = x + \sqrt[3]{xy}$. Find
   (a) $f(t, t^2)$ (b) $f(x, x^2)$ (c) $f(2y^2, 4y)$.

3. Let $f(x, y) = xy + 3$. Find
   (a) $f(x + y, x - y)$ (b) $f(xy, 3x^2y^3)$.

4. Let $g(x) = x \sin x$. Find
   (a) $g(x/y)$ (b) $g(xy)$ (c) $g(x - y)$.

5. Find $F(g(x), h(y))$ if $F(x, y) = xe^{xy}$, $g(x) = x^3$, and $h(y) = 3y + 1$.

6. Find $g(u(x, y), v(x, y))$ if $g(x, y) = y \sin(x^2y)$, $u(x, y) = x^2y^3$, and $v(x, y) = \pi xy$.

7. Let $f(x, y) = x + 3x^2y^2$, $x(t) = t^2$, and $y(t) = t^3$. Find
   (a) $f(x(t), y(t))$ (b) $f(x(0), y(0))$
   (c) $f(x(2), y(2))$.

8. Let $g(x, y) = ye^{-3x}$, $x(t) = \ln(t^2 + 1)$, and $y(t) = \sqrt{t}$. Find $g(x(t), y(t))$.

9. Let $f(x, y, z) = xy^2z^3 + 3$. Find
   (a) $f(2, 1, 2)$ (b) $f(-3, 2, 1)$
   (c) $f(0, 0, 0)$ (d) $f(a, a, a)$
   (e) $f(t, t^2, -t)$ (f) $f(a + b, a - b, b)$.

10. Let $f(x, y, z) = zxy + x$. Find
    (a) $f(x + y, x - y, x^2)$ (b) $f(xy, y/x, xz)$.

11. Find $F(f(x), g(y), h(z))$ if $F(x, y, z) = ye^{xyz}$, $f(x) = x^2$, $g(y) = y + 1$, and $h(z) = z^2$.

12. Find $g(u(x, y, z), v(x, y, z), w(x, y, z))$ if $g(x, y, z) = z \sin xy$, $u(x, y, z) = x^2z^3$, $v(x, y, z) = \pi xyz$, and $w(x, y, z) = xy/z$.

13. Let $f(x, y, z) = x^2y^2z^4$, $x(t) = t^3$, $y(t) = t^2$, and $z(t) = t$. Find
    (a) $f(x(t), y(t), z(t))$ (b) $f(x(0), y(0), z(0))$
    (c) $f(x(2), y(2), z(2))$.

In Exercises 14–19, sketch the domain of f. Use solid lines for portions of the boundary included in the domain and dashed lines for portions not included.

14. $f(x, y) = xy\sqrt{y - 1}$.

15. $f(x, y) = \ln(1 - x^2 - y^2)$.

16. $f(x, y) = \sqrt{x^2 + y^2 - 4}$.

17. $f(x, y) = \dfrac{1}{x - y^2}$.

18. $f(x, y) = \ln xy$. **19.** $f(x, y) = \sqrt{\dfrac{x^2 + y^2}{x^2 - y^2}}$.

In Exercises 20–27, describe the domain of f.

20. $f(x, y) = \sin^{-1}(x + y)$. **21.** $f(x, y) = xe^{-\sqrt{y+2}}$.

22. $f(x, y) = \dfrac{\sqrt{4 - x^2}}{y^2 + 3}$.

23. $f(x, y) = \ln(y - 2x)$.

24. $f(x, y, z) = \sqrt{25 - x^2 - y^2 - z^2}$.

25. $f(x, y, z) = \dfrac{xyz}{x + y + z}$. **26.** $f(x, y, z) = e^{xyz}$.

27. $f(x, y, z) = z + \ln(1 - x^2 - y^2)$.

In Exercises 28–39, sketch the graph of f.

28. $f(x, y) = 3$.

29. $f(x, y) = 4 - 2x - 4y$.

30. $f(x, y) = \sqrt{9 - x^2 - y^2}$.

31. $f(x, y) = \sqrt{x^2 + y^2}$. **32.** $f(x, y) = x^2 + y^2$.

33. $f(x, y) = x^2 - y^2$.

34. $f(x, y) = 4 - x^2 - y^2$.

35. $f(x, y) = -\sqrt{1 - x^2/4 - y^2/9}$.

36. $f(x, y) = \sqrt{x^2 + y^2 - 1}$.

37. $f(x, y) = \sqrt{x^2 + y^2 + 1}$.

38. $f(x, y) = x^2$. **39.** $f(x, y) = y + 1$.

In Exercises 40–47, sketch the level curve $z = k$ for the specified values of k.

40. $z = 3x + y$; $k = -2, -1, 0, 1, 2$.

41. $z = x^2 + y^2$; $k = 0, 1, 2, 3, 4$.

42. $z = y/x$; $k = -2, -1, 0, 1, 2$.

43. $z = x^2 + y$; $k = -2, -1, 0, 1, 2$.

44. $z = x^2 + 9y^2$; $k = 0, 1, 2, 3, 4$.

45. $z = x^2 - y^2$; $k = -2, -1, 0, 1, 2$.

46. $z = y \csc x$; $k = -2, -1, 0, 1, 2$.

47. $z = \sqrt{\dfrac{x + y}{x - y}}$; $k = 0, 1, 2, 3, 4$.

In Exercises 48–51, sketch the level surface $f(x, y, z) = k$.

48. $f(x, y, z) = 4x - 2y + z$; $k = 1$.

49. $f(x, y, z) = 4x^2 + y^2 + 4z^2$; $k = 16$.

50. $f(x, y, z) = x^2 + y^2 - z^2$; $k = 0$.

51. $f(x, y, z) = z - x^2 - y^2 + 4$; $k = 7$.

In Exercises 52–55, describe the level surfaces.

52. $f(x, y, z) = 3x - y + 2z$.

53. $f(x, y, z) = (x - 2)^2 + y^2 + z^2$.

54. $f(x, y, z) = z - x^2 - y^2$.

55. $f(x, y, z) = x^2 + z^2$.

56. Let $f(x, y) = yx^2 + 1$. Find an equation of the level curve that passes through the point
(a) $(1, 2)$ (b) $(-2, 4)$ (c) $(0, 0)$.

57. Let $f(x, y) = x^2 - 2x^3 + 3xy$. Find an equation of the level curve that passes through the point
(a) $(-1, 1)$ (b) $(0, 0)$ (c) $(2, -1)$.

58. Let $f(x, y) = ye^x$. Find an equation of the level curve that passes through the point
(a) $(\ln 2, 1)$ (b) $(0, 3)$ (c) $(1, -2)$.

59. Let $f(x, y, z) = x^2 + y^2 - z$. Find an equation of the level surface that passes through the point
(a) $(1, -2, 0)$ (b) $(1, 0, 3)$ (c) $(0, 0, 0)$.

60. Let $f(x, y, z) = xyz + 3$. Find an equation of the level surface that passes through the point
(a) $(1, 0, 2)$ (b) $(-2, 4, 1)$ (c) $(0, 0, 0)$.

61. If $V(x, y)$ is the voltage or potential at a point (x, y) in the xy-plane, then the level curves of V are called *equipotential curves*. Along such a curve, the voltage remains constant. Given that

$$V(x, y) = \frac{8}{\sqrt{16 + x^2 + y^2}}$$

sketch the equipotential curves at which $V = 2.0$, $V = 1.0$, and $V = 0.5$.

62. If $T(x, y)$ is the temperature at a point (x, y) on a thin metal plate in the xy-plane, then the level curves of T are called *isothermal curves*. All points on such a curve are at the same temperature. Suppose that a plate occupies the first quadrant and $T(x, y) = xy$.
  (a) Sketch the isothermal curves on which $T = 1$, $T = 2$, and $T = 3$.
  (b) An ant, initially at $(1, 4)$, wants to walk on the plate so that the temperature along its path remains constant. What path should the ant take?

In Exercises 63–65, use the contour map shown in the figure (all elevations in hundreds of feet).

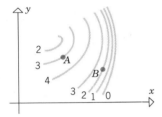

63. For points A and B,
  (a) which one is higher
  (b) which one is on the steeper slope?

64. Starting at point A, will the elevation begin to increase or decrease if we travel so that
  (a) y remains constant and x increases
  (b) y remains constant and x decreases
  (c) x remains constant and y increases
  (d) x remains constant and y decreases?

65. Starting at point B, will the elevation begin to increase or decrease if we travel so that

  (a) y remains constant and x increases
  (b) y remains constant and x decreases
  (c) x remains constant and y increases
  (d) x remains constant and y decreases?

In Exercises 66 and 67, classify the given set of points as open, closed, or neither.

66. The set of points (x, y) in the xy-plane that satisfy the inequality
  (a) $1 \leq x^2 + y^2 \leq 3$
  (b) $0 \leq y < 2x + 1$ and $0 \leq x \leq 1$
  (c) $\sqrt{y} < x < 1$
  (d) $0 < y < e^x$ and $0 < x < 2$.

67. The set of points (x, y, z) in 3-space that satisfy the inequality
  (a) $x^2 + y^2 + z^2 < 3$
  (b) $0 \leq z \leq x^2 + y^2$ for $x^2 + y^2 < 1$
  (c) $0 \leq z \leq 5 - 3x - 2y$ for $x \geq 0$ and $y \geq 0$
  (d) $y^2 + z^2 \leq 5$ for $1 \leq x \leq 2$.

In Exercises 68 and 69, classify the given set as bounded or unbounded.

68. The set of points (x, y) in the xy-plane that satisfy the inequality
  (a) $x^2 + y^2 < 100$ (b) $x^2 + y^2 \geq 100$
  (c) $-1 < y < 2$
  (d) $y \leq 3 - x$ and $-1 < x < 1$.

69. The set of points in 3-space that satisfy the inequality
  (a) $x^2 + y^2 + z^2 \leq 3$
  (b) $0 < z < 2 - y$ and $0 < x < 1$
  (c) $x^2 + y^2 < 4$ (d) $z \geq 2 - x - 3y$.

■ 16.2 LIMITS AND CONTINUITY

In this section we shall introduce the notions of limit and continuity for functions of two or more variables. We shall not go into great detail; our objective is to develop the basic ideas accurately, and to obtain results needed in later sections. A more extensive study of these topics is usually given in advanced calculus.

□ **LIMITS ALONG CURVES**

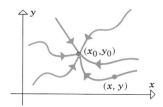

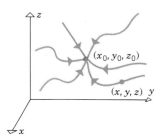

Figure 16.2.1

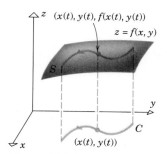

Figure 16.2.2

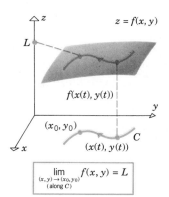

$$\lim_{\substack{(x,y)\to(x_0,y_0)\\ \text{(along } C)}} f(x, y) = L$$

Figure 16.2.3

For a function of one variable we considered the one-sided limits,

$$\lim_{x\to x_0^+} f(x) \quad \text{and} \quad \lim_{x\to x_0^-} f(x)$$

which reflected the fact that there are two directions from which x can approach x_0—the left or the right. For functions of two or three variables the situation is much more complicated because there are infinitely many different curves along which one point can approach another (Figure 16.2.1).

Suppose that C is a smooth curve in 2-space with parametric equations

$$x = x(t), \quad y = y(t)$$

and $z = f(x, y)$ is the equation of some surface S in 3-space. If we substitute the two parametric functions into the formula for f and graph the parametric equations

$$x = x(t), \quad y = y(t), \quad z = f(x(t), y(t))$$

we obtain a curve on the surface S whose projection onto the xy-plane is the curve C (Figure 16.2.2). Thus, as the parameter t increases, the point $(x(t), y(t))$ moves along the curve C in the xy-plane, and the "companion point" $(x(t), y(t), f(x(t), y(t)))$ moves along the surface S.

REMARK. Strictly speaking, in Figure 16.2.2 we should have denoted the point on the curve C by $(x(t), y(t), 0)$, since the curve is in 3-space. However, the omission of the zero z-coordinate of points in the xy-plane is common practice that we shall often follow.

Suppose that $f(x, y)$ is a function of two variables, C is a smooth parametric curve in the xy-plane with parametric equations

$$x = x(t), \quad y = y(t)$$

and $(x_0, y_0) = (x(t_0), y(t_0))$ is a point on the curve C. The limit of $f(x, y)$ as (x, y) approaches (x_0, y_0) along C is denoted by

$$\lim_{\substack{(x,y)\to(x_0,y_0)\\ \text{(along } C)}} f(x, y)$$

and is defined by

$$\lim_{\substack{(x,y)\to(x_0,y_0)\\ \text{(along } C)}} f(x, y) = \lim_{t\to t_0} f(x(t), y(t)) \tag{1}$$

Simply stated, the limit of $f(x, y)$ along C is obtained by substituting parametric equations for C into the formula for f and calculating the appropriate limit of the resulting function of one variable. A geometric interpretation of the limit along a curve is shown in Figure 16.2.3.

The limit of a function $f(x, y, z)$ along a smooth curve C in 3-space is defined similarly to (1):

$$\lim_{\substack{(x,y,z) \to (x_0, y_0, z_0) \\ \text{(along } C)}} f(x, y, z) = \lim_{t \to t_0} f(x(t), y(t), z(t)) \qquad (2)$$

In both (1) and (2), the limit of the function of t will have to be one-sided if (x_0, y_0) or (x_0, y_0, z_0) is an endpoint of C.

Example 1 Let

$$f(x, y) = \frac{xy}{x^2 + y^2}$$

(See Figure 16.2.4.) Find the limit of $f(x, y)$ as $(x, y) \to (0, 0)$ along

(a) the x-axis   (b) the y-axis

(c) the parabola $y = x^2$ (d) the line $y = x$

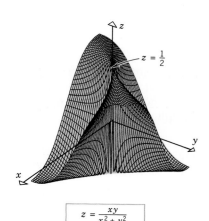

$$z = \frac{xy}{x^2 + y^2}$$

Figure 16.2.4

Solution (a). The x-axis has parametric equations $x = t$, $y = 0$, with $(0, 0)$ corresponding to $t = 0$, so

$$\lim_{\substack{(x,y) \to (0,0) \\ \text{(along } y=0)}} f(x, y) = \lim_{t \to 0} f(t, 0) = \lim_{t \to 0} \frac{0}{t^2} = \lim_{t \to 0} 0 = 0$$

Solution (b). The y-axis has parametric equations $x = 0$, $y = t$, with $(0, 0)$ corresponding to $t = 0$, so

$$\lim_{\substack{(x,y) \to (0,0) \\ \text{(along } x=0)}} f(x, y) = \lim_{t \to 0} f(0, t) = \lim_{t \to 0} \frac{0}{t^2} = \lim_{t \to 0} 0 = 0$$

Solution (c).   The parabola $y = x^2$ has parametric equations $x = t$, $y = t^2$, with $(0, 0)$ corresponding to $t = 0$, so

$$\lim_{\substack{(x,y)\to(0,0) \\ \text{(along } y=x^2)}} f(x, y) = \lim_{t\to 0} f(t, t^2) = \lim_{t\to 0} \frac{t^3}{t^2 + t^4} = \lim_{t\to 0} \frac{t}{1 + t^2} = 0$$

Solution (d).   The line $y = x$ has parametric equations $x = t$, $y = t$, with $(0, 0)$ corresponding to $t = 0$, so

$$\lim_{\substack{(x,y)\to(0,0) \\ \text{(along } y=x)}} f(x, y) = \lim_{t\to 0} f(t, t) = \lim_{t\to 0} \frac{t^2}{2t^2} = \lim_{t\to 0} \frac{1}{2} = \frac{1}{2} \quad \blacktriangleleft$$

Example 2   Find

$$\lim_{\substack{(x,y,z)\to(-1,0,\pi) \\ \text{(along } C)}} \frac{x^2 + y^2 + x}{z - \pi}$$

where C is the circular helix with parametric equations $x = \cos t$, $y = \sin t$, $z = t$.

Solution.   The point $(-1, 0, \pi)$ corresponds to $t = \pi$, so

$$\lim_{\substack{(x,y,z)\to(-1,0,\pi) \\ \text{(along } C)}} \frac{x^2 + y^2 + x}{z - \pi} = \lim_{t\to\pi} \frac{\cos^2 t + \sin^2 t + \cos t}{t - \pi}$$

$$= \lim_{t\to\pi} \frac{1 + \cos t}{t - \pi} = 0$$

where L'Hôpital's rule was applied to evaluate the last limit. $\blacktriangleleft$

☐ **GENERAL LIMITS OF FUNCTIONS OF TWO AND THREE VARIABLES**

Although limits along specific curves are useful for many purposes, they do not always tell the complete story about the limiting behavior of a function; what is required is a limit concept that accounts for the behavior of the function in an *entire vicinity* of a point, not just along smooth curves passing through the point. For motivation, recall that for a function of one variable, the statement

$$\lim_{x\to x_0} f(x) = L$$

means that the value of $f(x)$ can be made arbitrarily close to L by making x sufficiently close to (but different from) x_0. Similarly, the statement

$$\lim_{(x,y)\to(x_0,y_0)} f(x, y) = L$$

is intended to mean that the value of $f(x, y)$ can be made arbitrarily close to L by making the point (x, y) sufficiently close to (but different from) (x_0, y_0). To phrase this idea another way, for every positive number ϵ we want to be able to guarantee that $f(x, y)$ is within ϵ units of L by restricting (x, y) to some sufficiently small circle of radius δ centered at (x_0, y_0), with the possible exception that $f(x, y)$ need not be within ϵ units of L if (x, y) is at the center of the circle. The following definition makes this informal idea precise (Figure 16.2.5).

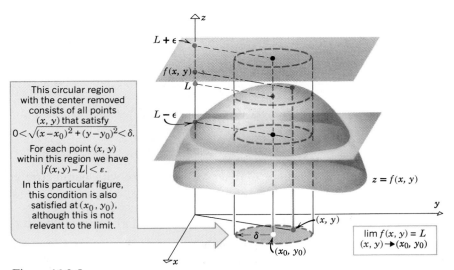

Figure 16.2.5

16.2.1   DEFINITION.   Let f be a function of two variables. We shall write

$$\lim_{(x,y)\to(x_0,y_0)} f(x, y) = L \tag{3}$$

if given any number $\epsilon > 0$, we can find a number $\delta > 0$ such that $f(x, y)$ satisfies

$$|f(x, y) - L| < \epsilon$$

whenever (x, y) lies in the domain of f and the distance between (x, y) and (x_0, y_0) satisfies

$$0 < \sqrt{(x - x_0)^2 + (y - y_0)^2} < \delta$$

The following definition extends the notion of a limit to functions of three variables. Note how closely it parallels Definition 16.2.1.

16.2.2 DEFINITION. Let f be a function of three variables. We shall write

$$\lim_{(x,y,z)\to(x_0,y_0,z_0)} f(x, y, z) = L \tag{4}$$

if given any number $\epsilon > 0$, we can find a number $\delta > 0$ such that $f(x, y, z)$ satisfies

$$|f(x, y, z) - L| < \epsilon$$

whenever (x, y, z) lies in the domain of f and the distance between (x, y, z) and (x_0, y_0, z_0) satisfies

$$0 < \sqrt{(x - x_0)^2 + (y - y_0)^2 + (z - z_0)^2} < \delta$$

When convenient, (3) and (4) can also be written in the alternative notations

$$f(x, y) \to L \quad \text{as} \quad (x, y) \to (x_0, y_0)$$

and

$$f(x, y, z) \to L \quad \text{as} \quad (x, y, z) \to (x_0, y_0, z_0)$$

REMARK.   Limits can also be defined for functions of more than three variables, but we shall leave this for more advanced courses.

☐ **RELATIONSHIPS BETWEEN GENERAL LIMITS AND LIMITS ALONG SMOOTH CURVES**

The following theorem, which we state without proof, establishes an important relationship between general limits and limits along smooth curves.

16.2.3 THEOREM.   *If a function $f(x, y)$ has a limit L as (x, y) approaches a point (x_0, y_0), then $f(x, y)$ approaches the same limit L as (x, y) approaches (x_0, y_0) along any smooth curve that lies in the domain of f. Similarly, for functions of three variables.*

It follows from this theorem that if one can find two different smooth curves containing (x_0, y_0) along which $f(x, y)$ has different limits as (x, y) approaches (x_0, y_0), or if one can find any single smooth curve containing (x_0, y_0) such that the limit of $f(x, y)$ does not exist as (x, y) approaches (x_0, y_0), then

$$\lim_{(x,y)\to(x_0,y_0)} f(x, y)$$

does not exist. Similarly, for functions of three variables.

Example 3   The limit

$$\lim_{(x,y)\to(0,0)} \frac{xy}{x^2 + y^2}$$

does not exist because in Example 1 we found two different smooth curves along which the stated limit had different values. For example, we saw that

$$\lim_{(x,y)\to(0,0) \atop \text{(along } x=0)} \frac{xy}{x^2+y^2} = 0 \quad \text{and} \quad \lim_{(x,y)\to(0,0) \atop \text{(along } y=x)} \frac{xy}{x^2+y^2} = \frac{1}{2} \quad \blacktriangleleft$$

☐ **PROPERTIES OF LIMITS**   We note without proof that the basic properties of limits hold for functions of two or three variables.

16.2.4   THEOREM.   *If*

$$\lim_{(x,y)\to(x_0,y_0)} f(x, y) = L_1 \quad and \quad \lim_{(x,y)\to(x_0,y_0)} g(x, y) = L_2$$

then

(*a*) $\displaystyle\lim_{(x,y)\to(x_0,y_0)} [cf(x, y)] = cL_1$ *(if c is constant)*

(*b*) $\displaystyle\lim_{(x,y)\to(x_0,y_0)} [f(x, y) + g(x, y)] = L_1 + L_2$

(*c*) $\displaystyle\lim_{(x,y)\to(x_0,y_0)} [f(x, y) - g(x, y)] = L_1 - L_2$

(*d*) $\displaystyle\lim_{(x,y)\to(x_0,y_0)} [f(x, y)g(x, y)] = L_1 L_2$

(*e*) $\displaystyle\lim_{(x,y)\to(x_0,y_0)} \frac{f(x, y)}{g(x, y)} = \frac{L_1}{L_2}$ *(if $L_2 \neq 0$)*

Similarly, for functions of three variables.

As with functions of one variable, results (*b*) and (*d*) can be extended to any finite number of functions; that is,

- the limit of a sum is the sum of the limits;
- the limit of a product is the product of the limits.

Example 4   It can be proved using Definitions 16.2.1 and 16.2.2 that

$$\lim_{(x,y)\to(x_0,y_0)} c = c \quad \text{(c a constant)}, \quad \lim_{(x,y)\to(x_0,y_0)} x = x_0, \quad \lim_{(x,y)\to(x_0,y_0)} y = y_0$$

with similar results for limits of functions with three variables. Thus,

$$\lim_{(x,y)\to(2,3)} x = 2, \quad \lim_{(x,y)\to(2,3)} y = 3, \quad \lim_{(x,y,z)\to(0,1,-2)} z = -2, \quad \lim_{(x,y)\to(0,0)} 7 = 7 \quad \blacktriangleleft$$

Example 5

$$\lim_{(x,y)\to(1,4)} [5x^3y^2 - 9] = \lim_{(x,y)\to(1,4)} [5x^3y^2] - \lim_{(x,y)\to(1,4)} 9$$

$$= 5\left[\lim_{(x,y)\to(1,4)} x\right]^3 \left[\lim_{(x,y)\to(1,4)} y\right]^2 - 9$$

$$= 5(1)^3(4)^2 - 9 = 71 \quad \blacktriangleleft$$

☐ **CONTINUITY**

Later in this section we shall describe some more techniques for evaluating limits, but most of the limits we shall need in this text will follow from the notion of continuity, which we shall discuss next.

The definition of continuity for functions of two or three variables is similar to the definition for functions of one variable (see Definition 2.7.1).

16.2.5 DEFINITION. A function f of two variables is called *continuous* at the point (x_0, y_0) if

1. $f(x_0, y_0)$ is defined;

2. $\displaystyle\lim_{(x,y)\to(x_0,y_0)} f(x, y)$ exists;

3. $\displaystyle\lim_{(x,y)\to(x_0,y_0)} f(x, y) = f(x_0, y_0)$.

16.2.6 DEFINITION. A function f of three variables is called *continuous at a point* (x_0, y_0, z_0) if

1. $f(x_0, y_0, z_0)$ is defined;

2. $\displaystyle\lim_{(x,y,z)\to(x_0,y_0,z_0)} f(x, y, z)$ exists;

3. $\displaystyle\lim_{(x,y,z)\to(x_0,y_0,z_0)} f(x, y, z) = f(x_0, y_0, z_0)$.

A function f that is continuous at each point of a region R in 2-space or 3-space is said to be *continuous on R*. A function that is continuous at every point in 2-space or 3-space is called *continuous everywhere* or simply *continuous*. As with functions of one variable, the first and second conditions in the above definitions are consequences of the third condition, so only the third condition needs to be verified to prove continuity.

Intuitively, we can imagine the graph of a continuous function of two variables to be constructed from a thin sheet of clay that has been hollowed and pinched into peaks and valleys without creating tears or pinholes: The requirement that $f(x_0, y_0)$ be defined eliminates the possibility of a hole in the surface $z = f(x, y)$ above the point (x_0, y_0); The requirement that $\displaystyle\lim_{(x,y)\to(x_0,y_0)} f(x, y)$ exists ensures that $z = f(x, y)$ does not become "infinite" at (x_0, y_0) or oscillate wildly; and the requirement that $\displaystyle\lim_{(x,y)\to(x_0,y_0)} f(x, y) = f(x_0, y_0)$ ensures that the surface does not have a vertical jump or step above the point (x_0, y_0). (See Figure 16.2.6.)

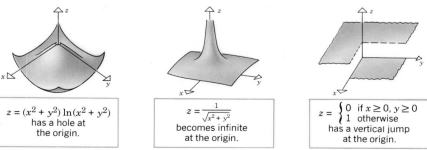

$z = (x^2 + y^2)\ln(x^2 + y^2)$
has a hole at
the origin.

$z = \dfrac{-1}{\sqrt{x^2 + y^2}}$
becomes infinite
at the origin.

$z = \begin{cases} 0 & \text{if } x \geq 0,\, y \geq 0 \\ 1 & \text{otherwise} \end{cases}$
has a vertical jump
at the origin.

Figure 16.2.6

☐ COMPOSITIONS OF
CONTINUOUS FUNCTIONS

The following theorem, which we state without proof, will help us to identify continuous functions of two variables.

16.2.7 THEOREM.

(a)   *If g and h are continuous functions of one variable, then $f(x, y) = g(x)h(y)$ is a continuous function of x and y.*

(b)   *If g is a continuous function of one variable and h is a continuous function of two variables, then their composition $f(x, y) = g(h(x, y))$ is a continuous function of x and y.*

Example 6   The function $f(x, y) = 3x^2y^5$ is continuous because it is the product of the continuous functions $g(x) = 3x^2$ and $h(y) = y^5$.

In general, any function of the form $f(x, y) = Ax^my^n$ (*m* and *n* nonnegative integers) is continuous because it is the product of the continuous functions Ax^m and y^n.   ◀

Example 7   The function $g(x) = \sin x$ is a continuous function of one variable and the function $h(x, y) = xy^2$ is a continuous function of two variables, so $g(h(x, y)) = g(xy^2) = \sin(xy^2)$ is a continuous function of *x* and *y*. By a similar argument, each of the following is continuous:

$$(x^4y^5)^{1/3}, \quad e^{xy}, \quad \cosh(x^3y) \quad ◀$$

Example 8   By Example 7, e^{xy} is continuous. Thus, $\cos(e^{xy})$ is continuous by part (*b*) of Theorem 16.2.7.   ◀

Theorem 16.2.7 is one of a whole class of theorems about continuity of functions in any number of variables. The content of these theorems can be summarized informally with three basic principles:

- A composition of continuous functions is continuous.
- A sum, difference, or product of continuous functions is continuous.
- A quotient of continuous functions is continuous, except where the denominator is zero.

Example 9   The following functions are continuous since they are sums, differences, products, and compositions of continuous functions:

$$3 - 2x^2yz + 9x^4y^8z^3, \quad e^{xy}\cos(xy^2 + 1), \quad (3z + ye^x)^{17} \quad ◀$$

Example 10   Since the function

$$f(x, y) = \frac{x^3y^2}{1 - xy}$$

is a quotient of continuous functions, it is continuous except where $1 - xy = 0$. Thus, $f(x, y)$ is continuous everywhere except on the hyperbola $xy = 1$.   ◀

Example 11 Evaluate

$$\lim_{(x,y)\to(-1,2)} \frac{xy}{x^2 + y^2}$$

Solution. Since $f(x, y) = xy/(x^2 + y^2)$ is continuous at $(-1, 2)$ (why?), it follows from part 3 of Definition 16.2.5 that

$$\lim_{(x,y)\to(-1,2)} \frac{xy}{x^2 + y^2} = \frac{(-1)(2)}{(-1)^2 + (2)^2} = -\frac{2}{5} \quad \blacktriangleleft$$

☐ **LIMITS AT POINTS OF DISCONTINUITY**

At a point of discontinuity the method of Example 11 cannot be used. However, sometimes such limits can be obtained by converting the given function to polar coordinates.

Example 12 Find

$$\lim_{(x,y)\to(0,0)} (x^2 + y^2) \ln (x^2 + y^2)$$

Solution. Let (r, θ) be the coordinates of the point (x, y) with $r \geq 0$. Then we have

$$x = r \cos \theta, \quad y = r \sin \theta, \quad r^2 = x^2 + y^2$$

Moreover, since $r \geq 0$ we have $r = \sqrt{x^2 + y^2}$, so that $r \to 0^+$ if and only if $(x, y) \to (0, 0)$. Thus, we can rewrite the given limit as

$$\lim_{(x,y)\to(0,0)} (x^2 + y^2) \ln (x^2 + y^2) = \lim_{r\to0^+} r^2 \ln r^2$$

$$= \lim_{r\to0^+} \frac{2 \ln r}{1/r^2} \quad \boxed{\text{This converts the limit to an indeterminate form of type } \infty/\infty.}$$

$$= \lim_{r\to0^+} \frac{2/r}{-2/r^3} \quad \boxed{\text{L'Hôpital's rule}}$$

$$= \lim_{r\to0^+} (-r^2) = 0 \quad \blacktriangleleft$$

REMARK. The function $f(x, y) = (x^2 + y^2) \ln (x^2 + y^2)$ from this example is graphed in Figure 16.2.6.

▶ **Exercise Set 16.2**

In Exercises 1–8, sketch the region where the function f is continuous.

1. $f(x, y) = y \ln (1 + x)$.

2. $f(x, y) = \sqrt{x - y}$.

3. $f(x, y) = \dfrac{x^2y}{\sqrt{25 - x^2 - y^2}}$.

4. $f(x, y) = \ln (2x - y + 1)$.

5. $f(x, y) = \cos \left(\dfrac{xy}{1 + x^2 + y^2} \right)$.

6. $f(x, y) = e^{(1-xy)}$.

7. $f(x, y) = \sin^{-1} (xy)$.

8. $f(x, y) = \tan^{-1} (y - x)$.

In Exercises 9–12, describe the region where f is continuous.

9. $f(x, y, z) = 3x^2 e^{yz} \cos(xyz)$.

10. $f(x, y, z) = \ln(4 - x^2 - y^2 - z^2)$.

11. $f(x, y, z) = \dfrac{y + 1}{x^2 + z^2 - 1}$.

12. $f(x, y, z) = \sin \sqrt{x^2 + y^2 + 3z^2}$.

In Exercises 13–35, find the limit, if it exists.

13. $\lim\limits_{(x,y) \to (1,3)} (4xy^2 - x)$.

14. $\lim\limits_{(x,y) \to (1/2,\pi)} (xy^2 \sin xy)$.

15. $\lim\limits_{(x,y) \to (-1,2)} \dfrac{xy^3}{x + y}$.

16. $\lim\limits_{(x,y) \to (1,-3)} e^{2x - y^2}$.

17. $\lim\limits_{(x,y) \to (0,0)} \ln(1 + x^2 y^3)$.

18. $\lim\limits_{(x,y) \to (4,-2)} x \sqrt[3]{y^3 + 2x}$.

19. $\lim\limits_{(x,y) \to (0,0)} \dfrac{x - y}{x^2 + y^2}$. [*Hint:* Let $(x, y) \to (0, 0)$ along the line $y = 0$.]

20. $\lim\limits_{(x,y) \to (0,0)} \dfrac{3}{x^2 + 2y^2}$.

21. $\lim\limits_{(x,y) \to (0,0)} \dfrac{\sin(x^2 + y^2)}{x^2 + y^2}$.

22. $\lim\limits_{(x,y) \to (0,0)} \dfrac{1 - \cos(x^2 + y^2)}{x^2 + y^2}$.

23. $\lim\limits_{(x,y) \to (0,0)} \dfrac{x^4 - y^4}{x^2 + y^2}$.

24. $\lim\limits_{(x,y) \to (0,0)} \dfrac{x^4 - 16y^4}{x^2 + 4y^2}$.

25. $\lim\limits_{(x,y) \to (0,0)} \dfrac{xy}{3x^2 + 2y^2}$.

26. $\lim\limits_{(x,y) \to (0,0)} \dfrac{1 - x^2 - y^2}{x^2 + y^2}$.

27. $\lim\limits_{(x,y) \to (0,0)} e^{-1/(x^2 + y^2)}$.

28. $\lim\limits_{(x,y) \to (0,0)} \dfrac{e^{-1/\sqrt{x^2 + y^2}}}{\sqrt{x^2 + y^2}}$.

29. $\lim\limits_{(x,y) \to (0,0)} y \ln(x^2 + y^2)$.

30. $\lim\limits_{(x,y) \to (0,0)} x \ln(|x| + |y|)$.

31. $\lim\limits_{(x,y,z) \to (2,-1,2)} \dfrac{xz^2}{\sqrt{x^2 + y^2 + z^2}}$.

32. $\lim\limits_{(x,y,z) \to (2,0,-1)} \ln(2x + y - z)$.

33. $\lim\limits_{(x,y,z) \to (0,0,0)} \dfrac{\sin(x^2 + y^2 + z^2)}{\sqrt{x^2 + y^2 + z^2}}$.

34. $\lim\limits_{(x,y,z) \to (0,0,0)} \dfrac{\sin \sqrt{x^2 + y^2 + z^2}}{x^2 + y^2 + z^2}$.

35. $\lim\limits_{(x,y,z) \to (0,0,0)} \dfrac{yz}{x^2 + y^2 + z^2}$.
[*Hint:* First let $(x, y, z) \to (0, 0, 0)$ along the z-axis and then along the line $x = t, y = t, z = t$.]

36. Show that $\dfrac{xy}{x^2 + y^2}$ can be made to approach any value in the interval $[-\frac{1}{2}, \frac{1}{2}]$ by letting $(x, y) \to (0, 0)$ along some line $y = mx$.

37. (a) Show that the value of $\dfrac{x^2 y}{x^4 + y^2}$ approaches zero as $(x, y) \to (0, 0)$ along any straight line $y = mx$.

(b) Show that $\lim\limits_{(x,y) \to (0,0)} \dfrac{x^2 y}{x^4 + y^2}$ does not exist by letting $(x, y) \to (0, 0)$ along the parabola $y = x^2$.

38. (a) Show that the value of $\dfrac{x^3 y}{2x^6 + y^2}$ approaches 0 as $(x, y) \to (0, 0)$ along any straight line $y = mx$, or along any parabola $y = kx^2$.

(b) Show that $\lim\limits_{(x,y) \to (0,0)} \dfrac{x^3 y}{2x^6 + y^2}$ does not exist by letting $(x, y) \to (0, 0)$ along the curve $y = x^3$.

39. (a) Show that the value of $\dfrac{xyz}{x^2 + y^4 + z^4}$ approaches 0 as $(x, y, z) \to (0, 0, 0)$ along any line $x = at, y = bt, z = ct$.

(b) Show that $\lim\limits_{(x,y,z) \to (0,0,0)} \dfrac{xyz}{x^2 + y^4 + z^4}$ does not exist by letting $(x, y, z) \to (0, 0, 0)$ along the curve $x = t^2, y = t, z = t$.

40. Find
$$\lim\limits_{(x,y) \to (0,1)} \tan^{-1} \left[\frac{x^2 + 1}{x^2 + (y - 1)^2} \right]$$

41. Find
$$\lim\limits_{(x,y) \to (0,1)} \tan^{-1} \left[\frac{x^2 - 1}{x^2 + (y - 1)^2} \right]$$

42. Let $f(x, y) = \begin{cases} \dfrac{\sin(x^2 + y^2)}{x^2 + y^2}, & (x, y) \neq (0, 0) \\ 1, & (x, y) = (0, 0). \end{cases}$

Show that f is continuous at $(0, 0)$.

43. Let $f(x, y) = \dfrac{x^2}{x^2 + y^2}$. Is it possible to define $f(0, 0)$ so that f will be continuous at $(0, 0)$?

44. Let $f(x, y) = xy \ln(x^2 + y^2)$. Is it possible to define $f(0, 0)$ so that f will be continuous at $(0, 0)$?

In Exercises 45 and 46, use Definition 16.2.1 to prove the given statement. [*Hint:* Let $r = \sqrt{x^2 + y^2}$.]

45. $\displaystyle\lim_{(x,y)\to(0,0)} (x^2 + y^2) = 0$.

46. $\displaystyle\lim_{(x,y)\to(0,0)} \frac{x^2 y^2}{\sqrt{x^2 + y^2}} = 0$.

In Exercises 47 and 48, use Definition 16.2.2 to prove the given statement. [*Hint:* Let $\rho = \sqrt{x^2 + y^2 + z^2}$.]

47. $\displaystyle\lim_{(x,y,z)\to(0,0,0)} (x^2 + y^2 + z^2) = 0$.

48. $\displaystyle\lim_{(x,y,z)\to(0,0,0)} e^{\sqrt{x^2+y^2+z^2}} = 1$.

■ 16.3 PARTIAL DERIVATIVES

*If f is a function of two or more independent variables and all but one of those variables are held fixed, then the derivative of f with respect to that one remaining independent variable is called a **partial derivative** of f. In this section we shall show how to compute partial derivatives and discuss their geometric significance.*

□ **PARTIAL DERIVATIVES OF FUNCTIONS OF TWO VARIABLES**

Let f be a function of x and y. If we hold y constant, say $y = y_0$, and view x as a variable, then $f(x, y_0)$ is a function of x alone. If this function is differentiable at $x = x_0$, then the value of this derivative is denoted by

$$f_x(x_0, y_0) \tag{1}$$

and is called the **partial derivative of f with respect to x** at the point (x_0, y_0). Similarly, if we hold x constant, say $x = x_0$, then $f(x_0, y)$ is a function of y alone. If this function is differentiable at $y = y_0$, then the value of this derivative is denoted by

$$f_y(x_0, y_0) \tag{2}$$

and is called the **partial derivative of f with respect to y** at (x_0, y_0).

The partial derivatives of $f(x, y)$ have a simple geometric interpretation. Let P be a point on the surface

$$z = f(x, y)$$

If y is held constant, say $y = y_0$, and x is allowed to vary, then the point P moves along the curve C_1 that is the intersection of the surface with the vertical plane $y = y_0$ (Figure 16.3.1a). Thus, the partial derivative $f_x(x_0, y_0)$ can be interpreted as the slope (change in z per unit increase in x) of the tangent line to the curve C_1 at the point (x_0, y_0). Similarly, if x is held constant, say $x = x_0$, and y is allowed to vary, then the point P moves along the curve C_2 that is the intersection of the surface with the vertical plane $x = x_0$. Thus, the partial derivative $f_y(x_0, y_0)$ can be interpreted as the slope of the tangent line (change in z per unit increase in y) to the curve C_2 at the point (x_0, y_0) (Figure 16.3.1b).

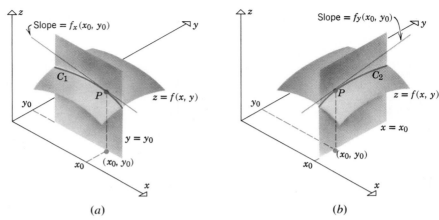

Figure 16.3.1

REMARK.   In this text we shall only consider partial derivatives at *interior points* of the domain of f. Partial derivatives at boundary points lead to complications that are best left for more advanced courses.

The values of $f_x(x_0, y_0)$ and $f_y(x_0, y_0)$ are usually obtained by finding expressions for $f_x(x, y)$ and $f_y(x, y)$ at a general point (x, y) and then substituting $x = x_0$ and $y = y_0$ in these expressions. To obtain $f_x(x, y)$ we differentiate $f(x, y)$ with respect to x, *treating y as a constant;* and to obtain $f_y(x, y)$ we differentiate $f(x, y)$ with respect to y, *treating x as a constant.*

Example 1   Find $f_x(1, 2)$ and $f_y(1, 2)$ if $f(x, y) = 2x^3y^2 + 2y + 4x$.

Solution.   Treating y as a constant and differentiating with respect to x, we obtain

$$f_x(x, y) = 6x^2y^2 + 4$$

Treating x as a constant and differentiating with respect to y, we obtain

$$f_y(x, y) = 4x^3y + 2$$

Substituting $x = 1$ and $y = 2$ in these partial-derivative formulas yields

$$f_x(1, 2) = 6(1)^2(2)^2 + 4 = 28$$
$$f_y(1, 2) = 4(1)^3(2) + 2 = 10 \quad \blacktriangleleft$$

The partial derivatives f_x and f_y are also denoted by the symbols*

$$\frac{\partial f}{\partial x} \quad \text{and} \quad \frac{\partial f}{\partial y}$$

_____

*The symbol ∂ is called a partial derivative sign. It is derived from the Cyrillic alphabet.

and if a dependent variable $z = f(x, y)$ is introduced, then the symbols

$$\frac{\partial z}{\partial x} \quad \text{and} \quad \frac{\partial z}{\partial y}$$

may be used. Some typical notations for the partial derivatives at a point (x_0, y_0) are

$$\left.\frac{\partial f}{\partial x}\right|_{x=x_0,\, y=y_0} \qquad \left.\frac{\partial z}{\partial y}\right|_{(x_0,\, y_0)} \qquad \left.\frac{\partial f}{\partial x}\right|_{(x_0,\, y_0)} \qquad \frac{\partial f}{\partial x}(x_0, y_0)$$

Example 2 Find $\partial z/\partial x$ and $\partial z/\partial y$ if $z = x^4 \sin(xy^3)$.

Solution.

$$\frac{\partial z}{\partial x} = \frac{\partial}{\partial x}[x^4 \sin(xy^3)] = x^4 \frac{\partial}{\partial x}[\sin(xy^3)] + \sin(xy^3) \cdot \frac{\partial}{\partial x}(x^4)$$

$$= x^4 \cos(xy^3) \cdot y^3 + \sin(xy^3) \cdot 4x^3 = x^4 y^3 \cos(xy^3) + 4x^3 \sin(xy^3)$$

$$\frac{\partial z}{\partial y} = \frac{\partial}{\partial y}[x^4 \sin(xy^3)] = x^4 \frac{\partial}{\partial y}[\sin(xy^3)] + \sin(xy^3) \cdot \frac{\partial}{\partial y}(x^4)$$

$$= x^4 \cos(xy^3) \cdot 3xy^2 + \sin(xy^3) \cdot 0 = 3x^5 y^2 \cos(xy^3) \quad \blacktriangleleft$$

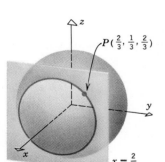

$P\left(\frac{2}{3}, \frac{1}{3}, \frac{2}{3}\right)$

$x = \frac{2}{3}$

Figure 16.3.2

Example 3 Suppose that a point Q moves along the intersection of the sphere $x^2 + y^2 + z^2 = 1$ with the plane $x = \frac{2}{3}$. At what rate is z changing with respect to y when the point is at $P(\frac{2}{3}, \frac{1}{3}, \frac{2}{3})$?

Solution. Since the z-coordinate of the point $P(\frac{2}{3}, \frac{1}{3}, \frac{2}{3})$ is positive, this point lies on the upper hemisphere

$$z = \sqrt{1 - x^2 - y^2} \tag{3}$$

(Figure 16.3.2). Since $x = \frac{2}{3}$, $y = \frac{1}{3}$ at P, the rate at which z changes with y at this point (as Q moves along the circle of intersection) is

$$\left.\frac{\partial z}{\partial y}\right|_{x=\frac{2}{3},\, y=\frac{1}{3}}$$

From (3) we obtain

$$\frac{\partial z}{\partial y} = \frac{\partial}{\partial y}[(1 - x^2 - y^2)^{1/2}] = \frac{1}{2}(1 - x^2 - y^2)^{-1/2}(-2y) = -\frac{y}{\sqrt{1 - x^2 - y^2}}$$

so that

$$\left.\frac{\partial z}{\partial y}\right|_{x=\frac{2}{3},\, y=\frac{1}{3}} = -\frac{\frac{1}{3}}{\sqrt{1 - (\frac{2}{3})^2 - (\frac{1}{3})^2}} = -\frac{1}{2}$$

Alternative Solution.   Instead of solving $x^2 + y^2 + z^2 = 1$ explicitly for z as a function of x and y, we can obtain $\partial z/\partial y$ by implicit differentiation. Differentiating both sides of $x^2 + y^2 + z^2 = 1$ with respect to y and treating z as a function of x and y yields

$$\frac{\partial}{\partial y}[x^2 + y^2 + z^2] = \frac{\partial}{\partial y}[1]$$

$$2y + 2z\frac{\partial z}{\partial y} = 0$$

$$\frac{\partial z}{\partial y} = -\frac{y}{z}$$

Substituting the y- and z-coordinates of the point $(\frac{2}{3}, \frac{1}{3}, \frac{2}{3})$ yields

$$\frac{\partial z}{\partial y} = -\frac{\frac{1}{3}}{\frac{2}{3}} = -\frac{1}{2}$$

which agrees with our previous result.   ◀

Example 4   According to the ideal gas law of physics, the pressure P exerted by a confined gas is related to its volume V and its temperature T by

$$P = k\frac{T}{V}$$

where k is a constant depending on the amount of gas present and the units of measurement. Show that if the temperature of such a gas remains constant, then the pressure and volume satisfy the condition

$$\frac{\partial P}{\partial V} = -\frac{P}{V} \tag{4}$$

Solution.   Treating T as a constant we obtain

$$\frac{\partial P}{\partial V} = \frac{\partial}{\partial V}\left(k\frac{T}{V}\right) = (kT)\left(-\frac{1}{V^2}\right) = -k\frac{T}{V}\cdot\frac{1}{V} = -\frac{P}{V}   ◀$$

Example 5   The diagonal D of a rectangle is given by $D = \sqrt{x^2 + y^2}$, where x and y are the lengths of the sides.

(a)   Find a formula for the instantaneous rate of change of D with respect to x if x varies while y remains constant.

(b)   Suppose $y = 4$ in. Find the rate of change of D with respect to x at the instant when $x = 3$ in.

Solution (a).

$$\frac{\partial D}{\partial x} = \frac{1}{2}(x^2 + y^2)^{-1/2}(2x) = \frac{x}{\sqrt{x^2 + y^2}}$$

Solution (b).

$$\frac{\partial D}{\partial x}\bigg|_{x=3, \, y=4} = \frac{3}{\sqrt{3^2 + 4^2}} = \frac{3}{5}$$

Thus, the diagonal D is increasing at a rate of $\frac{3}{5}$ inches per inch increase in x. ◀

◻ **HIGHER-ORDER PARTIAL DERIVATIVES**

Since the partial derivatives $\partial f/\partial x$ and $\partial f/\partial y$ are functions of x and y, each can have partial derivatives. This gives rise to four possible *second-order* partial derivatives of f, which are defined by

$$\frac{\partial^2 f}{\partial x^2} = \frac{\partial}{\partial x}\left(\frac{\partial f}{\partial x}\right), \quad \frac{\partial^2 f}{\partial y^2} = \frac{\partial}{\partial y}\left(\frac{\partial f}{\partial y}\right)$$

$$\frac{\partial^2 f}{\partial x \, \partial y} = \frac{\partial}{\partial x}\left(\frac{\partial f}{\partial y}\right), \quad \frac{\partial^2 f}{\partial y \, \partial x} = \frac{\partial}{\partial y}\left(\frac{\partial f}{\partial x}\right)$$

Henceforth, we shall call $\partial f/\partial x$ and $\partial f/\partial y$ the **first-order** partial derivatives of f.

Example 6 Find the second-order partial derivatives of the function $f(x, y) = x^2 y^3 + x^4 y$.

Solution. We have

$$\frac{\partial f}{\partial x} = 2xy^3 + 4x^3 y \quad \text{and} \quad \frac{\partial f}{\partial y} = 3x^2 y^2 + x^4$$

so that

$$\frac{\partial^2 f}{\partial x^2} = \frac{\partial}{\partial x}\left(\frac{\partial f}{\partial x}\right) = \frac{\partial}{\partial x}(2xy^3 + 4x^3 y) = 2y^3 + 12x^2 y$$

$$\frac{\partial^2 f}{\partial y^2} = \frac{\partial}{\partial y}\left(\frac{\partial f}{\partial y}\right) = \frac{\partial}{\partial y}(3x^2 y^2 + x^4) = 6x^2 y$$

$$\frac{\partial^2 f}{\partial x \, \partial y} = \frac{\partial}{\partial x}\left(\frac{\partial f}{\partial y}\right) = \frac{\partial}{\partial x}(3x^2 y^2 + x^4) = 6xy^2 + 4x^3$$

$$\frac{\partial^2 f}{\partial y \, \partial x} = \frac{\partial}{\partial y}\left(\frac{\partial f}{\partial x}\right) = \frac{\partial}{\partial y}(2xy^3 + 4x^3 y) = 6xy^2 + 4x^3 \quad ◀$$

REMARK. The derivatives

$$\frac{\partial^2 f}{\partial y \, \partial x} \quad \text{and} \quad \frac{\partial^2 f}{\partial x \, \partial y}$$

are called the *mixed second-order partial derivatives* or *mixed second-partials*. For most functions that arise in applications, these mixed partial derivatives are equal (as in the last example). In the next section we shall state precise conditions under which equality holds.

By successively differentiating, we can obtain third-order partial derivatives and higher. Some possibilities are

$$\frac{\partial^3 f}{\partial x^3} = \frac{\partial}{\partial x}\left(\frac{\partial^2 f}{\partial x^2}\right), \quad \frac{\partial^3 f}{\partial y^2\,\partial x} = \frac{\partial}{\partial y}\left(\frac{\partial^2 f}{\partial y\,\partial x}\right)$$

$$\frac{\partial^3 f}{\partial y\,\partial x^2} = \frac{\partial}{\partial y}\left(\frac{\partial^2 f}{\partial x^2}\right), \quad \frac{\partial^4 f}{\partial y^2\,\partial x^2} = \frac{\partial}{\partial y}\left(\frac{\partial^3 f}{\partial y\,\partial x^2}\right)$$

Higher-order partial derivatives can be denoted more compactly with subscript notation. For example,

$$\frac{\partial^2 f}{\partial y\,\partial x} = \frac{\partial}{\partial y}\left(\frac{\partial f}{\partial x}\right) = \frac{\partial}{\partial y}(f_x) = (f_x)_y$$

It is usual to drop the parentheses and write simply

$$\frac{\partial^2 f}{\partial y\,\partial x} = f_{xy}$$

Note that in "∂" notation the sequence of differentiations is obtained by reading from right to left, but in the subscript notation it is left to right. Some other examples are

$$f_{xx} = \frac{\partial^2 f}{\partial x^2}, \quad f_{yyx} = \frac{\partial^3 f}{\partial x\,\partial y^2}, \quad f_{xxyy} = \frac{\partial^4 f}{\partial y^2\,\partial x^2}$$

Example Let $f(x, y) = y^2 e^x + y$. Find f_{xyy}.

Solution.

$$f_{xyy} = \frac{\partial^3 f}{\partial y^2\,\partial x} = \frac{\partial^2}{\partial y^2}\left(\frac{\partial f}{\partial x}\right) = \frac{\partial^2}{\partial y^2}(y^2 e^x) = \frac{\partial}{\partial y}(2ye^x) = 2e^x \quad ◀$$

☐ **PARTIAL DERIVATIVES OF FUNCTIONS WITH MORE THAN TWO VARIABLES**

For a function $f(x, y, z)$ of three variables, there are three *partial derivatives:*

$$f_x(x, y, z), \quad f_y(x, y, z), \quad f_z(x, y, z)$$

The partial derivative f_x is calculated by holding y and z constant and differentiating with respect to x. For f_y the variables x and z are held constant, and for f_z the variables x and y are held constant. If a dependent variable

$$w = f(x, y, z)$$

is used, then the three partial derivatives of f may be denoted by

$$\frac{\partial w}{\partial x}, \quad \frac{\partial w}{\partial y}, \quad \text{and} \quad \frac{\partial w}{\partial z}$$

Example 8 If $f(x, y, z) = x^3 y^2 z^4 + 2xy + z$, then

$$f_x(x, y, z) = 3x^2 y^2 z^4 + 2y$$
$$f_y(x, y, z) = 2x^3 y z^4 + 2x$$
$$f_z(x, y, z) = 4x^3 y^2 z^3 + 1$$
$$f_z(-1, 1, 2) = 4(-1)^3(1)^2(2)^3 + 1 = -31 \quad \blacktriangleleft$$

Example 9 If $f(\rho, \theta, \phi) = \rho^2 \cos \phi \sin \theta$, then

$$f_\rho(\rho, \theta, \phi) = 2\rho \cos \phi \sin \theta$$
$$f_{\rho\phi}(\rho, \theta, \phi) = -2\rho \sin \phi \sin \theta$$
$$f_{\rho\phi\theta}(\rho, \theta, \phi) = -2\rho \sin \phi \cos \theta \quad \blacktriangleleft$$

In general, if $f(v_1, v_2, \ldots, v_n)$ is a function of n variables, there are n partial derivatives of f, each of which is obtained by holding $n - 1$ of the variables fixed and differentiating the function f with respect to the remaining variable. If $w = f(v_1, v_2, \ldots, v_n)$, then these partial derivatives are denoted by

$$\frac{\partial w}{\partial v_1}, \frac{\partial w}{\partial v_2}, \ldots, \frac{\partial w}{\partial v_n}$$

where $\partial w / \partial v_i$ is obtained by holding all variables except v_i fixed and differentiating with respect to v_i.

Example 10 Find

$$\frac{\partial}{\partial x_i} [\sqrt{x_1^2 + x_2^2 + \cdots + x_n^2}]$$

for $i = 1, 2, \ldots, n$.

Solution. For each $i = 1, 2, \ldots, n$ we obtain

$$\frac{\partial}{\partial x_i} [\sqrt{x_1^2 + x_2^2 + \cdots + x_n^2}] = \frac{1}{2\sqrt{x_1^2 + x_2^2 + \cdots + x_n^2}} \cdot \frac{\partial}{\partial x_i} [x_1^2 + x_2^2 + \cdots + x_n^2]$$

$$= \frac{1}{2\sqrt{x_1^2 + x_2^2 + \cdots + x_n^2}} [2x_i] \qquad \boxed{\text{All terms except } x_i^2 \text{ are constant.}}$$

$$= \frac{x_i}{\sqrt{x_1^2 + x_2^2 + \cdots + x_n^2}} \quad \blacktriangleleft$$

▶ Exercise Set 16.3 $\boxed{C}$ 59, 60

In Exercises 1–6, find $\partial z/\partial x$ and $\partial z/\partial y$.

1. $z = 3x^3y^2$.

2. $z = 4x^2 - 2y + 7x^4y^5$.

3. $z = 4e^{x^2y^3}$.   4. $z = \cos(x^5y^4)$.

5. $z = x^3 \ln(1 + xy^{-3/5})$.   6. $z = e^{xy} \sin 4y^2$.

In Exercises 7–12, find $f_x(x, y)$ and $f_y(x, y)$.

7. $f(x, y) = \sqrt{3x^5y - 7x^3y}$.

8. $f(x, y) = \dfrac{x + y}{x - y}$.

9. $f(x, y) = y^{-3/2} \tan^{-1}(x/y)$.

10. $f(x, y) = x^3 e^{-y} + y^3 \sec \sqrt{x}$.

11. $f(x, y) = (y^2 \tan x)^{-4/3}$.

12. $f(x, y) = \cosh(\sqrt{x}) \sinh^2(xy^2)$.

13. Given $f(x, y) = 9 - x^2 - 7y^3$, find
    (a) $f_x(3, 1)$ (b) $f_y(3, 1)$.

14. Given $f(x, y) = x^2ye^{xy}$, find
    (a) $\left. \dfrac{\partial f}{\partial x} \right|_{(1, 1)}$ (b) $\left. \dfrac{\partial f}{\partial y} \right|_{(1, 1)}$.

15. Given $z = \sqrt{x^2 + 4y^2}$, find
    (a) $\left. \dfrac{\partial z}{\partial x} \right|_{(1, 2)}$ (b) $\left. \dfrac{\partial z}{\partial y} \right|_{(1, 2)}$.

16. Given $w = x^2 \cos xy$, find
    (a) $\dfrac{\partial w}{\partial x}(\tfrac{1}{2}, \pi)$ (b) $\dfrac{\partial w}{\partial y}(\tfrac{1}{2}, \pi)$.

In Exercises 17–20, calculate $\partial z/\partial x$ and $\partial z/\partial y$ using implicit differentiation. Leave your answers in terms of x, y, and z.

17. $(x^2 + y^2 + z^2)^{3/2} = 1$.

18. $\ln(2x^2 + y - z^3) = x$.

19. $x^2 + z \sin xyz = 0$.

20. $e^{xy} \sinh z - z^2x + 1 = 0$.

In Exercises 21–26, find f_{xx}, f_{yy}, f_{xy}, and f_{yx}.

21. $f(x, y) = 4x^2 - 8xy^4 + 7y^5 - 3$.

22. $f(x, y) = \sqrt{x^2 + y^2}$.   23. $f(x, y) = e^x \cos y$.

24. $f(x, y) = e^{x-y^2}$.   25. $f(x, y) = \ln(4x - 5y)$.

26. $f(x, y) = (x^2 - y^2)/(x^2 + y^2)$.

27. Given $f(x, y) = x^3y^5 - 2x^2y + x$, find
    (a) f_{xxy} (b) f_{yxy} (c) f_{yyy}.

28. Given $z = (2x - y)^5$, find
    (a) $\dfrac{\partial^3 z}{\partial y \, \partial x \, \partial y}$ (b) $\dfrac{\partial^3 z}{\partial x^2 \, \partial y}$ (c) $\dfrac{\partial^4 z}{\partial x^2 \, \partial y^2}$.

29. Given $f(x, y) = y^3 e^{-5x}$, find
    (a) $f_{xyy}(0, 1)$ (b) $f_{xxx}(0, 1)$ (c) $f_{yyxx}(0, 1)$.

30. Given $w = e^y \cos x$, find
    (a) $\left. \dfrac{\partial^3 w}{\partial y^2 \, \partial x} \right|_{(\pi/4, 0)}$ (b) $\left. \dfrac{\partial^3 w}{\partial x^2 \, \partial y} \right|_{(\pi/4, 0)}$.

31. Express the following derivatives in "∂" notation.
    (a) f_{xxx} (b) f_{xyy} (c) f_{yyxx} (d) f_{xyyy}.

32. Express the following derivatives in "subscript" notation.
    (a) $\dfrac{\partial^3 f}{\partial y^2 \, \partial x}$ (b) $\dfrac{\partial^4 f}{\partial x^4}$
    (c) $\dfrac{\partial^4 f}{\partial y^2 \, \partial x^2}$ (d) $\dfrac{\partial^5 f}{\partial x^2 \, \partial y^3}$.

In Exercises 33–37, find $\partial w/\partial x$, $\partial w/\partial y$, and $\partial w/\partial z$.

33. $w = x^2y^4z^3 + xy + z^2 + 1$.

34. $w = ye^z \sin x$.   35. $w = \dfrac{x^2 - y^2}{y^2 + z^2}$.

36. $w = y^3 e^{2x+3z}$.   37. $w = \sqrt{x^2 + y^2 + z^2}$.

In Exercises 38–42, find f_x, f_y, and f_z.

38. $f(x, y, z) = z \ln(x^2y \cos z)$.

39. $f(x, y, z) = \tan^{-1}\left(\dfrac{1}{xy^2z^3}\right)$.

40. $f(x, y, z) = y^{-3/2} \sec\left(\dfrac{xz}{y}\right)$.

41. $f(x, y, z) = \cosh(\sqrt{z}) \sinh^2(x^2yz)$.

42. $f(x, y, z) = \left(\dfrac{xz}{1 - z^2 - y^2}\right)^{-3/4}$.

43. Let $f(x, y, z) = 5x^2yz^3$. Find
    (a) $f_x(1, -1, 2)$ (b) $f_y(1, -1, 2)$
    (c) $f_z(1, -1, 2)$.

44. Let $f(x, y, z) = y^2e^{xz}$. Find
    (a) $\partial f/\partial x|_{(1, 1, 1)}$ (b) $\partial f/\partial y|_{(1, 1, 1)}$
    (c) $\partial f/\partial z|_{(1, 1, 1)}$.

45. Let $w = \sqrt{x^2 + 4y^2 - z^2}$. Find

    (a) $\partial w/\partial x|_{(2, 1, -1)}$ (b) $\partial w/\partial y|_{(2, 1, -1)}$

    (c) $\partial w/\partial z|_{(2, 1, -1)}$.

46. Let $w = x \sin xyz$. Find

    (a) $\dfrac{\partial w}{\partial x}(1, \tfrac{1}{2}, \pi)$ (b) $\dfrac{\partial w}{\partial y}(1, \tfrac{1}{2}, \pi)$

    (c) $\dfrac{\partial w}{\partial z}(1, \tfrac{1}{2}, \pi)$.

In Exercises 47–50, find $\partial w/\partial x$, $\partial w/\partial y$, and $\partial w/\partial z$ using implicit differentiation. Leave your answers in terms of x, y, z, and w.

---

47. $(x^2 + y^2 + z^2 + w^2)^{3/2} = 4$.

48. $\ln(2x^2 + y - z^3 + 3w) = z$.

49. $w^2 + w \sin xyz = 0$.

50. $e^{xy} \sinh w - z^2 w + 1 = 0$.

51. Let $f(x, y, z) = x^3 y^5 z^7 + xy^2 + y^3 z$. Find

    (a) f_{xy} (b) f_{yz} (c) f_{xz} (d) f_{zz}

    (e) f_{zyy} (f) f_{xxy} (g) f_{zyx} (h) f_{xxyz}.

52. Let $w = (4x - 3y + 2z)^5$. Find

    (a) $\dfrac{\partial^2 w}{\partial x\, \partial z}$ (b) $\dfrac{\partial^3 w}{\partial x\, \partial y\, \partial z}$ (c) $\dfrac{\partial^4 w}{\partial z^2\, \partial y\, \partial x}$.

53. Show that the following functions satisfy

$$\frac{\partial^2 f}{\partial x^2} + \frac{\partial^2 f}{\partial y^2} = 0$$

(This is called *Laplace's equation*.)

    (a) $f(x, y) = e^x \sin y + e^y \cos x$

    (b) $f(x, y) = \ln(x^2 + y^2)$

    (c) $f(x, y) = \tan^{-1}\dfrac{2xy}{x^2 - y^2}$.

54. Show that $u(x, y)$ and $v(x, y)$ satisfy

$$\frac{\partial u}{\partial x} = \frac{\partial v}{\partial y} \quad \text{and} \quad \frac{\partial u}{\partial y} = -\frac{\partial v}{\partial x}$$

(These are called the *Cauchy-Riemann equations*.)

    (a) $u = x^2 - y^2$, $v = 2xy$

    (b) $u = e^x \cos y$, $v = e^x \sin y$

    (c) $u = \ln(x^2 + y^2)$, $v = 2 \tan^{-1}(y/x)$.

55. A point moves along the intersection of the elliptic paraboloid $z = x^2 + 3y^2$ and the plane $x = 2$. At what rate is z changing with y when the point is at $(2, 1, 7)$?

56. A point moves along the intersection of the surface $z = \sqrt{29 - x^2 - y^2}$ and the plane $y = 3$. At what rate is z changing with x when the point is at $(4, 3, 2)$?

57. Find the slope of the tangent line at $(-1, 1, 5)$ to the curve of intersection of the surface $z = x^2 + 4y^2$ and

    (a) the plane $x = -1$ (b) the plane $y = 1$.

58. Find the slope of the tangent line at $(2, 1, 2)$ to the curve of intersection of the surface $x^2 + y^2 + z^2 = 9$ and

    (a) the plane $x = 2$ (b) the plane $y = 1$.

59. The volume V of a right-circular cylinder is given by $V = \pi r^2 h$, where r is the radius and h is the height.

    (a) Find a formula for the instantaneous rate of change of V with respect to r if h remains constant.

    (b) Find a formula for the instantaneous rate of change of V with respect to h if r remains constant.

    (c) Suppose h has a constant value of 4 in., but r varies. Find the rate of change of V with respect to r at the instant when $r = 6$ in.

    (d) Suppose r has a constant value of 8 in., but h varies. Find the instantaneous rate of change of V with respect to h at the instant when $h = 10$ in.

60. The volume V of a right-circular cone is given by

$$V = \frac{\pi}{24} d^2 \sqrt{4s^2 - d^2}$$

where s is the slant height and d is the diameter of the base.

    (a) Find a formula for the instantaneous rate of change of V with respect to s if d remains constant.

    (b) Find a formula for the instantaneous rate of change of V with respect to d if s remains constant.

    (c) Suppose d has a constant value of 16 cm, but s varies. Find the rate of change of V with respect to s at the instant when $s = 10$ cm.

    (d) Suppose s has a constant value of 10 cm, but d varies. Find the rate of change of V with respect to d at the instant when $d = 16$ cm.

61. According to the ideal gas law (Example 4), the pressure, temperature, and volume of a gas are related by $P = kT/V$. Suppose that for a certain gas, $k = 10$.

    (a) Find the instantaneous rate of change of pressure (lb/in²) with respect to temperature if the temperature is 80 K and the volume remains fixed at 50 in³.

    (b) Find the instantaneous rate of change of volume with respect to pressure if the volume is 50 in³ and the temperature remains fixed at 80 K.

62. Find parametric equations for the tangent line at $(1, 3, 3)$ to the curve of intersection of the surface $z = x^2y$ and
(a) the plane $x = 1$ (b) the plane $y = 3$.

63. Suppose that $\sin(x + z) + \sin(x - y) = 1$. Use implicit differentiation to find $\partial z/\partial x$, $\partial z/\partial y$, and $\partial^2 z/\partial x \partial y$ in terms of x, y, and z.

64. The volume of a right-circular cone of radius r and height h is $V = \frac{1}{3}\pi r^2 h$. Show that if the height remains constant while the radius changes, then the volume satisfies

$$\frac{\partial V}{\partial r} = \frac{2V}{r}$$

65. The temperature at a point (x, y) on a metal plate in the xy-plane is $T(x, y) = x^3 + 2y^2 + x$ degrees. Assume that distance is measured in centimeters and find the rate at which temperature changes with distance if we start at the point $(1, 2)$ and move
(a)   to the right and parallel to the x-axis
(b)   upward and parallel to the y-axis.

66. When two resistors having resistances R_1 ohms and R_2 ohms are connected in parallel, their combined resistance R in ohms is $R = R_1R_2/(R_1 + R_2)$. Show:

$$\frac{\partial^2 R}{\partial R_1^2}\frac{\partial^2 R}{\partial R_2^2} = \frac{4R^2}{(R_1 + R_2)^4}$$

67. Prove: If $u(x, y)$ and $v(x, y)$ each have equal mixed second partials, and if u and v satisfy the Cauchy-Riemann equations (Exercise 54), then u and v both satisfy Laplace's equation (Exercise 53).

68. Recall that for a function of one variable the derivative $f'(x)$ can be expressed as the limit

$$f'(x) = \lim_{h \to 0}\frac{f(x + h) - f(x)}{h}$$

Express $f_x(x, y)$ and $f_y(x, y)$ as limits.

In Exercises 69 and 70, the figures show some of the isotherms (curves of constant temperature) of a temperature function $T(x, y)$ for a thin metal plate in the xy-plane. Based on the figures, determine at the point P
(a)   the signs of $\partial T/\partial x$ and $\partial T/\partial y$, and which is largest in absolute value
(b)   the signs of $\partial^2 T/\partial x^2$, $\partial^2 T/\partial y^2$, $\partial^2 T/\partial y\partial x$, and $\partial^2 T/\partial x\partial y$.

69.

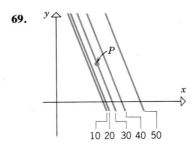

70.

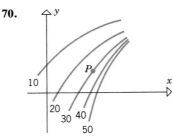

71. Let $f(x, y) = (x^2 + y^2)^{2/3}$. Show that

$$f_x(x, y) = \begin{cases} \dfrac{4x}{3(x^2 + y^2)^{1/3}}, & (x, y) \neq (0, 0) \\ 0, & (x, y) = (0, 0) \end{cases}$$

[This problem, due to Don Cohen, appeared in *Mathematics and Computer Education*, Vol. 25, No. 2, 1991, p. 179.]

72. Let $f(x, y) = (x^3 + y^3)^{1/3}$.
(a)   Show that $f_y(0, 0) = 1$.
(b)   At what points, if any, does $f_y(x, y)$ fail to exist?

■ **16.4** DIFFERENTIABILITY AND CHAIN RULES FOR FUNCTIONS OF TWO VARIABLES

In this section we shall extend the notion of differentiability to functions of two variables and derive versions of the chain rule for these functions. We have restricted the discussion in this section to functions of two variables because some of the results we shall discuss have geometric interpretations that only apply to such functions. In a later section we shall extend the concepts developed here to functions of three or more variables.

☐ **DIFFERENTIABILITY OF FUNCTIONS OF TWO VARIABLES**

Recall that a function f of one variable is called differentiable at x_0 if it has a derivative at x_0, or, in other words, if the limit

$$f'(x_0) = \lim_{\Delta x \to 0} \frac{f(x_0 + \Delta x) - f(x_0)}{\Delta x} \tag{1}$$

exists. A function f that is differentiable at a point x_0 enjoys two important properties:

- $f(x)$ is continuous at x_0.
- The curve $y = f(x)$ has a nonvertical tangent line at x_0.

Our primary objective in this section is to extend the notion of differentiability to functions of two variables in such a way that the natural analogs of these two properties hold. More precisely, when $f(x, y)$ is differentiable at (x_0, y_0), we shall want it to be the case that

- $f(x, y)$ is continuous at (x_0, y_0).
- The surface $z = f(x, y)$ has a nonvertical tangent plane at (x_0, y_0) (Figure 16.4.1). (A precise definition of a tangent plane will be given later.)

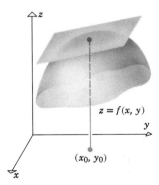

$z = f(x, y)$

(x_0, y_0)

Figure 16.4.1

It would not be unreasonable to guess that a function f of two variables should be called differentiable at (x_0, y_0) if the two partial derivatives $f_x(x_0, y_0)$ and $f_y(x_0, y_0)$ exist at (x_0, y_0). Unfortunately, this condition is not strong enough to meet our objectives, since there are functions that have partial derivatives at a point, but are not continuous at that point. For example, the function

$$f(x, y) = \begin{cases} -1 & \text{if } x > 0 \text{ and } y > 0 \\ 0 & \text{otherwise} \end{cases}$$

is discontinuous at $(0, 0)$, but has partial derivatives at $(0, 0)$; these derivatives are $f_x(0, 0) = 0$ and $f_y(0, 0) = 0$. The discontinuity and the values of the partial derivatives should be evident from Figure 16.4.2.

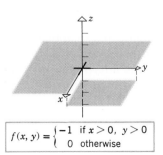

$f(x, y) = \begin{cases} -1 & \text{if } x > 0, \ y > 0 \\ 0 & \text{otherwise} \end{cases}$

Figure 16.4.2

To motivate an appropriate definition of differentiability for functions of two variables, it will be helpful to reexamine the concept of differentiability for functions of one variable. Assuming, for the moment, that f is a function of one variable that is differentiable at $x = x_0$, it follows from Formula (8) of Section 3.7 that (1) can be rewritten as

$$f'(x_0) = \lim_{\Delta x \to 0} \frac{\Delta f}{\Delta x} \tag{2}$$

or equivalently as

$$\lim_{\Delta x \to 0} \left[\frac{\Delta f}{\Delta x} - f'(x_0) \right] = 0 \tag{3}$$

where

$$\Delta f = f(x_0 + \Delta x) - f(x_0)$$

If we now define ϵ by

$$\epsilon = \frac{\Delta f}{\Delta x} - f'(x_0) \tag{4}$$

then it follows from this formula that

$$\Delta f = f'(x_0)\,\Delta x + \epsilon\,\Delta x \tag{5}$$

where ϵ is a function of Δx. Using (4), the limit in (3) can be rewritten as

$$\lim_{\Delta x \to 0} \epsilon = 0 \tag{6}$$

Formulas (5) and (6) suggest the following alternative definition of differentiability for functions of one variable.

16.4.1   DEFINITION.   A function f of one variable is said to be **differentiable** at x_0 if there exists a number $f'(x_0)$ such that Δf can be written in the form

$$\Delta f = f'(x_0)\,\Delta x + \epsilon\,\Delta x \tag{7}$$

where ϵ is a function of Δx such that $\epsilon \to 0$ as $\Delta x \to 0$.

Although this definition of differentiability is more complicated than that given earlier in the text, it provides the basis for extending the notion of differentiability to functions of two or more variables. A geometric interpretation of the terms appearing in (7) is shown in Figure 16.4.3. The term Δf represents the change in height that results when a point moves along the graph of f as the x-coordinate changes from x_0 to $x_0 + \Delta x$; the term $f'(x_0)\,\Delta x$ represents the change in height that results when a point moves along the tangent line at $(x_0, f(x_0))$ as the x-coordinate changes from x_0 to $x_0 + \Delta x$; finally, the term $\epsilon\,\Delta x$ represents the difference between Δf and $f'(x_0)\,\Delta x$. It is evident from Figure 16.4.3 that $\epsilon\,\Delta x \to 0$ as $\Delta x \to 0$. However, (7) actually makes the stronger

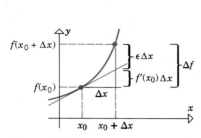

Figure 16.4.3

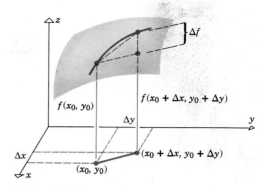

Figure 16.4.4

statement that $\epsilon \rightarrow 0$ as $\Delta x \rightarrow 0$. This is not at all evident from Figure 16.4.3, but it follows from (6).

If f is a function of x and y, then the symbol Δf, called the **increment** of f, denotes the change in the value of $f(x, y)$ that results when (x, y) varies from some initial position (x_0, y_0) to some new position $(x_0 + \Delta x, y_0 + \Delta y)$; thus,

$$\Delta f = f(x_0 + \Delta x, y_0 + \Delta y) - f(x_0, y_0) \tag{8}$$

(See Figure 16.4.4.) If a dependent variable $z = f(x, y)$ is used, then we shall sometimes write Δz rather than Δf.

Motivated by Definition 16.4.1, we now make the following definition of differentiability for functions of two variables.

> **16.4.2 DEFINITION.** A function f of two variables is said to be **differentiable** at (x_0, y_0) if $f_x(x_0, y_0)$ and $f_y(x_0, y_0)$ exist and Δf can be written in the form
>
> $$\Delta f = f_x(x_0, y_0)\, \Delta x + f_y(x_0, y_0)\, \Delta y + \epsilon_1\, \Delta x + \epsilon_2\, \Delta y \tag{9}$$
>
> where ϵ_1 and ϵ_2 are functions of Δx and Δy such that $\epsilon_1 \rightarrow 0$ and $\epsilon_2 \rightarrow 0$ as $(\Delta x, \Delta y) \rightarrow (0, 0)$.

A function is called **differentiable on a region R** of the xy-plane if it is differentiable at each point of R. A function that is differentiable on the entire xy-plane is called **everywhere differentiable** or simply **differentiable**.

REMARK. Before proceeding further, we note that for functions of one variable, the terms "differentiable" and "has a derivative" are synonymous. However, for functions of two variables, differentiability requires more than the existence of partial derivatives.

☐ **RELATIONSHIP BETWEEN DIFFERENTIABILITY AND CONTINUITY**

Earlier, we set two goals for our definition of differentiability: We wanted a function that is differentiable at (x_0, y_0) to be continuous at (x_0, y_0), and we wanted its graph to have a nonvertical tangent plane at (x_0, y_0). The next theorem shows that the continuity criterion is met; the existence of a nonvertical tangent plane will be demonstrated in the next section.

16.4.3 THEOREM. *If f is differentiable at (x_0, y_0), then f is continuous at (x_0, y_0).*

Proof. We must prove that

$$\lim_{(x,y)\to(x_0,y_0)} f(x, y) = f(x_0, y_0)$$

which, on letting $x = x_0 + \Delta x$ and $y = y_0 + \Delta y$, is equivalent to

$$\lim_{(\Delta x,\Delta y)\to(0,0)} f(x_0 + \Delta x, y_0 + \Delta y) = f(x_0, y_0)$$

which from (8) is equivalent to

$$\lim_{(\Delta x,\Delta y)\to(0,0)} \Delta f = 0$$

But f is assumed to be differentiable at (x_0, y_0), so it follows from (9) that

$$\Delta f = f_x(x_0, y_0)\, \Delta x + f_y(x_0, y_0)\, \Delta y + \epsilon_1\, \Delta x + \epsilon_2\, \Delta y$$

where $\epsilon_1 \to 0$, $\epsilon_2 \to 0$ as $(\Delta x, \Delta y) \to (0, 0)$. Thus,

$$\lim_{(\Delta x,\Delta y)\to(0,0)} \Delta f = \lim_{(\Delta x,\Delta y)\to(0,0)} [f_x(x_0, y_0)\, \Delta x + f_y(x_0, y_0)\, \Delta y + \epsilon_1\, \Delta x + \epsilon_2\, \Delta y] = 0$$

which completes the proof. ∎

The next theorem gives simple conditions for a function of two variables to be differentiable at a point.

16.4.4 THEOREM. *If f has first-order partial derivatives at each point in some circular region centered at (x_0, y_0), and if these partial derivatives are continuous at (x_0, y_0), then f is differentiable at (x_0, y_0).*

The proof is optional and can be found at the end of this section.

Example 1 Show that $f(x, y) = x^3 y^4$ is a differentiable function.

Solution. The partial derivatives $f_x = 3x^2 y^4$ and $f_y = 4x^3 y^3$ are defined and continuous everywhere in the xy-plane. Thus, the hypotheses of Theorem 16.4.4 are satisfied at each point (x_0, y_0) in the xy-plane, so $f(x, y) = x^3 y^4$ is everywhere differentiable. ◄

The following important corollary follows from Theorems 16.4.3 and 16.4.4.

16.4.5 COROLLARY. *If f has first-order partial derivatives at each point of some circular region centered at (x_0, y_0), and if these partial derivatives are continuous at (x_0, y_0), then f is continuous at (x_0, y_0).*

□ **EQUALITY OF MIXED PARTIALS**

In addition to guaranteeing differentiability and continuity of $f(x, y)$, continuity of the first-order partial derivatives also ensures that the mixed second-order partial derivatives of f are equal. This is the content of the following theorem, which we state without proof.

> **16.4.6** THEOREM. *Let f be a function of two variables. If f_x, f_y, f_{xy}, and f_{yx} are continuous on an open set, then $f_{xy} = f_{yx}$ at each point of the set.*

Example 2 Let $f(x, y) = 2e^{xy} \sin y$. It should be evident from the form of this function and from Theorem 16.2.7 that f and all its partial derivatives are continuous everywhere. Thus, Theorem 16.4.6 guarantees that $f_{xy} = f_{yx}$ everywhere. This is, in fact, the case since

$$f_x(x, y) = 2ye^{xy} \sin y = (2y \sin y)e^{xy}$$

$$f_{xy}(x, y) = (2y \sin y)(xe^{xy}) + e^{xy}(2y \cos y + 2 \sin y)$$
$$= 2e^{xy}(xy \sin y + y \cos y + \sin y)$$

$$f_y(x, y) = 2e^{xy} \cos y + 2xe^{xy} \sin y$$

$$f_{yx}(x, y) = 2ye^{xy} \cos y + 2xye^{xy} \sin y + 2e^{xy} \sin y$$
$$= 2e^{xy}(xy \sin y + y \cos y + \sin y)$$

so $f_{xy}(x, y) = f_{yx}(x, y)$ for all (x, y). ◄

In general, the order of differentiation in an nth order partial derivative can be changed without affecting the final result whenever the function and all its partial derivatives of order n or less are continuous. For example, if f and its partial derivatives of the first, second, and third orders are continuous on an open set, then at each point of that set,

$$f_{xyy} = f_{yxy} = f_{yyx}$$

or in another notation,

$$\frac{\partial^3 f}{\partial y^2\, \partial x} = \frac{\partial^3 f}{\partial y\, \partial x\, \partial y} = \frac{\partial^3 f}{\partial x\, \partial y^2}$$

□ **CHAIN RULES**

If y is a differentiable function of x and x is a differentiable function of t, then the chain rule for functions of one variable states that

$$\frac{dy}{dt} = \frac{dy}{dx}\frac{dx}{dt}$$

We shall now derive versions of the chain rule for functions of two variables. Assume that z is a function of x and y, say

$$z = f(x, y) \tag{10}$$

and suppose that x and y, in turn, are functions of a single variable t, say

$$x = x(t), \quad y = y(t)$$

On substituting these functions of t in (10), we obtain the relationship

$$z = f(x(t), y(t))$$

which expresses z as a function of the single variable t. Thus, we may ask for the derivative dz/dt, and we may inquire about its relationship to the derivatives $\partial z/\partial x$, $\partial z/\partial y$, dx/dt, and dy/dt.

16.4.7 THEOREM (Chain Rule). *If $x = x(t)$ and $y = y(t)$ are differentiable at t, and if $z = f(x, y)$ is differentiable at the point $(x(t), y(t))$, then $z = f(x(t), y(t))$ is differentiable at t, and*

$$\frac{dz}{dt} = \frac{\partial z}{\partial x}\frac{dx}{dt} + \frac{\partial z}{\partial y}\frac{dy}{dt} \tag{11}$$

Proof. From the derivative definition for functions of one variable

$$\frac{dz}{dt} = \lim_{\Delta t \to 0} \frac{\Delta z}{\Delta t} \tag{12}$$

Since $z = f(x, y)$ is differentiable at the point $(x, y) = (x(t), y(t))$, we can express Δz in the form

$$\Delta z = \frac{\partial z}{\partial x}\Delta x + \frac{\partial z}{\partial y}\Delta y + \epsilon_1 \Delta x + \epsilon_2 \Delta y \tag{13}$$

where the partial derivatives are evaluated at $(x(t), y(t))$ and $\epsilon_1 \to 0$, $\epsilon_2 \to 0$ as $(\Delta x, \Delta y) \to (0, 0)$. Thus, from (12) and (13),

$$\frac{dz}{dt} = \lim_{\Delta t \to 0} \frac{\Delta z}{\Delta t} = \lim_{\Delta t \to 0} \left[\frac{\partial z}{\partial x}\frac{\Delta x}{\Delta t} + \frac{\partial z}{\partial y}\frac{\Delta y}{\Delta t} + \epsilon_1 \frac{\Delta x}{\Delta t} + \epsilon_2 \frac{\Delta y}{\Delta t} \right] \tag{14}$$

But

$$\lim_{\Delta t \to 0} \frac{\Delta x}{\Delta t} = \frac{dx}{dt} \quad \text{and} \quad \lim_{\Delta t \to 0} \frac{\Delta y}{\Delta t} = \frac{dy}{dt}$$

Therefore, if we can show that $\epsilon_1 \to 0$, $\epsilon_2 \to 0$ as $\Delta t \to 0$, then the proof will be complete, since (14) will reduce to (11). But $\Delta x \to 0$ and $\Delta y \to 0$ as $\Delta t \to 0$, since

$$\lim_{\Delta t \to 0} \Delta x = \lim_{\Delta t \to 0} \frac{\Delta x}{\Delta t}\Delta t = \frac{dx}{dt} \cdot 0 = 0$$

and similarly for Δy. Thus, as Δt tends to zero, $(\Delta x, \Delta y) \to (0, 0)$, which implies that $\epsilon_1 \to 0$, $\epsilon_2 \to 0$. ∎

Example 3 Suppose that

$$z = x^2 y, \quad x = t^2, \quad y = t^3$$

Use the chain rule to find dz/dt, and check the result by expressing z as a function of t and differentiating directly.

Solution. By the chain rule

$$\frac{dz}{dt} = \frac{\partial z}{\partial x} \frac{dx}{dt} + \frac{\partial z}{\partial y} \frac{dy}{dt} = (2xy)(2t) + (x^2)(3t^2)$$

$$= (2t^5)(2t) + (t^4)(3t^2) = 7t^6$$

Alternatively, we may express z directly as a function of t,

$$z = x^2 y = (t^2)^2 (t^3) = t^7$$

and then differentiate to obtain $dz/dt = 7t^6$. However, this procedure is not always convenient. ◀

Example 4 Suppose that

$$z = \sqrt{xy + y}, \quad x = \cos \theta, \quad y = \sin \theta$$

Use the chain rule to find $dz/d\theta$ when $\theta = \pi/2$.

Solution. From the chain rule with θ in place of t,

$$\frac{dz}{d\theta} = \frac{\partial z}{\partial x} \frac{dx}{d\theta} + \frac{\partial z}{\partial y} \frac{dy}{d\theta}$$

we obtain

$$\frac{dz}{d\theta} = \frac{1}{2} (xy + y)^{-1/2}(y)(-\sin \theta) + \frac{1}{2} (xy + y)^{-1/2}(x + 1)(\cos \theta)$$

When $\theta = \pi/2$, we have

$$x = \cos \frac{\pi}{2} = 0, \quad y = \sin \frac{\pi}{2} = 1$$

Substituting $x = 0$, $y = 1$, $\theta = \pi/2$ in the formula for $dz/d\theta$ yields

$$\left. \frac{dz}{d\theta} \right|_{\theta = \pi/2} = \frac{1}{2} (1)(1)(-1) + \frac{1}{2} (1)(1)(0) = -\frac{1}{2} \quad ◀$$

REMARK. There are many variations in derivative notations, each of which gives the chain rule a different look. If $z = f(x, y)$, where x and y are functions of t, then some possibilities are

$$\frac{dz}{dt} = f_x \frac{dx}{dt} + f_y \frac{dy}{dt}$$

$$\frac{df}{dt} = \frac{\partial f}{\partial x}\frac{dx}{dt} + \frac{\partial f}{\partial y}\frac{dy}{dt}$$

$$\frac{df}{dt} = f_x x'(t) + f_y y'(t)$$

The reader may be able to construct other variations as well.

In the special case where $z = F(x, y)$ and y is a differentiable function of x, chain-rule formula (11) yields

$$\frac{dz}{dx} = \frac{\partial F}{\partial x}\frac{dx}{dx} + \frac{\partial F}{\partial y}\frac{dy}{dx} = \frac{\partial F}{\partial x} + \frac{\partial F}{\partial y}\frac{dy}{dx} \tag{15}$$

This result can be used to find derivatives of functions that are defined implicitly. Suppose that the equation

$$F(x, y) = 0 \tag{16}$$

defines y implicitly as a differentiable function of x, and we are interested in finding dy/dx. Differentiating both sides of (16) with respect to x and applying (15) yields

$$\frac{\partial F}{\partial x} + \frac{\partial F}{\partial y}\frac{dy}{dx} = 0$$

or

$$\frac{dy}{dx} = -\frac{\partial F/\partial x}{\partial F/\partial y} \tag{17}$$

provided $\partial F/\partial y \neq 0$.

Example 5 Given that

$$x^3 + y^2 x - 3 = 0$$

find dy/dx using (17) and check the result using implicit differentiation.

Solution. By (17) with $F(x, y) = x^3 + y^2 x - 3$

$$\frac{dy}{dx} = -\frac{\partial F/\partial x}{\partial F/\partial y} = -\frac{3x^2 + y^2}{2yx}$$

On the other hand, differentiating the given equation implicitly yields

$$3x^2 + y^2 + x\left(2y\frac{dy}{dx}\right) - 0 = 0 \quad \text{or} \quad \frac{dy}{dx} = -\frac{3x^2 + y^2}{2yx}$$

which agrees with the result obtained by (17). ◀

In Theorem 16.4.7 the variables x and y are each functions of a single variable t. We now consider the case where x and y are each functions of two variables. Let

$$z = f(x, y) \tag{18}$$

and suppose x and y are functions of u and v, say

$$x = x(u, v), \quad y = y(u, v)$$

On substituting these functions of u and v into (18), we obtain the relationship

$$z = f(x(u, v), y(u, v))$$

which expresses z as a function of the two variables u and v. Thus, we may ask for the partial derivatives $\partial z/\partial u$ and $\partial z/\partial v$; and we may inquire about the relationship between these derivatives and the derivatives $\partial z/\partial x$, $\partial z/\partial y$, $\partial x/\partial u$, $\partial x/\partial v$, $\partial y/\partial u$, and $\partial y/\partial v$.

16.4.8 THEOREM (***Chain Rule***). *If* $x = x(u, v)$ *and* $y = y(u, v)$ *have first-order partial derivatives at the point* (u, v), *and if* $z = f(x, y)$ *is differentiable at the point* $(x(u, v), y(u, v))$, *then* $z = f(x(u, v), y(u, v))$ *has first-order partial derivatives at* (u, v) *given by*

$$\frac{\partial z}{\partial u} = \frac{\partial z}{\partial x}\frac{\partial x}{\partial u} + \frac{\partial z}{\partial y}\frac{\partial y}{\partial u} \quad \text{and} \quad \frac{\partial z}{\partial v} = \frac{\partial z}{\partial x}\frac{\partial x}{\partial v} + \frac{\partial z}{\partial y}\frac{\partial y}{\partial v}$$

Proof. If v is held fixed, then $x = x(u, v)$ and $y = y(u, v)$ become functions of u alone. Thus, we are back to the case of Theorem 16.4.7. If we apply that theorem with u in place of t, and if we use ∂ rather than d to indicate that the variable v is fixed, we obtain

$$\frac{\partial z}{\partial u} = \frac{\partial z}{\partial x}\frac{\partial x}{\partial u} + \frac{\partial z}{\partial y}\frac{\partial y}{\partial u}$$

The formula for $\partial z/\partial v$ is derived similarly. ∎

Example 6 Given that

$$z = e^{xy}, \quad x = 2u + v, \quad y = u/v$$

find $\partial z/\partial u$ and $\partial z/\partial v$ using the chain rule.

Solution.

$$\frac{\partial z}{\partial u} = \frac{\partial z}{\partial x}\frac{\partial x}{\partial u} + \frac{\partial z}{\partial y}\frac{\partial y}{\partial u} = (ye^{xy})(2) + (xe^{xy})(1/v) = \left[2y + \frac{x}{v}\right]e^{xy}$$

$$= \left[\frac{2u}{v} + \frac{2u + v}{v}\right]e^{(2u+v)(u/v)} = \left[\frac{4u}{v} + 1\right]e^{(2u+v)(u/v)}$$

$$\frac{\partial z}{\partial v} = \frac{\partial z}{\partial x}\frac{\partial x}{\partial v} + \frac{\partial z}{\partial y}\frac{\partial y}{\partial v} = (ye^{xy})(1) + (xe^{xy})\left(-\frac{u}{v^2}\right)$$

$$= \left[y - x\left(\frac{u}{v^2}\right)\right]e^{xy} = \left[\frac{u}{v} - (2u + v)\left(\frac{u}{v^2}\right)\right]e^{(2u+v)(u/v)}$$

$$= -\frac{2u^2}{v^2}e^{(2u+v)(u/v)} \quad \blacktriangleleft$$

The chain rules are useful in related rates problems.

Example 7   At what rate is the area of a rectangle changing if its length is 15 ft and increasing at 3 ft/sec while its width is 6 ft and increasing at 2 ft/sec?

Solution.   Let

x = length of the rectangle in feet
y = width of the rectangle in feet
A = area of the rectangle in square feet
t = time in seconds

We are given that

$$\frac{dx}{dt} = 3 \quad \text{and} \quad \frac{dy}{dt} = 2 \tag{19}$$

at the instant when

$$x = 15, \quad y = 6 \tag{20}$$

We want to find dA/dt at that instant.
   From the area formula $A = xy$, we obtain

$$\frac{dA}{dt} = \frac{\partial A}{\partial x}\frac{dx}{dt} + \frac{\partial A}{\partial y}\frac{dy}{dt} = y\frac{dx}{dt} + x\frac{dy}{dt}$$

Substituting (19) and (20) in this equation yields

$$\frac{dA}{dt} = 6(3) + 15(2) = 48$$

Thus, the area is increasing at a rate of 48 ft^2/sec at the given instant. $\blacktriangleleft$

■ OPTIONAL

Proof of Theorem 16.4.4.   We must show that

$$\Delta f = f(x_0 + \Delta x, y_0 + \Delta y) - f(x_0, y_0)$$

can be expressed in the form

$$\Delta f = f_x(x_0, y_0)\,\Delta x + f_y(x_0, y_0)\,\Delta y + \epsilon_1\,\Delta x + \epsilon_2\,\Delta y$$

where $\epsilon_1 \to 0$, $\epsilon_2 \to 0$ as $(\Delta x, \Delta y) \to (0,0)$. In our discussion, we shall restrict Δx and Δy to be sufficiently small so that the points

Figure 16.4.5

$$B(x_0 + \Delta x, y_0) \quad \text{and} \quad C(x_0 + \Delta x, y_0 + \Delta y)$$

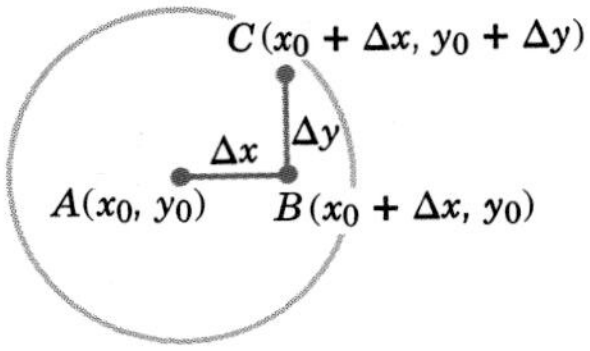

lie inside a circle where f_x and f_y exist (Figure 16.4.5). The increment Δf represents the change in the value of f as (x, y) varies from the initial point $A(x_0, y_0)$ to the final point $C(x_0 + \Delta x, y_0 + \Delta y)$. Let us resolve Δf into two terms

$$\Delta f = \Delta f_1 + \Delta f_2 \tag{21}$$

where Δf_1 is the change in the value of f as (x, y) varies from A to B and Δf_2 is the change in the value of f as (x, y) varies from B to C. Thus,

$$\Delta f_1 = f(x_0 + \Delta x, y_0) - f(x_0, y_0) \tag{22}$$
$$\Delta f_2 = f(x_0 + \Delta x, y_0 + \Delta y) - f(x_0 + \Delta x, y_0) \tag{23}$$

Along the line segment from A to B, y has a constant value of y_0, so that $f(x, y) = f(x, y_0)$ is a function of x alone. Let us denote this function by

$$f_1(x) = f(x, y_0)$$

Similarly, along the line segment from B to C, x has a constant value of $x = x_0 + \Delta x$, so that $f(x, y) = f(x_0 + \Delta x, y)$ is a function of y alone. Let us denote this function by

$$f_2(y) = f(x_0 + \Delta x, y)$$

With these definitions of f_1 and f_2, (22) and (23) can be written as

$$\Delta f_1 = f_1(x_0 + \Delta x) - f_1(x_0) \tag{24}$$
$$\Delta f_2 = f_2(y_0 + \Delta y) - f_2(y_0) \tag{25}$$

The reader should be able to show that f_1 and f_2 satisfy the hypotheses of the Mean-Value Theorem (4.10.2). Accepting this, we can rewrite (24) and (25) in the form

$$\Delta f_1 = f_1'(x^*) \, \Delta x \tag{26}$$
$$\Delta f_2 = f_2'(y^*) \, \Delta y \tag{27}$$

where x^* is between x_0 and $x_0 + \Delta x$ and y^* is between y_0 and $y_0 + \Delta y$. From the definitions of $f_1(x)$ and $f_2(y)$,

$$f_1'(x) = \frac{d}{dx} [f_1(x)] = f_x(x, y_0)$$

$$f_2'(y) = \frac{d}{dy} [f_2(y)] = f_y(x_0 + \Delta x, y)$$

so (26) and (27) can be rewritten as

$$\Delta f_1 = f_x(x^*, y_0) \, \Delta x \tag{28}$$
$$\Delta f_2 = f_y(x_0 + \Delta x, y^*) \, \Delta y \tag{29}$$

If we define

$$\epsilon_1 = f_x(x^*, y_0) - f_x(x_0, y_0) \tag{30}$$

$$\epsilon_2 = f_y(x_0 + \Delta x, y^*) - f_y(x_0, y_0) \tag{31}$$

then (28) and (29) can be rewritten as

$$\Delta f_1 = f_x(x_0, y_0)\,\Delta x + \epsilon_1\,\Delta x$$

$$\Delta f_2 = f_y(x_0, y_0)\,\Delta y + \epsilon_2\,\Delta y$$

and substituting these in (21) yields

$$\Delta f = f_x(x_0, y_0)\,\Delta x + f_y(x_0, y_0)\,\Delta y + \epsilon_1\,\Delta x + \epsilon_2\,\Delta y$$

To finish, we must show that $\epsilon_1 \to 0$, $\epsilon_2 \to 0$ as $(\Delta x, \Delta y) \to (0, 0)$.

Since x^* is between x_0 and $x_0 + \Delta x$ and y^* is between y_0 and $y_0 + \Delta y$, it follows that

$$x^* \to x_0 \quad \text{as} \quad \Delta x \to 0$$

$$y^* \to y_0 \quad \text{as} \quad \Delta y \to 0$$

But this, together with the hypothesis that f_x and f_y are continuous at (x_0, y_0), implies that

$$f_x(x^*, y_0) \to f_x(x_0, y_0)$$

$$f_y(x_0 + \Delta x, y^*) \to f_y(x_0, y_0)$$

as $(\Delta x, \Delta y) \to (0, 0)$. Thus, from (30) and (31) it follows that $\epsilon_1 \to 0$ and $\epsilon_2 \to 0$ as $(\Delta x, \Delta y) \to (0, 0)$. ∎

▶ **Exercise Set 16.4**   Ⓒ *44, 47, 48, 49*

1. Find Δf given that $f(x, y) = x^2 y$, $(x_0, y_0) = (1, 3)$, $\Delta x = 0.1$, and $\Delta y = 0.2$.

2. Find Δz given that $z = 3x^2 - 2y$, $(x_0, y_0) = (-2, 4)$, $\Delta x = 0.02$, and $\Delta y = -0.03$.

3. Find the increment of $f(x, y) = x/y$ as (x, y) varies from $(-1, 2)$ to $(3, 1)$.

4. Find the increment of $g(u, v) = 2uv - v^3$ as (u, v) varies from $(0, 1)$ to $(4, -2)$.

In Exercises 5–8, use Definition 16.4.2 to establish the differentiability of the given function. [*Remark:* ϵ_1 and ϵ_2 are not unique.]

5. $f(x, y) = xy$.

6. $f(x, y) = x^2 + y^2$.

7. $f(x, y) = x^2 y$.

8. $f(x, y) = 3x + y^2$.

9. Let $f(x, y) = \sqrt{x^2 + y^2}$.
   (a) Show that f is continuous at $(0, 0)$.

(b) Show that $f_x(0, 0)$ does not exist, and hence that f is not differentiable at $(0, 0)$. [*Hint:* Express $f_x(0, 0)$ as a limit (see Exercise 68, Section 16.3).]

10. Let
$$f(x, y) = \begin{cases} 5 - 3x - 2y, & x \ge 0 \text{ or } y \ge 0 \\ 0, & x < 0 \text{ and } y < 0 \end{cases}$$

Show that $f_x(0, 0)$ and $f_y(0, 0)$ exist, but f is not continuous at $(0, 0)$.

11. Let
$$f(x, y) = \begin{cases} \dfrac{xy}{x^2 + y^2}, & (x, y) \ne (0, 0) \\ 0, & (x, y) = (0, 0) \end{cases}$$

Prove: $f_x(0, 0)$ and $f_y(0, 0)$ exist, but f is not continuous at $(0, 0)$. [*Hint:* Show that $f_x(0, 0) = 0$ and $f_y(0, 0) = 0$ by expressing these derivatives as limits (see Exercise 68, Section 16.3). To prove that f is not continuous at $(0, 0)$, show that

$$\lim_{(x,y)\to(0,0)} f(x, y)$$

does not exist by letting $(x, y) \to (0, 0)$ along $y = 0$ and along $y = x$.]

In Exercises 12–16, find f_{xy} and f_{yx} and verify their equality.

12. $f(x, y) = 2xy - 3y^2$.
13. $f(x, y) = 4x^3y + 3x^2y$.
14. $f(x, y) = x^3/y$.
15. $f(x, y) = \sin(x^2 + y^3)$.
16. $f(x, y) = \sqrt{x^2 + y^2 - 1}$.
17. Let f be a function of two variables with continuous third- and fourth-order partial derivatives.
    (a) How many of the third-order partial derivatives can be distinct?
    (b) How many of the fourth order?
18. Let $f(x, y) = e^{xy^2}$. Find f_{xyx}, f_{xxy}, and f_{yxx} and verify their equality.

In Exercises 19–24, find dz/dt using the chain rule.

19. $z = 3x^2y^3$; $x = t^4$, $y = t^2$.
20. $z = \ln(2x^2 + y)$; $x = \sqrt{t}$, $y = t^{2/3}$.
21. $z = 3\cos x - \sin xy$; $x = 1/t$, $y = 3t$.
22. $z = \sqrt{1 + x - 2xy^4}$; $x = \ln t$, $y = t$.

23. $z = e^{1-xy}$; $x = t^{1/3}$, $y = t^3$.
24. $z = \cosh^2 xy$; $x = t/2$, $y = e^t$.

In Exercises 25–31, find $\partial z/\partial u$ and $\partial z/\partial v$ by the chain rule.

25. $z = 8x^2y - 2x + 3y$; $x = uv$, $y = u - v$.
26. $z = x^2 - y \tan x$; $x = u/v$, $y = u^2v^2$.
27. $z = x/y$; $x = 2\cos u$, $y = 3\sin v$.
28. $z = 3x - 2y$; $x = u + v \ln u$, $y = u^2 - v \ln v$.
29. $z = e^{x^2y}$; $x = \sqrt{uv}$, $y = 1/v$.
30. $z = \cos x \sin y$; $x = u - v$, $y = u^2 + v^2$.
31. $z = \tan^{-1}(x^2 + y^2)$; $x = e^u \sin v$; $y = e^u \cos v$.
32. Let $w = rs/(r^2 + s^2)$; $r = uv$, $s = u - 2v$. Use the chain rule to find $\partial w/\partial u$ and $\partial w/\partial v$.
33. Let $T = x^2y - xy^3 + 2$; $x = r \cos \theta$, $y = r \sin \theta$. Use the chain rule to find $\partial T/\partial r$ and $\partial T/\partial \theta$.
34. Let $R = e^{2s-t^2}$; $s = 3\phi$, $t = \phi^{1/2}$. Use the chain rule to find $dR/d\phi$.
35. Let $t = u/v$; $u = x^2 - y^2$, $v = 4xy^3$. Use the chain rule to find $\partial t/\partial x$ and $\partial t/\partial y$.
36. Use the chain rule to find $\dfrac{dz}{dt}\bigg|_{t=3}$ if $z = x^2y$; $x = t^2$, $y = t + 7$.
37. Use the chain rule to find the value of $\dfrac{dw}{ds}\bigg|_{s=1/4}$ if $w = r^2 - r \tan \theta$; $r = \sqrt{s}$, $\theta = \pi s$.
38. Use the chain rule to find the value of
$$\frac{\partial f}{\partial u}\bigg|_{u=1,\, v=-2} \quad \text{and} \quad \frac{\partial f}{\partial v}\bigg|_{u=1,\, v=-2}$$
if $f(x, y) = x^2y^2 - x + 2y$; $x = \sqrt{u}$, $y = uv^3$.
39. Use the chain rule to find the value of
$$\frac{\partial z}{\partial r}\bigg|_{r=2,\, \theta=\pi/6} \quad \text{and} \quad \frac{\partial z}{\partial \theta}\bigg|_{r=2,\, \theta=\pi/6}$$
if $z = xye^{x/y}$; $x = r \cos \theta$, $y = r \sin \theta$.

In Exercises 40–43, use (17) to find dy/dx and check your result using implicit differentiation.

40. $x^2y^3 + \cos y = 0$.　　41. $x^3 - 3xy^2 + y^3 = 5$.
42. $e^{xy} + ye^y = 1$.　　43. $x - \sqrt{xy} + 3y = 4$.

44. The portion of a tree that is usable for lumber may be viewed as a right-circular cylinder. If the height of a tree increases 2 ft per year and the diameter increases 3 in. per year, how fast is the volume of usable lumber increasing when the tree is 20 ft high and the diameter is 30 in.?

45. Two straight roads intersect at right angles. Car A, moving on one of the roads, approaches the intersection at 25 mi/hr and car B, moving on the other road, approaches the intersection at 30 mi/hr. At what rate is the distance between the cars changing when A is 0.3 mile from the intersection and B is 0.4 mile from the intersection?

46. Use the ideal gas law (Example 4, Section 16.3) with $k = 10$ to find the rate at which the temperature of a gas is changing when the volume is 200 in³ and increasing at the rate of 4 in³/sec, while the pressure is 5 lb/in² and decreasing at the rate of 1 lb/in²/sec.

47. Two sides of a triangle have lengths $a = 4$ cm and $b = 3$ cm, but are increasing at the rate of 1 cm/sec. If the area of the triangle remains constant, at what rate is the angle θ between a and b changing when $\theta = \pi/6$?

48. Two sides of a triangle have lengths $a = 5$ cm and $b = 10$ cm, and the included angle is $\theta = \pi/3$. If a is increasing at a rate of 2 cm/sec, b is increasing at a rate of 1 cm/sec, and θ remains constant, at what rate is the third side changing? Is it increasing or decreasing? [*Hint:* Use the law of cosines.]

49. Suppose that a particle moving along a metal plate in the xy-plane has velocity $\mathbf{v} = \mathbf{i} - 4\mathbf{j}$ (cm/sec) at the point $(3, 2)$. If the temperature of the plate at points in the xy-plane is $T(x, y) = y^2 \ln x$, $x \geq 1$, in degrees Celsius, find dT/dt at $(3, 2)$.

In Exercises 50–67, you may assume that the derivatives satisfy all continuity requirements needed to solve the problem.

50. Let $z = f(u)$ where $u = g(x, y)$. Express $\partial z/\partial x$ and $\partial z/\partial y$ in terms of dz/du, $\partial u/\partial x$, and $\partial u/\partial y$.

51. Let $z = f(x^2 - y^2)$. Show that $y \partial z/\partial x + x \partial z/\partial y = 0$.

52. Let $z = f(xy)$. Show that $x \, \partial z/\partial x - y \, \partial z/\partial y = 0$.

53. Let $z = f(u)$ where $u = g(x, y)$. Show that

(a) $\dfrac{\partial^2 z}{\partial x^2} = \dfrac{dz}{du} \dfrac{\partial^2 u}{\partial x^2} + \dfrac{d^2 z}{du^2} \left(\dfrac{\partial u}{\partial x} \right)^2$

and $\dfrac{\partial^2 z}{\partial y^2} = \dfrac{dz}{du} \dfrac{\partial^2 u}{\partial y^2} + \dfrac{d^2 z}{du^2} \left(\dfrac{\partial u}{\partial y} \right)^2$

(b) $\dfrac{\partial^2 z}{\partial y \, \partial x} = \dfrac{dz}{du} \dfrac{\partial^2 u}{\partial y \, \partial x} + \dfrac{d^2 z}{du^2} \dfrac{\partial u}{\partial x} \dfrac{\partial u}{\partial y}$.

54. Let $r = \sqrt{x^2 + y^2}$. Show that

(a) $\dfrac{\partial r}{\partial x} = \dfrac{x}{r}$ (b) $\dfrac{\partial r}{\partial y} = \dfrac{y}{r}$

(c) $\dfrac{\partial^2 r}{\partial x^2} = \dfrac{y^2}{r^3}$ (d) $\dfrac{\partial^2 r}{\partial y^2} = \dfrac{x^2}{r^3}$.

55. Let $z = f(r)$ where $r = \sqrt{x^2 + y^2}$, $r \neq 0$.

(a) Use the formulas in Exercises 53 and 54 to show that if z satisfies Laplace's equation, $\dfrac{\partial^2 z}{\partial x^2} + \dfrac{\partial^2 z}{\partial y^2} = 0$, then $r \dfrac{d^2 z}{dr^2} + \dfrac{dz}{dr} = 0$.

(b) Use the result of part (a) to find the general solution of Laplace's equation for $z = f(r)$.

$\left[\text{*Hint:* Write } r \dfrac{d^2 z}{dr^2} + \dfrac{dz}{dr} \text{ as the derivative of a product.} \right]$

56. Let $z = f(x - y, y - x)$. Show that $\partial z/\partial x + \partial z/\partial y = 0$.

57. Let $z = f(y + cx) + g(y - cx)$, where $c \neq 0$. Show that

$$\dfrac{\partial^2 z}{\partial x^2} = c^2 \dfrac{\partial^2 z}{\partial y^2}$$

58. Let $z = f(x, y)$ where $x = g(t)$ and $y = h(t)$.

(a) Show that

$$\dfrac{d}{dt}\left(\dfrac{\partial z}{\partial x} \right) = \dfrac{\partial^2 z}{\partial x^2} \dfrac{dx}{dt} + \dfrac{\partial^2 z}{\partial y \, \partial x} \dfrac{dy}{dt}$$

and

$$\dfrac{d}{dt}\left(\dfrac{\partial z}{\partial y} \right) = \dfrac{\partial^2 z}{\partial x \, \partial y} \dfrac{dx}{dt} + \dfrac{\partial^2 z}{\partial y^2} \dfrac{dy}{dt}$$

(b) Use the formulas in part (a) to help find a formula for $d^2 z/dt^2$.

59. Let $z = f(x, y)$ where $x = g(u, v)$ and $y = h(u, v)$. Show that

$$\dfrac{\partial}{\partial u}\left(\dfrac{\partial z}{\partial x} \right) = \dfrac{\partial^2 z}{\partial x^2} \dfrac{\partial x}{\partial u} + \dfrac{\partial^2 z}{\partial y \, \partial x} \dfrac{\partial y}{\partial u}$$

60. Let $z = f(x, y)$ where $x = u + v$ and $y = u - v$. Show that

$$\dfrac{\partial^2 z}{\partial v \, \partial u} = \dfrac{\partial^2 z}{\partial x^2} - \dfrac{\partial^2 z}{\partial y^2}$$

61. The equations $x = r \cos \theta$ and $y = r \sin \theta$, which relate Cartesian and polar coordinates, define r and θ implicitly as functions of x and y.

(a) Use implicit differentiation with respect to the variable x on both equations to show that

$$\frac{\partial r}{\partial x} = \cos\theta \quad \text{and} \quad \frac{\partial\theta}{\partial x} = -\frac{\sin\theta}{r}$$

(b) Use implicit differentiation with respect to y on both equations to show that

$$\frac{\partial r}{\partial y} = \sin\theta \quad \text{and} \quad \frac{\partial\theta}{\partial y} = \frac{\cos\theta}{r}$$

62. Let $z = f(r, \theta)$, where r and θ are defined implicitly as functions of x and y by the equations $x = r\cos\theta$ and $y = r\sin\theta$. Use the results in Exercise 61 to show that

(a) $\dfrac{\partial z}{\partial x} = \dfrac{\partial z}{\partial r}\cos\theta - \dfrac{1}{r}\dfrac{\partial z}{\partial\theta}\sin\theta$

(b) $\dfrac{\partial z}{\partial y} = \dfrac{\partial z}{\partial r}\sin\theta + \dfrac{1}{r}\dfrac{\partial z}{\partial\theta}\cos\theta.$

63. Use the formulas in Exercise 62 to show that

$$\left(\frac{\partial z}{\partial x}\right)^2 + \left(\frac{\partial z}{\partial y}\right)^2 = \left(\frac{\partial z}{\partial r}\right)^2 + \frac{1}{r^2}\left(\frac{\partial z}{\partial\theta}\right)^2$$

64. Use the formulas in Exercises 61 and 62 to show that Laplace's equation

$$\frac{\partial^2 z}{\partial x^2} + \frac{\partial^2 z}{\partial y^2} = 0$$

becomes

$$\frac{\partial^2 z}{\partial r^2} + \frac{1}{r^2}\frac{\partial^2 z}{\partial\theta^2} + \frac{1}{r}\frac{\partial z}{\partial r} = 0$$

in polar coordinates.

65. A function $f(x, y)$ is said to be **homogeneous of degree n** if $f(tx, ty) = t^n f(x, y)$ for $t > 0$. Show that the following functions are homogeneous and find the degree of each.

(a) $f(x, y) = 3x^2 + y^2$

(b) $f(x, y) = \sqrt{x^2 + y^2}$

(c) $f(x, y) = x^2 y - 2y^3$

(d) $f(x, y) = \dfrac{5}{(x^2 + 2y^2)^2}$, $(x, y) \neq (0, 0)$.

66. Show that for a homogeneous function $f(x, y)$ of degree n (see Exercise 65)

$$x\frac{\partial f}{\partial x} + y\frac{\partial f}{\partial y} = nf$$

[*Hint:* Let $u = tx$ and $v = ty$ in $f(tx, ty)$, and differentiate both sides of $f(u, v) = t^n f(x, y)$ with respect to t.]

67. Let $w = f(x, y)$ where $y = g(x, z)$. Taking x and z as the independent variables, express each of the following in terms of $\partial f/\partial x$, $\partial f/\partial y$, $\partial y/\partial x$, and $\partial y/\partial z$.

(a) $\dfrac{\partial w}{\partial x}$

(b) $\dfrac{\partial w}{\partial z}$.

68. Show that if $u(x, y)$ and $v(x, y)$ satisfy the Cauchy-Riemann equations (Exercise 54, Section 16.3), and if $x = r\cos\theta$ and $y = r\sin\theta$, then

$$\frac{\partial u}{\partial r} = \frac{1}{r}\frac{\partial v}{\partial\theta} \quad \text{and} \quad \frac{\partial v}{\partial r} = -\frac{1}{r}\frac{\partial u}{\partial\theta}$$

69. Prove: If f, f_x, and f_y are continuous on a circular region containing $A(x_0, y_0)$ and $B(x_1, y_1)$, then there is a point (x^*, y^*) on the line segment joining A and B such that

$$f(x_1, y_1) - f(x_0, y_0)$$
$$= f_x(x^*, y^*)(x_1 - x_0) + f_y(x^*, y^*)(y_1 - y_0)$$

This result is the two-dimensional version of the Mean-Value Theorem. [*Hint:* Express the line segment joining A and B in parametric form and use the Mean-Value Theorem for functions of one variable.]

70. Prove: If $f_x(x, y) = 0$ and $f_y(x, y) = 0$ throughout a circular region, then $f(x, y)$ is constant on that region. [*Hint:* Use the result of Exercise 69.]

■ 16.5 TANGENT PLANES; TOTAL DIFFERENTIALS FOR FUNCTIONS OF TWO VARIABLES

In this section we shall discuss tangent planes to surfaces in three-dimensional space. We are concerned with three main questions: What is a tangent plane? When do tangent planes exist? How do we find equations of tangent planes? We shall use our results on tangent planes to extend the concept of a differential to functions of two variables.

☐ **TANGENT PLANES**

Recall that if C is a smooth curve in 3-space, then the tangent line to C at a point P_0 is the line through P_0 along the unit tangent vector to C at P_0 (Figure 16.5.1). The concept of a *tangent plane* builds on this definition. If $P_0(x_0, y_0, z_0)$ is a point on a surface S, and if the tangent lines at P_0 to all smooth curves that pass through P_0 and lie on the surface S all lie in a common plane, then we shall regard that plane to be the *tangent plane* to the surface at P_0 (Figure 16.5.2).

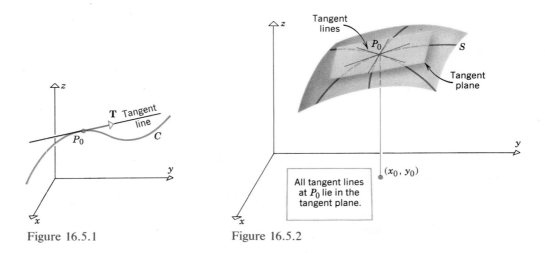

Figure 16.5.1   Figure 16.5.2

The following theorem states conditions that ensure the existence of a tangent plane and tell us how to find its equation.

16.5.1   THEOREM. *Let $P_0(x_0, y_0, z_0)$ be any point on the surface $z = f(x, y)$. If $f(x, y)$ is differentiable at (x_0, y_0), then the surface has a tangent plane at P_0, and this plane has the equation*

$$f_x(x_0, y_0)(x - x_0) + f_y(x_0, y_0)(y - y_0) - (z - z_0) = 0 \qquad (1)$$

Proof.   To prove the existence of a tangent plane at P_0, we must show that all smooth curves on the surface $z = f(x, y)$ that pass through P_0 have tangent lines that lie in a common plane. We shall do this by showing that these curves all have unit tangent vectors at P_0 that are perpendicular to the vector

$$\mathbf{n} = \langle f_x(x_0, y_0), f_y(x_0, y_0), -1 \rangle \tag{2}$$

This will force all the tangent lines at P_0 to lie in the plane through P_0 with $\mathbf{n}$ as a normal. Moreover, it will follow from (2) that the point-normal equation of this plane is (1), thereby completing the proof.

Assume that C is any smooth curve that lies on the surface $z = f(x, y)$ and passes through $P_0(x_0, y_0, z_0)$. Moreover, assume that C has parametric equations

$$x = x(s), \quad y = y(s), \quad z = z(s)$$

where s is an arc-length parameter, and assume that $P_0(x_0, y_0, z_0)$ is the point on C that corresponds to the parameter value $s = s_0$. Thus, $x_0 = x(s_0)$, $y_0 = y(s_0)$, and $z_0 = z(s_0)$. Because C lies on the surface $z = f(x, y)$, every point $(x(s), y(s), z(s))$ on C must satisfy this equation for all s, so

$$z(s) = f(x(s), y(s))$$

for all s. If we differentiate both sides of this equation and apply the version of the chain rule in Theorem 16.4.7 with s replacing t, we obtain

$$\frac{dz}{ds} = \frac{\partial f}{\partial x}\frac{dx}{ds} + \frac{\partial f}{\partial y}\frac{dy}{ds}$$

or equivalently

$$\frac{\partial f}{\partial x}\frac{dx}{ds} + \frac{\partial f}{\partial y}\frac{dy}{ds} - \frac{dz}{ds} = 0$$

The left side of this equation can be rewritten as a dot product of vectors:

$$\left\langle \frac{\partial f}{\partial x}, \frac{\partial f}{\partial y}, -1 \right\rangle \cdot \left\langle \frac{dx}{ds}, \frac{dy}{ds}, \frac{dz}{ds} \right\rangle = 0$$

or in an alternative notation as

$$\langle f_x(x, y), f_y(x, y), -1 \rangle \cdot \langle x'(s), y'(s), z'(s) \rangle = 0$$

In particular, if $s = s_0$, we have

$$\langle f_x(x_0, y_0), f_y(x_0, y_0), -1 \rangle \cdot \langle x'(s_0), y'(s_0), z'(s_0) \rangle = 0 \tag{3}$$

But the second vector in the dot product is the unit tangent vector to C at the point $P_0(x_0, y_0, z_0)$ [see Formula (7) of Section 15.5], so that from (3) the unit tangent vector to C at P_0 is perpendicular to the vector

$$\mathbf{n} = \langle f_x(x_0, y_0), f_y(x_0, y_0), -1 \rangle$$

which completes the proof.   ∎

If $f(x, y)$ is differentiable at (x_0, y_0), then the vector

$$\mathbf{n} = \langle f_x(x_0, y_0), f_y(x_0, y_0), -1 \rangle \tag{4}$$

is called a **normal vector** to the surface $z = f(x, y)$ at $P_0(x_0, y_0, z_0)$, and the line through P_0 parallel to $\mathbf{n}$ is called the **normal line** to the surface at P_0 (Figure 16.5.3). Parametric equations of the normal line are

$$x = x_0 + f_x(x_0, y_0)t, \quad y = y_0 + f_y(x_0, y_0)t, \quad z = z_0 - t \tag{5}$$

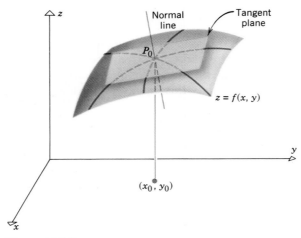

Figure 16.5.3

Example 1   Find equations of the tangent plane and normal line to the surface $z = x^2 y$ at the point $(2, 1, 4)$.

Solution.   Since $f(x, y) = x^2 y$, it follows that

$$f_x(x, y) = 2xy \quad \text{and} \quad f_y(x, y) = x^2$$

so that with $x = 2$, $y = 1$,

$$f_x(2, 1) = 4 \quad \text{and} \quad f_y(2, 1) = 4$$

Thus, a vector normal to the surface at $(2, 1, 4)$ is

$$\mathbf{n} = f_x(2, 1)\mathbf{i} + f_y(2, 1)\mathbf{j} - \mathbf{k} = 4\mathbf{i} + 4\mathbf{j} - \mathbf{k}$$

Therefore, the tangent plane has the equation

$$4(x - 2) + 4(y - 1) - (z - 4) = 0 \quad \text{or} \quad 4x + 4y - z = 8$$

and the normal line has equations

$$x = 2 + 4t, \quad y = 1 + 4t, \quad z = 4 - t \quad \blacktriangleleft$$

REMARK.   In the previous section we set two goals for the definition of differentiability of a function $f(x, y)$ of two variables at a point (x_0, y_0)—we wanted f to be continuous at (x_0, y_0) and we wanted the surface $z = f(x, y)$ to have a nonvertical tangent plane at (x_0, y_0). Both of these goals have now been achieved, for we showed in Theorem 16.4.3 that differentiability implies continuity and now Theorem 16.5.1 shows that differentiability implies the existence of a nonvertical tangent plane. [The tangent plane given by (1) is nonvertical because the third component of the normal vector **n** in (2) is not zero.]

☐ **DIFFERENTIALS**

Recall that if $y = f(x)$ is a function of one variable, then the differential

$$dy = f'(x_0)\, dx$$

represents the change in y along the *tangent line* at (x_0, y_0) produced by a change dx in x and

$$\Delta y = f(x_0 + \Delta x) - f(x_0)$$

represents the change in y along the *curve* $y = f(x)$ produced by a change Δx in x. Analogously, if $z = f(x, y)$ is a function of two variables, we shall define dz to be the change in z along the *tangent plane* at (x_0, y_0, z_0) to the surface $z = f(x, y)$ produced by changes dx and dy in x and y, respectively. This is in contrast to

$$\Delta z = f(x_0 + \Delta x, y_0 + \Delta y) - f(x_0, y_0)$$

which represents the change in z *along the surface* produced by changes Δx and Δy in x and y. A comparison of dz and Δz is shown in Figure 16.5.4 in the case where $dx = \Delta x$ and $dy = \Delta y$.

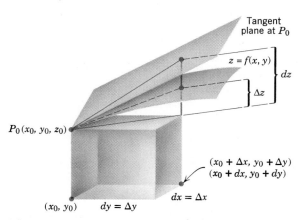

Figure 16.5.4

To derive a formula for dz, let $P_0(x_0, y_0, z_0)$ be a fixed point on the surface $z = f(x, y)$. If we assume f to be differentiable at (x_0, y_0), then the surface has a tangent plane at P_0, given by the equation

$$f_x(x_0, y_0)(x - x_0) + f_y(x_0, y_0)(y - y_0) - (z - z_0) = 0$$

or

$$z = z_0 + f_x(x_0, y_0)(x - x_0) + f_y(x_0, y_0)(y - y_0) \tag{6}$$

From (6) it follows that the tangent plane has height z_0 when $x = x_0$, $y = y_0$, and it has height

$$z_0 + f_x(x_0, y_0)\, dx + f_y(x_0, y_0)\, dy \tag{7}$$

when $x = x_0 + dx$, $y = y_0 + dy$. Thus, the change dz in the height of the tangent plane as (x, y) varies from (x_0, y_0) to $(x_0 + dx, y_0 + dy)$ is obtained by subtracting z_0 from expression (7). This yields

$$dz = f_x(x_0, y_0)\, dx + f_y(x_0, y_0)\, dy$$

This quantity is called the **total differential** of z at (x_0, y_0). Often, we omit the subscripts on x_0 and y_0 and write

$$dz = f_x(x, y)\, dx + f_y(x, y)\, dy \tag{8}$$

In this formula, dx and dy are usually viewed as variables and x and y as fixed numbers. Formula (8) may also be written using df in place of dz.

If $z = f(x, y)$ is differentiable at the point (x, y), then the increment Δz can be written as

$$\Delta z = f_x(x, y)\, \Delta x + f_y(x, y)\, \Delta y + \epsilon_1\, \Delta x + \epsilon_2\, \Delta y \tag{9}$$

where $\epsilon_1 \to 0$, $\epsilon_2 \to 0$ as $(\Delta x, \Delta y) \to (0, 0)$. In the case where $dx = \Delta x$ and $dy = \Delta y$, it follows from (8) and (9) that

$$\Delta z = dz + \epsilon_1\, \Delta x + \epsilon_2\, \Delta y$$

Thus, when $\Delta x = dx$ and $\Delta y = dy$ are small, we can approximate Δz by

$$\Delta z \approx dz \tag{10}$$

Geometrically, this approximation tells us that the change in z along the surface and the change in z along the tangent plane are approximately equal when $\Delta x = dx$ and $\Delta y = dy$ are small (see Figure 16.5.4).

Example 2 Let $z = 4x^3y^2$. Find dz.

Solution. Since $f(x, y) = 4x^3y^2$,

$$f_x(x, y) = 12x^2y^2 \quad \text{and} \quad f_y(x, y) = 8x^3y$$

so

$$dz = 12x^2y^2\, dx + 8x^3y\, dy \quad \blacktriangleleft$$

Example 3 Let $f(x, y) = \sqrt{x^2 + y^2}$. Use a total differential to approximate the change in $f(x, y)$ as (x, y) varies from the point $(3, 4)$ to the point $(3.04, 3.98)$.

Solution.   We shall approximate Δf (the change in f) by

$$df = f_x(x, y)\, dx + f_y(x, y)\, dy = \frac{x}{\sqrt{x^2 + y^2}}\, dx + \frac{y}{\sqrt{x^2 + y^2}}\, dy$$

Since $x = 3$, $y = 4$, $dx = 0.04$, and $dy = -0.02$, we obtain

$$\Delta f \approx df = \frac{3}{\sqrt{3^2 + 4^2}}\, (0.04) + \frac{4}{\sqrt{3^2 + 4^2}}\, (-0.02)$$

$$= \frac{3}{5}\, (0.04) - \frac{4}{5}\, (0.02) = 0.008$$

The reader may want to use a calculator to show that the true value of Δf to five decimal places is

$$\Delta f = \sqrt{(3.04)^2 + (3.98)^2} - \sqrt{3^2 + 4^2} \approx 0.00819 \qquad \blacktriangleleft$$

Example 4   The volume V of a right-circular cone of radius r and height h is given by $V = \frac{1}{3}\pi r^2 h$. Suppose the height increases from 10 cm to 10.01 cm, while the radius decreases from 12 cm to 11.95 cm. Use a total differential to approximate the change ΔV in volume.

Solution.   We shall approximate ΔV by

$$dV = \frac{\partial V}{\partial r}\, dr + \frac{\partial V}{\partial h}\, dh = \frac{2}{3}\pi r h\, dr + \frac{1}{3}\pi r^2\, dh$$

With $h = 10$, $dh = 0.01$, $r = 12$, and $dr = -0.05$, we obtain

$$\Delta V \approx dV = \frac{2}{3}\pi(12)(10)(-0.05) + \frac{1}{3}\pi(12)^2(0.01)$$

or

$$\Delta V \approx -3.52\pi \approx -11.06$$

Thus, the volume decreases by approximately 11.06 cm³. $\blacktriangleleft$

Example 5   The radius of a right-circular cylinder is measured with an error of at most 2%, and the height is measured with an error of at most 4%. Approximate the maximum possible percentage error in the volume V calculated from these measurements.

Solution.   Let r, h, and V be the true radius, height, and volume of the cylinder, and let Δr, Δh, and ΔV be the errors in these quantities. We are given that

$$\left| \frac{\Delta r}{r} \right| \le 0.02 \quad \text{and} \quad \left| \frac{\Delta h}{h} \right| \le 0.04$$

We want to find the maximum possible value of $|\Delta V / V|$. Since the volume of the cylinder is $V = \pi r^2 h$, it follows from (8) that

$$dV = \frac{\partial V}{\partial r}\, dr + \frac{\partial V}{\partial h}\, dh = 2\pi r h\, dr + \pi r^2\, dh$$

If we choose $dr = \Delta r$ and $dh = \Delta h$, then we can use the approximations

$$\Delta V \approx dV \quad \text{and} \quad \frac{\Delta V}{V} \approx \frac{dV}{V}$$

But

$$\frac{dV}{V} = \frac{2\pi rh \, dr + \pi r^2 \, dh}{\pi r^2 h} = 2\frac{dr}{r} + \frac{dh}{h}$$

so by the triangle inequality (Theorem 1.2.6)

$$\left|\frac{dV}{V}\right| = \left|2\frac{dr}{r} + \frac{dh}{h}\right| \le 2\left|\frac{dr}{r}\right| + \left|\frac{dh}{h}\right| \le 2(0.02) + (0.04) = 0.08$$

Thus, the maximum percentage error in V is approximately 8%.   ◀

▶ Exercise Set 16.5   Ⓒ 13, 14, 15, 16, 24, 32, 34

In Exercises 1–8, find equations for the tangent plane and normal line to the given surface at the point P.

1. $z = 4x^3 y^2 + 2y$; $P(1, -2, 12)$.

2. $z = \frac{1}{2}x^7 y^{-2}$; $P(2, 4, 4)$.

3. $z = xe^{-y}$; $P(1, 0, 1)$.

4. $z = \ln \sqrt{x^2 + y^2}$; $P(-1, 0, 0)$.

5. $z = e^{3y} \sin 3x$; $P(\pi/6, 0, 1)$.

6. $z = x^{1/2} + y^{1/2}$; $P(4, 9, 5)$.

7. $x^2 + y^2 + z^2 = 25$; $P(-3, 0, 4)$.

8. $x^2 y - 4z^2 = -7$; $P(-3, 1, -2)$.

In Exercises 9–12, find dz.

9. $z = 7x - 2y$.

10. $z = 5x^2 y^5 - 2x + 4y + 7$.

11. $z = \tan^{-1} xy$.   12. $z = \sec^2(x - 3y)$.

In Exercises 13–16, use a total differential to approximate the change in $f(x, y)$ as (x, y) varies from P to Q.

13. $f(x, y) = x^2 + 2xy - 4x$; $P(1, 2)$, $Q(1.01, 2.04)$.

14. $f(x, y) = x^{1/3} y^{1/2}$; $P(8, 9)$, $Q(7.78, 9.03)$.

15. $f(x, y) = \dfrac{x + y}{xy}$; $P(-1, -2)$, $Q(-1.02, -2.04)$.

16. $f(x, y) = \ln \sqrt{1 + xy}$; $P(0, 2)$, $Q(-0.09, 1.98)$.

17. Find all points on the surface at which the tangent plane is horizontal.
     (a) $z = x^3 y^2$
     (b) $z = x^2 - xy + y^2 - 2x + 4y$.

18. Find a point on the surface $z = 3x^2 - y^2$ at which the tangent plane is parallel to the plane $6x + 4y - z = 5$.

19. Find a point on the surface $z = 8 - 3x^2 - 2y^2$ at which the tangent plane is perpendicular to the line $x = 2 - 3t$, $y = 7 + 8t$, $z = 5 - t$.

20. Show that the surfaces $z = \sqrt{x^2 + y^2}$ and $z = \frac{1}{10}(x^2 + y^2) + \frac{5}{2}$ intersect at $(3, 4, 5)$ and have a common tangent plane at that point.

21. Show that the surfaces $z = \sqrt{16 - x^2 - y^2}$ and $z = \sqrt{x^2 + y^2}$ intersect at $(2, 2, 2\sqrt{2})$ and have tangent planes that are perpendicular at that point.

22. (a) Show that every line normal to the cone $z = \sqrt{x^2 + y^2}$ passes through the z-axis.
     (b) Show that every tangent plane intersects the cone in a line passing through the origin.

23. One leg of a right triangle increases from 3 cm to

3.2 cm, while the other leg decreases from 4 cm to 3.96 cm. Use a total differential to approximate the change in the length of the hypotenuse.

24. The volume V of a right-circular cone of radius r and height h is given by $V = \frac{1}{3}\pi r^2 h$. Suppose the height decreases from 20 in. to 19.95 in., while the radius increases from 4 in. to 4.05 in. Use a total differential to approximate the change in volume.

25. The length and width of a rectangle are measured with errors of at most 3% and 5%, respectively. Use differentials to approximate the maximum percentage error in the calculated area.

26. The radius and height of a right-circular cone are measured with errors of at most 1% and 4%, respectively. Use differentials to approximate the maximum percentage error in the calculated volume.

27. The length and width of a rectangle are measured with errors of at most $r\%$. Use differentials to approximate the maximum percentage error in the calculated length of the diagonal.

28. The legs of a right triangle are measured to be 3 cm and 4 cm, with a maximum error of 0.05 cm in each measurement. Use differentials to approximate the maximum possible error in the calculated value of (a) the hypotenuse and (b) the area of the triangle.

29. The total resistance R of two resistances R_1 and R_2, connected in parallel, is

$$R = \frac{R_1 R_2}{R_1 + R_2}$$

Suppose R_1 and R_2 are measured to be 200 ohms and 400 ohms, respectively, with a maximum error of 2% in each. Use differentials to approximate the maximum percentage error in the calculated value of R.

30. According to the ideal gas law, the pressure, temperature, and volume of a confined gas are related by $P = kT/V$, where k is a constant. Use differentials to approximate the percentage change in pressure if the temperature of a gas is increased 3% and the volume is increased 5%.

31. An angle θ of a right triangle is calculated by the formula

$$\theta = \sin^{-1}\frac{a}{c}$$

where a is the length of the side opposite to θ and c is the length of the hypotenuse. Suppose that the measurements $a = 3$ in. and $c = 5$ in. each have a maximum possible error of 0.01 in. Use differentials to approximate the maximum possible error in the calculated value of θ.

32. An open cylindrical can has an inside radius of 2 cm and an inside height of 5 cm. Use differentials to approximate the volume of metal in the can if it is 0.01 cm thick. [*Hint:* The volume of metal is the difference, ΔV, in the volumes of two cylinders.]

33. The period T of a simple pendulum with small oscillations is calculated from the formula $T = 2\pi\sqrt{L/g}$ where L is the length of the pendulum and g is the acceleration due to gravity. The values of L and g have errors of at most 0.5% and 0.1%, respectively. Use differentials to approximate the maximum percentage error in the calculated value of T.

34. The angle of elevation from a point on the ground to the top of a building is measured as 60° with a maximum possible error of 0.2°. The distance from the point to the building is measured as 100 ft with a maximum possible error of 2 in. Use differentials to approximate the maximum possible error in the calculated height of the building.

35. Suppose that x and y have errors of at most $r\%$ and $s\%$, respectively. For each of the following formulas in x and y, use differentials to approximate the maximum possible error in the calculated result.
(a) xy (b) x/y (c) $x^2 y^3$ (d) $x^3\sqrt{y}$.

36. Show that the volume of the solid bounded by the coordinate planes and the plane tangent to the portion of the surface $xyz = k$, $k > 0$ in the first octant does not depend on the point of tangency.

37. (a) Find all points of intersection of the line $x = -1 + t, y = 2 + t, z = 2t + 7$ and the surface $z = x^2 + y^2$.
(b) At each point of intersection, find the cosine of the acute angle between the given line and the line normal to the surface.

38. Show that if f is differentiable and $z = xf(x/y)$, then all tangent planes to the graph of this equation pass through a common point.

39. Show that the equation of the plane tangent to the ellipsoid

$$\frac{x^2}{a^2} + \frac{y^2}{b^2} + \frac{z^2}{c^2} = 1$$

at (x_0, y_0, z_0) can be written in the form

$$\frac{x_0 x}{a^2} + \frac{y_0 y}{b^2} + \frac{z_0 z}{c^2} = 1$$

40. Show that the equation of the plane tangent to the paraboloid

$$z = \frac{x^2}{a^2} + \frac{y^2}{b^2}$$

at (x_0, y_0, z_0) can be written in the form

$$z + z_0 = \frac{2x_0 x}{a^2} + \frac{2y_0 y}{b^2}$$

41. Prove: If $z = f(x, y)$ and $z = g(x, y)$ intersect at $P(x_0, y_0, z_0)$, and if f and g are differentiable at (x_0, y_0), then the normal lines at P are perpendicular if and only if

$$f_x(x_0, y_0)g_x(x_0, y_0) + f_y(x_0, y_0)g_y(x_0, y_0) = -1$$

16.6 DIRECTIONAL DERIVATIVES AND GRADIENTS FOR FUNCTIONS OF TWO VARIABLES

> *The partial derivatives $f_x(x, y)$ and $f_y(x, y)$ represent the rates of change of $f(x, y)$ in directions parallel to the x and y axes. In this section we shall investigate rates of change of $f(x, y)$ in other directions.*

☐ **DIRECTIONAL DERIVATIVES**

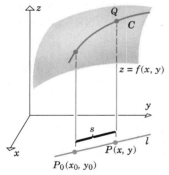

Figure 16.6.1

As shown in Figure 16.6.1, let $P_0(x_0, y_0)$ be a fixed point in the xy-plane, let l be a line in the xy-plane that passes through P_0, and let f be a function of two variables that is differentiable at (x_0, y_0). If a point $P(x, y)$ moves along the line l, then, as illustrated in the figure, there is a companion point Q directly above it, that moves along the surface $z = f(x, y)$, tracing out some curve C. We shall be interested in finding the rate at which the z-coordinate of Q changes with the distance s between P_0 and P. To make this idea more precise, consider the *unit vector*

$$\mathbf{u} = \langle u_1, u_2 \rangle$$

with initial point $P_0(x_0, y_0)$ and pointing in the direction of motion of $P(x, y)$. (See Figure 16.6.2.) Let s be the (signed) distance along l from $P_0(x_0, y_0)$ to $P(x, y)$, where s is nonnegative if P is in the direction of $\mathbf{u}$ from P_0 and is negative otherwise. The vector $\overrightarrow{P_0P}$ is related to the vector $\mathbf{u}$ by

$$\overrightarrow{P_0P} = s\mathbf{u} \quad \text{or} \quad \langle x - x_0, y - y_0 \rangle = \langle su_1, su_2 \rangle$$

(Figure 16.6.2). By equating corresponding components, we obtain the following parametric equations of l:

$$x = x_0 + su_1, \quad y = y_0 + su_2 \tag{1}$$

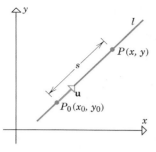

Figure 16.6.2

Since the point Q is directly above $P(x, y)$ and lies on the surface, the z-coordinate of Q is

$$z = f(x, y)$$

or, in terms of s,

$$z = f(x_0 + su_1, y_0 + su_2)$$

The rate at which z changes with respect to s can now be calculated using the chain rule. We obtain

$$\frac{dz}{ds} = f_x(x, y)\frac{dx}{ds} + f_y(x, y)\frac{dy}{ds}$$

or from (1)

$$\frac{dz}{ds} = f_x(x, y)u_1 + f_y(x, y)u_2 \tag{2}$$

where x and y in this formula are expressed in terms of s by the equations in (1). But $P_0(x_0, y_0)$ is the point on l that corresponds to $s = 0$ [see (1)], so setting $s = 0$ in (2) yields the instantaneous rate of change of z with respect to s at the point $P_0(x_0, y_0)$:

$$\left.\frac{dz}{ds}\right|_{s=0} = f_x(x_0, y_0)u_1 + f_y(x_0, y_0)u_2 \tag{3}$$

This quantity is commonly denoted by $D_{\mathbf{u}}z(x_0, y_0)$ or $D_{\mathbf{u}}f(x_0, y_0)$. We make the following definition.

16.6.1 DEFINITION. If $f(x, y)$ is differentiable at (x_0, y_0), and if $\mathbf{u} = \langle u_1, u_2 \rangle$ is a unit vector, then the *directional derivative* of f at (x_0, y_0) in the direction of $\mathbf{u}$ is defined by

$$D_{\mathbf{u}}f(x_0, y_0) = f_x(x_0, y_0)u_1 + f_y(x_0, y_0)u_2 \tag{4}$$

It should be kept in mind that there are infinitely many directional derivatives of $z = f(x, y)$ at a point (x_0, y_0), one for each possible choice of the direction vector $\mathbf{u}$ (Figure 16.6.3).

REMARK. The directional derivative $D_{\mathbf{u}}f(x_0, y_0)$ can be interpreted algebraically as the instantaneous rate of change in the direction of $\mathbf{u}$ at (x_0, y_0) of $z = f(x, y)$ with respect to the distance parameter s described above, or geometrically as the rise over the run of the tangent line to the curve C at the point Q_0 shown in Figure 16.6.4.

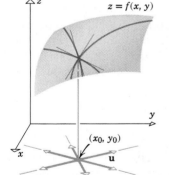

The rate at which z changes with s at (x_0, y_0) depends on the direction of $\mathbf{u}$.

Figure 16.6.3

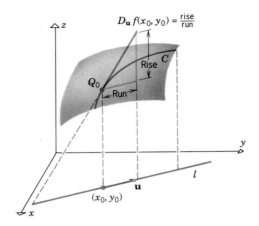

Figure 16.6.4

Example 1   Find the directional derivative of $f(x, y) = 3x^2y$ at the point $(1, 2)$ in the direction of the vector $\mathbf{a} = 3\mathbf{i} + 4\mathbf{j}$.

Solution.   The partial derivatives of f are

$$f_x(x, y) = 6xy, \quad f_y(x, y) = 3x^2$$

so that

$$f_x(1, 2) = 12, \quad f_y(1, 2) = 3$$

Thus, the directional derivative at $(1, 2)$ is

$$D_\mathbf{u} f(1, 2) = f_x(1, 2)u_1 + f_y(1, 2)u_2 = 12u_1 + 3u_2 \tag{5}$$

where u_1, u_2 are components of the *unit* vector in the direction of differentiation. We obtain these components by normalizing $\mathbf{a} = 3\mathbf{i} + 4\mathbf{j}$:

$$\mathbf{u} = \frac{\mathbf{a}}{\|\mathbf{a}\|} = \frac{1}{\sqrt{25}} (3\mathbf{i} + 4\mathbf{j}) = \frac{3}{5}\mathbf{i} + \frac{4}{5}\mathbf{j}$$

Thus, from (5)

$$D_\mathbf{u} f(1, 2) = 12\left(\frac{3}{5}\right) + 3\left(\frac{4}{5}\right) = \frac{48}{5} \quad \blacktriangleleft$$

If $\mathbf{u} = u_1\mathbf{i} + u_2\mathbf{j}$ is a unit vector making an angle θ with the positive x-axis, then

$$u_1 = \cos \theta \quad \text{and} \quad u_2 = \sin \theta$$

(see Figure 16.6.5). Thus, (4) can be written in the form

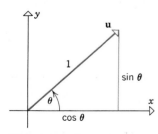

Figure 16.6.5

$$D_\mathbf{u} f(x_0, y_0) = f_x(x_0, y_0) \cos \theta + f_y(x_0, y_0) \sin \theta \tag{6}$$

Example 2 Find the directional derivative of e^{xy} at $(-2, 0)$ in the direction of the unit vector $\mathbf{u}$ that makes an angle of $\pi/3$ with the positive x-axis.

Solution. Let $f(x, y) = e^{xy}$ so that

$$f_x(x, y) = ye^{xy}, \quad f_y(x, y) = xe^{xy}$$

$$f_x(-2, 0) = 0, \quad f_y(-2, 0) = -2$$

From (6)

$$D_{\mathbf{u}}f(-2, 0) = f_x(-2, 0) \cos \frac{\pi}{3} + f_y(-2, 0) \sin \frac{\pi}{3}$$

$$= 0\left(\frac{1}{2}\right) + (-2)\left(\frac{\sqrt{3}}{2}\right) = -\sqrt{3} \quad \blacktriangleleft$$

☐ **GRADIENT**

As might be expected, directional derivatives in the directions of the positive coordinate axes reduce to partial derivatives:

$$D_{\mathbf{i}}f(x_0, y_0) = f_x(x_0, y_0)(1) + f_y(x_0, y_0)(0) = f_x(x_0, y_0)$$

$$D_{\mathbf{j}}f(x_0, y_0) = f_x(x_0, y_0)(0) + f_y(x_0, y_0)(1) = f_y(x_0, y_0)$$

Moreover, directional derivatives in the directions of the negative coordinate axes reduce to negatives of partial derivatives:

$$D_{-\mathbf{i}}f(x_0, y_0) = f_x(x_0, y_0)(-1) + f_y(x_0, y_0)(0) = -f_x(x_0, y_0)$$

$$D_{-\mathbf{j}}f(x_0, y_0) = f_x(x_0, y_0)(0) + f_y(x_0, y_0)(-1) = -f_y(x_0, y_0)$$

The directional derivative formula

$$D_{\mathbf{u}}f(x, y) = f_x(x, y)u_1 + f_y(x, y)u_2$$

can be expressed in the form of a dot product by writing

$$D_{\mathbf{u}}f(x, y) = (f_x(x, y)\mathbf{i} + f_y(x, y)\mathbf{j}) \cdot (u_1\mathbf{i} + u_2\mathbf{j}) \qquad (7)$$

The second vector in the dot product is $\mathbf{u}$. However, the first vector is something new; it is called the *gradient of f* and is denoted by the symbol ∇f or $\nabla f(x, y)$.*

16.6.2 DEFINITION. If f is a function of x and y, then the **gradient of f** is defined by

$$\nabla f(x, y) = f_x(x, y)\mathbf{i} + f_y(x, y)\mathbf{j} \qquad (8)$$

---

*The symbol ∇ (read, "del") is an inverted delta. In older books this symbol is sometimes called a "nabla" because of its similarity in form to an ancient Hebrew ten-stringed harp of that name.

With the gradient notation, Formula (7) for the directional derivative can be written in the following compact form:

$$D_{\mathbf{u}}f(x, y) = \nabla f(x, y) \cdot \mathbf{u} \tag{9}$$

In words, the dot product of the gradient of f with a unit vector $\mathbf{u}$ produces the directional derivative of f in the direction of $\mathbf{u}$.

Example 3 Find the gradient of $f(x, y) = 3x^2y$ at the point $(1, 2)$ and use it to calculate the directional derivative of f at $(1, 2)$ in the direction of the vector $\mathbf{a} = 3\mathbf{i} + 4\mathbf{j}$.

Solution. From (8)

$$\nabla f(x, y) = f_x(x, y)\mathbf{i} + f_y(x, y)\mathbf{j} = 6xy\mathbf{i} + 3x^2\mathbf{j}$$

so that the gradient of f at $(1, 2)$ is

$$\nabla f(1, 2) = 12\mathbf{i} + 3\mathbf{j}$$

The unit vector in the direction of $\mathbf{a}$ is

$$\mathbf{u} = \frac{\mathbf{a}}{\|\mathbf{a}\|} = \frac{1}{5}(3\mathbf{i} + 4\mathbf{j}) = \tfrac{3}{5}\mathbf{i} + \tfrac{4}{5}\mathbf{j}$$

Thus, from (9)

$$D_{\mathbf{u}}f(1, 2) = \nabla f(1, 2) \cdot \mathbf{u} = (12\mathbf{i} + 3\mathbf{j}) \cdot (\tfrac{3}{5}\mathbf{i} + \tfrac{4}{5}\mathbf{j}) = \tfrac{48}{5}$$

which agrees with the result obtained in Example 1.   ◄

☐ **PROPERTIES OF THE GRADIENT**

The gradient is not merely a notational device to simplify the formula for the directional derivative; the length and direction of the gradient ∇f provide important information about the function f.

16.6.3 THEOREM. *Let f be a function of two variables that is differentiable at (x_0, y_0).*

(a) *If $\nabla f(x_0, y_0) = \mathbf{0}$, then all directional derivatives of f at (x_0, y_0) are zero.*

(b) *If $\nabla f(x_0, y_0) \neq \mathbf{0}$, then among all possible directional derivatives of f at (x_0, y_0), the derivative in the direction of $\nabla f(x_0, y_0)$ has the largest value. The value of that directional derivative is $\|\nabla f(x_0, y_0)\|$.*

(c) *If $\nabla f(x_0, y_0) \neq \mathbf{0}$, then among all possible directional derivatives of f at (x_0, y_0), the derivative in the direction opposite to that of $\nabla f(x_0, y_0)$ has the smallest value. The value of that directional derivative is $-\|\nabla f(x_0, y_0)\|$.*

Proof (a). If $\nabla f(x_0, y_0) = \mathbf{0}$, then for all choices of $\mathbf{u}$ we have

$$D_{\mathbf{u}}f(x_0, y_0) = \nabla f(x_0, y_0) \cdot \mathbf{u} = \mathbf{0} \cdot \mathbf{u} = 0$$

Proofs (b) and (c). Assume that $\nabla f(x_0, y_0) \neq \mathbf{0}$ and let θ be the angle between $\nabla f(x_0, y_0)$ and an arbitrary unit vector $\mathbf{u}$. By the definition of dot product,

$$D_{\mathbf{u}}f(x_0, y_0) = \nabla f(x_0, y_0) \cdot \mathbf{u} = \|\nabla f(x_0, y_0)\| \, \|\mathbf{u}\| \cos \theta$$

or, since $\|\mathbf{u}\| = 1$,

$$D_{\mathbf{u}}f(x_0, y_0) = \|\nabla f(x_0, y_0)\| \cos \theta$$

Thus, the maximum value of $D_{\mathbf{u}}f(x_0, y_0)$ is $\|\nabla f(x_0, y_0)\|$, and this occurs when $\cos \theta = 1$, or when $\mathbf{u}$ has the same direction as $\nabla f(x_0, y_0)$ (since $\theta = 0$). The minimum value of $D_{\mathbf{u}}f(x_0, y_0)$ is $-\|\nabla f(x_0, y_0)\|$, and this occurs when $\cos \theta = -1$ or when $\mathbf{u}$ and $\nabla f(x_0, y_0)$ are oppositely directed (since $\theta = \pi$). ∎

Example 4 For the function $f(x, y) = x^2 e^y$, find the maximum value of a directional derivative at $(-2, 0)$, and give a unit vector in the direction in which the maximum value occurs.

Solution. Since

$$\nabla f(x, y) = f_x(x, y)\mathbf{i} + f_y(x, y)\mathbf{j} = 2xe^y\mathbf{i} + x^2 e^y\mathbf{j}$$

the gradient of f at $(-2, 0)$ is

$$\nabla f(-2, 0) = -4\mathbf{i} + 4\mathbf{j}$$

By Theorem 16.6.3, the maximum value of the directional derivative is

$$\|\nabla f(-2, 0)\| = \sqrt{(-4)^2 + 4^2} = \sqrt{32} = 4\sqrt{2}$$

This maximum occurs in the direction of $\nabla f(-2, 0)$. A unit vector in this direction is

$$\frac{\nabla f(-2, 0)}{\|\nabla f(-2, 0)\|} = \frac{1}{4\sqrt{2}}(-4\mathbf{i} + 4\mathbf{j}) = -\frac{1}{\sqrt{2}}\mathbf{i} + \frac{1}{\sqrt{2}}\mathbf{j} \quad \blacktriangleleft$$

There is an important geometric relationship between the gradient $\nabla f(x, y)$ and the level curves of $f(x, y)$. If (x_0, y_0) is any point in the domain of f, and if $f(x_0, y_0) = c$, then under appropriate conditions* the equation

$$f(x, y) = c$$

defines a level curve of f that passes through the point (x_0, y_0) (Figure 16.6.6a).

---

*Precise conditions under which the graph of $f(x, y) = c$ is a unique curve in the vicinity of (x_0, y_0) with a tangent line at (x_0, y_0) are specified in a theorem from advanced calculus called the "implicit function theorem." For simplicity, we shall assume that these conditions are satisfied.

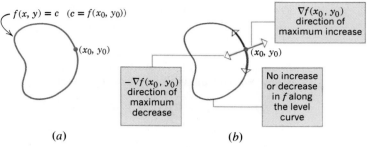

Figure 16.6.6

Let us recall that the function f has its maximum rate of increase at (x_0, y_0) in the direction of $\nabla f(x_0, y_0)$ and has its maximum rate of decrease in the direction of $-\nabla f(x_0, y_0)$. The level curve through (x_0, y_0) is the "dividing line" between these extremes in the sense that $f(x, y)$ neither increases nor decreases along this curve (Figure 16.6.6b). The following theorem shows that the gradient of f at (x_0, y_0) is perpendicular to the level curve of f through (x_0, y_0). (A vector is ***perpendicular*** or ***normal*** to a curve at a point if it is perpendicular to a nonzero tangent vector at that point.)

> **16.6.4**   THEOREM.   *If f is differentiable at (x_0, y_0) and $\nabla f(x_0, y_0) \neq \mathbf{0}$, then $\nabla f(x_0, y_0)$ is perpendicular to the level curve of f through (x_0, y_0).*

To see why this result is true, let

$$f(x, y) = c \tag{10}$$

be the level curve through (x_0, y_0). It is proved in advanced calculus that with the given hypotheses, this level curve can be represented by appropriate parametric equations

$$x = x(t), \quad y = y(t)$$

so that the level curve has a nonzero tangent vector at (x_0, y_0). More precisely,

$$x'(t_0)\mathbf{i} + y'(t_0)\mathbf{j} \neq \mathbf{0}$$

where t_0 is the value of the parameter corresponding to (x_0, y_0).

Differentiating both sides of (10) with respect to t and applying the chain rule yields

$$f_x(x(t), y(t))x'(t) + f_y(x(t), y(t))y'(t) = 0$$

Substituting $t = t_0$ and using the fact that $x(t_0) = x_0$ and $y(t_0) = y_0$ we obtain

$$f_x(x_0, y_0)x'(t_0) + f_y(x_0, y_0)y'(t_0) = 0$$

which can be written as

$$\nabla f(x_0, y_0) \cdot (x'(t_0)\mathbf{i} + y'(t_0)\mathbf{j}) = 0$$

This equation tells us that $\nabla f(x_0, y_0)$ is perpendicular to the tangent vector $x'(t_0)\mathbf{i} + y'(t_0)\mathbf{j}$ and, therefore, is perpendicular to the level curve through (x_0, y_0). ∎

Example 5 For the function $f(x, y) = x^2 + y^2$, sketch the level curve through the point $(3, 4)$, and draw the gradient vector at this point.

Solution. Since $f(3, 4) = 25$, the level curve through the point $(3, 4)$ has the equation $f(x, y) = 25$, which is the circle

$$x^2 + y^2 = 25$$

Since

$$\nabla f(x, y) = f_x(x, y)\mathbf{i} + f_y(x, y)\mathbf{j} = 2x\mathbf{i} + 2y\mathbf{j}$$

the gradient vector at $(3, 4)$ is

$$\nabla f(3, 4) = 6\mathbf{i} + 8\mathbf{j}$$

(Figure 16.6.7). Note that the gradient vector is perpendicular to the circle at $(3, 4)$, as guaranteed by Theorem 16.6.4. ◀

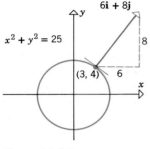

Figure 16.6.7

▶ **Exercise Set 16.6**   Ⓒ 48

In Exercises 1–4, find ∇z.

1. $z = 4x - 8y$.
2. $z = e^{-3y} \cos 4x$.
3. $z = \ln \sqrt{x^2 + y^2}$.
4. $z = e^{-5x} \sec x^2 y$.

In Exercises 5–8, find the gradient of f at the indicated point.

5. $f(x, y) = (x^2 + xy)^3$; $(-1, -1)$.
6. $f(x, y) = (x^2 + y^2)^{-1/2}$; $(3, 4)$.
7. $f(x, y) = y \ln (x + y)$; $(-3, 4)$.
8. $f(x, y) = y^2 \tan^3 x$; $(\pi/4, -3)$.

In Exercises 9–12, find $D_{\mathbf{u}} f$ at P.

9. $f(x, y) = (1 + xy)^{3/2}$; $P(3, 1)$; $\mathbf{u} = \dfrac{1}{\sqrt{2}}\mathbf{i} + \dfrac{1}{\sqrt{2}}\mathbf{j}$.

10. $f(x, y) = e^{2xy}$; $P(4, 0)$; $\mathbf{u} = -\tfrac{3}{5}\mathbf{i} + \tfrac{4}{5}\mathbf{j}$.

11. $f(x, y) = \ln (1 + x^2 + y)$; $P(0, 0)$;

  $\mathbf{u} = -\dfrac{1}{\sqrt{10}}\mathbf{i} - \dfrac{3}{\sqrt{10}}\mathbf{j}$.

12. $f(x, y) = \dfrac{cx + dy}{x - y}$; $P(3, 4)$; $\mathbf{u} = \dfrac{4}{5}\mathbf{i} + \dfrac{3}{5}\mathbf{j}$.

In Exercises 13–18, find the directional derivative of f at P in the direction of $\mathbf{a}$.

13. $f(x, y) = 4x^3y^2$; $P(2, 1)$; $\mathbf{a} = 4\mathbf{i} - 3\mathbf{j}$.
14. $f(x, y) = x^2 - 3xy + 4y^3$; $P(-2, 0)$; $\mathbf{a} = \mathbf{i} + 2\mathbf{j}$.
15. $f(x, y) = y^2 \ln x$; $P(1, 4)$; $\mathbf{a} = -3\mathbf{i} + 3\mathbf{j}$.
16. $f(x, y) = e^x \cos y$; $P(0, \pi/4)$; $\mathbf{a} = 5\mathbf{i} - 2\mathbf{j}$.
17. $f(x, y) = \tan^{-1}(y/x)$; $P(-2, 2)$; $\mathbf{a} = -\mathbf{i} - \mathbf{j}$.
18. $f(x, y) = xe^y - ye^x$; $P(0, 0)$; $\mathbf{a} = 5\mathbf{i} - 2\mathbf{j}$.

In Exercises 19–22, find the directional derivative of f at P in the direction of a vector making the angle θ with the positive x-axis.

19. $f(x, y) = \sqrt{xy}$; $P(1, 4)$; $\theta = \pi/3$.

20. $f(x, y) = \dfrac{x - y}{x + y}$; $P(-1, -2)$; $\theta = \pi/2$.

21. $f(x, y) = \tan (2x + y)$; $P(\pi/6, \pi/3)$; $\theta = 7\pi/4$.
22. $f(x, y) = \sinh x \cosh y$; $P(0, 0)$; $\theta = \pi$.

In Exercises 23–26, sketch the level curve of $f(x, y)$ that passes through P and draw the gradient vector at P.

23. $f(x, y) = 4x - 2y + 3$; $P(1, 2)$.

24. $f(x, y) = y/x^2$; $P(-2, 2)$.

25. $f(x, y) = x^2 + 4y^2$; $P(-2, 0)$.

26. $f(x, y) = x^2 - y^2$; $P(2, -1)$.

In Exercises 27–30, find a unit vector in the direction in which f increases most rapidly at P; and find the rate of change of f at P in that direction.

27. $f(x, y) = 4x^3y^2$; $P(-1, 1)$.

28. $f(x, y) = 3x - \ln y$; $P(2, 4)$.

29. $f(x, y) = \sqrt{x^2 + y^2}$; $P(4, -3)$.

30. $f(x, y) = \dfrac{x}{x + y}$; $P(0, 2)$.

In Exercises 31–34, find a unit vector in the direction in which f decreases most rapidly at P; and find the rate of change of f at P in that direction.

31. $f(x, y) = 20 - x^2 - y^2$; $P(-1, -3)$.

32. $f(x, y) = e^{xy}$; $P(2, 3)$.

33. $f(x, y) = \cos(3x - y)$; $P(\pi/6, \pi/4)$.

34. $f(x, y) = \sqrt{\dfrac{x - y}{x + y}}$; $P(3, 1)$.

35. Find the directional derivative of $f(x, y) = \dfrac{x}{x + y}$ at $P(1, 0)$ in the direction to $Q(-1, -1)$.

36. Find the directional derivative of $f(x, y) = e^{-x} \sec y$ at $P(0, \pi/4)$ in the direction of the origin.

37. Find the directional derivative of the function $f(x, y) = \sqrt{xy}\, e^y$ at $P(1, 1)$ in the direction of the negative y-axis.

38. Let $f(x, y) = \dfrac{y}{x + y}$. Find a unit vector $\mathbf{u}$ for which $D_{\mathbf{u}}f(2, 3) = 0$.

39. Find a unit vector $\mathbf{u}$ that is perpendicular at $P(1, -2)$ to the level curve of $f(x, y) = 4x^2y$ through P.

40. Find a unit vector $\mathbf{u}$ that is perpendicular at $P(2, -3)$ to the level curve of $f(x, y) = 3x^2y - xy$ through P.

41. Given that $D_{\mathbf{u}}f(1, 2) = -5$ if $\mathbf{u} = \frac{3}{5}\mathbf{i} - \frac{4}{5}\mathbf{j}$ and $D_{\mathbf{v}}f(1, 2) = 10$ if $\mathbf{v} = \frac{4}{5}\mathbf{i} + \frac{3}{5}\mathbf{j}$, find
  (a) $f_x(1, 2)$ (b) $f_y(1, 2)$
  (c) the directional derivative of f at $(1, 2)$ in the direction of the origin.

42. Given that $f_x(-5, 1) = -3$ and $f_y(-5, 1) = 2$, find the directional derivative of f at $P(-5, 1)$ in the direction to $Q(-4, 3)$.

43. Given that $\nabla f(4, -5) = 2\mathbf{i} - \mathbf{j}$, find the directional derivative of the function f at $(4, -5)$ in the direction of $\mathbf{a} = 5\mathbf{i} + 2\mathbf{j}$.

44. Given that $\nabla f(x_0, y_0) = \mathbf{i} - 2\mathbf{j}$ and $D_{\mathbf{u}}f(x_0, y_0) = -2$, find $\mathbf{u}$.

45. Let $z = 3x^2 - y^2$. Find all points where $\|\nabla z\| = 6$.

46. Given that $z = 3x + y^2$, find the maximum value of $d\|\nabla z\|/ds$ at the point $(5, 2)$ and a unit vector in the direction in which the maximum is attained.

47. A particle moves along a path C given by $x = t$ and $y = -t^2$. If $z = x^2 + y^2$, find dz/ds along C at the instant when the particle is at the point $(2, -4)$.

48. The temperature at a point (x, y) on a metal plate in the xy-plane is $T(x, y) = \dfrac{xy}{1 + x^2 + y^2}$ degrees Celsius.
  (a) Find the rate of change of temperature at $(1, 1)$ in the direction of $\mathbf{a} = 2\mathbf{i} - \mathbf{j}$.
  (b) An ant at $(1, 1)$ wants to walk in the direction in which the temperature drops most rapidly. Find a unit vector in that direction.

49. If the electric potential at a point (x, y) in the xy-plane is $V(x, y)$, then the **electric intensity vector** at (x, y) is $\mathbf{E} = -\nabla V(x, y)$. Suppose that $V(x, y) = e^{-2x} \cos 2y$.
  (a) Find the electric intensity vector at $(\pi/4, 0)$.
  (b) Show that at each point in the plane, the electric potential decreases most rapidly in the direction of $\mathbf{E}$.

50. On a certain mountain, the elevation z above a point (x, y) in a horizontal xy-plane that lies at sea level is $z = 2000 - 2x^2 - 4y^2$ ft. The positive x-axis points east, and the positive y-axis north. A climber is at the point $(-20, 5, 1100)$.
  (a) If the climber uses a compass reading to walk due west, will he begin to ascend or descend?
  (b) If the climber uses a compass reading to walk northeast, will he ascend or descend? At what rate?
  (c) In what compass direction should the climber walk to travel a level path?

51. Let $r = \sqrt{x^2 + y^2}$.
  (a) Show that $\nabla r = \dfrac{\mathbf{r}}{r}$, where $\mathbf{r} = x\mathbf{i} + y\mathbf{j}$.

(b) Show that $\nabla f(r) = f'(r)\nabla r = \dfrac{f'(r)}{r}\,\mathbf{r}$.

52. Use the formula of part (b) in Exercise 51 to find
(a) $\nabla f(r)$ if $f(r) = re^{-3r}$
(b) $f(r)$ if $\nabla f(r) = 3r^2\mathbf{r}$ and $f(2) = 1$.

53. Let $\mathbf{u}_r$ be a unit vector making an angle θ with the positive x-axis, and let $\mathbf{u}_\theta$ be a unit vector $90°$ counterclockwise from $\mathbf{u}_r$. Show that if $z = f(x, y)$, $x = r\cos\theta$, and $y = r\sin\theta$, then

$$\nabla z = \frac{\partial z}{\partial r}\mathbf{u}_r + \frac{1}{r}\frac{\partial z}{\partial \theta}\mathbf{u}_\theta$$

[*Hint:* Use parts (a) and (b) of Exercise 62, Section 16.4.]

54. Prove: If f and g are differentiable, then
(a) $\nabla(f + g) = \nabla f + \nabla g$
(b) $\nabla(cf) = c\nabla f$ (c constant)

(c) $\nabla(fg) = f\nabla g + g\nabla f$
(d) $\nabla\left(\dfrac{f}{g}\right) = \dfrac{g\nabla f - f\nabla g}{g^2}$
(e) $\nabla(f^p) = pf^{p-1}\nabla f$.

55. Prove: If $x = x(t)$ and $y = y(t)$ are differentiable at t, and if $z = f(x, y)$ is differentiable at the point $(x(t), y(t))$, then

$$\frac{dz}{dt} = \nabla z \cdot \mathbf{r}'(t)$$

where $\mathbf{r}(t) = x(t)\mathbf{i} + y(t)\mathbf{j}$.

56. Prove: If f, f_x, and f_y are continuous on a circular region, and if $\nabla f(x, y) = \mathbf{0}$ throughout the region, then $f(x, y)$ is constant on the region. [*Hint:* See Exercise 70, Section 16.4.]

57. Prove: If $D_\mathbf{u}f(x, y) = 0$ in two nonparallel directions, then $D_\mathbf{u}f(x, y) = 0$ in all directions.

■ **16.7** DIFFERENTIABILITY, DIRECTIONAL DERIVATIVES, AND GRADIENTS FOR FUNCTIONS OF THREE VARIABLES

In this section we shall extend most of the results obtained in the last two sections to functions of three variables. The main difference between functions of two and three variables is geometric: The graph of $z = f(x, y)$ represents a surface in 3-space, whereas $w = f(x, y, z)$ has no analogous interpretation.

☐ **DIFFERENTIABILITY**

The definition of differentiability for functions of three variables is a direct generalization of Definition 16.4.2 for functions of two variables.

16.7.1 DEFINITION. A function f of three variables is defined to be *differentiable* at the point (x_0, y_0, z_0) if the partial derivatives $f_x(x_0, y_0, z_0)$, $f_y(x_0, y_0, z_0)$, and $f_z(x_0, y_0, z_0)$ exist and

$$\Delta f = f(x_0 + \Delta x, y_0 + \Delta y, z_0 + \Delta z) - f(x_0, y_0, z_0)$$

can be written in the form

$$\Delta f = f_x(x_0, y_0, z_0)\,\Delta x + f_y(x_0, y_0, z_0)\,\Delta y$$
$$+ f_z(x_0, y_0, z_0)\,\Delta z + \epsilon_1\,\Delta x + \epsilon_2\,\Delta y + \epsilon_3\,\Delta z$$

where ϵ_1, ϵ_2, and ϵ_3 are functions of Δx, Δy, and Δz such that $\epsilon_1 \to 0$, $\epsilon_2 \to 0$, and $\epsilon_3 \to 0$ as $(\Delta x, \Delta y, \Delta z) \to (0, 0, 0)$.

The meaning of the terms **differentiable on a region R** and **differentiable** (i.e., **everywhere differentiable**) should be clear.

As in the cases of one and two variables, a function $f(x, y, z)$ that is differentiable at a point must also be continuous at that point. Moreover, the following generalization of Theorem 16.4.4, which we state without proof, gives simple conditions for $f(x, y, z)$ to be differentiable at a point.

16.7.2   THEOREM.   *If f has first-order partial derivatives at each point of some spherical region centered at (x_0, y_0, z_0), and if these partial derivatives are continuous at (x_0, y_0, z_0), then f is differentiable at (x_0, y_0, z_0).*

The reader should have no trouble modifying the proof of Theorem 16.4.7 to obtain the following version of the chain rule.

16.7.3   THEOREM (**Chain Rule**).   *If $x = x(t)$, $y = y(t)$, and $z = z(t)$ are differentiable at t and $w = f(x, y, z)$ is differentiable at the point $\big(x(t), y(t), z(t)\big)$, then $w = f\big(x(t), y(t), z(t)\big)$ is differentiable at t, and*

$$\frac{dw}{dt} = \frac{\partial w}{\partial x}\frac{dx}{dt} + \frac{\partial w}{\partial y}\frac{dy}{dt} + \frac{\partial w}{\partial z}\frac{dz}{dt}$$

Example 1   Suppose that

$$w = x^3 y^2 z, \quad x = t^2, \quad y = t^3, \quad z = t^4$$

Use the chain rule to find dw/dt.

Solution.   By the chain rule,

$$\begin{aligned}
\frac{dw}{dt} &= \frac{\partial w}{\partial x}\frac{dx}{dt} + \frac{\partial w}{\partial y}\frac{dy}{dt} + \frac{\partial w}{\partial z}\frac{dz}{dt} \\
&= (3x^2 y^2 z)(2t) + (2x^3 yz)(3t^2) + (x^3 y^2)(4t^3) \\
&= (3t^{14})(2t) + (2t^{13})(3t^2) + (t^{12})(4t^3) = 16t^{15} \quad \blacktriangleleft
\end{aligned}$$

Other variations of the chain rule for functions of three variables will be considered in the next section.

□ **DIRECTIONAL DERIVATIVES**

We shall now show how to define directional derivatives for functions of three variables. Suppose that $w = f(x, y, z)$, where f is differentiable at the point $P_0(x_0, y_0, z_0)$. Let l be a line passing through $P_0(x_0, y_0, z_0)$ and

$$\mathbf{u} = \langle u_1, u_2, u_3 \rangle$$

a unit vector along l. Let s be the (signed) distance from $P_0(x_0, y_0, z_0)$ to a point $P(x, y, z)$ on l, where s is nonnegative if P is in the direction of $\mathbf{u}$ from P_0 and negative otherwise (Figure 16.7.1).

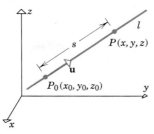

Figure 16.7.1

Since $\overrightarrow{P_0P} = s\mathbf{u}$, it follows that l is given parametrically by

$$x = x_0 + su_1, \quad y = y_0 + su_2, \quad z = z_0 + su_3 \tag{1}$$

As the parameter s varies, the point $P(x, y, z)$ moves along l and the value of $w = f(x, y, z)$ changes. The reader should have no trouble modifying the derivation of Formula (3) of Section 16.6 to show that the instantaneous rate of change of w with s in the direction of $\mathbf{u}$ at (x_0, y_0, z_0) (i.e., when $s = 0$) is

$$\left.\frac{dw}{ds}\right|_{s=0} = f_x(x_0, y_0, z_0)u_1 + f_y(x_0, y_0, z_0)u_2 + f_z(x_0, y_0, z_0)u_3$$

This quantity is denoted by $D_\mathbf{u}w(x_0, y_0, z_0)$ or $D_\mathbf{u}f(x_0, y_0, z_0)$.

16.7.4   DEFINITION.   If f is differentiable at (x_0, y_0, z_0), and if $\mathbf{u} = \langle u_1, u_2, u_3 \rangle$ is a unit vector, then the *directional derivative* of f at (x_0, y_0, z_0) in the direction of $\mathbf{u}$ is defined by

$$D_\mathbf{u}f(x_0, y_0, z_0) = f_x(x_0, y_0, z_0)u_1 + f_y(x_0, y_0, z_0)u_2 + f_z(x_0, y_0, z_0)u_3 \tag{2}$$

☐ **GRADIENT**

The directional derivative $D_\mathbf{u}f(x_0, y_0, z_0)$ represents the instantaneous rate at which $w = f(x, y, z)$ changes with distance as (x, y, z) moves from (x_0, y_0, z_0) in the direction of the unit vector $\mathbf{u}$.

Note the similarity between the formula for $D_\mathbf{u}f(x_0, y_0, z_0)$ and the formula for $D_\mathbf{u}f(x_0, y_0)$ in Definition 16.6.1. If we now define the *gradient* of f by

$$\nabla f(x, y, z) = f_x(x, y, z)\mathbf{i} + f_y(x, y, z)\mathbf{j} + f_z(x, y, z)\mathbf{k} \tag{3}$$

then the directional derivative at (x, y, z),

$$D_\mathbf{u}f(x, y, z) = f_x(x, y, z)u_1 + f_y(x, y, z)u_2 + f_z(x, y, z)u_3$$

can be expressed in the form

$$D_\mathbf{u}f(x, y, z) = \nabla f(x, y, z) \cdot \mathbf{u} \tag{4}$$

which parallels (9) of Section 16.6. Using this result, the reader should have no trouble proving the following extension of Theorem 16.6.3.

16.7.5 THEOREM.   *Let f be a function of three variables that is differentiable at (x_0, y_0, z_0).*

(a) *If $\nabla f(x_0, y_0, z_0) = \mathbf{0}$, then all directional derivatives of f at (x_0, y_0, z_0) are zero.*

(b) *If $\nabla f(x_0, y_0, z_0) \neq \mathbf{0}$, then among all possible directional derivatives of f at (x_0, y_0, z_0), the derivative in the direction of $\nabla f(x_0, y_0, z_0)$ has the largest value. The value of that directional derivative is $\|\nabla f(x_0, y_0, z_0)\|$.*

(c) *If $\nabla f(x_0, y_0, z_0) \neq \mathbf{0}$, then among all possible directional derivatives of f at (x_0, y_0, z_0), the derivative in the direction opposite to that of ∇f (x_0, y_0, z_0) has the smallest value. The value of that directional derivative is $-\|\nabla f(x_0, y_0, z_0)\|$.*

Example 2   Find the directional derivative of $f(x, y, z) = x^2y - yz^3 + z$ at the point $P(1, -2, 0)$ in the direction of the vector $\mathbf{a} = 2\mathbf{i} + \mathbf{j} - 2\mathbf{k}$, and find the maximum rate of increase of f at P.

Solution.   Since

$$f_x(x, y, z) = 2xy, \quad f_y(x, y, z) = x^2 - z^3, \quad f_z(x, y, z) = -3yz^2 + 1$$

it follows that

$$\nabla f(x, y, z) = 2xy\mathbf{i} + (x^2 - z^3)\mathbf{j} + (-3yz^2 + 1)\mathbf{k}$$
$$\nabla f(1, -2, 0) = -4\mathbf{i} + \mathbf{j} + \mathbf{k}$$

A unit vector in the direction of $\mathbf{a}$ is

$$\mathbf{u} = \frac{\mathbf{a}}{\|\mathbf{a}\|} = \frac{1}{\sqrt{9}}(2\mathbf{i} + \mathbf{j} - 2\mathbf{k}) = \tfrac{2}{3}\mathbf{i} + \tfrac{1}{3}\mathbf{j} - \tfrac{2}{3}\mathbf{k}$$

Therefore

$$D_\mathbf{u}f(1, -2, 0) = \nabla f(1, -2, 0) \cdot \mathbf{u} = (-4)(\tfrac{2}{3}) + (1)(\tfrac{1}{3}) + (1)(-\tfrac{2}{3}) = -3$$

The maximum rate of increase of f at P is

$$\|\nabla f(1, -2, 0)\| = \sqrt{(-4)^2 + (1)^2 + (1)^2} = 3\sqrt{2} \quad \blacktriangleleft$$

If $P_0(x_0, y_0, z_0)$ is a point in the domain of the function f and $f(x_0, y_0, z_0) = c$, then under appropriate conditions* the equation

$$f(x, y, z) = c$$

defines a level surface of f that passes through the point $P_0(x_0, y_0, z_0)$. The

---

*Precise conditions under which the graph of $f(x, y, z) = c$ is a unique surface in the vicinity of (x_0, y_0, z_0) with a tangent plane at (x_0, y_0, z_0) are specified in a theorem from advanced calculus called the "implicit function theorem." For simplicity, we shall assume that these conditions are satisfied.

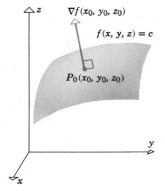

Figure 16.7.2

following theorem states that the gradient of f at P_0 is perpendicular to this surface at P_0 (Figure 16.7.2). (Recall that for functions of two variables, the gradient at a point is perpendicular to the level *curve* through the point.)

16.7.6 THEOREM.   *If f is differentiable at (x_0, y_0, z_0) and $\nabla f(x_0, y_0, z_0) \neq \mathbf{0}$, then $\nabla f(x_0, y_0, z_0)$ is perpendicular to the level surface of $f(x, y, z)$ through (x_0, y_0, z_0).*

We omit the proof.

Example 3   Find an equation of the plane that is tangent to the ellipsoid $x^2 + 4y^2 + z^2 = 18$ at the point $(1, 2, -1)$.

Solution.   If we let $F(x, y, z) = x^2 + 4y^2 + z^2$, then the given equation has the form $F(x, y, z) = 18$, which may be viewed as the equation of a level surface for F. Thus, the vector $\nabla F(1, 2, -1)$ is normal to the ellipsoid at the point $(1, 2, -1)$. To find this vector we write

$$\nabla F(x, y, z) = \frac{\partial F}{\partial x}\mathbf{i} + \frac{\partial F}{\partial y}\mathbf{j} + \frac{\partial F}{\partial z}\mathbf{k} = 2x\mathbf{i} + 8y\mathbf{j} + 2z\mathbf{k}$$

$$\nabla F(1, 2, -1) = 2\mathbf{i} + 16\mathbf{j} - 2\mathbf{k}$$

Using this normal and the point $(1, 2, -1)$, we obtain as the equation of the tangent plane

$$2(x - 1) + 16(y - 2) - 2(z + 1) = 0$$

or

$$x + 8y - z = 18$$

Alternative Solution.   Solving the given equation for the variable z, we obtain $z = \pm\sqrt{18 - x^2 - 4y^2}$. Since the point $(1, 2, -1)$ has a negative z-coordinate, this point lies on the lower half of the ellipsoid, which is given by

$$z = -\sqrt{18 - x^2 - 4y^2}$$

Using this expression and Theorem 16.5.1, we can obtain the equation of the tangent plane as follows:

$$\frac{\partial z}{\partial x} = \frac{x}{\sqrt{18 - x^2 - 4y^2}}, \quad \frac{\partial z}{\partial y} = \frac{4y}{\sqrt{18 - x^2 - 4y^2}}$$

so that

$$\frac{\partial z}{\partial x}\bigg|_{x=1,\, y=2} = 1, \quad \frac{\partial z}{\partial y}\bigg|_{x=1,\, y=2} = 8$$

Thus, from Formula (1) of Section 16.5 the tangent plane to the ellipsoid at $(1, 2, -1)$ is given by

$$(1)(x - 1) + 8(y - 2) - (z + 1) = 0$$

or

$$x + 8y - z = 18$$

which agrees with our previous solution. ◀

REMARK.   The first method in the last example is useful when it is inconvenient or impossible to express z explicitly as a function of x and y.

☐ **TOTAL**
DIFFERENTIALS

If $w = f(x, y, z)$, then we define the *increment* Δw (also written Δf) to be

$$\Delta w = f(x + \Delta x, y + \Delta y, z + \Delta z) - f(x, y, z)$$

and we define the *total differential* dw (also written df) to be

$$dw = f_x(x, y, z)\, dx + f_y(x, y, z)\, dy + f_z(x, y, z)\, dz$$

where Δx, Δy, Δz, dx, dy, and dz are all variables representing changes in the values of x, y, and z.

The increment Δw represents the change in the value of $w = f(x, y, z)$ when x, y, and z are changed by amounts Δx, Δy, and Δz, respectively. However, for functions of three variables, the total differential dw has no natural geometric interpretation.

If we let

$$dx = \Delta x, \quad dy = \Delta y, \quad dz = \Delta z$$

and if $w = f(x, y, z)$ is differentiable at (x, y, z), then it follows from Definition 16.7.1 that

$$\Delta w = dw + \epsilon_1 \Delta x + \epsilon_2 \Delta y + \epsilon_3 \Delta z \tag{5}$$

where $\epsilon_1 \to 0$, $\epsilon_2 \to 0$, $\epsilon_3 \to 0$ as $(\Delta x, \Delta y, \Delta z) \to (0, 0, 0)$. Thus, when Δx, Δy, and Δz are small, it follows from (5) that

$$\Delta w \approx dw$$

Example 4   The length, width, and height of a rectangular box are each measured with an error of at most 5%. Find an upper bound on the maximum possible percentage error that results if these quantities are used to calculate the diagonal of the box.

Solution.   Let x, y, z, and D be the true length, width, height, and diagonal of the box, respectively; and let Δx, Δy, Δz, and ΔD be the errors in these quantities. We are given that

$$\left| \frac{\Delta x}{x} \right| \leq 0.05, \quad \left| \frac{\Delta y}{y} \right| \leq 0.05, \quad \left| \frac{\Delta z}{z} \right| \leq 0.05$$

We want to estimate $|\Delta D/D|$. Since the diagonal D is related to the length, width, and height by

$$D = \sqrt{x^2 + y^2 + z^2}$$

it follows that

$$dD = \frac{\partial D}{\partial x}\, dx + \frac{\partial D}{\partial y}\, dy + \frac{\partial D}{\partial z}\, dz$$

$$= \frac{x}{\sqrt{x^2 + y^2 + z^2}}\, dx + \frac{y}{\sqrt{x^2 + y^2 + z^2}}\, dy + \frac{z}{\sqrt{x^2 + y^2 + z^2}}\, dz$$

If we choose $\Delta x = dx$, $\Delta y = dy$, $\Delta z = dz$, then we can use the approximation $\Delta D/D \approx dD/D$. But,

$$\frac{dD}{D} = \frac{x}{x^2 + y^2 + z^2}\, dx + \frac{y}{x^2 + y^2 + z^2}\, dy + \frac{z}{x^2 + y^2 + z^2}\, dz$$

or

$$\frac{dD}{D} = \frac{x^2}{x^2 + y^2 + z^2}\frac{dx}{x} + \frac{y^2}{x^2 + y^2 + z^2}\frac{dy}{y} + \frac{z^2}{x^2 + y^2 + z^2}\frac{dz}{z}$$

Thus,

$$\left|\frac{dD}{D}\right| = \left|\frac{x^2}{x^2 + y^2 + z^2}\frac{dx}{x} + \frac{y^2}{x^2 + y^2 + z^2}\frac{dy}{y} + \frac{z^2}{x^2 + y^2 + z^2}\frac{dz}{z}\right|$$

$$\leq \left|\frac{x^2}{x^2 + y^2 + z^2}\frac{dx}{x}\right| + \left|\frac{y^2}{x^2 + y^2 + z^2}\frac{dy}{y}\right| + \left|\frac{z^2}{x^2 + y^2 + z^2}\frac{dz}{z}\right|$$

$$\leq \frac{x^2}{x^2 + y^2 + z^2}(0.05) + \frac{y^2}{x^2 + y^2 + z^2}(0.05) + \frac{z^2}{x^2 + y^2 + z^2}(0.05)$$

$$= 0.05$$

Therefore, the maximum percentage error in D is at most 5%.   ◄

▶ Exercise Set 16.7   Ⓒ 24, 41

In Exercises 1–4, find dw/dt using the chain rule.

1. $w = 5x^2y^3z^4$; $x = t^2$, $y = t^3$, $z = t^5$.

2. $w = \ln(3x^2 - 2y + 4z^3)$; $x = t^{1/2}$, $y = t^{2/3}$, $z = t^{-2}$.

3. $w = 5\cos xy - \sin xz$; $x = 1/t$, $y = t$, $z = t^3$.

4. $w = \sqrt{1 + x - 2yz^4x}$; $x = \ln t$, $y = t$, $z = 4t$.

5. Use the chain rule to find $\left.\dfrac{dw}{dt}\right|_{t=1}$ if $w = x^3y^2z^4$; $x = t^2$, $y = t + 2$, $z = 2t^4$.

6. Use the chain rule to find $\left.\dfrac{dw}{dt}\right|_{t=0}$ if $w = x\sin yz^2$; $x = \cos t$, $y = t^2$, $z = e^t$.

In Exercises 7–10, find the gradient of f at P, and then use the gradient to calculate $D_{\mathbf{u}}f$ at P.

7. $f(x, y, z) = 4x^5y^2z^3$; $P(2, -1, 1)$; $\mathbf{u} = \frac{1}{3}\mathbf{i} + \frac{2}{3}\mathbf{j} - \frac{2}{3}\mathbf{k}$.

8. $f(x, y, z) = ye^{xz} + z^2$; $P(0, 2, 3)$; $\mathbf{u} = \frac{2}{7}\mathbf{i} - \frac{3}{7}\mathbf{j} + \frac{6}{7}\mathbf{k}$.

9. $f(x, y, z) = \ln(x^2 + 2y^2 + 3z^2);$ $P(-1, 2, 4);$
    $\mathbf{u} = -\frac{3}{13}\mathbf{i} - \frac{4}{13}\mathbf{j} - \frac{12}{13}\mathbf{k}.$

10. $f(x, y, z) = \sin xyz;$ $P(\frac{1}{2}, \frac{1}{3}, \pi);$
    $\mathbf{u} = \frac{1}{\sqrt{3}}\mathbf{i} - \frac{1}{\sqrt{3}}\mathbf{j} + \frac{1}{\sqrt{3}}\mathbf{k}.$

In Exercises 11–14, find the directional derivative of
f at P in the direction of $\mathbf{a}$.

11. $f(x, y, z) = x^3z - yx^2 + z^2;$ $P(2, -1, 1);$
    $\mathbf{a} = 3\mathbf{i} - \mathbf{j} + 2\mathbf{k}.$

12. $f(x, y, z) = y - \sqrt{x^2 + z^2};$ $P(-3, 1, 4);$
    $\mathbf{a} = 2\mathbf{i} - 2\mathbf{j} - \mathbf{k}.$

13. $f(x, y, z) = \dfrac{z - x}{z + y};$ $P(1, 0, -3);$
    $\mathbf{a} = -6\mathbf{i} + 3\mathbf{j} - 2\mathbf{k}.$

14. $f(x, y, z) = e^{x+y+3z};$ $P(-2, 2, -1);$
    $\mathbf{a} = 20\mathbf{i} - 4\mathbf{j} + 5\mathbf{k}.$

In Exercises 15–18, find a unit vector in the direction
in which f increases most rapidly at P, and find the
rate of increase of f in that direction.

15. $f(x, y, z) = x^3z^2 + y^3z + z - 1;$ $P(1, 1, -1).$

16. $f(x, y, z) = \sqrt{x - 3y + 4z};$ $P(0, -3, 0).$

17. $f(x, y, z) = \dfrac{x}{z} + \dfrac{z}{y^2};$ $P(1, 2, -2).$

18. $f(x, y, z) = \tan^{-1}\left(\dfrac{x}{y + z}\right);$ $P(4, 2, 2).$

In Exercises 19 and 20, find a unit vector in the direction in which f decreases most rapidly at P, and find
the rate of change of f in that direction.

19. $f(x, y, z) = \dfrac{x + z}{z - y};$ $P(5, 7, 6).$

20. $f(x, y, z) = 4e^{xy}\cos z;$ $P(0, 1, \pi/4).$

21. Find the directional derivative of the function
    $$f(x, y, z) = \frac{y}{x + z} \text{ at } P(2, 1, -1) \text{ in the direction from}$$
    P to $Q(-1, 2, 0).$

22. Find the directional derivative of the function
    $$f(x, y, z) = x^3y^2z^5 - 2xz + yz + 3x$$
    at $P(-1, -2, 1)$ in the direction of the negative z-
    axis.

23. Given that the directional derivative of $f(x, y, z)$ at

the point $(3, -2, 1)$ in the direction of $\mathbf{a} = 2\mathbf{i} - \mathbf{j} - 2\mathbf{k}$
is -5 and that $\|\nabla f(3, -2, 1)\| = 5$, find $\nabla f(3, -2, 1).$

24. The temperature (in degrees Celsius) at a point (x, y, z)
    in a metal solid is
    $$T(x, y, z) = \frac{xyz}{1 + x^2 + y^2 + z^2}$$

    (a) Find the rate of change of temperature at $(1, 1, 1)$
        in the direction of the origin.
    (b) Find the direction in which the temperature rises
        most rapidly at $(1, 1, 1)$. (Express your answer
        as a unit vector.)
    (c) Find the rate at which the temperature rises
        moving from $(1, 1, 1)$ in the direction obtained
        in part (b).

In Exercises 25–28, find equations for the tangent plane
and the line that is normal to the given surface at the
point P.

25. $x^2 + y^2 + z^2 = 49;$ $P(-3, 2, -6).$

26. $xz - yz^3 + yz^2 = 2;$ $P(2, -1, 1).$

27. $\sqrt{\dfrac{z + x}{y - 1}} = z^2;$ $P(3, 5, 1).$

28. $\sin xz - 4\cos yz = 4;$ $P(\pi, \pi, 1).$

29. Show that every line that is normal to the sphere
    $x^2 + y^2 + z^2 = 1$ passes through the origin.

30. Find all points on the ellipsoid $2x^2 + 3y^2 + 4z^2 = 9$
    at which the tangent plane is parallel to the plane
    $x - 2y + 3z = 5.$

31. Find all points on the surface $x^2 + y^2 - z^2 = 1$ at
    which the normal line is parallel to the line through
    $P(1, -2, 1)$ and $Q(4, 0, -1).$

32. Show that the ellipsoid $2x^2 + 3y^2 + z^2 = 9$ and the
    sphere
    $$x^2 + y^2 + z^2 - 6x - 8y - 8z + 24 = 0$$
    have a common tangent plane at the point $(1, 1, 2).$

In Exercises 33–36, find dw.

33. $w = 8x - 3y + 4z.$

34. $w = 4x^2y^3z^7 - 3xy + z + 5.$

35. $w = \tan^{-1}(xyz).$

36. $w = \sqrt{x} + \sqrt{y} + \sqrt{z}.$

37. Use a total differential to approximate the change in
    $f(x, y, z) = 2xy^2z^3$ as (x, y, z) varies from $P(1, -1, 2)$
    to $Q(0.99, -1.02, 2.02).$

38. Use a total differential to approximate the change in $f(x, y, z) = xyz/(x + y + z)$ as (x, y, z) varies from $P(-1, -2, 4)$ to $Q(-1.04, -1.98, 3.97)$.

39. The length, width, and height of a rectangular box are measured to be 3 cm, 4 cm, and 5 cm, respectively, with a maximum error of 0.05 cm in each measurement. Use differentials to approximate the maximum error in the calculated volume.

40. The total resistance R of three resistances R_1, R_2, and R_3, connected in parallel, is given by

$$\frac{1}{R} = \frac{1}{R_1} + \frac{1}{R_2} + \frac{1}{R_3}$$

Suppose that R_1, R_2, and R_3 are measured to be 100 ohms, 200 ohms, and 500 ohms, respectively, with a maximum error of 10% in each. Use differentials to approximate the maximum percentage error in the calculated value of R.

41. The area of a triangle is to be computed from the formula $A = \frac{1}{2}ab \sin \theta$, where a and b are the lengths of two sides and θ is the included angle. Suppose that a, b, and θ are measured to be 40 ft, 50 ft, and 30°, respectively. Use differentials to approximate the maximum error in the calculated value of A if the maximum errors in a, b, and θ are $\frac{1}{2}$ ft, $\frac{1}{4}$ ft, and 2°, respectively.

42. The length, width, and height of a rectangular box are measured with errors of at most $r\%$. Use differentials to approximate the maximum error in the computed value of the volume.

43. Use differentials to approximate the maximum percentage error in $w = xy^2z^3$ if x, y, and z have errors of at most 1%, 2%, and 3%, respectively.

44. Two surfaces are said to be **orthogonal** at a point of intersection if their normal lines are perpendicular at that point. Prove that the surfaces $f(x, y, z) = 0$ and $g(x, y, z) = 0$ are orthogonal at a point of intersection, (x_0, y_0, z_0), if and only if

$$f_x g_x + f_y g_y + f_z g_z = 0$$

at (x_0, y_0, z_0). [Assume that $\nabla f(x_0, y_0, z_0) \neq \mathbf{0}$ and $\nabla g(x_0, y_0, z_0) \neq \mathbf{0}$.]

45. Use the result of Exercise 44 to show that the sphere $x^2 + y^2 + z^2 = a^2$ and the cone $z^2 = x^2 + y^2$ are orthogonal at every point of intersection.

16.8 FUNCTIONS OF n VARIABLES; MORE ON THE CHAIN RULE

In this section we shall discuss functions involving more than three variables. Our main objective is to develop forms of the chain rule for such functions.

All the definitions and theorems we have stated for functions of two and three variables can be extended to functions of four or more variables. Recall that if

$$w = f(v_1, v_2, \ldots, v_n)$$

is a function of n variables, then there are n partial derivatives

$$\frac{\partial w}{\partial v_1}, \frac{\partial w}{\partial v_2}, \ldots, \frac{\partial w}{\partial v_n}$$

each of which is calculated by holding $n - 1$ of the variables fixed and differentiating with respect to the remaining variable. In order to define directional derivatives for a function of n variables, it is first necessary to define the notion of a vector in "n-dimensional space." This topic is studied in a branch of mathematics called **linear algebra,** and will not be considered in this text.

If $w = f(v_1, v_2, \ldots, v_n)$, we define the **increment** Δw and the **total differential** dw to be

$$\Delta w = f(v_1 + \Delta v_1, v_2 + \Delta v_2, \ldots, v_n + \Delta v_n) - f(v_1, v_2, \ldots, v_n) \qquad (1)$$

and

$$dw = \frac{\partial w}{\partial v_1}\, dv_1 + \frac{\partial w}{\partial v_2}\, dv_2 + \cdots + \frac{\partial w}{\partial v_n}\, dv_n \qquad (2)$$

where $\Delta v_1, \Delta v_2, \ldots, \Delta v_n$ and $dv_1, dv_2, \ldots, dv_n$ are variables representing changes in the values of $v_1, v_2, \ldots, v_n$.

If $v_1, v_2, \ldots, v_n$ are functions of a single variable t, then $w = f(v_1, v_2, \ldots, v_n)$ is a function of t, and a chain rule formula for dw/dt is

$$\frac{dw}{dt} = \frac{\partial w}{\partial v_1}\frac{dv_1}{dt} + \frac{\partial w}{\partial v_2}\frac{dv_2}{dt} + \cdots + \frac{\partial w}{\partial v_n}\frac{dv_n}{dt} \qquad (3)$$

which is a natural extension of the chain-rule formulas in Theorems 16.4.7 and 16.7.3. Observe that (3) results if we formally divide both sides of (2) by dt.

Other forms of the chain rule arise, depending on the number of variables involved. For example, in Theorem 16.4.8 we obtained the chain-rule formulas

$$\frac{\partial z}{\partial u} = \frac{\partial z}{\partial x}\frac{\partial x}{\partial u} + \frac{\partial z}{\partial y}\frac{\partial y}{\partial u} \qquad (4)$$

$$\frac{\partial z}{\partial v} = \frac{\partial z}{\partial x}\frac{\partial x}{\partial v} + \frac{\partial z}{\partial y}\frac{\partial y}{\partial v} \qquad (5)$$

for the case where z is a function of two variables $z = f(x, y)$; and x and y in turn are functions of two other variables, $x = x(u, v)$, $y = y(u, v)$.

Formulas (4) and (5) can be represented schematically by a "tree diagram" constructed as follows (Figure 16.8.1). Starting with z at the top of the diagram and moving downward, join each variable by lines (or branches) to those variables that depend *directly* on it. Thus, z is joined to x and y and these in turn are each joined to u and v. Next, label each branch with a derivative whose "numerator" contains the variable at the top end of that branch, and whose "denominator" contains the variable at the bottom end of that branch. This completes the "tree." To find the formula for $\partial z/\partial u$ trace all paths through the tree that start at z and end at u. Each such path produces one of the terms in the formula for $\partial z/\partial u$ (Figure 16.8.1a). Similarly, each term in the formula for $\partial z/\partial v$ corresponds to a path starting at z and ending at v (Figure 16.8.1b).

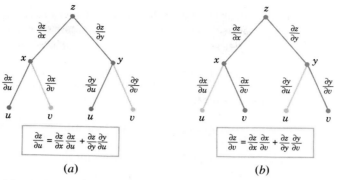

Figure 16.8.1

The following examples illustrate how tree diagrams can be used to construct other forms of the chain rule.

Example 1 Suppose that

$$w = e^{xyz}, \quad x = 3r + s, \quad y = 3r - s, \quad z = r^2 s$$

Use appropriate forms of the chain rule to find $\partial w/\partial r$ and $\partial w/\partial s$.

Solution. From the tree diagram on the left we obtain the chain-rule formulas on the right:

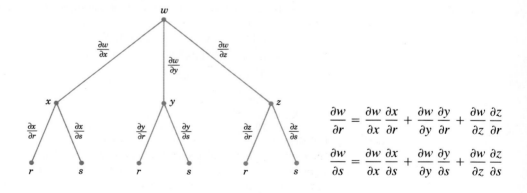

These formulas yield

$$\frac{\partial w}{\partial r} = yze^{xyz}(3) + xze^{xyz}(3) + xye^{xyz}(2rs) = e^{xyz}(3yz + 3xz + 2xyrs)$$

and

$$\frac{\partial w}{\partial s} = yze^{xyz}(1) + xze^{xyz}(-1) + xye^{xyz}(r^2) = e^{xyz}(yz - xz + xyr^2)$$

If desired, we can express $\partial w/\partial r$ and $\partial w/\partial s$ in terms of r and s alone by replacing x, y, and z by their expressions in terms of r and s. ◄

Example 2 Suppose that $w = x^2 + y^2 - z^2$ and

$$x = \rho \sin \phi \cos \theta, \quad y = \rho \sin \phi \sin \theta, \quad z = \rho \cos \phi$$

Use appropriate forms of the chain rule to find $\partial w/\partial \rho$ and $\partial w/\partial \theta$.

Solution. From the tree diagram on the left we obtain the chain-rule formulas on the right:

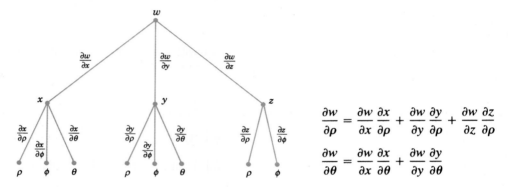

$$\frac{\partial w}{\partial \rho} = \frac{\partial w}{\partial x}\frac{\partial x}{\partial \rho} + \frac{\partial w}{\partial y}\frac{\partial y}{\partial \rho} + \frac{\partial w}{\partial z}\frac{\partial z}{\partial \rho}$$

$$\frac{\partial w}{\partial \theta} = \frac{\partial w}{\partial x}\frac{\partial x}{\partial \theta} + \frac{\partial w}{\partial y}\frac{\partial y}{\partial \theta}$$

From these formulas we obtain

$$\frac{\partial w}{\partial \rho} = (2x) \sin \phi \cos \theta + 2y \sin \phi \sin \theta - 2z \cos \phi$$

$$= 2\rho \sin^2 \phi \cos^2 \theta + 2\rho \sin^2 \phi \sin^2 \theta - 2\rho \cos^2 \phi$$
$$= 2\rho \sin^2 \phi (\cos^2 \theta + \sin^2 \theta) - 2\rho \cos^2 \phi$$
$$= 2\rho (\sin^2 \phi - \cos^2 \phi)$$
$$= -2\rho \cos 2\phi$$

$$\frac{\partial w}{\partial \theta} = (2x)(-\rho \sin \phi \sin \theta) + (2y) \rho \sin \phi \cos \theta$$

$$= -2\rho^2 \sin^2 \phi \sin \theta \cos \theta + 2\rho^2 \sin^2 \phi \sin \theta \cos \theta$$
$$= 0$$

This result is explained by the fact that w does not vary with θ. We may see this directly by expressing w in terms of ρ, ϕ, and θ. If this is done, the expressions involving θ will cancel, leaving w as a function of ρ and ϕ alone. (Verify that $w = -\rho^2 \cos 2\phi$.) ◄

In many applications of the chain rule, some of the variables in the function $w = f(v_1, v_2, \ldots, v_n)$ are functions of the remaining variables. Tree diagrams are especially helpful in such situations.

Example 3 **Suppose that**

$$w = xy + yz, \quad y = \sin x, \quad z = e^x$$

Use an appropriate form of the chain rule to find dw/dx.

Solution.   From the tree diagram on the left we obtain the chain-rule formula on the right:

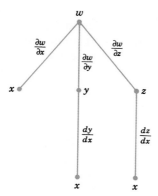

$$\frac{dw}{dx} = \frac{\partial w}{\partial x} + \frac{\partial w}{\partial y}\frac{dy}{dx} + \frac{\partial w}{\partial z}\frac{dz}{dx}$$

From this formula we obtain

$$\frac{dw}{dx} = y + (x + z)\cos x + ye^x$$

$$= \sin x + (x + e^x)\cos x + e^x \sin x$$

This same result can be obtained by first expressing w explicitly in terms of x,

$$w = x \sin x + e^x \sin x$$

and then differentiating with respect to x; however, such direct substitution is not always convenient.   ◄

REMARK.   Unlike the differential dz, a partial symbol ∂z has no meaning of its own. For example, if we were to "cancel" partial symbols in the chain-rule formula

$$\frac{\partial z}{\partial u} = \frac{\partial z}{\partial x}\frac{\partial x}{\partial u} + \frac{\partial z}{\partial y}\frac{\partial y}{\partial u}$$

we would obtain

$$\frac{\partial z}{\partial u} = \frac{\partial z}{\partial u} + \frac{\partial z}{\partial u}$$

which is nonsense in cases where $\partial z/\partial u \neq 0$.

In each of the expressions

$$z = \sin xy, \quad z = \frac{xy}{1 + xy}, \quad z = e^{xy}$$

the independent variables occur only in the combination xy, so the substitution $t = xy$ reduces the expression to a function of one variable:

$$z = \sin t, \quad z = \frac{t}{1 + t}, \quad z = e^t$$

Conversely, if we begin with a function of one variable $z = f(t)$ and substitute $t = xy$, we obtain a function $z = f(xy)$ in which the variables appear only in the combination xy. Functions whose variables occur in fixed combinations arise frequently in applications.

Example 4   Show that a function of the form $z = f(xy)$ satisfies the equation

$$x \frac{\partial z}{\partial x} - y \frac{\partial z}{\partial y} = 0$$

(assuming the derivatives exist).

Solution.   Let $t = xy$, so that $z = f(t)$. From the tree diagram on the left we obtain the chain-rule formulas on the right:

$$\frac{\partial z}{\partial x} = \frac{dz}{dt} \frac{\partial t}{\partial x} = y \frac{dz}{dt}$$

$$\frac{\partial z}{\partial y} = \frac{dz}{dt} \frac{\partial t}{\partial y} = x \frac{dz}{dt}$$

Thus,

$$x \frac{\partial z}{\partial x} - y \frac{\partial z}{\partial y} = xy \frac{dz}{dt} - yx \frac{dz}{dt} = 0 \quad \blacktriangleleft$$

▶ Exercise Set 16.8   [C] 15

1. Let $f(v, w, x, y) = 4v^2 w^3 x^4 y^5$. Find $\partial f/\partial v$, $\partial f/\partial w$, $\partial f/\partial x$, and $\partial f/\partial y$.

2. Let $w = r \cos st + e^u \sin ur$. Find $\partial w/\partial r$, $\partial w/\partial s$, $\partial w/\partial t$, and $\partial w/\partial u$.

3. Let $f(v_1, v_2, v_3, v_4) = \dfrac{v_1^2 - v_2^2}{v_3^2 + v_4^2}$. Find $\partial f/\partial v_1$, $\partial f/\partial v_2$, $\partial f/\partial v_3$, and $\partial f/\partial v_4$.

4. Let $V = xe^{2x-y} + we^{zw} + yw$. Find V_x, V_y, V_z, and V_w.

5. Let $f(v, w, x, y) = 2v^{1/2} w^4 x^{1/2} y^{2/3}$. Find $f_v(1, -2, 4, 8)$, $f_w(1, -2, 4, 8)$, $f_x(1, -2, 4, 8)$, and $f_y(1, -2, 4, 8)$.

6. Let $u(w, x, y, z) = xe^{yw} \sin^2 z$. Find

(a) $\dfrac{\partial u}{\partial x}(0, 0, 1, \pi)$

(b) $\dfrac{\partial u}{\partial y}(0, 0, 1, \pi)$

(c) $\dfrac{\partial u}{\partial w}(0, 0, 1, \pi)$

(d) $\dfrac{\partial u}{\partial z}(0, 0, 1, \pi)$

(e) $\dfrac{\partial^4 u}{\partial x \, \partial y \, \partial w \, \partial z}$

(f) $\dfrac{\partial^4 u}{\partial w \, \partial z \, \partial y^2}$.

In Exercises 7–13, use appropriate forms of the chain rule to find the derivatives.

7. Let $v = 7w^2x^3y^4z^5$, where $w = t^4$, $x = t^3$, $y = t^2$, $z = t$. Find dv/dt.

8. Let $z = u^7$, where $u = 5x^2 - 2y^3$. Find $\partial z/\partial x$ and $\partial z/\partial y$.

9. Let $z = \ln(x^2 + 1)$, where $x = r\cos\theta$. Find $\partial z/\partial r$ and $\partial z/\partial\theta$.

10. Let $u = rs^2 \ln t$, $r = x^2$, $s = 4y + 1$, $t = xy^3$. Find $\partial u/\partial x$ and $\partial u/\partial y$.

11. Let $w = 4x^2 + 4y^2 + z^2$, $x = \rho\sin\phi\cos\theta$, $y = \rho\sin\phi\sin\theta$, $z = \rho\cos\phi$. Find $\partial w/\partial\rho$, $\partial w/\partial\phi$, and $\partial w/\partial\theta$.

12. Let $w = 3xy^2z^3$, $y = 3x^2 + 2$, $z = \sqrt{x - 1}$. Find dw/dx.

13. Let $w = \sqrt{x^2 + y^2 + z^2}$, $x = \cos 2y$, $z = \sqrt{y}$. Find dw/dy.

14. The length, width, and height of a rectangular box are increasing at rates of 1 in./sec, 2 in./sec, and 3 in./sec, respectively.
(a) At what rate is the volume increasing when the length is 2 in., the width is 3 in., and the height is 6 in.?
(b) At what rate is the diagonal increasing at that instant?

15. Angle A of triangle ABC is increasing at a rate of $\pi/60$ rad/sec, side AB is increasing at a rate of 2 cm/sec, and side AC is increasing at a rate of 4 cm/sec. At what rate is the length of BC changing when angle A is $\pi/3$ rad, $AB = 20$ cm, and $AC = 10$ cm? Is the length of BC increasing or decreasing? [*Hint:* Use the law of cosines.]

16. The area A of a triangle is given by $A = \frac{1}{2}ab\sin\theta$, where a and b are the lengths of two sides and θ is the angle between these sides. Suppose $a = 5$, $b = 10$, and $\theta = \pi/3$. Find
(a) the rate at which A changes with a if b and θ are held constant
(b) the rate at which A changes with θ if a and b are held constant
(c) the rate at which b changes with a if A and θ are held constant.

17. Suppose $x^2 + 4xz + z^2 - 3yz + 5 = 0$. Find $\partial z/\partial x$ and $\partial z/\partial y$ by implicit differentiation.

18. Suppose $e^{xy}\cos yz - e^{yz}\sin xz + 2 = 0$. Find $\partial z/\partial x$ and $\partial z/\partial y$ by implicit differentiation.

19. Let $f(w, x, y, z) = wz\tan^{-1}\dfrac{x}{y} + 5w$. Show that
$$f_{ww} + f_{xx} + f_{yy} + f_{zz} = 0$$

In the remaining exercises, you may assume that all derivatives mentioned exist.

20. Let f be a function of one variable, and let $z = f(x + 2y)$. Show that
$$2\frac{\partial z}{\partial x} - \frac{\partial z}{\partial y} = 0$$

21. Let f be a function of one variable and let $z = f(x^2 + y^2)$. Show that
$$y\frac{\partial z}{\partial x} - x\frac{\partial z}{\partial y} = 0$$

22. Let f be a function of one variable, and let $w = f(\rho)$, where $\rho = (x^2 + y^2 + z^2)^{1/2}$. Show that
$$\left(\frac{\partial w}{\partial x}\right)^2 + \left(\frac{\partial w}{\partial y}\right)^2 + \left(\frac{\partial w}{\partial z}\right)^2 = \left(\frac{dw}{d\rho}\right)^2$$

23. Let f be a function of three variables and let $w = f(x - y, y - z, z - x)$. Show that
$$\frac{\partial w}{\partial x} + \frac{\partial w}{\partial y} + \frac{\partial w}{\partial z} = 0$$

24. Let $w = f(x, y, z)$, $x = \rho\sin\phi\cos\theta$, $y = \rho\sin\phi\sin\theta$, and $z = \rho\cos\phi$. Express $\partial w/\partial\rho$, $\partial w/\partial\phi$, and $\partial w/\partial\theta$ in terms of $\partial w/\partial x$, $\partial w/\partial y$, and $\partial w/\partial z$.

25. Assume that $F(x, y, z) = 0$ defines z implicitly as a function of x and y. Show that
$$\frac{\partial z}{\partial x} = -\frac{\partial F/\partial x}{\partial F/\partial z} \quad\text{and}\quad \frac{\partial z}{\partial y} = -\frac{\partial F/\partial y}{\partial F/\partial z}$$

In Exercises 26–28, use the formulas in Exercise 25 to find $\partial z/\partial x$ and $\partial z/\partial y$.

26. $x^2 - 3yz^2 + xyz - 2 = 0$.

27. $ye^x - 5\sin 3z = 3z$.

28. $\ln(1 + z) + xy^2 + z = 1$.

29. Given that $u = u(x, y, z)$, $v = v(x, y, z)$, and $w = w(x, y, z)$, show that
$$\nabla f(u, v, w) = \frac{\partial f}{\partial u}\nabla u + \frac{\partial f}{\partial v}\nabla v + \frac{\partial f}{\partial w}\nabla w$$

30. Let $w = f(x, y, z)$ where $z = g(x, y)$. Taking x and y as the independent variables, express each of the following in terms of $\partial f/\partial x$, $\partial f/\partial y$, $\partial f/\partial z$, $\partial z/\partial x$, and $\partial z/\partial y$.

(a) $\partial w/\partial x$ (b) $\partial w/\partial y$.

31. Let $w = \ln(e^r + e^s + e^t + e^u)$. Show that

$$w_{rstu} = -6e^{r+s+t+u-4w}$$

[*Hint:* Take advantage of the relationship $e^w = e^r + e^s + e^t + e^u$.]

32. Suppose w is a function of x_1, x_2, and x_3, and

$$x_1 = a_1y_1 + b_1y_2$$
$$x_2 = a_2y_1 + b_2y_2$$
$$x_3 = a_3y_1 + b_3y_2$$

where the a's and b's are constants. Express $\partial w/\partial y_1$ and $\partial w/\partial y_2$ in terms of $\partial w/\partial x_1$, $\partial w/\partial x_2$, and $\partial w/\partial x_3$.

33. (a) Let w be a function of x_1, x_2, x_3, and x_4, and let each x_i be a function of t. Find a chain rule formula for dw/dt.

(b) Let w be a function of x_1, x_2, x_3, and x_4, and let each x_i be a function of v_1, v_2, and v_3. Find chain-rule formulas for $\partial w/\partial v_1$, $\partial w/\partial v_2$, and $\partial w/\partial v_3$.

34. Let $w = (x_1^2 + x_2^2 + \cdots + x_n^2)^k$, where $n > 2$. For what values of k does

$$\frac{\partial^2 w}{\partial x_1^2} + \frac{\partial^2 w}{\partial x_2^2} + \cdots + \frac{\partial^2 w}{\partial x_n^2} = 0$$

hold?

35. Show that

$$\frac{d}{dx}\left[\int_{a(x)}^{b(x)} f(t)\, dt\right] = f(b(x))b'(x) - f(a(x))a'(x)$$

This result is called **Leibniz's rule**. [*Hint:* Let $u = a(x)$, $v = b(x)$, and

$$F(u, v) = \int_u^v f(t)\, dt$$

Then use a chain rule and Theorem 5.9.3.]

36. Use the result of Exercise 35 to compute the following derivatives without performing the integrations.

(a) $\dfrac{d}{dx}\displaystyle\int_x^{x^2} e^{t^2}\, dt$ (b) $\dfrac{d}{dx}\displaystyle\int_{\sin x}^{\cos x} (t^3 + 2)^{2/3}\, dt$

(c) $\dfrac{d}{dx}\displaystyle\int_{3x}^{x^3} \sin^5 t\, dt$ (d) $\dfrac{d}{dx}\displaystyle\int_{e^x}^{e^{2x}} (\ln t)^4\, dt$.

16.9 MAXIMA AND MINIMA OF FUNCTIONS OF TWO VARIABLES

Earlier in this text we learned how to find maximum and minimum values of a function of one variable. In this section we shall develop similar techniques for functions of two variables.

☐ **EXTREMA**

If we imagine the graph of a function f of two variables to be a portion of the earth's terrain (Figure 16.9.1), then the mountaintops, which are the high points in their immediate vicinity, are called *relative maxima* of f, and the valley bottoms, which are the low points in their immediate vicinity, are called *relative minima* of f.

Just as a geologist might be interested in finding the highest mountain and deepest valley in an entire mountain range, so a mathematician might be interested in finding the largest and smallest values of $f(x, y)$ over the *entire* domain of f. These are called the *absolute maximum* and *absolute minimum* values of f. The following definitions make these informal ideas precise.

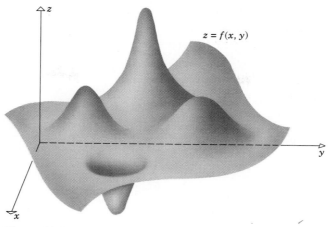

Figure 16.9.1

16.9.1 DEFINITION. A function f of two variables is said to have a **relative maximum** at a point (x_0, y_0) if there is a circle centered at (x_0, y_0) such that $f(x_0, y_0) \geq f(x, y)$ for all points (x, y) in the domain of f that lie inside the circle, and f is said to have an **absolute maximum** at (x_0, y_0) if $f(x_0, y_0) \geq f(x, y)$ for all points (x, y) in the domain of f.

16.9.2 DEFINITION. A function f of two variables is said to have a **relative minimum** at a point (x_0, y_0) if there is a circle centered at (x_0, y_0) such that $f(x_0, y_0) \leq f(x, y)$ for all points (x, y) in the domain of f that lie inside the circle, and f is said to have an **absolute minimum** at (x_0, y_0) if $f(x_0, y_0) \leq f(x, y)$ for all points (x, y) in the domain of f.

If f has a relative maximum or a relative minimum at (x_0, y_0), then we say that f has a **relative extremum** at (x_0, y_0), and if f has an absolute maximum or absolute minimum at (x_0, y_0), then we say that f has an **absolute extremum** at (x_0, y_0).

In Figure 16.9.2 we have sketched the graph of a function f whose domain is the closed square region in the xy-plane whose points satisfy the inequalities $0 \leq x \leq 1, 0 \leq y \leq 1$. The function f has relative minima at the points A and C and a relative maximum at B. There is an absolute minimum at A and an absolute maximum at D.

For functions of two variables we shall be concerned with two important questions:

- Are there any relative or absolute extrema?
- If so, where are they located?

For functions of one variable that are continuous on a closed interval, the Extreme-Value Theorem (4.6.4) answered the existence question for absolute extrema. The following theorem, which we state without proof, is the corresponding result for functions of two variables.

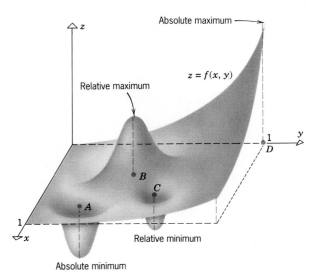

Figure 16.9.2

16.9.3   THEOREM (***Extreme-Value Theorem***).   *If $f(x, y)$ is continuous on a closed and bounded set R, then f has both an absolute maximum and an absolute minimum on R.*

Example 1   The square region R whose points satisfy the inequalities

$$0 \le x \le 1 \quad \text{and} \quad 0 \le y \le 1$$

is a closed and bounded set in the xy-plane. The function f whose graph is sketched in Figure 16.9.2 is continuous on R; thus, it is guaranteed to have an absolute maximum and minimum on R by the foregoing theorem. These occur at points D and A in the figure.   ◀

REMARK.   If any of the conditions in the Extreme-Value Theorem fail to hold, then there is no guarantee that an absolute maximum or absolute minimum exists on the region R. Thus, a discontinuous function on a closed and bounded set need not have any absolute extrema, and a continuous function on a set that is not closed and bounded also need not have any absolute extrema. (See Exercise 49.)

□ **FINDING RELATIVE EXTREMA**

Recall that if a function g of one variable has a relative extremum at a point x_0 where g is differentiable, then $g'(x_0) = 0$ (Theorem 4.3.4). To obtain the analog of this result for functions of two variables, suppose that $f(x, y)$ has a relative maximum at a point (x_0, y_0) and that the partial derivatives of f exist at (x_0, y_0). It seems plausible geometrically that the traces of the surface $z = f(x, y)$ on the planes $x = x_0$ and $y = y_0$ have horizontal tangent lines at (x_0, y_0) (Figure 16.9.3), so

$$f_x(x_0, y_0) = 0 \quad \text{and} \quad f_y(x_0, y_0) = 0$$

The same conclusion holds if f has a relative minimum at (x_0, y_0), all of which suggests the following result.

16.9.4 THEOREM. *If f has a relative extremum at a point (x_0, y_0), and if the first-order partial derivatives of f exist at this point, then*

$$f_x(x_0, y_0) = 0 \quad and \quad f_y(x_0, y_0) = 0$$

Proof. Assume that f has a relative maximum at (x_0, y_0) and that both partial derivatives of f exist at (x_0, y_0). We leave it as an exercise to show that

$$G(x) = f(x, y_0)$$

has a relative maximum at $x = x_0$ and

$$H(y) = f(x_0, y)$$

has a relative maximum at $y = y_0$ (Figure 16.9.3). It follows from these results that

$$G'(x_0) = f_x(x_0, y_0) = 0$$

and

$$H'(y_0) = f_y(x_0, y_0) = 0$$

(Exercise 48). The proof for a relative minimum is similar.   ▮

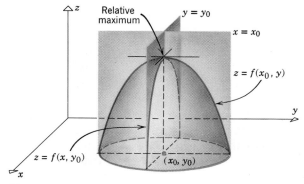

Figure 16.9.3

Recall that for a function f of one variable we defined a critical point to be a point x_0 at which $f'(x_0) = 0$ or at which $f'(x_0)$ does not exist. Analogously, for a function f of two variables we define a **critical point** of f to be a point (x_0, y_0) in the interior of the domain of f at which

$$f_x(x_0, y_0) = 0 \quad and \quad f_y(x_0, y_0) = 0$$

or at which one or both of the partial derivatives does not exist.

For a function of one variable, the condition $f'(x_0) = 0$ is *not* sufficient to guarantee that f has a relative extremum at x_0. (The graph of f may have an inflection point with a horizontal tangent line at x_0.) Similarly, the conditions $f_x(x_0, y_0) = 0$ and $f_y(x_0, y_0) = 0$ are not sufficient to guarantee that a function f of two variables has a relative extremum at (x_0, y_0).

Example 2   The graphs of the functions

$$z = f(x, y) = x^2 + y^2 \qquad \boxed{\text{Paraboloid}}$$

$$z = g(x, y) = 1 - x^2 - y^2 \qquad \boxed{\text{Paraboloid}}$$

$$z = h(x, y) = y^2 - x^2 \qquad \boxed{\text{Hyperbolic paraboloid}}$$

are the quadric surfaces graphed in Figure 16.9.4. We have

$$f_x(x, y) = 2x, \qquad f_y(x, y) = 2y$$
$$g_x(x, y) = -2x, \qquad g_y(x, y) = -2y$$
$$h_x(x, y) = -2x, \qquad h_y(x, y) = 2y$$

so in all three cases the partial derivatives are zero at $(0, 0)$, which means that $(0, 0)$ is a critical point for all three functions. The function f has a relative minimum at $(0, 0)$ and the function g a relative maximum. However, the function h has neither. To see this, observe that inside any circle in the xy-plane centered at $(0, 0)$, there exist points where $h(x, y)$ is positive (points on the y-axis) and there exist points where $h(x, y)$ is negative (points on the x-axis). Thus, $h(0, 0) = 0$ is neither the largest nor the smallest value of $h(x, y)$ in the circle.   ◀

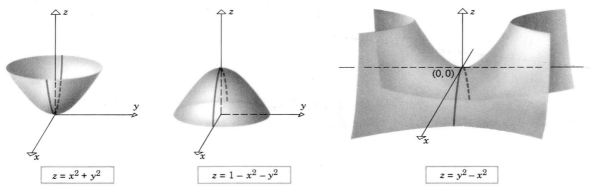

$z = x^2 + y^2$

$z = 1 - x^2 - y^2$

$z = y^2 - x^2$

Figure 16.9.4

A critical point at which a function does not have a relative extremum is called a **saddle point** of the function. Thus, the point $(0, 0)$ is a saddle point of the function $h(x, y) = y^2 - x^2$. (More advanced books distinguish between various types of saddle points. We shall not do this, however.)

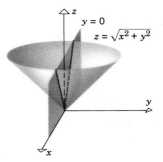

Figure 16.9.5

Example 3 The graph of the function

$$f(x, y) = \sqrt{x^2 + y^2}$$

is the circular cone shown in Figure 16.9.5. The point $(0, 0)$ is a critical point of f because the partial derivatives do not both exist there. It is evident geometrically that $f_x(0, 0)$ does not exist because the trace of the cone in the plane $y = 0$ has a corner at the origin. The fact that $f_x(0, 0)$ does not exist can also be seen algebraically by noting that $f_x(0, 0)$ can be interpreted as the derivative with respect to x of the function

$$f(x, 0) = \sqrt{x^2 + 0} = |x|$$

at $x = 0$. But $|x|$ is not differentiable at $x = 0$, so $f_x(0, 0)$ does not exist. Similarly, $f_y(0, 0)$ does not exist. The function f has a relative minimum at the critical point $(0, 0)$. ◄

For functions of one variable the second derivative test (Theorem 4.3.7) was used to determine the behavior of a function at a critical point. The following theorem, which is usually proved in advanced calculus, is the analog of that theorem for functions of two variables.

> **16.9.5** THEOREM (***The Second Partials Test***). *Let f be a function of two variables with continuous second-order partial derivatives in some circle centered at a critical point (x_0, y_0), and let*
>
> $$D = f_{xx}(x_0, y_0)f_{yy}(x_0, y_0) - f_{xy}^2(x_0, y_0)$$
>
> *(a) If $D > 0$ and $f_{xx}(x_0, y_0) > 0$, then f has a relative minimum at (x_0, y_0).*
> *(b) If $D > 0$ and $f_{xx}(x_0, y_0) < 0$, then f has a relative maximum at (x_0, y_0).*
> *(c) If $D < 0$, then f has a saddle point at (x_0, y_0).*
> *(d) If $D = 0$, then no conclusion can be drawn.*

Example 4 Locate all relative extrema and saddle points of

$$f(x, y) = 3x^2 - 2xy + y^2 - 8y$$

Solution. Since $f_x(x, y) = 6x - 2y$ and $f_y(x, y) = -2x + 2y - 8$, the critical points of f satisfy the equations

$$6x - 2y = 0$$
$$-2x + 2y - 8 = 0$$

Solving these for x and y yields $x = 2, y = 6$ (verify), so $(2, 6)$ is the only critical point. To apply Theorem 16.9.5 we need the second-order partial derivatives

$$f_{xx}(x, y) = 6, \quad f_{yy}(x, y) = 2, \quad f_{xy}(x, y) = -2$$

At the point $(2, 6)$ we have

$$D = f_{xx}(2, 6)f_{yy}(2, 6) - f_{xy}^2(2, 6) = (6)(2) - (-2)^2 = 8 > 0$$

and

$$f_{xx}(2, 6) = 6 > 0$$

so that f has a relative minimum at $(2, 6)$ by part (a) of the second partials test.   ◄

Example 5   Locate all relative extrema and saddle points of

$$f(x, y) = 4xy - x^4 - y^4$$

Solution.   Since

$$f_x(x, y) = 4y - 4x^3 \tag{1}$$
$$f_y(x, y) = 4x - 4y^3$$

the critical points of f have coordinates satisfying the equations

$$\begin{matrix} 4y - 4x^3 = 0 \\ 4x - 4y^3 = 0 \end{matrix} \quad \text{or} \quad \begin{matrix} y = x^3 \\ x = y^3 \end{matrix} \tag{2}$$

Substituting the top equation in the bottom yields $x = (x^3)^3$ or $x^9 - x = 0$ or $x(x^8 - 1) = 0$, which has solutions $x = 0$, $x = 1$, $x = -1$. Substituting these values in the top equation of (2) we obtain the corresponding y values $y = 0$, $y = 1$, $y = -1$. Thus, the critical points of f are $(0, 0)$, $(1, 1)$, and $(-1, -1)$.
   From (1),

$$f_{xx}(x, y) = -12x^2, \quad f_{yy}(x, y) = -12y^2, \quad f_{xy}(x, y) = 4$$

which yields the following table:

CRITICAL POINT (x_0, y_0)	$f_{xx}(x_0, y_0)$	$f_{yy}(x_0, y_0)$	$f_{xy}(x_0, y_0)$	$D = f_{xx}f_{yy} - f_{xy}^2$
$(0, 0)$	0	0	4	-16
$(1, 1)$	-12	-12	4	128
$(-1, -1)$	-12	-12	4	128

At the points $(1, 1)$ and $(-1, -1)$, we have $D > 0$ and $f_{xx} < 0$, so relative maxima occur at these critical points. At $(0, 0)$ there is a saddle point since $D < 0$. The surface is shown in Figure 16.9.6.   ◄

   The following analog of Theorem 4.6.5 is a key result for locating the absolute extreme values of a function of two variables.

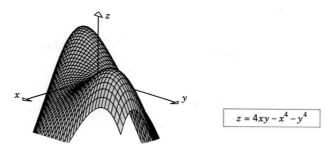

$$z = 4xy - x^4 - y^4$$

Figure 16.9.6

16.9.6   THEOREM.   *If a function f of two variables has an absolute extremum (either an absolute maximum or an absolute minimum) at an interior point of its domain, then this extremum occurs at a critical point.*

Proof.   We shall prove the result for an absolute maximum. The proof for an absolute minimum is similar.

   If f has an absolute maximum at the point (x_0, y_0) in the interior of the domain of f, then f has a relative maximum at (x_0, y_0) (why?). If both partial derivatives exist at (x_0, y_0), then

$$f_x(x_0, y_0) = 0 \quad \text{and} \quad f_y(x_0, y_0) = 0$$

by Theorem 16.9.4, so (x_0, y_0) is a critical point of f. If either partial derivative does not exist, then again (x_0, y_0) is a critical point, so (x_0, y_0) is a critical point in all cases.   ∎

☐ **FINDING ABSOLUTE EXTREMA ON CLOSED AND BOUNDED SETS**

If $f(x, y)$ is continuous on a closed and bounded set R, then the Extreme-Value Theorem (16.9.3) guarantees the existence of an absolute maximum and an absolute minimum of f on R. These absolute extrema can occur either on the boundary of R or in the interior of R, but if an absolute extremum occurs in the interior, then it occurs at a critical point by Theorem 16.9.6. Thus, we are led to the following procedure for finding absolute extrema:

> *How to Find the Absolute Extrema of a Continuous Function f of Two Variables on a Closed and Bounded Set R*
>
> **Step 1.**   Find the critical points of f that lie in the interior of R.
>
> **Step 2.**   Find all boundary points at which the absolute extrema can occur.
>
> **Step 3.**   Evaluate $f(x, y)$ at the points obtained in the previous steps. The largest of these values is the absolute maximum and the smallest the absolute minimum.

Example 6   Find the absolute maximum and minimum values of

$$f(x, y) = 3xy - 6x - 3y + 7 \tag{3}$$

on the closed triangular region R with vertices $(0, 0)$, $(3, 0)$, and $(0, 5)$.

Solution.   The region R is shown in Figure 16.9.7. We have

$$\frac{\partial f}{\partial x} = 3y - 6 \quad \text{and} \quad \frac{\partial f}{\partial y} = 3x - 3$$

so all critical points occur where

$$3y - 6 = 0 \quad \text{and} \quad 3x - 3 = 0$$

Solving these equations yields $x = 1$ and $y = 2$, so $(1, 2)$ is the only critical point. As shown in Figure 16.9.7, this critical point is in the interior of R.

Next, we want to determine the location of the points on the boundary of R at which the absolute extrema might occur. The boundary of R consists of three line segments, each of which we shall treat separately:

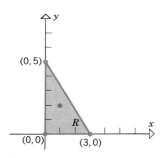

(0, 5)

R

(0, 0)   (3, 0)

Figure 16.9.7

The line segment between $(0, 0)$ *and* $(3, 0)$:   On this line segment we have $y = 0$, so (3) simplifies to a function of the single variable x,

$$u(x) = f(x, 0) = -6x + 7, \quad 0 \le x \le 3$$

This function has no critical points because $u'(x) = -6$ is nonzero for all x. Thus the extreme values of $u(x)$ occur at the endpoints $x = 0$ and $x = 3$, which correspond to the points $(0, 0)$ and $(3, 0)$ of R.

The line segment between $(0, 0)$ *and* $(0, 5)$:   On this line segment we have $x = 0$, so (3) simplifies to a function of the single variable y,

$$v(y) = f(0, y) = -3y + 7, \quad 0 \le y \le 5$$

This function has no critical points because $v'(y) = -3$ is nonzero for all y. Thus, the extreme values of $v(y)$ occur at the endpoints $y = 0$ and $y = 5$, which correspond to the points $(0, 0)$ and $(0, 5)$ of R.

The line segment between $(3, 0)$ *and* $(0, 5)$:   In the xy-plane, an equation for this line segment is

$$y = -\tfrac{5}{3}x + 5, \quad 0 \le x \le 3 \tag{4}$$

so (3) simplifies to a function of the single variable x,

$$w(x) = f(x, -\tfrac{5}{3}x + 5) = 3x(-\tfrac{5}{3}x + 5) - 6x - 3(-\tfrac{5}{3}x + 5) + 7$$
$$= -5x^2 + 14x - 8, \quad 0 \le x \le 3$$

Since $w'(x) = -10x + 14$, the equation $w'(x) = 0$ yields $x = \tfrac{7}{5}$ as the only critical point of w. Thus, the extreme values of w occur either at the critical point $x = \tfrac{7}{5}$ or at the endpoints $x = 0$ and $x = 3$. The endpoints correspond to the points $(0, 5)$ and $(3, 0)$ of R, and from (4) the critical point corresponds to $(\tfrac{7}{5}, \tfrac{8}{3})$.

Finally, Table 16.9.1 lists the values of $f(x, y)$ at the interior critical point and at the points on the boundary where an absolute extremum can occur. From the table we conclude that the absolute maximum value of f is $f(0, 0) = 7$ and the absolute minimum value is $f(3, 0) = -11$. ◄

(x, y)	$(0, 0)$	$(3, 0)$	$(0, 5)$	$(\frac{7}{5}, \frac{8}{3})$	$(1, 2)$
$f(x, y)$	7	-11	-8	$\frac{9}{5}$	1

Example 7 Determine the dimensions of a rectangular box, open at the top, having a volume of 32 ft³, and requiring the least amount of material for its construction.

Solution. Let

x = length of the box (in feet)
y = width of the box (in feet)
z = height of the box (in feet)
S = surface area of the box (in square feet)

We may reasonably assume that the box with least surface area requires the least amount of material, so our objective is to minimize the surface area

$$S = xy + 2xz + 2yz \tag{5}$$

(see Figure 16.9.8) subject to the volume requirement

$$xyz = 32 \tag{6}$$

From (6) we obtain $z = 32/xy$, so (5) can be rewritten as

$$S = xy + \frac{64}{y} + \frac{64}{x} \tag{7}$$

which expresses S as a function of two variables. The dimensions x and y in this formula must be positive, but otherwise have no limitation, so our problem reduces to finding the absolute minimum value of S over the region for which $x > 0$ and $y > 0$ (Figure 16.9.9). Because this region is not bounded, we have no mathematical guarantee at this stage that an absolute minimum exists. However, if it does, then it occurs at a critical point of S, so we shall begin by finding the critical points. Differentiating (7) we obtain

$$\frac{\partial S}{\partial x} = y - \frac{64}{x^2}, \quad \frac{\partial S}{\partial y} = x - \frac{64}{y^2} \tag{8}$$

so the coordinates of the critical points of S satisfy

$$y - \frac{64}{x^2} = 0, \quad x - \frac{64}{y^2} = 0$$

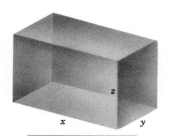

Two sides have area xy.
Two sides have area yz.
The base has area xy.

Figure 16.9.8

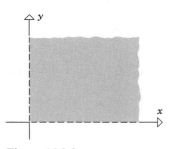

Figure 16.9.9

Solving the first equation for y yields

$$y = \frac{64}{x^2} \tag{9}$$

and substituting this expression in the second equation yields

$$x - \frac{64}{(64/x^2)^2} = 0$$

which can be rewritten as

$$x\left(1 - \frac{x^3}{64}\right) = 0$$

The solutions of this equation are $x = 0$ and $x = 4$. Since we require $x > 0$, the only solution of significance is $x = 4$. Substituting this value in (9) yields $y = 4$. To see that we have located a relative minimum, we use the second partials test. From (8),

$$\frac{\partial^2 S}{\partial x^2} = \frac{128}{x^3}, \quad \frac{\partial^2 S}{\partial y^2} = \frac{128}{y^3}, \quad \frac{\partial^2 S}{\partial y \, \partial x} = 1$$

Thus, when $x = 4$ and $y = 4$, we have

$$\frac{\partial^2 S}{\partial x^2} = 2, \quad \frac{\partial^2 S}{\partial y^2} = 2, \quad \frac{\partial^2 S}{\partial y \, \partial x} = 1$$

and

$$D = \frac{\partial^2 S}{\partial x^2}\frac{\partial^2 S}{\partial y^2} - \left(\frac{\partial^2 S}{\partial y \, \partial x}\right)^2 = (2)(2) - (1)^2 = 3$$

Since $\partial^2 S/\partial x^2 > 0$ and $D > 0$, it follows from the second partials test that a relative minimum occurs when $x = y = 4$. Substituting these values in (6) yields $z = 2$, so the box using least material has a height of 2 ft and a square base whose edges are 4 ft long. ◄

REMARK.   Strictly speaking, the solution in the last example is incomplete since we have not shown that an *absolute minimum* for S occurs when $x = y = 4$ and $z = 2$, only a relative minimum. The problem of showing that a relative extremum is also an absolute extremum can be difficult for functions of two or more variables and will not be considered in this text. However, in applied problems such as this it is often obvious from physical or geometric considerations that an absolute extremum has been found.

☐ **RELATIVE EXTREMA FOR FUNCTIONS OF THREE OR MORE VARIABLES**

Definitions of relative extrema can be given for functions of three or more variables. For example, if f is a function of three variables, then f has a *relative maximum* at (x_0, y_0, z_0) if $f(x_0, y_0, z_0) \geq f(x, y, z)$ for all points (x, y, z) in the domain of f that lie within some sphere centered at (x_0, y_0, z_0), and for a *relative minimum* the inequality is reversed. If $f(x, y, z)$ has first partial derivatives, then the relative extrema occur at **critical points,** that is, points where

$$f_x(x_0, y_0, z_0) = f_y(x_0, y_0, z_0) = f_z(x_0, y_0, z_0) = 0$$

The extension of the second partials test (Theorem 16.9.5) to functions of three or more variables is given in advanced calculus texts.

▶ Exercise Set 16.9

In Exercises 1–20, locate all relative maxima, relative minima, and saddle points.

1. $f(x, y) = 3x^2 + 2xy + y^2$.

2. $f(x, y) = x^3 - 3xy - y^3$.

3. $f(x, y) = y^2 + xy + 3y + 2x + 3$.

4. $f(x, y) = x^2 + xy - 2y - 2x + 1$.

5. $f(x, y) = x^2 + xy + y^2 - 3x$.

6. $f(x, y) = xy - x^3 - y^2$.

7. $f(x, y) = x^2 + 2y^2 - x^2y$.

8. $f(x, y) = 2x^2 - 4xy + y^4 + 2$.

9. $f(x, y) = x^2 + y^2 + 2/(xy)$.

10. $f(x, y) = x^3 + y^3 - 3x - 3y$.

11. $f(x, y) = x^2 + y - e^y$.

12. $f(x, y) = xe^y$.

13. $f(x, y) = e^x \sin y$.

14. $f(x, y) = xy + 2/x + 4/y$.

15. $f(x, y) = 2y^2x - yx^2 + 4xy$.

16. $f(x, y) = y \sin x$.

17. $f(x, y) = e^{-(x^2+y^2+2x)}$.

18. $f(x, y) = xy + \dfrac{a^3}{x} + \dfrac{b^3}{y}$ $(a \neq 0, b \neq 0)$.

19. $f(x, y) = \sin x + \sin y$, $0 < x < \pi, 0 < y < \pi$.

20. $f(x, y) = \sin x + \sin y + \sin (x + y)$,
    $0 < x < \pi/2, 0 < y < \pi/2$.

21. (a) Show that the second partials test provides no information about the critical points of
    $f(x, y) = x^4 + y^4$.
    (b) Classify all critical points of f as relative maxima, relative minima, or saddle points.

22. (a) Show that the second partials test provides no information about the critical points of
    $f(x, y) = x^4 - y^4$.

(b) Classify all critical points of f as relative maxima, relative minima, or saddle points.

23. If f is a function of one variable, and f is continuous on an interval I and has exactly one relative extremum on I, say at x_0, then f has an absolute extremum at x_0 (Theorem 4.6.6). This exercise shows that a similar result does not hold for functions of two variables.

(a) Show that $f(x, y) = 3xe^y - x^3 - e^{3y}$ has only one critical point and that a relative maximum occurs there (Figure 16.9.10).

(b) Show that f does not have an absolute maximum.

[This exercise is based on the article "The Only Critical Point in Town Test" by Ira Rosenholtz and Lowell Smylie, *Mathematics Magazine*, Vol. 58, No. 3, May 1985, pp. 149–150.]

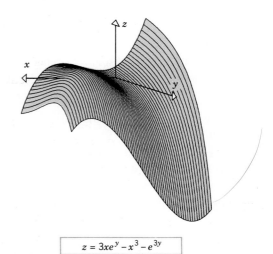

$$z = 3xe^y - x^3 - e^{3y}$$

Figure 16.9.10

24. Let f be a function of two variables that is continuous everywhere. One might think that if f has relative maxima at two points, then f must have another critical point because it is impossible to have two mountains without some sort of valley in between. This exercise shows that this is not true. Let $f(x, y) = 4x^2e^y - 2x^4 - e^{4y}$.

(a) Show that f has exactly two critical points and that a relative maximum occurs at each one (Figure 16.9.11).

(b) Show that f does not have an absolute maximum.

[This exercise is based on the problem Two Mountains Without a Valley, proposed and solved by Ira Rosenholtz, *Mathematics Magazine*, Vol. 60, No. 1, February 1987, p. 48.]

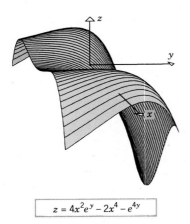

$$z = 4x^2e^y - 2x^4 - e^{4y}$$

Figure 16.9.11

In Exercises 25–30, find the absolute extrema of the given function on the indicated closed and bounded set R.

25. $f(x, y) = xy - x - 3y$; R is the triangular region with vertices $(0, 0)$, $(0, 4)$, and $(5, 0)$.

26. $f(x, y) = xy - 2x$; R is the triangular region with vertices $(0, 0)$, $(0, 4)$, and $(4, 0)$.

27. $f(x, y) = x^2 - 3y^2 - 2x + 6y$; R is the square region with vertices $(0, 0)$, $(0, 2)$, $(2, 2)$, and $(2, 0)$.

28. $f(x, y) = xe^y - x^2 - e^y$; R is the rectangular region with vertices $(0, 0)$, $(0, 1)$, $(2, 1)$, and $(2, 0)$.

29. $f(x, y) = x^2 + 2y^2 - x$; R is the circular region $x^2 + y^2 \leq 4$.

30. $f(x, y) = xy^2$; R is the region that satisfies the inequalities $x \geq 0$, $y \geq 0$, and $x^2 + y^2 \leq 1$.

31. Show that $f(x, y) = y^2 - 2xy + x^2$ has an absolute minimum at each point on the line $y = x$.

32. Find three positive numbers whose sum is 48 and such that their product is as large as possible.

33. Find three positive numbers whose sum is 27 and such that the sum of their squares is as small as possible.

34. Find all points on the plane $x + y + z = 5$ in the first octant at which $f(x, y, z) = xy^2z^2$ has a maximum value.

35. Find the points on the surface $x^2 - yz = 5$ that are closest to the origin.

36. Find the dimensions of the rectangular box of maximum volume that can be inscribed in a sphere of radius a.

37. Find the maximum volume of a rectangular box with three faces in the coordinate planes and a vertex in the first octant on the plane $x + y + z = 1$.

38. A manufacturer makes two models of an item, standard and deluxe. It costs \$40 to manufacture the standard model and \$60 for the deluxe. A market research firm estimates that if the standard model is priced at x dollars and the deluxe at y dollars, then the manufacturer will sell $500(y - x)$ of the standard items and $45,000 + 500(x - 2y)$ of the deluxe each year. How should the items be priced to maximize the profit?

39. A closed rectangular box with a volume of 16 ft³ is made from two kinds of materials. The top and bottom are made of material costing 10¢ per square foot and the sides from material costing 5¢ per square foot. Find the dimensions of the box so that the cost of materials is minimized.

40. Use the methods of this section to find the distance between the lines

$$\begin{matrix} x = 3t \\ y = 2t \\ z = t \end{matrix} \quad \text{and} \quad \begin{matrix} x = 2t \\ y = 2t + 3 \\ z = 2t \end{matrix}$$

41. Use the methods of this section to find the distance from the point $(-1, 3, 2)$ to the plane $x - 2y + z = 4$.

42. Show that among all parallelograms with perimeter l, a square with sides of length $l/4$ has maximum area. [*Hint:* The area of a parallelogram is given by the formula $A = ab \sin \alpha$, where a and b are the lengths of two adjacent sides and α is the angle between them.]

43. Determine the dimensions of a rectangular box, open at the top, having volume V, and requiring the least amount of material for its construction.

44. A length of sheet metal 27 in. wide is to be made into a water trough by bending up two sides as shown in Figure 16.9.12. Find x and ϕ so that the trapezoid-shaped cross section has a maximum area.

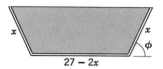
$$27 - 2x$$ Figure 16.9.12

45. A common problem in experimental work is to obtain a mathematical relationship $y = f(x)$ between two variables x and y by "fitting" a curve to points in the plane corresponding to various experimentally determined values of x and y, say

$$(x_1, y_1), (x_2, y_2), \ldots, (x_n, y_n)$$

Based on theoretical considerations, or simply on the pattern of the points, one decides on the general form of the curve $y = f(x)$ to be fitted. Often, the "curve" to be fitted is a straight line, $y = ax + b$. However, because of experimental errors in the data it is often impossible to find a line that passes through all of the points (Figure 16.9.13), so one looks for a line that "best fits" the data. One criterion for selecting a line of "best fit" is to choose a and b to minimize the function

$$f(a, b) = \sum_{k=1}^{n} (ax_k + b - y_k)^2$$

Geometrically, $|ax_k + b - y_k|$ is the vertical distance between the data point (x_k, y_k) and the line $y = ax + b$, so in effect, minimizing $f(a, b)$ minimizes the sum of the squares of these vertical distances (Figure 16.9.13). This procedure is called the method of *least squares*.

Figure 16.9.13

(a) Show that the conditions $\dfrac{\partial f}{\partial a} = 0$ and $\dfrac{\partial f}{\partial b} = 0$ result in the equations

$$(\Sigma x_k^2)a + (\Sigma x_k)b = \Sigma x_k y_k$$
$$(\Sigma x_k)a + nb = \Sigma y_k$$

where $\Sigma = \displaystyle\sum_{k=1}^{n}$.

(b) Solve the equations in part (a) for a and b to show that

$$a = \frac{n(\Sigma x_k y_k) - (\Sigma x_k)(\Sigma y_k)}{n(\Sigma x_k^2) - (\Sigma x_k)^2}$$

and

$$b = \frac{(\Sigma x_k^2)(\Sigma y_k) - (\Sigma x_k)(\Sigma x_k y_k)}{n(\Sigma x_k^2) - (\Sigma x_k)^2}$$

46. This exercise shows that the values of a and b obtained in Exercise 45 result in the absolute minimum value of $f(a, b)$.

(a) Given that the arithmetic average $\bar{x} = \dfrac{1}{n}\Sigma x_k$ minimizes $\Sigma(x_k - \bar{x})^2$ (see Exercise 24, Section 4.8), show that $n(\Sigma x_k^2) - (\Sigma x_k)^2 > 0$. [*Note:* $\Sigma(x_k - \bar{x})^2 > 0$ if the x_k's are not all the same.]

(b) Find $f_{aa}(a, b)$, $f_{bb}(a, b)$, and $f_{ab}(a, b)$.

(c) Use Theorem 16.9.5 and the results of parts (a) and (b) to show that f has a relative minimum at the critical point found in Exercise 45.

(d) Based on the result of part (c) and the nature of the function $f(a, b)$, show that f takes on its absolute minimum value at the critical point.

47. Use the formulas in part (b) of Exercise 45 to find the equation of the least squares line $y = ax + b$ for the following data:

x	1	2	3	4
y	1.5	1.6	2.1	3.0

48. Suppose that f has a relative maximum at (x_0, y_0), and both $f_x(x_0, y_0)$ and $f_y(x_0, y_0)$ exist. Prove that $G(x) = f(x, y_0)$ has a relative maximum at $x = x_0$ and $H(y) = f(x_0, y)$ has a relative maximum at $y = y_0$.

49. Find an example to show that a function that has a discontinuity on a closed and bounded set need not have any absolute extrema and give an example to show that a continuous function on a set that is not closed and bounded also need not have any absolute extrema.

■ **16.10**   LAGRANGE MULTIPLIERS

> *In this section we shall study a powerful method for solving certain types of optimization problems that is sometimes simpler to apply than the methods studied in the last section.*

□ **EXTREMUM PROBLEMS WITH CONSTRAINTS**

In the previous section, we considered the problem of minimizing

$$S = xy + 2xz + 2yz$$

subject to the constraint

$$xyz - 32 = 0$$

This is a special case of the following general problem, which we shall study in this section:

> **Three-Variable Extremum Problem with One Constraint**
> Maximize or minimize the function $f(x, y, z)$ subject to the constraint $g(x, y, z) = 0$.

We shall also be interested in the two-variable version of this problem.

> **Two-Variable Extremum Problem with One Constraint**
> Maximize or minimize the function $f(x, y)$ subject to the constraint $g(x, y) = 0$.

□ **LAGRANGE MULTIPLIERS**

One way to attack these problems is to solve the constraint equation for one of the variables in terms of the rest and substitute the result into f. The resulting function of one or two variables can then be maximized or minimized by finding its critical points. (See the solution of Example 7 in the preceding section.) However, if the constraint equation is too complicated to solve for one of the variables in terms of the rest, then other techniques must be used. We shall discuss one such technique, called *the method of Lagrange multipliers*. (See p. 723 for biography.) Since a rigorous discussion of this topic requires results from advanced calculus, we shall not be too concerned about all the technical details; we shall emphasize computational techniques.

Let us begin with the two-variable problem of maximizing or minimizing $f(x, y)$ subject to the constraint $g(x, y) = 0$. The graph of $g(x, y) = 0$ is usually some curve C in the xy-plane. Geometrically, we are concerned with finding the maximum or minimum value of $f(x, y)$ as (x, y) varies over the constraint curve C. If (x_0, y_0) is a point on the constraint curve C, then we shall say that $f(x, y)$ has a *constrained relative maximum* at (x_0, y_0) if there is a circle centered at (x_0, y_0) such that

$$f(x_0, y_0) \geq f(x, y) \tag{1}$$

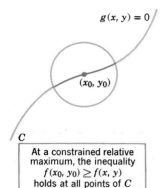

$g(x, y) = 0$

(x_0, y_0)

C

At a constrained relative maximum, the inequality
$f(x_0, y_0) \geq f(x, y)$
holds at all points of C within some circle centered at (x_0, y_0)

Figure 16.10.1

for all points (x, y) on C within the circle (Figure 16.10.1). For a *constrained relative minimum* at (x_0, y_0), the inequality in (1) is reversed. We shall say that f has a *constrained relative extremum* at (x_0, y_0) if f has either a constrained relative maximum or a constrained relative minimum at (x_0, y_0). The following result is the key to finding constrained relative extrema.

16.10.1 THEOREM (*The Constrained-Extremum Principle for Two Variables*). *Let f and g be functions of two variables with continuous first partial derivatives on some open set containing the constraint curve* $g(x, y) = 0$, *and assume that* $\nabla g \neq \mathbf{0}$ *at any point on this curve. If f has a constrained relative extremum, then this extremum occurs at a point* (x_0, y_0) *on the constraint curve where the gradient vectors* $\nabla f(x_0, y_0)$ *and* $\nabla g(x_0, y_0)$ *are parallel, that is, there is some number* λ *such that*

$$\nabla f(x_0, y_0) = \lambda \nabla g(x_0, y_0) \tag{2}$$

The number λ in (2) is called a *Lagrange multiplier.*

To see why the constrained-extremum principle is true, let C denote the constraint curve $g(x, y) = 0$, and suppose that a constrained relative extremum occurs at the point (x_0, y_0) on C. It is proved in advanced calculus that under the hypotheses of this theorem, the curve C can be represented parametrically by the equations

$$x = x(t), \quad y = y(t)$$

such that C has a nonzero tangent vector at (x_0, y_0). More precisely,

$$x'(t_0)\mathbf{i} + y'(t_0)\mathbf{j} \neq \mathbf{0}$$

where t_0 is the value of the parameter corresponding to (x_0, y_0). In terms of the parameter t, the value of f at a point (x, y) on the curve C is $f(x, y) = f(x(t), y(t))$. Since f has a constrained relative extremum at the point (x_0, y_0), and since this point corresponds to $t = t_0$, it follows that the function

$$F(t) = f(x(t), y(t))$$

has a relative extremum at $t = t_0$, that is,

$$F'(t_0) = 0 \tag{3}$$

By the chain rule

$$F'(t) = \frac{d}{dt} [f(x(t), y(t))] = f_x(x, y) \frac{dx}{dt} + f_y(x, y) \frac{dy}{dt}$$

so that condition (3) yields the equation

$$f_x(x_0, y_0)x'(t_0) + f_y(x_0, y_0)y'(t_0) = 0$$

which may be expressed in the form

$$\nabla f(x_0, y_0) \cdot [x'(t_0)\mathbf{i} + y'(t_0)\mathbf{j}] = 0 \tag{4}$$

Since the vector $x'(t_0)\mathbf{i} + y'(t_0)\mathbf{j}$ is tangent to the curve C at (x_0, y_0), it follows from (4) that $\nabla f(x_0, y_0)$ is perpendicular to C at (x_0, y_0). But $\nabla g(x_0, y_0)$ is also perpendicular to C at (x_0, y_0) since C is the level curve $g(x, y) = 0$ for the function g (see Theorem 16.6.4). Since $\nabla f(x_0, y_0)$ and $\nabla g(x_0, y_0)$ are both perpendicular to C at (x_0, y_0), these vectors are parallel.

Example 1   At what point on the circle $x^2 + y^2 = 1$ does the product xy have a maximum?

Solution.   We want to maximize $f(x, y) = xy$ subject to the constraint

$$g(x, y) = x^2 + y^2 - 1 = 0 \tag{5}$$

We have

$$\nabla f = y\mathbf{i} + x\mathbf{j} \quad \text{and} \quad \nabla g = 2x\mathbf{i} + 2y\mathbf{j}$$

From the formula for ∇g we see that $\nabla g = \mathbf{0}$ if and only if $x = 0$ and $y = 0$, so $\nabla g \neq \mathbf{0}$ at any point on the circle $x^2 + y^2 = 1$. Thus, it follows from Theorem 16.10.1 that at a constrained relative extremum we must have

$$\nabla f = \lambda \nabla g \quad \text{or} \quad y\mathbf{i} + x\mathbf{j} = \lambda(2x\mathbf{i} + 2y\mathbf{j})$$

which is equivalent to the pair of equations

$$y = 2x\lambda \quad \text{and} \quad x = 2y\lambda \tag{6}$$

Since the maximum value of xy on the circle $x^2 + y^2 = 1$ is obviously greater than zero, we must have $x \neq 0$ and $y \neq 0$ at a constrained maximum. Thus, the equations in (6) can be rewritten as

$$\lambda = \frac{y}{2x} \quad \text{and} \quad \lambda = \frac{x}{2y}$$

from which we obtain

$$\frac{y}{2x} = \frac{x}{2y}$$

or

$$y^2 = x^2 \tag{7}$$

Substituting this in (5) yields

$$2x^2 - 1 = 0$$

or

$$x = 1/\sqrt{2} \quad \text{and} \quad x = -1/\sqrt{2}$$

Substituting $x = 1/\sqrt{2}$ in (7) yields $y = \pm 1/\sqrt{2}$ and substituting $x = -1/\sqrt{2}$ in (7) yields $y = \pm 1/\sqrt{2}$, so there are four candidates for the location of a maximum:

$$(1/\sqrt{2}, 1/\sqrt{2}), \quad (1/\sqrt{2}, -1/\sqrt{2}), \quad (-1/\sqrt{2}, 1/\sqrt{2}), \quad (-1/\sqrt{2}, -1/\sqrt{2})$$

At the first and fourth points the function $f(x, y) = xy$ has value $\frac{1}{2}$, while at the second and third points the value is $-\frac{1}{2}$. Thus, the constrained maximum value of $\frac{1}{2}$ occurs at $(1/\sqrt{2}, 1/\sqrt{2})$ and $(-1/\sqrt{2}, -1/\sqrt{2})$. ◀

REMARK. If c is a constant, then the functions $g(x, y)$ and $g(x, y) - c$ have the same gradient since the constant c drops out when we differentiate. Consequently, it is *not* essential to rewrite a constraint of the form $g(x, y) = c$ as $g(x, y) - c = 0$ in order to apply the constrained-extremum principle. Thus, in the last example, we could have kept the constraint in the form $x^2 + y^2 = 1$ and taken $g(x, y) = x^2 + y^2$ rather than $g(x, y) = x^2 + y^2 - 1$.

In Exercise 6 of Section 4.7, it was stated that among all rectangles of perimeter p, a square of side $p/4$ has maximum area. This result can be obtained using Lagrange multipliers.

Example 2 Find the dimensions of a rectangle having perimeter p and maximum area.

Solution. Let

> $x = $ length of the rectangle
> $y = $ width of the rectangle
> $A = $ area of the rectangle

We want to maximize $A = xy$ subject to the perimeter constraint

$$2x + 2y = p \tag{8}$$

If we let $f(x, y) = xy$ and $g(x, y) = 2x + 2y$, then we have

$$\nabla f = y\mathbf{i} + x\mathbf{j} \quad \text{and} \quad \nabla g = 2\mathbf{i} + 2\mathbf{j}$$

Noting that $\nabla g \neq \mathbf{0}$, it follows from the constrained-extremum principle (Theorem 16.10.1) that

$$\nabla f = \lambda \nabla g \quad \text{or} \quad y\mathbf{i} + x\mathbf{j} = \lambda(2\mathbf{i} + 2\mathbf{j})$$

at a constrained relative maximum. This is equivalent to the two equations

$$y = 2\lambda \quad \text{and} \quad x = 2\lambda$$

Eliminating λ from these equations we obtain

$$x = y$$

Using this condition and constraint (8), we obtain $x = p/4$, $y = p/4$. ◀

Lagrange multipliers can also be used in the three-variable problem of maximizing or minimizing $f(x, y, z)$ subject to the constraint $g(x, y, z) = 0$. The graph of $g(x, y, z) = 0$ is generally some surface S in 3-space. Geometrically we are concerned with finding the maximum or minimum of the function $f(x, y, z)$ as (x, y, z) varies over the surface S. We shall say that $f(x, y, z)$ has a **constrained relative maximum** at (x_0, y_0, z_0) if there is a sphere centered at (x_0, y_0, z_0) such that

$$f(x_0, y_0, z_0) \geq f(x, y, z)$$

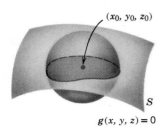

At a constrained relative maximum, the inequality
$f(x_0, y_0, z_0) \geq f(x, y, z)$
holds at all points of S inside some sphere centered at (x_0, y_0, z_0).

Figure 16.10.2

for all points (x, y, z) on S within the sphere (Figure 16.10.2). The meaning of the terms **constrained relative minimum** and **constrained relative extremum** should be clear. It can be shown that if f and g have continuous first partial derivatives and $\nabla g \neq \mathbf{0}$ on the surface $g(x, y, z) = 0$, then a constrained relative extremum can only occur at a point (x_0, y_0, z_0) where $\nabla f(x_0, y_0, z_0)$ and $\nabla g(x_0, y_0, z_0)$ are parallel, that is,

$$\nabla f(x_0, y_0, z_0) = \lambda \nabla g(x_0, y_0, z_0)$$

for some number λ.

Example 3 Find the points on the sphere $x^2 + y^2 + z^2 = 36$ that are closest to and farthest from $(1, 2, 2)$.

Solution. To avoid radicals, we shall find points on the sphere that minimize and maximize the *square* of the distance to $(1, 2, 2)$. Thus, we want to extremize

$$f(x, y, z) = (x - 1)^2 + (y - 2)^2 + (z - 2)^2$$

subject to the constraint

$$x^2 + y^2 + z^2 = 36 \tag{9}$$

Therefore, with $g(x, y, z) = x^2 + y^2 + z^2$, we must have

$$\nabla f(x, y, z) = \lambda \nabla g(x, y, z)$$

at a constrained relative extremum; that is,

$$2(x - 1)\mathbf{i} + 2(y - 2)\mathbf{j} + 2(z - 2)\mathbf{k} = \lambda(2x\mathbf{i} + 2y\mathbf{j} + 2z\mathbf{k})$$

which leads to the equations

$$2(x - 1) = 2x\lambda, \quad 2(y - 2) = 2y\lambda, \quad 2(z - 2) = 2z\lambda \tag{10}$$

We may assume that x, y, and z are nonzero since $x = 0$ does not satisfy the first equation, $y = 0$ does not satisfy the second, and $z = 0$ does not satisfy the third. Thus, we can rewrite (10) as

$$\frac{x - 1}{x} = \lambda, \quad \frac{y - 2}{y} = \lambda, \quad \frac{z - 2}{z} = \lambda$$

The first two equations imply that

$$\frac{x - 1}{x} = \frac{y - 2}{y}$$

from which it follows that

$$y = 2x \tag{11}$$

Similarly, the first and third equations imply that

$$z = 2x \tag{12}$$

Substituting (11) and (12) in the constraint equation (9), we obtain

$$9x^2 = 36, \quad \text{or} \quad x = \pm 2$$

Substituting these values in (11) and (12) yields two points

$$(2, 4, 4) \quad \text{and} \quad (-2, -4, -4)$$

Since $f(2, 4, 4) = 9$ and $f(-2, -4, -4) = 81$, it follows that $(2, 4, 4)$ is the point on the sphere closest to $(1, 2, 2)$, and $(-2, -4, -4)$ is the point that is farthest (Figure 16.10.3). ◀

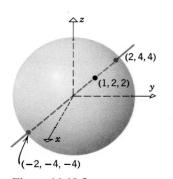

Figure 16.10.3

Next we shall use Lagrange multipliers to solve the problem of Example 7 in the preceding section.

Example 4 Use Lagrange multipliers to determine the dimensions of a rectangular box, open at the top, having a volume of 32 ft³, and requiring the least amount of material for its construction.

Solution. With the notation of Example 7 in Section 16.9, the problem is to minimize the surface area

$$S = xy + 2xz + 2yz$$

subject to the volume constraint

$$xyz = 32 \tag{13}$$

If we let $f(x, y, z) = xy + 2xz + 2yz$ and $g(x, y, z) = xyz$, then

$$\nabla f = (y + 2z)\mathbf{i} + (x + 2z)\mathbf{j} + (2x + 2y)\mathbf{k} \quad \text{and} \quad \nabla g = yz\mathbf{i} + xz\mathbf{j} + xy\mathbf{k}$$

It follows from the formula for ∇g that $\nabla g = \mathbf{0}$ if and only if $x = 0$, $y = 0$, and $z = 0$ (verify); thus $\nabla g \neq \mathbf{0}$ at any point on the surface $xyz = 32$. Thus, at a constrained relative extremum we must have $\nabla f = \lambda \Delta g$, that is,

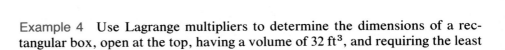

$$(y + 2z)\mathbf{i} + (x + 2z)\mathbf{j} + (2x + 2y)\mathbf{k} = \lambda(yz\mathbf{i} + xz\mathbf{j} + xy\mathbf{k})$$

This condition yields the three equations

$$y + 2z = \lambda yz, \quad x + 2z = \lambda xz, \quad 2x + 2y = \lambda xy$$

Because of the volume constraint $xyz = 32$, we know that x, y, and z are nonzero, so these equations can be rewritten as

$$\frac{1}{z} + \frac{2}{y} = \lambda, \quad \frac{1}{z} + \frac{2}{x} = \lambda, \quad \frac{2}{y} + \frac{2}{x} = \lambda$$

From the first two equations,

$$y = x \tag{14}$$

and from the second and third equations, $z = \frac{1}{2}y$. This and (14) imply that

$$z = \frac{1}{2}x \tag{15}$$

Substituting (14) and (15) in the volume constraint (13) yields

$$\tfrac{1}{2}x^3 = 32$$

This equation, together with (14) and (15), yields

$$x = 4, \quad y = 4, \quad z = 2$$

which agrees with the result that was obtained in Example 7 of the preceding section. ◄

Lagrange multipliers can also be used in problems involving two or more constraints. However, we shall not pursue this topic here.

▶ Exercise Set 16.10

In Exercises 1–7, use Lagrange multipliers to find the maximum and minimum values of f subject to the given constraint. Also, find the points at which these extreme values occur.

1. $f(x, y) = xy$; $4x^2 + 8y^2 = 16$.

2. $f(x, y) = x^2 - y$; $x^2 + y^2 = 25$.

3. $f(x, y) = 4x^3 + y^2$; $2x^2 + y^2 = 1$.

4. $f(x, y) = x - 3y - 1$; $x^2 + 3y^2 = 16$.

5. $f(x, y, z) = 2x + y - 2z$; $x^2 + y^2 + z^2 = 4$.

6. $f(x, y, z) = 3x + 6y + 2z$; $2x^2 + 4y^2 + z^2 = 70$.

7. $f(x, y, z) = xyz$; $x^2 + y^2 + z^2 = 1$.

In Exercises 8–16, solve using Lagrange multipliers.

8. Find the point on the line $2x - 4y = 3$ that is closest to the origin.

9. Find the point on the line $y = 2x + 3$ that is closest to $(4, 2)$.

10. Find the point on the plane $x + 2y + z = 1$ that is closest to the origin.

11. Find the point on the plane $4x + 3y + z = 2$ that is closest to $(1, -1, 1)$.

12. Find the points on the surface $xy - z^2 = 1$ that are closest to the origin.

13. Find the maximum value of $\sin x \sin y$, where x and y denote the acute angles of a right triangle.

14. Find a vector in 3-space whose length is 5 and whose components have the largest possible sum.

15. Find the points on the circle $x^2 + y^2 = 45$ that are closest to and farthest from $(1, 2)$.

16. The temperature at a point (x, y) on a metal plate is $T(x, y) = 4x^2 - 4xy + y^2$. An ant, walking on the plate, traverses a circle of radius 5 centered at the origin. What are the highest and lowest temperatures encountered by the ant?

In Exercises 17–24, use Lagrange multipliers to solve the indicated problems from Section 16.9.

17. Exercise 33.   **18.** Exercise 34.

19. Exercise 35.   **20.** Exercise 36.

21. Exercise 39.   **22.** Exercise 41.

23. Exercise 42.   **24.** Exercise 43.

▶ SUPPLEMENTARY EXERCISES C 42

1. Let $f(x, y) = e^x \ln y$. Find
 (a) $f(0, e)$ (b) $f(\ln y, e^x)$ (c) $f(r + s, rs)$.

2. Sketch the domain of f using solid lines for portions of the boundary included in the domain and dashed lines for portions not included.
 (a) $f(x, y) = \sqrt{x - y}/(2x - y)$
 (b) $f(x, y) = \ln(xy - 1)$
 (c) $f(x, y) = (\sin^{-1} x)/e^y$.

3. Describe the graph of f.
 (a) $f(x, y) = \sqrt{x^2 + 4y^2}$
 (b) $f(x, y) = 1 - x/a - y/b$.

4. Find f_x, f_y, and f_z if $f(x, y, z) = x^2/(y^2 + z^2)$.

5. Find $\partial w/\partial r$ if $w = \ln(xy)/\sin yz$, $x = r + s$, $y = s$, $z = 3r - s$.

6. Find $\partial f/\partial x$, $\partial f/\partial y$, and $\partial f/\partial u$ if
$$f(x, y, z, u) = (e^{yz}/x) + \ln(u - x)$$

7. Find f_x, f_y, f_{yx}, and f_{yzx} at $(0, \pi/2, 1)$ if
$$f(x, y, z) = e^{xy} \sin yz$$

8. Find $g_{xy}(0, 3)$ and $g_{yy}(2, 0)$ if
$$g(x, y) = \sin(xy) + xe^y$$

9. Find $\partial w/\partial\theta|_{r=1, \theta=0}$ if $w = \ln(x^2 + y^2)$, $x = re^\theta$, $y = \tan(r\theta)$.

10. Find $\partial w/\partial r$ if $w = x \cos y + y \sin x$, $x = rs^2$, $y = r + s$.

11. Find $\partial w/\partial s$ if $w = \ln(x^2 + y^2 + 2z)$, $x = r + s$, $y = r - s$, $z = 2rs$.

In Exercises 12–15, verify the assertion.

12. If $w = \tan(x^2 + y^2) + x\sqrt{y}$, then $w_{xy} = w_{yx}$.

13. If $w = \ln(3x - 3y) + \cos(x + y)$, then $\partial^2 w/\partial x^2 = \partial^2 w/\partial y^2$.

14. If $F(x, y, z) = 2z^3 - 3(x^2 + y^2)z$, then $F_{xx} + F_{yy} + F_{zz} = 0$.

15. If $f(x, y, z) = xyz + x^2 + \ln(y/z)$, then $f_{xyzx} = f_{zxxy}$.

16. Find the slope of the tangent line at $(1, -2, -3)$ to the curve of intersection of the surface $z = 5 - 4x^2 - y^2$ with
 (a) the plane $x = 1$ (b) the plane $y = -2$.

17. The pressure in newtons/meter^2 of a gas in a cylinder is given by $P = 10T/V$ with T in kelvins (K) and V in meters^3.
 (a) If T is increasing at a rate of 3 K/min with V held fixed at 2.5 m^3, find the rate at which the pressure is changing when $T = 50$ K.
 (b) If T is held fixed at 50 K while V is decreasing at the rate of 3 m^3/min, find the rate at which the pressure is changing when $V = 2.5$ m^3.

In Exercises 18 and 19,
 (a) find the limit of $f(x, y)$ as $(x, y) \to (0, 0)$ if it exists;
 (b) determine whether f is continuous at $(0, 0)$.

18. $f(x, y) = \dfrac{x^4 - x + y - x^3y}{x - y}$.

19. $f(x, y) = \begin{cases} \dfrac{x^4 - y^4}{x^2 + y^2} & \text{if } (x, y) \neq (0, 0) \\ 0 & \text{if } (x, y) = (0, 0). \end{cases}$

20. Find dw/dt using the chain rule.
 (a) $w = \sin xy + y\ln xz + z$, $x = e^t$, $y = t^2$, $z = 1$
 (b) $w = \sqrt{xy - e^z}$, $x = \sin t$, $y = 3t$, $z = \cos t$.

21. Use Formula (17) of Section 16.4 to find dy/dx.
 (a) $3x^2 - 5xy + \tan xy = 0$
 (b) $x \ln y + \sin(x - y) = \pi$.

22. If $F(x, y) = 0$, find a formula for d^2y/dx^2 in terms of partial derivatives of F. [*Hint:* Use Formula (17) of Section 16.4.]

23. The voltage V across a fixed resistance R in series with a variable resistance r is $V = RE/(r + R)$, where E is the source voltage. Express dV/dt in terms of dE/dt and dr/dt.

24. Let $f(x, y, z) = 1/(z - x^2 - 4y^2)$.
   (a) Describe the domain of f.
   (b) Describe the level surface $f(x, y, z) = 2$.
   (c) Find $f(3t, uv, e^{3t})$.

In Exercises 25–29, find
(a) the gradient of f at P_0;
(b) the directional derivative at P_0 in the indicated direction.

25. $f(x, y) = x^2y^5$, $P_0(3, 1)$; from P_0 toward $P_1(4, -3)$.

26. $f(x, y, z) = ye^x \sin z$, $P_0(\ln 2, 2, \pi/4)$; in the direction of $\mathbf{a} = \langle 1, -2, 2 \rangle$.

27. $f(x, y, z) = \ln(xyz)$, $P_0(3, 2, 6)$; $\mathbf{u} = \langle -1, 1, 1 \rangle / \sqrt{3}$.

28. $f(x, y) = x^2y + 2xy^2$, $P_0(1, 2)$; $\mathbf{u}$ makes an angle of $60°$ with the positive x-axis.

29. $f(x, y, z) = xy + yz + zx$, $P_0(1, -1, 2)$; from P_0 toward $P_1(11, 10, 0)$.

30. Let $f(x, y, z) = (x + y)^2 + (y + z)^2 + (z + x)^2$. Find the maximum rate of decrease of f at $P_0(2, -1, 2)$ and the direction in which this rate of decrease occurs.

31. Find all unit vectors $\mathbf{u}$ such that $D_{\mathbf{u}}f = 0$ at P_0.
   (a) $f(x, y) = x^3y^3 - xy$, $P_0(1, -1)$
   (b) $f(x, y) = xe^y$, $P_0(-2, 0)$.

32. The directional derivative $D_{\mathbf{u}}f$ at (x_0, y_0) is known to be 2 when $\mathbf{u}$ makes an angle of $30°$ with the positive x-axis, and 8 when this angle is $150°$. Find $D_{\mathbf{u}}f(x_0, y_0)$ in the direction of the vector $\sqrt{3}\mathbf{i} + 2\mathbf{j}$.

33. At the point $(1, 2)$, the directional derivative $D_{\mathbf{u}}f$ is $2\sqrt{2}$ toward $P_1(2, 3)$ and -3 toward $P_2(1, 0)$. Find $D_{\mathbf{u}}f(1, 2)$ toward the origin.

In Exercises 34 and 35,
(a) find a normal vector $\mathbf{N}$ at $P_0(x_0, y_0, f(x_0, y_0))$;
(b) find an equation for the tangent plane at P_0.

34. $f(x, y) = 4x^2 + y^2 + 1$; $P_0(1, 2, 9)$.

35. $f(x, y) = 2\sqrt{x^2 + y^2}$; $P_0(4, -3, 10)$.

36. Find equations for the tangent plane and normal line to the given surface at P_0.
   (a) $z = x^2e^{2y}$; $P_0(1, \ln 2, 4)$
   (b) $x^2y^3z^4 + xyz = 2$; $P_0(2, 1, -1)$.

37. Find all points P_0 on the surface $z = 2 - xy$ at which the normal line passes through the origin.

38. Show that for all tangent planes to the surface $x^{2/3} + y^{2/3} + z^{2/3} = 1$, the sum of the squares of the x-, y-, and z-intercepts is 1.

39. Find all points on the elliptic paraboloid $z = 9x^2 + 4y^2$ at which the normal line is parallel to the line through the points $P(4, -2, 5)$ and $Q(-2, -6, 4)$.

40. If $w = x^2y - 2xy + y^2x$, find the increment Δw and the differential dw if (x, y) varies from $(1, 0)$ to $(1.1, -0.1)$.

41. Use differentials to approximate the change in the volume $V = \frac{1}{3}x^2h$ of a pyramid with a square base when its height h is increased from 2 to 2.2 m, while its base dimension x is decreased from 1 to 0.9 m. Compare this to ΔV.

42. If $f(x, y, z) = x^2y^4/(1 + z^2)$, use differentials to approximate $f(4.996, 1.003, 1.995)$.

In Exercises 43–45, locate all relative minima, relative maxima, and saddle points.

43. $f(x, y) = x^2 + 3xy + 3y^2 - 6x + 3y$.

44. $f(x, y) = x^2y - 6y^2 - 3x^2$.

45. $f(x, y) = x^3 - 3xy + \frac{1}{2}y^2$.

Solve Exercises 46 and 47 two ways:
(a) Use the constraint to eliminate a variable.
(b) Use Lagrange multipliers.

46. Find all relative extrema of x^2y^2 subject to the constraint $4x^2 + y^2 = 8$.

47. Find the dimensions of the rectangular box of maximum volume that can be inscribed in the ellipsoid $(x/a)^2 + (y/b)^2 + (z/c)^2 = 1$.

In Exercises 48 and 49, use Lagrange multipliers.

48. Find the points on the curve $5x^2 - 6xy + 5y^2 = 8$ whose distance from the origin is (i) minimum and (ii) maximum.

49. A current I branches into currents I_1, I_2, and I_3 through resistors with resistances R_1, R_2, and R_3 (see figure) in such a way that the total energy to the three resistors is a minimum. If the energy delivered to R_i is $I_i^2R_i$ ($i = 1, 2, 3$), find the ratios $I_1:I_2:I_3$.

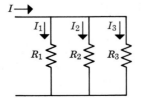

17
Multiple Integrals

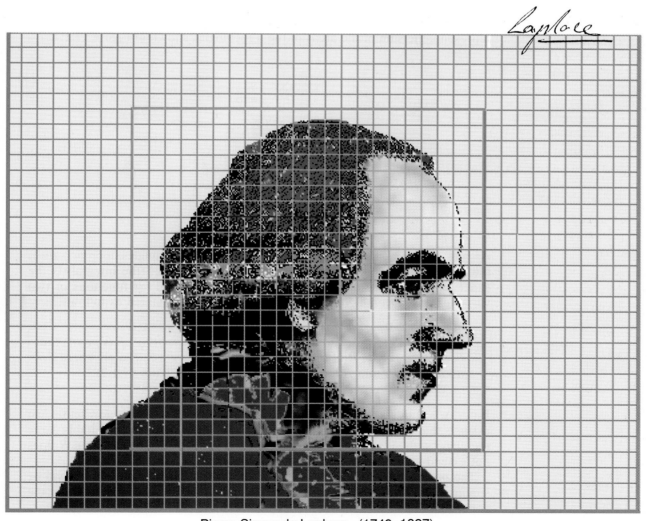

Pierre-Simon de Laplace   (1749 -1827)

■ 17.1   DOUBLE INTEGRALS

The notion of a definite integral can be extended to functions of two or more variables. In this section we shall discuss the double integral, which is the extension to functions of two variables.

☐ **DEFINITION OF A DOUBLE INTEGRAL**

Recall that the definite integral

$$\int_a^b f(x)\,dx = \lim_{n\to+\infty} \sum_{k=1}^n f(x_k^*)\,\Delta x_k \tag{1}$$

arose from the problem of finding areas under curves. For a continuous function with nonnegative values on $[a,b]$, the Riemann sum on the right side of (1) represents the sum of the areas of the n rectangles shown in Figure 17.1.1.

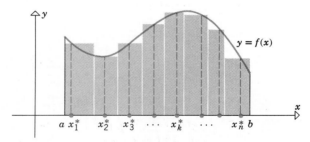

Figure 17.1.1

As n increases in such a way that the widths of the rectangles approach zero, the errors in the approximations diminish, and the limit (the integral) is the exact area under the curve. The concept of an integral for functions of two variables arises from a volume problem, which can be stated as follows:

> **17.1.1**   PROBLEM.   *Find the volume of the solid consisting of all points that lie between a region R in the xy-plane and a surface $z = f(x,y)$, where f is continuous on R and $f(x,y) \geq 0$ for all (x,y) in R (Figure 17.1.2).*

Later, we shall place more restrictions on the region R, but for now let us just assume that the entire region R can be enclosed within some suitably large rectangle with sides parallel to the coordinate axes. This ensures that R does not extend indefinitely in any direction (Figure 17.1.3).

The procedure for finding the volume V of the solid in Figure 17.1.2 will be similar to the limiting process used for finding areas, except that now the approximating elements will be rectangular parallelepipeds rather than rectangles. We proceed as follows:

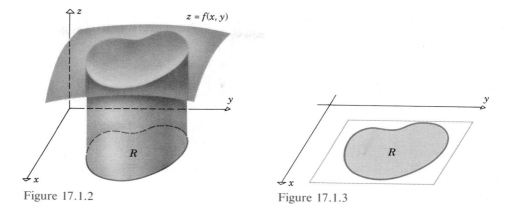

Figure 17.1.2   Figure 17.1.3

Step 1. Using lines parallel to the coordinate axes, divide the rectangle enclosing the region R into subrectangles, and exclude from consideration all those subrectangles that contain any points outside of R. This leaves only rectangles that are subsets of R (Figure 17.1.4). Denote the areas of these rectangles by

$$\Delta A_1, \Delta A_2, \ldots, \Delta A_n$$

Step 2. Choose an arbitrary point in each of these subrectangles, and denote them by

$$(x_1^*, y_1^*), (x_2^*, y_2^*), \ldots, (x_n^*, y_n^*)$$

As shown in Figure 17.1.5, the product $f(x_k^*, y_k^*) \, \Delta A_k$ is the volume of a rectangular parallelepiped with base area ΔA_k and height $f(x_k^*, y_k^*)$, so the sum

$$\sum_{k=1}^{n} f(x_k^*, y_k^*) \, \Delta A_k$$

can be viewed as an approximation to the volume V of the entire solid.

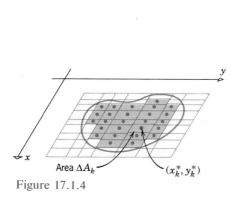

Area ΔA_k (x_k^*, y_k^*)

Figure 17.1.4

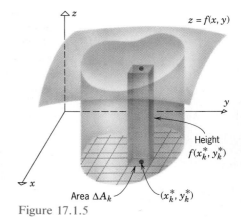

Height $f(x_k^*, y_k^*)$

Area ΔA_k (x_k^*, y_k^*)

Figure 17.1.5

Step 3. There are two sources of error in the approximation: First, the parallelepipeds have flat tops, whereas the surface $z = f(x, y)$ may be curved, and, second, the rectangles that form the bases of the parallelepipeds do not completely cover the region R. However, if we repeat the above process with more and more subdivisions, so that the lengths and widths of the base rectangles approach zero, then it is plausible that the errors of both types approach zero, and the exact volume of the solid is

$$V = \lim_{n \to +\infty} \sum_{k=1}^{n} f(x_k^*, y_k^*) \, \Delta A_k \tag{2}$$

The sums in (2) are called **Riemann sums,** and the limit of the Riemann sums is denoted by

$$\iint_R f(x, y) \, dA = \lim_{n \to +\infty} \sum_{k=1}^{n} f(x_k^*, y_k^*) \, \Delta A_k \tag{3}$$

which is called the **double integral** of $f(x, y)$ over R. With this notation, the volume V of the solid can be expressed as

$$V = \iint_R f(x, y) \, dA \tag{4}$$

This result was derived under the assumption that f is continuous and non-negative on the region R. In the case where $f(x, y)$ can have both positive and negative values on R, the double integral over R represents a difference of two volumes: the volume of the solid that is above R but below $z = f(x, y)$ minus the volume of the solid that is below R but above $z = f(x, y)$. We call this difference the **net signed volume** between $z = f(x, y)$ and R; it is analogous to the concept of net signed area defined in Section 5.6. Thus, a positive value for a double integral means that there is more volume above R than below, a negative value means that there is more volume below than above, and a value of zero means that the two volumes are the same.

A precise definition of expression (3) (in terms of ϵ's and δ's) and conditions under which the double integral exists are studied in advanced calculus. However, for our purposes it suffices to say that existence is ensured when f is continuous on R and the region R is not too "complicated."

Observe the similarity between (1) and (3). Because of this it should not be surprising that double integrals have many of the same properties as definite integrals (which we now also call **single integrals**):

$$\iint_R cf(x, y) \, dA = c \iint_R f(x, y) \, dA \quad (c \text{ a constant}) \tag{5}$$

properties of double integrals are the ~~similar to~~ same as single integrals

$$\iint\limits_{R} [f(x, y) + g(x, y)] \, dA = \iint\limits_{R} f(x, y) \, dA + \iint\limits_{R} g(x, y) \, dA \qquad (6)$$

$$\iint\limits_{R} [f(x, y) - g(x, y)] \, dA = \iint\limits_{R} f(x, y) \, dA - \iint\limits_{R} g(x, y) \, dA \qquad (7)$$

It is evident intuitively that if $f(x, y)$ is nonnegative on a region R, then subdividing R into two regions R_1 and R_2 has the effect of subdividing the solid between R and $z = f(x, y)$ into two solids, the sum of whose volumes is the volume of the entire solid (Figure 17.1.6). This suggests the following result, which holds even if f has negative values:

$$\iint\limits_{R} f(x, y) \, dA = \iint\limits_{R_1} f(x, y) \, dA + \iint\limits_{R_2} f(x, y) \, dA \qquad (8)$$

The proofs of the properties above will be omitted.

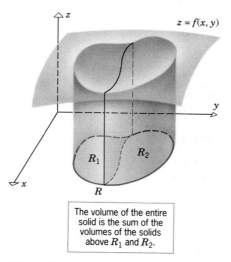

The volume of the entire solid is the sum of the volumes of the solids above R_1 and R_2.

Figure 17.1.6

□ EVALUATING DOUBLE
INTEGRALS

Except in the simplest cases, it is impractical to obtain the value of a double integral from the limit in (3). However, we shall now show how to evaluate double integrals by calculating two successive single integrals. For the remainder of this section, we shall limit our discussion to the case where R is a rectangle. In the next section we shall consider double integrals over more complicated regions.

The partial derivatives of a function $f(x, y)$ are calculated by holding one of the variables fixed and differentiating with respect to the other variable. Let us consider the reverse of this process, *partial integration*. The symbol

$$\int_{a}^{b} f(x, y) \, dx$$

is a *partial definite integral with respect to x.* It is evaluated by holding y fixed and integrating with respect to x. Similarly, the *partial definite integral with respect to y,*

$$\int_c^d f(x, y)\, dy$$

is evaluated by holding x fixed and integrating with respect to y.

Example 1

$$\int_0^1 xy^2\, dx = y^2 \int_0^1 x\, dx = \frac{y^2 x^2}{2}\bigg]_{x=0}^1 = \frac{y^2}{2}$$

$$\int_0^1 xy^2\, dy = x \int_0^1 y^2\, dy = \frac{xy^3}{3}\bigg]_{y=0}^1 = \frac{x}{3} \qquad \blacktriangleleft$$

As this example shows, an integral of the form $\int_a^b f(x, y)\, dx$ produces a function of y as the result, while an integral of the form $\int_c^d f(x, y)\, dy$ produces a function of x. This being the case, we can consider the following types of calculations:

$$\int_c^d \left[\int_a^b f(x, y)\, dx \right] dy \tag{9a}$$

$$\int_a^b \left[\int_c^d f(x, y)\, dy \right] dx \tag{9b}$$

In (9a), the inside integration, $\int_a^b f(x, y)\, dx$, yields a function of y, which is then integrated over the interval $c \le y \le d$. In (9b), the integration, $\int_c^d f(x, y)\, dy$, yields a function of x, which is then integrated over the interval $a \le x \le b$. Expressions (9a) and (9b) are called *iterated* (or *repeated*) *integrals.* Often the brackets are omitted and these expressions are written as

$$\int_c^d \int_a^b f(x, y)\, dx\, dy = \int_c^d \left[\int_a^b f(x, y)\, dx \right] dy \tag{10a}$$

$$\int_a^b \int_c^d f(x, y)\, dy\, dx = \int_a^b \left[\int_c^d f(x, y)\, dy \right] dx \tag{10b}$$

Example 2   Evaluate

(a) $\displaystyle\int_0^3 \int_1^2 (1 + 8xy)\, dy\, dx$ (b) $\displaystyle\int_1^2 \int_0^3 (1 + 8xy)\, dx\, dy$

Solution (a).

$$\int_0^3 \int_1^2 (1 + 8xy) \, dy \, dx = \int_0^3 \left[\int_1^2 (1 + 8xy) \, dy \right] dx$$

$$= \int_0^3 \left[y + 4xy^2 \right]_{y=1}^2 dx$$

$$= \int_0^3 [(2 + 16x) - (1 + 4x)] \, dx$$

$$= \int_0^3 (1 + 12x) \, dx$$

$$= x + 6x^2 \Big]_0^3 = 57$$

Solution (b).

$$\int_1^2 \int_0^3 (1 + 8xy) \, dx \, dy = \int_1^2 \left[\int_0^3 (1 + 8xy) \, dx \right] dy$$

$$= \int_1^2 \left[x + 4x^2 y \right]_{x=0}^3 dy$$

$$= \int_1^2 (3 + 36y) \, dy$$

$$= 3y + 18y^2 \Big]_1^2 = 57 \quad \blacktriangleleft$$

It is no accident that the two iterated integrals in the last example have the same value; it is a consequence of the following theorem.

Fubini's Theorem

> **17.1.2 THEOREM.** *Let R be the rectangle defined by the inequalities*
>
> $$a \le x \le b, \quad c \le y \le d$$
>
> *If $f(x, y)$ is continuous on this rectangle, then*
>
> $$\iint_R f(x, y) \, dA = \int_c^d \int_a^b f(x, y) \, dx \, dy = \int_a^b \int_c^d f(x, y) \, dy \, dx$$

This major theorem enables us to evaluate a double integral over a rectangle by calculating an iterated integral. Moreover, the theorem tells us that the order of integration in the iterated integral does not matter. We shall not formally prove this result. However, we offer the following geometric argument for the case where $f(x, y)$ is nonnegative on R.

If $f(x, y)$ is nonnegative on R, the double integral

$$\iint_R f(x, y)\ dA$$

represents the volume of the solid S bounded above by the surface $z = f(x, y)$ and below by the rectangle R. However, as discussed in Section 6.2 (Formula 6.2.2), the volume of the solid S is also given by

$$\text{Vol}\,(S) = \int_c^d A(y)\ dy \tag{11}$$

where $A(y)$ represents the area of the cross section perpendicular to the y-axis taken at the point y (Figure 17.1.7). But consider how we might compute the cross-sectional area $A(y)$. For each *fixed y* in the interval $c \le y \le d$, the function $f(x, y)$ is a function of x alone, and $A(y)$ may be viewed as the area under the graph of this function along the interval $a \le x \le b$ (see the blue shaded area in Figure 17.1.7). Thus,

$$A(y) = \int_a^b f(x, y)\ dx$$

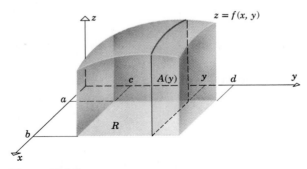

Figure 17.1.7

Substituting this expression in (11) yields

$$\text{Vol}\,(S) = \int_c^d \left[\int_a^b f(x, y)\ dx \right] dy = \int_c^d \int_a^b f(x, y)\ dx\ dy \tag{12}$$

The volume of the solid S can also be obtained using cross sections perpendicular to the x-axis. By Formula 6.2.1,

$$\text{Vol}\,(S) = \int_a^b A(x)\ dx \tag{13}$$

where $A(x)$ is the area of the cross section perpendicular to the x-axis taken at the point x (Figure 17.1.8). For each *fixed x* in the interval $a \le x \le b$, the

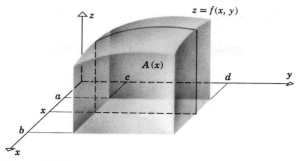

Figure 17.1.8

function $f(x, y)$ is a function of y alone, so that the area $A(x)$ is given by

$$A(x) = \int_c^d f(x, y)\, dy$$

Substituting this expression in (13) yields

$$\text{Vol}\,(S) = \int_a^b \left[\int_c^d f(x, y)\, dy \right] dx = \int_a^b \int_c^d f(x, y)\, dy\, dx \qquad (14)$$

Since the volume of S is also given by the double integral $\iint\limits_R f(x, y)\, dA$, it

follows from (12) and (14) that

$$\iint\limits_R f(x, y)\, dA = \int_c^d \int_a^b f(x, y)\, dx\, dy = \int_a^b \int_c^d f(x, y)\, dy\, dx$$

which is what we intended to show.

Example 3 Evaluate the double integral

$$\iint\limits_R y^2 x\, dA$$

over the rectangle $R = \{(x, y): -3 \le x \le 2, 0 \le y \le 1\}$.

Solution. In view of Theorem 17.1.2, the value of the double integral may be obtained from either of the iterated integrals

$$\int_{-3}^2 \int_0^1 y^2 x\, dy\, dx \quad \text{or} \quad \int_0^1 \int_{-3}^2 y^2 x\, dx\, dy \qquad (15)$$

Using the first of these, we obtain

$$\iint\limits_{R} y^2x \; dA = \int_{-3}^{2} \int_{0}^{1} y^2x \; dy \; dx = \int_{-3}^{2} \left[\frac{1}{3} y^3 x \right]_{y=0}^{1} dx$$

$$= \int_{-3}^{2} \frac{1}{3} x \; dx = \frac{x^2}{6} \Bigg]_{-3}^{2} = -\frac{5}{6}$$

The reader can check this result by evaluating the second integral in (15). ◀

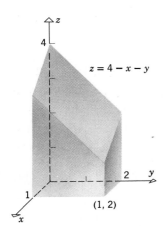

Figure 17.1.9

Example 4   Use a double integral to find the volume of the solid that is bounded above by the plane $z = 4 - x - y$ and below by the rectangle $R = \{(x, y): 0 \le x \le 1, 0 \le y \le 2\}$ (Figure 17.1.9).

Solution.

$$V = \iint\limits_{R} (4 - x - y) \; dA = \int_{0}^{2} \int_{0}^{1} (4 - x - y) \; dx \; dy$$

$$= \int_{0}^{2} \left[4x - \frac{x^2}{2} - xy \right]_{x=0}^{1} dy = \int_{0}^{2} \left(\frac{7}{2} - y \right) dy$$

$$= \left[\frac{7}{2} y - \frac{y^2}{2} \right]_{0}^{2} = 5$$

The volume can also be obtained by first integrating with respect to y and then with respect to x.   ◀

▶ Exercise Set 17.1

In Exercises 1–14, evaluate the iterated integrals.

1. $\int_{0}^{1} \int_{0}^{2} (x + 3) \; dy \; dx.$

2. $\int_{1}^{3} \int_{-1}^{1} (2x - 4y) \; dy \; dx.$

3. $\int_{2}^{4} \int_{0}^{1} x^2y \; dx \; dy.$

4. $\int_{-2}^{0} \int_{-1}^{2} (x^2 + y^2) \; dx \; dy.$

5. $\int_{0}^{\ln 3} \int_{0}^{\ln 2} e^{x+y} \; dy \; dx.$

6. $\int_{0}^{2} \int_{0}^{1} y \sin x \; dy \; dx.$

7. $\int_{0}^{3} \int_{0}^{1} x(x^2 + y)^{1/2} \; dx \; dy.$

8. $\int_{-1}^{2} \int_{2}^{4} (2x^2y + 3xy^2) \; dx \; dy.$

9. $\int_{-1}^{0} \int_{2}^{5} dx \; dy.$

10. $\displaystyle\int_4^6 \int_{-3}^7 dy\, dx.$

11. $\displaystyle\int_0^1 \int_0^1 \frac{x}{(xy+1)^2}\, dy\, dx.$

12. $\displaystyle\int_{\pi/2}^\pi \int_1^2 x \cos xy\, dy\, dx.$

13. $\displaystyle\int_0^{\ln 2} \int_0^1 xy\, e^{y^2 x}\, dy\, dx.$

14. $\displaystyle\int_3^4 \int_1^2 \frac{1}{(x+y)^2}\, dy\, dx.$

In Exercises 15–19, evaluate the double integral over the rectangular region R.

15. $\displaystyle\iint_R 4xy^3\, dA;$

$R = \{(x,y): -1 \leq x \leq 1,\, -2 \leq y \leq 2\}.$

16. $\displaystyle\iint_R \frac{xy}{\sqrt{x^2+y^2+1}}\, dA;$

$R = \{(x,y): 0 \leq x \leq 1,\, 0 \leq y \leq 1\}.$

17. $\displaystyle\iint_R x\sqrt{1-x^2}\, dA;$

$R = \{(x,y): 0 \leq x \leq 1,\, 2 \leq y \leq 3\}.$

18. $\displaystyle\iint_R (x \sin y - y \sin x)\, dA;$

$R = \{(x,y): 0 \leq x \leq \pi/2,\, 0 \leq y \leq \pi/3\}.$

19. $\displaystyle\iint_R \cos(x+y)\, dA;$

$R = \{(x,y): -\pi/4 \leq x \leq \pi/4,\, 0 \leq y \leq \pi/4\}.$

In Exercises 20–25, the iterated integral represents the volume of a solid. Make an accurate sketch of the solid. (You do *not* have to find the volume.)

20. $\displaystyle\int_0^5 \int_1^2 4\, dx\, dy.$

21. $\displaystyle\int_0^1 \int_0^1 (2 - x - y)\, dy\, dx.$

22. $\displaystyle\int_2^3 \int_3^4 y\, dx\, dy.$

23. $\displaystyle\int_0^3 \int_0^4 \sqrt{25 - x^2 - y^2}\, dy\, dx.$

24. $\displaystyle\int_{-2}^2 \int_{-2}^2 (x^2 + y^2)\, dx\, dy.$

25. $\displaystyle\int_0^1 \int_{-1}^1 \sqrt{4 - x^2}\, dy\, dx.$

In Exercises 26–30, use a double integral to find the volume.

26. The volume under the plane $z = 2x + y$ and over the rectangle $R = \{(x,y): 3 \leq x \leq 5,\, 1 \leq y \leq 2\}.$

27. The volume under the surface $z = 3x^3 + 3x^2y$ and over the rectangle $R = \{(x,y): 1 \leq x \leq 3,\, 0 \leq y \leq 2\}.$

28. The volume in the first octant bounded by the co-ordinate planes, the plane $y = 4$, and the plane $x/3 + z/5 = 1.$

29. The volume of the solid in the first octant enclosed by the surface $z = x^2$ and the planes $x = 2$, $y = 3$, $y = 0$, and $z = 0.$

30. The volume of the solid in the first octant that is enclosed by the planes $x = 0$, $z = 0$, $x = 5$, $z - y = 0$, and $z = -2y + 6$. [*Hint:* Break the solid into two parts.]

31. Suppose that $f(x,y) = g(x)h(y)$ and $R = \{(x,y): a \leq x \leq b,\, c \leq y \leq d\}$. Show that

$$\iint_R f(x,y)\, dA = \left[\int_a^b g(x)\, dx\right]\left[\int_c^d h(y)\, dy\right]$$

32. Evaluate

$$\iint_R x \cos(xy) \cos^2 \pi x\, dA$$

where $R = \{(x,y): 0 \leq x \leq \tfrac{1}{2},\, 0 \leq y \leq \pi\}.$
[*Hint:* One order of integration leads to a simpler solution than the other.]

■ 17.2 DOUBLE INTEGRALS OVER NONRECTANGULAR REGIONS

> *In this section we shall show how to evaluate double integrals over regions other than rectangles.*

☐ ITERATED INTEGRALS
WITH NONCONSTANT
LIMITS OF INTEGRATION

In this section we shall see that evaluating double integrals over nonrectangular regions can often be reduced to evaluating iterated integrals of the following types:

$$\int_a^b \int_{g_1(x)}^{g_2(x)} f(x, y) \, dy \, dx = \int_a^b \left[\int_{g_1(x)}^{g_2(x)} f(x, y) \, dy \right] dx \qquad \text{(1a)}$$

$$\int_c^d \int_{h_1(y)}^{h_2(y)} f(x, y) \, dx \, dy = \int_c^d \left[\int_{h_1(y)}^{h_2(y)} f(x, y) \, dx \right] dy \qquad \text{(1b)}$$

We begin with some examples that illustrate methods for evaluating such iterated integrals.

Example 1   Evaluate $\displaystyle\int_0^2 \int_{x^2}^x y^2 x \, dy \, dx$.

Solution.

$$\int_0^2 \int_{x^2}^x y^2 x \, dy \, dx = \int_0^2 \left[\int_{x^2}^x y^2 x \, dy \right] dx = \int_0^2 \left[\frac{y^3 x}{3} \right]_{y=x^2}^x dx$$

$$= \int_0^2 \left(\frac{x^4}{3} - \frac{x^7}{3} \right) dx = \left[\frac{x^5}{15} - \frac{x^8}{24} \right]_0^2$$

$$= \frac{32}{15} - \frac{256}{24} = -\frac{128}{15} \qquad \blacktriangleleft$$

Example 2   Evaluate $\displaystyle\int_0^\pi \int_0^{\cos y} x \sin y \, dx \, dy$.

Solution.

$$\int_0^\pi \int_0^{\cos y} x \sin y \, dx \, dy = \int_0^\pi \left[\int_0^{\cos y} x \sin y \, dx \right] dy$$

$$= \int_0^\pi \left[\frac{x^2}{2} \sin y \right]_{x=0}^{\cos y} dy$$

$$= \int_0^\pi \frac{1}{2} \cos^2 y \, \sin y \, dy$$

$$= \left[-\frac{1}{6} \cos^3 y \right]_0^\pi = \frac{1}{3} \quad \blacktriangleleft$$

☐ **DOUBLE INTEGRALS OVER NONRECTANGULAR REGIONS**

Plane regions can be extremely complex, and the theory of double integrals over very general regions is a topic for advanced courses in mathematics. We shall limit our study of double integrals to two basic types of regions, which we shall call *type I* and *type II*; they are defined as follows:

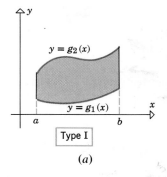

Type I

(a)

Type II

(b)

Figure 17.2.1

17.2.1   DEFINITION.

(a) A *type I region* is bounded on the left and right by vertical lines $x = a$ and $x = b$ and is bounded below and above by continuous curves $y = g_1(x)$ and $y = g_2(x)$, where $g_1(x) \leq g_2(x)$ for $a \leq x \leq b$ (Figure 17.2.1a).

(b) A *type II region* is bounded below and above by horizontal lines $y = c$ and $y = d$ and is bounded on the left and right by continuous curves $x = h_1(y)$ and $x = h_2(y)$ satisfying $h_1(y) \leq h_2(y)$ for $c \leq y \leq d$ (Figure 17.2.1b).

The following theorem will enable us to evaluate double integrals over type I and type II regions using iterated integrals.

17.2.2   THEOREM.

(a) *If R is a type I region on which $f(x, y)$ is continuous, then*

$$\iint_R f(x, y) \, dA = \int_a^b \int_{g_1(x)}^{g_2(x)} f(x, y) \, dy \, dx \qquad (2a)$$

(b) *If R is a type II region on which $f(x, y)$ is continuous, then*

$$\iint_R f(x, y) \, dA = \int_c^d \int_{h_1(y)}^{h_2(y)} f(x, y) \, dx \, dy \qquad (2b)$$

Although we shall not prove this theorem, it is easy to visualize geometrically if $f(x, y)$ is nonnegative, in which case the double integral

$$\iint_R f(x, y) \, dA \qquad (3)$$

represents the volume of the solid S bounded above by the surface $z = f(x, y)$ and below by the region R. By the method of cross sections, the volume of S is also given by

$$\text{Vol}(S) = \int_a^b A(x) \, dx \qquad (4)$$

where $A(x)$ is the area of the cross section at the fixed point x. As shown in Figure 17.2.2, this cross-sectional area extends from $g_1(x)$ to $g_2(x)$ in the y-direction, so

$$A(x) = \int_{g_1(x)}^{g_2(x)} f(x, y) \, dy$$

Substituting this in (4) we obtain

$$\text{Vol}(S) = \int_a^b \int_{g_1(x)}^{g_2(x)} f(x, y) \, dy \, dx$$

Since the volume of S is also given by (3), we obtain

$$\iint_R f(x, y) \, dA = \int_a^b \int_{g_1(x)}^{g_2(x)} f(x, y) \, dy \, dx$$

Part (*b*) of Theorem 17.2.2 can be motivated in a similar way by using cross sections parallel to the x-axis.

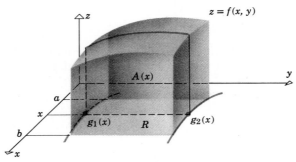

Figure 17.2.2

□ **SETTING UP LIMITS OF INTEGRATION FOR EVALUATING DOUBLE INTEGRALS**

To apply Theorem 17.2.2, it is usual to start with a two-dimensional sketch of the region R. [It is not necessary to graph $f(x, y)$.] For a type I region, the limits of integration in Formula (2a) can be obtained as follows:

Step 1. Since x is held fixed for the first integration, we draw a vertical line through the region R at an arbitrary fixed point x (Figure 17.2.3). This line crosses the boundary of R twice. The lower point of intersection is on the curve $y = g_1(x)$ and the higher point is on the curve $y = g_2(x)$. These two intersections determine the lower and upper y-limits of integration in Formula (2a).

Step 2. Imagine moving the line drawn in Step 1 first to the left and then to the right (Figure 17.2.3). The leftmost position where the line intersects the region R is $x = a$ and the rightmost position where the line intersects the region R is $x = b$. This yields the limits for the x-integration in Formula (2a).

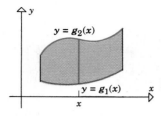
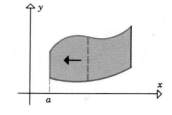
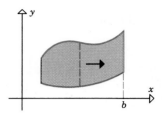

Figure 17.2.3

Example 3 Evaluate

$$\iint\limits_{R} xy \, dA$$

over the region R enclosed between $y = \frac{1}{2}x$, $y = \sqrt{x}$, $x = 2$, and $x = 4$.

Solution. We view R as a type I region. The region R and a vertical line corresponding to a fixed x are shown in Figure 17.2.4. This line meets the region R at the lower boundary $y = \frac{1}{2}x$ and the upper boundary $y = \sqrt{x}$. These are the y-limits of integration. Moving this line first left and then right yields the x-limits of integration, $x = 2$ and $x = 4$. Thus,

$$\iint\limits_{R} xy \, dA = \int_{2}^{4} \int_{x/2}^{\sqrt{x}} xy \, dy \, dx = \int_{2}^{4} \left[\frac{xy^2}{2} \right]_{y=x/2}^{\sqrt{x}} dx = \int_{2}^{4} \left(\frac{x^2}{2} - \frac{x^3}{8} \right) dx$$

$$= \left[\frac{x^3}{6} - \frac{x^4}{32} \right]_{2}^{4} = \left(\frac{64}{6} - \frac{256}{32} \right) - \left(\frac{8}{6} - \frac{16}{32} \right) = \frac{11}{6} \qquad \blacktriangleleft$$

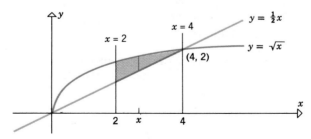

Figure 17.2.4

If R is a type II region, then the limits of integration in Formula (2b) can be obtained as follows:

> **Step 1.** Since y is held fixed for the first integration, we draw a horizontal line through the region R at a fixed point y (Figure 17.2.5). This line crosses the boundary of R twice. The leftmost point of intersection is on the curve $x = h_1(y)$ and the rightmost point is on the curve $x = h_2(y)$. These intersections determine x-limits of integration in (2b).

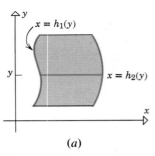
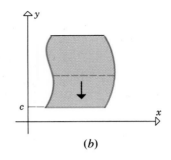
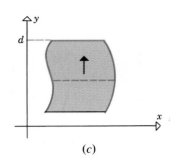

(a)   (b)   (c)

Figure 17.2.5

> **Step 2.** Imagine moving the line drawn in Step 1 first down and then up (Figure 17.2.5). The lowest position where the line intersects the region R is $y = c$, and the highest position where the line intersects the region R is $y = d$. This yields the y-limits of integration in (2b).

Example 4   Evaluate

$$\iint_R (2x - y^2)\, dA$$

over the triangular region R enclosed between the lines $y = -x + 1$, $y = x + 1$, and $y = 3$.

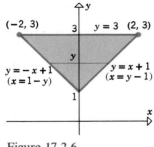

Figure 17.2.6

Solution.   We view R as a type II region. The region R and a horizontal line corresponding to a fixed y are shown in Figure 17.2.6. This line meets the region R at its left-hand boundary $x = 1 - y$ and its right-hand boundary $x = y - 1$. These are the x-limits of integration. Moving this line first down and then up yields the y-limits, $y = 1$ and $y = 3$. Thus,

$$\iint_R (2x - y^2)\, dA = \int_1^3 \int_{1-y}^{y-1} (2x - y^2)\, dx\, dy = \int_1^3 \left[x^2 - y^2 x \right]_{x=1-y}^{y-1} dy$$

$$= \int_1^3 [(1 - 2y + 2y^2 - y^3) - (1 - 2y + y^3)]\, dy$$

$$= \int_1^3 (2y^2 - 2y^3)\, dy = \left[\frac{2y^3}{3} - \frac{y^4}{2} \right]_1^3 = -\frac{68}{3} \blacktriangleleft$$

REMARK.   To integrate over a type II region, the left- and right-hand boundaries must be expressed in the form $x = h_1(y)$ and $x = h_2(y)$. This is why we rewrote the boundary equations $y = -x + 1$ and $y = x + 1$ as $x = 1 - y$ and $x = y - 1$ in the last example.

In Example 4 we could have treated R as a type I region, but with an added complication. Viewed as a type I region, the upper boundary of R is the line $y = 3$ (Figure 17.2.7) and the lower boundary consists of two parts, the line

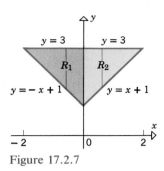

Figure 17.2.7

$y = -x + 1$ to the left of the origin and the line $y = x + 1$ to the right of the origin. To carry out the integration it is necessary to decompose the region R into two parts, R_1 and R_2, as shown in Figure 17.2.7, and write

$$\iint_R (2x - y^2)\, dA = \iint_{R_1} (2x - y^2)\, dA + \iint_{R_2} (2x - y^2)\, dA$$

$$= \int_{-2}^{0} \int_{-x+1}^{3} (2x - y^2)\, dy\, dx + \int_{0}^{2} \int_{x+1}^{3} (2x - y^2)\, dy\, dx$$

This will yield the same result that was obtained in Example 4.

□ **REVERSING THE ORDER OF INTEGRATION**

Sometimes the evaluation of an iterated integral can be simplified by reversing the order of integration. The next example illustrates how this is done.

Example 5   Since there is no elementary antiderivative of e^{x^2}, the integral

$$\int_0^2 \int_{y/2}^1 e^{x^2}\, dx\, dy$$

cannot be evaluated by performing the x-integration first. Evaluate this integral by expressing it as an equivalent iterated integral with the order of integration reversed.

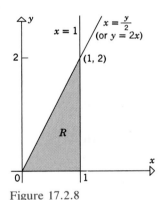

Figure 17.2.8

Solution.   For the inside integration, y is fixed and x varies from the line $x = y/2$ to the line $x = 1$ (Figure 17.2.8). For the outside integration, y varies from 0 to 2, so the given iterated integral is equal to a double integral over the triangular region R in Figure 17.2.8.

To reverse the order of integration, we treat R as a type I region, which enables us to write the given integral as

$$\int_0^2 \int_{y/2}^1 e^{x^2}\, dx\, dy = \iint_R e^{x^2}\, dA = \int_0^1 \int_0^{2x} e^{x^2}\, dy\, dx = \int_0^1 \left[e^{x^2} y \right]_{y=0}^{2x} dx$$

$$= \int_0^1 2x e^{x^2}\, dx = e^{x^2} \Big]_0^1 = e - 1 \quad \blacktriangleleft$$

□ **AREA CALCULATED AS A DOUBLE INTEGRAL**

Although double integrals arose in the context of calculating volumes, they can also be used to calculate areas. For this purpose, we consider the solid consisting of the points between the plane $z = 1$ and a region R in the xy-plane (Figure 17.2.9). The volume V of this solid is

$$V = \iint_R 1\, dA = \iint_R dA \tag{5}$$

However, the solid has congruent cross sections taken parallel to the xy-plane, so that

$$V = \text{area of base} \cdot \text{height} = \text{area of } R \cdot 1 = \text{area of } R$$

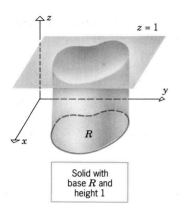

Solid with
base R and
height 1

Figure 17.2.9

Combining this with (5) yields the area formula

$$\text{area of } R = \iint_R dA \tag{6}$$

REMARK. Formula (6) is sometimes confusing because it equates an area and a volume; the formula is intended to equate only the *numerical values* of the area and volume and not the units, which must, of course, be different.

Example 6 Use a double integral to find the area of the region R enclosed between the parabola $y = \frac{1}{2}x^2$ and the line $y = 2x$.

Solution. The region R may be treated equally well as type I (Figure 17.2.10a) or type II (Figure 17.2.10b). Treating R as type I yields

$$\text{area of } R = \iint_R dA = \int_0^4 \int_{x^2/2}^{2x} dy\, dx = \int_0^4 \left[y \right]_{y=x^2/2}^{2x} dx$$

$$= \int_0^4 \left(2x - \frac{1}{2}x^2 \right) dx = \left[x^2 - \frac{x^3}{6} \right]_0^4 = \frac{16}{3}$$

Treating R as type II yields

$$\text{area of } R = \iint_R dA = \int_0^8 \int_{y/2}^{\sqrt{2y}} dx\, dy = \int_0^8 \left[x \right]_{x=y/2}^{\sqrt{2y}} dy$$

$$= \int_0^8 \left(\sqrt{2y} - \frac{1}{2}y \right) dy = \left[\frac{2\sqrt{2}}{3}y^{3/2} - \frac{y^2}{4} \right]_0^8 = \frac{16}{3} \quad \blacktriangleleft$$

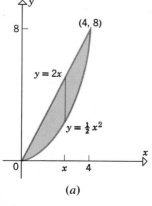

(a)

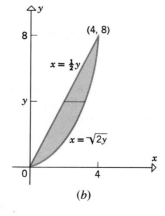

(b)

Figure 17.2.10

Volume

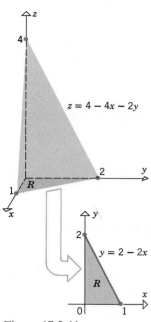

Figure 17.2.11

Example 7   Use a double integral to find the volume of the tetrahedron bounded by the coordinate planes and the plane $z = 4 - 4x - 2y$.

Solution.   The tetrahedron in question is bounded above by the plane

$$z = 4 - 4x - 2y \qquad (7)$$

and below by the triangular region R shown in Figure 17.2.11. Thus, the volume is given by

$$V = \iint\limits_{R} (4 - 4x - 2y)\, dA$$

The region R is bounded by the x-axis, the y-axis, and the line $y = 2 - 2x$ [set $z = 0$ in (7)], so that treating R as a type I region yields

$$V = \iint\limits_{R} (4 - 4x - 2y)\, dA = \int_0^1 \int_0^{2-2x} (4 - 4x - 2y)\, dy\, dx$$

$$= \int_0^1 \Big[4y - 4xy - y^2 \Big]_{y=0}^{2-2x} dx = \int_0^1 (4 - 8x + 4x^2)\, dx = \frac{4}{3} \qquad \blacktriangleleft$$

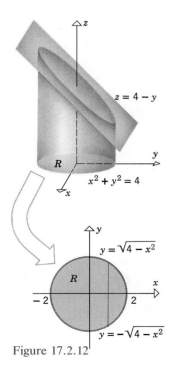

Figure 17.2.12

Example 8   Find the volume of the solid bounded by the cylinder $x^2 + y^2 = 4$ and the planes $y + z = 4$ and $z = 0$.

Solution.   The solid shown in Figure 17.2.12 is bounded above by the plane $z = 4 - y$ and below by the region R within the circle $x^2 + y^2 = 4$. The volume is given by

$$V = \iint\limits_{R} (4 - y)\, dA$$

Treating R as a type I region we obtain

$$V = \int_{-2}^{2} \int_{-\sqrt{4-x^2}}^{\sqrt{4-x^2}} (4 - y)\, dy\, dx = \int_{-2}^{2} \Big[4y - \frac{1}{2} y^2 \Big]_{y=-\sqrt{4-x^2}}^{\sqrt{4-x^2}} dx$$

$$= \int_{-2}^{2} 8\sqrt{4 - x^2}\, dx = 8(2\pi) = 16\pi \qquad \boxed{\text{See Example 4 of Section 9.5.}} \qquad \blacktriangleleft$$

▶ Exercise Set 17.2

In Exercises 1–12, evaluate the iterated integral.

1. $\int_0^1 \int_{x^2}^x xy^2 \, dy \, dx.$

2. $\int_1^2 \int_y^{3-y} y \, dx \, dy.$

3. $\int_0^3 \int_0^{\sqrt{9-y^2}} y \, dx \, dy.$

4. $\int_{1/4}^1 \int_{x^2}^x \sqrt{\dfrac{x}{y}} \, dy \, dx.$

5. $\int_{\sqrt{\pi}}^{\sqrt{2\pi}} \int_0^{x^3} \sin \dfrac{y}{x} \, dy \, dx.$

6. $\int_{-1}^1 \int_{-x^2}^{x^2} (x^2 - y) \, dy \, dx.$

7. $\int_{\pi/2}^\pi \int_0^{x^2} \dfrac{1}{x} \cos \dfrac{y}{x} \, dy \, dx.$

8. $\int_0^{\pi/2} \int_0^{\sin y} e^x \cos y \, dx \, dy.$

9. $\int_0^a \int_0^{\sqrt{a^2-x^2}} (x + y) \, dy \, dx \qquad (a > 0).$

10. $\int_1^2 \int_0^{y^2} e^{x/y^2} \, dx \, dy.$

11. $\int_0^1 \int_0^x y\sqrt{x^2 - y^2} \, dy \, dx.$

12. $\int_0^1 \int_0^x e^{x^2} \, dy \, dx.$

In Exercises 13–28, evaluate the double integral.

13. $\displaystyle\iint_R 6xy \, dA$; R is the region bounded by $y = 0$, $x = 2$, and $y = x^2$.

14. $\displaystyle\iint_R xy \, dA$; R is the region bounded by the trapezoid with vertices $(1, 3)$, $(5, 3)$, $(2, 1)$, and $(4, 1)$.

15. $\displaystyle\iint_R x \cos xy \, dA$; R is the region enclosed by $x = 1$, $x = 2$, $y = \pi/2$, and $y = 2\pi/x$.

16. $\displaystyle\iint_R (x + y) \, dA$; R is the region enclosed between the curves $y = x^2$ and $y = \sqrt{x}$.

17. $\displaystyle\iint_R x^2 \, dA$; R is the region bounded by $y = 16/x$, $y = x$, and $x = 8$.

18. $\displaystyle\iint_R xy^2 \, dA$; R is the region enclosed by $y = 1$, $y = 2$, $x = 0$, and $y = x$.

19. $\displaystyle\iint_R x(1 + y^2)^{-1/2} \, dA$; R is the region in the first quadrant enclosed by $y = x^2$, $y = 4$, and $x = 0$. [*Hint:* Choose your order of integration carefully.]

20. $\displaystyle\iint_R x \cos y \, dA$; R is the triangular region bounded by $y = x$, $y = 0$, and $x = \pi$.

21. $\displaystyle\iint_R (3x - 2y) \, dA$; R is the region enclosed by the circle $x^2 + y^2 = 1$.

22. $\displaystyle\iint_R y \, dA$; R is the region in the first quadrant enclosed between the circle $x^2 + y^2 = 25$ and the line $x + y = 5$.

23. $\displaystyle\iint_R \dfrac{1}{1 + x^2} \, dA$; R is the triangular region with vertices $(0, 0)$, $(1, 1)$, and $(0, 1)$.

24. $\displaystyle\iint_R (x^2 - xy) \, dA$; R is the region enclosed by $y = x$ and $y = 3x - x^2$.

25. $\displaystyle\iint_R xy \, dA$; R is the region enclosed by $y = \sqrt{x}$, $y = 6 - x$, and $y = 0$.

26. $\displaystyle\iint_R x \, dA$; R is the region enclosed by $y = \sin^{-1} x$, $x = 1/\sqrt{2}$, and $y = 0$.

27. $\displaystyle\iint_R (x - 1) \, dA$; R is the region enclosed between $y = x$ and $y = x^3$.

28. $\displaystyle\iint_R x^2 \, dA$; R is the region in the first quadrant enclosed by $xy = 1$, $y = x$, and $y = 2x$.

In Exercises 29–34, use double integration to find the area of the plane region enclosed by the given curves.

29. $x + y = 5$, $x = 0$, and $y = 0$.

30. $y = x^2$ and $y = 4x$.

31. $y = \sin x$ and $y = \cos x$, for $0 \le x \le \pi/4$.

32. $y^2 = -x$ and $3y - x = 4$.

33. $y^2 = 9 - x$ and $y^2 = 9 - 9x$.

34. $y = \cosh x$, $y = \sinh x$, $x = 0$, and $x = 1$.

In Exercises 35–48, use double integration to find the volume of each solid.

35. The tetrahedron that lies in the first octant that is bounded by the three coordinate planes and the plane $z = 5 - 2x - y$.

36. The solid bounded by the cylinder $x^2 + y^2 = 9$ and the planes $z = 0$ and $z = 3 - x$.

37. The solid that is bounded above by the plane $z = x + 2y + 2$, below by the xy-plane, and laterally by $y = 0$ and $y = 1 - x^2$.

38. The solid in the first octant bounded above by the paraboloid $z = x^2 + 3y^2$, below by the plane $z = 0$, and laterally by $y = x^2$ and $y = x$.

39. The solid that is bounded above by the paraboloid $z = 9x^2 + y^2$, below by the plane $z = 0$, and laterally by the planes $x = 0$, $y = 0$, $x = 3$, and $y = 2$.

40. The solid enclosed by $y^2 = x$, $z = 0$, and $x + z = 1$.

41. The wedge cut from the cylinder $4x^2 + y^2 = 9$ by the planes $z = 0$ and $z = y + 3$.

42. The solid in the first octant bounded above by $z = 9 - x^2$, below by $z = 0$, and laterally by $y^2 = 3x$.

43. The solid in the first octant bounded by the three coordinate planes and the planes $x + 2y = 4$ and $x + 8y - 4z = 0$.

44. The solid in the first octant bounded by the surface $z = e^{y-x}$, the plane $x + y = 1$, and the coordinate planes.

45. The solid bounded above by the paraboloid $z = 1 - x^2 - y^2$ and below by the xy-plane. [*Hint:* Use a trigonometric substitution to evaluate the integral.]

46. The solid in the first octant bounded by the paraboloid $z = x^2 + y^2$ and the cylinder $x^2 + y^2 = 4$. [*Hint:* Use a trigonometric substitution to evaluate the integral.]

47. The solid common to the cylinders $x^2 + y^2 = 25$ and $x^2 + z^2 = 25$.

48. The solid that is bounded above by the paraboloid $z = x^2 + y^2$, bounded laterally by the circular cylinder $x^2 + (y - 1)^2 = 1$, and bounded below by the xy-plane.

In Exercises 49–56, express the integral as an equivalent integral with the order of integration reversed.

49. $\displaystyle\int_0^2 \int_0^{\sqrt{x}} f(x, y)\, dy\, dx$.

50. $\displaystyle\int_0^4 \int_{2y}^8 f(x, y)\, dx\, dy$.

51. $\displaystyle\int_0^2 \int_1^{e^y} f(x, y)\, dx\, dy$.

52. $\displaystyle\int_1^e \int_0^{\ln x} f(x, y)\, dy\, dx$.

53. $\displaystyle\int_{-2}^2 \int_{-\sqrt{1-(x^2/4)}}^{\sqrt{1-(x^2/4)}} f(x, y)\, dy\, dx$.

54. $\displaystyle\int_0^1 \int_{y^2}^{\sqrt{y}} f(x, y)\, dx\, dy$.

55. $\displaystyle\int_0^1 \int_{\sin^{-1}y}^{\pi/2} f(x, y)\, dx\, dy$.

56. $\displaystyle\int_{-3}^1 \int_{x^2+6x}^{4x+3} f(x, y)\, dy\, dx$.

In Exercises 57–62, evaluate the integral by first reversing the order of integration.

57. $\displaystyle\int_0^1 \int_{4x}^4 e^{-y^2}\, dy\, dx$.

58. $\displaystyle\int_0^2 \int_{y/2}^1 \cos(x^2)\, dx\, dy$.

59. $\displaystyle\int_0^4 \int_{\sqrt{y}}^2 e^{x^3}\, dx\, dy$.

60. $\displaystyle\int_1^3 \int_0^{\ln x} x\, dy\, dx$.

61. $\displaystyle\int_0^1 \int_0^{\cos^{-1}x} x\, dy\, dx$.

62. $\displaystyle\int_0^1 \int_{\sin^{-1}y}^{\pi/2} \sec^2(\cos x)\, dx\, dy$.

63. Evaluate $\displaystyle\iint_R \sin(y^3)\, dA$, where R is the region bounded by $y = \sqrt{x}$, $y = 2$, and $x = 0$. [*Hint:* Choose the order of integration carefully.]

64. Evaluate $\displaystyle\iint_R x\, dA$, where R is the region bounded by $x = \ln y$, $x = 0$, and $y = e$. [*Hint:* Choose the order of integration carefully.]

65. In each part evaluate $\displaystyle\iint_R xy^2 \, dA$.

(a)

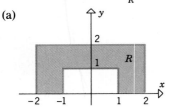

(b)

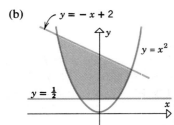

66. Assume that the region R shown in the following figure is symmetric about the y-axis and that the y-axis divides R into the subregions R_1 and R_2 shown.

Suppose also that $\displaystyle\iint_{R_1} f(x, y) \, dA = 3$.

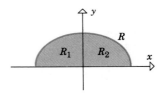

(a) Find $\displaystyle\iint_R f(x, y) \, dA$ if $f(-x, y) = f(x, y)$.

(b) Find $\displaystyle\iint_R f(x, y) \, dA$ if $f(-x, y) = -f(x, y)$.

In Exercises 67–69, use symmetry to evaluate the integral without performing an integration. [*Hint:* See Exercise 66.]

67. $\displaystyle\int_{-1}^{1} \int_{0}^{\sqrt{1-x^2}} (2 + x\sqrt{9 - y^2}) \, dy \, dx$.

68. $\displaystyle\int_{0}^{2} \int_{x-2}^{2-x} \sin(xy^3) \, dy \, dx$.

69. $\displaystyle\iint_R x^3 y \, dA$, where R is the semicircular region bounded by $y = \sqrt{4 - x^2}$ and $x = 0$.

■ **17.3 DOUBLE INTEGRALS IN POLAR COORDINATES**

In this section we shall study double integrals in which the integrand and the region of integration are expressed in polar coordinates. Such integrals are important for two reasons: First, they arise naturally in many applications, and second, many double integrals in rectangular coordinates are more easily evaluated if they are converted to polar coordinates.

☐ **DEFINITION OF DOUBLE INTEGRALS IN POLAR COORDINATES**

Double integrals whose integrands are in polar coordinates arise naturally in volume problems in which the base region R can be expressed more easily in polar coordinates than in rectangular coordinates. This is generally the case when R is of the form shown in Figure 17.3.1a. This region consists of all points enclosed between two rays, $\theta = \alpha$ and $\theta = \beta$, and two continuous polar curves, $r = r_1(\theta)$ and $r = r_2(\theta)$, where $\alpha < \beta$, $0 \le r_1(\theta) \le r_2(\theta)$ for $\alpha \le \theta \le \beta$,

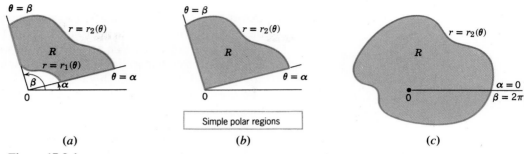

Figure 17.3.1

and $\beta - \alpha \leq 2\pi$. We shall call regions of this form *simple polar regions*. In the case where $r_1(\theta)$ is identically zero, the inner boundary of a simple polar region reduces to a single point (the origin) and the region looks like that in Figure 17.3.1*b*. If, in addition, $\alpha = 0$ and $\beta = 2\pi$, then the side boundaries coincide and the region looks like that in Figure 17.3.1*c*.

The concept of a double integral in polar coordinates can be motivated by the following volume problem:

17.3.1 PROBLEM. *Find the volume of the solid consisting of all points that lie between a simple polar region R in the xy-plane and a surface whose equation in cylindrical coordinates is $z = f(r, \theta)$, where f is continuous on R and $f(r, \theta) \geq 0$ for all (r, θ) in R (Figure 17.3.2).*

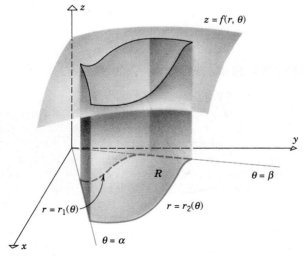

Figure 17.3.2

The method for finding the volume V of the solid in Figure 17.3.2 will be a limiting process similar to that used to derive Formula (2) in Section 17.1, except that here the region R will be subdivided by circular arcs and rays rather than lines parallel to the x and y axes. The procedure is as follows:

Step 1. Cover the region R with a grid of circular arcs centered at the origin and rays emanating from the origin (Figure 17.3.3). The blocks formed by this grid are called **polar rectangles**. Exclude from consideration all polar rectangles that contain any points outside of R. This leaves only polar rectangles that are subsets of R. Denote the areas of these polar rectangles by

$$\Delta A_1, \Delta A_2, \ldots, \Delta A_n$$

Step 2. Choose an arbitrary point in each of these polar rectangles, and denote them by

$$(r_1^*, \theta_1^*), (r_2^*, \theta_2^*), \ldots, (r_n^*, \theta_n^*)$$

As shown in Figure 17.3.4, the product $f(r_k^*, \theta_k^*) \, \Delta A_k$ is the volume of a solid with base area ΔA_k and height $f(r_k^*, \theta_k^*)$, so the sum

$$\sum_{k=1}^{n} f(r_k^*, \theta_k^*) \, \Delta A_k$$

can be viewed as an approximation to the volume V of the entire solid.

Step 3. If we repeat the process using more and more subdivisions in such a way that the dimensions of the polar rectangles approach zero, then it is plausible that the errors in the approximations approach zero, and the exact volume of the solid is

$$V = \lim_{n \to +\infty} \sum_{k=1}^{n} f(r_k^*, \theta_k^*) \, \Delta A_k \tag{1}$$

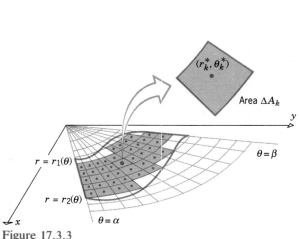

Figure 17.3.3

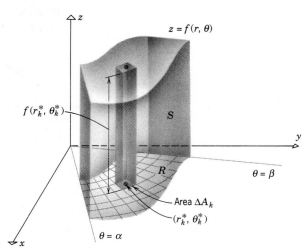

Figure 17.3.4

The sums in (1) are called *polar Riemann sums,* and the limit of the polar Riemann sums is denoted by

$$\iint\limits_{R} f(r, \theta)\, dA = \lim_{n \to +\infty} \sum_{k=1}^{n} f(r_k^*, \theta_k^*)\, \Delta A_k \tag{2}$$

which is called the *polar double integral* of $f(r, \theta)$ over R. With this notation, the volume V of the solid can be expressed as

$$V = \iint\limits_{R} f(r, \theta)\, dA \tag{3}$$

If $f(r, \theta)$ can have both positive and negative values on R, then, in keeping with past experience, the polar double integral over R represents the *net signed volume* between $z = f(r, \theta)$ and R, that is, the difference between the volume of the solid above R but below $z = f(r, \theta)$ and the volume of the solid that is below R but above $z = f(r, \theta)$.

It can be shown that polar double integrals satisfy integral properties (5), (6), (7), and (8) of Section 17.1 with $f(r, \theta)$ replacing $f(x, y)$, and the argument used to derive area formula (6) of Section 17.2 can be used to show that the area of a simple polar region R can be expressed as

$$\text{area of } R = \iint\limits_{R} dA \tag{4}$$

☐ **EVALUATION OF POLAR DOUBLE INTEGRALS**

The following theorem will enable us to evaluate double polar integrals by using iterated integrals.

17.3.2 THEOREM. *If R is a region of the type shown in* Figure 17.3.1, *and if $f(r, \theta)$ is continuous on R, then*

$$\iint\limits_{R} f(r, \theta)\, dA = \int_{\alpha}^{\beta} \int_{r_1(\theta)}^{r_2(\theta)} f(r, \theta) r\, dr\, d\theta \tag{5}$$

REMARK. In (5) note that the dA in the double integral becomes $r\, dr\, d\theta$ in the iterated integral.

We shall not formally prove this theorem. However, the appearance of the factor r in the iterated integral may be explained by returning to the definition

$$\iint\limits_{R} f(r, \theta)\, dA = \lim_{n \to +\infty} \sum_{k=1}^{n} f(r_k^*, \theta_k^*)\, \Delta A_k \tag{6}$$

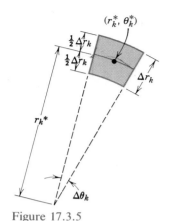

Figure 17.3.5

In the polar Riemann sum $\sum_{k=1}^{n} f(r_k^*, \theta_k^*) \Delta A_k$, suppose that the arbitrary point (r_k^*, θ_k^*) is chosen at the "center" of the kth polar rectangle, that is, at the point halfway between the bounding circular arcs on the ray that bisects them (Figure 17.3.5). Suppose also that this polar rectangle has a central angle $\Delta \theta_k$ and a "radial thickness" Δr_k. Thus, the inner radius of this polar rectangle is $r_k^* - \frac{1}{2} \Delta r_k$ and the outer radius is $r_k^* + \frac{1}{2} \Delta r_k$. Treating the area ΔA_k of this polar rectangle as the difference in area of two sectors, we obtain

$$\Delta A_k = \frac{1}{2} \left(r_k^* + \frac{1}{2} \Delta r_k \right)^2 \Delta \theta_k - \frac{1}{2} \left(r_k^* - \frac{1}{2} \Delta r_k \right)^2 \Delta \theta_k$$

which simplifies to

$$\Delta A_k = r_k^* \, \Delta r_k \, \Delta \theta_k \tag{7}$$

(Verify.) Substituting this expression in (6) yields

$$\iint\limits_R f(r, \theta) \, dA = \lim_{n \to +\infty} \sum_{k=1}^{n} f(r_k^*, \theta_k^*) r_k^* \, \Delta r_k \, \Delta \theta_k$$

which explains the form of the integrand in (5).

To apply Theorem 17.3.2, we start with a sketch of the region R. From this sketch the limits of integration in the formula

$$\iint\limits_R f(r, \theta) \, dA = \int_\alpha^\beta \int_{r_1(\theta)}^{r_2(\theta)} f(r, \theta) r \, dr \, d\theta \tag{8}$$

can be obtained as follows:

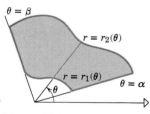

Figure 17.3.6

Step 1. Since θ is held fixed for the first integration, draw a radial line from the origin through the region R at a fixed angle θ (Figure 17.3.6). This line crosses the boundary of R at most twice. The innermost point of intersection is on the curve $r = r_1(\theta)$ and the outermost point is on the curve $r = r_2(\theta)$. These intersections determine the r-limits of integration in (8).

Step 2. Imagine rotating a ray along the positive x-axis one revolution counterclockwise about the origin. The smallest angle at which this ray intersects the region R is $\theta = \alpha$ and the largest angle is $\theta = \beta$. This yields the θ-limits of integration.

Example 1   Evaluate

$$\iint\limits_R \sin \theta \, dA$$

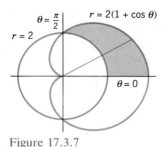

Figure 17.3.7

where R is the region in the first quadrant that is outside the circle $r = 2$ and inside the cardioid $r = 2(1 + \cos \theta)$.

Solution. The region R is sketched in Figure 17.3.7. Following the two steps outlined above we obtain

$$\iint_R \sin \theta \, dA = \int_0^{\pi/2} \int_2^{2(1+\cos \theta)} (\sin \theta) r \, dr \, d\theta$$

$$= \int_0^{\pi/2} \frac{1}{2} r^2 \sin \theta \bigg]_{r=2}^{2(1+\cos \theta)} d\theta$$

$$= 2 \int_0^{\pi/2} [(1 + \cos \theta)^2 \sin \theta - \sin \theta] \, d\theta$$

$$= 2 \left[-\frac{1}{3}(1 + \cos \theta)^3 + \cos \theta \right]_0^{\pi/2}$$

$$= 2 \left[-\frac{1}{3} - \left(-\frac{5}{3} \right) \right] = \frac{8}{3} \quad \blacktriangleleft$$

Example 2   In cylindrical coordinates, the equation $r^2 + z^2 = a^2$ represents a sphere of radius a centered at the origin. (It is the sphere whose equation in rectangular coordinates is $x^2 + y^2 + z^2 = a^2$.) Use a double polar integral to calculate the volume of this sphere.

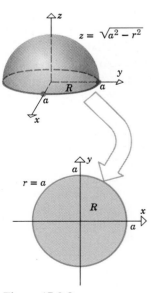

Figure 17.3.8

Solution.   The upper hemisphere is given by the (cylindrical) equation

$$z = \sqrt{a^2 - r^2}$$

so the volume enclosed by the entire sphere is

$$V = 2 \iint_R \sqrt{a^2 - r^2} \, dA$$

where R is the circular region shown in Figure 17.3.8. Thus,

$$V = 2 \iint_R \sqrt{a^2 - r^2} \, dA = \int_0^{2\pi} \int_0^a \sqrt{a^2 - r^2} \, (2r) \, dr \, d\theta$$

$$= \int_0^{2\pi} \left[-\frac{2}{3}(a^2 - r^2)^{3/2} \right]_{r=0}^a d\theta = \int_0^{2\pi} \frac{2}{3} a^3 \, d\theta$$

$$= \left[\frac{2}{3} a^3 \theta \right]_0^{2\pi} = \frac{4}{3} \pi a^3 \quad \blacktriangleleft$$

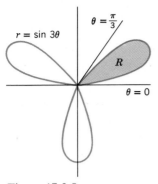

$\theta = \frac{\pi}{3}$

$r = \sin 3\theta$

R

$\theta = 0$

Figure 17.3.9

Example 3   Use a double polar integral to find the area enclosed by the three-petaled rose $r = \sin 3\theta$.

Solution.   The rose is sketched in Figure 17.3.9. We shall use Formula (4) to calculate the area of the petal R in the first quadrant and multiply by three.

$$A = 3 \iint\limits_{R} dA = 3 \int_{0}^{\pi/3} \int_{0}^{\sin 3\theta} r \, dr \, d\theta$$

$$= \frac{3}{2} \int_{0}^{\pi/3} \sin^2 3\theta \, d\theta = \frac{3}{4} \int_{0}^{\pi/3} (1 - \cos 6\theta) \, d\theta$$

$$= \left[\frac{3}{4} \theta - \frac{3}{24} \sin 6\theta \right]_{0}^{\pi/3} = \frac{1}{4} \pi \quad \blacktriangleleft$$

☐ **CONVERTING DOUBLE INTEGRALS FROM RECTANGULAR TO POLAR COORDINATES**

Polar double integrals are sometimes called *double integrals in polar coordinates* and integrals of the type studied in the preceding two sections *double integrals in rectangular coordinates*. Sometimes a double integral in rectangular coordinates that is difficult to evaluate can be evaluated more easily by converting it to an equivalent integral in polar coordinates. This is often true when the integrand or the boundary of R involves expressions of the form $x^2 + y^2$ or $\sqrt{x^2 + y^2}$, since these simplify to r^2 and r in polar coordinates.

Example 4   Use polar coordinates to evaluate $\displaystyle\int_{-1}^{1} \int_{0}^{\sqrt{1-x^2}} (x^2 + y^2)^{3/2} \, dy \, dx$.

Solution.   The first step is to express this iterated integral as a double integral in rectangular coordinates. For fixed x, the y-integration runs from the lower boundary $y = 0$ up to the semicircle $y = \sqrt{1 - x^2}$. The fixed x can extend from -1 on the left to $+1$ on the right, so

$$\int_{-1}^{1} \int_{0}^{\sqrt{1-x^2}} (x^2 + y^2)^{3/2} \, dy \, dx = \iint\limits_{R} (x^2 + y^2)^{3/2} \, dA$$

where R is the region shown in Figure 17.3.10. Expressing this double integral as a double integral in polar coordinates yields

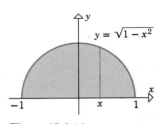

y

$y = \sqrt{1 - x^2}$

x

-1 x 1

Figure 17.3.10

$$\int_{-1}^{1} \int_{0}^{\sqrt{1-x^2}} (x^2 + y^2)^{3/2} \, dy \, dx = \iint\limits_{R} (r^2)^{3/2} \, dA$$

$$= \int_{0}^{\pi} \int_{0}^{1} (r^3) r \, dr \, d\theta$$

$$= \int_{0}^{\pi} \frac{1}{5} \, d\theta = \frac{\pi}{5} \quad \blacktriangleleft$$

▶ Exercise Set 17.3

In Exercises 1–6, evaluate the iterated integral.

1. $\displaystyle\int_0^{\pi/2} \int_0^{\sin\theta} r\cos\theta \, dr \, d\theta.$

2. $\displaystyle\int_0^{\pi} \int_0^{1+\cos\theta} r \, dr \, d\theta.$ **3.** $\displaystyle\int_{-\pi/2}^{\pi/2} \int_0^{a\sin\theta} r^2 \, dr \, d\theta.$

4. $\displaystyle\int_0^{\pi/3} \int_0^{\cos 3\theta} r \, dr \, d\theta.$

5. $\displaystyle\int_0^{\pi} \int_0^{1-\sin\theta} r^2 \cos\theta \, dr \, d\theta.$

6. $\displaystyle\int_0^{\pi} \int_0^{\cos\theta} r^3 \, dr \, d\theta.$

In Exercises 7–12, use a double integral in polar coordinates to find the area of the region described.

7. The region enclosed by the cardioid $r = 1 - \cos\theta$.

8. The region enclosed by the rose $r = \sin 2\theta$.

9. The region in the first quadrant bounded by $r = 1$ and $r = \sin 2\theta$, with $\pi/4 \le \theta \le \pi/2$.

10. The region inside the circle $x^2 + y^2 = 4$ and to the right of the line $x = 1$.

11. The region inside the circle $r = 4\sin\theta$ and outside the circle $r = 2$.

12. The region inside the circle $r = 1$ and outside the cardioid $r = 1 + \cos\theta$.

In Exercises 13–18, use a double integral in polar coordinates to find the volume of the solid.

13. The solid enclosed by the sphere $x^2 + y^2 + z^2 = 9$ and the cylinder $x^2 + y^2 = 1$.

14. The solid enclosed by the sphere $r^2 + z^2 = 4$ and the cylinder $r = 2\cos\theta$.

15. The solid that is bounded above by the cone $z = \sqrt{x^2 + y^2}$, below by the xy-plane, and laterally by the cylinder $x^2 + y^2 = 2y$.

16. The solid that is bounded above by the surface $z = (x^2 + y^2)^{-1/2}$, below by the xy-plane, and enclosed between the cylinders $x^2 + y^2 = 1$ and $x^2 + y^2 = 9$.

17. The solid bounded above by the paraboloid $z = 1 - x^2 - y^2$, below by the xy-plane, and laterally by the cylinder $x^2 + y^2 - x = 0$.

18. The solid in the first octant bounded above by the plane $z = r\sin\theta$, below by the xy-plane, and laterally by the plane $x = 0$ and the cylinder $r = 3\sin\theta$.

In Exercises 19–22, use polar coordinates to evaluate the double integral.

19. $\displaystyle\iint_R e^{-(x^2+y^2)} \, dA$, where R is the region enclosed by the circle $x^2 + y^2 = 1$.

20. $\displaystyle\iint_R \sqrt{9 - x^2 - y^2} \, dA$, where R is the region in the first quadrant within the circle $x^2 + y^2 = 9$.

21. $\displaystyle\iint_R \frac{1}{1 + x^2 + y^2} \, dA$, where R is the sector in the first quadrant that is bounded by $y = 0$, $y = x$, and $x^2 + y^2 = 4$.

22. $\displaystyle\iint_R 2y \, dA$, where R is the region in the first quadrant bounded above by the circle $(x - 1)^2 + y^2 = 1$ and below by the line $y = x$.

In Exercises 23–30, evaluate the iterated integral by converting to polar coordinates.

23. $\displaystyle\int_0^1 \int_0^{\sqrt{1-x^2}} (x^2 + y^2) \, dy \, dx.$

24. $\displaystyle\int_{-2}^2 \int_{-\sqrt{4-y^2}}^{\sqrt{4-y^2}} e^{-(x^2+y^2)} \, dx \, dy.$

25. $\displaystyle\int_0^2 \int_0^{\sqrt{2x-x^2}} \sqrt{x^2 + y^2} \, dy \, dx.$

26. $\displaystyle\int_0^1 \int_0^{\sqrt{1-y^2}} \cos(x^2 + y^2) \, dx \, dy.$

27. $\displaystyle\int_0^a \int_0^{\sqrt{a^2-x^2}} \frac{dy \, dx}{(1 + x^2 + y^2)^{3/2}} \quad (a > 0).$

28. $\displaystyle\int_0^1 \int_y^{\sqrt{y}} \sqrt{x^2 + y^2} \, dx \, dy.$

29. $\displaystyle\int_0^{\sqrt{2}} \int_y^{\sqrt{4-y^2}} \frac{1}{\sqrt{1 + x^2 + y^2}} \, dx \, dy.$

30. $\displaystyle\int_0^4 \int_3^{\sqrt{25-x^2}} dy \, dx.$

31. Use polar coordinates to find the volume of the solid that is bounded above by the ellipsoid $x^2/a^2 + y^2/a^2 + z^2/c^2 = 1$, below by the xy-plane, and laterally by the cylinder $x^2 + y^2 - ay = 0$.

32. Find the area of the region enclosed by the lemniscate $r^2 = 2a^2 \cos 2\theta$.

33. Find the area in the first quadrant inside the circle $r = 4 \sin \theta$ and outside the lemniscate $r^2 = 8 \cos 2\theta$.

34. Show that the shaded area in the following figure is $a^2\phi - \frac{1}{2}a^2 \sin 2\phi$.

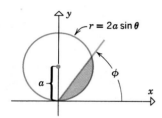

35. The integral $\displaystyle\int_0^{+\infty} e^{-x^2}\, dx$, which arises in probability theory, can be evaluated using a trick. Let the value of the integral be I. Thus,

$$I = \int_0^{+\infty} e^{-x^2}\, dx = \int_0^{+\infty} e^{-y^2}\, dy$$

since the letter used for the variable of integration in a definite integral does not matter.

(a) Show that

$$I^2 = \int_0^{+\infty} \int_0^{+\infty} e^{-(x^2+y^2)}\, dx\, dy$$

(b) Evaluate the iterated integral in part (a) by converting to polar coordinates.

(c) Use the result in part (b) to find I.

36. In each part evaluate $\displaystyle\iint_R x^2\, dA$.

(a)

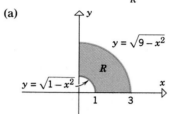

(b)

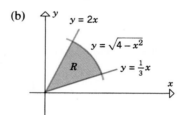

17.4 SURFACE AREA

In Section 6.5 we showed how to find the surface area of a surface of revolution. In this section we shall consider more general surface area problems.

☐ **DERIVATION OF THE SURFACE AREA FORMULA**

The surface area problem of concern in this section can be posed as follows:

17.4.1 PROBLEM. *Let f be a function defined on a closed region R of the xy-plane. Find the area of that portion of the surface z = f(x, y) whose projection on the xy-plane is the region R (Figure 17.4.1).*

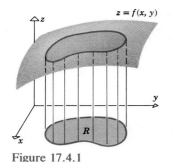

Figure 17.4.1

We will attack this problem in stages, first solving it in the special case where the surface is a plane and the region R is a rectangle, then using the special case to solve the general problem.

> **17.4.2** THEOREM. *Let R be a closed rectangular region in the xy-plane. If R has sides of length l and w, then the surface area S of that portion of the plane $z = ax + by + c$ that projects onto the region R is given by*
>
> $$S = \sqrt{a^2 + b^2 + 1}\, lw \tag{1}$$

Proof. The portion of the plane that projects onto the region R is a parallelogram (Figure 17.4.2). Thus, if we can find vectors **a** and **b** forming adjacent sides of this parallelogram, then we can obtain its surface area S from the formula

$$S = \|\mathbf{a} \times \mathbf{b}\| \tag{2}$$

(see Section 14.4).

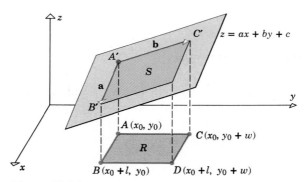

Figure 17.4.2

To find the vectors **a** and **b**, suppose that the four corners of the rectangle R are

$$A(x_0, y_0), \quad B(x_0 + l, y_0), \quad C(x_0, y_0 + w), \quad D(x_0 + l, y_0 + w)$$

(Figure 17.4.2). Then those points in the plane $z = ax + by + c$ that lie above the corners A, B, and C are

$$A'(x_0, y_0, ax_0 + by_0 + c)$$
$$B'(x_0 + l, y_0, a[x_0 + l] + by_0 + c)$$
$$C'(x_0, y_0 + w, ax_0 + b[y_0 + w] + c)$$

Thus, the vectors

$$\mathbf{a} = \overrightarrow{A'B'} = l\,\mathbf{i} + 0\,\mathbf{j} + al\,\mathbf{k}$$
$$\mathbf{b} = \overrightarrow{A'C'} = 0\,\mathbf{i} + w\,\mathbf{j} + bw\,\mathbf{k}$$

form adjacent sides of the parallelogram in question. Since

$$\mathbf{a} \times \mathbf{b} = \begin{vmatrix} \mathbf{i} & \mathbf{j} & \mathbf{k} \\ l & 0 & al \\ 0 & w & bw \end{vmatrix} = -alw\,\mathbf{i} - lbw\,\mathbf{j} + lw\,\mathbf{k}$$

it follows from (2) that

$$S = \|\mathbf{a} \times \mathbf{b}\| = \sqrt{(-alw)^2 + (-lbw)^2 + (lw)^2} = \sqrt{a^2 + b^2 + 1}\;lw \qquad ∎$$

We shall now show how Theorem 17.4.2 can be used to derive the following more general result.

17.4.3 SURFACE AREA FORMULA. If f has continuous first partial derivatives on a closed region R of the xy-plane, then the area S of that portion of the surface $z = f(x, y)$ that projects onto R is

$$S = \iint\limits_{R} \sqrt{\left(\frac{\partial z}{\partial x}\right)^2 + \left(\frac{\partial z}{\partial y}\right)^2 + 1}\; dA \qquad (3)$$

We proceed as follows:

Step 1. Enclose R within a rectangle whose sides are parallel to the x and y axes. Using lines parallel to the x and y axes, divide this rectangle into subrectangles and exclude from consideration all those subrectangles that contain any points outside of R. This leaves only rectangles that are subsets of R (Figure 17.4.3a). Let these rectangles be denoted by

$$R_1, R_2, \ldots, R_n$$

and suppose that the sides of rectangle R_k have lengths Δx_k and Δy_k.

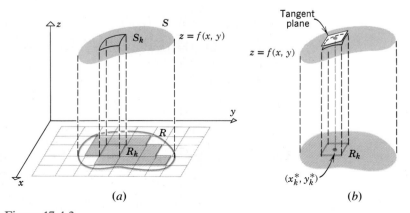

Figure 17.4.3

(a)   (b)

Step 2. When projected up to the surface $z = f(x, y)$, each subrectangle determines a patch of area on the surface (Figure 17.4.3a). If we denote the areas of these patches by

$$S_1, S_2, \ldots, S_n$$

then the entire surface area S can be approximated by adding the areas of these patches:

$$S \approx S_1 + S_2 + \cdots + S_n \tag{4}$$

(This is only an approximation since the subrectangles may not completely fill out the region R.)

Step 3. We shall now approximate each of the areas

$$S_1, S_2, \ldots, S_n$$

Let

$$(x_k^*, y_k^*)$$

be an arbitrary point in the kth subrectangle and above this point construct the tangent plane to the surface $z = f(x, y)$ (Figure 17.4.3b). It follows from Theorem 16.5.1 that the equation of this tangent plane can be written in the form

$$z = f_x(x_k^*, y_k^*)x + f_y(x_k^*, y_k^*)y + c$$

where c is an appropriate constant. If the rectangle R_k is small, then we can reasonably approximate the area S_k of the kth patch on the surface by the portion of area on the tangent plane over R_k (Figure 17.4.3b). Thus, by Theorem 17.4.2

$$S_k \approx \sqrt{f_x(x_k^*, y_k^*)^2 + f_y(x_k^*, y_k^*)^2 + 1} \, \Delta x_k \, \Delta y_k$$

Substituting this expression in (4) and writing the area of the kth subrectangle as $\Delta A_k = \Delta x_k \, \Delta y_k$, we obtain the Riemann sum

$$S \approx \sum_{k=1}^{n} \sqrt{f_x(x_k^*, y_k^*)^2 + f_y(x_k^*, y_k^*)^2 + 1} \, \Delta A_k \tag{5}$$

Step 4. There are two sources of error in approximation (5). First, the rectangles $R_1, R_2, \ldots, R_n$ may not fill up the region R completely, and second, we have approximated patches of area on the surface by areas on tangent planes. However, let us repeat this approximation process using more and more rectangles of decreasing dimensions. It is intuitively plausible that both kinds of errors diminish and the exact surface area S is given by

$$S = \lim_{n \to +\infty} \sum_{k=1}^{n} \sqrt{f_x(x_k^*, y_k^*)^2 + f_y(x_k^*, y_k^*)^2 + 1} \ \Delta A_k$$

or equivalently

$$S = \iint_R \sqrt{f_x(x, y)^2 + f_y(x, y)^2 + 1} \ dA$$

which is just Formula (3) with a different notation for the partial derivatives.

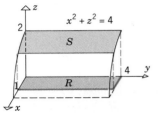

Figure 17.4.4

Example 1   Find the surface area of the portion of the cylinder $x^2 + z^2 = 4$ above the rectangle $R = \{(x, y): 0 \le x \le 1, 0 \le y \le 4\}$ in the xy-plane.

Solution.   The surface is shown in Figure 17.4.4. The portion of the cylinder $x^2 + z^2 = 4$ that lies above the xy-plane has the equation $z = \sqrt{4 - x^2}$. Thus, from (3)

$$S = \iint_R \sqrt{\left(\frac{\partial z}{\partial x}\right)^2 + \left(\frac{\partial z}{\partial y}\right)^2 + 1} \ dA$$

$$= \iint_R \sqrt{\left(-\frac{x}{\sqrt{4 - x^2}}\right)^2 + 0 + 1} \ dA = \int_0^4 \int_0^1 \frac{2}{\sqrt{4 - x^2}} \ dx \ dy$$

$$= 2 \int_0^4 \left[\sin^{-1}\left(\frac{1}{2} x\right) \right]_{x=0}^{1} dy = 2 \int_0^4 \frac{\pi}{6} \ dy = \frac{4}{3}\pi \quad \blacktriangleleft$$

Formula (17)
of Section 8.2

Example 2   Find the surface area of the portion of the paraboloid $z = x^2 + y^2$ below the plane $z = 1$.

Solution.   The surface is shown in Figure 17.4.5. From the equations $z = 1$ and $z = x^2 + y^2$, we see that the plane and the paraboloid intersect in a circle

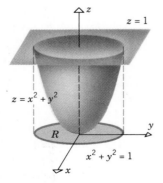

Figure 17.4.5

whose projection on the xy-plane has the equation $x^2 + y^2 = 1$. Consequently, the surface whose area we seek projects onto the region R enclosed by this circle. Since the surface has the equation

$$z = x^2 + y^2$$

it follows from (3) that

$$S = \iint_R \sqrt{4x^2 + 4y^2 + 1} \; dA$$

This integral is best evaluated in polar coordinates. Replacing $x^2 + y^2$ by r^2 and substituting $r \, dr \, d\theta$ for dA we obtain

$$S = \int_0^{2\pi} \int_0^1 \sqrt{4r^2 + 1} \; r \, dr \, d\theta = \int_0^{2\pi} \left[\frac{1}{12} (4r^2 + 1)^{3/2} \right]_{r=0}^1 d\theta$$

$$= \int_0^{2\pi} \frac{1}{12} (5\sqrt{5} - 1) \, d\theta = \frac{1}{6} \pi (5\sqrt{5} - 1) \quad \blacktriangleleft$$

▶ Exercise Set 17.4

1. Find the surface area of the portion of the cylinder $y^2 + z^2 = 9$ that is above the rectangle $R = \{(x, y) : 0 \le x \le 2, -3 \le y \le 3\}$.

2. By integration, find the surface area of the portion of the plane $2x + 2y + z = 8$ in the first octant that is cut off by the three coordinate planes.

3. Find the surface area of the portion of the cone $z^2 = 4x^2 + 4y^2$ that is above the region in the first quadrant bounded by the line $y = x$ and the parabola $y = x^2$.

4. Find the surface area of the portion of the cone $z = \sqrt{x^2 + y^2}$ that lies inside the cylinder $x^2 + y^2 = 2x$.

5. Find the surface area of the portion of the paraboloid $z = 1 - x^2 - y^2$ that is above the xy-plane.

6. Find the area of the portion of the surface $z = 2x + y^2$ that is above the triangular region with vertices $(0, 0)$, $(0, 1)$, and $(1, 1)$.

7. Find the area of the portion of the surface $z = xy$ that is above the sector in the first quadrant bounded by the lines $y = x/\sqrt{3}$, $y = 0$, and the circle $x^2 + y^2 = 9$.

8. Find the surface area of the portion of the paraboloid $2z = x^2 + y^2$ that is inside the cylinder $x^2 + y^2 = 8$.

9. Find the surface area of the portion of the sphere $x^2 + y^2 + z^2 = 16$ between the planes $z = 1$ and $z = 2$.

10. Find the surface area of the portion of the sphere $x^2 + y^2 + z^2 = 8$ that is cut out by the cone $z = \sqrt{x^2 + y^2}$.

11. Find the total surface area of the portion of the sphere $x^2 + y^2 + z^2 = a^2$ inside the cylinder $x^2 + y^2 = ay$.

12. Use a double integral to derive the formula for the surface area of a sphere of radius a.

13. Find the surface area of the portion of the cylinder $x^2 + z^2 = 16$ that lies inside the circular cylinder $x^2 + y^2 = 16$. [Hint: Find the area in the first octant and use symmetry.]

14. Find the surface area of the portion of the cylinder $x^2 + z^2 = 5x$ that lies inside the sphere $x^2 + y^2 + z^2 = 25$.

15. The portion of the surface

$$z = \frac{h}{a} \sqrt{x^2 + y^2} \qquad (a, h > 0)$$

between the xy-plane and the plane $z = h$ is a right-circular cone of height h and radius a. Use a double integral to show that the lateral surface area of this cone is $S = \pi a \sqrt{a^2 + h^2}$.

▪ 17.5 TRIPLE INTEGRALS

In the preceding sections we defined and discussed properties of double integrals for functions of two variables. In this section we shall define triple integrals for functions of three variables.

□ **DEFINITION OF A TRIPLE INTEGRAL**

Whereas a double integral of a function $f(x, y)$ is defined over a closed region R in the xy-plane, a triple integral of a function $f(x, y, z)$ is defined over a closed three-dimensional solid region G. To ensure that G does not extend indefinitely in any direction, we shall assume that G can be enclosed within some suitably large box (rectangular parallelepiped) with sides parallel to the coordinate planes (Figure 17.5.1).

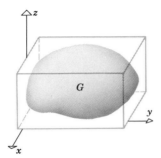

Figure 17.5.1

Volume $= \Delta V_k$

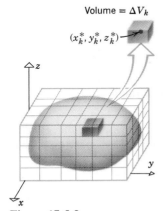

Figure 17.5.2

To define the triple integral of $f(x, y, z)$ over G, we first divide G into smaller "subboxes" by planes parallel to the coordinate planes. We then discard those subboxes that contain any points outside of G and choose an arbitrary point in each of the remaining subboxes. As shown in Figure 17.5.2, we denote the volume of the kth remaining subbox by ΔV_k and the point selected in the kth subbox by (x_k^*, y_k^*, z_k^*). Next, we form the product

$$f(x_k^*, y_k^*, z_k^*) \, \Delta V_k$$

for each subbox, then add the products for all of the subboxes to obtain the **Riemann sum**

$$\sum_{k=1}^{n} f(x_k^*, y_k^*, z_k^*) \, \Delta V_k$$

Finally, we repeat this process with more and more subdivisions in such a way that the length, width, and height of each subbox approaches zero, and n (the number of subboxes) approaches $+\infty$. The limit

$$\iiint_G f(x, y, z) \, dV = \lim_{n \to +\infty} \sum_{k=1}^{n} f(x_k^*, y_k^*, z_k^*) \, \Delta V_k \tag{1}$$

is called the *triple integral* of $f(x, y, z)$ over the region G. Conditions under which the triple integral exists are studied in advanced calculus. However, for our purposes it suffices to say that existence is ensured when f is continuous on G and the region G is not too "complicated."

☐ VOLUME CALCULATED
AS A TRIPLE INTEGRAL

Triple integrals have a number of physical interpretations, some of which we shall consider in the next section. In the special case where $f(x, y, z) = 1$ the triple integral over G represents the volume of the solid G, that is,

$$\text{volume of } G = \iiint_G dV \tag{2}$$

[Compare this to (6) of Section 17.2.] To obtain (2), let $f(x, y, z) = 1$ in (1). This yields

$$\iiint_G dV = \lim_{n \to +\infty} \sum_{k=1}^{n} \Delta V_k$$

As suggested by Figure 17.5.2, the sum $\Sigma \, \Delta V_k$ represents the total volume of the boxes interior to the solid G. As n increases and the dimensions of these boxes approach zero, the boxes tend to fill up the solid G, so that their total volume approaches the volume of G, that is,

$$\iiint_G dV = \lim_{n \to +\infty} \sum_{k=1}^{n} \Delta V_k = \text{volume of } G$$

☐ PROPERTIES OF TRIPLE
INTEGRALS

Triple integrals enjoy many properties of single and double integrals:

$$\iiint_G cf(x, y, z) \, dV = c \iiint_G f(x, y, z) \, dV \quad (c \text{ a constant})$$

$$\iiint_G [f(x, y, z) + g(x, y, z)] \, dV = \iiint_G f(x, y, z) \, dV + \iiint_G g(x, y, z) \, dV$$

$$\iiint_G [f(x, y, z) - g(x, y, z)] \, dV = \iiint_G f(x, y, z) \, dV - \iiint_G g(x, y, z) \, dV$$

Moreover, if the region G is subdivided into two subregions G_1 and G_2 (Figure 17.5.3), then

$$\iiint_G f(x, y, z) \, dV = \iiint_{G_1} f(x, y, z) \, dV + \iiint_{G_2} f(x, y, z) \, dV$$

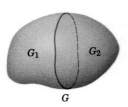

G_1 G_2

G

Figure 17.5.3

We omit the proofs.

☐ **EVALUATING TRIPLE INTEGRALS OVER RECTANGULAR BOXES**

Just as a double integral can be evaluated by two single integrations, a triple integral can be evaluated by three single integrations.

17.5.1   THEOREM.   *Let G be the rectangular box defined by the inequalities*

$$a \leq x \leq b, \quad c \leq y \leq d, \quad k \leq z \leq l$$

If f is continuous on the region G, then

$$\iiint_G f(x, y, z) \, dV = \int_a^b \int_c^d \int_k^l f(x, y, z) \, dz \, dy \, dx \tag{3}$$

Moreover, the iterated integral on the right can be replaced with any of the five other iterated integrals that result by altering the order of integration.

We omit the proof.

Example 1   Evaluate the triple integral

$$\iiint_G 12xy^2z^3 \, dV$$

over the rectangular box G defined by the inequalities $-1 \leq x \leq 2, 0 \leq y \leq 3$, $0 \leq z \leq 2$.

Solution.   Of the six possible iterated integrals we might use, we shall choose the one in (3). Thus, we shall first integrate with respect to z, holding x and y fixed, then with respect to y holding x fixed, and finally with respect to x.

$$\iiint_G 12xy^2z^3 \, dV = \int_{-1}^2 \int_0^3 \int_0^2 12xy^2z^3 \, dz \, dy \, dx$$

$$= \int_{-1}^2 \int_0^3 \left[3xy^2z^4 \right]_{z=0}^2 dy \, dx = \int_{-1}^2 \int_0^3 48xy^2 \, dy \, dx$$

$$= \int_{-1}^2 \left[16xy^3 \right]_{y=0}^3 dx = \int_{-1}^2 432x \, dx$$

$$= 216x^2 \Big]_{-1}^2 = 648 \quad \blacktriangleleft$$

☐ **EVALUATING TRIPLE INTEGRALS OVER MORE GENERAL REGIONS**

We shall also be concerned with evaluating triple integrals over solid regions other than rectangular boxes. For simplicity, we shall restrict our discussion to solid regions constructed as follows. Let R be a closed region in the xy-plane and let $g_1(x, y)$ and $g_2(x, y)$ be continuous functions satisfying

$$g_1(x, y) \leq g_2(x, y)$$

for all (x, y) in R. Geometrically, this condition states that the surface $z = g_2(x, y)$ does not dip below the surface $z = g_1(x, y)$ over R (Figure 17.5.4a). We shall call $z = g_1(x, y)$ the *lower surface* and $z = g_2(x, y)$ the *upper surface*. Let G be the solid consisting of all points above or below the region R that lie between the upper surface and the lower surface (Figure 17.5.4b). A solid G constructed in this way will be called a *simple solid,* and the region R will be called the *projection* of G on the xy-plane.

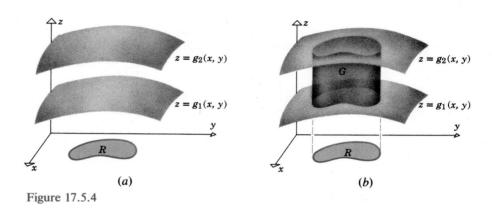

Figure 17.5.4

The following theorem, which we state without proof, will enable us to evaluate triple integrals over simple solids.

17.5.2 THEOREM. *Let G be a simple solid with upper surface $z = g_2(x, y)$ and lower surface $z = g_1(x, y)$, and let R be the projection of G on the xy-plane. If $f(x, y, z)$ is continuous on G, then*

$$\iiint_G f(x, y, z)\, dV = \iint_R \left[\int_{g_1(x, y)}^{g_2(x,y)} f(x, y, z)\, dz \right] dA \tag{4}$$

In (4), the first integration is with respect to z, after which a function of x and y remains. This function of x and y is then integrated over the region R in the xy-plane. To apply (4), it is usual to begin with a three-dimensional sketch of the solid G, from which the limits of integration can be obtained as follows:

Step 1. Find an equation $z = g_2(x, y)$ for the upper surface and an equation $z = g_1(x, y)$ for the lower surface of G. The functions $g_1(x, y)$ and $g_2(x, y)$ determine the lower and upper z-limits of integration.

Step 2. Make a two-dimensional sketch of the projection R of the solid on the xy-plane. From this sketch determine the limits of integration for the double integral over R in (4).

Example 2 Let G be the wedge in the first octant cut from the cylindrical solid $y^2 + z^2 \leq 1$ by the planes $y = x$ and $x = 0$. Evaluate

$$\iiint_G z \, dV$$

Solution. The solid G and its projection R on the xy-plane are shown in Figure 17.5.5. The upper surface of the solid is formed by the cylinder and the lower surface by the xy-plane. Since the portion of the cylinder $y^2 + z^2 = 1$ that lies above the xy-plane has the equation $z = \sqrt{1 - y^2}$, and the xy-plane has the equation $z = 0$, it follows from (4) that

$$\iiint_G z \, dV = \iint_R \left[\int_0^{\sqrt{1-y^2}} z \, dz \right] dA \tag{5}$$

For the double integral over R, the x and y integrations can be performed in either order, since R is both a type I and type II region. We shall integrate with respect to x first. With this choice, (5) yields

$$\iiint_G z \, dV = \int_0^1 \int_0^y \int_0^{\sqrt{1-y^2}} z \, dz \, dx \, dy = \int_0^1 \int_0^y \frac{1}{2} z^2 \Big]_{z=0}^{\sqrt{1-y^2}} dx \, dy$$

$$= \int_0^1 \int_0^y \frac{1}{2} (1 - y^2) \, dx \, dy = \frac{1}{2} \int_0^1 (1 - y^2) x \Big]_{x=0}^{y} dy$$

$$= \frac{1}{2} \int_0^1 (y - y^3) \, dy = \frac{1}{2} \left[\frac{1}{2} y^2 - \frac{1}{4} y^4 \right]_0^1 = \frac{1}{8} \quad \blacktriangleleft$$

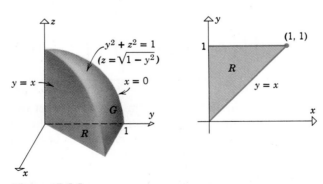

Figure 17.5.5

Example 3 Use a triple integral to find the volume of the solid enclosed between the cylinder $x^2 + y^2 = 9$ and the planes $z = 1$ and $x + z = 5$.

Solution. The solid G and its projection R on the xy-plane are shown in Figure 17.5.6. The lower surface of the solid is the plane $z = 1$, and the upper surface

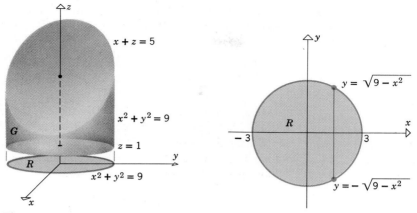

Figure 17.5.6

is the plane $x + z = 5$, or equivalently, $z = 5 - x$. Thus, from (2) and (4)

$$\text{volume of } G = \iiint_G dV = \iint_R \left[\int_1^{5-x} dz \right] dA \tag{6}$$

For the double integral over R, we shall integrate with respect to y first. Thus, (6) yields

$$\text{volume of } G = \int_{-3}^{3} \int_{-\sqrt{9-x^2}}^{\sqrt{9-x^2}} \int_1^{5-x} dz \, dy \, dx = \int_{-3}^{3} \int_{-\sqrt{9-x^2}}^{\sqrt{9-x^2}} z \Big]_{z=1}^{5-x} dy \, dx$$

$$= \int_{-3}^{3} \int_{-\sqrt{9-x^2}}^{\sqrt{9-x^2}} (4 - x) \, dy \, dx = \int_{-3}^{3} (8 - 2x)\sqrt{9 - x^2} \, dx$$

$$= 8 \int_{-3}^{3} \sqrt{9 - x^2} \, dx - \int_{-3}^{3} 2x\sqrt{9 - x^2} \, dx \qquad \boxed{\text{For the first integral, see Example 4 of Section 9.5.}}$$

$$= 8\left(\frac{9}{2}\pi\right) - \int_{-3}^{3} 2x\sqrt{9 - x^2} \, dx \qquad \boxed{\text{Let } u = 9 - x^2, \text{ or better, apply the result in Exercise 42(a) of Section 5.8.}}$$

$$= 8\left(\frac{9}{2}\pi\right) - 0 = 36\pi \qquad \blacktriangleleft$$

Example 4 Find the volume of the solid enclosed by the paraboloids

$$z = 5x^2 + 5y^2 \quad \text{and} \quad z = 6 - 7x^2 - y^2$$

Solution. The solid G and its projection R on the xy-plane are shown in Figure 17.5.7. The projection R is obtained by solving the given equations simultaneously to determine where the paraboloids intersect. We obtain

$$5x^2 + 5y^2 = 6 - 7x^2 - y^2$$

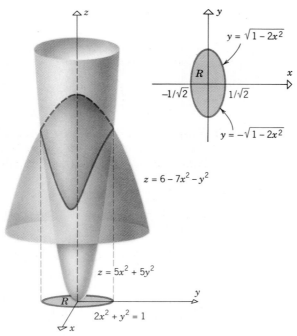

Figure 17.5.7

or

$$2x^2 + y^2 = 1 \tag{7}$$

which tells us that the paraboloids intersect in a curve on the elliptic cylinder given by (7).

The projection of this intersection on the xy-plane is an ellipse with this same equation. Therefore,

$$\text{volume of } G = \iiint\limits_G dV = \iint\limits_R \left[\int_{5x^2+5y^2}^{6-7x^2-y^2} dz \right] dA$$

$$\text{volume of } G = \int_{-1/\sqrt{2}}^{1/\sqrt{2}} \int_{-\sqrt{1-2x^2}}^{\sqrt{1-2x^2}} \int_{5x^2+5y^2}^{6-7x^2-y^2} dz \, dy \, dx$$

$$= \int_{-1/\sqrt{2}}^{1/\sqrt{2}} \int_{-\sqrt{1-2x^2}}^{\sqrt{1-2x^2}} (6 - 12x^2 - 6y^2) \, dy \, dx$$

$$= \int_{-1/\sqrt{2}}^{1/\sqrt{2}} \left[6(1 - 2x^2)y - 2y^3 \right]_{y=-\sqrt{1-2x^2}}^{\sqrt{1-2x^2}} dx$$

$$= 8 \int_{-1/\sqrt{2}}^{1/\sqrt{2}} (1 - 2x^2)^{3/2} \, dx = \frac{8}{\sqrt{2}} \int_{-\pi/2}^{\pi/2} \cos^4 \theta \, d\theta = \frac{3\pi}{\sqrt{2}} \quad \blacktriangleleft$$

Let $x = \dfrac{1}{\sqrt{2}} \sin \theta$.

Formula (5), Section 9.3

☐ **INTEGRATION IN OTHER ORDERS**

For certain regions, triple integrals are best evaluated by integrating first with respect to x or y rather than z. For example, if the solid G is bounded on the left and right by the surfaces $y = g_1(x, z)$ and $y = g_2(x, z)$ and bounded laterally by a cylinder extending in the y-direction (Figure 17.5.8a), then

$$\iiint_G f(x, y, z)\, dV = \iint_R \left[\int_{g_1(x,z)}^{g_2(x,z)} f(x, y, z)\, dy \right] dA$$

where R is the projection of the solid G on the xz-plane.

Similarly, if the solid G is bounded in the back and front by the surfaces $x = g_1(y, z)$ and $x = g_2(y, z)$ and bounded laterally by a cylinder extending in the x-direction, then

$$\iiint_G f(x, y, z)\, dV = \iint_R \left[\int_{g_1(y,z)}^{g_2(y,z)} f(x, y, z)\, dx \right] dA$$

where R is the projection of the solid on the yz-plane (Figure 17.5.8b).

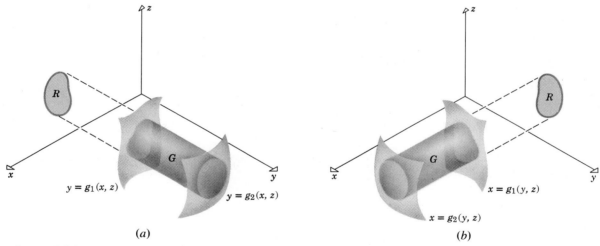

(a) (b)

Figure 17.5.8

Example 5 In Example 2, we evaluated

$$\iiint_G z\, dV$$

over the wedge in Figure 17.5.5 by integrating first with respect to z. Evaluate this integral by integrating first with respect to x.

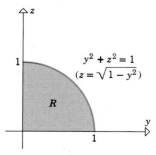

Figure 17.5.9

Solution. The solid is bounded in the back by the plane $x = 0$ and in the front by the plane $x = y$, so

$$\iiint_G z\, dV = \iint_R \left[\int_0^y z\, dx \right] dA$$

where R is the projection of G on the yz-plane (Figure 17.5.9). The integration over R can be performed first with respect to z and then y or vice versa. Performing the z-integration first yields

$$\iiint_G z\, dV = \int_0^1 \int_0^{\sqrt{1-y^2}} \int_0^y z\, dx\, dz\, dy = \int_0^1 \int_0^{\sqrt{1-y^2}} zx \Big]_{x=0}^y dz\, dy$$

$$= \int_0^1 \int_0^{\sqrt{1-y^2}} zy\, dz\, dy = \int_0^1 \frac{1}{2} z^2 y \Big]_{z=0}^{\sqrt{1-y^2}} = \int_0^1 \frac{1}{2}(1-y^2)y\, dy = \frac{1}{8}$$

which agrees with the result in Example 2. ◀

▶ Exercise Set 17.5

In Exercises 1–8, evaluate the iterated integral.

1. $\displaystyle\int_{-1}^1 \int_0^2 \int_0^1 (x^2 + y^2 + z^2)\, dx\, dy\, dz.$

2. $\displaystyle\int_{1/3}^{1/2} \int_0^\pi \int_0^1 zx \sin xy\, dz\, dy\, dx.$

3. $\displaystyle\int_0^2 \int_{-1}^{y^2} \int_1^z yz\, dx\, dz\, dy.$

4. $\displaystyle\int_0^{\pi/4} \int_0^1 \int_0^{x^2} x \cos y\, dz\, dx\, dy.$

5. $\displaystyle\int_0^3 \int_0^{\sqrt{9-z^2}} \int_0^x xy\, dy\, dx\, dz.$

6. $\displaystyle\int_1^3 \int_x^{x^2} \int_0^{\ln z} xe^y\, dy\, dz\, dx.$

7. $\displaystyle\int_0^2 \int_0^{\sqrt{4-x^2}} \int_{-5+x^2+y^2}^{3-x^2-y^2} x\, dz\, dy\, dx.$

8. $\displaystyle\int_1^2 \int_z^2 \int_0^{\sqrt{3}y} \frac{y}{x^2+y^2}\, dx\, dy\, dz.$

In Exercises 9–12, evaluate the triple integral.

9. $\displaystyle\iiint_G xy \sin yz\, dV$, where G is the rectangular box defined by the inequalities $0 \leq x \leq \pi$, $0 \leq y \leq 1$, $0 \leq z \leq \pi/6$.

10. $\displaystyle\iiint_G y\, dV$, where G is the solid enclosed by the

plane $z = y$, the xy-plane, and the parabolic cylinder $y = 1 - x^2$.

11. $\displaystyle\iiint_G xyz\, dV$, where G is the solid in the first octant that is bounded by the parabolic cylinder $z = 2 - x^2$ and the planes $z = 0$, $y = x$, and $y = 0$.

12. $\displaystyle\iiint_G \cos (z/y)\, dV$, where G is the solid defined by the inequalities, $\pi/6 \leq y \leq \pi/2$, $y \leq x \leq \pi/2$, $0 \leq z \leq xy$.

In Exercises 13–21, use a triple integral to find the volume of the solid.

13. The solid in the first octant bounded by the coordinate planes and the plane $3x + 6y + 4z = 12$.

14. The solid bounded by the surface $z = \sqrt{y}$ and the planes $x + y = 1$, $x = 0$, and $z = 0$.

15. The solid bounded by the surface $y = x^2$ and the planes $y + z = 4$ and $z = 0$.

16. The wedge in the first octant cut from the cylinder $y^2 + z^2 \leq 1$ by the planes $y = x$ and $x = 0$.

17. The solid enclosed between the elliptic cylinder $x^2 + 9y^2 = 9$ and the planes $z = 0$ and $z = x + 3$.

18. The solid enclosed by the cylinders $x^2 + y^2 = 1$ and $x^2 + z^2 = 1$.

19. The solid bounded by the paraboloid $z = 4x^2 + y^2$ and the parabolic cylinder $z = 4 - 3y^2$.

20. The solid that is enclosed between the paraboloids $z = 8 - x^2 - y^2$ and $z = 3x^2 + y^2$.

21. The solid that is enclosed between the sphere $x^2 + y^2 + z^2 = 2a^2$ and the paraboloid $az = x^2 + y^2$ $(a > 0)$.

22. In each part sketch the solid whose volume is given by the integral.

(a) $\displaystyle\int_0^3 \int_{x^2}^9 \int_0^2 dz\, dy\, dx$

(b) $\displaystyle\int_0^2 \int_0^{2-y} \int_0^{2-x-y} dz\, dx\, dy$.

23. In each part sketch the solid whose volume is given by the integral.

(a) $\displaystyle\int_{-1}^1 \int_{-\sqrt{1-x^2}}^{\sqrt{1-x^2}} \int_0^{y+1} dz\, dy\, dx$

(b) $\displaystyle\int_0^9 \int_0^{y/3} \int_0^{\sqrt{y^2-9x^2}} dz\, dx\, dy$.

24. In each part sketch the solid whose volume is given by the integral.

(a) $\displaystyle\int_0^1 \int_0^{\sqrt{1-x^2}} \int_0^2 dy\, dz\, dx$

(b) $\displaystyle\int_{-2}^2 \int_0^{4-y^2} \int_0^2 dx\, dz\, dy$.

25. Let G be the tetrahedron in the first octant bounded by the coordinate planes and the plane

$$\frac{x}{a} + \frac{y}{b} + \frac{z}{c} = 1 \qquad (a > 0,\, b > 0,\, c > 0)$$

(a) List six different iterated integrals that represent the volume of G.

(b) Evaluate any one of the six to show that the volume of G is $\frac{1}{6}abc$.

26. In parts (a)–(c), express the integral as an equivalent integral in which the z-integration is performed first, the y-integration second, and the x-integration last.

(a) $\displaystyle\int_0^3 \int_0^{\sqrt{9-z^2}} \int_0^{\sqrt{9-y^2-z^2}} f(x, y, z)\, dx\, dy\, dz$

(b) $\displaystyle\int_0^4 \int_0^2 \int_0^{x/2} f(x, y, z)\, dy\, dz\, dx$

(c) $\displaystyle\int_0^4 \int_0^{4-y} \int_0^{\sqrt{z}} f(x, y, z)\, dx\, dz\, dy$.

27. Use a triple integral to find the volume of the tetrahedron with vertices $(0, 0, 0)$, $(a, a, 0)$, $(a, 0, 0)$, and $(a, 0, a)$.

28. Let G be the rectangular box defined by the inequalities $a \le x \le b,\ c \le y \le d,\ k \le z \le l$. Show that

$$\iiint_G f(x)g(y)h(z)\, dV$$

$$= \left[\int_a^b f(x)\, dx\right]\left[\int_c^d g(y)\, dy\right]\left[\int_k^l h(z)\, dz\right]$$

29. Use the result of Exercise 28 to evaluate

(a) $\displaystyle\iiint_G xy^2 \sin z\, dV$, where G is the set of points satisfying

$-1 \le x \le 1,\ 0 \le y \le 1,\ 0 \le z \le \pi/2$

(b) $\displaystyle\iiint_G e^{2x+y-z}\, dV$, where G is the set of points satisfying

$0 \le x \le 1,\ 0 \le y \le \ln 3,\ 0 \le z \le \ln 2$.

30. Use a triple integral to find the volume of the ellipsoid

$$\frac{x^2}{a^2} + \frac{y^2}{b^2} + \frac{z^2}{c^2} = 1$$

31. Let G be a region that is symmetric about the xy-plane, and let G_1 and G_2 be the portions of G above and below the xy-plane, respectively. Suppose that

$$\iiint_{G_1} f(x, y, z)\, dV = 5.$$

(a) Find $\displaystyle\iiint_G f(x, y, z)\, dV$ given that

$f(x, y, -z) = f(x, y, z)$.

(b) Find $\displaystyle\iiint_G f(x, y, z)\, dV$ given that

$f(x, y, -z) = -f(x, y, z)$.

In Exercises 32–34, use symmetry and Exercise 31 to evaluate the integral without integrating.

32. $\displaystyle\int_{-2}^2 \int_0^{\sqrt{4-x^2}} \int_{-\sqrt{4-x^2-y^2}}^{\sqrt{4-x^2-y^2}} [z^3 \cos(xyz) - 3]\, dz\, dy\, dx$.

33. $\displaystyle\int_{-3}^3 \int_{-\sqrt{9-y^2}}^{\sqrt{9-y^2}} \int_{x^2+y^2}^9 x^2 y^3 e^z\, dz\, dx\, dy$.

34. $\displaystyle\iiint_G e^{xy} \sin z\, dV$, where G is the spherical solid enclosed by $x^2 + y^2 + z^2 = R^2$.

■ **17.6**   CENTROIDS, CENTERS OF GRAVITY, THEOREM OF PAPPUS

*Suppose that a physical body is acted on by a gravitational field. Because the body is composed of many particles, each of which is affected by gravity, the action of the gravitational field on the body consists of a large number of forces distributed over the entire body. However, it is a physical fact that a body subjected to gravity behaves as if the entire force of gravity acts at a single point. This point is called the **center of gravity** or **center of mass** of the body. In this section we shall show how double and triple integrals can be used to locate centers of gravity.*

□ **DENSITY OF A LAMINA**

The thickness of a lamina is negligible.

Figure 17.6.1

To begin, let us consider an idealized flat object that is sufficiently thin to be viewed as a two-dimensional solid (Figure 17.6.1). Such a solid is called a *lamina*. A lamina is called *homogeneous* if its composition and structure are uniform throughout, and is called *inhomogeneous* otherwise.

For a *homogeneous* lamina of mass M and area A, we define the *density* δ of the lamina by

$$\delta = \frac{M}{A}$$

Thus, δ measures the mass per unit area.

For an inhomogeneous lamina the composition may vary from point to point. If the lamina is placed in the xy-plane, then the "density" of the lamina at a general point (x, y) can be specified by a function $\delta(x, y)$, called the *density function* for the lamina. Informally, the density function can be visualized as follows. Construct a small rectangle centered at (x, y) and let ΔM and ΔA be the mass and area of the portion of lamina enclosed by this rectangle (Figure 17.6.2). If the ratio $\Delta M / \Delta A$ tends toward a limiting value as the dimensions of the rectangle tend to zero, then this limit is considered to be the density of the lamina at (x, y). Symbolically,

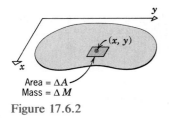

Area = ΔA
Mass = ΔM

Figure 17.6.2

$$\delta(x, y) = \lim \frac{\Delta M}{\Delta A} \tag{1}$$

From this relationship we obtain the approximation

$$\Delta M \approx \delta(x, y)\, \Delta A \tag{2}$$

which relates the mass and area of a small rectangular portion of lamina centered at (x, y). It is assumed that as the dimensions of the rectangle tend to zero, the error in this approximation also tends to zero.

☐ MASS OF A LAMINA

We can now obtain a formula for the mass of a lamina.

17.6.1 DEFINITION.   If a lamina with a continuous density function $\delta(x, y)$ occupies a region R in the xy-plane, then its total mass M is defined to be

$$M = \iint\limits_{R} \delta(x, y) \, dA \tag{3}$$

To motivate this result, let a rectangle enclosing the region R be divided into subrectangles by lines parallel to the coordinate axes, and exclude from consideration all those rectangles that contain any points outside of R. Let

$$\Delta A_1, \Delta A_2, \ldots, \Delta A_n$$

be the areas of the remaining rectangles and let

$$\Delta M_1, \Delta M_2, \ldots, \Delta M_n$$

be the masses of the portions of lamina covered by these rectangles. We can approximate the total mass of the lamina by adding the masses of these pieces:

$$M = \text{total mass} \approx \sum_{k=1}^{n} \Delta M_k \tag{4}$$

If we now let (x_k^*, y_k^*) be the center of the kth rectangle, then it follows from (2) that

$$\Delta M_k \approx \delta(x_k^*, y_k^*) \, \Delta A_k$$

Substituting this in (4) yields

$$M \approx \sum_{k=1}^{n} \delta(x_k^*, y_k^*) \, \Delta A_k \tag{5}$$

If we now increase n in such a way that the dimensions of the rectangles tend to zero, then the errors in our approximations will diminish so that

$$M = \lim_{n \to +\infty} \sum_{k=1}^{n} \delta(x_k^*, y_k^*) \, \Delta A_k = \iint\limits_{R} \delta(x, y) \, dA$$

Example 1   A triangular lamina with vertices $(0, 0)$, $(0, 1)$, and $(1, 0)$ has density function $\delta(x, y) = xy$. Find its total mass.

Solution.   Referring to (3) and Figure 17.6.3, the mass M of the lamina is

$$M = \iint\limits_{R} \delta(x, y) \, dA = \iint\limits_{R} xy \, dA = \int_0^1 \int_0^{-x+1} xy \, dy \, dx$$

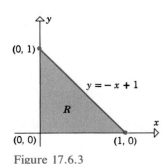

Figure 17.6.3

$$= \int_0^1 \left[\frac{1}{2} xy^2 \right]_{y=0}^{-x+1} dx = \int_0^1 \left[\frac{1}{2} x^3 - x^2 + \frac{1}{2} x \right] dx$$

$$= \frac{1}{24} \text{ (units of mass)} \quad \blacktriangleleft$$

☐ **CENTER OF GRAVITY OF A LAMINA**

Let us now consider the following problem.

17.6.2 PROBLEM. *Suppose that a lamina with density function $\delta(x, y)$ occupies a region R in a horizontal xy-plane. Find the coordinates $(\bar{x}, \bar{y})$ of the center of gravity.*

To motivate the solution, consider what happens if we try to balance the lamina on a knife-edge parallel to the x-axis. Suppose the lamina in Figure 17.6.4 is placed on a knife-edge along a line $y = c$ that does not pass through the center of gravity. Because the lamina behaves as if its entire mass is concentrated at the center of gravity $(\bar{x}, \bar{y})$, the lamina will be rotationally unstable and the force of gravity will cause a rotation about $y = c$. Similarly, the lamina will undergo a rotation if placed on a knife-edge along $y = d$. However, if the knife-edge runs along the line $y = \bar{y}$ through the center of gravity, the lamina will be in perfect balance. Similarly, the lamina will be in perfect balance on a knife-edge along the line $x = \bar{x}$ through the center of gravity.

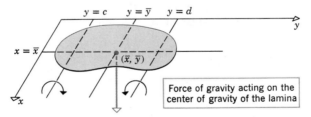

Figure 17.6.4

This discussion suggests that the center of gravity of a lamina can be determined as the intersection of two lines of balance, one parallel to the x-axis and the other parallel to the y-axis. In order to find these lines of balance, we shall need some preliminary results about rotations.

Children on a seesaw learn by experience that a lighter child can balance a heavier one by sitting farther from the fulcrum or pivot point. This is because the tendency for a mass to produce rotation is proportional not only to the magnitude of the mass but also to the distance between the mass and the fulcrum. To be precise, if a point-mass m is located on a coordinate axis at a point x, then the tendency for that mass to produce a rotation about a point a on the axis is measured by the following quantity, called the **moment of m about a:**

$$\begin{bmatrix} \text{moment of } m \\ \text{about } a \end{bmatrix} = m(x - a)$$

The number $x - a$ is called the ***lever arm***. Depending on whether the mass is to the right or left of a, the lever arm is either the distance between x and a or the negative of this distance (Figure 17.6.5). Positive lever arms result in positive moments and clockwise rotations, while negative lever arms result in negative moments and counterclockwise rotations.

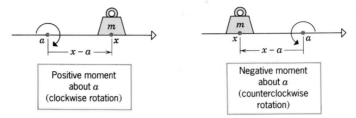

Figure 17.6.5

Suppose that masses $m_1, m_2, \ldots, m_n$ are located at points $x_1, x_2, \ldots, x_n$ on a coordinate axis and a fulcrum is positioned at the point a (Figure 17.6.6). Depending on whether the sum of the moments about a,

$$\sum_{k=1}^{n} m_k(x_k - a) = m_1(x_1 - a) + m_2(x_2 - a) + \cdots + m_n(x_n - a)$$

is positive, negative, or zero, the axis will rotate clockwise about a, rotate counterclockwise about a, or balance perfectly. In the last case, the system of masses is said to be in ***equilibrium***.

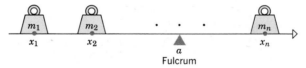

Figure 17.6.6

Example 2   Where should a fulcrum be placed so that the following system of masses is in equilibrium?

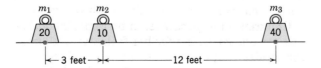

Solution.   Introduce an x-axis with the origin at mass m_1, and let a be the unknown coordinate of the fulcrum (Figure 17.6.7). (Our choice of location for the origin is arbitrary. Any convenient location will suffice.) We shall determine a so that the total moment about a is zero. The sum of the moments about a is

$$\text{total moment} = 20(0 - a) + 10(3 - a) + 40(15 - a) = 630 - 70a$$

Figure 17.6.7

For equilibrium, a should be chosen so that

$$630 - 70a = 0$$

or $a = 9$. Thus, the fulcrum should be located 9 feet to the right of mass m_1.

◀

The ideas we have discussed can be extended to masses distributed in two-dimensional space. If a mass m is located at a point (x, y) in a horizontal xy-plane, then we define the **moment of m about the line x = a** and the **moment of m about the line y = c** by

$$\begin{bmatrix} \text{moment of } m \\ \text{about the} \\ \text{line } x = a \end{bmatrix} = m(x - a) \quad \left(\begin{matrix} \text{mass times} \\ \text{lever arm} \end{matrix}\right) \tag{6}$$

$$\begin{bmatrix} \text{moment of } m \\ \text{about the} \\ \text{line } y = c \end{bmatrix} = m(y - c) \quad \left(\begin{matrix} \text{mass times} \\ \text{lever arm} \end{matrix}\right) \tag{7}$$

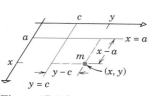

Figure 17.6.8

(See Figure 17.6.8.)

If a number of masses are distributed in the xy-plane, and if we imagine the xy-plane to be a weightless sheet, then this sheet will balance on the line $x = a$ (or $y = c$) if the sum of the moments about that line is zero.

Suppose now that we are dealing not with isolated point masses but with a lamina in the xy-plane. How should we define the moments about the lines $x = a$ and $y = c$? To motivate the appropriate definitions, assume that the lamina occupies a region R in the xy-plane and has a continuous density function $\delta(x, y)$. Following a now familiar procedure, let R be partitioned into subregions and let R_k be a typical rectangle interior to R. Let ΔA_k be the area of R_k, let ΔM_k be the mass of that portion of the lamina covered by R_k, and let (x_k^*, y_k^*) be the center of the rectangle R_k. From (2), the mass ΔM_k may be approximated by

$$\Delta M_k \approx \delta(x_k^*, y_k^*) \, \Delta A_k$$

Moreover, if the rectangle R_k is small, there should be little error in assuming that the entire mass ΔM_k is concentrated at the center (x_k^*, y_k^*) of the rectangle. Thus, from (6) and (7), the moment of the mass ΔM_k about the line $x = a$ is approximately

$$(x_k^* - a) \, \Delta M_k \approx (x_k^* - a) \, \delta(x_k^*, y_k^*) \, \Delta A_k$$

and the moment about the line $y = c$ is approximately

$$(y_k^* - c) \, \Delta M_k \approx (y_k^* - c) \, \delta(x_k^*, y_k^*) \, \Delta A_k$$

Therefore, the total moment about $x = a$ produced by all the rectangular regions interior to R is approximately

$$\sum_{k=1}^{n} (x_k^* - a) \, \delta(x_k^*, y_k^*) \, \Delta A_k \tag{8}$$

and the total moment about $y = c$ is approximately

$$\sum_{k=1}^{n} (y_k^* - c) \, \delta(x_k^*, y_k^*) \, \Delta A_k \tag{9}$$

If we assume that the errors in our approximations diminish as the dimensions of the rectangles tend to zero, then (8) and (9) suggest the following definitions:

$$\begin{bmatrix} \text{moment of} \\ \text{lamina } R \\ \text{about the} \\ \text{line } x = a \end{bmatrix} = \lim_{n \to +\infty} \sum_{k=1}^{n} (x_k^* - a) \, \delta(x_k^*, y_k^*) \, \Delta A_k = \iint_R (x - a) \, \delta(x, y) \, dA \tag{10}$$

$$\begin{bmatrix} \text{moment of} \\ \text{lamina } R \\ \text{about the} \\ \text{line } y = c \end{bmatrix} = \lim_{n \to +\infty} \sum_{k=1}^{n} (y_k^* - c) \, \delta(x_k^*, y_k^*) \, \Delta A_k = \iint_R (y - c) \, \delta(x, y) \, dA \tag{11}$$

Of special importance are the moments about the y-axis ($a = 0$) and about the x-axis ($c = 0$). These moments are denoted by

$$M_y = \begin{bmatrix} \text{moment of lamina } R \\ \text{about the } y\text{-axis} \end{bmatrix} = \iint_R x \, \delta(x, y) \, dA \tag{12}$$

$$M_x = \begin{bmatrix} \text{moment of lamina } R \\ \text{about the } x\text{-axis} \end{bmatrix} = \iint_R y \, \delta(x, y) \, dA \tag{13}$$

We are now in a position to obtain formulas for the center of gravity $(\bar{x}, \bar{y})$ of a lamina. Since a lamina balances on the lines $x = \bar{x}$ and $y = \bar{y}$ through the center of gravity, the moments of the lamina about these lines must be zero. Thus, from (10) and (11)

$$\iint_R (x - \bar{x}) \, \delta(x, y) \, dA = 0$$

$$\iint_R (y - \bar{y}) \, \delta(x, y) \, dA = 0$$

Since $\bar{x}$ and $\bar{y}$ are constant, these equations can be rewritten as

$$\iint\limits_R x\delta(x, y)\, dA = \bar{x} \iint\limits_R \delta(x, y)\, dA$$

$$\iint\limits_R y\delta(x, y)\, dA = \bar{y} \iint\limits_R \delta(x, y)\, dA$$

from which we obtain the following formulas:

Center of Gravity of a Lamina

$$\bar{x} = \frac{\displaystyle\iint\limits_R x\delta(x, y)\, dA}{\displaystyle\iint\limits_R \delta(x, y)\, dA}, \qquad \bar{y} = \frac{\displaystyle\iint\limits_R y\delta(x, y)\, dA}{\displaystyle\iint\limits_R \delta(x, y)\, dA} \qquad (14a)$$

By virtue of (3), (12), and (13), these formulas can be written as

$$\bar{x} = \frac{\begin{bmatrix}\text{moment about}\\\text{the } y\text{-axis}\end{bmatrix}}{\begin{bmatrix}\text{mass of the}\\\text{lamina}\end{bmatrix}} = \frac{M_y}{M}, \qquad \bar{y} = \frac{\begin{bmatrix}\text{moment about}\\\text{the } x\text{-axis}\end{bmatrix}}{\begin{bmatrix}\text{mass of the}\\\text{lamina}\end{bmatrix}} = \frac{M_x}{M} \qquad (14b)$$

Example 3   Find the center of gravity of the triangular lamina with vertices $(0, 0)$, $(0, 1)$, and $(1, 0)$ and density function $\delta(x, y) = xy$.

Solution.   The lamina is shown in Figure 17.6.3. In Example 1 we found the mass of the lamina to be

$$M = \iint\limits_R \delta(x, y)\, dA = \iint\limits_R xy\, dA = \frac{1}{24}$$

The moment of the lamina about the y-axis is

$$M_y = \iint\limits_R x\delta(x, y)\, dA = \iint\limits_R x^2 y\, dA = \int_0^1 \int_0^{-x+1} x^2 y\, dy\, dx$$

$$= \int_0^1 \left[\frac{1}{2} x^2 y^2\right]_{y=0}^{-x+1} dx = \int_0^1 \left(\frac{1}{2} x^4 - x^3 + \frac{1}{2} x^2\right) dx = \frac{1}{60}$$

and the moment about the x-axis is

$$M_x = \iint_R y\delta(x, y)\, dA = \iint_R xy^2\, dA = \int_0^1 \int_0^{-x+1} xy^2\, dy\, dx$$

$$= \int_0^1 \left[\frac{1}{3} xy^3 \right]_{y=0}^{-x+1} dx = \int_0^1 \left(-\frac{1}{3} x^4 + x^3 - x^2 + \frac{1}{3} x \right) dx = \frac{1}{60}$$

From (14b),

$$\bar{x} = \frac{M_y}{M} = \frac{1/60}{1/24} = \frac{2}{5}, \quad \bar{y} = \frac{M_x}{M} = \frac{1/60}{1/24} = \frac{2}{5}$$

so the center of gravity is $(\frac{2}{5}, \frac{2}{5})$. ◀

☐ **CENTROIDS**

In the special case of a *homogeneous* lamina, the center of gravity is called the ***centroid of the lamina*** or sometimes the ***centroid of the region R***. Because the density function δ is constant for a homogeneous lamina, the factor δ may be moved through the integral signs in (14a) and canceled. Thus, the centroid $(\bar{x}, \bar{y})$ of a region R is given by the following formulas:

Centroid of a Region R

$$\bar{x} = \frac{\displaystyle\iint_R x\, dA}{\displaystyle\iint_R dA} = \frac{1}{\text{area of } R} \iint_R x\, dA \qquad (15a)$$

$$\bar{y} = \frac{\displaystyle\iint_R y\, dA}{\displaystyle\iint_R dA} = \frac{1}{\text{area of } R} \iint_R y\, dA \qquad (15b)$$

Example 4 Find the centroid of the following semicircular region:

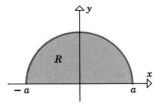

Solution. By symmetry, $\bar{x} = 0$ since the y-axis is obviously a line of balance. From (15b),

$$\bar{y} = \frac{\displaystyle\iint_R y\, dA}{\text{area of } R} = \frac{\displaystyle\iint_R y\, dA}{\frac{1}{2}\pi a^2} \qquad (16)$$

The integral in (16) is best evaluated in polar coordinates:

$$\iint\limits_{R} y \, dA = \int_0^\pi \int_0^a (r \sin \theta) r \, dr \, d\theta = \int_0^\pi \left[\frac{1}{3} r^3 \sin \theta \right]_{r=0}^a d\theta$$

$$= \frac{1}{3} a^3 \int_0^\pi \sin \theta \, d\theta = \frac{2}{3} a^3$$

Thus, from (16)

$$\bar{y} = \frac{\frac{2}{3} a^3}{\frac{1}{2} \pi a^2} = \frac{4a}{3\pi}$$

so the centroid is $\left(0, \dfrac{4a}{3\pi} \right)$.   ◄

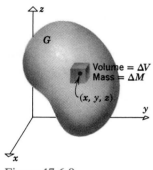

Figure 17.6.9

☐ **CENTER OF GRAVITY OF A SOLID**

For three-dimensional solids the formulas for moments, centers of gravity, and centroids are similar to the formulas for laminas. If a solid G of mass M and volume V is homogeneous in composition, we define the **density** δ of the solid to be

$$\delta = \frac{M}{V}$$

which measures the mass per unit volume. If the solid G is not homogeneous, then the density at a general point (x, y, z) is specified by a **density function** $\delta(x, y, z)$ that may be viewed as a limit

$$\delta(x, y, z) = \lim \frac{\Delta M}{\Delta V}$$

where ΔM and ΔV represent the mass and volume of a rectangular parallel-epiped, centered at (x, y, z), whose dimensions tend to zero (Figure 17.6.9). The following formulas generalize the corresponding results for laminas:

$$\text{mass of } G = \iiint\limits_{G} \delta(x, y, z) \, dV \tag{17}$$

Center of Gravity $(\bar{x}, \bar{y}, \bar{z})$ of a Solid G

$$\bar{x} = \frac{1}{\text{mass of } G} \iiint\limits_{G} x \delta(x, y, z) \, dV \tag{18a}$$

$$\bar{y} = \frac{1}{\text{mass of } G} \iiint\limits_{G} y \delta(x, y, z) \, dV \tag{18b}$$

$$\bar{z} = \frac{1}{\text{mass of } G} \iiint\limits_{G} z \delta(x, y, z) \, dV \tag{18c}$$

Centroid $(\bar{x}, \bar{y}, \bar{z})$ *of a Solid G*

$$\bar{x} = \frac{1}{\text{volume of } G} \iiint\limits_{G} x \, dV \qquad (19a)$$

$$\bar{y} = \frac{1}{\text{volume of } G} \iiint\limits_{G} y \, dV \qquad (19b)$$

$$\bar{z} = \frac{1}{\text{volume of } G} \iiint\limits_{G} z \, dV \qquad (19c)$$

Example 5   Find the mass and the center of gravity of a cylindrical solid of height h and radius a (Figure 17.6.10), assuming that the density at each point is proportional to the distance between the point and the base of the solid.

Solution.   Since the density is proportional to the distance z from the base, the density function has the form $\delta(x, y, z) = kz$ where k is some (unknown) positive constant of proportionality. From (17) the mass of the solid is

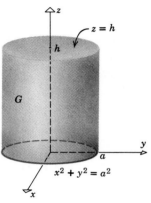

Figure 17.6.10

$$M = \iiint\limits_{G} \delta(x, y, z) \, dV = \int_{-a}^{a} \int_{-\sqrt{a^2-x^2}}^{\sqrt{a^2-x^2}} \int_{0}^{h} kz \, dz \, dy \, dx$$

$$= k \int_{-a}^{a} \int_{-\sqrt{a^2-x^2}}^{\sqrt{a^2-x^2}} \tfrac{1}{2}h^2 \, dy \, dx$$

$$= kh^2 \int_{-a}^{a} \sqrt{a^2 - x^2} \, dx$$

$$= \tfrac{1}{2}kh^2\pi a^2 \qquad \boxed{\begin{array}{l}\text{See Example 4}\\ \text{of Section 9.5.}\end{array}}$$

Without additional information, the constant k cannot be determined. However, as we shall now see, the value of k does not affect the center of gravity.
From (18c),

$$\bar{z} = \frac{\displaystyle\iiint\limits_{G} z\delta(x, y, z) \, dV}{\text{mass of } G} = \frac{\displaystyle\iiint\limits_{G} z\delta(x, y, z) \, dV}{\tfrac{1}{2}kh^2\pi a^2}$$

But

$$\iiint\limits_{G} z\delta(x, y, z) \, dV = \int_{-a}^{a} \int_{-\sqrt{a^2-x^2}}^{\sqrt{a^2-x^2}} \int_{0}^{h} z(kz) \, dz \, dy \, dx$$

$$= k \int_{-a}^{a} \int_{-\sqrt{a^2-x^2}}^{\sqrt{a^2-x^2}} \tfrac{1}{3}h^3 \, dy \, dx$$

$$= \tfrac{2}{3}kh^3 \int_{-a}^{a} \sqrt{a^2 - x^2} \, dx$$

$$= \tfrac{1}{3}kh^3\pi a^2$$

so that

$$\overline{z} = \frac{\frac{1}{3}kh^3\pi a^2}{\frac{1}{2}kh^2\pi a^2} = \frac{2}{3}h$$

Similar calculations using (18) will yield $\overline{x} = \overline{y} = 0$. However, this is evident by inspection, since it follows from the symmetry of the solid and the form of its density function that the center of gravity is on the z-axis. Thus, the center of gravity is $(0, 0, \frac{2}{3}h)$.   ◄

☐ **THEOREM OF PAPPUS**

The next theorem, due to the Greek mathematician Pappus,* gives an interesting and useful relationship between the centroid of a plane region R and the volume of the solid generated when the region is revolved about a line.

17.6.3 THEOREM. *If R is a plane region and L is a line that lies in the plane of R, but does not intersect R, then the volume of the solid formed by revolving R about L is given by*

$$volume = (area\ of\ R) \cdot \left(\begin{array}{c} distance\ traveled \\ by\ the\ centroid \end{array}\right)$$

Proof. Introduce an xy-coordinate system so that L is along the y-axis and the region R is in the first quadrant (Figure 17.6.11). Let R be partitioned into subregions in the usual way and let R_k be a typical rectangle interior to R. If (x_k^*, y_k^*) is the center of R_k, and if the area of R_k is $\Delta A_k = \Delta x_k \, \Delta y_k$, then the

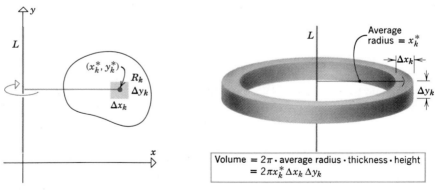

Figure 17.6.11

---

*PAPPUS OF ALEXANDRIA (4th century A.D.). Greek mathematician. Pappus lived during the early Christian era when mathematical activity was in a period of decline. His main contributions to mathematics appeared in a series of eight books called *The Collection* (written about 340 A.D.). This work, which survives only partially, contained some original results, but was devoted mostly to statements, refinements, and proofs of results by earlier mathematicians. Pappus's Theorem, stated without proof in Book VII of *The Collection*, was probably known and proved in earlier times. This result is sometimes called Guldin's Theorem in recognition of the Swiss mathematician, Paul Guldin (1577–1643), who rediscovered it independently.

volume generated by R_k as it revolves about L is

$$2\pi x_k^* \, \Delta x_k \, \Delta y_k = 2\pi x_k^* \, \Delta A_k$$

[See Figure 17.6.11 and Formula (1) of Section 6.3.] Therefore, the total volume of the solid is approximately

$$V \approx \sum_{k=1}^{n} 2\pi x_k^* \, \Delta A_k$$

from which it follows that the exact volume is

$$V = \iint_R 2\pi x \, dA = 2\pi \iint_R x \, dA$$

Thus, from (15a)

$$V = 2\pi \cdot \bar{x} \cdot [\text{area of } R]$$

This completes the proof since $2\pi\bar{x}$ is the distance traveled by the centroid when R is revolved about the y-axis. ∎

Example 6   Use Pappus's Theorem to find the volume V of the torus generated by revolving a circular region of radius r about a line at a distance b (greater than r) from the center of the circle (Figure 17.6.12).

Solution.   By symmetry, the centroid of a circular region is its center. Thus, the distance traveled by the centroid is $2\pi b$. Since the area of a circle is πr^2, it follows from Pappus's Theorem that the volume of the torus is

$$V = (2\pi b)(\pi r^2) = 2\pi^2 b r^2 \quad \blacktriangleleft$$

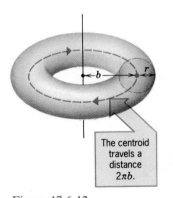

The centroid travels a distance $2\pi b$.

Figure 17.6.12

▶ Exercise Set 17.6

1. Where should the fulcrum be placed so that the seesaw is in equilibrium?

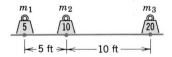

m_1 m_2 m_3
⟨5⟩   ⟨10⟩   ⟨20⟩
|← 5 ft →|←—— 10 ft ——→|

2. A rectangular lamina with density $\delta(x, y) = xy^2$ has vertices $(0, 0)$, $(0, 2)$, $(3, 0)$, $(3, 2)$. Find its mass and center of gravity.

3. A lamina with density $\delta(x, y) = x + y$ is bounded by the x-axis, the line $x = 1$, and the curve $y = \sqrt{x}$. Find its mass and center of gravity.

4. A lamina with density $\delta(x, y) = y$ is bounded by

$y = \sin x$, $y = 0$, $x = 0$, and $x = \pi$. Find its mass and center of gravity.

5. A lamina with density $\delta(x, y) = xy$ is in the first quadrant and is bounded by the circle $x^2 + y^2 = a^2$ and the coordinate axes. Find its mass and center of gravity.

6. A lamina with density $\delta(x, y) = x^2 + y^2$ is bounded by the x-axis and the upper half of the circle $x^2 + y^2 = 1$. Find its mass and center of gravity.

In Exercises 7–13, find the centroid of the region.

7. The triangular region bounded by $y = x$, $x = 1$, and the x-axis.

8. The region bounded by $y = x^2$, $x = 1$, and the x-axis.

9. The region that is enclosed between $y = x$ and $y = 2 - x^2$.

10. The region enclosed by the parabolas $x = 3y^2 - 6y$ and $x = 2y - y^2$.

11. The region above the x-axis and between the circles $x^2 + y^2 = a^2$ and $x^2 + y^2 = b^2$ $(a < b)$.

12. The region enclosed between the y-axis and the right half of the circle $x^2 + y^2 = a^2$.

13. The region enclosed between $y = |x|$ and the line $y = 4$.

14. A rectangular box that is defined by the three inequalities $0 \leq x \leq 1, 0 \leq y \leq 1, 0 \leq z \leq 1$ has density $\delta(x, y, z) = 3xyz$. Find its mass and center of gravity.

15. A cube that is defined by the three inequalities $0 \leq x \leq a$, $0 \leq y \leq a$, $0 \leq z \leq a$ has density $\delta(x, y, z) = a - x$. Find its mass and center of gravity.

16. The cylindrical solid enclosed by $x^2 + y^2 = a^2$, $z = 0$, and $z = h$ has density $\delta(x, y, z) = h - z$. Find its mass and center of gravity.

17. The solid enclosed by the surface $z = 1 - y^2$ (where $y \geq 0$) and the planes $z = 0$, $x = -1$, $x = 1$ has density $\delta(x, y, z) = yz$. Find its mass and center of gravity.

18. The solid enclosed by the surface $y = 9 - x^2$ (where $x \geq 0$) and the planes $y = 0$, $z = 0$, and $z = 1$ has density $\delta(x, y, z) = xz$. Find its mass and center of gravity.

In Exercises 19–23, find the centroid of the solid.

19. The tetrahedron in the first octant enclosed by the coordinate planes and the plane $x + y + z = 1$.

20. The solid enclosed by the xy-plane and the hemisphere $z = \sqrt{a^2 - x^2 - y^2}$.

21. The solid bounded by the surface $z = y^2$, and the planes $x = 0$, $x = 1$, and $z = 1$.

22. The solid in the first octant bounded by the surface $z = xy$ and the planes $z = 0$, $x = 2$, and $y = 2$.

23. The solid in the first octant bounded by the sphere $x^2 + y^2 + z^2 = a^2$ and the coordinate planes.

24. Find the center of gravity of the square lamina with vertices $(0, 0)$, $(1, 0)$, $(0, 1)$, and $(1, 1)$ if
   (a) the density is proportional to the square of the distance from the origin

(b) the density is proportional to the distance from the y-axis.

25. Find the mass of a circular lamina of radius a whose density at a point is proportional to the distance from the center. (Let k denote the constant of proportionality.)

26. Find the center of gravity of the cube that is determined by the inequalities $0 \leq x \leq 1$, $0 \leq y \leq 1$, $0 \leq z \leq 1$ if
   (a) the density is proportional to the square of the distance to the origin
   (b) the density is proportional to the sum of the distances to the faces that lie in the coordinate planes.

27. Show that in polar coordinates the formulas for the centroid $(\bar{x}, \bar{y})$ of a region R are
$$\bar{x} = \frac{1}{\text{area of } R} \iint_R r^2 \cos \theta \, dr \, d\theta$$
$$\bar{y} = \frac{1}{\text{area of } R} \iint_R r^2 \sin \theta \, dr \, d\theta$$

28. Use the result of Exercise 27 to find the centroid $(\bar{x}, \bar{y})$ of the region that is enclosed by the cardioid $r = a(1 + \sin \theta)$.

29. Use the result of Exercise 27 to find the centroid $(\bar{x}, \bar{y})$ of the petal of the rose $r = \sin 2\theta$ that lies in the first quadrant.

30. Let R be the rectangle bounded by the lines $x = 0$, $x = 3$, $y = 0$, and $y = 2$. By inspection, find the centroid of R and use it to evaluate
$$\iint_R x \, dA \quad \text{and} \quad \iint_R y \, dA$$

31. Use the Theorem of Pappus and the fact that the volume of a sphere of radius a is $V = \frac{4}{3}\pi a^3$ to show that the centroid of the lamina that is bounded by the x-axis and the semicircle $y = \sqrt{a^2 - x^2}$ is $(0, 4a/(3\pi))$. (This problem was solved directly in Example 4.)

32. Use the Theorem of Pappus and the result of Exercise 31 to find the volume of the solid generated when the region bounded by the x-axis and the semicircle $y = \sqrt{a^2 - x^2}$ is revolved about
   (a) the line $y = -a$ (b) the line $y = x - a$.

33. Use the Theorem of Pappus and the fact that the area of an ellipse with semiaxes a and b is πab to find the volume of the elliptical torus generated by

revolving the ellipse

$$\frac{(x - k)^2}{a^2} + \frac{y^2}{b^2} = 1$$

about the y-axis. Assume that $k > a$.

34. Use the Theorem of Pappus to find the volume of the solid generated when the region enclosed by $y = x^2$ and $y = 8 - x^2$ is revolved about the x-axis.

35. Use the Theorem of Pappus to find the centroid of the triangular region with vertices $(0, 0)$, $(a, 0)$, and $(0, b)$, where $a > 0$ and $b > 0$. [*Hint:* Revolve the region about the x-axis to obtain $\bar{y}$ and about the y-axis to obtain $\bar{x}$.]

36. If a lamina with a continuous density function $\delta(x, y)$ occupies a region R in the xy-plane, then the *second moments* (or *moments of inertia*) of the lamina about the x-axis, y-axis, and z-axis, respectively, are defined by

$$I_x = \iint\limits_{R} y^2 \delta \, dA, \qquad I_y = \iint\limits_{R} x^2 \delta \, dA,$$

$$I_z = \iint\limits_{R} (x^2 + y^2) \delta \, dA$$

The corresponding definitions for a solid with density function $\delta(x, y, z)$ that occupies a region G are

$$I_x = \iiint\limits_{G} (y^2 + z^2) \delta(x, y, z) \, dV$$

$$I_y = \iiint\limits_{G} (x^2 + z^2) \delta(x, y, z) \, dV$$

$$I_z = \iiint\limits_{G} (x^2 + y^2) \delta(x, y, z) \, dV$$

Find I_x, I_y, and I_z for the homogeneous (constant δ) rectangular lamina described by the inequalities $0 \le x \le a$ and $0 \le y \le b$. Express your answers in terms of the mass M of the lamina. [*Note:* Moments of inertia are important in problems involving rotational motion.]

In Exercises 37–44, a homogeneous lamina or solid is given. Use the formulas in Exercise 36 to find the indicated moment of inertia. Express your answers in terms of the mass M of the lamina or solid.

37. The moment I_z for the rectangular lamina $-a/2 \le x \le a/2$, $-b/2 \le y \le b/2$.

38. The moment I_x for the rectangular lamina $-a/2 \le x \le a/2$, $-b/2 \le y \le b/2$.

39. The circular lamina $x^2 + y^2 \le R^2$; I_y.

40. The circular lamina $x^2 + y^2 \le R^2$; I_z.

41. The circular lamina $x^2 + (y - R)^2 \le R^2$; I_z.

42. The triangular lamina $0 \le y \le 2 - 2x$, $x \ge 0$; I_y.

43. The solid $0 \le x \le a$, $0 \le y \le a$, $0 \le z \le a$; I_z.

44. The rectangular solid $0 \le x \le a$, $-b/2 \le y \le b/2$, $-c/2 \le z \le c/2$; I_x.

◼ 17.7 TRIPLE INTEGRALS IN CYLINDRICAL AND SPHERICAL COORDINATES

In Section 17.3 we saw that some double integrals are easier to evaluate in polar coordinates. Similarly, some triple integrals are easier to evaluate in cylindrical or spherical coordinates. In this section we shall study triple integrals in these coordinate systems.

☐ **TRIPLE INTEGRALS IN CYLINDRICAL COORDINATES**

In rectangular coordinates the triple integral of a continuous function f over a solid region G is defined as

$$\iiint\limits_{G} f(x, y, z) \, dV = \lim_{n \to +\infty} \sum_{k=1}^{n} f(x_k^*, y_k^*, z_k^*) \, \Delta V_k$$

where ΔV_k denotes the volume of a rectangular parallelepiped interior to G

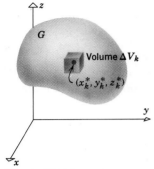

Figure 17.7.1

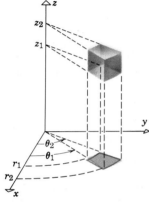

Figure 17.7.2

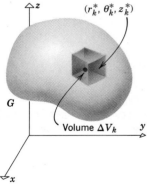

Figure 17.7.3

and (x_k^*, y_k^*, z_k^*) is a point in this parallelepiped (Figure 17.7.1). Triple integrals in cylindrical and spherical coordinates are defined similarly, except that the region G is divided not into rectangular parallelepipeds but into regions more appropriate to these coordinate systems.

In cylindrical coordinates, the simplest equations are of the form

$$r = \text{constant}, \quad \theta = \text{constant}, \quad z = \text{constant}$$

As indicated in Figure 14.8.2b, the first equation represents a right-circular cylinder centered on the z-axis, the second a vertical half-plane hinged on the z-axis, and the third a horizontal plane. These surfaces can be paired up to determine solids called **cylindrical wedges** or **cylindrical elements of volume.** To be precise, a cylindrical wedge is a solid enclosed between six surfaces of the following type:

two cylinders	$r = r_1, r = r_2$	$(r_1 < r_2)$
two half-planes	$\theta = \theta_1, \theta = \theta_2$	$(\theta_1 < \theta_2)$
two planes	$z = z_1, z = z_2$	$(z_1 < z_2)$

(Figure 17.7.2). The dimensions $\theta_2 - \theta_1$, $r_2 - r_1$, and $z_2 - z_1$ are called the **central angle, thickness,** and **height** of the wedge.

To define the triple integral over G of a function $f(r, \theta, z)$ in cylindrical coordinates, we proceed as follows:

Step 1. Construct a three-dimensional grid consisting of right-circular cylinders centered on the z-axis, half-planes hinged on the z-axis, and horizontal planes. The blocks determined by this grid will be cylindrical wedges. Using this grid, divide the region G into subregions and exclude from consideration subregions that contain any portion of the surface of the solid G. This leaves only cylindrical wedges interior to G (Figure 17.7.3). Denote the volumes of these interior cylindrical wedges by

$$\Delta V_1, \Delta V_2, \ldots, \Delta V_n$$

Step 2. Choose an arbitrary point in each interior cylindrical wedge, and let these be denoted by

$$(r_1^*, \theta_1^*, z_1^*), (r_2^*, \theta_2^*, z_2^*), \ldots, (r_n^*, \theta_n^*, z_n^*)$$

(See Figure 17.7.3.)

Step 3. Form the sum

$$\sum_{k=1}^{n} f(r_k^*, \theta_k^*, z_k^*) \, \Delta V_k$$

> **Step 4.** Repeat this process with more and more subdivisions so that the height, thickness, and central angle of each interior cylindrical wedge approaches zero, and n, the number of such wedges, approaches $+\infty$. Define
>
> $$\iiint\limits_{G} f(r, \theta, z)\, dV = \lim_{n \to +\infty} \sum_{k=1}^{n} f(r_k^*, \theta_k^*, z_k^*)\, \Delta V_k$$

In Section 17.5, we defined simple solids. The following theorem will enable us to evaluate triple integrals in cylindrical coordinates over such solids by repeated integration.

17.7.1 THEOREM. *Let G be a simple solid whose upper surface has the equation $z = g_2(r, \theta)$ and whose lower surface has the equation $z = g_1(r, \theta)$ in cylindrical coordinates. If R is the projection of the solid on the xy-plane, and if $f(r, \theta, z)$ is continuous on G, then*

$$\iiint\limits_{G} f(r, \theta, z)\, dV = \iint\limits_{R} \left[\int_{g_1(r,\theta)}^{g_2(r,\theta)} f(r, \theta, z)\, dz \right] dA \tag{1}$$

where the double integral over R is evaluated in polar coordinates. In particular, if the projection R is as shown in Figure 17.7.4, then (1) can be written as

$$\iiint\limits_{G} f(r, \theta, z)\, dV = \int_{\theta_1}^{\theta_2} \int_{r_1(\theta)}^{r_2(\theta)} \int_{g_1(r,\theta)}^{g_2(r,\theta)} f(r, \theta, z) r\, dz\, dr\, d\theta \tag{2}$$

The type of solid to which formula (2) applies is illustrated in Figure 17.7.4.

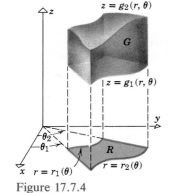

$z = g_2(r, \theta)$

G

$z = g_1(r, \theta)$

θ_2
θ_1

R

$r = r_1(\theta)$ $r = r_2(\theta)$

Figure 17.7.4

We shall not formally prove Theorem 17.7.1. However, the appearance of the factor r in the integrand of (2) can be explained by returning to the definition

$$\iiint\limits_{G} f(r, \theta, z)\, dV = \lim_{n \to +\infty} \sum_{k=1}^{n} f(r_k^*, \theta_k^*, z_k^*)\, \Delta V_k \tag{3}$$

In this formula, the volume ΔV_k of the kth cylindrical wedge can be written as

$$\Delta V_k = [\text{area of base}] \cdot [\text{height}] \tag{4}$$

If we denote the thickness, central angle, and height of this wedge by Δr_k, $\Delta\theta_k$, and Δz_k, and if we choose the arbitrary point $(r_k^*, \theta_k^*, z_k^*)$ to lie above the

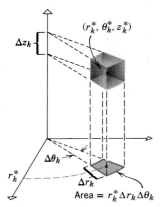

Figure 17.7.5

"center" of the base (Figures 17.3.5 and 17.7.5), then it follows from (7) of Section 17.3 that the base has area $r_k^* \, \Delta r_k \, \Delta \theta_k$. Thus, (4) can be written as

$$\Delta V_k = r_k^* \, \Delta r_k \, \Delta \theta_k \, \Delta z_k$$

Substituting this expression in (3) yields

$$\iiint\limits_G f(r, \theta, z) \, dV = \lim_{n \to +\infty} \sum_{k=1}^{n} f(r_k^*, \theta_k^*, z_k^*) r_k^* \, \Delta z_k \, \Delta r_k \, \Delta \theta_k$$

which explains the form of the integrand in (2).

To apply (1) and (2) it is best to begin with a three-dimensional sketch of the solid G, from which the limits of integration can be obtained as follows:

> **Step 1.** Identify the upper surface $z = g_2(r, \theta)$ and the lower surface $z = g_1(r, \theta)$ of the solid. The functions $g_1(r, \theta)$ and $g_2(r, \theta)$ determine the z-limits of integration. (If the upper and lower surfaces are given in rectangular coordinates, convert them to cylindrical coordinates.)
>
> **Step 2.** Make a two-dimensional sketch of the projection R of the solid on the xy-plane. From this sketch the r- and θ-limits of integration may be obtained exactly as with double integrals in polar coordinates.

Example 1 Use triple integration in cylindrical coordinates to find the volume and the centroid of the solid G that is bounded above by the hemisphere $z = \sqrt{25 - x^2 - y^2}$, below by the xy-plane, and laterally by the cylinder $x^2 + y^2 = 9$.

Solution. The solid G and its projection R on the xy-plane are shown in Figure 17.7.6. In cylindrical coordinates, the upper surface of G is the hemisphere $z = \sqrt{25 - r^2}$ and the lower surface is the plane $z = 0$. Thus, from (1), the volume of G is

$$V = \iiint\limits_G dV = \iint\limits_R \left[\int_0^{\sqrt{25-r^2}} dz \right] dA$$

For the double integral over R, we use polar coordinates:

$$V = \int_0^{2\pi} \int_0^3 \int_0^{\sqrt{25-r^2}} r \, dz \, dr \, d\theta = \int_0^{2\pi} \int_0^3 \left[rz \right]_{z=0}^{\sqrt{25-r^2}} dr \, d\theta$$

$$= \int_0^{2\pi} \int_0^3 r\sqrt{25 - r^2} \, dr \, d\theta = \int_0^{2\pi} \left[-\frac{1}{3}(25 - r^2)^{3/2} \right]_{r=0}^3 d\theta$$

$$= \int_0^{2\pi} \frac{61}{3} \, d\theta = \frac{122}{3} \pi$$

$$\boxed{\begin{array}{l} u = 25 - r^2 \\ du = -2r \, dr \end{array}}$$

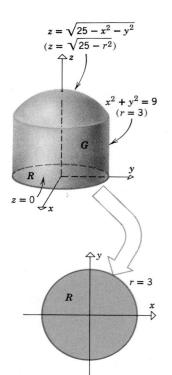

Figure 17.7.6

From this result and Formula (19c) of Section 17.6,

$$\bar{z} = \frac{1}{V} \iiint_G z \, dV = \frac{3}{122\pi} \iiint_G z \, dV = \frac{3}{122\pi} \iint_R \left[\int_0^{\sqrt{25-r^2}} z \, dz \right] dA$$

$$= \frac{3}{122\pi} \int_0^{2\pi} \int_0^3 \int_0^{\sqrt{25-r^2}} zr \, dz \, dr \, d\theta = \frac{3}{122\pi} \int_0^{2\pi} \int_0^3 \left[\frac{1}{2} rz^2 \right]_{z=0}^{\sqrt{25-r^2}} dr \, d\theta$$

$$= \frac{3}{244\pi} \int_0^{2\pi} \int_0^3 (25r - r^3) \, dr \, d\theta = \frac{3}{244\pi} \int_0^{2\pi} \frac{369}{4} \, d\theta = \frac{1107}{488}$$

By symmetry, the centroid $(\bar{x}, \bar{y}, \bar{z})$ of G lies on the z-axis, so $\bar{x} = \bar{y} = 0$. Thus, the centroid is at the point $(0, 0, 1107/488)$.   ◄

Frequently, a triple integral that is hard to evaluate in rectangular coordinates is easy to evaluate when the integrand and the limits of integration are expressed in cylindrical coordinates. This is especially true when the integrand or the limits of integration involve expressions of the form

$$x^2 + y^2 \quad \text{or} \quad \sqrt{x^2 + y^2}$$

since these simplify to

$$r^2 \quad \text{and} \quad r$$

in cylindrical coordinates.

Example 2   Use cylindrical coordinates to evaluate

$$\int_{-3}^3 \int_{-\sqrt{9-x^2}}^{\sqrt{9-x^2}} \int_0^{9-x^2-y^2} x^2 \, dz \, dy \, dx$$

Solution.   In problems of this type, it is helpful to sketch the region of integration G and its projection R on the xy-plane. From the z-limits of integration, the upper surface of G is the paraboloid $z = 9 - x^2 - y^2$ and the lower surface is the xy-plane $z = 0$. From the x- and y-limits of integration, the projection R is the region in the xy-plane enclosed by the circle $x^2 + y^2 = 9$ (Figure 17.7.7). Thus,

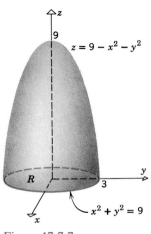

Figure 17.7.7

$$\int_{-3}^3 \int_{-\sqrt{9-x^2}}^{\sqrt{9-x^2}} \int_0^{9-x^2-y^2} x^2 \, dz \, dy \, dx = \iiint_G x^2 \, dV$$

$$= \iiint_G (r \cos \theta)^2 \, dV = \iint_R \left[\int_0^{9-r^2} r^2 \cos^2 \theta \, dz \right] dA$$

$$= \int_0^{2\pi} \int_0^3 \int_0^{9-r^2} r^3 \cos^2 \theta \, dz \, dr \, d\theta = \int_0^{2\pi} \int_0^3 \left[zr^3 \cos^2 \theta \right]_{z=0}^{9-r^2} dr \, d\theta$$

$$= \int_0^{2\pi} \int_0^3 (9r^3 - r^5) \cos^2 \theta \, dr \, d\theta = \int_0^{2\pi} \left[\left(\frac{9r^4}{4} - \frac{r^6}{6} \right) \cos^2 \theta \right]_{r=0}^3 d\theta$$

$$= \frac{243}{4} \int_0^{2\pi} \cos^2 \theta \, d\theta = \frac{243}{4} \int_0^{2\pi} \frac{1}{2} (1 + \cos 2\theta) \, d\theta = \frac{243\pi}{4} \quad \blacktriangleleft$$

☐ **TRIPLE INTEGRALS IN SPHERICAL COORDINATES**

In spherical coordinates, the simplest equations are of the form

$$\rho = \text{constant}, \quad \theta = \text{constant}, \quad \phi = \text{constant}$$

As indicated in Figure 14.8.2c, the first equation represents a sphere centered at the origin, the second a half-plane hinged on the z-axis, and the third a right-circular cone with its vertex at the origin and its line of symmetry along the z-axis. By a **spherical wedge** or **spherical element of volume** we mean a solid enclosed between six surfaces of the following form:

two spheres	$\rho = \rho_1, \rho = \rho_2$	$(\rho_1 < \rho_2)$
two half-planes	$\theta = \theta_1, \theta = \theta_2$	$(\theta_1 < \theta_2)$
two right-circular cones	$\phi = \phi_1, \phi = \phi_2$	$(\phi_1 < \phi_2)$

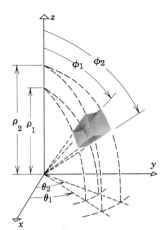

Figure 17.7.8

(Figure 17.7.8). We shall refer to the numbers $\rho_2 - \rho_1$, $\theta_2 - \theta_1$, and $\phi_2 - \phi_1$ as the **dimensions** of a spherical wedge.

If G is a solid region in three-dimensional space, then the triple integral over G of a continuous function $f(\rho, \theta, \phi)$ in spherical coordinates is similar in definition to the triple integral in cylindrical coordinates, except that the solid G is partitioned into *spherical* wedges by a three-dimensional grid consisting of spheres centered at the origin, half-planes hinged on the z-axis, and right-circular cones with vertices at the origin and lines of symmetry along the z-axis (Figure 17.7.9).

The defining equation of a triple integral in spherical coordinates is

$$\iiint_G f(\rho, \theta, \phi) \, dV = \lim_{n \to +\infty} \sum_{k=1}^n f(\rho_k^*, \theta_k^*, \phi_k^*) \, \Delta V_k \tag{5}$$

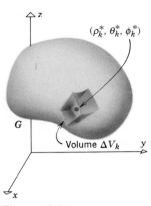

Figure 17.7.9

where ΔV_k is the volume of the kth spherical wedge that is interior to G, $(\rho_k^*, \theta_k^*, \phi_k^*)$ is an arbitrary point in this wedge, and n increases in such a way that the dimensions of each interior spherical wedge tend to zero.

In the exercises we shall help the reader to show that if the point $(\rho_k^*, \theta_k^*, \phi_k^*)$ is suitably chosen, then the volume ΔV_k in (5) can be written as

$$\Delta V_k = \rho_k^{*2} \sin \phi_k^* \, \Delta \rho_k \, \Delta \phi_k \, \Delta \theta_k$$

where $\Delta \rho_k$, $\Delta \phi_k$, and $\Delta \theta_k$ are the dimensions of the wedge. Substituting this in (5) we obtain

$$\iiint_G f(\rho, \theta, \phi) \, dV = \lim_{n \to +\infty} \sum_{k=1}^n f(\rho_k^*, \theta_k^*, \phi_k^*) \rho_k^{*2} \sin \phi_k^* \, \Delta \rho_k \, \Delta \phi_k \, \Delta \theta_k$$

which suggests the following formula for evaluating a triple integral in spherical coordinates by repeated integration:

$$\iiint\limits_G f(\rho, \theta, \phi) \, dV = \iiint f(\rho, \theta, \phi)\rho^2 \sin\phi \, d\rho \, d\phi \, d\theta \qquad (6)$$

Appropriate
limits of
integration

REMARK.   Note the extra factor of $\rho^2 \sin\phi$ that appears in the integrand of the iterated integral. This is analogous to the extra factor of r that appeared when we integrated in cylindrical coordinates.

In (6) we omitted the limits of integration; instead, we shall give some examples that explain how to obtain such limits. In all our examples we shall use the same order of integration: first with respect to ρ, then ϕ, and then θ.

Suppose that we want to integrate $f(\rho, \theta, \phi)$ over the spherical solid G enclosed by the sphere $\rho = \rho_0$. The basic idea is to choose the limits of integration so that every point of the solid is accounted for in the integration process. Figure 17.7.10 illustrates one way of doing this.

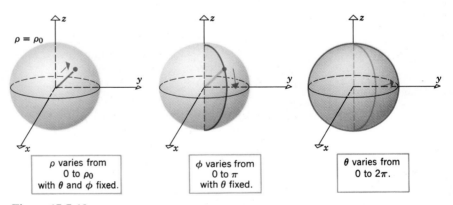

ρ varies from
0 to ρ_0
with θ and ϕ fixed.

ϕ varies from
0 to π
with θ fixed.

θ varies from
0 to 2π.

Figure 17.7.10

Holding θ and ϕ fixed for the first integration, we let ρ vary from 0 to ρ_0. This covers a radial line from the origin to the surface of the sphere. Next, keeping θ fixed, we let ϕ vary from 0 to π so that the radial line sweeps out a fan-shaped region. Finally, we let θ vary from 0 to 2π so that the fan-shaped region makes a complete revolution, thereby sweeping out the entire sphere. Thus, the triple integral of $f(\rho, \theta, \phi)$ over the spherical solid G may be evaluated by writing

$$\iiint\limits_G f(\rho, \theta, \phi) \, dV = \int_0^{2\pi} \int_0^{\pi} \int_0^{\rho_0} f(\rho, \theta, \phi)\rho^2 \sin\phi \, d\rho \, d\phi \, d\theta$$

Table 17.7.1 (pages 1140–1141) suggests how the limits of integration in spherical coordinates can be obtained for some other common solids.

Table 17.7.1

DETERMINATION OF LIMITS   INTEGRAL

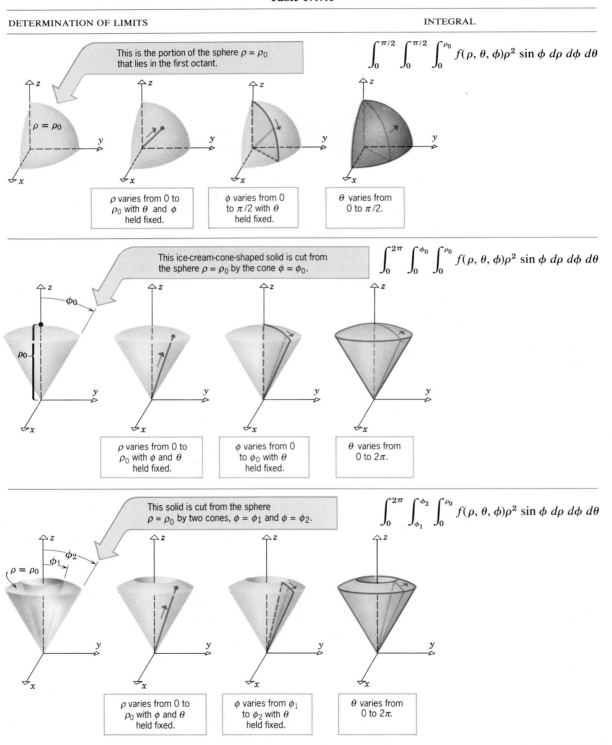

This is the portion of the sphere $\rho = \rho_0$ that lies in the first octant.

$$\int_0^{\pi/2} \int_0^{\pi/2} \int_0^{\rho_0} f(\rho, \theta, \phi)\rho^2 \sin\phi \, d\rho \, d\phi \, d\theta$$

ρ varies from 0 to ρ_0 with θ and ϕ held fixed.

ϕ varies from 0 to $\pi/2$ with θ held fixed.

θ varies from 0 to $\pi/2$.

This ice-cream-cone-shaped solid is cut from the sphere $\rho = \rho_0$ by the cone $\phi = \phi_0$.

$$\int_0^{2\pi} \int_0^{\phi_0} \int_0^{\rho_0} f(\rho, \theta, \phi)\rho^2 \sin\phi \, d\rho \, d\phi \, d\theta$$

ρ varies from 0 to ρ_0 with ϕ and θ held fixed.

ϕ varies from 0 to ϕ_0 with θ held fixed.

θ varies from 0 to 2π.

This solid is cut from the sphere $\rho = \rho_0$ by two cones, $\phi = \phi_1$ and $\phi = \phi_2$.

$$\int_0^{2\pi} \int_{\phi_1}^{\phi_2} \int_0^{\rho_0} f(\rho, \theta, \phi)\rho^2 \sin\phi \, d\rho \, d\phi \, d\theta$$

ρ varies from 0 to ρ_0 with ϕ and θ held fixed.

ϕ varies from ϕ_1 to ϕ_2 with θ held fixed.

θ varies from 0 to 2π.

DETERMINATION OF LIMITS	INTEGRAL

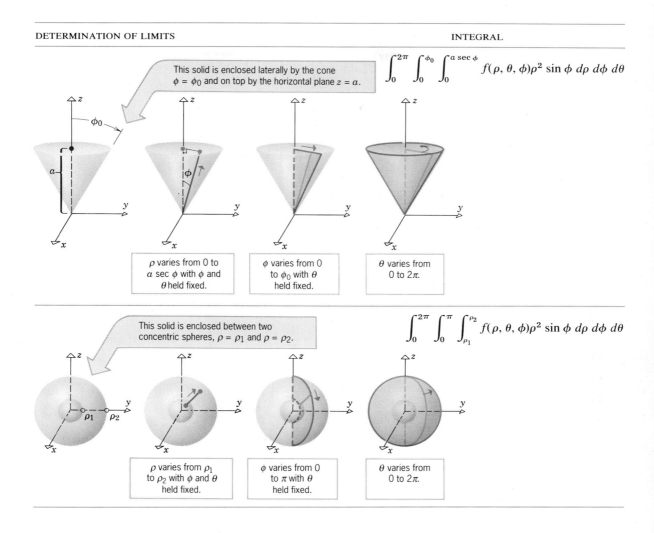

This solid is enclosed laterally by the cone $\phi = \phi_0$ and on top by the horizontal plane $z = a$.

$$\int_0^{2\pi} \int_0^{\phi_0} \int_0^{a \sec \phi} f(\rho, \theta, \phi) \rho^2 \sin \phi \, d\rho \, d\phi \, d\theta$$

ρ varies from 0 to $a \sec \phi$ with ϕ and θ held fixed.

ϕ varies from 0 to ϕ_0 with θ held fixed.

θ varies from 0 to 2π.

This solid is enclosed between two concentric spheres, $\rho = \rho_1$ and $\rho = \rho_2$.

$$\int_0^{2\pi} \int_0^{\pi} \int_{\rho_1}^{\rho_2} f(\rho, \theta, \phi) \rho^2 \sin \phi \, d\rho \, d\phi \, d\theta$$

ρ varies from ρ_1 to ρ_2 with ϕ and θ held fixed.

ϕ varies from 0 to π with θ held fixed.

θ varies from 0 to 2π.

Example 3 Use spherical coordinates to find the volume and the centroid of the solid G bounded above by the sphere $x^2 + y^2 + z^2 = 16$ and below by the cone $z = \sqrt{x^2 + y^2}$.

Solution. The solid G is sketched in Figure 17.7.11.

In spherical coordinates, the equation of the sphere $x^2 + y^2 + z^2 = 16$ is $\rho = 4$ and the equation of the cone $z = \sqrt{x^2 + y^2}$ is

$$\rho \cos \phi = \sqrt{\rho^2 \sin^2 \phi \cos^2 \theta + \rho^2 \sin^2 \phi \sin^2 \theta}$$

which simplifies to

$$\rho \cos \phi = \rho \sin \phi$$

or, on dividing both sides by $\rho \cos \phi$,

$$\tan \phi = 1$$

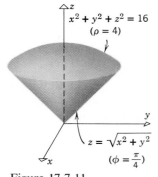

Figure 17.7.11

Thus, $\phi = \pi/4$, so the volume of G is

$$
V = \iiint_G dV = \int_0^{2\pi} \int_0^{\pi/4} \int_0^4 \rho^2 \sin\phi \, d\rho \, d\phi \, d\theta
$$

$$
= \int_0^{2\pi} \int_0^{\pi/4} \left[\frac{\rho^3}{3} \sin\phi \right]_{\rho=0}^4 d\phi \, d\theta
$$

$$
= \int_0^{2\pi} \int_0^{\pi/4} \frac{64}{3} \sin\phi \, d\phi \, d\theta
$$

$$
= \frac{64}{3} \int_0^{2\pi} \left[-\cos\phi \right]_{\phi=0}^{\pi/4} d\theta = \frac{64}{3} \int_0^{2\pi} \left(1 - \frac{\sqrt{2}}{2} \right) d\theta
$$

$$
= \frac{64\pi}{3} (2 - \sqrt{2})
$$

By symmetry, the centroid $(\bar{x}, \bar{y}, \bar{z})$ is on the z-axis, so $\bar{x} = \bar{y} = 0$. From Formula (19c) of Section 17.6 and the volume calculated above,

$$
\bar{z} = \frac{1}{V} \iiint_G z \, dV = \frac{1}{V} \int_0^{2\pi} \int_0^{\pi/4} \int_0^4 (\rho \cos\phi) \rho^2 \sin\phi \, d\rho \, d\phi \, d\theta
$$

$$
= \frac{1}{V} \int_0^{2\pi} \int_0^{\pi/4} \left[\frac{\rho^4}{4} \cos\phi \sin\phi \right]_{\rho=0}^4 d\phi \, d\theta
$$

$$
= \frac{64}{V} \int_0^{2\pi} \int_0^{\pi/4} \sin\phi \cos\phi \, d\phi \, d\theta = \frac{64}{V} \int_0^{2\pi} \left[\frac{1}{2} \sin^2\phi \right]_{\phi=0}^{\pi/4} d\theta
$$

$$
= \frac{16}{V} \int_0^{2\pi} d\theta = \frac{32\pi}{V} = \frac{3}{2(2 - \sqrt{2})}
$$

With the help of a calculator, $\bar{z} \approx 2.56$ (to two decimal places), so the approximate location of the centroid in the xyz-coordinate system is $(0, 0, 2.56)$.   ◀

Example 4   Use spherical coordinates to evaluate

$$
\int_{-2}^2 \int_{-\sqrt{4-x^2}}^{\sqrt{4-x^2}} \int_0^{\sqrt{4-x^2-y^2}} z^2 \sqrt{x^2 + y^2 + z^2} \, dz \, dy \, dx
$$

Solution.   In problems like this, it is helpful to begin (when possible) with a sketch of the region G of integration. From the z-limits of integration, the upper surface of G is the hemisphere $z = \sqrt{4 - x^2 - y^2}$ and the lower surface is the xy-plane $z = 0$. From the x- and y-limits of integration, the projection of the solid G on the xy-plane is the region enclosed by the circle $x^2 + y^2 = 4$. From this information we obtain the sketch of G in Figure 17.7.12.

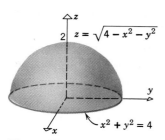

Figure 17.7.12

Thus,

$$\int_{-2}^{2} \int_{-\sqrt{4-x^2}}^{\sqrt{4-x^2}} \int_{0}^{\sqrt{4-x^2-y^2}} z^2 \sqrt{x^2 + y^2 + z^2} \, dz \, dy \, dx$$

$$= \iiint_{G} z^2 \sqrt{x^2 + y^2 + z^2} \, dV$$

$$= \iiint_{G} (\rho^2 \cos^2 \phi) \rho \, dV$$

$$= \int_{0}^{2\pi} \int_{0}^{\pi/2} \int_{0}^{2} \rho^5 \cos^2 \phi \sin \phi \, d\rho \, d\phi \, d\theta$$

$$= \int_{0}^{2\pi} \int_{0}^{\pi/2} \frac{32}{3} \cos^2 \phi \sin \phi \, d\phi \, d\theta$$

$$= \frac{32}{3} \int_{0}^{2\pi} \left[-\frac{1}{3} \cos^3 \phi \right]_{\phi=0}^{\pi/2} d\theta = \frac{32}{9} \int_{0}^{2\pi} d\theta = \frac{64}{9} \pi \quad \blacktriangleleft$$

▶ Exercise Set 17.7

In Exercises 1–4, evaluate the iterated integral.

1. $\int_{0}^{2\pi} \int_{0}^{1} \int_{0}^{\sqrt{1-r^2}} zr \, dz \, dr \, d\theta$.

2. $\int_{0}^{\pi/2} \int_{0}^{\cos \theta} \int_{0}^{r^2} r \sin \theta \, dz \, dr \, d\theta$.

3. $\int_{0}^{\pi/2} \int_{0}^{\pi/2} \int_{0}^{1} \rho^3 \sin \phi \cos \phi \, d\rho \, d\phi \, d\theta$.

4. $\int_{0}^{2\pi} \int_{0}^{\pi/4} \int_{0}^{a \sec \phi} \rho^2 \sin \phi \, d\rho \, d\phi \, d\theta \quad (a > 0)$.

In Exercises 5–9, use cylindrical coordinates to find the volume of the solid.

5. The solid bounded by the paraboloid $z = x^2 + y^2$ and the plane $z = 9$.

6. The solid bounded above and below by the sphere $x^2 + y^2 + z^2 = 9$ and laterally by the cylinder $x^2 + y^2 = 4$.

7. The solid that is enclosed between the surface $r^2 + z^2 = 20$ and the surface $z = r^2$.

8. The solid that is enclosed between the cone $z = hr/a$ and the plane $z = h$.

9. The solid in the first octant that is below the sphere $x^2 + y^2 + z^2 = 16$ and inside the cylinder $x^2 + y^2 = 4x$.

In Exercises 10–14, use spherical coordinates to find the volume of the solid.

10. The solid bounded above by the sphere $\rho = 4$ and below by the cone $\phi = \pi/3$.

11. The solid in the first octant bounded by the sphere $\rho = 2$, the coordinate planes, and the cones $\phi = \pi/6$ and $\phi = \pi/3$.

12. The solid within the cone $\phi = \pi/4$ and between the spheres $\rho = 1$ and $\rho = 2$.

13. The solid within the sphere $x^2 + y^2 + z^2 = 9$, outside the cone $z = \sqrt{x^2 + y^2}$ and above the xy-plane.

14. The solid bounded by the sphere $x^2 + y^2 + z^2 = 4a^2$ and the planes $z = 0$ and $z = a$.

15. Find the volume of $x^2 + y^2 + z^2 = a^2$ using
  (a) cylindrical coordinates
  (b) spherical coordinates.

In Exercises 16–18, use cylindrical coordinates.

16. Find the mass of the solid that is bounded by the cone $z = \sqrt{x^2 + y^2}$ and $z = 3$ if the density is $\delta(x, y, z) = 3 - z$.

17. Find the mass of the solid in the first octant bounded above by the paraboloid $z = 4 - x^2 - y^2$, below by the plane $z = 0$, and laterally by the cylinder $x^2 + y^2 = 2x$ and the planes $x = 0$ and $y = 0$, assuming the density to be $\delta(x, y, z) = z$. [*Hint:* The Wallis formulas in Exercises 34 and 36 of Section 9.3 will help with the integration.]

18. Find the mass of a right-circular cylinder of radius a and height h if the density is proportional to the distance from the base. (Let k be the constant of proportionality.)

In Exercises 19–21, use spherical coordinates.

19. Find the mass of the solid enclosed by the sphere $x^2 + y^2 + z^2 = 1$ and the cone $z = \sqrt{x^2 + y^2}$ if the density is $\delta(x, y, z) = \sqrt{x^2 + y^2 + z^2}$.

20. Find the mass of the solid enclosed between the spheres $x^2 + y^2 + z^2 = 1$ and $x^2 + y^2 + z^2 = 4$ if the density is $\delta(x, y, z) = (x^2 + y^2 + z^2)^{-1/2}$.

21. Find the mass of a spherical solid of radius a if the density is proportional to the distance from the center. (Let k be the constant of proportionality.)

In Exercises 22–24, use cylindrical coordinates to find the centroid of the solid.

22. The solid bounded by the cone $z = \sqrt{x^2 + y^2}$ and the plane $z = 2$.

23. The solid that is bounded above by the sphere $x^2 + y^2 + z^2 = 2$ and below by the paraboloid $z = x^2 + y^2$.

24. The solid bounded above by the paraboloid $z = x^2 + y^2$, below by the plane $z = 0$, and laterally by the cylinder $(x - 1)^2 + y^2 = 1$. [*Hint:* The Wallis formulas in Exercises 34 and 36 of Section 9.3 will help with the integration.]

In Exercises 25–27, use spherical coordinates to find the centroid.

25. The solid in the first octant bounded by the coordinate planes and the sphere $x^2 + y^2 + z^2 = a^2$.

26. The solid bounded above by the sphere $\rho = 4$ and below by the cone $\phi = \pi/3$.

27. The solid that is enclosed by the hemispheres $y = \sqrt{9 - x^2 - z^2}$, $y = \sqrt{4 - x^2 - z^2}$, and the plane $y = 0$.

In Exercises 28–31, use cylindrical or spherical coordinates to evaluate the integral.

28. $\displaystyle\int_0^a \int_0^{\sqrt{a^2 - x^2}} \int_0^{a^2 - x^2 - y^2} x^2 \, dz \, dy \, dx$ $(a > 0)$.

29. $\displaystyle\int_{-1}^1 \int_0^{\sqrt{1 - x^2}} \int_0^{\sqrt{1 - x^2 - y^2}} e^{-(x^2 + y^2 + z^2)^{3/2}} \, dz \, dy \, dx$.

30. $\displaystyle\int_0^2 \int_0^{\sqrt{4 - y^2}} \int_{\sqrt{x^2 + y^2}}^{\sqrt{8 - x^2 - y^2}} z^2 \, dz \, dx \, dy$.

31. $\displaystyle\int_{-3}^3 \int_{-\sqrt{9 - y^2}}^{\sqrt{9 - y^2}} \int_{-\sqrt{9 - x^2 - y^2}}^{\sqrt{9 - x^2 - y^2}} \sqrt{x^2 + y^2 + z^2} \, dz \, dx \, dy$.

32. Let G be the solid in the first octant bounded by the sphere $x^2 + y^2 + z^2 = 4$ and the coordinate planes.

Evaluate $\displaystyle\iiint_G xyz \, dV$ using

(a) rectangular coordinates
(b) cylindrical coordinates
(c) spherical coordinates.

Solve Exercises 33–38 using either cylindrical or spherical coordinates, whichever seems appropriate.

33. Find the center of gravity of the solid hemisphere bounded by $z = \sqrt{a^2 - x^2 - y^2}$ and $z = 0$ if the density is proportional to the distance from the origin.

34. Find the center of gravity of the solid in the first octant bounded by the cylinder $x^2 + y^2 = a^2$, the coordinate planes, and the plane $z = a$ if the density is $\delta(x, y, z) = xyz$.

35. Find the center of gravity of the solid bounded by the paraboloid $z = 1 - x^2 - y^2$ and the xy-plane if the density is $\delta(x, y, z) = x^2 + y^2 + z^2$.

36. Find the center of gravity of the solid that is bounded by the cylinder $x^2 + y^2 = 1$, the cone $z = \sqrt{x^2 + y^2}$, and the xy-plane if the density is $\delta(x, y, z) = z$.

37. Find the volume that is enclosed by the spheres $x^2 + y^2 + z^2 = 9$ and $x^2 + y^2 + (z - 2)^2 = 4$.

38. Suppose that the density at a point on a spherical planet is assumed to be

$$\delta = \delta_0 e^{[(\rho/R^3) - 1]}$$

where δ_0 is a positive constant, R is the radius of the planet, and ρ is the distance from the point to the planet's center. Calculate the mass of the planet.

39. In this exercise we shall obtain a formula for the volume of the spherical wedge in Figure 17.7.8.
    (a) Use a triple integral in cylindrical coordinates to show that the volume of the solid bounded above by a sphere $\rho = \rho_0$, below by a cone $\phi = \phi_0$, and on the sides by $\theta = \theta_1$ and $\theta = \theta_2$ $(\theta_1 < \theta_2)$ is

    $$V = \tfrac{1}{3}\rho_0{}^3(1 - \cos \phi_0)(\theta_2 - \theta_1)$$

    [*Hint:* In cylindrical coordinates, the sphere has the equation $r^2 + z^2 = \rho_0{}^2$ and the cone has the equation $z = r \cot \phi_0$. For simplicity, consider only the case $0 < \phi_0 < \pi/2$.]
    (b) Subtract appropriate volumes and use the result in part (a) to deduce that the volume ΔV of the spherical wedge is

    $$\Delta V = \frac{\rho_2{}^3 - \rho_1{}^3}{3}(\cos \phi_1 - \cos \phi_2)(\theta_2 - \theta_1)$$

    (c) Apply the Mean-Value Theorem to the functions $\cos \phi$ and ρ^3 to deduce that the formula in part (b) can be written as

    $$\Delta V = \rho^{*2} \sin \phi^* \, \Delta\rho \, \Delta\phi \, \Delta\theta$$

    where ρ^* is between ρ_1 and ρ_2, ϕ^* is between ϕ_1 and ϕ_2, and $\Delta\rho = \rho_2 - \rho_1$, $\Delta\phi = \phi_2 - \phi_1$, $\Delta\theta = \theta_2 - \theta_1$.

In Exercises 40–44, use the formulas in Exercise 36, Section 17.6, to find the indicated moment of inertia for the given homogeneous solid. Express your answer in terms of the mass M of the solid.

40. I_z for the solid cylinder $x^2 + y^2 \le R^2$, $0 \le z \le h$.

41. I_y for the solid cylinder $x^2 + y^2 \le R^2$, $0 \le z \le h$.

42. I_z for the hollow cylinder $R_1{}^2 \le x^2 + y^2 \le R_2{}^2$, $0 \le z \le h$.

43. I_z for the solid sphere $x^2 + y^2 + z^2 \le R^2$.

44. I_z for the solid cone $\dfrac{h}{R}\sqrt{x^2 + y^2} \le z \le h$.

45. If d is the distance between two point masses m and m', then according to Newton's *inverse-square law of gravitational attraction* each point mass attracts the other with a force of magnitude kmm'/d^2, where k is a constant. Because force is a vector, the attraction between a lamina or solid and a point mass can be found by considering the components of the force. For example, assume that a solid G with mass den-

sity δ attracts a unit point mass at the origin (see figure). The component in the z-direction of the force that G exerts on the unit mass can be found by approximating the corresponding components exerted by small subregions, summing, and taking the limit. This leads to the formula, using the spherical coordinates ρ and ϕ,

$$\iiint\limits_{G} \frac{k\delta \cos \phi}{\rho^2}\, dV$$

for the component of force in the z-direction.
(a) Use cylindrical coordinates to find the component in the z-direction of the force of attraction on a unit mass at the origin by the homogeneous solid cylinder $x^2 + y^2 \le R^2$, $0 < a \le z \le a + h$.
(b) Explain why the force in part (a) is the total force that acts on the unit mass.

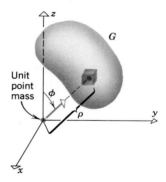

In Exercises 46–48, use the formula in Exercise 45 to find the magnitude of the force of attraction on a unit mass at the origin by the given homogeneous lamina or solid. (The regions are all such that the force in the z-direction is the only force of attraction.)

46. The circular lamina $x^2 + y^2 \le R^2$, $z = a > 0$. [*Hint:* If S is a lamina with area A, and δ is the *surface density*, then the formula is

$$\iint\limits_{S} \frac{k\delta \cos \phi}{\rho^2}\, dA$$

Use polar coordinates to do the integration.]

47. The solid cone $\dfrac{h}{R}r \le z \le h$. (Use spherical coordinates.)

48. The solid sphere of radius R with its center on the z-axis at $z = a > R$. (Use spherical coordinates.) Show that the force is the same as if the total mass of the sphere were concentrated at its center.

1146   MULTIPLE INTEGRALS

► SUPPLEMENTARY EXERCISES

In Exercises 1–4, evaluate the iterated integrals.

1. $\displaystyle\int_{1/2}^{1}\int_{0}^{2x}\cos(\pi x^2)\,dy\,dx.$

2. $\displaystyle\int_{0}^{2}\int_{-y}^{2y}xe^{y^3}\,dx\,dy.$

3. $\displaystyle\int_{-1}^{0}\int_{0}^{y^2}\int_{xy}^{1}2y\,dz\,dx\,dy.$

4. $\displaystyle\int_{0}^{1}\int_{0}^{z}\int_{0}^{\sqrt{yz}}x\,dx\,dy\,dz.$

In Exercises 5 and 6, express the iterated integral as an equivalent integral with the order of integration reversed.

5. $\displaystyle\int_{0}^{2}\int_{0}^{x/2}e^x e^y\,dy\,dx.$ **6.** $\displaystyle\int_{0}^{\pi}\int_{y}^{\pi}\frac{\sin x}{x}\,dx\,dy.$

7. Use a double integral to find the area of the region bounded by $y = 2x^3$, $2x + y = 4$, and the x-axis.

8. Use a double integral to find the area of the region bounded by $x = y^2$ and $x = 4y - y^2$.

9. Sketch the region R whose area is given by the iterated integral.

(a) $\displaystyle\int_{0}^{\pi/2}\int_{\tan(x/2)}^{\sin x}dy\,dx$

(b) $\displaystyle\int_{\pi/6}^{\pi/2}\int_{a}^{a(1+\cos\theta)}r\,dr\,d\theta$ $(a > 0)$.

In Exercises 10–12, evaluate the double integral over R using either rectangular or polar coordinates.

10. $\displaystyle\iint_{R}xy\,dA$; R is the region bounded by $y = \sqrt{x}$, $y = 2 - \sqrt{x}$, and the y-axis.

11. $\displaystyle\iint_{R}x^2\sin y^2\,dA$; R is the region bounded by $y = x^3$, $y = -x^3$, and $y = 8$.

12. $\displaystyle\iint_{R}(4 - x^2 - y^2)\,dA$; R is the sector in the first quadrant bounded by the circle $x^2 + y^2 = 4$ and the coordinate axes.

In Exercises 13–15, use a double integral in rectangular or polar coordinates to find the volume of the solid.

13. The solid in the first octant bounded by the coordinate planes and the plane $3x + 2y + z = 6$.

14. The solid enclosed by the cylinders $y = 3x + 4$ and $y = x^2$, and such that $0 \leq z \leq \sqrt{y}$.

15. The solid $G = \{(x, y, z) : 2 \leq x^2 + y^2 \leq 4$ and $0 \leq z \leq 1/(x^2 + y^2)^2\}$.

16. Convert to polar coordinates and evaluate:

$$\int_{0}^{\sqrt{2}}\int_{x}^{\sqrt{4-x^2}}4xy\,dy\,dx$$

17. Convert to rectangular coordinates and evaluate:

$$\int_{0}^{\pi/2}\int_{0}^{2a\sin\theta}r\sin 2\theta\,dr\,d\theta \qquad (a > 0)$$

In Exercises 18–20, find the area of the region using a double integral in polar coordinates.

18. The region outside the circle $r = \sqrt{2}a$ and inside the lemniscate $r^2 = 4a^2\cos 2\theta$.

19. The region enclosed by the rose $r = \cos 3\theta$.

20. The region inside the circle $r = 2\sqrt{3}\sin\theta$ and outside the circle $r = 3$.

In Exercises 21–23, find the area of the surface described.

21. The part of the paraboloid $z = 3x^2 + 3y^2 - 3$ below the xy-plane.

22. The part of the plane $2x + 2y + z = 7$ in the first octant.

23. The part of the cone $z^2 = x^2 + y^2$ between the planes $z = 1$ and $z = 4$.

24. Evaluate $\displaystyle\iiint_{G}x^2yz\,dV$, where G is the set of points satisfying the inequalities $0 \leq x \leq 2$, $-x \leq y \leq x^2$, and $0 \leq z \leq x + y$.

25. Evaluate $\displaystyle\iiint_{G}\sqrt{x^2 + y^2}\,dV$, where G is the set of points satisfying $x^2 + y^2 \leq 16$, $0 \leq z \leq 4 - y$.

26. If $G = \{(x, y, z) : x^2 + y^2 \leq z \leq 4x\}$, express the volume of G as a triple integral in (a) rectangular coordinates and (b) cylindrical coordinates.

27. In each part find an equivalent integral of the form

$$\iiint\limits_{G} (\quad) \, dx \, dz \, dy.$$

(a) $\displaystyle\int_{0}^{1} \int_{0}^{(1-x)/2} \int_{0}^{1-x-2y} z \, dz \, dy \, dx$

(b) $\displaystyle\int_{0}^{2} \int_{x^2}^{4} \int_{0}^{4-y} 3 \, dz \, dy \, dx$.

28. (a) Change to cylindrical coordinates and then evaluate:

$$\int_{-2}^{2} \int_{-\sqrt{4-x^2}}^{\sqrt{4-x^2}} \int_{(x^2+y^2)^2}^{16} x^2 \, dz \, dy \, dx$$

(b) Change to spherical coordinates and evaluate:

$$\int_{0}^{1} \int_{0}^{\sqrt{1-x^2}} \int_{0}^{\sqrt{1-x^2-y^2}} \frac{1}{1 + x^2 + y^2 + z^2} \, dz \, dy \, dx$$

29. If G is the region bounded above by the sphere $\rho = a$ and bounded below by the cone $\phi = \pi/3$, express $\iiint\limits_{G}(x^2 + y^2) \, dV$ as an iterated integral in (a) spherical coordinates, (b) cylindrical coordinates, and (c) Cartesian coordinates.

In Exercises 30–33, find the volume of G.

30. $G = \{(r, \theta, z) : 0 \leq r \leq 2 \sin \theta, \ 0 \leq z \leq r \sin \theta\}$.

31. G is the solid enclosed by the "inverted apple" $\rho = a(1 + \cos \phi)$.

32. G is the solid that is enclosed between the surfaces $x = y^2 + z^2$ and $x = 1 - y^2$.

33. G is the solid bounded below by the upper nappe of the cone $\phi = \pi/6$ and above by the plane $z = a$.

In Exercises 34–36, find the centroid $(\bar{x}, \bar{y})$ of the plane region R.

34. The region R is the upper half of the ellipse $(x/a)^2 + (y/b)^2 = 1$.

35. The region R is enclosed by the cardioid $r = a(1 + \sin \theta)$.

36. The region R is bounded by $y^2 = 4x$ and $y^2 = 8(x - 2)$.

In Exercises 37 and 38, find the center of gravity of the lamina with density δ.

37. The triangular lamina with vertices $(a, 0)$, $(-a, 0)$, and $(0, b)$, where $a > 0$ and $b > 0$; and $\delta(x, y)$ is proportional to the distance from (x, y) to the y-axis.

38. The lamina enclosed by the circle $r = 3 \cos \theta$, but outside the cardioid $r = 1 + \cos \theta$, and with $\delta(r, \theta)$ proportional to the distance from (r, θ) to the x-axis.

In Exercises 39 and 40, find the mass of the solid G if its density is δ.

39. The solid G is the part of the first octant under the plane $x/a + y/b + z/c = 1$, where a, b, c are positive, and $\delta(x, y, z) = kz$.

40. The spherical solid G is bounded by $\rho = a$ and $\delta(x, y, z)$ is twice the distance from (x, y, z) to the origin.

In Exercises 41–43, find the centroid of G.

41. The solid G bounded by $y = x^2$, $y = 4$, $z = 0$, and $y + z = 4$.

42. The solid G is the part of the sphere $\rho \leq a$ lying within the cone $\phi \leq \phi_0$, where $\phi_0 \leq \pi/2$.

43. The solid G is bounded by the cone with vertex $(0, 0, h)$ and base $x^2 + y^2 \leq R^2$ in the xy-plane.

18
Topics in
Vector Calculus

Johann Friedrich Carl Gauss   (1777 -1855)

■ **18.1** LINE INTEGRALS

In previous chapters we considered three kinds of integrals in rectangular coordinates: single integrals over intervals, double integrals over two-dimensional regions, and triple integrals over three-dimensional regions. In this section we shall discuss line integrals, which are integrals over curves in two- or three-dimensional space.

□ LINE INTEGRALS ALONG
CURVES IN 2-SPACE

18.1.1 DEFINITION.   Let C be a smooth curve in 2-space given parametrically by the equations

$$x = x(t), \quad y = y(t) \qquad (a \leq t \leq b)$$

and let $f(x, y)$ and $g(x, y)$ be continuous on some open region containing the curve C. Then we define

$$\int_C f(x, y)\, dx = \int_a^b f(x(t), y(t))x'(t)\, dt \tag{1}$$

$$\int_C g(x, y)\, dy = \int_a^b g(x(t), y(t))y'(t)\, dt \tag{2}$$

The symbol $\int_C f(x, y)\, dx$ is called the **line integral of f along C with respect to x,** and $\int_C g(x, y)\, dy$ is called the **line integral of g along C with respect to y.**

REMARK.   In words, the line integrals $\int_C f(x, y)\, dx$ and $\int_C g(x, y)\, dy$ are evaluated by replacing x and y by their expressions in terms of t, writing dx and dy as $dx = x'(t)\, dt$ and $dy = y'(t)\, dt$, and then integrating in the usual way over the interval $a \leq t \leq b$.

Frequently, the line integrals in (1) and (2) occur in combination, in which case we dispense with one of the integral signs and write

$$\int_C f(x, y)\, dx + g(x, y)\, dy = \int_C f(x, y)\, dx + \int_C g(x, y)\, dy \tag{3}$$

The expression $f(x, y)\, dx + g(x, y)\, dy$ is called a **differential form.**

Example 1 Evaluate

$$\int_C 2xy\,dx + (x^2 + y^2)\,dy$$

over the circular arc C given by $x = \cos t$, $y = \sin t$ $(0 \le t \le \pi/2)$ (Figure 18.1.1).

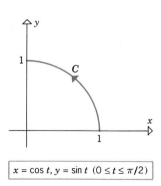

$x = \cos t, y = \sin t \ (0 \le t \le \pi/2)$

Figure 18.1.1

Solution. From (1) and (2)

$$\int_C 2xy\,dx = \int_0^{\pi/2} (2 \cos t \sin t)\left[\frac{d}{dt}(\cos t)\right] dt$$

$$= -2\int_0^{\pi/2} \sin^2 t \cos t\,dt = -\frac{2}{3}\sin^3 t\Big]_0^{\pi/2} = -\frac{2}{3}$$

$$\int_C (x^2 + y^2)\,dy = \int_0^{\pi/2} (\cos^2 t + \sin^2 t)\left[\frac{d}{dt}(\sin t)\right] dt$$

$$= \int_0^{\pi/2} \cos t\,dt = \sin t\Big]_0^{\pi/2} = 1$$

Thus, from (3)

$$\int_C 2xy\,dx + (x^2 + y^2)\,dy = \int_C 2xy\,dx + \int_C (x^2 + y^2)\,dy$$

$$= -\frac{2}{3} + 1 = \frac{1}{3} \quad \blacktriangleleft$$

□ INDEPENDENCE OF
PARAMETRIZATION

Since the parametric equations of a curve are used to evaluate line integrals over that curve, it seems possible that two different parametrizations of a curve C might produce different values for the same line integral over C. The following theorems deal with this question.

18.1.2 THEOREM (*Independence of Parametrization*).   *If C is a smooth parametric curve, and if a smooth change of parameter is made that does not change the orientation of the curve, then both sets of parametric equations produce the same value for any line integral over C.*

18.1.3 THEOREM (*Reversal of Orientation*).   *If C is a smooth parametric curve, and if a smooth change of parameter is made that reverses the orientation of the curve, then for any line integral over C the two sets of parametric equations produce values that are negatives of one another.*

We omit the proofs.

Example 2 Evaluate

$$\int_C 2xy\, dx + (x^2 + y^2)\, dy$$

over the curve C given parametrically by

(a) $x = \cos 2t$, $y = \sin 2t$, where $0 \leq t \leq \pi/4$
(b) $x = \cos(t^2)$, $y = \sin(t^2)$, where $0 \leq t \leq \sqrt{\pi/2}$
(c) $x = \cos(-t)$, $y = \sin(-t)$, where $3\pi/2 \leq t \leq 2\pi$
(d) $x = t$, $y = \sqrt{1 - t^2}$, where $0 \leq t \leq 1$

Solution. In parts (a) and (b), the curve C is the quarter-circle shown in Figure 18.1.1, traced in the counterclockwise direction. Thus, from Example 1 and Theorem 18.1.2,

$$\int_C 2xy\, dx + (x^2 + y^2)\, dy = \frac{1}{3}$$

in both cases. In parts (c) and (d), the curve is the same quarter-circle, but traced in the clockwise direction. Thus,

$$\int_C 2xy\, dx + (x^2 + y^2)\, dy = -\frac{1}{3}$$

These results should be verified using (1), (2), and (3) to evaluate the integrals directly. ◄

If a curve C is traced in a certain direction, then the same curve traced in the opposite direction is often denoted by the symbol $-C$. Thus, it follows from Theorem 18.1.3 that

$$\int_{-C} f(x, y)\, dx + g(x, y)\, dy = -\int_C f(x, y)\, dx + g(x, y)\, dy \qquad (4)$$

☐ **LINE INTEGRALS OVER PIECEWISE SMOOTH CURVES**

In the definition of a line integral, we required the curve C to be smooth. However, the definition can be extended to curves formed by finitely many smooth curves $C_1, C_2, \ldots, C_n$ joined end to end. Such a curve is called *piecewise smooth* (Figure 18.1.2). We define a line integral over a piecewise smooth curve C to be the sum of the integrals over the pieces:

$$\int_C = \int_{C_1} + \int_{C_2} + \cdots + \int_{C_n}$$

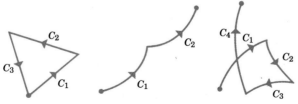

Figure 18.1.2

Example 3 Evaluate

$$\int_C x^2 y \, dx + x \, dy$$

in a counterclockwise direction around the triangular path shown in Figure 18.1.3.

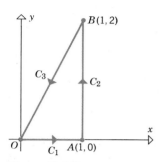

Figure 18.1.3

Solution. We shall integrate over C_1, C_2, and C_3 separately and add the results. For each of the three integrals we must find parametric equations that trace the path of integration in the correct direction.

The horizontal line segment C_1 can be represented as follows using $x = t$ as the parameter:

$$x = t, \quad y = 0 \qquad (0 \le t \le 1)$$

The vertical line segment C_2 can be represented as follows using $y = t$ as the parameter:

$$x = 1, \quad y = t \qquad (0 \le t \le 2)$$

To find parametric equations for C_3 we shall use Formula (5) of Section 14.5 with $\mathbf{v} = \vec{BO} = \langle -1, -2 \rangle$ and $\mathbf{r_0} = \langle 1, 2 \rangle$ to obtain the vector equation

$$\langle x, y \rangle = \langle 1, 2 \rangle + t \langle -1, -2 \rangle$$

for the line through O and B traced in the direction of C_3. Equating corresponding components and using the fact that the line segment C_3 is traced as t varies from 0 to 1 yields the parametric equations

$$x = 1 - t, \quad y = 2 - 2t \qquad (0 \le t \le 1)$$

We can now carry out the calculations.

$$\int_{C_1} x^2 y \, dx + x \, dy = \int_0^1 (t^2)(0) \frac{d}{dt} [t] \, dt + \int_0^1 (t) \frac{d}{dt} [0] \, dt = 0$$

$$\int_{C_2} x^2 y \, dx + x \, dy = \int_0^2 (1^2)(t) \frac{d}{dt} [1] \, dt + \int_0^2 (1) \frac{d}{dt} [t] \, dt = 0 + 2 = 2$$

$$\int_{C_3} x^2 y \, dx + x \, dy = \int_0^1 (1 - t)^2 (2 - 2t) \frac{d}{dt} [1 - t] \, dt$$

$$+ \int_0^1 (1 - t) \frac{d}{dt} [2 - 2t] \, dt$$

$$= 2 \int_0^1 (t - 1)^3 \, dt + 2 \int_0^1 (t - 1) \, dt = -\frac{1}{2} - 1 = -\frac{3}{2}$$

Thus,

$$\int_C x^2 y \, dx + x \, dy = 0 + 2 + \left(-\frac{3}{2} \right) = \frac{1}{2} \qquad \blacktriangleleft$$

☐ **VECTOR NOTATION FOR LINE INTEGRALS**

For many applications it is helpful to express the line integral

$$\int_C f(x, y)\, dx + g(x, y)\, dy \tag{5}$$

in vector notation. To do this, we introduce the vector-valued function

$$\mathbf{F}(x, y) = f(x, y)\mathbf{i} + g(x, y)\mathbf{j} \tag{6}$$

and express the curve $C : x = x(t)$, $y = y(t)$ in terms of its position vector

$$\mathbf{r}(t) = x(t)\mathbf{i} + y(t)\mathbf{j}$$

If we formally multiply both sides of the equation

$$\frac{d\mathbf{r}}{dt} = \frac{dx}{dt}\mathbf{i} + \frac{dy}{dt}\mathbf{j}$$

by dt, we are led to the following convenient notation:

$$d\mathbf{r} = \frac{d\mathbf{r}}{dt}\, dt = dx\mathbf{i} + dy\mathbf{j} \tag{7}$$

If we now calculate the dot product of (6) and (7), we obtain

$$\mathbf{F}(x, y) \cdot d\mathbf{r} = f(x, y)\, dx + g(x, y)\, dy$$

which is precisely the integrand in (5). This suggests the notation

$$\int_C f(x, y)\, dx + g(x, y)\, dy = \int_C \mathbf{F}(x, y) \cdot d\mathbf{r} \tag{8}$$

Moreover, the equation

$$\int_C f(x, y)\, dx + g(x, y)\, dy = \int_a^b \left[f(x(t), y(t)) \frac{dx}{dt} + g(x(t), y(t)) \frac{dy}{dt} \right] dt$$

can be written as

$$\int_C \mathbf{F}(x, y) \cdot d\mathbf{r} = \int_a^b \left[\mathbf{F}(x(t), y(t)) \cdot \frac{d\mathbf{r}}{dt} \right] dt \tag{9}$$

☐ **WORK**

One of the most important applications of line integrals occurs in the study of work. Recall that if a constant force $\mathbf{F}$ acts on a particle moving along a straight line from a point P to a point Q, then the work done by $\mathbf{F}$ is

$$W = \mathbf{F} \cdot \overrightarrow{PQ} = (\|\mathbf{F}\| \cos \theta)\|\overrightarrow{PQ}\| \tag{10}$$

where $\|\mathbf{F}\| \cos \theta$ is the component of force in the direction of motion and $\|\overrightarrow{PQ}\|$ is the distance traveled by the particle [see (13), Section 14.3].

This elementary notion of work is inadequate for many applications. For example, if a rocket moves through the earth's gravitational field, the path is usually a curve rather than a straight line; and the force of gravity on the rocket does not remain constant, but changes with the position of the rocket relative to the earth's center. Thus, we must consider the more general notion of work done on a particle moving along a curve C while subject to a force $\mathbf{F}(x, y)$ that varies from point to point along the curve (Figure 18.1.4).

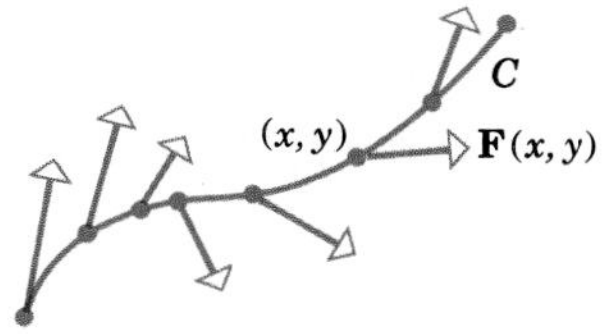

Let us assume that C is a smooth curve given parametrically by the equations

$$x = x(t), \quad y = y(t) \quad (a \le t \le b)$$

and that the force vector at a point (x, y) on the curve is

$$\mathbf{F}(x, y) = f(x, y)\mathbf{i} + g(x, y)\mathbf{j}$$

where $f(x, y)$ and $g(x, y)$ are continuous functions. Let $\mathbf{r}(t) = x(t)\mathbf{i} + y(t)\mathbf{j}$ $(a \le t \le b)$ denote the position vector for C, and let us define the function $W(t)$ by

$$W(t) = \left[\begin{array}{c} \text{the work done as the particle moves from} \\ \text{the initial point } \mathbf{r}(a) \text{ to the point } \mathbf{r}(t) \end{array} \right]$$

(See Figure 18.1.5.) Thus, the work done as the particle moves from $\mathbf{r}(a)$ to $\mathbf{r}(t + \Delta t)$ is $W(t + \Delta t)$, and the work done as the particle moves from $\mathbf{r}(t)$ to $\mathbf{r}(t + \Delta t)$ is $W(t + \Delta t) - W(t)$.

If Δt is small, then we may reasonably approximate the portion of curve between $\mathbf{r}(t)$ and $\mathbf{r}(t + \Delta t)$ by a straight line. If the force vector $\mathbf{F}$ does not vary much over this small section of curve, we may assume the force to be constant between $\mathbf{r}(t)$ and $\mathbf{r}(t + \Delta t)$ and equal to its value at $\mathbf{r}(t)$. Thus, (10) implies that the work done as the particle moves from $\mathbf{r}(t)$ to $\mathbf{r}(t + \Delta t)$ can be approximated in the following way if Δt is small:

$$W(t + \Delta t) - W(t) \approx \mathbf{F}\big(x(t), y(t)\big) \cdot [\mathbf{r}(t + \Delta t) - \mathbf{r}(t)]$$

If we assume that the error in this approximation diminishes to zero as $\Delta t \to 0$, then it follows on dividing by Δt and taking the limit that

$$\lim_{\Delta t \to 0} \frac{W(t + \Delta t) - W(t)}{\Delta t} = \lim_{\Delta t \to 0} \mathbf{F}\big(x(t), y(t)\big) \cdot \frac{\mathbf{r}(t + \Delta t) - \mathbf{r}(t)}{\Delta t}$$

or

$$W'(t) = \mathbf{F}\big(x(t), y(t)\big) \cdot \mathbf{r}'(t) \tag{11}$$

Figure 18.1.4

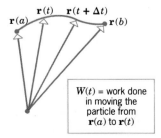

Figure 18.1.5

Since $W(t)$ represents the total work done as the particle moves from $\mathbf{r}(a)$ to $\mathbf{r}(t)$, it follows that

$$W(a) = 0$$

Thus, the total work W done as the particle moves from $\mathbf{r}(a)$ to $\mathbf{r}(b)$ can be written as

$$W = W(b) = W(b) - W(a) = \int_a^b W'(t)\, dt$$

or from (11)

$$W = \int_a^b \mathbf{F}(x(t), y(t)) \cdot \mathbf{r}'(t)\, dt$$

Since this is precisely the line integral over C of the force $\mathbf{F}$ [see (9)], we are led to the following definition of the work W done by a force $\mathbf{F}(x, y)$ acting on a particle as it moves along a curve C:

$$W = \int_C \mathbf{F}(x, y) \cdot d\mathbf{r} \tag{12}$$

Example 4 Find the work done if a particle moves from $(-2, 4)$ to $(1, 1)$ along the parabola $y = x^2$, while subject to the force

$$\mathbf{F}(x, y) = x^3 y\mathbf{i} + (x - y)\mathbf{j}$$

Solution. If we use $x = t$ as the parameter, the path of the particle is represented by

$$x = t, \quad y = t^2 \quad (-2 \le t \le 1)$$

or in vector notation as

$$\mathbf{r}(t) = t\mathbf{i} + t^2\mathbf{j} \quad (-2 \le t \le 1)$$

Since $\mathbf{F}(x, y) = x^3 y\mathbf{i} + (x - y)\mathbf{j}$, it follows that

$$\mathbf{F}(x(t), y(t)) = (t^3)(t^2)\mathbf{i} + (t - t^2)\mathbf{j} = t^5\mathbf{i} + (t - t^2)\mathbf{j}$$

Thus, from (12) and (9)

$$W = \int_C \mathbf{F}(x, y) \cdot d\mathbf{r} = \int_a^b \left[\mathbf{F}(x(t), y(t)) \cdot \frac{d\mathbf{r}}{dt} \right] dt$$

$$= \int_{-2}^1 [t^5\mathbf{i} + (t - t^2)\mathbf{j}] \cdot [\mathbf{i} + 2t\mathbf{j}]\, dt = \int_{-2}^1 (t^5 + 2t^2 - 2t^3)\, dt$$

$$= \frac{1}{6}t^6 + \frac{2}{3}t^3 - \frac{1}{2}t^4 \Big]_{-2}^1 = 3 \quad \text{(units of work)} \quad \blacktriangleleft$$

☐ **LINE INTEGRALS ALONG CURVES IN 3-SPACE**

The notion of a line integral can be extended to three dimensions. If C is a smooth curve in three dimensions given parametrically by

$$x = x(t), \quad y = y(t), \quad z = z(t) \qquad (a \leq t \leq b)$$

and if $f(x, y, z)$, $g(x, y, z)$, and $h(x, y, z)$ are continuous on some region containing C, then we define

$$\int_C f(x, y, z) \, dx = \int_a^b f(x(t), y(t), z(t))x'(t) \, dt$$

$$\int_C g(x, y, z) \, dy = \int_a^b g(x(t), y(t), z(t))y'(t) \, dt \qquad (13)$$

$$\int_C h(x, y, z) \, dz = \int_a^b h(x(t), y(t), z(t))z'(t) \, dt$$

$$\int_C f(x, y, z) \, dx + g(x, y, z) \, dy + h(x, y, z) \, dz$$

$$= \int_C f(x, y, z) \, dx + \int_C g(x, y, z) \, dy + \int_C h(x, y, z) \, dz \qquad (14)$$

If we let

$$\mathbf{F}(x, y, z) = f(x, y, z)\mathbf{i} + g(x, y, z)\mathbf{j} + h(x, y, z)\mathbf{k}$$

$$\mathbf{r}(t) = x(t)\mathbf{i} + y(t)\mathbf{j} + z(t)\mathbf{k}$$

$$d\mathbf{r} = \frac{d\mathbf{r}}{dt} \, dt = dx\mathbf{i} + dy\mathbf{j} + dz\mathbf{k}$$

then (14) can be written as

$$\int_C f(x, y, z) \, dx + g(x, y, z) \, dy + h(x, y, z) \, dz = \int_C \mathbf{F}(x, y, z) \cdot d\mathbf{r} \qquad (15)$$

Moreover, it follows from (13) that

$$\int_C \mathbf{F}(x, y, z) \cdot d\mathbf{r} = \int_a^b \left[\mathbf{F}(x(t), y(t), z(t)) \cdot \frac{d\mathbf{r}}{dt} \right] dt \qquad (16)$$

Finally, if a particle moves along a curve C in three dimensions while subject to a force $\mathbf{F}(x, y, z) = f(x, y, z)\mathbf{i} + g(x, y, z)\mathbf{j} + h(x, y, z)\mathbf{k}$, then we define the work W done by the force on the particle to be

$$W = \int_C \mathbf{F}(x, y, z) \cdot d\mathbf{r} \qquad (17)$$

Example 5   Evaluate $\int_C \mathbf{F} \cdot d\mathbf{r}$ if $\mathbf{F}(x, y, z) = yz\mathbf{i} + xz\mathbf{j} + xy\mathbf{k}$ and C is the curve $\mathbf{r}(t) = t\mathbf{i} + t^2\mathbf{j} + t^3\mathbf{k}$ $(0 \leq t \leq 1)$.

Solution.   From (16)

$$\int_C \mathbf{F} \cdot d\mathbf{r} = \int_0^1 \left[\mathbf{F}(x(t), y(t), z(t)) \cdot \frac{d\mathbf{r}}{dt} \right] dt$$

$$= \int_0^1 (t^2 t^3 \mathbf{i} + t t^3 \mathbf{j} + t t^2 \mathbf{k}) \cdot (\mathbf{i} + 2t\mathbf{j} + 3t^2\mathbf{k}) \, dt$$

$$= \int_0^1 (t^5 + 2t^5 + 3t^5) \, dt = \int_0^1 6t^5 \, dt = 1 \quad \blacktriangleleft$$

We conclude by noting that Theorems 18.1.2 and 18.1.3 hold for line integrals in 3-space.

▶ **Exercise Set 18.1**   ⓒ 28

1.   Let C be the curve $x = t$, $y = t^2$, $0 \leq t \leq 1$. Find
   (a) $\int_C (2x + y) \, dx$ (b) $\int_C (x^2 - y) \, dy$
   (c) $\int_C (2x + y) \, dx + (x^2 - y) \, dy$.

2.   Let $\mathbf{F} = (3x + 2y)\mathbf{i} + (2x - y)\mathbf{j}$. Evaluate $\int_C \mathbf{F} \cdot d\mathbf{r}$, where C is
   (a)   the line segment from $(0, 0)$ to $(1, 1)$.
   (b)   the parabolic arc $y = x^2$ from $(0, 0)$ to $(1, 1)$.
   (c)   the curve $y = \sin(\pi x/2)$ from $(0, 0)$ to $(1, 1)$.
   (d)   the curve $x = y^3$ from $(0, 0)$ to $(1, 1)$.

In Exercises 3–10, evaluate the line integral.

3. $\int_C y \, dx - x^2 \, dy$, where C is the curve $x = t$, $y = \frac{1}{2}t^2$, $0 \leq t \leq 2$.

4. $\int_C e^{y-x} \, dx + xy \, dy$, where C is the curve $x = 2 + t$, $y = 3 - t$, $0 \leq t \leq 1$.

5. $\int_C (x + 2y) \, dx + (x - y) \, dy$, where C is the curve $x = 2 \cos t$, $y = 4 \sin t$, $0 \leq t \leq \pi/4$.

6. $\int_C (x^2 - y^2) \, dx + x \, dy$, where C is the curve $x = t^{2/3}$, $y = t$, $-1 \leq t \leq 1$.

7. $\int_C -y \, dx + x \, dy$ along $y^2 = 3x$ from the point $(3, 3)$ to the point $(0, 0)$.

8. $\int_C (y - x) \, dx + x^2 y \, dy$ along $y^2 = x^3$ from the point $(1, -1)$ to the point $(1, 1)$.

9. $\int_C (x^2 + y^2) \, dx - x \, dy$ along the quarter-circle $x^2 + y^2 = 1$ from $(1, 0)$ to $(0, 1)$.

10. $\int_C (y - x) \, dx + xy \, dy$ along the line segment from $(3, 4)$ to $(2, 1)$.

11.   Find $\int_C \mathbf{F} \cdot d\mathbf{r}$, where $\mathbf{F}(x, y) = x^2\mathbf{i} + xy\mathbf{j}$ and C is the semicircle given by $\mathbf{r}(t) = 2 \cos t\mathbf{i} + 2 \sin t\mathbf{j}$, $0 \leq t \leq \pi$.

12.   Find $\int_C \mathbf{F} \cdot d\mathbf{r}$, where $\mathbf{F}(x, y) = x^2 y\mathbf{i} + 4\mathbf{j}$ and C is the curve $\mathbf{r}(t) = e^t\mathbf{i} + e^{-t}\mathbf{j}$, $0 \leq t \leq 1$.

13.   Find $\int_C \mathbf{F} \cdot d\mathbf{r}$, where $\mathbf{F}(x, y) = (x^2 + y^2)^{-3/2}(x\mathbf{i} + y\mathbf{j})$ and C is the curve $\mathbf{r}(t) = e^t \sin t\mathbf{i} + e^t \cos t\mathbf{j}$, $0 \leq t \leq 1$.

14. Evaluate $\int_C y\, dx - x\, dy$ over each of the following curves:

(a)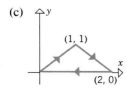

(b)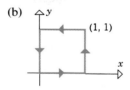

(c)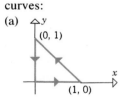

(d)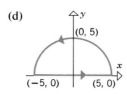

In Exercises 15–18, evaluate the line integral in 3-space.

15. $\int_C yz\, dx - xz\, dy + xy\, dz$, where C is the curve $x = e^t$, $y = e^{3t}$, $z = e^{-t}$, $0 \le t \le 1$.

16. $\int_C y\, dx + z\, dy - x\, dz$, where C is the line segment from $(1, 1, 1)$ to $(-3, 2, 0)$.

17. $\int_C \mathbf{F} \cdot d\mathbf{r}$, where $\mathbf{F}(x, y, z) = z\mathbf{i} + x\mathbf{j} + y\mathbf{k}$ and C is the curve $\mathbf{r}(t) = \sin t\,\mathbf{i} + 3\sin t\,\mathbf{j} + \sin^2 t\,\mathbf{k}$, $0 \le t \le \pi/2$.

18. $\int_C x^2 z\, dx - yx^2\, dy + 3xz\, dz$, where C is the triangular path from $(0,0,0)$ to $(1,1,0)$ to $(1,1,1)$ to $(0,0,0)$.

19. Find the work done if a particle moves along the parabolic arc $x = y^2$ from $(0, 0)$ to $(1, 1)$ while subject to the force $\mathbf{F}(x, y) = xy\mathbf{i} + x^2\mathbf{j}$.

20. Find the work done if a particle moves along the curve $x = t$, $y = 1/t$, $1 \le t \le 3$ while subject to the force $\mathbf{F}(x, y) = (x^2 + xy)\mathbf{i} + (y - x^2 y)\mathbf{j}$.

21. Find the work done by the force

$$\mathbf{F}(x, y) = \frac{1}{x^2 + y^2}\mathbf{i} + \frac{4}{x^2 + y^2}\mathbf{j}$$

acting on a particle that moves along the curve C shown.

(a)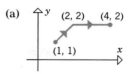

(b)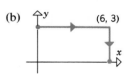

(c)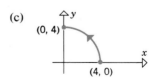

22. Find the work that is done by a force

$$\mathbf{F}(x, y, z) = xy\mathbf{i} + yz\mathbf{j} + xz\mathbf{k}$$

acting on a particle that moves along the curve $\mathbf{r}(t) = t\mathbf{i} + t^2\mathbf{j} + t^3\mathbf{k}$, $0 \le t \le 1$.

23. Find the work that is done by a force

$$\mathbf{F}(x, y, z) = (x + y)\mathbf{i} + xy\mathbf{j} - z^2\mathbf{k}$$

acting on a particle that moves along the line segment from $(0, 0, 0)$ to $(1, 3, 1)$ and then along the line segment from $(1, 3, 1)$ to $(2, -1, 4)$.

24. Evaluate $\int_{-C} \dfrac{x\, dy - y\, dx}{x^2 + y^2}$, where C is the circle $x^2 + y^2 = a^2$ traversed counterclockwise.

25. A particle moves from the point $(0, 0)$ to the point $(1, 0)$ along the curve $x = t$, $y = \lambda t(1 - t)$ while subject to the force $\mathbf{F}(x, y) = xy\mathbf{i} + (x - y)\mathbf{j}$. For what value of λ is the work done equal to 1?

26. (a) Show that $\int_C f(x, y)\, dx = 0$ if C is a vertical line segment in the xy-plane.

   (b) Show that $\int_C f(x, y)\, dy = 0$ if C is a horizontal line segment in the xy-plane.

27. A particle moves counterclockwise along the upper half of the circle $x^2 + y^2 = a^2$ from $(a, 0)$ to $(-a, 0)$ while subject to the force $\mathbf{F}(x, y) = k\mathbf{r}/\|\mathbf{r}\|^3$, where k is a constant and $\mathbf{r} = x\mathbf{i} + y\mathbf{j}$. Find the work done.

28. A farmer carries a sack of grain weighing 20 lb up a circular helical staircase encircling a silo of radius 25 ft. As the farmer climbs, grain leaks from the sack at a rate of 1 lb per 10 ft of ascent. How much work is performed by the farmer on the sack of grain if he climbs 60 ft in exactly four revolutions?

■ 18.2 LINE INTEGRALS INDEPENDENT OF PATH

In general, the value of a line integral $\int_C \mathbf{F} \cdot d\mathbf{r}$ depends on the curve C. However, we shall show in this section that when the integrand satisfies appropriate conditions, the value of the integral depends only on the location of the endpoints of the curve C and not on the shape of the curve joining the endpoints. In such cases the evaluation of the line integral is greatly simplified.

☐ **INDEPENDENCE OF PATH**   Let us begin with an example.

Example 1   Let $\mathbf{F}(x, y) = y\mathbf{i} + x\mathbf{j}$. Evaluate the line integral

$$\int_C \mathbf{F} \cdot d\mathbf{r} = \int_C y\,dx + x\,dy$$

over the following curves (Figure 18.2.1):

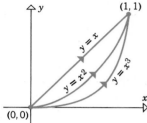

Figure 18.2.1

(a)   The line segment $y = x$ from $(0, 0)$ to $(1, 1)$
(b)   The parabola $y = x^2$ from $(0, 0)$ to $(1, 1)$
(c)   The cubic $y = x^3$ from $(0, 0)$ to $(1, 1)$

Solution (a).   With $x = t$ as the parameter, the path of integration is given by $\mathbf{r}(t) = t\mathbf{i} + t\mathbf{j}$ $(0 \le t \le 1)$, or in parametric form, $x = t$, $y = t$ $(0 \le t \le 1)$. Since $\mathbf{F}(x, y) = y\mathbf{i} + x\mathbf{j}$, it follows that $\mathbf{F}(x, y) = \mathbf{F}(x(t), y(t)) = t\mathbf{i} + t\mathbf{j}$. Thus,

$$\int_C \mathbf{F} \cdot d\mathbf{r} = \int_0^1 \left[\mathbf{F}(x(t), y(t)) \cdot \frac{d\mathbf{r}}{dt} \right] dt = \int_0^1 (t\mathbf{i} + t\mathbf{j}) \cdot (\mathbf{i} + \mathbf{j})\,dt$$

$$= \int_0^1 2t\,dt = 1$$

Solution (b).   With $x = t$ as the parameter, $\mathbf{r}(t) = t\mathbf{i} + t^2\mathbf{j}$ $(0 \le t \le 1)$, and $\mathbf{F}(x(t), y(t)) = t^2\mathbf{i} + t\mathbf{j}$. Thus,

$$\int_C \mathbf{F} \cdot d\mathbf{r} = \int_0^1 \left[\mathbf{F}(x(t), y(t)) \cdot \frac{d\mathbf{r}}{dt} \right] dt$$

$$= \int_0^1 (t^2\mathbf{i} + t\mathbf{j}) \cdot (\mathbf{i} + 2t\mathbf{j})\,dt$$

$$= \int_0^1 3t^2\,dt = 1$$

Solution (c). With $x = t$ as the parameter, $\mathbf{r}(t) = t\mathbf{i} + t^3\mathbf{j}$ $(0 \leq t \leq 1)$, and $\mathbf{F}(x(t), y(t)) = t^3\mathbf{i} + t\mathbf{j}$. Thus,

$$\int_C \mathbf{F} \cdot d\mathbf{r} = \int_0^1 \left[\mathbf{F}(x(t), y(t)) \cdot \frac{d\mathbf{r}}{dt} \right] dt$$

$$= \int_0^1 (t^3\mathbf{i} + t\mathbf{j}) \cdot (\mathbf{i} + 3t^2\mathbf{j}) \, dt$$

$$= \int_0^1 4t^3 \, dt = 1 \qquad \blacktriangleleft$$

In this example we obtained the same value for the line integral even though we integrated over three different paths joining $(0, 0)$ to $(1, 1)$. This is not accidental. The following theorem, called the **Fundamental Theorem of Line Integrals,** shows that this is the case because $\mathbf{F}(x, y) = y\mathbf{i} + x\mathbf{j}$ is the gradient of some function ϕ [specifically $\mathbf{F}(x, y) = \nabla\phi$, where $\phi(x, y) = xy$].

18.2.1 THEOREM (*The Fundamental Theorem of Line Integrals*). *Suppose that* $\mathbf{F}(x, y) = f(x, y)\mathbf{i} + g(x, y)\mathbf{j}$, *where* f *and* g *are continuous in some open region containing the points* (x_0, y_0) *and* (x_1, y_1). *If*

$$\mathbf{F}(x, y) = \nabla\phi(x, y)$$

at each point of this region, then for any piecewise smooth curve C *that starts at* (x_0, y_0), *ends at* (x_1, y_1), *and lies entirely in the region,*

$$\int_C \mathbf{F}(x, y) \cdot d\mathbf{r} = \phi(x_1, y_1) - \phi(x_0, y_0) \tag{1}$$

We shall give the proof for a smooth curve C. The proof for piecewise smooth curves is obtained by considering each piece separately. The details are left for the exercises.

Proof. If C is given parametrically by $x = x(t)$, $y = y(t)$ $(a \leq t \leq b)$, then the initial and final points of the curve C are

$$(x_0, y_0) = (x(a), y(a)) \quad \text{and} \quad (x_1, y_1) = (x(b), y(b))$$

Since $\mathbf{F}(x, y) = \nabla\phi$, it follows that

$$\mathbf{F}(x, y) = \frac{\partial\phi}{\partial x}\mathbf{i} + \frac{\partial\phi}{\partial y}\mathbf{j}$$

so

$$\int_C \mathbf{F}(x, y) \cdot d\mathbf{r} = \int_C \frac{\partial \phi}{\partial x} \, dx + \frac{\partial \phi}{\partial y} \, dy = \int_a^b \left[\frac{\partial \phi}{\partial x} \frac{dx}{dt} + \frac{\partial \phi}{\partial y} \frac{dy}{dt} \right] dt$$

$$= \int_a^b \frac{d}{dt} \left[\phi(x(t), y(t)) \right] dt = \phi[x(t), y(t)] \Big]_{t=a}^b$$

$$= \phi(x(b), y(b)) - \phi(x(a), y(a))$$

$$= \phi(x_1, y_1) - \phi(x_0, y_0) \qquad \blacksquare$$

☐ **CONSERVATIVE FUNCTIONS**

Since the right side of (1) involves only the values of ϕ at the endpoints (x_0, y_0) and (x_1, y_1), it follows that the integral on the left has the same value for *every* piecewise smooth curve C joining these endpoints. We say that the integral is *independent of the path C*. If $\mathbf{F}(x, y)$ is the gradient of a function $\phi(x, y)$ in some open region, then $\mathbf{F}$ is said to be *conservative* in that region, and ϕ is called a *potential function* for $\mathbf{F}$ in that region. In brief, Theorem 18.2.1 tells us that if the function $\mathbf{F}$ is conservative in a region, and C is a path (curve) in that region, then the line integral of $\mathbf{F} \cdot d\mathbf{r}$ is independent of the path and the value of the integral can be determined from the values of a potential function at the endpoints of the path.

Example 2 The function $\mathbf{F}(x, y) = y\mathbf{i} + x\mathbf{j}$ is conservative in the entire *xy*-plane since it is the gradient of $\phi(x, y) = xy$. Thus, for any piecewise smooth curve C running from $(0, 0)$ to $(1, 1)$, it follows from (1) that

$$\int_C y \, dx + x \, dy = \int_C \mathbf{F} \cdot d\mathbf{r} = \phi(1, 1) - \phi(0, 0) = 1 - 0 = 1$$

which explains the results obtained in Example 1.   ◄

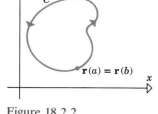

Figure 18.2.2

A curve C given by $\mathbf{r}(t) = x(t)\mathbf{i} + y(t)\mathbf{j}$ ($a \le t \le b$) is called *closed* if the initial point $\mathbf{r}(a) = \langle x_0, y_0 \rangle$ and the terminal point $\mathbf{r}(b) = \langle x_1, y_1 \rangle$ coincide (Figure 18.2.2). Line integrals over smooth closed curves require no new computational techniques. However, if $\mathbf{F}(x, y) = f(x, y)\mathbf{i} + g(x, y)\mathbf{j}$ is the gradient of $\phi(x, y)$, then it follows from the Fundamental Theorem of Line Integrals (18.2.1) that

$$\int_C \mathbf{F} \cdot d\mathbf{r} = \int_C f(x, y) \, dx + g(x, y) \, dy = \phi(x_1, y_1) - \phi(x_0, y_0) = 0$$

Thus, we have the following result.

18.2.2   THEOREM.   *Suppose that* $\mathbf{F}(x, y) = f(x, y)\mathbf{i} + g(x, y)\mathbf{j}$, *where f and g are continuous in some open region. If* $\mathbf{F}$ *is conservative in that region, then*

$$\int_C \mathbf{F} \cdot d\mathbf{r} = 0$$

for every piecewise smooth closed curve C in the region.

Theorem 18.2.1 raises two basic questions that must be answered if Formula (1) is to be useful for calculations.

- How can we tell whether a continuous function $\mathbf{F}(x, y) = f(x, y)\mathbf{i} + g(x, y)\mathbf{j}$ is the gradient of some function $\phi(x, y)$?
- If $\mathbf{F}(x, y)$ is the gradient of some function $\phi(x, y)$, how do we find ϕ?

To answer these questions, we shall need some terminology. A plane curve $\mathbf{r} = \mathbf{r}(t)$ ($a \le t \le b$) is said to be *simple* if it does not intersect itself anywhere between its endpoints. A simple curve may or may not be closed (Figure 18.2.3).

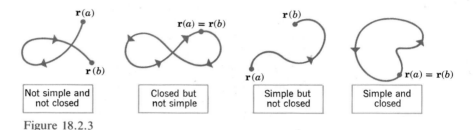

| Not simple and not closed | Closed but not simple | Simple but not closed | Simple and closed |

Figure 18.2.3

A plane region whose boundary consists of *one* simple closed curve is called *simply connected*. As illustrated in Figure 18.2.4, a simply connected region cannot consist of two or more separated pieces and cannot contain any holes.

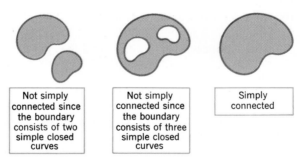

| Not simply connected since the boundary consists of two simple closed curves | Not simply connected since the boundary consists of three simple closed curves | Simply connected |

Figure 18.2.4

The entire plane (which has no boundary) is also regarded to be simply connected. The following theorem will help us to determine whether a given function $\mathbf{F}(x, y)$ is conservative.

18.2.3 THEOREM. *Let* $\mathbf{F}(x, y) = f(x, y)\mathbf{i} + g(x, y)\mathbf{j}$, *where f and g have continuous first partial derivatives in an open simply connected region. Then* $\mathbf{F}(x, y)$ *is the gradient of some function* $\phi(x, y)$ *on that region if and only if*

$$\frac{\partial f}{\partial y} = \frac{\partial g}{\partial x} \tag{2}$$

at each point of the region.

We shall prove that if $\mathbf{F}$ is the gradient of some function, then condition (2) holds. The proof of the converse is omitted since it requires results from advanced calculus.

Proof.   Suppose that $\mathbf{F}$ is the gradient of some function ϕ, that is,

$$\mathbf{F}(x, y) = f(x, y)\mathbf{i} + g(x, y)\mathbf{j} = \frac{\partial \phi}{\partial x}\mathbf{i} + \frac{\partial \phi}{\partial y}\mathbf{j} = \nabla\phi$$

It follows that

$$f(x, y) = \frac{\partial \phi}{\partial x} \quad \text{and} \quad g(x, y) = \frac{\partial \phi}{\partial y}$$

so

$$\frac{\partial f}{\partial y} = \frac{\partial^2 \phi}{\partial y \, \partial x} \quad \text{and} \quad \frac{\partial g}{\partial x} = \frac{\partial^2 \phi}{\partial x \, \partial y}$$

Since $\partial f/\partial y$ and $\partial g/\partial x$ were assumed continuous, it follows from Theorem 16.4.6 that the two mixed partial derivatives of ϕ are equal. Thus, $\partial f/\partial y = \partial g/\partial x$.   ∎

WARNING.   In (2), the $\mathbf{i}$-component of $\mathbf{F}$ is differentiated with respect to y and the $\mathbf{j}$-component with respect to x. It is easy to get this backwards by mistake.

Example 3   Show that $\mathbf{F}(x, y) = 2xy^3\mathbf{i} + (1 + 3x^2y^2)\mathbf{j}$ is the gradient of some function ϕ on the entire xy-plane, and find ϕ.

Solution.   Since $f(x, y) = 2xy^3$ and $g(x, y) = 1 + 3x^2y^2$ we have

$$\frac{\partial f}{\partial y} = 6xy^2 = \frac{\partial g}{\partial x}$$

so that (2) holds for all (x, y). Thus, $\mathbf{F}$ is the gradient of some function ϕ on the entire xy-plane, that is,

$$\mathbf{F}(x, y) = 2xy^3\mathbf{i} + (1 + 3x^2y^2)\mathbf{j} = \frac{\partial \phi}{\partial x}\mathbf{i} + \frac{\partial \phi}{\partial y}\mathbf{j} = \nabla\phi \tag{3}$$

To find ϕ, we begin by equating components in (3):

$$\frac{\partial \phi}{\partial x} = 2xy^3 \tag{4}$$

$$\frac{\partial \phi}{\partial y} = 1 + 3x^2y^2 \tag{5}$$

Integrating (4) with respect to x (and treating y as constant), we obtain

$$\phi = \int 2xy^3 \, dx = x^2y^3 + k(y) \tag{6}$$

where $k(y)$ represents the "constant" of integration. We are justified in treating the constant of integration as a function of y, since y is held constant in the differentiation process, so

$$\frac{\partial}{\partial x}[k(y)] = 0$$

To find $k(y)$, we differentiate (6) with respect to y:

$$\frac{\partial \phi}{\partial y} = 3x^2y^2 + k'(y)$$

Equating this with (5) yields

$$1 + 3x^2y^2 = 3x^2y^2 + k'(y)$$

Thus,

$$k'(y) = 1 \quad \text{or} \quad k(y) = \int 1\, dy = y + K$$

where K is a (numerical) constant of integration. Substituting in (6) we obtain

$$\phi = x^2y^3 + y + K$$

The appearance of the arbitrary constant K tells us that ϕ is not unique. The reader may want to verify that $\nabla\phi = \mathbf{F}$.

Alternative Solution. Instead of integrating (4) with respect to x, as above, we can begin by integrating (5) with respect to y (treating x as a constant in the integration process). We obtain

$$\phi = \int (1 + 3x^2y^2)\, dy = y + x^2y^3 + k(x) \tag{7}$$

where $k(x)$ is the "constant" of integration. Differentiating this expression with respect to x yields

$$\frac{\partial \phi}{\partial x} = 2xy^3 + k'(x)$$

and equating this with (4) gives

$$2xy^3 = 2xy^3 + k'(x)$$

so $k'(x) = 0$ and $k(x) = K$, where K is a (numerical) constant of integration. Substituting this in (7) gives

$$\phi = y + x^2y^3 + K$$

which agrees with our previous solution.   ◀

If a curve C runs from the initial point (x_0, y_0) to the terminal point (x_1, y_1), and if the line integral $\int_C \mathbf{F} \cdot d\mathbf{r}$ is independent of the path joining these points, then we write

$$\int_C \mathbf{F} \cdot d\mathbf{r} = \int_{(x_0, y_0)}^{(x_1, y_1)} \mathbf{F} \cdot d\mathbf{r}$$

or, equivalently,

$$\int_C f(x, y)\, dx + g(x, y)\, dy = \int_{(x_0, y_0)}^{(x_1, y_1)} f(x, y)\, dx + g(x, y)\, dy$$

Example 4   Evaluate $\displaystyle\int_{(1,4)}^{(3,1)} 2xy^3\, dx + (1 + 3x^2y^2)\, dy$.

Solution.   From Example 3, $\mathbf{F}(x, y) = 2xy^3\mathbf{i} + (1 + 3x^2y^2)\mathbf{j}$ is the gradient of $\phi(x, y) = y + x^2y^3 + K$. Thus, from Theorem 18.2.1,

$$\int_{(1,4)}^{(3,1)} 2xy^3\, dx + (1 + 3x^2y^2)\, dy = \int_{(1,4)}^{(3,1)} \mathbf{F} \cdot d\mathbf{r} = \phi(3, 1) - \phi(1, 4)$$

$$= (10 + K) - (68 + K) = -58 \qquad \blacktriangleleft$$

REMARK.   Note that the constant K drops out. In future integration problems we shall omit K from the computations.

Example 5   A particle moves over the semicircle $C : \mathbf{r}(t) = \cos t\mathbf{i} + \sin t\mathbf{j}$ $(0 \leq t \leq \pi)$ while subject to the force $\mathbf{F}(x, y) = e^y\mathbf{i} + xe^y\mathbf{j}$. Find the work done.

Solution.   The work done is

$$W = \int_C \mathbf{F} \cdot d\mathbf{r} = \int_C e^y\, dx + xe^y\, dy \tag{8}$$

If we try to evaluate this integral directly, we must deal with the complicated integral

$$W = \int_C \mathbf{F} \cdot d\mathbf{r} = \int_0^\pi \left[\mathbf{F}(x(t), y(t)) \cdot \frac{d\mathbf{r}}{dt} \right] dt$$

$$= \int_0^\pi [e^{\sin t}\mathbf{i} + (\cos t)e^{\sin t}\mathbf{j}] \cdot [-\sin t\mathbf{i} + \cos t\mathbf{j}]\, dt$$

$$= \int_0^\pi [(-\sin t)e^{\sin t} + (\cos^2 t)e^{\sin t}]\, dt$$

To avoid this difficulty, we shall try an alternative approach: The force $\mathbf{F}$ is the gradient of some function ϕ on the entire xy-plane since

$$\frac{\partial}{\partial y}(e^y) = e^y = \frac{\partial}{\partial x}(xe^y)$$

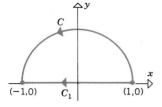

Figure 18.2.5

Thus, the integral in (8) is independent of the path and we are free to replace C by a straight-line segment C_1 joining the endpoint $(1, 0)$ to the endpoint $(-1, 0)$ (Figure 18.2.5). This will simplify the computations.

The straight-line segment C_1 can be represented parametrically by

$$x = 1 - 2t, \quad y = 0 \qquad (0 \le t \le 1)$$

(Verify.) Thus,

$$\mathbf{r}(t) = (1 - 2t)\mathbf{i} + 0\mathbf{j} = (1 - 2t)\mathbf{i} \quad (0 \le t \le 1) \quad \text{and} \quad \frac{d\mathbf{r}}{dt} = -2\mathbf{i}$$

Therefore, the work done is

$$
\begin{aligned}
W = \int_{C_1} \mathbf{F} \cdot d\mathbf{r} &= \int_{C_1} \left[(e^y \mathbf{i} + x e^y \mathbf{j}) \cdot \frac{d\mathbf{r}}{dt} \right] dt \\
&= \int_0^1 [e^0 \mathbf{i} + (1 - 2t)e^0 \mathbf{j}] \cdot (-2\mathbf{i}) \, dt \\
&= \int_0^1 [\mathbf{i} + (1 - 2t)\mathbf{j}] \cdot (-2\mathbf{i}) \, dt \\
&= \int_0^1 -2 \, dt = -2 \quad \text{(units of work)}
\end{aligned}
$$

Second Solution. Since the line segment C_1 has initial point $(1, 0)$ and terminal point $(-1, 0)$, the line segment $-C_1$ has initial point $(-1, 0)$ and terminal point $(1, 0)$. Thus, $-C_1$ can be represented parametrically by

$$x = t, \quad y = 0 \qquad (-1 \le t \le 1)$$

It follows that

$$\mathbf{r}(t) = t\mathbf{i} + 0\mathbf{j} = t\mathbf{i} \quad (-1 \le t \le 1) \quad \text{and} \quad \frac{d\mathbf{r}}{dt} = \mathbf{i}$$

From (4) of Section 18.1,

$$
\begin{aligned}
W = \int_{C_1} \mathbf{F} \cdot d\mathbf{r} &= -\int_{-C_1} \mathbf{F} \cdot d\mathbf{r} = -\int_{-C_1} \left[(e^y \mathbf{i} + x e^y \mathbf{j}) \cdot \frac{d\mathbf{r}}{dt} \right] dt \\
&= -\int_{-1}^1 [(e^0 \mathbf{i} + t e^0 \mathbf{j}) \cdot \mathbf{i}] \, dt = -\int_{-1}^1 [(\mathbf{i} + t\mathbf{j}) \cdot \mathbf{i}] \, dt = -\int_{-1}^1 dt = -2
\end{aligned}
$$

Third Solution. Since $\mathbf{F}(x, y)$ is the gradient of some function ϕ, we can proceed, as we did in Example 4, by finding ϕ and applying (1). Since $\nabla \phi = \mathbf{F}(x, y) = e^y \mathbf{i} + x e^y \mathbf{j}$,

$$\frac{\partial \phi}{\partial x} = e^y \quad \text{and} \quad \frac{\partial \phi}{\partial y} = x e^y \tag{9}$$

Thus,

$$\phi = \int e^y \, dx = xe^y + k(y) \tag{10}$$

and consequently

$$\frac{\partial \phi}{\partial y} = xe^y + k'(y)$$

Equating this with the expression for $\partial \phi / \partial y$ in (9), we obtain $k'(y) = 0$ or $k(y) = K$. Thus, from (10),

$$\phi = xe^y + K$$

Since the curve C begins at $(1,0)$ and ends at $(-1,0)$, it follows that

$$W = \int_C \mathbf{F} \cdot d\mathbf{r} = \int_{(1,0)}^{(-1,0)} e^y \, dx + xe^y \, dy = \phi(-1,0) - \phi(1,0)$$

$$= (-1)e^0 - (1)e^0 = -2 \quad \blacktriangleleft$$

Example 6   Suppose that a particle moves around the circle C given by

$$\mathbf{r}(t) = \cos t\mathbf{i} + \sin t\mathbf{j} \quad (0 \le t \le 2\pi)$$

while it is subject to the force $\mathbf{F}(x, y) = e^y\mathbf{i} + xe^y\mathbf{j}$. Find the work done.

Solution.   In Example 5 we showed that $\mathbf{F} \cdot d\mathbf{r} = e^y \, dx + xe^y \, dy$ is conservative. Since C is a closed curve, it follows from Theorem 18.2.2 that

$$W = \int_C \mathbf{F} \cdot d\mathbf{r} = \int_C e^y \, dx + xe^y \, dy = 0 \quad \blacktriangleleft$$

▶ Exercise Set 18.2

In Exercises 1–6, determine whether **F** is conservative. If it is, find a potential function for it.

1. $\mathbf{F}(x, y) = x\mathbf{i} + y\mathbf{j}$.

2. $\mathbf{F}(x, y) = 3y^2\mathbf{i} + 6xy\mathbf{j}$.

3. $\mathbf{F}(x, y) = x^2y\mathbf{i} + 5xy^2\mathbf{j}$.

4. $\mathbf{F}(x, y) = e^x \cos y\mathbf{i} - e^x \sin y\mathbf{j}$.

5. $\mathbf{F}(x, y) = (\cos y + y \cos x)\mathbf{i} + (\sin x - x \sin y)\mathbf{j}$.

6. $\mathbf{F}(x, y) = x \ln y\mathbf{i} + y \ln x\mathbf{j}$.

7. Show that $\displaystyle\int_{(-1, 2)}^{(1, 3)} y^2 \, dx + 2xy \, dy$ is independent of the path and evaluate the integral by
  (a)   using Theorem 18.2.1
  (b)   integrating along the line segment from $(-1, 2)$ to $(1, 3)$.

8. Show that $\displaystyle\int_{(0,\,1)}^{(\pi,\,-1)} y \sin x \, dx - \cos x \, dy$ is indepen-

dent of the path, and evaluate the integral by
(a) using Theorem 18.2.1
(b) integrating along the line segment from $(0, 1)$ to $(\pi, -1)$.

In Exercises 9–14, show that the integral is independent of the path, and find its value by any method.

9. $\displaystyle\int_{(1,\,2)}^{(4,\,0)} 3y \, dx + 3x \, dy.$

10. $\displaystyle\int_{(0,\,0)}^{(1,\,\pi/2)} e^x \sin y \, dx + e^x \cos y \, dy.$

11. $\displaystyle\int_{(0,\,0)}^{(3,\,2)} 2xe^y \, dx + x^2 e^y \, dy.$

12. $\displaystyle\int_{(-1,\,2)}^{(0,\,1)} (3x - y + 1) \, dx - (x + 4y + 2) \, dy.$

13. $\displaystyle\int_{(2,\,-2)}^{(-1,\,0)} 2xy^3 \, dx + 3y^2 x^2 \, dy.$

14. $\displaystyle\int_{(1,\,1)}^{(3,\,3)} \left(e^x \ln y - \frac{e^y}{x} \right) dx + \left(\frac{e^x}{y} - e^y \ln x \right) dy,$
where x and y are positive.

In Exercises 15–18, find the work done by the conservative force $\mathbf{F}$ as it acts on a particle moving from
P to Q.

15. $\mathbf{F}(x, y) = xy^2\mathbf{i} + x^2y\mathbf{j}; \ P(1, 1), \ Q(0, 0).$

16. $\mathbf{F}(x, y) = ye^{xy}\mathbf{i} + xe^{xy}\mathbf{j}; \ P(-1, 1), \ Q(2, 0).$

17. $\mathbf{F}(x, y) = \dfrac{y}{x^2 + y^2}\mathbf{i} - \dfrac{x}{x^2 + y^2}\mathbf{j}; \ P(0, 2), \ Q(3, 3).$
(Assume that $y > 0$.)

18. $\mathbf{F}(x, y) = e^{-y} \cos x\mathbf{i} - e^{-y} \sin x\mathbf{j}; \ P(\pi/2, 1),$
$Q(-\pi/2, 0).$

19. Let $\mathbf{F}(x, y) = (e^y + ye^x)\mathbf{i} + (xe^y + e^x)\mathbf{j}.$ Find the
work done by the force $\mathbf{F}$ if it acts on a particle that
moves
(a) from $(a, 0)$ to $(-a, 0)$ along the x-axis, where
$a > 0$

(b) from $(a, 0)$ to $(-a, 0)$ along the upper half of the
circle $x^2 + y^2 = a^2$
(c) from $(a, 0)$ to $(-a, 0)$ along the upper half of the
ellipse

$$\frac{x^2}{a^2} + \frac{y^2}{b^2} = 1$$

(d) once around the circle $x^2 + y^2 = a^2$.

20. Evaluate

$$\int_C (\sin y \sinh x + \cos y \cosh x) \, dx$$
$$+ (\cos y \cosh x - \sin y \sinh x) \, dy$$

where C is the line segment

$$x = t, \quad y = \tfrac{1}{2}\pi(t - 1) \qquad (1 \le t \le 2)$$

21. Let

$$\mathbf{F}(x, y) = \frac{k}{\|\mathbf{r}\|^3} \mathbf{r}$$

where $\mathbf{r} = x\mathbf{i} + y\mathbf{j}$. Show that $\mathbf{F}$ is conservative on
any region that does not contain the origin.

22. Find a function h for which

$$\mathbf{F}(x, y) = h(x)[x \sin y + y \cos y]\mathbf{i}$$
$$+ h(x)[x \cos y - y \sin y]\mathbf{j}$$

is conservative.

23. (a) Prove: $\mathbf{F}(x, y) = f(x, y)\mathbf{i} + g(x, y)\mathbf{j}$ is conservative in a region if and only if there is a function
$\phi(x, y)$ such that

$$d\phi = f(x, y) \, dx + g(x, y) \, dy$$

on the region.

(b) Prove: If $\mathbf{F}(x, y)$ is conservative on a simply
connected open region containing (x_0, y_0) and
(x_1, y_1), then

$$\int_{(x_0, y_0)}^{(x_1, y_1)} d\phi = \phi(x_1, y_1) - \phi(x_0, y_0)$$

where ϕ is the function in part (a).

24. Prove Theorem 18.2.1 in the case where C is a
piecewise smooth curve composed of smooth curves
$C_1, C_2, \ldots, C_n$.

■ 18.3 GREEN'S THEOREM

In this section we shall discuss a remarkable and beautiful theorem that expresses the double integral over a plane region in terms of a line integral over its boundary.

18.3.1 THEOREM *(Green's* Theorem).* *Let R be a simply connected plane region whose boundary is a simple, closed piecewise smooth curve C oriented counterclockwise. If f(x, y) and g(x, y) have continuous first partial derivatives on some open set containing R, then*

$$\int_C f(x, y)\, dx + g(x, y)\, dy = \iint_R \left(\frac{\partial g}{\partial x} - \frac{\partial f}{\partial y} \right) dA \tag{1}$$

Proof. For simplicity, we shall prove the theorem only for regions that are simultaneously type I and type II (see Section 17.2). Such a region is shown in Figure 18.3.1. The crux of the proof is to show that

$$\int_C f(x, y)\, dx = - \iint_R \frac{\partial f}{\partial y}\, dA \tag{2a}$$

and

$$\int_C g(x, y)\, dy = \iint_R \frac{\partial g}{\partial x}\, dA \tag{2b}$$

Formula (1) will then follow by adding (2a) and (2b).

---

*GEORGE GREEN (1793–1841). English mathematician and physicist. Green left school at an early age to work in his father's bakery and consequently had little early formal education. When his father opened a mill, the boy used the top room as a study in which he taught himself physics and mathematics from library books. In 1828 Green published his most important work, *An Essay on the Application of Mathematical Analysis to the Theories of Electricity and Magnetism.* Although Green's Theorem appeared in that paper, the result went virtually unnoticed because of the small pressrun and local distribution. Following the death of his father in 1829, Green was urged by friends to seek a college education. In 1833, after four years of self-study to close the gaps in his elementary education, Green was admitted to Caius College, Cambridge. He graduated four years later, but with a disappointing performance on his final examinations—possibly because he was more interested in his own research. After a succession of works on light and sound, he was named to be Perse Fellow at Caius College. Two years later he died. In 1845, four years after his death, his paper of 1828 was published and the theories developed therein by this obscure, self-taught baker's son helped pave the way to the modern theories of electricity and magnetism.

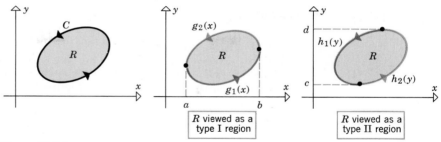

Figure 18.3.1

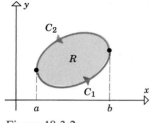

Figure 18.3.2

To prove (2a), view R as a type I region and let C_1 and C_2 be the lower and upper boundary curves, oriented as in Figure 18.3.2. Then

$$\int_C f(x, y)\, dx = \int_{C_1} f(x, y)\, dx + \int_{C_2} f(x, y)\, dx$$

or, equivalently,

$$\int_C f(x, y)\, dx = \int_{C_1} f(x, y)\, dx - \int_{-C_2} f(x, y)\, dx \qquad (3)$$

(This step will help simplify our calculations since C_1 and $-C_2$ are both oriented left to right.) With $x = t$ as the parameter, C_1 and $-C_2$ are represented by

$$C_1: x = t,\ y = g_1(t) \quad (a \le t \le b)$$
$$-C_2: x = t,\ y = g_2(t) \quad (a \le t \le b)$$

Thus, (3) yields

$$\int_C f(x, y)\, dx = \int_a^b f(t, g_1(t)) \left(\frac{dx}{dt}\right) dt - \int_a^b f(t, g_2(t)) \left(\frac{dx}{dt}\right) dt$$

$$= \int_a^b f(t, g_1(t))\, dt - \int_a^b f(t, g_2(t))\, dt$$

$$= -\int_a^b [f(t, g_2(t)) - f(t, g_1(t))]\, dt$$

$$= -\int_a^b \left[f(t, y) \right]_{y=g_1(t)}^{y=g_2(t)} dt = -\int_a^b \left[\int_{g_1(t)}^{g_2(t)} \frac{\partial f}{\partial y}\, dy \right] dt$$

$$= -\int_a^b \int_{g_1(x)}^{g_2(x)} \frac{\partial f}{\partial y}\, dy\, dx = -\iint_R \frac{\partial f}{\partial y}\, dA$$

Since $x = t$

The proof of (2b) is obtained similarly by treating R as a type II region. The details are omitted. ∎

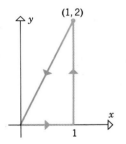

(1, 2)

Figure 18.3.3

Example 1 Use Green's Theorem to evaluate

$$\int_C x^2 y \, dx + x \, dy$$

over the triangular path shown in Figure 18.3.3.

Solution. Since $f(x, y) = x^2 y$ and $g(x, y) = x$, it follows from (1) that

$$\int_C x^2 y \, dx + x \, dy = \iint_R \left[\frac{\partial}{\partial x} (x) - \frac{\partial}{\partial y} (x^2 y) \right] dA$$

$$= \int_0^1 \int_0^{2x} (1 - x^2) \, dy \, dx$$

$$= \int_0^1 (2x - 2x^3) \, dx$$

$$= \left[x^2 - \frac{x^4}{2} \right]_0^1 = \frac{1}{2}$$

This agrees with the result obtained in Example 3 of Section 18.1, where we evaluated the line integral directly. Note how much simpler this solution is. ◀

Example 2 Suppose that a particle travels once around the unit circle in the counterclockwise direction and that at the point (x, y) it is subject to the force $\mathbf{F}(x, y) = (e^x - y^3)\mathbf{i} + (\cos y + x^3)\mathbf{j}$. Use Green's Theorem to find the work done by the force.

Solution.

$$W = \int_C (e^x - y^3) \, dx + (\cos y + x^3) \, dy$$

$$= \iint_R \left[\frac{\partial}{\partial x} (\cos y + x^3) - \frac{\partial}{\partial y} (e^x - y^3) \right] dA$$

$$= \iint_R (3x^2 + 3y^2) \, dA = 3 \iint_R (x^2 + y^2) \, dA$$

$$= 3 \int_0^{2\pi} \int_0^1 (r^2) r \, dr \, d\theta = \frac{3}{4} \int_0^{2\pi} d\theta = \frac{3\pi}{2} \quad ◀$$

We converted to polar coordinates.

☐ **COMPUTING AREAS USING GREEN'S THEOREM**

Green's Theorem leads to some new formulas for area. Letting $f(x, y) = 0$ and $g(x, y) = x$ in (1) yields

$$\int_C x \, dy = \iint_R dA = \text{area of } R \tag{4a}$$

Letting $f(x, y) = -y$ and $g(x, y) = 0$ in (1) yields

$$\int_C -y \, dx = \iint_R dA = \text{area of } R \tag{4b}$$

Adding (4a) and (4b) and dividing the result by 2 yields

$$\frac{1}{2} \int_C -y \, dx + x \, dy = \text{area of } R \tag{4c}$$

Formulas (4a), (4b), and (4c) each express the area of R in terms of a line integral taken counterclockwise around the boundary of R. These formulas apply to those regions that satisfy the hypotheses of Green's Theorem. Although (4c) is more complicated than (4a) or (4b), it frequently results in an easier integral to evaluate.

Example 3 Use Formula (4c) to find the area enclosed by the ellipse $x^2/a^2 + y^2/b^2 = 1$.

Solution. The ellipse, with counterclockwise orientation, can be represented parametrically by

$$x = a \cos t, \quad y = b \sin t \qquad (0 \le t \le 2\pi)$$

If we denote this curve by C, then the area enclosed by the ellipse is

$$A = \frac{1}{2} \int_C -y \, dx + x \, dy$$

$$= \frac{1}{2} \int_0^{2\pi} [(-b \sin t)(-a \sin t) + (a \cos t)(b \cos t)] \, dt$$

$$= \frac{1}{2} ab \int_0^{2\pi} (\sin^2 t + \cos^2 t) \, dt$$

$$= \frac{1}{2} ab \int_0^{2\pi} dt = \pi ab$$

This result can also be obtained from Formula (4a) or (4b). ◀

REMARK. Green's Theorem provides another viewpoint about some earlier results. If $\partial f/\partial x$ and $\partial f/\partial y$ are continuous and if $\partial f/\partial y = \partial g/\partial x$ throughout a simply connected open region R with boundary C, then it follows from Green's Theorem that

$$\int_C f\,dx + g\,dy = \iint_R \left(\frac{\partial g}{\partial x} - \frac{\partial f}{\partial y} \right) dA = 0$$

However, this is to be expected since the condition $\partial f/\partial y = \partial g/\partial x$ implies that $\mathbf{F}(x, y) = f(x, y)\mathbf{i} + g(x, y)\mathbf{j}$ is conservative in R (Theorem 18.2.3) and we already know that a line integral around a closed curve has value zero in this case (Theorem 18.2.2).

▶ Exercise Set 18.3

In Exercises 1 and 2, evaluate the line integral using Green's Theorem and check the answer by evaluating directly.

1. $\int_C y^2\,dx + x^2\,dy$, where C is the square with vertices $(0, 0)$, $(1, 0)$, $(1, 1)$, and $(0, 1)$ oriented counterclockwise.

2. $\int_C y\,dx + x\,dy$, where C is the unit circle oriented counterclockwise.

In Exercises 3–13, use Green's Theorem to evaluate the integral. In each exercise, assume that the curve C is oriented counterclockwise.

3. $\int_C 3xy\,dx + 2xy\,dy$, where C is the rectangle bounded by $x = -2$, $x = 4$, $y = 1$, and $y = 2$.

4. $\int_C (x^2 - y^2)\,dx + x\,dy$, where C is the circle $x^2 + y^2 = 9$.

5. $\int_C x\cos y\,dx - y\sin x\,dy$, where C is the square with vertices $(0, 0)$, $(0, \pi/2)$, $(\pi/2, \pi/2)$, and $(\pi/2, 0)$.

6. $\int_C y\tan^2 x\,dx + \tan x\,dy$, where C is the circle $x^2 + (y + 1)^2 = 1$.

7. $\int_C (x^2 - y)\,dx + x\,dy$, where C is the circle $x^2 + y^2 = 4$.

8. $\int_C (e^x + y^2)\,dx + (e^y + x^2)\,dy$, where C is the boundary of the region between $y = x^2$ and $y = x$.

9. $\int_C \ln(1 + y)\,dx - \frac{xy}{1 + y}\,dy$, where C is the triangle with vertices $(0, 0)$, $(2, 0)$, and $(0, 4)$.

10. $\int_C x^2 y\,dx - y^2 x\,dy$, where C is the boundary of the region in the first quadrant, enclosed between the coordinate axes and the circle $x^2 + y^2 = 16$.

11. $\int_C \tan^{-1} y\,dx - \frac{y^2 x}{1 + y^2}\,dy$, where C is the square with vertices $(0, 0)$, $(0, 1)$, $(1, 1)$, and $(1, 0)$.

12. $\int_C \cos x\sin y\,dx + \sin x\cos y\,dy$, where C is the triangle with vertices $(0, 0)$, $(3, 3)$, and $(0, 3)$.

13. $\int_C x^2 y\,dx + (y + xy^2)\,dy$, where C is the boundary of the region enclosed by $y = x^2$ and $x = y^2$.

14. Let C be the boundary of the region enclosed between $y = x^2$ and $y = 2x$. Assuming that C is oriented counterclockwise, evaluate the following integrals by Green's Theorem:

(a) $\int_C (6xy - y^2)\,dx$ (b) $\int_C (6xy - y^2)\,dy$.

15. Find the area of the ellipse in Example 3 using
(a)   Formula (4a)   (b)   Formula (4b).

16. Use a line integral to find the area of the region enclosed by

$$x = a \cos^3 t, \quad y = a \sin^3 t \quad (0 \le t \le 2\pi)$$

17. Use a line integral to find the area of the triangle with vertices $(0, 0)$, $(a, 0)$, and $(0, b)$, where a and b are positive.

18. Use a line integral to find the area of the region in the first quadrant enclosed by $y = x$, $y = 1/x$, and $y = x/9$.

19. A particle, starting at $(5, 0)$, traverses the upper semicircle $x^2 + y^2 = 25$ and returns to its starting point along the x-axis. Use Green's Theorem to find the work done on the particle by a force
$\mathbf{F}(x, y) = xy\mathbf{i} + (\frac{1}{2}x^2 + yx)\mathbf{j}$.

20. A particle moves counterclockwise one time around the closed curve formed by $y = 0, x = 2$, and $y = x^3/4$. Use Green's Theorem to find the work done by a force $\mathbf{F}(x, y) = \sqrt{y}\,\mathbf{i} + \sqrt{x}\,\mathbf{j}$.

21. (a) Let R be a plane region with area A whose boundary is a piecewise smooth simple closed curve C. Use Green's Theorem to prove that the centroid $(\bar{x}, \bar{y})$ of R is given by

$$\bar{x} = \frac{1}{2A} \int_C x^2\, dy, \qquad \bar{y} = -\frac{1}{2A} \int_C y^2\, dx$$

(b) Use the result in part (a) to find the centroid of the region enclosed between the x-axis and the upper half of the circle $x^2 + y^2 = a^2$.

22. Evaluate $\displaystyle\int_C y\, dx - x\, dy$, where C is the cardioid

$$r = a(1 + \cos\theta), \ 0 \le \theta \le 2\pi$$

23. (a) Let C be the line segment from a point (a, b) to a point (c, d). Show that

$$\int_C x\, dy - y\, dx = ad - bc$$

(b) Use the result in part (a) to show that the area A of a triangle with successive vertices (x_1, y_1), (x_2, y_2), and (x_3, y_3) going counterclockwise is

$$A = \frac{1}{2}[(x_1 y_2 - x_2 y_1)$$
$$+ (x_2 y_3 - x_3 y_2) + (x_3 y_1 - x_1 y_3)]$$

(c) Find a formula for the area of a polygon with successive vertices (x_1, y_1), (x_2, y_2), . . . , (x_n, y_n) going counterclockwise.

(d) Use the result in part (c) to find the area of a quadrilateral with vertices $(0, 0)$, $(3, 4)$, $(-2, 2)$, $(-1, 0)$.

24. Find a simple closed curve C that maximizes the value of

$$\int_C \frac{1}{3} y^3\, dx + \left(x - \frac{1}{3}x^3\right) dy$$

25. (a) Let R be the region enclosed by the circle $C : x = \cos t, \ y = \sin t, \ 0 \le t \le 2\pi$. By evaluating directly, show that

$$\int_C -\frac{y}{x^2 + y^2}\, dx + \frac{x}{x^2 + y^2}\, dy = 2\pi$$

(b) Let f and g be the functions

$$f(x, y) = -\frac{y}{x^2 + y^2}, \quad g(x, y) = \frac{x}{x^2 + y^2}$$

Show that $\partial g/\partial x = \partial f/\partial y$.

(c) Find the error in the following argument: By Green's Theorem

$$\int_C -\frac{y}{x^2 + y^2}\, dx + \frac{x}{x^2 + y^2}\, dy$$
$$= \int_C f(x, y)\, dx + g(x, y)\, dy$$
$$= \iint_R \left(\frac{\partial g}{\partial x} - \frac{\partial f}{\partial y}\right) dA = 0$$

[which contradicts the result in part (a)].

■ 18.4 INTRODUCTION TO SURFACE INTEGRALS

In previous sections we considered four kinds of integrals—integrals over intervals, double integrals over two-dimensional regions, triple integrals over three-dimensional solids, and line integrals over curves in two- or three-dimensional space. In this section we shall discuss integrals over surfaces in three-dimensional space. Such integrals occur in problems of fluid and heat flow, electricity, magnetism, mass, and center of gravity.

☐ MASS OF A
CURVED LAMINA

The thickness of a curved lamina is negligible.

Figure 18.4.1

Figure 18.4.2

We shall motivate the definition of a surface integral by considering a problem about mass. In Section 17.6 we defined a lamina to be an idealized flat object that is sufficiently thin to be viewed as a two-dimensional solid. If we imagine a flat lamina to be bent into some curved shape, the result is a **curved lamina** (Figure 18.4.1). The density of a curved lamina at a point (x, y, z) can be specified by a function $\delta(x, y, z)$, called the **density function** for the lamina. Informally, the density function can be visualized as follows: Consider a small section of the lamina containing the point (x, y, z), and let ΔM and ΔS be the mass and surface area of this section (Figure 18.4.2). If the ratio $\Delta M / \Delta S$ tends toward a limiting value when the small section of lamina is allowed to shrink down to the point (x, y, z), then this limit is the density of the lamina at (x, y, z). Symbolically,

$$\delta(x, y, z) = \lim \frac{\Delta M}{\Delta S}$$

From this equation we obtain the approximation

$$\Delta M \approx \delta(x, y, z) \, \Delta S \tag{1}$$

which relates the mass and surface area of a small section of lamina containing the point (x, y, z).

We shall now use (1) to obtain a formula for the mass of a curved lamina.

18.4.1 DEFINITION.   If a curved lamina with density $\delta(x, y, z)$ has the equation $z = f(x, y)$, and if the projection of this lamina on the xy-plane is the region R (Figure 18.4.3), then the mass M of the lamina is defined to be

$$M = \iint_R \delta[x, y, f(x, y)]\sqrt{f_x(x, y)^2 + f_y(x, y)^2 + 1} \, dA \tag{2a}$$

or in an alternative notation

$$M = \iint_R \delta[x, y, f(x, y)] \sqrt{\left(\frac{\partial z}{\partial x}\right)^2 + \left(\frac{\partial z}{\partial y}\right)^2 + 1} \, dA \tag{2b}$$

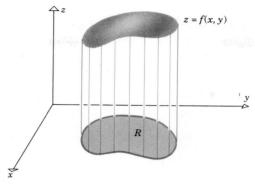

Figure 18.4.3

We shall motivate this formula in four steps that closely parallel the four steps in the motivation of the formula for surface area (see 17.4.3). (The reader should review the discussion following that definition before continuing.)

Step 1. As shown in Figure 18.4.4, subdivide a rectangle containing R into subrectangles and discard all those subrectangles that contain any points outside of R. This leaves only rectangles $R_1, R_2, \ldots, R_n$ that are subsets of R. We shall denote the area of R_k by ΔA_k.

Step 2. As illustrated in Figure 18.4.4, project each subrectangle onto the lamina, let the surface areas of the resulting sections of lamina be

$$\Delta S_1, \Delta S_2, \ldots, \Delta S_n$$

and let $(x_k^*, y_k^*, z_k^*) = [x_k^*, y_k^*, f(x_k^*, y_k^*)]$ be an arbitrary point in the kth section.

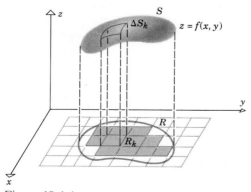

Figure 18.4.4

Step 3. As in Step 3 of the surface area discussion, the surface area ΔS_k can be approximated by

$$\Delta S_k \approx \sqrt{f_x(x_k^*, y_k^*)^2 + f_y(x_k^*, y_k^*)^2 + 1}\ \Delta A_k \qquad (3)$$

It follows from (1) that the mass ΔM_k of the kth section of lamina is given by the approximation

$$\Delta M_k \approx \delta(x_k^*, y_k^*, z_k^*)\ \Delta S_k = \delta[x_k^*, y_k^*, f(x_k^*, y_k^*)]\ \Delta S_k \quad (4)$$

Substituting (3) into (4) and adding the masses of the individual sections yields the following approximation to the total mass M of the lamina

$$M \approx \sum_{k=1}^{n} \delta[x_k^*, y_k^*, f(x_k^*, y_k^*)]\sqrt{f_x(x_k^*, y_k^*)^2 + f_y(x_k^*, y_k^*)^2 + 1}\ \Delta A_k$$

Step 4. If we repeat the subdivision process using more and more rectangles with dimensions that decrease to zero as $n \to +\infty$, then it is reasonable to expect that the errors in the approximations will diminish to zero and the exact mass of the lamina will be

$$M = \lim_{n \to +\infty} \sum_{k=1}^{n} \delta[x_k^*, y_k^*, f(x_k^*, y_k^*)]\sqrt{f_x(x_k^*, y_k^*)^2 + f_y(x_k^*, y_k^*)^2 + 1}\ \Delta A_k$$

$$= \iint\limits_{R} \delta[x, y, f(x, y)]\sqrt{f_x(x, y)^2 + f_y(x, y)^2 + 1}\ dA$$

which establishes (2a).

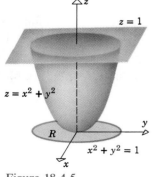

Figure 18.4.5

Example 1   A curved lamina is the portion of the paraboloid $z = x^2 + y^2$ below the plane $z = 1$ (Figure 18.4.5) and has constant density $\delta(x, y, z) = \delta_0$. Find the mass of the lamina.

Solution.   Since $z = f(x, y) = x^2 + y^2$, it follows that

$$\frac{\partial z}{\partial x} = 2x \quad \text{and} \quad \frac{\partial z}{\partial y} = 2y$$

Substituting these expressions and $\delta(x, y, z) = \delta(x, y, f(x, y)) = \delta_0$ into (2b) yields

$$M = \iint\limits_{R} \delta_0 \sqrt{(2x)^2 + (2y)^2 + 1}\ dA = \delta_0 \iint\limits_{R} \sqrt{4x^2 + 4y^2 + 1}\ dA \qquad (5)$$

where R is the circular region enclosed by $x^2 + y^2 = 1$. To evaluate (5) we use polar coordinates:

$$M = \delta_0 \int_0^{2\pi} \int_0^1 \sqrt{4r^2 + 1} \; r \, dr \, d\theta = \frac{\delta_0}{12} \int_0^{2\pi} (4r^2 + 1)^{3/2} \Big]_{r=0}^1 d\theta$$

$$= \frac{\delta_0}{12} \int_0^{2\pi} (5^{3/2} - 1) \, d\theta = \frac{\pi\delta_0}{6} (5\sqrt{5} - 1) \qquad \blacktriangleleft$$

☐ **SURFACE INTEGRALS**

We shall now generalize the procedure used to derive Formulas (2a) and (2b) to develop the concept of a *surface integral*.

Let σ be a surface with finite surface area and $g(x, y, z)$ a continuous function defined on σ. We shall *not* assume in this discussion that σ can be represented by an equation of the form $z = f(x, y)$. Thus, the surface might be a sphere $x^2 + y^2 + z^2 = a^2$ or perhaps a surface represented by an equation of the form $y = f(x, z)$ or $x = f(y, z)$.

Subdivide σ into n parts $\sigma_1, \sigma_2, \ldots, \sigma_n$ with surface areas $\Delta S_1, \Delta S_2, \ldots, \Delta S_n$ and form the sum

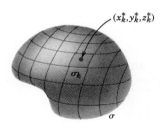

$$\sum_{k=1}^{n} g(x_k^*, y_k^*, z_k^*) \, \Delta S_k \tag{6}$$

where (x_k^*, y_k^*, z_k^*) is an arbitrary point in the kth part σ_k (Figure 18.4.6).

Now, repeat the subdivision process, dividing σ into more and more parts so that each part shrinks to a point as $n \to +\infty$. If (6) tends toward a limit as $n \to +\infty$, and if this limit does not depend on the way the subdivisions are made or how the points (x_k^*, y_k^*, z_k^*) are chosen, then this limit is called the *surface integral* of $g(x, y, z)$ over σ, and is denoted by

Figure 18.4.6

$$\iint_{\sigma} g(x, y, z) \, dS = \lim_{n \to +\infty} \sum_{k=1}^{n} g(x_k^*, y_k^*, z_k^*) \, \Delta S_k \tag{7}$$

Example 2 In the special case where $g(x, y, z) = 1$, the integral in (7) is the surface area of σ, that is,

$$S = \text{surface area of } \sigma = \iint_{\sigma} dS \tag{8}$$

To see this, observe that for every subdivision of σ into parts with surface areas $\Delta S_1, \Delta S_2, \ldots, \Delta S_n$, these surface areas add up to the total surface area

S, that is,

$$S = \sum_{k=1}^{n} \Delta S_k$$

Since the right side has the value S for all n, it follows that

$$S = \lim_{n \to +\infty} \sum_{k=1}^{n} \Delta S_k$$

Formula (8) now follows from this and (7). ◄

Example 3   If σ is a curved lamina with density $\delta(x, y, z)$ and mass M, then

$$M = \iint_{\sigma} \delta(x, y, z)\, dS \qquad (9)$$

To see this, let σ be subdivided into n parts with surface areas $\Delta S_1, \Delta S_2, \ldots,$ ΔS_n and for $k = 1, 2, \ldots, n$ let (x_k^*, y_k^*, z_k^*) be an arbitrary point in the kth part. It follows from (1) that

$$\delta(x_k^*, y_k^*, z_k^*)\, \Delta S_k$$

is approximately the mass of the kth part, and so the entire mass M can be approximated by

$$M \approx \sum_{k=1}^{n} \delta(x_k^*, y_k^*, z_k^*)\, \Delta S_k$$

If we assume that the error in this approximation tends to zero as the number of subdivisions tends to infinity and the parts shrink to points, then the exact mass is

$$M = \lim_{n \to +\infty} \sum_{k=1}^{n} \delta(x_k^*, y_k^*, z_k^*)\, \Delta S_k = \iint_{\sigma} \delta(x, y, z)\, dS \qquad ◄$$

If σ is a curved lamina with density $\delta(x, y, z)$ and equation $z = f(x, y)$, then (2b) expresses the mass of the lamina as a double integral and (9) as a surface integral. Equating (2b) and (9) yields the following relationship between the surface integral and the double integral:

$$\iint_{\sigma} \delta(x, y, z)\, dS = \iint_{R} \delta[x, y, f(x, y)] \sqrt{\left(\frac{\partial z}{\partial x}\right)^2 + \left(\frac{\partial z}{\partial y}\right)^2 + 1}\; dA$$

This is a special case of the following general result that can be used to evaluate many surface integrals.

18.4.2 THEOREM.

(a) *Let σ be a surface with equation $z = f(x, y)$ and let R be its projection on the xy-plane. If f has continuous first partial derivatives on R and $g(x, y, z)$ is continuous on σ, then*

$$\iint_\sigma g(x, y, z)\, dS = \iint_R g[x, y, f(x, y)] \sqrt{\left(\frac{\partial z}{\partial x}\right)^2 + \left(\frac{\partial z}{\partial y}\right)^2 + 1}\, dA \qquad (10)$$

(b) *Let σ be a surface with equation $y = f(x, z)$ and let R be its projection on the xz-plane. If f has continuous first partial derivatives on R and $g(x, y, z)$ is continuous on σ, then*

$$\iint_\sigma g(x, y, z)\, dS = \iint_R g[x, f(x, z), z] \sqrt{\left(\frac{\partial y}{\partial x}\right)^2 + \left(\frac{\partial y}{\partial z}\right)^2 + 1}\, dA \qquad (11)$$

(c) *Let σ be a surface with equation $x = f(y, z)$ and let R be its projection on the yz-plane. If f has continuous first partial derivatives on R and $g(x, y, z)$ is continuous on σ, then*

$$\iint_\sigma g(x, y, z)\, dS = \iint_R g[f(y, z), y, z] \sqrt{\left(\frac{\partial x}{\partial y}\right)^2 + \left(\frac{\partial x}{\partial z}\right)^2 + 1}\, dA \qquad (12)$$

We omit the proof.

Example 4   Evaluate the surface integral

$$\iint_\sigma xz\, dS$$

where σ is the part of the plane $x + y + z = 1$ that lies in the first octant.

Solution.   The equation of the plane can be written as

$$z = 1 - x - y$$

which is of the form $z = f(x, y)$. Consequently, we can apply Formula (10) with $z = f(x, y) = 1 - x - y$ and $g(x, y, z) = xz$. Thus,

$$\frac{\partial z}{\partial x} = -1 \quad \text{and} \quad \frac{\partial z}{\partial y} = -1$$

so (10) becomes

$$\iint_\sigma xz \, dS = \iint_R x(1 - x - y)\sqrt{(-1)^2 + (-1)^2 + 1} \, dA \tag{13}$$

where R is the projection of σ on the xy-plane (Figure 18.4.7a). Rewriting the double integral in (13) as an iterated integral yields

$$\iint_\sigma xz \, dS = \sqrt{3} \int_0^1 \int_0^{1-x} (x - x^2 - xy) \, dy \, dx$$

$$= \sqrt{3} \int_0^1 xy - x^2y - \frac{xy^2}{2} \Bigg]_{y=0}^{1-x} dx$$

$$= \sqrt{3} \int_0^1 \left(\frac{x}{2} - x^2 + \frac{x^3}{2} \right) dx$$

$$= \sqrt{3} \left[\frac{x^2}{4} - \frac{x^3}{3} + \frac{x^4}{8} \right]_0^1 = \frac{\sqrt{3}}{24}$$

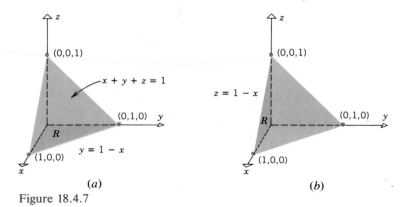

(a)   (b)

Figure 18.4.7

Alternative Solution.   The equation of the plane can be written as

$$y = 1 - x - z$$

which is of the form $y = f(x, z)$. Consequently, we can apply Formula (11) with $y = f(x, z) = 1 - x - z$. Thus,

$$\frac{\partial y}{\partial x} = -1 \quad \text{and} \quad \frac{\partial y}{\partial z} = -1$$

so (11) yields

$$\iint_\sigma xz \, dS = \iint_R xz\sqrt{(-1)^2 + (-1)^2 + 1} \, dA \tag{14}$$

where R is the projection of σ on the xz-plane (Figure 18.4.7b). Rewriting the double integral in (14) as an iterated integral yields

$$\iint_\sigma xz \, dS = \sqrt{3} \int_0^1 \int_0^{1-x} xz \, dz \, dx = \sqrt{3} \int_0^1 \frac{xz^2}{2} \Big]_{z=0}^{1-x} dx$$

$$= \frac{\sqrt{3}}{2} \int_0^1 (x - 2x^2 + x^3) \, dx$$

$$= \frac{\sqrt{3}}{2} \left[\frac{x^2}{2} - \frac{2x^3}{3} + \frac{x^4}{4} \right]_0^1 = \frac{\sqrt{3}}{24} \quad \blacktriangleleft$$

Example 5   Evaluate the surface integral

$$\iint_\sigma y^2 z^2 \, dS$$

where σ is the part of the cone $z = \sqrt{x^2 + y^2}$ between the planes $z = 1$ and $z = 2$ (Figure 18.4.8).

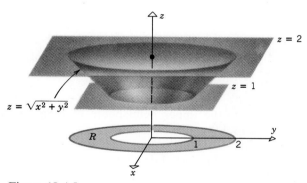

Figure 18.4.8

Solution.   We shall apply Formula (10) with $z = f(x, y) = \sqrt{x^2 + y^2}$ and $g(x, y, z) = y^2 z^2$. Thus,

$$\frac{\partial z}{\partial x} = \frac{x}{\sqrt{x^2 + y^2}} \quad \text{and} \quad \frac{\partial z}{\partial y} = \frac{y}{\sqrt{x^2 + y^2}}$$

so

$$\sqrt{\left(\frac{\partial z}{\partial x}\right)^2 + \left(\frac{\partial z}{\partial y}\right)^2 + 1} = \sqrt{2}$$

(verify), and (10) yields

$$\iint_\sigma y^2 z^2 \, dS = \iint_R y^2 (\sqrt{x^2 + y^2})^2 \sqrt{2} \, dA = \sqrt{2} \iint_R y^2 (x^2 + y^2) \, dA$$

where R is the annulus enclosed between $x^2 + y^2 = 1$ and $x^2 + y^2 = 4$ (Figure 18.4.8). Using polar coordinates to evaluate this double integral over the annulus R yields

$$\iint_\sigma y^2 z^2 \, dS = \sqrt{2} \int_0^{2\pi} \int_1^2 (r \sin \theta)^2 (r^2) r \, dr \, d\theta$$

$$= \sqrt{2} \int_0^{2\pi} \int_1^2 r^5 \sin^2 \theta \, dr \, d\theta$$

$$= \sqrt{2} \int_0^{2\pi} \frac{r^6}{6} \sin^2 \theta \Big]_{r=1}^2 d\theta = \frac{21}{\sqrt{2}} \int_0^{2\pi} \sin^2 \theta \, d\theta$$

$$= \frac{21}{\sqrt{2}} \left[\frac{1}{2} \theta - \frac{1}{4} \sin 2\theta \right]_0^{2\pi} = \frac{21\pi}{\sqrt{2}} \quad \boxed{\text{See Example 1, Section 9.3.}} \quad \blacktriangleleft$$

Sometimes Formulas (10), (11), and (12) cannot be applied because the partial derivatives in these formulas do not exist at every point of the region R. However, it may still be possible to compute surface integrals in such cases by means of a limiting procedure. The next example illustrates this.

Example 6   Evaluate the surface integral

$$\iint_\sigma dS \tag{15}$$

where σ is the upper hemisphere of radius a given by $z = \sqrt{a^2 - x^2 - y^2}$ (Figure 18.4.9).

Solution.   The partial derivatives

$$\frac{\partial z}{\partial x} = \frac{-x}{\sqrt{a^2 - x^2 - y^2}} \quad \text{and} \quad \frac{\partial z}{\partial y} = \frac{-y}{\sqrt{a^2 - x^2 - y^2}}$$

do not exist on the boundary of the region R since $x^2 + y^2 = a^2$ there. In order to overcome this problem we evaluate the double integral over a slightly smaller circular region R_ρ of radius ρ, and then let ρ approach a. The computations are as follows:

$$\iint_\sigma dS = \lim_{\rho \to a} \iint_{R_\rho} \sqrt{\left(\frac{\partial z}{\partial x} \right)^2 + \left(\frac{\partial z}{\partial y} \right)^2 + 1} \, dA$$

$$= \lim_{\rho \to a} \iint_{R_\rho} \sqrt{\frac{x^2}{a^2 - x^2 - y^2} + \frac{y^2}{a^2 - x^2 - y^2} + 1} \, dA$$

$$= \lim_{\rho \to a} \iint_{R_\rho} \frac{a}{\sqrt{a^2 - x^2 - y^2}} \, dA = \lim_{\rho \to a} \int_0^{2\pi} \int_0^\rho \frac{a}{\sqrt{a^2 - r^2}} r \, dr \, d\theta$$

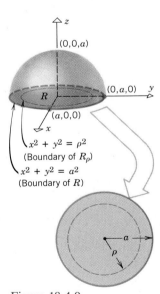

Figure 18.4.9

$$= \lim_{\rho \to a} \int_0^{2\pi} -a\sqrt{a^2 - r^2} \Big]_{r=0}^{\rho} \, d\theta = \lim_{\rho \to a} \int_0^{2\pi} (a^2 - a\sqrt{a^2 - \rho^2}) \, d\theta$$

$$= \lim_{\rho \to a} 2\pi(a^2 - a\sqrt{a^2 - \rho^2}) = 2\pi a^2 \quad \blacktriangleleft$$

Note that (15) is the surface area of the hemisphere σ [see (8)], so we have shown that the surface area of a *sphere* of radius a is $4\pi a^2$.

☐ **PROPERTIES OF SURFACE INTEGRALS**

We conclude this section by noting without proof that the standard properties of integrals also hold for surface integrals. More precisely,

$$\iint_\sigma (f + g) \, dS = \iint_\sigma f \, dS + \iint_\sigma g \, dS$$

$$\iint_\sigma (f - g) \, dS = \iint_\sigma f \, dS - \iint_\sigma g \, dS$$

$$\iint_\sigma kf \, dS = k \iint_\sigma f \, dS \quad (k \text{ constant})$$

and, finally, if a surface σ is subdivided into finitely many parts, then a surface integral over σ can be computed by computing it over each part and adding the results. Thus, if σ is subdivided into two parts, σ_1 and σ_2 (Figure 18.4.10), then

Figure 18.4.10

$$\iint_\sigma f \, dS = \iint_{\sigma_1} f \, dS + \iint_{\sigma_2} f \, dS$$

▶ **Exercise Set 18.4**

In Exercises 1–4, find the mass of the given lamina assuming the density to be a constant δ_0.

1. The lamina that is the portion of the paraboloid $z = 1 - x^2 - y^2$ above the xy-plane.

2. The lamina that is the portion of the plane $2x + 2y + z = 8$ in the first octant.

3. The lamina that is the portion of the circular cylinder $x^2 + z^2 = 4$ that lies directly above the rectangle $R = \{(x, y) : 0 \le x \le 1, 0 \le y \le 4\}$ in the xy-plane.

4. The lamina that is the portion of the paraboloid $2z = x^2 + y^2$ inside the cylinder $x^2 + y^2 = 8$.

In Exercises 5–10, evaluate the surface integrals.

5. $\displaystyle\iint_\sigma z^2 \, dS$, where σ is the portion of the cone $z = \sqrt{x^2 + y^2}$ between the planes $z = 1$ and $z = 2$.

6. $\displaystyle\iint_\sigma xyz \, dS$, where σ is the portion of the plane $x + y + z = 1$ lying in the first octant.

7. $\displaystyle\iint_\sigma x^2 y \, dS$, where σ is the portion of the cylinder $x^2 + z^2 = 1$ between the planes $y = 0$, $y = 1$, and above the xy-plane.

8. $\iint_\sigma (x^2 + y^2)z \, dS$, where σ is the portion of the sphere $x^2 + y^2 + z^2 = 4$ above the plane $z = 1$.

9. $\iint_\sigma (x + y + z) \, dS$, where σ is the portion of the plane $x + y = 1$ in the first octant between $z = 0$ and $z = 1$.

10. $\iint_\sigma (x + y) \, dS$, where σ is the portion of the plane $z = 6 - 2x - 3y$ in the first octant.

11. Evaluate

$$\iint_\sigma (x + y + z) \, dS$$

over the surface of the cube defined by the inequalities $0 \le x \le 1$, $0 \le y \le 1$, $0 \le z \le 1$. [*Hint:* Integrate over each face separately.]

12. Evaluate

$$\iint_\sigma zx^2 \, dS$$

over the portion of the cylinder $x^2 + y^2 = 1$ between $z = 0$ and $z = 1$. [*Hint:* Divide the surface into two parts and evaluate by projecting on the xz-plane.]

13. Evaluate

$$\iint_\sigma \sqrt{x^2 + y^2 + z^2} \, dS$$

over the portion of the cone $z = \sqrt{x^2 + y^2}$ below the plane $z = 1$.

14. Evaluate

$$\iint_\sigma (z + 1) \, dS$$

where σ is the upper hemisphere

$$z = \sqrt{1 - x^2 - y^2}$$

15. Evaluate

$$\iint_\sigma (x^2 + y^2) \, dS$$

over the surface of the sphere $x^2 + y^2 + z^2 = a^2$. [*Hint:* Divide the sphere into two parts and integrate over each part separately.]

16. Find the mass of the lamina that is the portion of the cone

$$z = \sqrt{x^2 + y^2}$$

between $z = 1$ and $z = 4$ if the density is

$$\delta(x, y, z) = x^2 z$$

17. Find the mass of the lamina that is the portion of the surface $y^2 = 4 - z$ between the planes $x = 0$, $x = 3$, $y = 0$, and $y = 3$ if the density is $\delta(x, y, z) = y$.

18. Find the mass of the lamina that is the portion of the paraboloid $z = x^2 + y^2$ below the plane $z = 1$ if the density is $\delta(x, y, z) = \sqrt{x^2 + y^2}$.

In Exercises 19 and 20, set up, but do not evaluate, an iterated integral equal to the given surface integral by projecting σ on (a) the xy-plane, (b) the yz-plane, and (c) the xz-plane.

19. $\iint_\sigma xyz \, dS$, where σ is the portion of the plane $2x + 3y + 4z = 12$ in the first octant.

20. $\iint_\sigma xz \, dS$, where σ is the portion of the sphere $x^2 + y^2 + z^2 = a^2$ in the first octant.

In Exercises 21 and 22, set up, but do not evaluate, two different iterated integrals equal to the given surface integral.

21. $\iint_\sigma xyz \, dS$, where σ is the portion of the surface $y^2 = x$ between the planes $z = 0$, $z = 4$, $y = 1$, and $y = 2$.

22. $\iint_\sigma x^2 y \, dS$, where σ is the portion of the cylinder $y^2 + z^2 = a^2$ in the first octant between the planes $x = 0$, $x = 9$, $z = y$, and $z = 2y$.

23. Prove: If a curved lamina has constant density δ_0, then its mass is the density times the surface area.

24. Show that the mass of the spherical lamina $x^2 + y^2 + z^2 = a^2$ is $2\pi a^3$ if the density at any point is equal to the distance from the point to the xy-plane.

■ 18.5 SURFACE INTEGRALS OF VECTOR FUNCTIONS

In this section we shall study surface integrals with integrands that involve vector functions. Applications of the results we obtain here will be given in later sections.

□ **ORIENTED SURFACES**

Figure 18.5.1

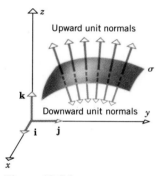

Figure 18.5.2

If a surface σ has a nonzero normal vector at a point (x, y, z), then there are two oppositely directed *unit normal* vectors at that point (Figure 18.5.1). These vectors are described by various names, depending on the signs of their components. For example, if the unit normal vector $\mathbf{n}$ has a positive $\mathbf{k}$ component, then $\mathbf{n}$ points roughly in the upward direction and is called an **upward unit normal** (Figure 18.5.2). If $\mathbf{n}$ has a negative $\mathbf{k}$ component, then $\mathbf{n}$ points roughly downward and is called a **downward unit normal**.

Table 18.5.1 explains the terminology used to describe unit normal vectors.

Example 1 If the vector

$$\mathbf{n} = \frac{1}{\sqrt{3}}\,\mathbf{i} - \frac{1}{\sqrt{3}}\,\mathbf{j} + \frac{1}{\sqrt{3}}\,\mathbf{k}$$

is normal to a surface σ, then $\mathbf{n}$ is an upward unit normal (positive $\mathbf{k}$ component), a left unit normal (negative $\mathbf{j}$ component), and a forward unit normal (positive $\mathbf{i}$ component). ◀

To compute the unit normals to a surface σ given by an equation $z = z(x, y)$, we first rewrite this equation as

$$z - z(x, y) = 0$$

which is a level surface for the function

$$G(x, y, z) = z - z(x, y)$$

Table 18.5.1

	Upward unit normal	Downward unit normal	Right unit normal	Left unit normal	Forward unit normal	Backward unit normal
TERMINOLOGY						
PICTURE						
MATHEMATICAL DESCRIPTION	Positive $\mathbf{k}$ component	Negative $\mathbf{k}$ component	Positive $\mathbf{j}$ component	Negative $\mathbf{j}$ component	Positive $\mathbf{i}$ component	Negative $\mathbf{i}$ component

By Theorem 16.7.6, if $z(x, y)$ has continuous first partial derivatives, then the gradient

$$\nabla G = -\frac{\partial z}{\partial x}\mathbf{i} - \frac{\partial z}{\partial y}\mathbf{j} + \mathbf{k} \tag{1}$$

is normal to this surface. To obtain a *unit* normal vector, we must normalize (1). Since

$$\|\nabla G\| = \sqrt{\left(\frac{\partial z}{\partial x}\right)^2 + \left(\frac{\partial z}{\partial y}\right)^2 + 1}$$

it follows that

$$\frac{\nabla G}{\|\nabla G\|} = \frac{-\dfrac{\partial z}{\partial x}\mathbf{i} - \dfrac{\partial z}{\partial y}\mathbf{j} + \mathbf{k}}{\sqrt{\left(\dfrac{\partial z}{\partial x}\right)^2 + \left(\dfrac{\partial z}{\partial y}\right)^2 + 1}} \tag{2}$$

is a unit normal vector to σ at the point (x, y, z). Moreover, because the $\mathbf{k}$ component is positive, (2) is the upward unit normal. To find the downward unit normal, we must multiply this vector by -1. This yields

$$-\frac{\nabla G}{\|\nabla G\|} = \frac{\dfrac{\partial z}{\partial x}\mathbf{i} + \dfrac{\partial z}{\partial y}\mathbf{j} - \mathbf{k}}{\sqrt{\left(\dfrac{\partial z}{\partial x}\right)^2 + \left(\dfrac{\partial z}{\partial y}\right)^2 + 1}} \tag{3}$$

For surfaces expressed in the form $y = y(x, z)$ or $x = x(y, z)$, the unit normal vectors are obtained by normalizing the gradients of the functions $y - y(x, z)$ and $x - x(y, z)$, respectively. The resulting formulas are listed in Table 18.5.2.

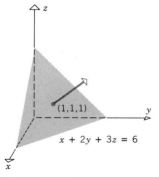

Figure 18.5.3

Example 2   Find the unit normal vector at the point $(1, 1, 1)$ to $x + 2y + 3z = 6$ shown in Figure 18.5.3.

Solution.   Since the vector points above the xy-plane, to the front of the yz-plane, and to the right of the xz-plane, it can be calculated from the formula for an upward normal, a right normal, or a forward normal. For illustrative purposes we shall calculate the vector two ways—as an upward normal and as a forward normal.

To use the formula for an upward normal we rewrite the equation of the plane as

$$z = 2 - \frac{1}{3}x - \frac{2}{3}y$$

from which we obtain

$$\frac{\partial z}{\partial x} = -\frac{1}{3} \quad \text{and} \quad \frac{\partial z}{\partial y} = -\frac{2}{3}$$

Table 18.5.2

EQUATION OF σ	NORMALS TO σ	
	Upward unit normal	Downward unit normal
$z = z(x, y)$	$\dfrac{-\dfrac{\partial z}{\partial x}\mathbf{i} - \dfrac{\partial z}{\partial y}\mathbf{j} + \mathbf{k}}{\sqrt{\left(\dfrac{\partial z}{\partial x}\right)^2 + \left(\dfrac{\partial z}{\partial y}\right)^2 + 1}}$	$\dfrac{\dfrac{\partial z}{\partial x}\mathbf{i} + \dfrac{\partial z}{\partial y}\mathbf{j} - \mathbf{k}}{\sqrt{\left(\dfrac{\partial z}{\partial x}\right)^2 + \left(\dfrac{\partial z}{\partial y}\right)^2 + 1}}$
	Right unit normal	Left unit normal
$y = y(x, z)$	$\dfrac{-\dfrac{\partial y}{\partial x}\mathbf{i} + \mathbf{j} - \dfrac{\partial y}{\partial z}\mathbf{k}}{\sqrt{\left(\dfrac{\partial y}{\partial x}\right)^2 + \left(\dfrac{\partial y}{\partial z}\right)^2 + 1}}$	$\dfrac{\dfrac{\partial y}{\partial x}\mathbf{i} - \mathbf{j} + \dfrac{\partial y}{\partial z}\mathbf{k}}{\sqrt{\left(\dfrac{\partial y}{\partial x}\right)^2 + \left(\dfrac{\partial y}{\partial z}\right)^2 + 1}}$
	Forward unit normal	Backward unit normal
$x = x(y, z)$	$\dfrac{\mathbf{i} - \dfrac{\partial x}{\partial y}\mathbf{j} - \dfrac{\partial x}{\partial z}\mathbf{k}}{\sqrt{\left(\dfrac{\partial x}{\partial y}\right)^2 + \left(\dfrac{\partial x}{\partial z}\right)^2 + 1}}$	$\dfrac{-\mathbf{i} + \dfrac{\partial x}{\partial y}\mathbf{j} + \dfrac{\partial x}{\partial z}\mathbf{k}}{\sqrt{\left(\dfrac{\partial x}{\partial y}\right)^2 + \left(\dfrac{\partial x}{\partial z}\right)^2 + 1}}$

Thus, from Table 18.5.2, the upward unit normal is

$$\mathbf{n} = \frac{\dfrac{1}{3}\mathbf{i} + \dfrac{2}{3}\mathbf{j} + \mathbf{k}}{\sqrt{\left(-\dfrac{1}{3}\right)^2 + \left(-\dfrac{2}{3}\right)^2 + 1}} = \frac{3}{\sqrt{14}}\left(\frac{1}{3}\mathbf{i} + \frac{2}{3}\mathbf{j} + \mathbf{k}\right)$$

$$= \frac{1}{\sqrt{14}}\mathbf{i} + \frac{2}{\sqrt{14}}\mathbf{j} + \frac{3}{\sqrt{14}}\mathbf{k}$$

To use the formula for a forward unit normal we rewrite the equation of the plane as

$$x = 6 - 2y - 3z$$

from which we obtain

$$\frac{\partial x}{\partial y} = -2 \quad \text{and} \quad \frac{\partial x}{\partial z} = -3$$

Thus, from Table 18.5.2, the forward unit normal is

$$\mathbf{n} = \frac{\mathbf{i} + 2\mathbf{j} + 3\mathbf{k}}{\sqrt{(-2)^2 + (-3)^2 + 1}} = \frac{1}{\sqrt{14}}\mathbf{i} + \frac{2}{\sqrt{14}}\mathbf{j} + \frac{3}{\sqrt{14}}\mathbf{k}$$

which agrees with the result obtained above. ◄

A surface σ is said to be **oriented** if a unit normal vector is constructed at each point of the surface in such a way that the vectors vary continuously (have no abrupt changes in direction) as we traverse any curve on the surface. The unit normal vectors are then said to form an **orientation** of the surface. For example, the surface σ in Figure 18.5.4*a* can be oriented by constructing an upward unit normal at each point (Figure 18.5.4*b*) or a downward unit normal at each point (Figure 18.5.4*c*). However, the unit normals in Figure 18.5.4*d* do not form an orientation because the directions of these vectors change abruptly as we cross the curve drawn on the surface.

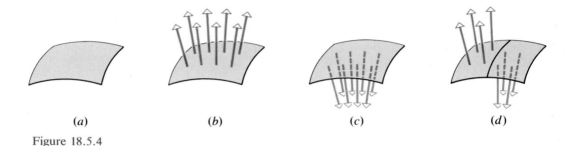

(*a*)   (*b*)   (*c*)   (*d*)

Figure 18.5.4

Most common surfaces can be oriented by constructing the unit normal vectors appropriately; such surfaces are called **orientable**. (A famous non-orientable surface, called a Möbius strip, is discussed in Exercise 22.) It is proved in advanced courses that an orientable surface has only two possible orientations. For example, the surface σ in Figure 18.5.4 can be oriented by upward normals or by downward normals, but any mixture of the two is not an orientation since abrupt direction changes in the normals are introduced.

Example 3 The two possible orientations of a sphere are by **inward normals** or **outward normals** (Figure 18.5.5). ◄

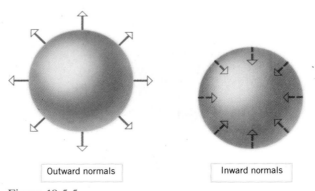

Outward normals   Inward normals

Figure 18.5.5

☐ **SURFACE INTEGRALS INVOLVING VECTOR-VALUED FUNCTIONS**

We now turn to the main topic of this section, surface integrals of vector-valued functions.

18.5.1 DEFINITION.   If $\mathbf{F}(x, y, z) = f(x, y, z)\mathbf{i} + g(x, y, z)\mathbf{j} + h(x, y, z)\mathbf{k}$ has continuous components on the oriented surface σ, and if $\mathbf{n} = \mathbf{n}(x, y, z)$ is the unit normal vector of the orientation at (x, y, z), then

$$\iint_\sigma \mathbf{F} \cdot \mathbf{n}\, dS$$

is called the **flux integral of F over σ,** or the **surface integral of $\mathbf{F} \cdot \mathbf{n}$ over σ,** or the **surface integral of the normal component of F over σ.**

In Section 18.8, we shall see that surface integrals of vector-valued functions have important applications and physical interpretations. However, for now we shall concentrate on computing them.

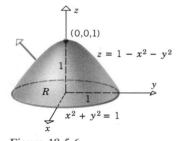

Figure 18.5.6

Example 4   Suppose that σ is the portion of the surface $z = 1 - x^2 - y^2$ above the xy-plane. Let σ be oriented by upward normals (Figure 18.5.6), and let $\mathbf{F}(x, y, z) = x\mathbf{i} + y\mathbf{j} + z\mathbf{k}$. Evaluate

$$\iint_\sigma \mathbf{F} \cdot \mathbf{n}\, dS$$

Solution.   From the first entry in Table 18.5.2, the upward unit normal to the surface

$$z = 1 - x^2 - y^2 \tag{4}$$

is given by

$$\mathbf{n} = \frac{-\dfrac{\partial z}{\partial x}\mathbf{i} - \dfrac{\partial z}{\partial y}\mathbf{j} + \mathbf{k}}{\sqrt{\left(\dfrac{\partial z}{\partial x}\right)^2 + \left(\dfrac{\partial z}{\partial y}\right)^2 + 1}}$$

Thus, from Formula (10) in Theorem 18.4.2,

$$\iint_\sigma \mathbf{F} \cdot \mathbf{n}\, dS = \iint_R \mathbf{F} \cdot \left[\frac{-\dfrac{\partial z}{\partial x}\mathbf{i} - \dfrac{\partial z}{\partial y}\mathbf{j} + \mathbf{k}}{\sqrt{\left(\dfrac{\partial z}{\partial x}\right)^2 + \left(\dfrac{\partial z}{\partial y}\right)^2 + 1}}\right] \sqrt{\left(\frac{\partial z}{\partial x}\right)^2 + \left(\frac{\partial z}{\partial y}\right)^2 + 1}\, dA$$

or, on canceling square roots,

$$\iint_\sigma \mathbf{F} \cdot \mathbf{n}\, dS = \iint_R \mathbf{F} \cdot \left(-\frac{\partial z}{\partial x}\, \mathbf{i} - \frac{\partial z}{\partial y}\, \mathbf{j} + \mathbf{k} \right) dA \tag{5}$$

where R is the region shown in Figure 18.5.6. From (4) and the given formula for $\mathbf{F}$, it follows that

$$\mathbf{F} \cdot \left(-\frac{\partial z}{\partial x}\, \mathbf{i} - \frac{\partial z}{\partial y}\, \mathbf{j} + \mathbf{k} \right) = (x\mathbf{i} + y\mathbf{j} + z\mathbf{k}) \cdot (2x\mathbf{i} + 2y\mathbf{j} + \mathbf{k})$$

$$= 2x^2 + 2y^2 + z$$

$$= 2x^2 + 2y^2 + (1 - x^2 - y^2)$$

$$= x^2 + y^2 + 1$$

Substituting this expression in (5) yields

$$\iint_\sigma \mathbf{F} \cdot \mathbf{n}\, dS = \iint_R (x^2 + y^2 + 1)\, dA$$

$$= \int_0^{2\pi} \int_0^1 (r^2 + 1) r\, dr\, d\theta \qquad \boxed{\text{Using polar coordinates to evaluate the integral}}$$

$$= \int_0^{2\pi} \left(\frac{3}{4} \right) d\theta = \frac{3\pi}{2} \qquad \blacktriangleleft$$

☐ **EVALUATING SURFACE INTEGRALS**

The procedure used to derive Formula (5) in the last example can be used to obtain the following general results.

18.5.2   THEOREM.   *If the surface σ is given by an equation of the form $z = z(x, y)$ where z has continuous first partial derivatives, and if R is the projection of the surface on the xy-plane, then*

(a) $$\iint_\sigma \mathbf{F} \cdot \mathbf{n}\, dS = \iint_R \mathbf{F} \cdot \left(-\frac{\partial z}{\partial x}\, \mathbf{i} - \frac{\partial z}{\partial y}\, \mathbf{j} + \mathbf{k} \right) dA \tag{6}$$

if σ is oriented by upward normals.

(b) $$\iint_\sigma \mathbf{F} \cdot \mathbf{n}\, dS = \iint_R \mathbf{F} \cdot \left(\frac{\partial z}{\partial x}\, \mathbf{i} + \frac{\partial z}{\partial y}\, \mathbf{j} - \mathbf{k} \right) dA \tag{7}$$

if σ is oriented by downward normals.

Similar formulas that apply to surfaces that are expressed in the form $x = x(y, z)$ and $y = y(x, z)$ are discussed in the exercises.

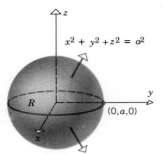

Figure 18.5.7

Example 5 Let σ be the sphere $x^2 + y^2 + z^2 = a^2$ oriented by outward normals (Figure 18.5.7), and let $\mathbf{F}(x, y, z) = z\mathbf{k}$. Evaluate

$$\iint_\sigma \mathbf{F} \cdot \mathbf{n} \, dS$$

Solution.   On the upper hemisphere the outward unit normal is upward, while on the lower hemisphere it is downward. Since different formulas for these normals apply on the two hemispheres, it is desirable to write

$$\iint_\sigma \mathbf{F} \cdot \mathbf{n} \, dS = \iint_{\sigma_1} \mathbf{F} \cdot \mathbf{n} \, dS + \iint_{\sigma_2} \mathbf{F} \cdot \mathbf{n} \, dS \tag{8}$$

where σ_1 is the upper hemisphere and σ_2 is the lower hemisphere.

The upper hemisphere σ_1 has the equation

$$z = \sqrt{a^2 - x^2 - y^2} \tag{9}$$

so that (6) yields

$$\iint_{\sigma_1} \mathbf{F} \cdot \mathbf{n} \, dS = \iint_R (z\mathbf{k}) \cdot \left(\frac{x}{\sqrt{a^2 - x^2 - y^2}} \mathbf{i} + \frac{y}{\sqrt{a^2 - x^2 - y^2}} \mathbf{j} + \mathbf{k} \right) dA$$

$$= \iint_R z \, dA = \iint_R \sqrt{a^2 - x^2 - y^2} \, dA \qquad \boxed{\begin{array}{c} R \text{ is the region shown} \\ \text{in Figure 18.5.7.} \end{array}}$$

$$= \int_0^{2\pi} \int_0^a \sqrt{a^2 - r^2} \, r \, dr \, d\theta$$

$$= \int_0^{2\pi} -\frac{1}{3} (a^2 - r^2)^{3/2} \Big]_0^a \, d\theta$$

$$= \int_0^{2\pi} \frac{1}{3} a^3 \, d\theta = \frac{2\pi a^3}{3}$$

The lower hemisphere σ_2 has the equation

$$z = -\sqrt{a^2 - x^2 - y^2}$$

so that (7) yields

$$\iint_{\sigma_2} \mathbf{F} \cdot \mathbf{n} \, dS = \iint_R (z\mathbf{k}) \cdot \left(\frac{x}{\sqrt{a^2 - x^2 - y^2}} \mathbf{i} + \frac{y}{\sqrt{a^2 - x^2 - y^2}} \mathbf{j} - \mathbf{k} \right) dA$$

$$= \iint_R -z \, dA = \iint_R \sqrt{a^2 - x^2 - y^2} \, dA$$

$$= \frac{2\pi a^3}{3} \qquad \boxed{\begin{array}{c} \text{The computations} \\ \text{are identical to} \\ \text{those above.} \end{array}}$$

Thus, from (8)

$$\iint_\sigma \mathbf{F} \cdot \mathbf{n} \, dS = \frac{2\pi a^3}{3} + \frac{2\pi a^3}{3} = \frac{4\pi a^3}{3} \quad \blacktriangleleft$$

▶ **Exercise Set 18.5**

1. Use three different formulas in Table 18.5.2 to calculate the unit normal to $2x + 3y + 4z = 9$ at $(1, 1, 1)$ that has positive components.

2. Use three different formulas in Table 18.5.2 to calculate the unit normal to $x^2 + y^2 + z^2 = 9$ at $(2, 1, -2)$ that points below the xy-plane.

3. In parts (a)–(c), use any appropriate formula to calculate the indicated unit normal.
   (a) The unit normal to $z = x^2 + y^2$ at $(1, 2, 5)$ that points toward the z-axis
   (b) The unit normal to $z = \sqrt{x^2 + y^2}$ at $(-3, 4, 5)$ that points toward the xz-plane
   (c) The unit normal to the cylinder $x^2 + z^2 = 25$ at $(3, 2, -4)$ that points away from the xy-plane.

4. In each part, use any appropriate formula to calculate the indicated unit normal.
   (a) The unit normal to the surface $y^2 = x$ at $(1, 1, 2)$ that points toward the xz-plane
   (b) The unit normal to the hyperbolic paraboloid $y = z^2 - x^2$ at $(1, 3, 2)$ that points toward the yz-plane
   (c) The unit normal to the cone $x^2 = y^2 + z^2$ at $(\sqrt{2}, -1, -1)$ that points away from the xy-plane.

In Exercises 5–15, evaluate $\iint_\sigma \mathbf{F} \cdot \mathbf{n} \, dS$.

5. $\mathbf{F}(x, y, z) = x\mathbf{i} + y\mathbf{j} + 2z\mathbf{k}$; σ is the portion of the surface $z = 1 - x^2 - y^2$ above the xy-plane, oriented by upward normals.

6. $\mathbf{F}(x, y, z) = (x + y)\mathbf{i} + (y + z)\mathbf{j} + (z + x)\mathbf{k}$; σ is the portion of the plane $x + y + z = 1$ in the first octant, oriented by unit normals with positive components.

7. $\mathbf{F}(x, y, z) = z^2\mathbf{k}$; σ is the upper hemisphere given by $z = \sqrt{1 - x^2 - y^2}$, oriented by upward unit normals.

8. $\mathbf{F}(x, y, z) = x^2\mathbf{i} + yx\mathbf{j} + zx\mathbf{k}$; σ is the portion of the plane $6x + 3y + 2z = 6$ in the first octant, oriented by unit normals with positive components.

9. $\mathbf{F}(x, y, z) = x\mathbf{i} + y\mathbf{j} + z\mathbf{k}$; σ is the upper hemisphere $z = \sqrt{9 - x^2 - y^2}$, oriented by upward unit normals.

10. $\mathbf{F}(x, y, z) = \mathbf{i} + \mathbf{j} + \mathbf{k}$; σ is the portion of the cone $z = \sqrt{x^2 + y^2}$ below the plane $z = 1$, oriented by downward unit normals.

11. $\mathbf{F}(x, y, z) = x\mathbf{i} + y\mathbf{j} + 2z\mathbf{k}$; σ is the portion of the cone $z^2 = x^2 + y^2$ between the planes $z = 1$ and $z = 2$, oriented by upward unit normals.

12. $\mathbf{F}(x, y, z) = y\mathbf{j} + \mathbf{k}$; σ is the portion of the paraboloid $z = x^2 + y^2$ below the plane $z = 4$, oriented by downward unit normals.

13. $\mathbf{F}(x, y, z) = x\mathbf{k}$; σ is the portion of the paraboloid $z = x^2 + y^2$ below the plane $z = y$, oriented by downward unit normals.

14. $\mathbf{F}(x, y, z) = x\mathbf{i} + y\mathbf{j} + z\mathbf{k}$; σ is the portion of the cylinder $z^2 = 1 - x^2$ between the planes $y = 1$ and $y = -2$, oriented by outward unit normals.

15. $\mathbf{F}(x, y, z) = x\mathbf{i} + y\mathbf{j} + z\mathbf{k}$, where σ is the sphere $x^2 + y^2 + z^2 = a^2$ oriented by outward unit normals.

16. Let σ be the surface of the cube bounded by the planes $x = \pm 1$, $y = \pm 1$, $z = \pm 1$, oriented by outward unit normals. In each part, evaluate the flux integral of $\mathbf{F}$ over σ.
    (a) $\mathbf{F}(x, y, z) = x\mathbf{i}$
    (b) $\mathbf{F}(x, y, z) = x\mathbf{i} + y\mathbf{j} + z\mathbf{k}$
    (c) $\mathbf{F}(x, y, z) = x^2\mathbf{i} + y^2\mathbf{j} + z^2\mathbf{k}$.

17. Show that reversing the orientation of σ reverses the sign of

$$\iint_\sigma \mathbf{F} \cdot \mathbf{n} \, dS$$

18. Derive Formulas (6) and (7) in Theorem 18.5.2. [*Hint:* Review the derivation of (5) in Example 4.]

19. Obtain the analogs of Formulas (6) and (7) for
    (a) surfaces of the form $x = x(y, z)$
    (b) surfaces of the form $y = y(x, z)$.

20. Evaluate $\iint\limits_{\sigma} \mathbf{F} \cdot \mathbf{n} \, dS$ where σ is the portion of the paraboloid $x = y^2 + z^2$ with $x \le 1$ and $z \ge 0$ oriented by backward unit normals and

$$\mathbf{F}(x, y, z) = y\mathbf{i} - z\mathbf{j} + 8\mathbf{k}$$

[*Hint:* Exercise 19(a).]

21. Evaluate $\iint\limits_{\sigma} \mathbf{F} \cdot \mathbf{n} \, dS$ if σ is the hemisphere $y = \sqrt{1 - x^2 - z^2}$ oriented by right unit normals and $\mathbf{F}(x, y, z) = x\mathbf{i} + y\mathbf{j} + z\mathbf{k}$. [*Hint:* Exercise 19(b).]

22. The best-known example of a nonorientable surface is the **Möbius strip**, which can be visualized by taking a band of paper, twisting it once, and gluing the ends together (Figures 18.5.8a and 18.5.8b). In Figure 18.5.8c, we have tried to construct unit normal vectors whose directions vary continuously moving counterclockwise around the dashed curve starting and finishing on the vertical line.

(a) Explain why these vectors are not part of an orientation of the surface.

(b) Explain why the surface is not orientable.

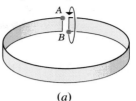

(a)

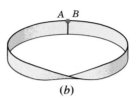

(b)

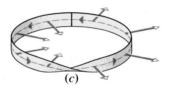

(c)

Figure 18.5.8

18.6 THE DIVERGENCE THEOREM

In this section we shall discuss an important three-dimensional analog of Green's Theorem.

We begin with some terminology.

☐ DIVERGENCE

18.6.1 DEFINITION. If $\mathbf{F}(x, y, z) = f(x, y, z)\mathbf{i} + g(x, y, z)\mathbf{j} + h(x, y, z)\mathbf{k}$, then we define the *divergence of F*, written div F, by

$$\text{div } \mathbf{F} = \frac{\partial f}{\partial x} + \frac{\partial g}{\partial y} + \frac{\partial h}{\partial z}$$

Example 1 The divergence of

$$\mathbf{F}(x, y, z) = xy^2z^2\mathbf{i} + (3xz^2 + y^4)\mathbf{j} + x^3y^2z\mathbf{k}$$

is

$$\text{div } \mathbf{F} = \frac{\partial}{\partial x}(xy^2z^2) + \frac{\partial}{\partial y}(3xz^2 + y^4) + \frac{\partial}{\partial z}(x^3y^2z)$$

$$= y^2z^2 + 4y^3 + x^3y^2 \quad \blacktriangleleft$$

The following result, called the **Divergence Theorem** or **Gauss's* Theorem,** expresses a triple integral over a solid in terms of a surface integral over its boundary.

18.6.2   THEOREM (*Divergence Theorem*).   *Let G be a solid with surface σ oriented by outward unit normals. If*

$$\mathbf{F}(x, y, z) = f(x, y, z)\mathbf{i} + g(x, y, z)\mathbf{j} + h(x, y, z)\mathbf{k}$$

where f, g, and h have continuous first partial derivatives on some open set containing G, then

$$\iint_{\sigma} \mathbf{F} \cdot \mathbf{n}\, dS = \iiint_{G} \text{div } \mathbf{F}\, dV \tag{1}$$

The proof of this theorem for a general solid G is too difficult to present here. However, we can give a proof for the special case where G is a simple solid (see Figure 17.5.4 and the discussion preceding it).

---

*KARL FRIEDRICH GAUSS (1777–1855). German mathematician and scientist. Sometimes called the "prince of mathematicians," Gauss ranks with Newton and Archimedes as one of the three greatest mathematicians who ever lived. His father, a laborer, was an uncouth but honest man who would have liked Gauss to take up a trade such as gardening or bricklaying; but the boy's genius for mathematics was not to be denied. In the entire history of mathematics there may never have been a child so precocious as Gauss—by his own account he worked out the rudiments of arithmetic before he could talk. One day, before he was even three years old, his genius became apparent to his parents in a very dramatic way. His father was preparing the weekly payroll for the laborers under his charge while the boy watched quietly from a corner. At the end of the long and tedious calculation, Gauss informed his father that there was an error in the result and stated the answer, which he had worked out in his head. To the astonishment of his parents, a check of the computations showed Gauss to be correct!

For his elementary education Gauss was enrolled in a squalid school run by a man named Büttner whose main teaching technique was thrashing. Büttner was in the habit of assigning long addition problems which, unknown to his students, were arithmetic progressions that he could sum up using formulas. On the first day that Gauss entered the arithmetic class, the students were asked to sum the numbers from 1 to 100. But no sooner had Büttner stated the problem than Gauss turned over his slate and exclaimed in his peasant dialect, "Ligget se'." (Here it lies.) For nearly an hour Büttner glared at Gauss, who sat with folded hands while his classmates toiled away. When Büttner examined the slates at the end of the period, Gauss's slate contained a single number, 5050—the only correct solution in the class.

To his credit, Büttner recognized the genius of Gauss and with the help of his assistant, John Bartels, had him brought to the attention of Karl Wilhelm Ferdinand, Duke of Brunswick. The shy and awkward boy, who was then fourteen, so captivated the Duke that he subsidized him through preparatory school, college, and the early part of his career.

From 1795 to 1798 Gauss studied mathematics at the University of Göttingen, receiving his degree in absentia from the University of Helmstadt. For his dissertation, he gave the first complete proof of the fundamental theorem of algebra, which states that every polynomial equation has as many solutions as its degree. At age 19 he solved a problem that baffled Euclid, inscribing a regular polygon of seventeen sides in a circle using straightedge and compass; and in 1801, at age 24, he

(*continued on next page*)

Proof (for simple solids).   Let G be a simple solid with upper surface $z = g_2(x, y)$, lower surface $z = g_1(x, y)$, and projection R on the xy-plane. Let σ_1 denote the lower surface, σ_2 the upper surface, and σ_3 the lateral surface (Figure 18.6.1a). If the upper surface and lower surface meet as in Figure 18.6.1b, then there is no lateral surface σ_3. Our proof will allow for both cases shown in those figures.

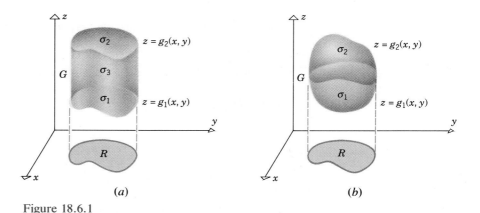

(a)   (b)

Figure 18.6.1

(continued)

published his first masterpiece, *Disquisitiones Arithmeticae,* considered by many to be one of the most brilliant achievements in mathematics. In that book Gauss systematized the study of number theory (properties of the integers) and formulated the basic concepts that form the foundation of that subject.

In the same year that the *Disquisitiones* was published, Gauss again applied his phenomenal computational skills in a dramatic way. The astronomer Giuseppi Piazzi had observed the asteroid Ceres for $\frac{1}{40}$ of its orbit, but lost it in the sun. Using only three observations and the "method of least squares" that he had developed in 1795, Gauss computed the orbit with such accuracy that astronomers had no trouble relocating it the following year. This achievement brought him instant recognition as the premier mathematician in Europe, and in 1807 he was made Professor of Astronomy and head of the astronomical observatory at Göttingen.

In the years that followed, Gauss revolutionized mathematics by bringing to it standards of precision and rigor undreamed of by his predecessors. He had a passion for perfection that drove him to polish and rework his papers rather than publish less finished work in greater numbers— his favorite saying was "Pauca, sed matura" (Few, but ripe). As a result, many of his important discoveries were squirreled away in diaries that remained unpublished until years after his death.

Among his myriad achievements, Gauss discovered the Gaussian or "bell-shaped" error curve fundamental in probability, gave the first geometric interpretation of complex numbers and established their fundamental role in mathematics, developed methods of characterizing surfaces intrinsically by means of the curves that they contain, developed the theory of conformal (angle-preserving) maps, and discovered non-Euclidean geometry 30 years before the ideas were published by others. In physics he made major contributions to the theory of lenses and capillary action, and with Wilhelm Weber he did fundamental work in electromagnetism. Gauss invented the heliotrope, bifilar magnetometer, and an electrotelegraph.

Gauss was deeply religious and aristocratic in demeanor. He mastered foreign languages with ease, read extensively, and enjoyed minerology and botany as hobbies. He disliked teaching and was usually cool and discouraging to other mathematicians, possibly because he had already anticipated their work. It has been said that if Gauss had published all of his discoveries, the current state of mathematics would be advanced by 50 years. He was without a doubt the greatest mathematician of the modern era.

We want to show that

$$\iint_\sigma \mathbf{F} \cdot \mathbf{n} \, dS = \iiint_G \operatorname{div} \mathbf{F} \, dV$$

or equivalently

$$\iint_\sigma [f(x, y, z)\mathbf{i} + g(x, y, z)\mathbf{j} + h(x, y, z)\mathbf{k}] \cdot \mathbf{n} \, dS = \iiint_G \left(\frac{\partial f}{\partial x} + \frac{\partial g}{\partial y} + \frac{\partial h}{\partial z} \right) dV$$

To prove this, it suffices to prove the following three equalities:

$$\iint_\sigma [f(x, y, z)\mathbf{i} \cdot \mathbf{n}] \, dS = \iiint_G \frac{\partial f}{\partial x} \, dV \tag{2a}$$

$$\iint_\sigma [g(x, y, z)\mathbf{j} \cdot \mathbf{n}] \, dS = \iiint_G \frac{\partial g}{\partial y} \, dV \tag{2b}$$

$$\iint_\sigma [h(x, y, z)\mathbf{k} \cdot \mathbf{n}] \, dS = \iiint_G \frac{\partial h}{\partial z} \, dV \tag{2c}$$

Since the proofs of all three formulas are similar, we shall prove only the third. It follows from Theorem 17.5.2 that

$$\iiint_G \frac{\partial h}{\partial z} \, dV = \iint_R \left[\int_{g_1(x,y)}^{g_2(x,y)} \frac{\partial h}{\partial z} \, dz \right] dA = \iint_R \left[h(x, y, z) \right]_{z=g_1(x,y)}^{g_2(x,y)} dA$$

so

$$\iiint_G \frac{\partial h}{\partial z} \, dV = \iint_R [h(x, y, g_2(x, y)) - h(x, y, g_1(x, y))] \, dA \tag{3}$$

We shall evaluate the surface integral in (2c) by integrating over each surface of G separately. If there is a lateral surface σ_3, then at each point of this surface $\mathbf{n} \cdot \mathbf{k} = 0$ since $\mathbf{n}$ is horizontal and $\mathbf{k}$ is vertical. Thus,

$$\iint_{\sigma_3} [h(x, y, z)\mathbf{k} \cdot \mathbf{n}] \, dS = 0$$

Therefore, regardless of whether G has a lateral surface or not, we can write

$$\iint_\sigma [h(x, y, z)\mathbf{k} \cdot \mathbf{n}] \, dS = \iint_{\sigma_1} [h(x, y, z)\mathbf{k} \cdot \mathbf{n}] \, dS + \iint_{\sigma_2} [h(x, y, z)\mathbf{k} \cdot \mathbf{n}] \, dS \tag{4}$$

On the upper surface σ_2, the outer normal is an upward normal, and on the lower surface σ_1, the outer normal is a downward normal. Thus, Theorem 18.5.2 implies that

$$\iint_{\sigma_2} [h(x, y, z)\mathbf{k} \cdot \mathbf{n}] \, dS = \iint_R \left[h(x, y, g_2(x, y))\mathbf{k} \cdot \left(-\frac{\partial z}{\partial x}\mathbf{i} - \frac{\partial z}{\partial y}\mathbf{j} + \mathbf{k} \right) \right] dA$$

$$= \iint_R [h(x, y, g_2(x, y))] \, dA \tag{5}$$

and

$$\iint_{\sigma_1} [h(x, y, z)\mathbf{k} \cdot \mathbf{n}] \, dS = \iint_R \left[h(x, y, g_1(x, y))\mathbf{k} \cdot \left(\frac{\partial z}{\partial x}\mathbf{i} + \frac{\partial z}{\partial y}\mathbf{j} - \mathbf{k} \right) \right] dA$$

$$= -\iint_R [h(x, y, g_1(x, y))] \, dA \tag{6}$$

Substituting (5) and (6) into (4) and combining the terms into a single integral yields

$$\iint_{\sigma} [h(x, y, z)\mathbf{k} \cdot \mathbf{n}] \, dS = \iint_R [h(x, y, g_2(x, y)) - h(x, y, g_1(x, y))] \, dA \tag{7}$$

Equation (2c) now follows from (3) and (7).   ∎

Example 2 Let σ be the sphere $x^2 + y^2 + z^2 = a^2$ oriented by outward normals and let $\mathbf{F}(x, y, z) = z\mathbf{k}$. Use the Divergence Theorem to evaluate

$$\iint_{\sigma} \mathbf{F} \cdot \mathbf{n} \, dS$$

Solution. Let G be the spherical solid enclosed by σ. Since

$$\text{div } \mathbf{F} = \frac{\partial z}{\partial z} = 1$$

it follows from (1) that

$$\iint_{\sigma} \mathbf{F} \cdot \mathbf{n} \, dS = \iiint_G dV = \text{volume of } G = \frac{4\pi a^3}{3} \quad \blacktriangleleft$$

REMARK. Notice how much simpler this solution is than the direct calculation of $\iint_{\sigma} \mathbf{F} \cdot \mathbf{n} \, dS$ in Example 5 of Section 18.5.

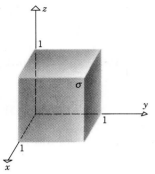

Figure 18.6.2

Example 3 Let G be the cube in the first octant shown in Figure 18.6.2, and let σ be its surface. Use the Divergence Theorem to compute

$$\iint_\sigma \mathbf{F} \cdot \mathbf{n}\ dS$$

where $\mathbf{F}(x, y, z) = 2x\mathbf{i} + 3y\mathbf{j} + z^2\mathbf{k}$, and $\mathbf{n}$ is the outward unit normal.

Solution. Since

$$\operatorname{div} \mathbf{F} = \frac{\partial}{\partial x}(2x) + \frac{\partial}{\partial y}(3y) + \frac{\partial}{\partial z}(z^2) = 5 + 2z$$

it follows from (1) that

$$\iint_\sigma \mathbf{F} \cdot \mathbf{n}\ dS = \iiint_G (5 + 2z)\ dV = \int_0^1 \int_0^1 \int_0^1 (5 + 2z)\ dz\ dy\ dx$$

$$= \int_0^1 \int_0^1 [5z + z^2]_{z=0}^1\ dy\ dx = \int_0^1 \int_0^1 6\ dy\ dx = 6 \quad \blacktriangleleft$$

Example 4 Let G be the cylindrical solid bounded by $x^2 + y^2 = 9$, the xy-plane, and the plane $z = 2$ (Figure 18.6.3), and let σ be its surface. Use the Divergence Theorem to evaluate

$$\iint_\sigma \mathbf{F} \cdot \mathbf{n}\ dS$$

where $\mathbf{F}(x, y, z) = x^3\mathbf{i} + y^3\mathbf{j} + z^2\mathbf{k}$, and $\mathbf{n}$ is the outward unit normal to σ.

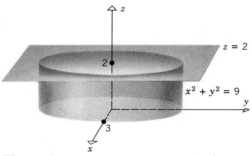

Figure 18.6.3

Solution. Since

$$\operatorname{div} \mathbf{F} = \frac{\partial}{\partial x}(x^3) + \frac{\partial}{\partial y}(y^3) + \frac{\partial}{\partial z}(z^2) = 3x^2 + 3y^2 + 2z$$

it follows from (1) that

$$\iint_{\sigma} \mathbf{F} \cdot \mathbf{n} \, dS = \iiint_{G} (3x^2 + 3y^2 + 2z) \, dV$$

We shall use cylindrical coordinates to evaluate the triple integral. We obtain

$$\iint_{\sigma} \mathbf{F} \cdot \mathbf{n} \, dS = \int_0^{2\pi} \int_0^3 \int_0^2 (3r^2 + 2z)r \, dz \, dr \, d\theta$$

$$= \int_0^{2\pi} \int_0^3 \left[3r^3 z + z^2 r \right]_{z=0}^2 \, dr \, d\theta$$

$$= \int_0^{2\pi} \int_0^3 (6r^3 + 4r) \, dr \, d\theta$$

$$= \int_0^{2\pi} \left[\frac{3r^4}{2} + 2r^2 \right]_0^3 \, d\theta$$

$$= \int_0^{2\pi} \frac{279}{2} \, d\theta = 279\pi \quad \blacktriangleleft$$

Example 5 Let *G* be the solid that is bounded above by the hemisphere $z = \sqrt{a^2 - x^2 - y^2}$ and below by the *xy*-plane (Figure 18.6.4). Find

$$\iint_{\sigma} \mathbf{F} \cdot \mathbf{n} \, dS$$

where $\mathbf{F}(x, y, z) = x^3 \mathbf{i} + y^3 \mathbf{j} + z^3 \mathbf{k}$, σ is the surface of *G*, and **n** is the outer unit normal.

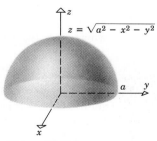

$z = \sqrt{a^2 - x^2 - y^2}$

Figure 18.6.4

Solution. Since

$$\text{div } \mathbf{F} = \frac{\partial}{\partial x} (x^3) + \frac{\partial}{\partial y} (y^3) + \frac{\partial}{\partial z} (z^3) = 3x^2 + 3y^2 + 3z^2$$

it follows from (1) that

$$\iint_{\sigma} \mathbf{F} \cdot \mathbf{n} \, dS = \iiint_{G} (3x^2 + 3y^2 + 3z^2) \, dV$$

We shall use spherical coordinates to evaluate the triple integral. We obtain

$$\iint\limits_{\sigma} \mathbf{F} \cdot \mathbf{n}\, dS = \int_0^{2\pi} \int_0^{\pi/2} \int_0^a (3\rho^2)\rho^2 \sin\phi\, d\rho\, d\phi\, d\theta$$

$$= 3 \int_0^{2\pi} \int_0^{\pi/2} \int_0^a \rho^4 \sin\phi\, d\rho\, d\phi\, d\theta$$

$$= 3 \int_0^{2\pi} \int_0^{\pi/2} \left[\frac{\rho^5}{5} \sin\phi \right]_{\rho=0}^a d\phi\, d\theta$$

$$= \frac{3a^5}{5} \int_0^{2\pi} \int_0^{\pi/2} \sin\phi\, d\phi\, d\theta$$

$$= \frac{3a^5}{5} \int_0^{2\pi} \left[-\cos\phi \right]_0^{\pi/2} d\theta$$

$$= \frac{3a^5}{5} \int_0^{2\pi} d\theta = \frac{6\pi a^5}{5} \quad \blacktriangleleft$$

▶ Exercise Set 18.6

In Exercises 1–5, find the divergence of **F**.

1. $\mathbf{F}(x, y, z) = xz^3\mathbf{i} + 2y^4x^2\mathbf{j} + 5z^2y\mathbf{k}$.

2. $\mathbf{F}(x, y, z) = 7y^3z^2\mathbf{i} - 8x^2z^5\mathbf{j} - 3xy^4\mathbf{k}$.

3. $\mathbf{F}(x, y, z) = e^{xy}\mathbf{i} - \cos y\mathbf{j} + \sin^2 z\mathbf{k}$.

4. $\mathbf{F}(x, y, z) = \dfrac{1}{\sqrt{x^2 + y^2 + z^2}}(x\mathbf{i} + y\mathbf{j} + z\mathbf{k})$.

5. $\mathbf{F}(x, y, z) = \ln x\mathbf{i} + e^{xyz}\mathbf{j} + \tan^{-1}\left(\dfrac{z}{x}\right)\mathbf{k}$.

In Exercises 6–17, use the Divergence Theorem to evaluate $\iint\limits_{\sigma} \mathbf{F} \cdot \mathbf{n}\, dS$, where **n** is the outer unit normal to σ.

6. $\mathbf{F}(x, y, z) = 2x\mathbf{i} + 2y\mathbf{j} + 2z\mathbf{k}$, where σ is the sphere $x^2 + y^2 + z^2 = 9$.

7. $\mathbf{F}(x, y, z) = 4x\mathbf{i} - 3y\mathbf{j} + 7z\mathbf{k}$; σ is the surface of the cube bounded by the coordinate planes and the planes $x = 1$, $y = 1$, and $z = 1$.

8. $\mathbf{F}(x, y, z) = z^3\mathbf{i} - x^3\mathbf{j} + y^3\mathbf{k}$, where σ is the sphere $x^2 + y^2 + z^2 = a^2$.

9. $\mathbf{F}(x, y, z) = (x - z)\mathbf{i} + (y - x)\mathbf{j} + (z - y)\mathbf{k}$; σ is the surface of the cylindrical solid bounded by $x^2 + y^2 = a^2$, $z = 0$, and $z = 1$.

10. $\mathbf{F}(x, y, z) = x\mathbf{i} + y\mathbf{j} + z\mathbf{k}$; σ is the surface of the solid bounded by the paraboloid $z = 1 - x^2 - y^2$ and the xy-plane.

11. $\mathbf{F}(x, y, z) = x^3\mathbf{i} + y^3\mathbf{j} + z^3\mathbf{k}$; σ is the surface of the cylindrical solid bounded by $x^2 + y^2 = 4$, $z = 0$, and $z = 3$.

12. $\mathbf{F}(x, y, z) = (x^3 - e^y)\mathbf{i} + (y^3 + \sin z)\mathbf{j} + (z^3 - xy)\mathbf{k}$, where σ is the surface of the solid bounded by $z = \sqrt{4 - x^2 - y^2}$ and the xy-plane. [*Hint:* Use spherical coordinates.]

13. $\mathbf{F}(x, y, z) = (x^2 + y)\mathbf{i} + xy\mathbf{j} - (2xz + y)\mathbf{k}$; σ is the surface of the tetrahedron in the first octant bounded by $x + y + z = 1$ and the coordinate planes.

14. $\mathbf{F}(x, y, z) = 2xz\mathbf{i} + yz\mathbf{j} + z^2\mathbf{k}$, where σ is the surface of the hemispherical solid bounded above by $z = \sqrt{a^2 - x^2 - y^2}$ and below by the xy-plane.

15. $\mathbf{F}(x, y, z) = x^2\mathbf{i} + y^2\mathbf{j} + z^2\mathbf{k}$; σ is the surface of the conical solid bounded by $z = \sqrt{x^2 + y^2}$ and $z = 1$.

16. $\mathbf{F}(x, y, z) = x^2y\mathbf{i} - xy^2\mathbf{j} + (z + 2)\mathbf{k}$; σ is the surface of the solid bounded above by the plane $z = 2x$ and below by the paraboloid $z = x^2 + y^2$.

17. $\mathbf{F}(x, y, z) = x^3\mathbf{i} + x^2y\mathbf{j} + xy\mathbf{k}$; σ is the surface of the solid bounded by $z = 4 - x^2$, $y + z = 5$, $z = 0$, and $y = 0$.

In Exercises 18–20, verify the Divergence Theorem by computing the integrals in (1) separately and verifying that they are equal.

18. $F(x, y, z) = x^2\mathbf{i} + y^2\mathbf{j} + z^2\mathbf{k}$; σ is the surface of the cube bounded by $x = 1$, $y = 1$, $z = 1$, and the coordinate planes.

19. $F(x, y, z) = x\mathbf{i} + y\mathbf{j} + z\mathbf{k}$, where σ is the sphere $x^2 + y^2 + z^2 = a^2$.

20. $F(x, y, z) = (x + y)\mathbf{i} + (y + z)\mathbf{j} + (z + x)\mathbf{k}$; σ is the surface of the cylindrical solid bounded by $x^2 + y^2 = 9$, $z = 0$, and $z = 5$.

21. Prove the following properties of divergence.
   (a) $\operatorname{div}(\mathbf{F} + \mathbf{G}) = \operatorname{div}\mathbf{F} + \operatorname{div}\mathbf{G}$
   (b) $\operatorname{div}(f\mathbf{F}) = f\operatorname{div}\mathbf{F} + \nabla f \cdot \mathbf{F}$, where $f(x, y, z)$ is a scalar-valued function and $\mathbf{F}(x, y, z)$ a vector-valued function.

22. Prove that if σ is the surface of a solid G oriented by outward unit normals, then the volume of G is given by

$$\operatorname{vol}(G) = \frac{1}{3} \iint\limits_{\sigma} \mathbf{F} \cdot \mathbf{n} \, dS$$

where $\mathbf{F}(x, y, z) = x\mathbf{i} + y\mathbf{j} + z\mathbf{k}$ and $\mathbf{n}$ is an outer unit normal to σ.

23. Use the formula in Exercise 22 to find the volume of a right-circular cylinder of radius a and height h.

24. Prove: If $\mathbf{F}(x, y, z) = a\mathbf{i} + b\mathbf{j} + c\mathbf{k}$, where a, b, and c are constant, and σ is the surface of a solid G, then

$$\iint\limits_{\sigma} \mathbf{F} \cdot \mathbf{n} \, dS = 0$$

where $\mathbf{n}$ is an outer unit normal to σ.

■ 18.7 STOKES' THEOREM

*In this section we shall discuss a generalization of Green's Theorem to three dimensions, called **Stokes'* Theorem**. In two dimensions, Green's Theorem expresses a double integral over a plane region in terms of a line integral over the boundary of the region. In three dimensions, Stokes' Theorem expresses a surface integral in terms of a line integral over the boundary of the surface.*

---

* GEORGE GABRIEL STOKES (1819–1903). Irish mathematician and physicist. Born in Skreen, Ireland, Stokes came from a family deeply rooted in the Church of Ireland. His father was a rector, his mother the daughter of a rector, and three of his brothers took holy orders. He received his early education from his father and a local parish clerk. In 1837, he entered Pembroke College and after graduating with top honors accepted a fellowship at the college. In 1847 he was appointed Lucasian professor of mathematics at Cambridge, a position once held by Isaac Newton, but one that had lost its esteem through the years. By virtue of his accomplishments, Stokes ultimately restored the position to the eminence it once held. Unfortunately, the position paid very little and Stokes was forced to teach at the Government School of Mines during the 1850s to supplement his income.

Stokes was one of several outstanding nineteenth century scientists who helped turn the physical sciences in a more empirical direction. He systematically studied hydrodynamics, elasticity of solids, behavior of waves in elastic solids, and diffraction of light. For Stokes, mathematics was a tool for his physical studies. He wrote classic papers on the motion of viscous fluids that laid the foundation for modern hydrodynamics; he elaborated on the wave theory of light; and he wrote papers on gravitational variation that established him as a founder of the modern science of geodesy.

Stokes was honored in his later years with degrees, medals, and memberships in foreign societies. He was knighted in 1889. Throughout his life, Stokes gave generously of his time to learned societies and readily assisted those who sought his help in solving problems. He was deeply religious and vitally concerned with the relationship between science and religion.

We begin with some terminology.

18.7.1 DEFINITION. If $\mathbf{F}(x, y, z) = f(x, y, z)\mathbf{i} + g(x, y, z)\mathbf{j} + h(x, y, z)\mathbf{k}$, then we define the *curl of* $\mathbf{F}$ by

$$\text{curl } \mathbf{F} = \left(\frac{\partial h}{\partial y} - \frac{\partial g}{\partial z}\right)\mathbf{i} + \left(\frac{\partial f}{\partial z} - \frac{\partial h}{\partial x}\right)\mathbf{j} + \left(\frac{\partial g}{\partial x} - \frac{\partial f}{\partial y}\right)\mathbf{k} \tag{1}$$

REMARK. This formula can be remembered by writing it in the determinant form

$$\text{curl } \mathbf{F} = \begin{vmatrix} \mathbf{i} & \mathbf{j} & \mathbf{k} \\ \dfrac{\partial}{\partial x} & \dfrac{\partial}{\partial y} & \dfrac{\partial}{\partial z} \\ f & g & h \end{vmatrix} \tag{2}$$

The reader should verify that Formula (1) results if we compute this determinant by interpreting a "product" such as $(\partial/\partial x)(f)$ to mean $\partial f/\partial x$. We note, however, that (2) is just a mnemonic device and not a true determinant, since the entries in a determinant must be numbers, not vectors and partial derivative symbols.

Example 1   If $\mathbf{F}(x, y, z) = x^2 y\mathbf{i} + 2y^3 z\mathbf{j} + 3z\mathbf{k}$, then

$$\text{curl } \mathbf{F} = \begin{vmatrix} \mathbf{i} & \mathbf{j} & \mathbf{k} \\ \dfrac{\partial}{\partial x} & \dfrac{\partial}{\partial y} & \dfrac{\partial}{\partial z} \\ x^2 y & 2y^3 z & 3z \end{vmatrix}$$

$$= \left[\frac{\partial}{\partial y}(3z) - \frac{\partial}{\partial z}(2y^3 z)\right]\mathbf{i} + \left[\frac{\partial}{\partial z}(x^2 y) - \frac{\partial}{\partial x}(3z)\right]\mathbf{j}$$

$$+ \left[\frac{\partial}{\partial x}(2y^3 z) - \frac{\partial}{\partial y}(x^2 y)\right]\mathbf{k} = -2y^3\mathbf{i} - x^2\mathbf{k} \quad \blacktriangleleft$$

To state Stokes' Theorem we need some terminology about oriented surfaces. If σ is an oriented surface bounded by a smooth curve C (Figure 18.7.1a), then the orientation of σ induces an orientation or direction along C as follows: If

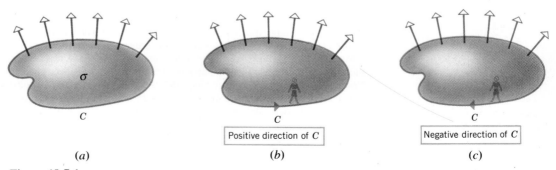

(a)	(b)	(c)
σ ... C	C — Positive direction of C	C — Negative direction of C

Figure 18.7.1

we imagine a woman walking along the curve C with her head in the (general) direction of the normals that orient σ, then the woman is walking in the *positive direction* of C if the surface σ is on the woman's left and in the *negative direction* if σ is on her right (Figures 18.7.1*b* and 18.7.1*c*).

18.7.2 THEOREM (*Stokes' Theorem*). *Let σ be an oriented surface, bounded by a single smooth parametric curve C. If the components of*

$$\mathbf{F}(x, y, z) = f(x, y, z)\mathbf{i} + g(x, y, z)\mathbf{j} + h(x, y, z)\mathbf{k}$$

have continuous first partial derivatives on some open set containing σ, then

$$\int_C \mathbf{F} \cdot d\mathbf{r} = \iint_\sigma (\text{curl } \mathbf{F}) \cdot \mathbf{n} \, dS \tag{3}$$

where the line integral is taken in the positive direction of C.

The proof of Stokes' Theorem is too difficult for a first course and will be omitted.

Recall that if s is an arc-length parameter for the curve C, and if the positive direction of C is in the direction of increasing s, then

$$\frac{d\mathbf{r}}{ds} = \mathbf{T}$$

where $\mathbf{T}$ is the unit tangent vector to C that points in the positive direction of C. This suggests the notation

$$d\mathbf{r} = \mathbf{T} \, ds$$

which enables us to write (3) as

$$\int_C \mathbf{F} \cdot \mathbf{T} \, ds = \iint_\sigma (\text{curl } \mathbf{F}) \cdot \mathbf{n} \, dS \tag{4}$$

Example 2 Verify Stokes' Theorem if σ is the portion of the paraboloid $z = 4 - x^2 - y^2$ for which $z \geq 0$ and $\mathbf{F}(x, y, z) = 2z\mathbf{i} + 3x\mathbf{j} + 5y\mathbf{k}$.

Solution. If σ is oriented by outward normals, then the positive direction for the boundary curve C is as shown in Figure 18.7.2. Since the curve C is a circle in the xy-plane with radius 2 and traversed counterclockwise looking down the z-axis, it is represented (with the proper direction) by the parametric equations

$$x = 2 \cos t, \quad y = 2 \sin t, \quad z = 0 \quad (0 \leq t \leq 2\pi) \tag{5}$$

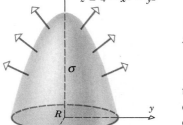
Figure 18.7.2

or equivalently by the position function

$$\mathbf{r}(t) = 2 \cos t\mathbf{i} + 2 \sin t\mathbf{j} + 0\mathbf{k} = 2 \cos t\mathbf{i} + 2 \sin t\mathbf{j}$$

where $0 \le t \le 2\pi$. Since $\mathbf{F}(x, y, z) = 2z\mathbf{i} + 3x\mathbf{j} + 5y\mathbf{k}$, it follows from (5) that

$$\mathbf{F}(x(t), y(t), z(t)) = 2(0)\mathbf{i} + 3(2 \cos t)\mathbf{j} + 5(2 \sin t)\mathbf{k} = 6 \cos t\mathbf{j} + 10 \sin t\mathbf{k}$$

Thus,

$$\int_C \mathbf{F} \cdot d\mathbf{r} = \int_0^{2\pi} \left[\mathbf{F}(x(t), y(t), z(t)) \cdot \frac{d\mathbf{r}}{dt} \right] dt$$

$$= \int_0^{2\pi} (6 \cos t\mathbf{j} + 10 \sin t\mathbf{k}) \cdot (-2 \sin t\mathbf{i} + 2 \cos t\mathbf{j}) \, dt$$

$$= \int_0^{2\pi} 12 \cos^2 t \, dt = 12 \left[\frac{1}{2} t + \frac{1}{4} \sin 2t \right]_0^{2\pi} = 12\pi$$

On the other hand,

$$\text{curl } \mathbf{F} = \begin{vmatrix} \mathbf{i} & \mathbf{j} & \mathbf{k} \\ \dfrac{\partial}{\partial x} & \dfrac{\partial}{\partial y} & \dfrac{\partial}{\partial z} \\ 2z & 3x & 5y \end{vmatrix} = 5\mathbf{i} + 2\mathbf{j} + 3\mathbf{k}$$

Since the outward normals are upward normals, and the paraboloid is given by $z = z(x, y) = 4 - x^2 - y^2$, it follows from Formula (6) of Section 18.5 (with curl F replacing F) that

$$\iint_\sigma (\text{curl } \mathbf{F}) \cdot \mathbf{n} \, dS = \iint_R (\text{curl } \mathbf{F}) \cdot \left(-\frac{\partial z}{\partial x} \mathbf{i} - \frac{\partial z}{\partial y} \mathbf{j} + \mathbf{k} \right) dA$$

$$= \iint_R (5\mathbf{i} + 2\mathbf{j} + 3\mathbf{k}) \cdot (2x\mathbf{i} + 2y\mathbf{j} + \mathbf{k}) \, dA$$

$$= \iint_R (10x + 4y + 3) \, dA$$

$$= \int_0^{2\pi} \int_0^2 (10r \cos \theta + 4r \sin \theta + 3) r \, dr \, d\theta$$

$$= \int_0^{2\pi} \left[\frac{10r^3}{3} \cos \theta + \frac{4r^3}{3} \sin \theta + \frac{3r^2}{2} \right]_{r=0}^2 d\theta$$

$$= \int_0^{2\pi} \left(\frac{80}{3} \cos \theta + \frac{32}{3} \sin \theta + 6 \right) d\theta$$

$$= \left[\frac{80}{3} \sin \theta - \frac{32}{3} \cos \theta + 6\theta \right]_0^{2\pi} = 12\pi$$

which is the same as the value of the line integral.   ◄

REMARK. Had we oriented σ by inward normals in this example, then the surface integral and the line integral would have had value -12π. (Why?)

Example 3 Let C be the rectangle in the plane $z = y$ oriented as in Figure 18.7.3, and let $\mathbf{F}(x, y, z) = x^2\mathbf{i} + 4xy^3\mathbf{j} + y^2x\mathbf{k}$. Find

$$\int_C \mathbf{F} \cdot d\mathbf{r}$$

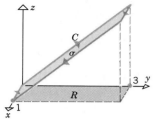

Figure 18.7.3

Solution. To evaluate the integral directly would require four separate integrations, one over each side of the rectangle. Instead, we shall apply Stokes' Theorem, so we need only evaluate a single surface integral over the rectangular surface σ bounded by C. For the positive direction of C to be as shown in Figure 18.7.3, the surface σ must be oriented by downward normals.

Since the surface σ has equation $z = y$ and

$$\text{curl } \mathbf{F} = \begin{vmatrix} \mathbf{i} & \mathbf{j} & \mathbf{k} \\ \dfrac{\partial}{\partial x} & \dfrac{\partial}{\partial y} & \dfrac{\partial}{\partial z} \\ x^2 & 4xy^3 & y^2x \end{vmatrix} = 2yx\mathbf{i} - y^2\mathbf{j} + 4y^3\mathbf{k}$$

it follows from Formula (7) of Theorem 18.5.2 that

$$\iint_\sigma (\text{curl } \mathbf{F}) \cdot \mathbf{n} \, dS = \iint_R (\text{curl } \mathbf{F}) \cdot \left(\frac{\partial z}{\partial x} \mathbf{i} + \frac{\partial z}{\partial y} \mathbf{j} - \mathbf{k} \right) dA$$

$$= \iint_R (2yx\mathbf{i} - y^2\mathbf{j} + 4y^3\mathbf{k}) \cdot (0\mathbf{i} + \mathbf{j} - \mathbf{k}) \, dA$$

$$= \int_0^1 \int_0^3 (-y^2 - 4y^3) \, dy \, dx$$

$$= -\int_0^1 \left[\frac{y^3}{3} + y^4 \right]_{y=0}^3 dx$$

$$= -\int_0^1 90 \, dx = -90 \quad \blacktriangleleft$$

▶ **Exercise Set 18.7**

In Exercises 1–5, find the curl of $\mathbf{F}$.

1. $\mathbf{F}(x, y, z) = x\mathbf{i} + y\mathbf{j} + z\mathbf{k}$.

2. $\mathbf{F}(x, y, z) = xz^3\mathbf{i} + 2y^4x^2\mathbf{j} + 5z^2y\mathbf{k}$.

3. $\mathbf{F}(x, y, z) = e^{xy}\mathbf{i} - \cos y\mathbf{j} + \sin^2 z\mathbf{k}$.

4. $\mathbf{F}(x, y, z) = 7y^3z^2\mathbf{i} - 8x^2z^5\mathbf{j} - 3xy^4\mathbf{k}$.

5. $\mathbf{F}(x, y, z) = \ln x\mathbf{i} + e^{xyz}\mathbf{j} + \tan^{-1}\left(\dfrac{z}{x}\right)\mathbf{k}$.

In Exercises 6–12, use Stokes' Theorem to evaluate the integral $\displaystyle\int_C \mathbf{F} \cdot d\mathbf{r}$.

6. $\mathbf{F}(x, y, z) = xz\mathbf{i} + 3x^2y^2\mathbf{j} + yx\mathbf{k}$; C is the rectangle in the plane $z = y$ shown in Figure 18.7.3.

7. $\mathbf{F}(x, y, z) = 3z\mathbf{i} + 4x\mathbf{j} + 2y\mathbf{k}$; C is the boundary of the paraboloid shown in Figure 18.7.2.

8. $\mathbf{F}(x, y, z) = -3y^2\mathbf{i} + 4z\mathbf{j} + 6x\mathbf{k}$; C is the triangle in the plane $z = \frac{1}{2}y$ with vertices $(2, 0, 0)$, $(0, 2, 1)$, and $(0, 0, 0)$ with a counterclockwise orientation looking down the positive z-axis.

9. $\mathbf{F}(x, y, z) = xy\mathbf{i} + x^2\mathbf{j} + z^2\mathbf{k}$; C is the intersection of the paraboloid $z = x^2 + y^2$ and the plane $z = y$ with a counterclockwise orientation looking down the positive z-axis.

10. $\mathbf{F}(x, y, z) = xy\mathbf{i} + yz\mathbf{j} + zx\mathbf{k}$; C is the triangle in the plane $x + y + z = 1$ with vertices $(1, 0, 0)$, $(0, 1, 0)$, and $(0, 0, 1)$ with a counterclockwise orientation looking from the first octant toward the origin.

11. $\mathbf{F}(x, y, z) = (x - y)\mathbf{i} + (y - z)\mathbf{j} + (z - x)\mathbf{k}$; C is the circle $x^2 + y^2 = a^2$ in the xy-plane with counterclockwise orientation looking down the positive z-axis.

12. $\mathbf{F}(x, y, z) = (z + \sin x)\mathbf{i} + (x + y^2)\mathbf{j} + (y + e^z)\mathbf{k}$; C is the intersection of the sphere $x^2 + y^2 + z^2 = 1$ and the cone $z = \sqrt{x^2 + y^2}$ with counterclockwise orientation looking down the positive z-axis.

In Exercises 13–16, verify Stokes' Theorem by computing the line and surface integrals in (3) and showing that they are equal.

13. $\mathbf{F}(x, y, z) = (x - y)\mathbf{i} + (y - z)\mathbf{j} + (z - x)\mathbf{k}$; σ is the portion of the plane $x + y + z = 1$ in the first octant.

14. $\mathbf{F}(x, y, z) = x^2\mathbf{i} + y^2\mathbf{j} + z^2\mathbf{k}$; σ is the portion of the cone $z = \sqrt{x^2 + y^2}$ below the plane $z = 1$.

15. $\mathbf{F}(x, y, z) = x\mathbf{i} + y\mathbf{j} + z\mathbf{k}$; σ is the upper hemisphere $z = \sqrt{a^2 - x^2 - y^2}$.

16. $\mathbf{F}(x, y, z) = (z - y)\mathbf{i} + (z + x)\mathbf{j} - (x + y)\mathbf{k}$; σ is the portion of the paraboloid $z = 9 - x^2 - y^2$ above the xy-plane.

17. Use Stokes' Theorem to evaluate

$$\int_C z^2\, dx + 2x\, dy - y^3\, dz$$

over the circle $x^2 + y^2 = 1$ in the xy-plane traversed counterclockwise looking down the positive z-axis. [*Hint:* Find a surface σ whose boundary is C.]

18. Use Stokes' Theorem to evaluate

$$\int_C y^2\, dx + z^2\, dy + x^2\, dz$$

over the triangle with vertices $P(a, 0, 0)$, $Q(0, a, 0)$, $R(0, 0, a)$ traversed counterclockwise looking toward the origin from the first octant.

19. Prove: If the components of $\mathbf{F}$ and their first- and second-order partial derivatives are continuous, then

$$\text{div (curl } \mathbf{F}) = 0$$

20. Use Exercise 19 and the Divergence Theorem to prove that

$$\iint_\sigma (\text{curl } \mathbf{F}) \cdot \mathbf{n}\, dS = 0$$

if σ is a sphere, $\mathbf{n}$ is the unit outer normal to σ, and the components of $\mathbf{F}$ satisfy the conditions stated in Exercise 19 at all points on or inside the sphere.

21. Prove: If f and the components of $\mathbf{F}$ are continuous and have continuous second-order partial derivatives, then

(a) $\text{curl } (\nabla f) = \mathbf{0}$

(b) $\text{curl } (\nabla f + \text{curl } \mathbf{F}) = \text{curl (curl } \mathbf{F})$.
[*Hint:* You may use the property $\text{curl } (\mathbf{F}_1 + \mathbf{F}_2) = \text{curl } \mathbf{F}_1 + \text{curl } \mathbf{F}_2$ without proof.]

22. (a) Use Stokes' Theorem to prove: If σ_1 and σ_2 are oriented surfaces with the same boundary C, and if C has the same positive orientation relative to each surface, then

$$\iint_{\sigma_1} (\text{curl } \mathbf{F}) \cdot \mathbf{n}\, dS = \iint_{\sigma_2} (\text{curl } \mathbf{F}) \cdot \mathbf{n}\, dS$$

(b) Use the result in part (a) to evaluate

$$\iint_\sigma (\text{curl } \mathbf{F}) \cdot \mathbf{n}\, dS$$

where $\mathbf{F}(x, y, z) = x^3\mathbf{i} + y\mathbf{j} + z^2e^{xy}\mathbf{k}$ and the surface σ is the upper hemisphere given by $z = \sqrt{a^2 - x^2 - y^2}$ and oriented by upward unit normals. [*Hint:* Find a circular disk in the xy-plane that shares a common boundary with the hemisphere.]

23. Use Stokes' Theorem to prove the result in Exercise 20. [*Hint:* See Exercise 22(a) and note that the upper and lower hemispheres share a common boundary.]

18.8 APPLICATIONS

> *In this section we shall discuss applications of Stokes' Theorem and the Divergence Theorem to problems of fluid flow. These theorems also have applications to problems in electricity, magnetism, and mechanics which we shall not discuss here.*

□ **FLUX**

Consider a body of fluid flowing in a region of three-dimensional space, and suppose that at each point in the region the velocity of the fluid is given by

$$\mathbf{F}(x, y, z) = f(x, y, z)\mathbf{i} + g(x, y, z)\mathbf{j} + h(x, y, z)\mathbf{k}$$

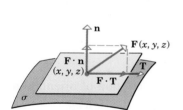

Figure 18.8.1

The function $\mathbf{F}(x, y, z)$ is called the *flow field* for the fluid motion (Figure 18.8.1). For simplicity, we shall make two assumptions:

1. The velocity of the fluid at each point of the region does not vary with time. Such a flow is said to be *steady* over the region.

2. The density (mass per unit volume) of the fluid is constant throughout the region. Such a fluid is called *incompressible* over the region. To keep our formulas simple, we shall assume this density to be 1.

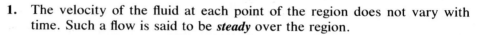

Figure 18.8.2

If σ is an oriented surface within such a three-dimensional body of fluid, then at each point (x, y, z) on the surface the velocity vector $\mathbf{F}(x, y, z)$ can be resolved into two components, a component $\mathbf{F} \cdot \mathbf{T}$ tangent to the surface, and a component $\mathbf{F} \cdot \mathbf{n}$ along the unit normal vector $\mathbf{n}$ in the orientation of σ (Figure 18.8.2). The component $\mathbf{F} \cdot \mathbf{T}$ is the component of flow *along* the surface, and $\mathbf{F} \cdot \mathbf{n}$ is the component *across* the surface. Fluid crosses σ in the direction of orientation where $\mathbf{F} \cdot \mathbf{n}$ is positive and opposite to that direction where $\mathbf{F} \cdot \mathbf{n}$ is negative. (Why?) In a specified time interval, the volume of fluid crossing σ in the direction of orientation minus the volume crossing in the opposite direction is called the *net volume* of fluid crossing σ during the time interval, and the surface integral

$$\iint_{\sigma} \mathbf{F} \cdot \mathbf{n} \, dS$$

is called the *flux of F across σ*; it represents the net volume of fluid that crosses the oriented surface σ in one unit of time. To see this, subdivide σ into n parts $\sigma_1, \sigma_2, \ldots, \sigma_n$ with areas

$$\Delta S_1, \Delta S_2, \ldots, \Delta S_n$$

If the parts are small and the flow is not too erratic, it is reasonable to assume that the velocity does not vary much on each part. Thus, if (x_k^*, y_k^*, z_k^*) is any point in the kth part, we can assume that $\mathbf{F}(x, y, z)$ is constant and equal to

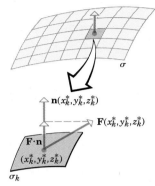

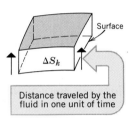

Figure 18.8.3

Surface

ΔS_k

Distance traveled by the fluid in one unit of time

Figure 18.8.4

$\mathbf{F}(x_k^*, y_k^*, z_k^*)$ throughout the part, and that the component of flow across the surface σ_k is

$$\mathbf{F}(x_k^*, y_k^*, z_k^*) \cdot \mathbf{n}(x_k^*, y_k^*, z_k^*) \tag{1}$$

where $\mathbf{n}(x_k^*, y_k^*, z_k^*)$ is the unit normal at (x_k^*, y_k^*, z_k^*) in the orientation of σ (Figure 18.8.3).

If we observe the flow across σ_k for one unit of time, then the section of fluid initially on the surface will move to some new position and the volume of fluid that crosses σ_k will form a solid (Figure 18.8.4) with a volume that can be approximated by

$$\text{volume} \approx \begin{bmatrix} \text{distance traveled by} \\ \text{the fluid in one unit} \\ \text{of time} \end{bmatrix} \Delta S_k \tag{2}$$

But (1) is the component of fluid velocity across the surface and therefore represents either the distance traveled by the fluid in one unit of time or the negative of that distance, depending on whether $\mathbf{F} \cdot \mathbf{n}$ is positive or negative. Thus, from (2)

$$\mathbf{F}(x_k^*, y_k^*, z_k^*) \cdot \mathbf{n}(x_k^*, y_k^*, z_k^*) \, \Delta S_k$$

is either the approximate volume of fluid crossing σ_k in one unit of time or the negative of that approximate volume. It is positive if the fluid is crossing in the same direction as the orienting vector $\mathbf{n}(x_k^*, y_k^*, z_k^*)$ and negative if it is crossing in the opposite direction. Thus,

$$\sum_{k=1}^{n} \mathbf{F}(x_k^*, y_k^*, z_k^*) \cdot \mathbf{n}(x_k^*, y_k^*, z_k^*) \, \Delta S_k \tag{3}$$

is approximately the volume of fluid crossing σ in the direction of orientation minus the volume crossing in the opposite direction.

If we now repeat the subdivision process, dividing σ into more and more parts in such a way that each part shrinks to a point, then we can expect that the errors in our approximations diminish to zero, and

$$\lim_{n \to +\infty} \sum_{k=1}^{n} \mathbf{F}(x_k^*, y_k^*, z_k^*) \cdot \mathbf{n}(x_k^*, y_k^*, z_k^*) \, \Delta S_k = \iint_{\sigma} \mathbf{F}(x, y, z) \cdot \mathbf{n}(x, y, z) \, dS$$

represents the *exact* net volume of fluid crossing σ in one unit of time.

In summary, we have the following result:

Formula for Flux

$$\begin{bmatrix} \text{flux of } \mathbf{F} \\ \text{across } \sigma \end{bmatrix} = \begin{bmatrix} \text{net volume of fluid crossing} \\ \sigma \text{ in the direction of orien-} \\ \text{tation in one unit of time.} \end{bmatrix} = \iint_{\sigma} \mathbf{F} \cdot \mathbf{n} \, dS \tag{4}$$

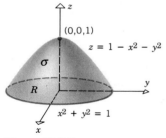

Figure 18.8.5

Example 1 Let σ be the portion of the surface $z = 1 - x^2 - y^2$ above the xy-plane oriented by upward normals (Figure 18.8.5), and suppose that the flow field is

$$\mathbf{F}(x, y, z) = x\mathbf{i} + y\mathbf{j} + z\mathbf{k}$$

Find the net volume of fluid that crosses σ in the direction of orientation in one unit of time.

Solution. From (4), the net volume of fluid crossing σ in one unit of time is

$$\iint_\sigma \mathbf{F} \cdot \mathbf{n} \, dS$$

We showed in Example 4 of Section 18.5 that

$$\iint_\sigma \mathbf{F} \cdot \mathbf{n} \, dS = \iint_\sigma (x^2 + y^2 + 1) \, dA = \frac{3\pi}{2}$$

so there is a net volume of $3\pi/2$ cubic units of fluid crossing σ in one unit of time. ◄

□ **PHYSICAL INTERPRETATION OF DIVERGENCE**

We now have the background to give a physical interpretation of the divergence of a flow field

$$\mathbf{F}(x, y, z) = f(x, y, z)\mathbf{i} + g(x, y, z)\mathbf{j} + h(x, y, z)\mathbf{k}$$

If G is a *small* spherical region centered at (x, y, z) and div $\mathbf{F}$ is continuous on G, then div $\mathbf{F}$ will not vary much from its value at the center (x, y, z); and we can assume that div $\mathbf{F}$ is constant over G. If σ is the surface of G oriented by outward normals, then the Divergence Theorem (18.6.2) yields

$$\iint_\sigma \mathbf{F} \cdot \mathbf{n} \, dS = \iiint_G \operatorname{div} \mathbf{F} \, dV \approx \operatorname{div} \mathbf{F}(x, y, z) \iiint_G dV$$

where div $\mathbf{F}(x, y, z)$ is the divergence at the center of the sphere. Therefore,

$$\iint_\sigma \mathbf{F} \cdot \mathbf{n} \, dS \approx \operatorname{div} \mathbf{F}(x, y, z) \, [\text{volume } (G)] \tag{5}$$

or

$$\operatorname{div} \mathbf{F}(x, y, z) \approx \frac{1}{\operatorname{vol}(G)} \iint_\sigma \mathbf{F} \cdot \mathbf{n} \, dS = \frac{\text{flux of } \mathbf{F} \text{ across } \sigma}{\operatorname{vol}(G)} \tag{6}$$

If we assume that the errors in our approximations diminish to zero over smaller and smaller spheres, then it follows from (6) that

$$\text{div } \mathbf{F}(x, y, z) = \lim_{\text{vol}\,(G) \to 0} \frac{\text{flux of } \mathbf{F} \text{ across } \sigma}{\text{vol}\,(G)}$$

The quantity

$$\frac{\text{flux of } \mathbf{F} \text{ across } \sigma}{\text{vol}\,(G)}$$

is called the ***flux density of F over the sphere G,*** and its limit as $\text{vol}\,(G) \to 0$ is called the ***flux density of F*** at the point (x, y, z). Thus, we have the following physical description of the divergence:

$$\text{div } \mathbf{F}(x, y, z) = \text{flux density of } \mathbf{F} \text{ at } (x, y, z)$$

The flux density at (x, y, z) is a measure of the rate at which fluid is approaching or departing from the point (x, y, z). To see why, consider a sphere G centered at (x, y, z) with its surface σ oriented by outward unit normals. If the sphere is small, then the volume of fluid approaching or departing from (x, y, z) in one unit of time is approximately the volume of fluid crossing the surface σ in that time. But (4) and (5) imply that the net volume of fluid crossing σ in one unit of time is directly proportional to the divergence of $\mathbf{F}$ at (x, y, z).

☐ SOURCES AND SINKS

We leave it for the reader to show that the net fluid flow is away from (x, y, z) if $\text{div } \mathbf{F}(x, y, z) > 0$, and toward (x, y, z) if $\text{div } \mathbf{F}(x, y, z) < 0$. Points where $\text{div } \mathbf{F}(x, y, z) > 0$ are called ***sources,*** and points where $\text{div } \mathbf{F}(x, y, z) < 0$ are called ***sinks.*** At a source, fluid is entering the flow, and at a sink it is leaving the flow. If $\text{div } \mathbf{F}(x, y, z) = 0$ throughout the flow, then there are no sources or sinks.

Example 2   The fluid with flow field $\mathbf{F}(x, y, z) = yz\mathbf{i} + xz\mathbf{j} + xy\mathbf{k}$ has no sources or sinks since

$$\text{div } \mathbf{F}(x, y, z) = \frac{\partial}{\partial x}\,(yz) + \frac{\partial}{\partial y}\,(xz) + \frac{\partial}{\partial z}\,(xy) = 0 \qquad ◀$$

We are now in a position to give an important physical interpretation of the Divergence Theorem. Let σ be an oriented surface that encloses a finite three-dimensional region G inside a moving body of fluid (Figure 18.8.6). If $\mathbf{F}(x, y, z)$ is the flow field, then it follows from the Divergence Theorem [Formula (1), Section 18.6] that

$$\iint_\sigma \mathbf{F} \cdot \mathbf{n}\, dS = \iiint_G \text{div } \mathbf{F}\, dV$$

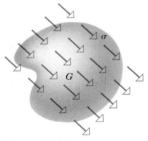

Figure 18.8.6

But from (4) and (7) this equation states that

$$\begin{bmatrix} \text{flux of } \mathbf{F} \\ \text{across } \sigma \end{bmatrix} = \iiint\limits_{G} [\text{flux density}] \, dV$$

In other words, the Divergence Theorem asserts that *the net volume of fluid that crosses σ in one unit of time is obtained by integrating the flux density over the region enclosed by σ.*

□ **PHYSICAL INTERPRETATION OF CURL**

Figure 18.8.7

We conclude this section by obtaining a physical interpretation of the curl of a flow field

$$\mathbf{F}(x, y, z) = f(x, y, z)\mathbf{i} + g(x, y, z)\mathbf{j} + h(x, y, z)\mathbf{k}$$

Let σ_a be a *small* oriented disk-shaped region of radius a, centered at (x, y, z), and having boundary C_a (Figure 18.8.7a). If the components of curl $\mathbf{F}$ are continuous on σ_a, then curl $\mathbf{F}$ will not vary much from its "value" at the center (x, y, z), and we can assume curl $\mathbf{F}$ to be approximately constant on σ_a. Since σ_a is a flat surface, the unit normal vectors that orient σ_a are all the same as the unit normal $\mathbf{n} = \mathbf{n}(x, y, z)$ at the center (x, y, z) (Figure 18.8.7b). Thus, curl $\mathbf{F} \cdot \mathbf{n}$ is also constant on σ_a, so Equation (4) of Section 18.7 and Stokes' Theorem (18.7.2) yield

$$\int_{C_a} \mathbf{F} \cdot \mathbf{T} \, ds = \iint\limits_{\sigma_a} (\text{curl } \mathbf{F}) \cdot \mathbf{n} \, dS \approx \text{curl } \mathbf{F}(x, y, z) \cdot \mathbf{n} \iint\limits_{\sigma_a} dS$$

where the line integral is taken in the positive direction of C_a, and $\mathbf{T}$ is the unit tangent vector to C_a pointing in the positive direction. Therefore,

$$\int_{C_a} \mathbf{F} \cdot \mathbf{T} \, ds \approx [\text{curl } \mathbf{F}(x, y, z) \cdot \mathbf{n}] \, [\text{surface area of } \sigma_a]$$

or

$$\int_{C_a} \mathbf{F} \cdot \mathbf{T} \, ds \approx \pi a^2 [\text{curl } \mathbf{F}(x, y, z) \cdot \mathbf{n}]$$

or

$$\text{curl } \mathbf{F}(x, y, z) \cdot \mathbf{n} \approx \frac{1}{\pi a^2} \int_{C_a} \mathbf{F} \cdot \mathbf{T} \, ds \tag{8}$$

If we assume that the errors in our approximations diminish to zero over smaller and smaller disks, then it follows from (8) that

$$\text{curl } \mathbf{F}(x, y, z) \cdot \mathbf{n} = \lim_{a \to 0} \frac{1}{\pi a^2} \int_{C_a} \mathbf{F} \cdot \mathbf{T} \, ds \tag{9}$$

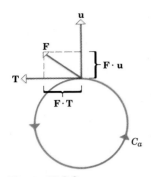

Figure 18.8.8

The integral in (9) has a physical interpretation that we shall now explain. At each point of C_a, the flow field $\mathbf{F}$ has a component $\mathbf{F} \cdot \mathbf{u}$ along the outward normal to C_a and a component $\mathbf{F} \cdot \mathbf{T}$ tangent to C_a (Figure 18.8.8). When $\mathbf{F}$ is closely aligned with $\mathbf{T}$, the magnitude of $\mathbf{F} \cdot \mathbf{T}$ is large compared with the magnitude of $\mathbf{F} \cdot \mathbf{u}$, so the fluid tends to move along the circle C_a rather than pass through it. Thus, physicists call the integral

$$\int_{C_a} \mathbf{F} \cdot \mathbf{T} \, ds$$

the *circulation of $\mathbf{F}$ around C_a* and use it as a measure of the tendency for fluid to circulate in the positive direction of C_a.

If the flow is normal to the circle at each point of C_a, then $\mathbf{F} \cdot \mathbf{T} = 0$, so

$$\int_{C_a} \mathbf{F} \cdot \mathbf{T} \, ds = 0$$

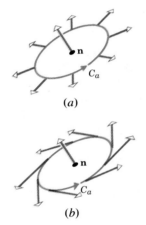

(a)

(b)

Figure 18.8.9

and there is no tendency for the fluid to circulate around C_a (Figure 18.8.9a). However, the more closely $\mathbf{F}$ is aligned with $\mathbf{T}$, the larger the value of $\mathbf{F} \cdot \mathbf{T}$ and the larger the value of the circulation. The maximum circulation around C_a occurs when $\mathbf{F}$ is aligned with $\mathbf{T}$ at each point of C_a (Figure 18.8.9b).

The quantity

$$\frac{1}{\pi a^2} \int_{C_a} \mathbf{F} \cdot \mathbf{T} \, ds = \frac{\text{circulation of } \mathbf{F} \text{ around } C_a}{\text{area of the disk } \sigma_a}$$

is called the *specific circulation* of $\mathbf{F}$ over the disk σ_a, and its limit as $a \to 0$ is called the *rotation of $\mathbf{F}$ about $\mathbf{n}$* at (x, y, z). Thus, from (9),

> curl $\mathbf{F}(x, y, z) \cdot \mathbf{n}$ = rotation of $\mathbf{F}$ about $\mathbf{n}$ at (x, y, z)

Since curl $\mathbf{F} \cdot \mathbf{n}$ has its maximum value when $\mathbf{n}$ is in the same direction as curl $\mathbf{F}(x, y, z)$, it follows that the rotation of $\mathbf{F}$ at (x, y, z) has its maximum value about this $\mathbf{n}$. Thus, in the vicinity of the point (x, y, z) the maximum circulation occurs around small circles in the plane normal to curl $\mathbf{F}$. Physically, if a paddle wheel is immersed in the fluid so that the pivot point is at (x, y, z), then the paddles will turn most quickly when the spindle is aligned with curl $\mathbf{F}(x, y, z)$ (Figure 18.8.10).

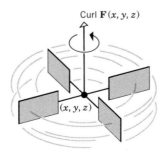

Figure 18.8.10

▶ Exercise Set 18.8

In Exercises 1–4, a flow field $F(x, y, z)$ is given. Find the net volume of fluid that crosses the surface σ in the direction of orientation in one unit of time.

1. $F(x, y, z) = x\mathbf{i} + y\mathbf{j} + 2z\mathbf{k}$; σ is the portion of the surface $z = 1 - x^2 - y^2$ oriented by upward unit normals.

2. $F(x, y, z) = x^2\mathbf{i} + yx\mathbf{j} + zx\mathbf{k}$; σ is the portion of the plane $6x + 3y + 2z = 6$ in the first octant oriented by upward unit normals.

3. $F(x, y, z) = (x + y)\mathbf{i} + (y + z)\mathbf{j} + (z + x)\mathbf{k}$; σ is the upper hemisphere $z = \sqrt{9 - x^2 - y^2}$ oriented by upward unit normals.

4. $F(x, y, z) = x\mathbf{i} + y\mathbf{j} + 2z\mathbf{k}$; σ is the portion of the cone $z^2 = x^2 + y^2$ between the planes $z = 1$ and $z = 2$, oriented by upward unit normals.

In Exercises 5–8, a flow field $F(x, y, z)$ is given. Use the Divergence Theorem to find the net volume of fluid that crosses the surface σ in the direction of orientation in one unit of time.

5. $F(x, y, z) = x\mathbf{i} + y\mathbf{j} + z\mathbf{k}$, where σ is the sphere $x^2 + y^2 + z^2 = a^2$ oriented by outward unit normals.

6. $F(x, y, z) = xy\mathbf{i} + yz\mathbf{j} + xz\mathbf{k}$; σ is the cube bounded by the coordinate planes and the planes $x = 2$, $y = 2$, and $z = 2$, oriented by outward unit normals.

7. $F(x, y, z) = (x^3 - y)\mathbf{i} + (y^3 + z)\mathbf{j} + (z^3 - 2xy)\mathbf{k}$; σ is the surface of the cylinder (including top and bottom) bounded by $x^2 + y^2 = 4$, $z = 0$, and $z = 3$, oriented by outward unit normals.

8. $F(x, y, z) = x^2\mathbf{i} + xy\mathbf{j} - 2xz\mathbf{k}$; σ is the tetrahedron bounded by the coordinate planes and the plane $x + y + z = 1$, oriented by outward unit normals.

In Exercises 9–12, determine whether the flow field $F(x, y, z)$ is free of sources and sinks. If it is not, find the location of all sources and sinks.

9. $F(x, y, z) = (y + z)\mathbf{i} - xz^3\mathbf{j} + (x^2 \sin y)\mathbf{k}$.

10. $F(x, y, z) = xy\mathbf{i} - xy\mathbf{j} + y^2\mathbf{k}$.

11. $F(x, y, z) = x^3\mathbf{i} + y^3\mathbf{j} + z^3\mathbf{k}$.

12. $F(x, y, z) = (x^3 - x)\mathbf{i} + (y^3 - y)\mathbf{j} + (z^3 - z)\mathbf{k}$.

13. For the flow field given by the formula
$F(x, y, z) = (x - z)\mathbf{i} + (y - x)\mathbf{j} + (z - xy)\mathbf{k}$,
(a) use Stokes' Theorem to find the circulation around the triangle with vertices $A(1, 0, 0)$, $B(0, 2, 0)$, and $C(0, 0, 1)$ oriented counterclockwise looking from the origin toward the first octant
(b) find the rotation of F about the vector $\mathbf{k}$ at the origin
(c) find the unit vector $\mathbf{n}$ such that the rotation of F about $\mathbf{n}$ at the origin is maximum.

▶ SUPPLEMENTARY EXERCISES

In Exercises 1–6, evaluate the line integral by any method.

1. $\int_C 2y\, dx + 3\, dy$ along $y = \sin x$ from the point $(0, 0)$ to the point $(\pi, 0)$.

2. $\int_C x^5\, dy$ along the curve C given by $x = 1/t$, $y = 4t^2$, $1 \le t \le 2$.

3. $\int_C \langle 2e^y, -x \rangle \cdot d\mathbf{r}$ along $y = \ln x$ from the point $(1, 0)$ to the point $(3, \ln 3)$.

4. $\int_C (y\mathbf{i} + z\mathbf{j} + x\mathbf{k}) \cdot d\mathbf{r}$ along the curve C given by $\mathbf{r} = \langle t^2 - 2t, -2t, t - 2 \rangle$, $0 \le t \le 2$.

5. $\int_C z\, dx + y\, dy - x\, dz$ along the line segment from $(0, 0, 0)$ to $(1, 2, 3)$.

6. $\int_C x \sin xy\, dx - y \sin xy\, dy$ along the line segment from $(0, 0)$ to $(1, \pi)$.

In Exercises 7–9, determine whether F is conservative. If it is, find a potential function for it.

7. $F(x, y) = y \sin xy\mathbf{i} - x \cos xy\mathbf{j}$.

8. $F(x, y) = 2x(\ln y - 1)\mathbf{i} + \left(\dfrac{x^2}{y} - 3y^2\right)\mathbf{j}$.

9. $F(x, y) = \left(3x^2 - \dfrac{y^2}{x^2}\right)\mathbf{i} + \left(\dfrac{2y}{x} + 4y\right)\mathbf{j}$.

10. Let $F(x, y) = y\mathbf{i} - 2x\mathbf{j}$. Find functions $h(x)$ and $g(y)$ such that $h(x)F(x, y)$ and $g(y)F(x, y)$ are each conservative.

In Exercises 11–14, determine whether the integral is independent of the path. If it is, find its value by any method.

11. $\displaystyle\int_{(1,\pi/4)}^{(2,\pi/4)} (\cos 2y - 3x^2y^2)\, dx$
$$+ (\cos 2y - 2x\sin 2y - 2x^3y)\, dy.$$

12. $\displaystyle\int_{(0,0)}^{(2,-2)} x^2y^4\, dx + y^2x^4\, dy.$

13. $\displaystyle\int_{(1,2)}^{(2,1)} \frac{y\, dx - x\, dy}{y^2}.$

14. $\displaystyle\int_{(0,1)}^{(1,0)} (ye^{xy} - 1)\, dx + (xe^{xy})\, dy.$

In Exercises 15–18, evaluate the line integral counterclockwise over C using Green's Theorem.

15. $\displaystyle\int_C 2xy\, dx + xy^2\, dy$ along the triangle with vertices $(0, 0)$, $(2, 0)$, $(2, 4)$.

16. $\displaystyle\int_C 3y\, dx - 4x\, dy$ along the boundary of the semicircular region $\{(x, y): x^2 + y^2 \le 2 \text{ and } 0 \le x\}$.

17. $\displaystyle\int_C -2y\, dx + (x - 4)\, dy$; C runs from $(0, 2)$ to $(0, -2)$ on the y-axis, then to $(0, 2)$ along the curve $x = 4 - y^2$.

18. $\displaystyle\int_C 5xy\, dx + x^3\, dy$ along the curve $y = x^2$ from $(0, 0)$ to $(2, 4)$, then back to $(0, 0)$ along $y = 2x$.

19. Use Green's Theorem to evaluate $\displaystyle\int_C y^3\, dx - x^3\, dy$ counterclockwise around the boundary of the region $R = \{(x, y): 1 \le x^2 + y^2 \le 4 \text{ and } 0 \le y\}$.

20. Find the area of the circle $r = 2\cos\theta$ using Formula (4c) of Section 18.3.

In Exercises 21–23, use any method to find the work done by the force F as it acts on a particle moving along C.

21. $F = (2x + 3y)\mathbf{i} + (xy)\mathbf{j}$ along $y = x^2$ from $(0, 0)$ to $(1, 1)$.

22. $F = 3x^2y^3\mathbf{i} + 3x^3y^2\mathbf{j}$ clockwise once around the cardioid $r = 3 + 3\cos\theta$.

23. $F = z\mathbf{i} + y\mathbf{j} - x\mathbf{k}$ on the line segment from $(0, 0, 0)$ to $(1, 2, 3)$.

24. Assuming that $\phi(x, y)$ has continuous first- and second-order partial derivatives, show that
(a) $\text{curl}(\nabla\phi) = \mathbf{0}$ (b) $\text{div}(\nabla\phi) = \phi_{xx} + \phi_{yy}$.

The centroid of a curved lamina σ is defined by
$$\bar{x} = \frac{\displaystyle\iint_\sigma x\, dS}{\text{area of }\sigma}, \quad \bar{y} = \frac{\displaystyle\iint_\sigma y\, dS}{\text{area of }\sigma}, \quad \bar{z} = \frac{\displaystyle\iint_\sigma z\, dS}{\text{area of }\sigma}$$
In Exercises 25 and 26, find the centroid of the lamina.

25. The lamina that is the portion of the paraboloid $z = \frac{1}{2}(x^2 + y^2)$ below the plane $z = 4$.

26. The lamina that is the portion of the sphere $x^2 + y^2 + z^2 = 4$ above the plane $z = 1$.

27. Let σ be the portion of the sphere
$$x^2 + y^2 + z^2 = 1$$
that lies in the first octant, $\mathbf{n}$ an outward unit normal (with respect to the sphere), and $D_{\mathbf{n}}f$ the directional derivative of
$$f(x, y, z) = \ln\sqrt{x^2 + y^2 + z^2}$$
Evaluate the surface integral
$$\iint_\sigma D_{\mathbf{n}}f\, dS$$

28. Let σ be the sphere $x^2 + y^2 + z^2 = 1$, $\mathbf{n}$ an inward unit normal, and $D_{\mathbf{n}}f$ the directional derivative of $f(x, y, z) = x^2 + y^2 + z^2$. Evaluate the surface integral
$$\iint_\sigma D_{\mathbf{n}}f\, dS$$

29. Let G be a solid with surface σ oriented by outward unit normals, suppose that ϕ has continuous first and second partial derivatives in some open set containing G, and let $D_{\mathbf{n}}\phi$ be the directional derivative of ϕ. Show that
$$\iint_\sigma D_{\mathbf{n}}\phi\, dS = \iiint_G \left[\frac{\partial^2\phi}{\partial x^2} + \frac{\partial^2\phi}{\partial y^2} + \frac{\partial^2\phi}{\partial z^2}\right] dV$$

30. Show that Green's Theorem is a special case of Stokes' Theorem by applying Stokes' Theorem to $F(x, y) = f(x, y)\mathbf{i} + g(x, y)\mathbf{j}$.

19

Second-Order Differential Equations

Gabriel Cramer   (1704-1752)

▪ 19.1 SECOND-ORDER LINEAR HOMOGENEOUS DIFFERENTIAL EQUATIONS WITH CONSTANT COEFFICIENTS

> *In Section 7.7 we showed how to solve first-order linear differential equations. In this section we shall show how to solve certain second-order linear differential equations.*

☐ **SECOND-ORDER LINEAR DIFFERENTIAL EQUATIONS**

Recall from Section 7.7 that a first-order linear differential equation has the form

$$\frac{dy}{dx} + p(x)y = q(x)$$

In this section we shall be concerned with the general **second-order linear differential equation**

$$\frac{d^2y}{dx^2} + p(x)\frac{dy}{dx} + q(x)y = r(x) \tag{1}$$

or in an alternative notation

$$y'' + p(x)y' + q(x)y = r(x)$$

where $p(x)$, $q(x)$, and $r(x)$ are continuous functions. If $r(x) = 0$ for all x, then (1) reduces to

$$\frac{d^2y}{dx^2} + p(x)\frac{dy}{dx} + q(x)y = 0$$

which is called the general second-order linear **homogeneous** differential equation. If $r(x)$ is not identically zero, then (1) is said to be **nonhomogeneous**. Some examples of second-order linear differential equations are

$$\frac{d^2y}{dx^2} + x^2\frac{dy}{dx} - xy = e^x \qquad \boxed{p(x) = x^2, \quad q(x) = -x, \quad r(x) = e^x}$$

$$y'' + y' - 3y = \sin x \qquad \boxed{p(x) = 1, \quad q(x) = -3, \quad r(x) = \sin x}$$

$$y'' + e^x y = 3 \qquad \boxed{p(x) = 0, \quad q(x) = e^x, \quad r(x) = 3}$$

$$\frac{d^2y}{dx^2} - \frac{dy}{dx} + 2y = 0 \qquad \boxed{p(x) = -1, \quad q(x) = 2, \quad r(x) = 0}$$

The last equation is homogeneous and the first three are nonhomogeneous.

In Section 7.7 we gave a general procedure for solving first-order linear differential equations by integration. For second-order linear differential equations the situation is more complicated and simple general procedures for solving such equations can only be given in special cases. Before we can pursue this matter further, we shall need some preliminary results.

First, some terminology. Two continuous functions f and g are said to be *linearly dependent* if one is a constant multiple of the other. If neither is a constant multiple of the other, then they are called *linearly independent*. Thus,

$$f(x) = \sin x \quad \text{and} \quad g(x) = 3 \sin x$$

are linearly dependent, but

$$f(x) = x \quad \text{and} \quad g(x) = x^2$$

are linearly independent.

The following theorem is central to the study of second-order linear differential equations.

19.1.1   THEOREM.   *If $y_1 = y_1(x)$ and $y_2 = y_2(x)$ are linearly independent solutions of the homogeneous equation*

$$\frac{d^2y}{dx^2} + p(x)\frac{dy}{dx} + q(x)y = 0 \tag{2}$$

then

$$y(x) = c_1 y_1(x) + c_2 y_2(x) \tag{3}$$

is the general solution of (2) in the sense that every solution of (2) can be obtained from (3) by choosing appropriate values for the arbitrary constants c_1 and c_2; conversely, (3) is a solution of (2) for all choices of c_1 and c_2.

(Readers interested in a proof of this theorem are referred to *Elementary Differential Equations and Boundary Value Problems*, John Wiley & Sons, New York, 1988, by William E. Boyce and Richard C. DiPrima.)

REMARK.   The expression on the right side of (3) is called a *linear combination* of $y_1(x)$ and $y_2(x)$. Thus, Theorem 19.1.1 tells us that once we find two linearly independent solutions of (2), we essentially know all the solutions because every other solution can be expressed as a linear combination of those two.

☐ CONSTANT
COEFFICIENTS

For the remainder of this section we shall restrict our attention to second-order linear homogeneous equations in which $p(x)$ and $q(x)$ are constants, p and q. Such equations have the form

$$\frac{d^2y}{dx^2} + p\frac{dy}{dx} + qy = 0 \tag{4}$$

Our objective is to find two linearly independent solutions of this equation. To start, we note that the function e^{mx} has the property that its derivatives are constant multiples of the function itself. This suggests the possibility that

$$y = e^{mx} \tag{5}$$

might be a solution of (4) if the constant m is suitably chosen. Since

$$\frac{dy}{dx} = me^{mx}, \quad \frac{d^2y}{dx^2} = m^2e^{mx} \tag{6}$$

substituting (5) and (6) into (4) yields

$$(m^2 + pm + q)e^{mx} = 0 \tag{7}$$

which is satisfied if and only if

$$m^2 + pm + q = 0 \tag{8}$$

since $e^{mx} \neq 0$ for every x.

Equation (8), which is called the **auxiliary equation** for (4), can be obtained from (4) by replacing d^2y/dx^2 by m^2, dy/dx by m (= m^1), and y by 1 (= m^0). The solutions, m_1 and m_2, of the auxiliary equation can be obtained by factoring or by the quadratic formula. These solutions are

$$m_1 = \frac{-p + \sqrt{p^2 - 4q}}{2}, \quad m_2 = \frac{-p - \sqrt{p^2 - 4q}}{2} \tag{9}$$

Depending on whether $p^2 - 4q$ is positive, zero, or negative, these roots will be distinct and real, equal and real, or complex conjugates.* We shall consider each of these cases separately.

☐ **DISTINCT REAL ROOTS**   If m_1 and m_2 are distinct real roots, then (4) has the two solutions

$$y_1 = e^{m_1x}, \quad y_2 = e^{m_2x}$$

Neither of the functions e^{m_1x} and e^{m_2x} is a constant multiple of the other (Exercise 29), so the general solution of (4) in this case is

$$y(x) = c_1e^{m_1x} + c_2e^{m_2x} \tag{10}$$

Example 1   Find the general solution of $y'' - y' - 6y = 0$.

Solution.   The auxiliary equation

$$m^2 - m - 6 = 0$$

can be rewritten as

$$(m + 2)(m - 3) = 0$$

---

*Recall that the complex solutions of a polynomial equation, and in particular of a quadratic equation, occur as conjugate pairs $a + bi$ and $a - bi$.

so its roots are $m = -2$, $m = 3$. Thus, from (10) the general solution of the differential equation is

$$y = c_1 e^{-2x} + c_2 e^{3x}$$

where c_1 and c_2 are arbitrary constants. ◄

☐ **EQUAL REAL ROOTS**   If m_1 and m_2 are equal real roots, say $m_1 = m_2$ $(= m)$, then the auxiliary equation yields only one solution of (4):

$$y_1(x) = e^{mx}$$

We shall now show that

$$y_2(x) = xe^{mx} \tag{11}$$

is a second linearly independent solution. To see that this is so, note that $p^2 - 4q = 0$ in (9) since the roots are equal. Thus,

$$m = m_1 = m_2 = -p/2$$

and (11) becomes

$$y_2(x) = xe^{(-p/2)x}$$

Differentiating yields

$$y_2'(x) = \left(1 - \frac{p}{2}x\right)e^{(-p/2)x} \quad \text{and} \quad y_2''(x) = \left(\frac{p^2}{4}x - p\right)e^{-(p/2)x}$$

so

$$
\begin{aligned}
y_2''(x) + py_2'(x) + qy_2(x) &= \left[\left(\frac{p^2}{4}x - p\right) + p\left(1 - \frac{p}{2}x\right) + qx\right]e^{(-p/2)x} \\
&= \left[-\frac{p^2}{4} + q\right]xe^{(-p/2)x}
\end{aligned}
\tag{12}
$$

But $p^2 - 4q = 0$ implies that $(-p^2/4) + q = 0$, so (12) becomes

$$y_2''(x) + py_2'(x) + qy_2(x) = 0$$

which tells us that $y_2(x)$ is a solution of (4). It can be shown that

$$y_1(x) = e^{mx} \quad \text{and} \quad y_2(x) = xe^{mx}$$

are linearly independent (Exercise 29), so the general solution of (4) in this case is

$$y = c_1 e^{mx} + c_2 xe^{mx} \tag{13}$$

Example 2 Find the general solution of $y'' - 8y' + 16y = 0$.

Solution. The auxiliary equation

$$m^2 - 8m + 16 = 0$$

can be rewritten as

$$(m - 4)^2 = 0$$

so $m = 4$ is the only root. Thus, from (13) the general solution of the differential equation is

$$y = c_1 e^{4x} + c_2 x e^{4x} \quad \blacktriangleleft$$

☐ **COMPLEX ROOTS**

If the auxiliary equation has complex roots $m_1 = a + bi$ and $m_2 = a - bi$, then $y_1(x) = e^{ax} \cos bx$ and $y_2(x) = e^{ax} \sin bx$ are linearly independent solutions of (4) and

$$y = e^{ax}(c_1 \cos bx + c_2 \sin bx) \tag{14}$$

is the general solution. The proof is discussed in the exercises (Exercise 30).

Example 3 Find the general solution of $y'' + y' + y = 0$.

Solution. The auxiliary equation $m^2 + m + 1 = 0$ has roots

$$m_1 = \frac{-1 + \sqrt{1 - 4}}{2} = -\frac{1}{2} + \frac{\sqrt{3}}{2} i$$

$$m_2 = \frac{-1 - \sqrt{1 - 4}}{2} = -\frac{1}{2} - \frac{\sqrt{3}}{2} i$$

Thus, from (14) with $a = -1/2$ and $b = \sqrt{3}/2$, the general solution of the differential equation is

$$y = e^{-x/2} \left(c_1 \cos \frac{\sqrt{3}}{2} x + c_2 \sin \frac{\sqrt{3}}{2} x \right) \quad \blacktriangleleft$$

☐ **INITIAL-VALUE PROBLEMS**

When a physical problem leads to a second-order differential equation, there are usually two conditions in the problem that determine specific values for the two arbitrary constants in the general solution of the equation. Conditions that specify the value of the solution $y(x)$ and its derivative $y'(x)$ at some point $x = x_0$ are called *initial conditions*. A second-order differential equation with initial conditions is called a *second-order initial-value problem*.

Example 4 Solve the initial-value problem

$$y'' - y = 0, \quad y(0) = 1, \quad y'(0) = 0$$

Solution. We must first solve the differential equation. The auxiliary equation

$$m^2 - 1 = 0$$

has distinct real roots $m_1 = 1$, $m_2 = -1$, so from (10) the general solution is

$$y(x) = c_1 e^x + c_2 e^{-x} \tag{15}$$

and the derivative of this solution is

$$y'(x) = c_1 e^x - c_2 e^{-x} \tag{16}$$

Substituting $x = 0$ in (15) and (16) and using the initial conditions $y(0) = 1$ and $y'(0) = 0$ yields the system of equations

$$c_1 + c_2 = 1$$
$$c_1 - c_2 = 0$$

Solving this system yields $c_1 = \frac{1}{2}$, $c_2 = \frac{1}{2}$, so from (15) the solution of the initial-value problem is

$$y(x) = \tfrac{1}{2} e^x + \tfrac{1}{2} e^{-x} \qquad \blacktriangleleft$$

The following summary is included as a ready reference for the solution of second-order homogeneous linear differential equations with constant coefficients.

Summary

EQUATION: $y'' + py' + qy = 0$	
AUXILIARY EQUATION: $m^2 + pm + q = 0$	
CASE	GENERAL SOLUTION
Distinct real roots m_1, m_2 to the auxiliary equation	$y = c_1 e^{m_1 x} + c_2 e^{m_2 x}$
Equal real roots $m_1 = m_2 \, (= m)$ to the auxiliary equation	$y = c_1 e^{mx} + c_2 x e^{mx}$
Complex roots $m_1 = a + bi$, $m_2 = a - bi$ to the auxiliary equation	$y = e^{ax}(c_1 \cos bx + c_2 \sin bx)$

▶ Exercise Set 19.1

1. Verify that the following are solutions of the differential equation $y'' - y' - 2y = 0$ by substituting these functions into the equation:
    (a) e^{2x} and e^{-x}
    (b) $c_1 e^{2x} + c_2 e^{-x}$ (c_1, c_2 constants).

2. Verify that the following are solutions of the differential equation $y'' + 4y' + 4y = 0$ by substituting these functions into the equation:
    (a) e^{-2x} and $x e^{-2x}$
    (b) $c_1 e^{-2x} + c_2 x e^{-2x}$ (c_1, c_2 constants).

In Exercises 3–16, find the general solution of the differential equation.

3. $y'' + 3y' - 4y = 0$.   **4.** $y'' + 6y' + 5y = 0$.

5. $y'' - 2y' + y = 0$.   **6.** $y'' + 6y' + 9y = 0$.

7. $y'' + 5y = 0$.   **8.** $y'' + y = 0$.

9. $\dfrac{d^2y}{dx^2} - \dfrac{dy}{dx} = 0$.   **10.** $\dfrac{d^2y}{dx^2} + 3\dfrac{dy}{dx} = 0$.

11. $\dfrac{d^2y}{dt^2} + 4\dfrac{dy}{dt} + 4y = 0$.

12. $\dfrac{d^2y}{dt^2} - 10\dfrac{dy}{dt} + 25y = 0$.

13. $\dfrac{d^2y}{dx^2} - 4\dfrac{dy}{dx} + 13y = 0$.

14. $\dfrac{d^2y}{dx^2} - 6\dfrac{dy}{dx} + 25y = 0$.

15. $8y'' - 2y' - 1 = 0$.   **16.** $9y'' - 6y' + 1 = 0$.

In Exercises 17–22, solve the initial-value problem.

17. $y'' + 2y' - 3y = 0$; $y(0) = 1$, $y'(0) = 5$.

18. $y'' - 6y' - 7y = 0$; $y(0) = 5$, $y'(0) = 3$.

19. $y'' - 6y' + 9y = 0$; $y(0) = 2$, $y'(0) = 1$.

20. $y'' + 4y' + y = 0$; $y(0) = 5$, $y'(0) = 4$.

21. $y'' + 4y' + 5y = 0$; $y(0) = -3$, $y'(0) = 0$.

22. $y'' - 6y' + 13y = 0$; $y(0) = -1$, $y'(0) = 1$.

23. In each part find a second-order linear homogeneous differential equation with constant coefficients that has the given functions as solutions.
(a) $y_1 = e^{5x}$, $y_2 = e^{-2x}$
(b) $y_1 = e^{4x}$, $y_2 = xe^{4x}$
(c) $y_1 = e^{-x}\cos 4x$, $y_2 = e^{-x}\sin 4x$.

24. Show that if e^x and e^{-x} are solutions of a second-order linear homogeneous differential equation, then so are $\cosh x$ and $\sinh x$.

25. Find all values of k for which the differential equation $y'' + ky' + ky = 0$ has a general solution of the given form.
(a) $y = c_1 e^{ax} + c_2 e^{bx}$ (b) $y = c_1 e^{ax} + c_2 x e^{ax}$
(c) $y = c_1 e^{ax}\cos bx + c_2 e^{ax}\sin bx$.

26. The equation
$$x^2\frac{d^2y}{dx^2} + px\frac{dy}{dx} + qy = 0 \qquad (x > 0)$$
where p and q are constants, is called *Euler's equidimensional equation*. Show that the substitution

$x = e^z$ transforms this equation into the equation
$$\frac{d^2y}{dz^2} + (p - 1)\frac{dy}{dz} + qy = 0$$

27. Use the result in Exercise 26 to find the general solution of
(a) $x^2\dfrac{d^2y}{dx^2} + 3x\dfrac{dy}{dx} + 2y = 0$ $(x > 0)$
(b) $x^2\dfrac{d^2y}{dx^2} - x\dfrac{dy}{dx} - 2y = 0$ $(x > 0)$.

28. Let $y(x)$ be a solution of $y'' + py' + qy = 0$. Prove: If p and q are positive constants, then $\lim\limits_{x \to +\infty} y(x) = 0$.

29. The *Wronskian* of two differentiable functions y_1 and y_2 is denoted by $W(y_1, y_2)$ and is defined to be the function
$$W(y_1, y_2) = y_1 y_2' - y_1' y_2 = \begin{vmatrix} y_1 & y_2 \\ y_1' & y_2' \end{vmatrix}$$
The value of $W(y_1, y_2)$ at a point x is denoted by $W(y_1, y_2)(x)$ or often more simply by $W(x)$. It can be proved that two solutions, $y_1 = y_1(x)$ and $y_2 = y_2(x)$, of Equation (2) are linearly dependent if and only if $W(x) = 0$ for all x. Equivalently, the functions are linearly independent if and only if $W(x) \neq 0$ for at least one value of x. Use this result to prove that the following solutions of Equation (4) are linearly independent:
(a) $y_1 = e^{m_1 x}$, $y_2 = e^{m_2 x}$ $(m_1 \neq m_2)$
(b) $y_1 = e^{mx}$, $y_2 = xe^{mx}$.

30. Prove: If the auxiliary equation of
$$y'' + py' + qy = 0$$
has complex roots $a + bi$ and $a - bi$, then the general solution of this differential equation is
$$y(x) = e^{ax}(c_1\cos bx + c_2\sin bx)$$
[*Hint:* By substitution, verify that $y_1 = e^{ax}\cos bx$ and $y_2 = e^{ax}\sin bx$ are solutions of the differential equation. Then use Exercise 29 to prove that y_1 and y_2 are linearly independent.]

31. Suppose that the auxiliary equation of the differential equation $y'' + py' + qy = 0$ has distinct real roots μ and m.
(a) Show that the function
$$g_\mu(x) = \frac{e^{\mu x} - e^{mx}}{\mu - m}$$
is a solution of the differential equation.

(b) Use L'Hôpital's rule to show that

$$\lim_{\mu \to m} g_\mu(x) = xe^{mx}$$

[*Note:* Can you see how the result in part (b) makes it plausible that the function $y(x) = xe^{mx}$ is a solution of $y'' + py' + qy = 0$ when m is a repeated root of the auxiliary equation?]

32. Consider the problem of solving the differential equation

$$y'' + \lambda y = 0$$

subject to the conditions $y(0) = 0$, $y(\pi) = 0$.

(a) Show that if $\lambda \le 0$, then $y = 0$ is the only solution.

(b) Show that if $\lambda > 0$, then the solution is

$$y = c \sin \sqrt{\lambda} x$$

where c is an arbitrary constant, if

$$\lambda = 1, 2^2, 3^2, 4^2, \ldots$$

and the only solution is $y = 0$ otherwise.

■ **19.2** SECOND-ORDER LINEAR NONHOMOGENEOUS DIFFERENTIAL EQUATIONS WITH CONSTANT COEFFICIENTS; UNDETERMINED COEFFICIENTS

In this section we shall study techniques for solving second-order linear differential equations that are not homogeneous.

☐ **THE COMPLEMENTARY EQUATION**

The following theorem is the key result for solving a second-order *nonhomogeneous* linear differential equation with constant coefficients

$$y'' + py' + qy = r(x) \tag{1}$$

where p and q are constants and $r(x)$ is a continuous function of x.

19.2.1 THEOREM. *The general solution of* (1) *is*

$$y(x) = c_1 y_1(x) + c_2 y_2(x) + y_p(x)$$

where $c_1 y_1(x) + c_2 y_2(x)$ *is the general solution of the homogeneous equation*

$$y'' + py' + qy = 0 \tag{2}$$

and $y_p(x)$ *is any solution of* (1).

The proof is deferred to the end of the section.

REMARK. Equation (2) is called the ***complementary equation*** to (1) and $y_p(x)$ is called a ***particular solution*** of (1). Thus, Theorem 1 states that *the general solution of* (1) *is obtained by adding a particular solution of* (1) *to the general solution of the complementary equation.*

☐ **THE METHOD OF UNDETERMINED COEFFICIENTS**

Since we already know how to obtain the general solution of (2), we shall focus our attention on the problem of obtaining a particular solution of (1). In this section we shall discuss a procedure for doing this called the method of *undetermined coefficients,* and in the next section we shall discuss a second procedure called *variation of parameters.*

The first step in the method of undetermined coefficients is to make a reasonable guess about the form of the particular solution. This guess will involve one or more unknown coefficients which we shall then determine from conditions obtained by substituting the proposed solution into the differential equation.

We shall begin with some examples that illustrate the basic idea.

Example 1   Find a particular solution of the differential equation

$$y'' + 2y' - 8y = e^{3x} \tag{3}$$

Solution.   It should be clear that such functions as $\sin x$, $\cos x$, $\ln x$, or x^3 are not reasonable possibilities for $y_p(x)$ because the derivatives of such functions do not yield expressions involving e^{3x}. Since the obvious choice for $y_p(x)$ is an expression involving e^{3x}, we shall *guess* that $y_p(x)$ has the form

$$y_p(x) = Ae^{3x} \tag{4}$$

where A is an unknown constant to be determined. It follows that

$$y_p'(x) = 3Ae^{3x} \tag{5}$$
$$y_p''(x) = 9Ae^{3x} \tag{6}$$

Substituting expressions (4), (5), and (6) for y, y', and y'' in (3) yields

$$9Ae^{3x} + 2(3Ae^{3x}) - 8(Ae^{3x}) = e^{3x}$$

or

$$7Ae^{3x} = e^{3x}$$

Thus, $A = \frac{1}{7}$. From this result and (4), a particular solution of (3) is

$$y_p(x) = \tfrac{1}{7}e^{3x} \qquad \blacktriangleleft$$

Example 2   Find a particular solution of the differential equation

$$y'' - y' - 6y = e^{3x} \tag{7}$$

Solution.   As in Example 1, a reasonable guess is $y_p(x) = Ae^{3x}$. However, if we substitute (4), (5), and (6) in (7), we obtain

$$9Ae^{3x} - 3Ae^{3x} - 6Ae^{3x} = e^{3x} \quad \text{or} \quad 0 = e^{3x}$$

which is contradictory. The problem here is that $y_p(x) = Ae^{3x}$ is a solution of the complementary equation

$$y'' - y' - 6y = 0$$

which makes it impossible for y_p to satisfy (7), no matter how A is selected. How should we proceed in this case? Experience has shown that multiplying (4) by x produces the correct form for the solution. Thus, we shall try

$$y_p(x) = Axe^{3x} \tag{8}$$

It follows that

$$y_p'(x) = 3Axe^{3x} + Ae^{3x} \tag{9}$$
$$y_p''(x) = 9Axe^{3x} + 6Ae^{3x} \tag{10}$$

Substituting expressions (8), (9), and (10) for y, y', and y'' in (7) yields

$$(9Axe^{3x} + 6Ae^{3x}) - (3Axe^{3x} + Ae^{3x}) - 6(Axe^{3x}) = e^{3x}$$

or

$$5Ae^{3x} = e^{3x}$$

Thus, $A = \frac{1}{5}$, so (8) implies that a particular solution of (7) is

$$y_p(x) = \tfrac{1}{5}xe^{3x} \qquad \blacktriangleleft$$

REMARK.   Had it turned out that Axe^{3x} was also a solution of the complementary equation, then we would have tried $y_p(x) = Ax^2e^{3x}$ as our candidate for a particular solution.

 SUMMARY

In summary, a particular solution of an equation of the form

$$y'' + py' + qy = ke^{ax}$$

can be obtained as follows:

> **Step 1.** Start with $y_p = Ae^{ax}$ as an initial guess.
>
> **Step 2.** Determine if the initial guess is a solution of the complementary equation $y'' + py' + qy = 0$.
>
> **Step 3.** If the initial guess is not a solution of the complementary equation, then $y_p = Ae^{ax}$ is the correct form of a particular solution.
>
> **Step 4.** If the initial guess is a solution of the complementary equation, then multiply it by the smallest positive integer power of x required to produce a function that is not a solution of the complementary equation. This will yield either $y_p = Axe^{ax}$ or $y_p = Ax^2e^{ax}$.

Table 19.2.1, which we provide without proof, restates the preceding procedure more compactly and also explains how to find a particular solution of (1) when $r(x)$ is a polynomial or a combination of sine and cosine functions.

Table 19.2.1

EQUATION	INITIAL GUESS FOR y_p
$y'' + py' + qy = ke^{ax}$	$y_p = Ae^{ax}$
$y'' + py' + qy = a_0 + a_1x + \cdots + a_nx^n$	$y_p = A_0 + A_1x + \cdots + A_nx^n$
$y'' + py' + qy = a_1\cos bx + a_2\sin bx$	$y_p = A_1\cos bx + A_2\sin bx$

MODIFICATION RULE
If any term in the initial guess is a solution of the complementary equation, then the correct form for y_p is obtained by multiplying the initial guess by the smallest positive integer power of x required so that no term is a solution of the complementary equation.

Example 3   Find the general solution of

$$y'' + y' = 4x^2 \tag{11}$$

Solution.   We begin by finding the general solution of the complementary equation

$$y'' + y' = 0$$

The auxiliary equation for this homogeneous equation is

$$m^2 + m = 0$$

which has roots $m_1 = 0$, $m_2 = -1$. Thus, the general solution, $y_c(x)$, of the complementary equation is

$$y_c(x) = c_1e^{0x} + c_2e^{-x} = c_1 + c_2e^{-x}$$

Since the right-hand side of (11) is a second-degree polynomial, our initial guess for y_p is

$$y_p = A_0 + A_1x + A_2x^2$$

But A_0 is a solution of the complementary equation (take $c_1 = A_0$, $c_2 = 0$), so our second guess for y_p is

$$y_p = x(A_0 + A_1x + A_2x^2) = A_0x + A_1x^2 + A_2x^3 \tag{12}$$

Since no term of this function is a solution of the complementary equation, this is the correct form for y_p. Differentiating (12) we obtain

$$y_p'(x) = A_0 + 2A_1x + 3A_2x^2 \tag{13}$$
$$y_p''(x) = 2A_1 + 6A_2x \tag{14}$$

Substituting (13) and (14) in (11) yields

$$(2A_1 + 6A_2x) + (A_0 + 2A_1x + 3A_2x^2) = 4x^2$$

or

$$(A_0 + 2A_1) + (2A_1 + 6A_2)x + 3A_2x^2 = 4x^2$$

To solve for A_0, A_1, and A_2, we equate corresponding coefficients on the two sides of this equation; this yields

$$A_0 + 2A_1 = 0$$
$$2A_1 + 6A_2 = 0$$
$$3A_2 = 4$$

from which we obtain

$$A_2 = \tfrac{4}{3}, \quad A_1 = -4, \quad A_0 = 8$$

Substituting these values in (12) yields the particular solution

$$y_p(x) = 8x - 4x^2 + \tfrac{4}{3}x^3$$

Thus, the general solution of (11) is

$$y(x) = y_c(x) + y_p(x) = c_1 + c_2e^{-x} + 8x - 4x^2 + \tfrac{4}{3}x^3 \quad \blacktriangleleft$$

Example 4 Find the general solution of

$$y'' - 2y' + y = \sin 2x \tag{15}$$

Solution. We begin by finding the general solution of the complementary equation

$$y'' - 2y' + y = 0$$

The auxiliary equation for this homogeneous equation is

$$m^2 - 2m + 1 = 0$$

which has the repeated root $m_1 = 1$, $m_2 = 1$. Thus, the general solution of the complementary equation is

$$y_c(x) = c_1e^x + c_2xe^x \tag{16}$$

Since $\sin 2x$ has the form $a_1 \cos bx + a_2 \sin bx$ ($a_1 = 0$, $a_2 = 1$, $b = 2$), our initial guess for y_p is

$$y_p(x) = A_1 \cos 2x + A_2 \sin 2x \tag{17}$$

Since no term of y_p is a solution of the complementary equation [see (16)], this is the correct form for a particular solution. Differentiating (17) we obtain

$$y_p'(x) = -2A_1 \sin 2x + 2A_2 \cos 2x \tag{18}$$
$$y_p''(x) = -4A_1 \cos 2x - 4A_2 \sin 2x \tag{19}$$

Substituting (17), (18), and (19) in (15) yields

$$(-4A_1 \cos 2x - 4A_2 \sin 2x) - 2(-2A_1 \sin 2x + 2A_2 \cos 2x)$$
$$+ (A_1 \cos 2x + A_2 \sin 2x) = \sin 2x$$

or

$$(-3A_1 - 4A_2) \cos 2x + (4A_1 - 3A_2) \sin 2x = \sin 2x$$

To solve for A_1 and A_2 we equate the coefficients of $\sin 2x$ and $\cos 2x$ on the two sides of this equation; this yields

$$-3A_1 - 4A_2 = 0$$
$$4A_1 - 3A_2 = 1$$

from which we obtain

$$A_1 = \tfrac{4}{25}, \quad A_2 = -\tfrac{3}{25}$$

(Verify.) From this result and (17), a particular solution of (15) is

$$y_p(x) = \tfrac{4}{25} \cos 2x - \tfrac{3}{25} \sin 2x$$

Thus, the general solution of (15) is

$$y(x) = y_c(x) + y_p(x) = c_1 e^x + c_2 x e^x + \tfrac{4}{25} \cos 2x - \tfrac{3}{25} \sin 2x \qquad \blacktriangleleft$$

We conclude this section with a proof of Theorem 19.2.1.

■ OPTIONAL

Proof of Theorem 19.2.1. **To prove that**

$$y(x) = c_1 y_1(x) + c_2 y_2(x) + y_p(x) \tag{20}$$

is the general solution of (1) we must prove two results: first, that for all choices of c_1 and c_2 the function $y(x)$ defined by (20) satisfies (1); and second, that every solution of (1) can be obtained from (20) by choosing appropriate values for the constants c_1 and c_2.

To prove the first statement, let c_1 and c_2 have any real values. Then differentiating (20) yields

$$y'(x) = c_1 y_1'(x) + c_2 y_2'(x) + y_p'(x)$$
$$y''(x) = c_1 y_1''(x) + c_2 y_2''(x) + y_p''(x)$$

so

$$y''(x) + py'(x) + qy(x) = c_1(y_1''(x) + py_1'(x) + qy_1(x))$$
$$+ c_2(y_2''(x) + py_2'(x) + qy_2(x))$$
$$+ y_p''(x) + py_p'(x) + qy_p(x) \tag{21}$$

But $y_1(x)$ and $y_2(x)$ satisfy (2) and $y_p(x)$ satisfies (1), so (21) reduces to

$$y''(x) + py'(x) + qy(x) = 0 + 0 + r(x) = r(x)$$

which proves that $y(x)$ satisfies (1).

To prove the second statement, let $y(x)$ be any solution of (1). We must find values of c_1 and c_2 such that

$$y(x) = c_1 y_1(x) + c_2 y_2(x) + y_p(x) \tag{22}$$

But $y(x) - y_p(x)$ satisfies (2) since

$$
\begin{aligned}
(y(x) - y_p(x))'' &+ p(y(x) - y_p(x))' + q(y(x) - y_p(x)) \\
&= [y''(x) + py'(x) + qy(x)] - [y_p''(x) + py_p'(x) + qy_p(x)] \\
&= r(x) - r(x) = 0
\end{aligned}
$$

Thus, since $c_1 y_1(x) + c_2 y_2(x)$ is the general solution of (2), there exist values of c_1 and c_2 such that the solution $y(x) - y_p(x)$ can be written as

$$y(x) - y_p(x) = c_1 y_1(x) + c_2 y_2(x)$$

from which (22) follows.   ∎

▶ Exercise Set 19.2

In Exercises 1–24, use the method of undetermined coefficients to find the general solution of the differential equation.

1. $y'' + 6y' + 5y = 2e^{3x}$.   2. $y'' + 3y' - 4y = 5e^{7x}$.

3. $y'' - 9y' + 20y = -3e^{5x}$.

4. $y'' + 7y' - 8y = 7e^x$.

5. $y'' + 2y' + y = e^{-x}$.

6. $y'' + 4y' + 4y = 4e^{-2x}$.

7. $y'' + y' - 12y = 4x^2$.

8. $y'' - 4y' - 5y = -6x^2$.

9. $y'' - 6y' = x - 1$.   10. $y'' + 3y' = 2x + 2$.

11. $y'' - x^3 + 1 = 0$.   12. $y'' + 3x^3 + x = 0$.

13. $y'' - y' - 2y = 10 \cos x$.

14. $y'' - 3y' - 4y = 2 \sin x$.

15. $y'' - 4y = 2 \sin 2x + 3 \cos 2x$.

16. $y'' - 9y = \cos 3x - \sin 3x$.

17. $y'' + y = \sin x$.

18. $y'' + 4y = \cos 2x$.

19. $y'' - 3y' + 2y = x$.

20. $y'' + 4y' + 4y = 3x + 3$.

21. $y'' + 4y' + 9y = x^2 + 3x$.

22. $y'' - y = 1 + x + x^2$.

23. $y'' + 4y = \sin x \cos x$.

24. $y'' + 4y = \cos^2 x - \sin^2 x$.

25. (a) Prove: If $y_1(x)$ is a solution of

$$y'' + p(x)y' + q(x)y = r_1(x)$$

and $y_2(x)$ is a solution of

$$y'' + p(x)y' + q(x)y = r_2(x)$$

then $y_1(x) + y_2(x)$ is a solution of

$$y'' + p(x)y' + q(x)y = r_1(x) + r_2(x)$$

(b) Use the result in part (a) to find a particular solution of the equation

$$y'' + 3y' - 4y = x + e^x$$

(c) State a generalization of the result in part (a) that is applicable to the equation

$$
\begin{aligned}
y'' + p(x)y' &+ q(x)y \\
&= r_1(x) + r_2(x) + \cdots + r_n(x)
\end{aligned}
$$

In Exercises 26–34, use the results in Exercise 25 to find the general solution of the differential equation.

26. $y'' - y' - 2y = x + e^{-x}$.

27. $y'' - y = 1 + e^x$.

28. $y'' - 4y' + 3y = 2 \cos x + 4 \sin x$.

29. $y'' + 4y = 1 + x + \sin x$.

30. $y'' + 2y' + y = 2 + 3x + 3e^x + 2 \cos 2x$.

31. $y'' - 2y' + y = \sinh x$. [*Hint:* $\sinh x = \frac{1}{2}(e^x - e^{-x})$.]

32. $y'' + 4y' - 5y = \cosh x$.
[*Hint:* $\cosh x = \frac{1}{2}(e^x + e^{-x})$.]

33. $y'' + y = 12 \cos^2 x$. [*Hint:* $\cos^2 x = \frac{1}{2}(1 + \cos 2x)$.]

34. $y'' + 2y' + y = \sin^2 x$.
[*Hint:* $\sin^2 x = \frac{1}{2}(1 - \cos 2x)$.]

35. (a)   Find the general solution of

$$y'' + \mu^2 y = a \sin bx$$

where a is an arbitrary constant and μ and b are positive constants such that $\mu \neq b$.

(b)   Use part (a) and Exercise 25 to find the general solution of

$$y'' + \mu^2 y = \sum_{k=1}^{n} a_k \sin k\pi x$$

where $\mu > 0$ and $\mu \neq k\pi$, $k = 1, 2, \ldots, n$.

36.   Find the general solution of

$$y'' + \lambda^2 y = \sum_{k=1}^{n} a_k \cos k\pi x$$

where $\lambda > 0$ and $\lambda \neq k\pi$, $k = 1, 2, \ldots, n$. [*Hint:* See Exercise 35.]

37.   Find all solutions of the equation

$$y'' - y' = 4 - 4x$$

(if any) with the property that $y'(x_0) = y''(x_0) = 0$ at some point x_0.

■ 19.3   VARIATION OF PARAMETERS

In this section* we shall consider an alternative method for finding a particular solution of a second-order linear nonhomogeneous equation that often works when the method of undetermined coefficients cannot be applied.

As in the previous section, we shall be concerned with equations of the form

$$y'' + py' + qy = r(x) \tag{1}$$

where p and q are constants and $r(x)$ is a continuous function of x. The method we shall discuss, called *variation of parameters,* is based on the (not very obvious) fact that if

$$y_c(x) = c_1 y_1(x) + c_2 y_2(x) \tag{2}$$

---

*It is assumed in this section that the reader is familiar with Cramer's rule, which is discussed in Section VI of Appendix C.

is the general solution of the complementary equation

$$y'' + py' + qy = 0$$

then it is possible to find functions $u(x)$ and $v(x)$ such that

$$y_p(x) = u(x)y_1(x) + v(x)y_2(x) \tag{3}$$

is a particular solution of (1).

To see how the functions $u(x)$ and $v(x)$ can be found, consider the derivative of (3):

$$y_p' = (uy_1' + vy_2') + (u'y_1 + v'y_2) \tag{4}$$

If we now differentiate y_p', we shall introduce second derivatives of the unknown functions $u(x)$ and $v(x)$. To avoid this complication we shall require that $u(x)$ and $v(x)$ satisfy the condition

$$u'y_1 + v'y_2 = 0 \tag{5}$$

so (4) simplifies to

$$y_p' = uy_1' + vy_2'$$

Thus,

$$y_p'' = uy_1'' + u'y_1' + vy_2'' + v'y_2' \tag{6}$$

On substituting (3), (4), and (6) in (1) and rearranging terms we obtain

$$u(y_1'' + py_1' + qy_1) + v(y_2'' + py_2' + qy_2) + u'y_1' + v'y_2' = r(x) \tag{7}$$

Since y_1 and y_2 are solutions of the complementary equation, we have

$$y_1'' + py_1' + qy_1 = 0 \quad \text{and} \quad y_2'' + py_2' + qy_2 = 0$$

so (7) simplifies to

$$u'y_1' + v'y_2' = r(x) \tag{8}$$

Equations (5) and (8) together yield two equations in the two unknowns $u'(x)$ and $v'(x)$:

$$\begin{aligned} u'y_1 + v'y_2 &= 0 \\ u'y_1' + v'y_2' &= r(x) \end{aligned} \tag{9}$$

It can be shown (see Boyce and DiPrima, *Elementary Differential Equations and Boundary Value Problems*, John Wiley & Sons, New York, 1988) that this system has a unique solution for u' and v'. Once this solution is found, u and v can be obtained by integration.

Example 1 Find the general solution of

$$y'' + y = \sec x \qquad \left(-\frac{\pi}{2} < x < \frac{\pi}{2} \right) \tag{10}$$

Solution. We begin by finding the general solution of the complementary equation

$$y'' + y = 0$$

The auxiliary equation for this homogeneous equation is

$$m^2 + 1 = 0$$

which has roots $m_1 = i$ and $m_2 = -i$. Thus, the general solution of the complementary equation is

$$y_c(x) = c_1 \cos x + c_2 \sin x$$

Comparing this with (2) yields $y_1(x) = \cos x$ and $y_2(x) = \sin x$, so we shall look for a particular solution of the form

$$y_p(x) = u(x) \cos x + v(x) \sin x \tag{11}$$

Substituting $y_1 = \cos x$, $y_2 = \sin x$, and $r(x) = \sec x$ in (9) yields

$$\begin{aligned} u' \cos x + v' \sin x &= 0 \\ -u' \sin x + v' \cos x &= \sec x \end{aligned} \tag{12}$$

Solving this system for u' and v' by Cramer's rule we obtain

$$u' = \frac{\begin{vmatrix} 0 & \sin x \\ \sec x & \cos x \end{vmatrix}}{\begin{vmatrix} \cos x & \sin x \\ -\sin x & \cos x \end{vmatrix}} = \frac{-\sec x \sin x}{\cos^2 x + \sin^2 x} = \frac{-\tan x}{1} = -\tan x$$

$$v' = \frac{\begin{vmatrix} \cos x & 0 \\ -\sin x & \sec x \end{vmatrix}}{\begin{vmatrix} \cos x & \sin x \\ -\sin x & \cos x \end{vmatrix}} = \frac{\cos x \sec x}{\cos^2 x + \sin^2 x} = \frac{1}{1} = 1$$

Integrating u' and v' yields

$$u(x) = \int -\tan x \, dx = \ln|\cos x| \quad \text{and} \quad v(x) = \int 1 \, dx = x$$

[We set the constants of integration equal to zero because any functions $u(x)$ and $v(x)$ satisfying (12) will suffice.] Substituting these functions in (11) yields

$$y_p(x) = (\ln|\cos x|) \cos x + x \sin x$$

Thus, the general solution of (10) is

$$y(x) = y_c(x) + y_p(x) = c_1 \cos x + c_2 \sin x + (\ln|\cos x|) \cos x + x \sin x$$

◀

▶ Exercise Set 19.3

In Exercises 1–6, use variation of parameters to find the general solution of the differential equation and check your result using undetermined coefficients.

1. $y'' + y = x^2$.   2. $y'' + 9y = 3x$.

3. $y'' + y' - 2y = 2e^x$.   4. $y'' + 5y' + 6y = e^{-x}$.

5. $y'' + 4y = \sin 2x$.   6. $y'' + 9y = \cos 3x$.

In Exercises 7–28, use variation of parameters to find the general solution of the differential equation.

7. $y'' + y = \tan x$.   8. $y'' + y = \cot x$.

9. $y'' - 2y' + y = e^x/x$.

10. $y'' - 4y' + 4y = e^{2x}/x$.

11. $y'' + y = 3\sin^2 x$.   12. $y'' + y = 6\cos^2 x$.

13. $y'' + y = \csc x$.   14. $y'' + 9y = 6\sec 3x$.

15. $y'' + y = \sec x \tan x$.   16. $y'' + y = \csc x \cot x$.

17. $y'' + 2y' + y = e^{-x}/x^2$.

18. $y'' - y = x^2 e^x$.

19. $y'' + 4y' + 4y = xe^{-x}$.

20. $y'' + 4y' + 4y = xe^{2x}$.

21. $y'' + y = \sec^2 x$.   22. $y'' + y = \sec^3 x$.

23. $y'' - 2y' + y = e^x/x^2$.

24. $y'' - 2y' + y = x^3 e^x$.

25. $y'' - y = e^x \cos x$.

26. $y'' - 2y' + 2y = e^{2x} \sin x$.

27. $y'' + 2y' + y = e^{-x} \ln|x|$.

28. $y'' - 3y' + 2y = \dfrac{e^x}{1 + e^x}$.

29. Use the method of variation of parameters to show that the general solution of $y'' + y = r(x)$ is
$$y = c_1 \cos x + c_2 \sin x + g(x) \cos x + h(x) \sin x$$
where
$$g(x) = -\int r(x) \sin x \, dx$$
$$h(x) = \int r(x) \cos x \, dx$$

30. Use the method of variation of parameters to show that if y_1 and y_2 are linearly independent solutions of $y'' + py' + qy = 0$, then a particular solution of $y'' + py' + qy = r(x)$ is given by
$$y_p(x) = -y_1(x) \int \frac{y_2(x)r(x)}{W(x)} \, dx + y_2(x) \int \frac{y_1(x)r(x)}{W(x)} \, dx$$
where $W(x) = y_1(x)y_2'(x) - y_1'(x)y_2(x)$.

■ **19.4 VIBRATION OF A SPRING**

> *In this section we shall use second-order linear differential equations with constant coefficients to study the vibration of a spring.*

We shall be interested in solving the following problem:

> **19.4.1 THE VIBRATING SPRING PROBLEM.** *As shown in* Figure 19.4.1, *let a mass attached to a vertical spring be allowed to settle into an equilibrium position. Then, let the spring be stretched (or compressed) by pulling (or pushing) the mass, and finally, let the mass be released, thereby causing it to undergo a vibratory motion. We shall be interested in finding a formula for the position of the mass at any time t.*

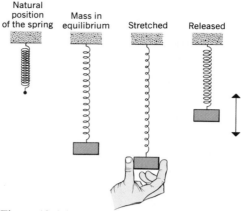

Figure 19.4.1

To solve this problem we shall need three results from physics:

HOOKE'S LAW.   *If a spring is stretched (or compressed) l units beyond its natural position, then it pulls back (or pushes) with a force of magnitude*

$$F = kl$$

*where k is a positive constant, called the **spring constant**. This constant depends on such factors as the thickness of the spring, the material from which it is made, and the units of force and length; it is measured in units of force per unit of length.*

NEWTON'S SECOND LAW OF MOTION.   *If an object with mass **M** is subjected to a force **F**, then it undergoes an acceleration **a** satisfying*

$$\mathbf{F} = M\mathbf{a}$$

WEIGHT.   *The gravitational force exerted by the earth on an object is called the **weight** (more precisely, **earth weight**) of the object. It follows from Newton's Second Law of Motion that an object with mass M has a weight whose magnitude w is given by*

$$w = Mg \tag{1}$$

*where g is a constant, called the **acceleration due to gravity**. Near the surface of the earth an approximate value of g is $g \approx 32$ ft/sec² if length is measured in feet and time in seconds, or $g \approx 980$ cm/sec² if length is measured in centimeters and time in seconds.*

In any problem involving length, mass, time, and force it is important that the units of measurement be consistent. The most important systems of measurement are summarized in Table 19.4.1.

Table 19.4.1

SYSTEM OF MEASUREMENT	LENGTH	TIME	MASS	FORCE
Engineering system	foot (ft)	second (sec)	slug	pound (lb)
Mks system	meter (m)	second (sec)	kilogram (kg)	newton (N)
cgs system	centimeter (cm)	second (sec)	gram (g)	dyne

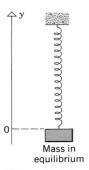

Figure 19.4.2

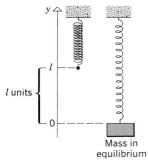

Figure 19.4.3

To solve the vibrating spring problem posed above, we introduce a coordinate axis (a y-axis) with the positive direction up and the origin at the bottom end of the spring when the mass is in its equilibrium position (Figure 19.4.2). If we take $t = 0$ to be the time at which the mass is released, then at each subsequent time t, the end of the spring has a position $y(t)$, a velocity $y'(t)$, and an acceleration $y''(t)$.* Because the positive direction of the y-axis is up, force, velocity, and acceleration are positive when directed up and negative when directed down.

Before we can solve the spring problem, it will be necessary to derive a preliminary result from the equilibrium conditions for the mass. Let us assume that the spring constant is k, the mass of the object is M, and the spring is stretched l units beyond its natural length when the mass is in equilibrium (Figure 19.4.3). When the mass is in its equilibrium position its downward weight, $-Mg$, is balanced exactly by the upward force, kl, of the spring, so the sum of these forces must be zero†; that is,

$$kl - Mg = 0$$

or

$$kl = Mg \tag{2}$$

The basic strategy for solving the spring problem is to find the total force $F(t)$ that acts on the mass at time t. Then, since the acceleration of the mass at time t is $y''(t)$, it will follow from Newton's Second Law of Motion that

$$My''(t) = F(t)$$

or

$$y''(t) = F(t)/M \tag{3}$$

which is a second-order linear differential equation that can be solved for the position function $y(t)$.

---

*For convenience of terminology we shall refer to $y(t)$, $y'(t)$, and $y''(t)$ as the position, velocity, and acceleration of the mass, respectively.

†We shall assume that the weight of the spring is small relative to the weight of the mass and can be neglected.

At each instant, the force $F(t)$ in (3) consists of four possible components:

$F_g(t)$ = the force of gravity (weight of the mass)

$F_s(t)$ = the force of the spring

$F_d(t)$ = the **damping force** or frictional force exerted on the mass by the surrounding medium (air, water, oil, etc.)

$F_e(t)$ = external forces due to such factors as movement in the spring support, magnetic forces acting on the mass, and so on.

☐ **UNDAMPED FREE VIBRATIONS**

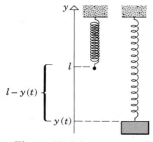

Figure 19.4.4

The simplest case occurs when there is no damping ($F_d = 0$) and the mass is *free* of external forces ($F_e = 0$). In this case the only forces acting on the mass are the force of gravity, F_g, and the spring force, F_s. Let us try to calculate these forces when the mass is at an arbitrary point $y(t)$.

When the mass is at the point $y(t)$, the spring is stretched (or compressed) $l - y(t)$ units from its natural length (Figure 19.4.4), so by Hooke's law the spring exerts a force of

$$F_s(t) = k(l - y(t))$$

on the mass. Adding this to the force of gravity,

$$F_g(t) = -Mg$$

acting on the mass yields the total force acting on the mass:

$$F_s(t) + F_g(t) = k(l - y(t)) - Mg$$

But $kl = Mg$ from (2), so

$$F_s(t) + F_g(t) = -ky(t) \qquad (4)$$

Substituting this expression for $F(t)$ in (3) yields

$$y''(t) = -\frac{k}{M} y(t)$$

or

$$y''(t) + \frac{k}{M} y(t) = 0 \qquad (5)$$

which is a second-order linear differential equation with constant coefficients.

Because the mass is *released* (i.e., has zero initial velocity) at time $t = 0$, we have $y'(0) = 0$. If we assume, in addition, that the position of the mass at time $t = 0$ is $y(0) = y_0$, then we have two initial conditions that can be combined with (5) to yield an initial-value problem for $y(t)$:

$$y'' + \frac{k}{M} y = 0$$

$$y(0) = y_0, \quad y'(0) = 0 \qquad (6)$$

The auxiliary equation for the differential equation in (6) is

$$m^2 + \frac{k}{M} = 0$$

which has roots $m_1 = \sqrt{k/M}\ i$, $m_2 = -\sqrt{k/M}\ i$ (since k and M are positive), so the general solution of the differential equation is

$$y(t) = c_1 \cos(\sqrt{k/M}\ t) + c_2 \sin(\sqrt{k/M}\ t)$$

From the initial conditions $y(0) = y_0$ and $y'(0) = 0$ it follows that $c_1 = y_0$, $c_2 = 0$ (verify), so the solution of (6) is

$$y(t) = y_0 \cos(\sqrt{k/M}\ t) \tag{7}$$

This formula describes a periodic vibration with an *amplitude* of $|y_0|$ and *period* T given by

$$T = \frac{2\pi}{\sqrt{k/M}} = 2\pi\sqrt{M/k} \tag{8}$$

(Figure 19.4.5). The *frequency* f of the vibration is the number of cycles per unit time, so

$$f = \frac{1}{T} = \frac{1}{2\pi}\sqrt{k/M} \tag{9}$$

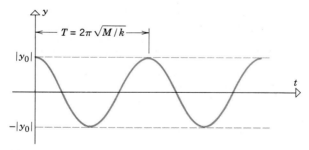

Figure 19.4.5

REMARK. In the preceding problem, the spring oscillates indefinitely with constant amplitude because there are no frictional forces to dissipate the energy in the mass–spring system.

Example 1 Suppose that the top of a spring is fixed to a ceiling and a mass attached to the bottom end stretches the spring $\frac{1}{2}$ foot. If the mass is pulled 2 feet below its equilibrium position and released, find

(a) a formula for the position of the mass at any time t;

(b) the amplitude, period, and frequency of the motion.

Solution (a).   The appropriate formula is (7). Although we are not given the mass M or the spring constant k, it does not matter because we are given that the mass stretches the spring $l = \frac{1}{2}$ ft and we know that $g = 32$ ft/sec^2, so (2) implies that

$$\frac{k}{M} = \frac{g}{l} = \frac{32}{\frac{1}{2}} = 64 \text{ lb/ft}$$

Also, the mass is initially 2 units *below* its equilibrium position, so $y_0 = -2$. Therefore, from (7) the formula for the position of the mass at time t is

$$y(t) = -2\cos 8t$$

Solution (b).   From (8) and (9)

amplitude $= |y_0| = |-2| = 2$ ft

period $= T = 2\pi\sqrt{M/k} = 2\pi\sqrt{1/64} = \pi/4$ sec/cycle

frequency $= f = 1/T = 4/\pi$ cycles/sec   ◄

☐ **DAMPED FREE VIBRATIONS**

We shall now consider the solution of the spring problem in the case where damping cannot be neglected; that is, $F_d \neq 0$.

Physicists have shown that under appropriate conditions the damping force F_d is opposite to the direction of motion of the mass (i.e., tends to slow the mass down) and has a magnitude that is proportional to the speed of the mass (the greater the speed, the greater the effect of friction). This type of damping force is described by an equation of the form

$$F_d(t) = -cy'(t) \tag{10}$$

where c is a positive constant, called the ***damping constant.*** The damping constant depends on the viscosity of the surrounding medium; it is measured in units of force per unit of velocity (lb/ft/sec or equivalently lb · sec/ft, for example).

It follows from (4) and (10) that

$$F_s(t) + F_g(t) + F_d(t) = -ky(t) - cy'(t)$$

If we substitute this expression in (3) for $F(t)$, we obtain

$$y''(t) = -\frac{k}{M}y(t) - \frac{c}{M}y'(t)$$

or

$$y''(t) + \frac{c}{M}y'(t) + \frac{k}{M}y(t) = 0$$

Combining this equation with the initial conditions $y(0) = y_0$, $y'(0) = 0$ yields the following initial-value problem whose solution describes the motion of the mass subject to damping:

$$y'' + \frac{c}{M}y' + \frac{k}{M}y = 0$$
$$y(0) = y_0, \quad y'(0) = 0 \tag{11}$$

The form of the solution to (11) will depend on whether the auxiliary equation

$$m^2 + \frac{c}{M} m + \frac{k}{M} = 0 \tag{12}$$

has distinct real roots, equal real roots, or complex roots. We leave it for the reader to show that the roots are distinct and real if $c^2 > 4kM$, equal and real if $c^2 = 4kM$, and complex if $c^2 < 4kM$ (Exercise 22).

☐ UNDERDAMPED
VIBRATIONS

The cases $c^2 > 4kM$ and $c^2 = 4kM$ are called **overdamped** and **critically damped,** respectively. In these cases vibration in the usual sense does not occur: the mass simply drifts slowly toward its equilibrium position without ever passing through it. We shall leave the analysis of these cases for the exercises and concentrate on the case $c^2 < 4kM$ in which true vibratory motion occurs. This is called the **underdamped case**.

In the underdamped case the roots of the auxiliary equation in (12) are

$$m_1 = -\frac{c}{2M} + \frac{\sqrt{4Mk - c^2}}{2M} i, \quad m_2 = -\frac{c}{2M} - \frac{\sqrt{4Mk - c^2}}{2M} i$$

(Verify.) For convenience, let

$$\alpha = \frac{c}{2M}, \quad \beta = \frac{\sqrt{4Mk - c^2}}{2M} \tag{13}$$

so the solution of the differential equation in (11) is

$$y(t) = e^{-\alpha t}(c_1 \cos \beta t + c_2 \sin \beta t) \tag{14}$$

It follows that

$$y'(t) = -\alpha e^{-\alpha t}(c_1 \cos \beta t + c_2 \sin \beta t) + \beta e^{-\alpha t}(-c_1 \sin \beta t + c_2 \cos \beta t)$$

so the initial conditions $y(0) = y_0$, $y'(0) = 0$ yield the equations

$$c_1 = y_0$$
$$-\alpha c_1 + \beta c_2 = 0$$

which can be solved to obtain

$$c_1 = y_0, \quad c_2 = \alpha y_0 / \beta$$

Substituting these values in (14) yields the solution of (11):

$$y(t) = \frac{y_0}{\beta} e^{-\alpha t}(\beta \cos \beta t + \alpha \sin \beta t) \tag{15}$$

In the exercises (Exercise 23) we ask the reader to use the cosine addition formula to show that this solution can be rewritten in the alternative form

$$y(t) = \frac{y_0\sqrt{\alpha^2 + \beta^2}}{\beta} e^{-\alpha t} \cos(\beta t - \omega) \tag{16a}$$

where

$$\omega = \tan^{-1}\left(\frac{\alpha}{\beta}\right), \quad 0 < \omega < \pi/2 \tag{16b}$$

Since $\cos(\beta t - \omega)$ has values between $+1$ and -1, it follows from (16a) and (16b) that the graph of $y(t)$ oscillates between the curves

$$y = \frac{y_0\sqrt{\alpha^2 + \beta^2}}{\beta} e^{-\alpha t} \quad \text{and} \quad y = -\frac{y_0\sqrt{\alpha^2 + \beta^2}}{\beta} e^{-\alpha t}$$

Thus, the graph of $y(t)$ resembles a cosine curve, but with decreasing amplitude (Figure 19.4.6). Strictly speaking, the function $y(t)$ is not periodic. However, $\cos(\beta t - \omega)$ has a period of $2\pi/\beta$, so the displacement $y(t)$ reaches a relative maximum at times spaced $2\pi/\beta$ units apart. Thus, we shall define the *period T* of this motion as

$$T = \frac{2\pi}{\beta} = \frac{4M\pi}{\sqrt{4Mk - c^2}} \tag{17}$$

(Figure 19.4.6) and the *frequency* to be

$$f = \frac{1}{T} = \frac{\sqrt{4Mk - c^2}}{4M\pi} \tag{18}$$

If we think of a relative maximum as marking the end of one cycle of motion and the start of the next, then the period T is the time required for the completion of one cycle, and the frequency is the number of cycles that occur per unit time. The physically significant fact is that the frequency and period of the

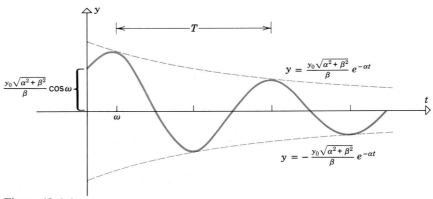

Figure 19.4.6

mass–spring system do not change as the amplitudes of the vibrations get smaller (why?). Galileo, who first observed this fact, used it to help design clocks.

Example 2 A spring with a spring constant of $k = 4$ lb/ft is attached to a ceiling and a 64-lb weight is attached to the bottom end. The weight is pushed 2 ft above its equilibrium position and released with an initial velocity of 0. Assuming that the damping constant due to air resistance is $c = 4$ lb · sec/ft, find

(a) a formula for the position of the weight at any time t;

(b) the period and frequency of the vibration.

Solution (a). We shall first use (13) to find α and β, after which we can use either (15) or (16a) and (16b). The attached weight is 64 lb, so from (1) its mass is

$$M = \frac{w}{g} = \frac{64}{32} = 2 \text{ slugs}$$

Thus,

$$\alpha = \frac{c}{2M} = \frac{4}{4} = 1, \quad \beta = \frac{\sqrt{4Mk - c^2}}{2M} = \frac{4}{4} = 1$$

Since the weight is initially 2 ft above its equilibrium position, we have $y_0 = 2$, so from (15) the position function of the weight is

$$y(t) = 2e^{-t}(\cos t + \sin t)$$

Alternatively, we can apply (16a) and (16b), but we must first calculate ω. Since

$$\omega = \tan^{-1}\left(\frac{\alpha}{\beta}\right) = \tan^{-1}(1) = \frac{\pi}{4}$$

(16a) and (16b) yield the alternative formula

$$y(t) = 2\sqrt{2}e^{-t} \cos\left(t - \frac{\pi}{4}\right)$$

Solution (b). From (17) and (18)

$$T = \frac{2\pi}{\beta} = 2\pi \quad \text{sec/cycle}$$

$$f = \frac{1}{T} = \frac{1}{2\pi} \quad \text{cycles/sec} \quad \blacktriangleleft$$

☐ **FORCED VIBRATIONS**

If there are external forces such as movement of the spring support or magnetic forces acting on the mass, then $F_e \neq 0$ and the vibrations are said to be *forced*. The study of forced vibrations leads to important phenomena such as *resonance* and *beats*, but this is beyond the scope of this section. Readers interested in this topic are referred to *Elementary Differential Equations and Boundary Value Problems*, John Wiley & Sons, New York, 1988, by William E. Boyce and Richard C. DiPrima.

▶ Exercise Set 19.4

In this exercise set assume that the y-axis is oriented
as in Figure 19.4.2.

1. A weight of 64 lb is attached to a vertical spring with
    spring constant $k = 8$ lb/ft. The weight is pushed 1 ft
    above its equilibrium position and released.
    (a) Find an initial-value problem whose solution
         $y(t)$ is the position function of the weight, as-
         suming that there is no damping.
    (b) Solve the initial-value problem.
    (c) Check the solution in (b) using Formula (7).

2. A mass of 1000 g is attached to a vertical spring with
    spring constant $k = 25$ dyne/cm. The mass is pushed
    50 cm above its equilibrium position and released.
    (a) Find an initial-value problem whose solution
         $y(t)$ is the position function of the mass, assum-
         ing that there is no damping.
    (b) Solve the initial-value problem.
    (c) Check the solution in (b) using Formula (7).

3. A mass attached to a vertical spring stretches the
    spring 5 cm. The mass is pulled 10 cm below its
    equilibrium position and released.
    (a) Find an initial-value problem whose solution
         $y(t)$ is the position function of the mass, assum-
         ing that there is no damping.
    (b) Solve the initial-value problem.
    (c) Check the solution in (b) using Formula (7).

4. A mass attached to a vertical spring stretches the
    spring 2 ft. The mass is pulled 4 ft below its equilib-
    rium position and released.
    (a) Find an initial-value problem whose solution
         $y(t)$ is the position function of the mass, assum-
         ing that there is no damping.
    (b) Solve the initial-value problem.
    (c) Check the solution in (b) using Formula (7).

5. A weight of $\frac{1}{2}$ lb is attached to a vertical spring with
    spring constant $k = 1$ lb/ft. The weight is pushed 2 ft
    above its equilibrium position and released. Use For-
    mulas (7), (8), and (9) to find
    (a) the position function of the weight
    (b) the amplitude of the vibration
    (c) the period of the vibration
    (d) the frequency of the vibration.

6. A mass of 2 slugs is attached to a vertical spring
    with spring constant $k = 4$ lb/ft. The mass is pushed

1 ft above its equilibrium position and released. Use
Formulas (7), (8), and (9) to find
    (a) the position function of the mass
    (b) the amplitude of the vibration
    (c) the period of the vibration
    (d) the frequency of the vibration.

7. A mass attached to a vertical spring stretches the
    spring 1 in. The mass is pulled 3 in. below its equi-
    librium position and released. Use Formulas (7), (8),
    and (9) to find
    (a) the position function of the mass
    (b) the amplitude of the vibration
    (c) the period of the vibration
    (d) the frequency of the vibration.

8. A mass attached to a vertical spring stretches the
    spring 8 m. The mass is pulled 2 m below its equi-
    librium position and released. Use Formulas (7), (8),
    and (9) to find
    (a) the position function of the mass
    (b) the amplitude of the vibration
    (c) the period of the vibration
    (d) the frequency of the vibration.

9. A weight of 32 lb is attached to a vertical spring with
    spring constant $k = 8$ lb/ft. The surrounding medium
    has a damping constant of $c = 4$ lb · sec/ft. The weight
    is pulled 3 ft below its equilibrium position and re-
    leased.
    (a) Find an initial-value problem whose solution
         $y(t)$ is the position function of the weight.
    (b) Solve the initial-value problem.
    (c) Check the solution in (b) using Formula (15).
    (d) Express the solution in the form of (16a).
    (e) Find the period of the vibration.
    (f) Find the frequency of the vibration.

10. A mass of 3 slugs is attached to a vertical spring
    with spring constant $k = 9$ lb/ft. The surrounding
    medium has a damping constant of $c = 6$ lb · sec/ft.
    The mass is pulled 1 ft below its equilibrium position
    and released.
    (a) Find an initial-value problem whose solution
         $y(t)$ is the position function of the mass.
    (b) Solve the initial-value problem.
    (c) Check the solution in (b) using Formula (15).
    (d) Express the solution in the form of (16a).
    (e) Find the period of the vibration.
    (f) Find the frequency of the vibration.

11. A mass of 25 g is attached to a vertical spring with spring constant $k = 3$ dyne/cm. The surrounding medium has a damping constant of $c = 10$ dyne · sec/cm. The mass is pushed 5 cm above its equilibrium position and released.
    (a) Use Formula (16a) to find the position function of the mass.
    (b) Find the period of the vibration.
    (c) Find the frequency of the vibration.

12. A mass of 490 g is attached to a vertical spring with spring constant $k = 40$ dyne/cm. The surrounding medium has a damping constant of $c = 140$ dyne · sec/cm. The mass is pushed 20 cm above its equilibrium position and released.
    (a) Use Formula (16a) to find the position function of the mass.
    (b) Find the period of the vibration.
    (c) Find the frequency of the vibration.

> If the object in Figure 19.4.1 is given an initial velocity v_0 rather than being released with initial velocity 0, then in Formulas (6) and (11) the initial conditions become $y(0) = y_0$, $y'(0) = v_0$. Use this fact in Exercises 13–16.

13. A weight of 3 lb attached to a vertical spring stretches the spring $\frac{1}{2}$ ft. While in its equilibrium position, the weight is struck to give it a downward initial velocity of 2 ft/sec. Assuming that there is no damping, find
    (a) the position function of the weight
    (b) the amplitude of the vibration
    (c) the period of the vibration
    (d) the frequency of the vibration.

14. A mass of 64 slugs attached to a vertical spring stretches the spring $1\frac{1}{2}$ in. The mass is pulled 4 in. below its equilibrium position and struck to give it a downward initial velocity of 8 ft/sec. Assuming that there is no damping, find
    (a) the position function of the weight
    (b) the amplitude of the vibration
    (c) the period of the vibration
    (d) the frequency of the vibration.

15. A weight of 4 lb is attached to a vertical spring with spring constant $k = 6\frac{1}{4}$ lb/ft. The weight is pushed $\frac{1}{3}$ ft above its equilibrium position and struck to give it a downward initial velocity of 5 ft/sec. Assuming that the damping constant is $c = \frac{1}{4}$ lb · sec/ft, find the position function of the weight.

16. A weight of 49 dynes is attached to a vertical spring with spring constant $k = \frac{1}{4}$ dyne/cm. The weight is pushed 1 cm above its equilibrium position and struck to give it an upward initial velocity of 2 cm/sec. Assuming that the damping constant is $c = \frac{1}{5}$ dyne · sec/cm, find the position function of the weight.

17. A weight of w pounds is attached to a spring, then pulled below its equilibrium position and released, thereby causing it to vibrate with a period of 3 sec. When 4 additional pounds are added, the period becomes 5 sec. Assuming there is no damping, find
    (a) the spring constant k (b) the weight w.

18. As illustrated in the following figure, let a toy cart of mass M be attached to a wall by a spring with spring constant k, and let an x-axis be introduced as shown with its origin at the point where the cart is in equilibrium. Suppose that the cart is pulled or pushed horizontally, then released. Find a differential equation for the position function of the cart if
    (a) there is no damping
    (b) there is a damping constant of c.

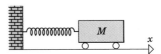

19. A cylindrical buoy of height h and radius r floats in the water with its axis vertical. By *Archimedes' principle* the water exerts an upward force on the buoy (the buoyancy force) with magnitude equal to the weight of the water displaced. Neglecting all forces except the buoyancy force and the force of gravity, determine the period with which the buoy will vibrate vertically if it is depressed slightly from its equilibrium position and released. Take ρ lb/in³ to be the density of water and δ lb/in³ to be the density of the buoy material (density = weight per unit of volume).

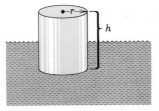

20. Suppose that the mass of the object in Figure 19.4.1 is M and the spring constant is k. Show that if the mass is displaced y_0 units from its equilibrium position and given an initial velocity of v_0 rather than being released with velocity 0, then its position function is given by

$$y(t) = y_0 \cos\left(\sqrt{\frac{k}{M}}\, t\right) + v_0 \sqrt{\frac{M}{k}} \sin\left(\sqrt{\frac{k}{M}}\, t\right)$$

21. A mass attached to a vertical spring is displaced from its equilibrium position and released, thereby causing it to vibrate with amplitude $|y_0|$ and period T (no damping).
    (a) Show that the velocity of the mass has maximum magnitude $2\pi|y_0|/T$ and that the maximum occurs when the mass is at its equilibrium position.
    (b) Show that the acceleration of the mass has maximum magnitude $4\pi^2|y_0|/T^2$ and that the maximum occurs when the mass is at a top or bottom point of its motion.

22. Prove that the roots of Equation (12) are distinct and real, equal and real, or complex according to whether $c^2 > 4kM$, $c^2 = 4kM$, or $c^2 < 4kM$.

23. Use the addition formula for cosine to show that Formula (15) can be rewritten as (16a).

24. **Overdamped Motion**
    (a) Prove that in the case of overdamped motion $(c^2 > 4kM)$ the solution of (11) is
    $$y(t) = \frac{y_0}{m_2 - m_1}(m_2 e^{m_1 t} - m_1 e^{m_2 t})$$
    (b) Prove that $\lim\limits_{t \to +\infty} y(t) = 0$. [*Hint:* Show that m_1 and m_2 are negative.]

25. **Critically Damped Motion**
    (a) Prove that in the case of critically damped motion $(c^2 = 4kM)$ the solution of (11) is
    $$y(t) = y_0 e^{-\alpha t}(1 + \alpha t)$$
    (b) Prove that $\lim\limits_{t \to +\infty} y(t) = 0$.
    (c) Prove that $y(t) \neq 0$ for any positive value of t (assuming that the initial displacement y_0 is not zero).

A

Appendix

CRAMER'S RULE

■ **A.1** CRAMER'S RULE

In this section we shall show how determinants can be used to solve systems of two linear equations in two unknowns and three linear equations in three unknowns. We assume that the reader has read the text through Section 14.6.

Recall that a **solution** of a system of two linear equations in two unknowns

$$a_1x + b_1y = k_1 \tag{1}$$
$$a_2x + b_2y = k_2$$

is a value for x and a value for y that satisfy *both* equations. The graphs of these equations are lines, which we shall denote by L_1 and L_2. Since a point (x, y) lies on a line if and only if the numbers x and y satisfy the equation of the line, the solutions of the system will correspond to points of intersection of L_1 and L_2. There are three possibilities:

- The lines L_1 and L_2 are parallel and distinct (Figure A.1.1a), in which case there is no intersection and, consequently, no solution to the system.
- The lines L_1 and L_2 intersect at only one point (Figure A.1.1b), in which case the system has exactly one solution.
- The lines L_1 and L_2 coincide (Figure A.1.1c), in which case there are infinitely many points of intersection, and consequently infinitely many solutions to the system.

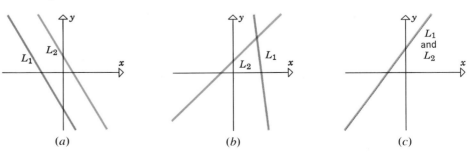

Figure A.1.1   (a)   (b)   (c)

Example 1   If we multiply the second equation of the system

$$x + y = 4$$
$$2x + 2y = 6$$

by $\frac{1}{2}$, it becomes evident that there is no solution since the two equations in the resulting system

$$x + y = 4$$
$$x + y = 3$$

contradict each other. Geometrically, the lines $x + y = 4$ and $2x + 2y = 6$ are distinct and parallel.   ◀

Example 2   Since the second equation of the system

$$x + y = 4$$
$$2x + 2y = 8$$

is a multiple of the first equation, it is evident that any solution of the first equation will satisfy the second equation automatically. But the first equation, $x + y = 4$, has infinitely many solutions since we may assign x an arbitrary value and determine y from the relationship $y = 4 - x$. (Some possibilities are $x = 0$, $y = 4$; $x = 1$, $y = 3$; $x = -1$, $y = 5$.) Thus, the system has infinitely many solutions. Geometrically, the lines $x + y = 4$ and $2x + 2y = 8$ coincide.   ◀

Example 3   If we add the equations of the system

$$x + y = 3$$
$$2x - y = 6$$

we obtain $3x = 9$ or $x = 3$, and if we substitute this value in the first equation we obtain $y = 0$, so that this system has the unique solution $x = 3$, $y = 0$. Geometrically, the lines $x + y = 3$ and $2x - y = 6$ intersect only at the point $(3, 0)$.   ◀

A *solution* of a system of three linear equations in three unknowns

$$a_1x + b_1y + c_1z = k_1$$
$$a_2x + b_2y + c_2z = k_2 \qquad (2)$$
$$a_3x + b_3y + c_3z = k_3$$

consists of values of x, y, and z that satisfy all three equations. The graphs of these equations are planes in 3-space, and the solutions of this system correspond to points where all three planes intersect. It follows that system (2), like system (1), has zero, one, or infinitely many solutions, depending on the relative positions of the planes (Figure A.1.2).

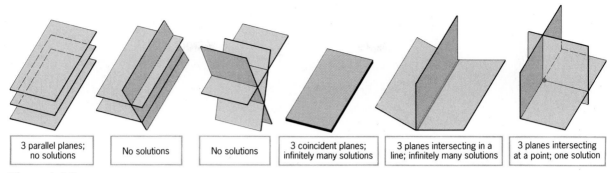

| 3 parallel planes; no solutions | No solutions | No solutions | 3 coincident planes; infinitely many solutions | 3 planes intersecting in a line; infinitely many solutions | 3 planes intersecting at a point; one solution |

Figure A.1.2

If a system of two linear equations in two unknowns or three linear equations in three unknowns has a unique solution, the solution can be expressed as a ratio of determinants by using the following theorems, which are cases of a general result called *Cramer's* * *rule.*

THEOREM **A.1.1** (*Cramer's Rule for Two Unknowns*). *If the system of equations*

$$a_1 x + b_1 y = k_1$$
$$a_2 x + b_2 y = k_2$$

is such that $\begin{vmatrix} a_1 & b_1 \\ a_2 & b_2 \end{vmatrix} \neq 0$

then the system has a unique solution. This solution is

$$x = \frac{\begin{vmatrix} k_1 & b_1 \\ k_2 & b_2 \end{vmatrix}}{\begin{vmatrix} a_1 & b_1 \\ a_2 & b_2 \end{vmatrix}}, \quad y = \frac{\begin{vmatrix} a_1 & k_1 \\ a_2 & k_2 \end{vmatrix}}{\begin{vmatrix} a_1 & b_1 \\ a_2 & b_2 \end{vmatrix}}$$

---

*GABRIEL CRAMER (1704–1752). Swiss mathematician. Although Cramer does not rank with the great mathematicians of his time, his contributions as a disseminator of mathematical ideas have earned him a well-deserved place in the history of mathematics. The son of a physician, Cramer was born and educated in Geneva, Switzerland. At age 20 he competed for, but failed to secure, the chair of philosophy at the Académie de Calvin at Geneva. However, the awarding magistrates were sufficiently impressed with Cramer and a fellow competitor to create a new chair of mathematics for both men to share. Alternately, each assumed the full responsibility and salary associated with the chair for two or three years while the other traveled. During his travels Cramer met many of the great mathematicians and scientists of his day: the Bernoullis, Euler, Halley, D'Alembert, and others. Many of these contacts and friendships led to extensive correspondence in which information about new mathematical discoveries was transmitted. Eventually, Cramer became sole possessor of the mathematics chair and the chair of philosophy as well.

Cramer's mathematical work was primarily in geometry and probability; he had relatively little knowledge of calculus and did not use it to any great extent in his work. In 1730 he finished second to Johann I Bernoulli in a competition for a prize offered by the Paris Academy to explain properties of planetary orbits.

THEOREM **A.1.2 (*Cramer's Rule for Three Unknowns*).**   *If the system of equations*

$$a_1 x + b_1 y + c_1 z = k_1$$
$$a_2 x + b_2 y + c_2 z = k_2$$
$$a_3 x + b_3 y + c_3 z = k_3$$

is such that $\begin{vmatrix} a_1 & b_1 & c_1 \\ a_2 & b_2 & c_2 \\ a_3 & b_3 & c_3 \end{vmatrix} \neq 0$

then the system has a unique solution. This solution is

$$x = \frac{\begin{vmatrix} k_1 & b_1 & c_1 \\ k_2 & b_2 & c_2 \\ k_3 & b_3 & c_3 \end{vmatrix}}{\begin{vmatrix} a_1 & b_1 & c_1 \\ a_2 & b_2 & c_2 \\ a_3 & b_3 & c_3 \end{vmatrix}}, \quad y = \frac{\begin{vmatrix} a_1 & k_1 & c_1 \\ a_2 & k_2 & c_2 \\ a_3 & k_3 & c_3 \end{vmatrix}}{\begin{vmatrix} a_1 & b_1 & c_1 \\ a_2 & b_2 & c_2 \\ a_3 & b_3 & c_3 \end{vmatrix}}, \quad z = \frac{\begin{vmatrix} a_1 & b_1 & k_1 \\ a_2 & b_2 & k_2 \\ a_3 & b_3 & k_3 \end{vmatrix}}{\begin{vmatrix} a_1 & b_1 & c_1 \\ a_2 & b_2 & c_2 \\ a_3 & b_3 & c_3 \end{vmatrix}}$$

REMARK.   There is a pattern to the formulas in these theorems. In each formula the determinant in the denominator is formed from the coefficients of the unknowns; and the determinant in the numerator differs from the determinant in the denominator in that the coefficients of the unknown being calculated are replaced by the k's. It is assumed in these theorems that the system is written so that like unknowns are aligned vertically and the constants appear by themselves on the right side of each equation.

We shall prove Cramer's rule for the case of three unknowns. However, let us first look at some examples.

(continued)

Cramer's most widely known work, *Introduction à l'analyse des lignes courbes algébriques* (1750), was a study and classification of algebraic curves; Cramer's rule appeared in the appendix. Although the rule bears his name, variations of the basic idea were formulated earlier by Leibniz (and even earlier by Chinese mathematicians). However, Cramer's superior notation helped clarify and popularize the technique.

Perhaps Cramer's most important contributions stemmed from his work as an editor of the mathematical creations of others. He edited and published the works of Jacob I Bernoulli and Leibniz.

Overwork combined with a fall from a carriage eventually led to his death in 1752. Cramer was apparently a good-natured and pleasant person, though he never married. His interests were broad. He wrote on philosophy of law and government and the history of mathematics. He served in public office, participated in artillery and fortifications activities for the government, instructed workers on techniques of cathedral repair, and undertook excavations of cathedral archives. Cramer received numerous honors for his activities.

Example 4   Use Cramer's rule to solve

$$5x - 2y = -1$$
$$2x + 3y = 3$$

Solution.

$$x = \frac{\begin{vmatrix} -1 & -2 \\ 3 & 3 \end{vmatrix}}{\begin{vmatrix} 5 & -2 \\ 2 & 3 \end{vmatrix}} = \frac{3}{19}; \quad y = \frac{\begin{vmatrix} 5 & -1 \\ 2 & 3 \end{vmatrix}}{\begin{vmatrix} 5 & -2 \\ 2 & 3 \end{vmatrix}} = \frac{17}{19} \quad \blacktriangleleft$$

Example 5   Use Cramer's rule to solve

$$x \qquad\quad + 2z = 6$$
$$-3x + 4y + 6z = 30$$
$$- x - 2y + 3z = 8$$

Solution.

$$x = \frac{\begin{vmatrix} 6 & 0 & 2 \\ 30 & 4 & 6 \\ 8 & -2 & 3 \end{vmatrix}}{\begin{vmatrix} 1 & 0 & 2 \\ -3 & 4 & 6 \\ -1 & -2 & 3 \end{vmatrix}} = \frac{-10}{11}; \quad y = \frac{\begin{vmatrix} 1 & 6 & 2 \\ -3 & 30 & 6 \\ -1 & 8 & 3 \end{vmatrix}}{\begin{vmatrix} 1 & 0 & 2 \\ -3 & 4 & 6 \\ -1 & -2 & 3 \end{vmatrix}} = \frac{18}{11};$$

$$z = \frac{\begin{vmatrix} 1 & 0 & 6 \\ -3 & 4 & 30 \\ -1 & -2 & 8 \end{vmatrix}}{\begin{vmatrix} 1 & 0 & 2 \\ -3 & 4 & 6 \\ -1 & -2 & 3 \end{vmatrix}} = \frac{38}{11} \quad \blacktriangleleft$$

Proof of Cramer's Rule for Three Unknowns.   Since two vectors are equal if and only if their corresponding components are equal, the system

$$a_1x + b_1y + c_1z = k_1$$
$$a_2x + b_2y + c_2z = k_2 \qquad\qquad (3)$$
$$a_3x + b_3y + c_3z = k_3$$

is equivalent to the single vector equation

$$\langle a_1x + b_1y + c_1z, a_2x + b_2y + c_2z, a_3x + b_3y + c_3z \rangle = \langle k_1, k_2, k_3 \rangle$$

or, equivalently,

$$x\langle a_1, a_2, a_3 \rangle + y\langle b_1, b_2, b_3 \rangle + z\langle c_1, c_2, c_3 \rangle = \langle k_1, k_2, k_3 \rangle \tag{4}$$

If we define

$$\mathbf{a} = \langle a_1, a_2, a_3 \rangle, \quad \mathbf{b} = \langle b_1, b_2, b_3 \rangle, \quad \mathbf{c} = \langle c_1, c_2, c_3 \rangle, \quad \mathbf{k} = \langle k_1, k_2, k_3 \rangle$$

then (4) can be written as

$$x\mathbf{a} + y\mathbf{b} + z\mathbf{c} = \mathbf{k} \tag{5}$$

Therefore, solving system (3) is equivalent to solving the vector equation (5) for the unknown scalars x, y, and z. If we take the dot product of both sides of (5) with $\mathbf{b} \times \mathbf{c}$, we obtain

$$(x\mathbf{a} + y\mathbf{b} + z\mathbf{c}) \cdot (\mathbf{b} \times \mathbf{c}) = \mathbf{k} \cdot (\mathbf{b} \times \mathbf{c})$$

or

$$x(\mathbf{a} \cdot \mathbf{b} \times \mathbf{c}) + y(\mathbf{b} \cdot \mathbf{b} \times \mathbf{c}) + z(\mathbf{c} \cdot \mathbf{b} \times \mathbf{c}) = \mathbf{k} \cdot (\mathbf{b} \times \mathbf{c}) \tag{6}$$

By Theorem 14.4.2,

$$\mathbf{b} \cdot (\mathbf{b} \times \mathbf{c}) = 0 \quad \text{and} \quad \mathbf{c} \cdot (\mathbf{b} \times \mathbf{c}) = 0$$

so that (6) reduces to

$$x(\mathbf{a} \cdot \mathbf{b} \times \mathbf{c}) = \mathbf{k} \cdot (\mathbf{b} \times \mathbf{c})$$

Expressing these triple scalar products as determinants yields

$$x \begin{vmatrix} a_1 & a_2 & a_3 \\ b_1 & b_2 & b_3 \\ c_1 & c_2 & c_3 \end{vmatrix} = \begin{vmatrix} k_1 & k_2 & k_3 \\ b_1 & b_2 & b_3 \\ c_1 & c_2 & c_3 \end{vmatrix}$$

Solving for x and applying Theorem 14.4.5(b) yields

$$x = \frac{\begin{vmatrix} k_1 & b_1 & c_1 \\ k_2 & b_2 & c_2 \\ k_3 & b_3 & c_3 \end{vmatrix}}{\begin{vmatrix} a_1 & b_1 & c_1 \\ a_2 & b_2 & c_2 \\ a_3 & b_3 & c_3 \end{vmatrix}}$$

The formulas for y and z have similar derivations.

As a matter of logic, we have shown that *if there exists a solution* of system (3), then the solution is unique and is given by the formulas in Cramer's rule. However, it is logically possible that the system has no solution, even though Cramer's rule generates values of x, y, and z. This can be ruled out by substituting these values into (3) and verifying that the equations are satisfied. The details are tedious and will be omitted. ∎

▶ Exercises A.1

In Exercises 1–8, solve the system using Cramer's rule.

1. $3x - 4y = -5$
    $2x + y = 4.$

2. $-x + 3y = 8$
    $2x + 5y = 7.$

3. $2x_1 - 5x_2 = -2$
    $4x_1 + 6x_2 = 1.$

4. $3a + 2b = 4$
    $-a + b = 7.$

5. $x + 2y + z = 3$
    $2x + y - z = 0$
    $x - y + z = 6.$

6. $x - 3y + z = 4$
    $2x - y = -2$
    $4x - 3z = 0.$

7. $x_1 + x_2 - 2x_3 = 1$
    $2x_1 - x_2 + x_3 = 2$
    $x_1 - 2x_2 - 4x_3 = -4.$

8. $r + s + t = 2$
    $r - s - 2t = 0$
    $-r + 2s + t = 4.$

9. Use Cramer's rule to solve the rotation equations

$$x = x' \cos \theta - y' \sin \theta$$
$$y = x' \sin \theta + y' \cos \theta$$

for x' and y' in terms of x and y.

10. Solve the following system of equations for the unknown angles α, β, and γ, where $0 \le \alpha \le 2\pi$, $0 \le \beta \le 2\pi$, and $0 \le \gamma < \pi$:

$$2 \sin \alpha - \cos \beta + 3 \tan \gamma = 3$$
$$4 \sin \alpha + 2 \cos \beta - 2 \tan \gamma = 2$$
$$6 \sin \alpha - 3 \cos \beta + \tan \gamma = 9$$

[*Hint:* First solve for $\sin \alpha$, $\cos \beta$, and $\tan \gamma$.]

B

Appendix

COMPLEX NUMBERS

B.1 COMPLEX NUMBERS

Complex numbers arise naturally in the course of solving equations. For example, the solutions of the quadratic equation $ax^2 + bx + c = 0$, which are given by the quadratic formula

$$x = \frac{-b \pm \sqrt{b^2 - 4ac}}{2a}$$

are complex numbers if the expression $b^2 - 4ac$ inside the radical is negative. In this section we shall develop some of the basic properties of complex numbers.

Since $x^2 \geq 0$ for every real number x, the equation

$$x^2 = -1$$

has no real solutions. To deal with this problem, mathematicians of the eighteenth century introduced the "imaginary" number,

$$i = \sqrt{-1}$$

which they assumed had the property

$$i^2 = (\sqrt{-1})^2 = -1$$

but which otherwise could be treated as a real number. Expressions of the form

$$a + bi$$

where a and b are real numbers were called "complex numbers," and these were manipulated according to the standard rules of arithmetic with the added property that $i^2 = -1$.

By the beginning of the nineteenth century it was recognized that a complex number,

$$a + bi$$

could be regarded as an alternative symbol for the ordered pair

$$(a, b)$$

of real numbers, and that operations of addition, subtraction, multiplication, and division could be defined on these ordered pairs so that the familiar laws of arithmetic hold and $i^2 = -1$. This is the approach we will follow.

Definition B.1.1. A *complex number* is an ordered pair of real numbers, denoted either by (a, b) or $a + bi$.

Example 1 Some examples of complex numbers in both notations are as follows:

Ordered Pair	Equivalent Notation
$(3, 4)$	$3 + 4i$
$(-1, 2)$	$-1 + 2i$
$(0, 1)$	$0 + i$
$(2, 0)$	$2 + 0i$
$(4, -2)$	$4 + (-2)i$

For simplicity, the last three complex numbers would usually be abbreviated as

$$0 + i = i \qquad 2 + 0i = 2 \qquad 4 + (-2)i = 4 - 2i \qquad \blacktriangleleft$$

Geometrically, a complex number can be viewed either as a point or a vector in the xy-plane (Figure B.1.1).

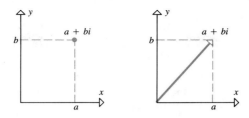

Figure B.1.1 | A complex number can be viewed as a point or a vector. |

Example 2   Some complex numbers are shown as points in Figure B.1.2*a* and as vectors in Figure B.1.2*b*.   ◀

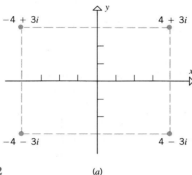 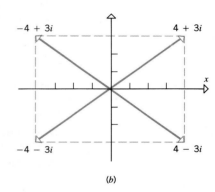

Figure B.1.2   (*a*)   (*b*)

THE COMPLEX PLANE

Sometimes it is convenient to use a single letter, such as z, to denote a complex number. Thus, we might write

$$z = a + bi$$

The real number a is called the ***real part of*** z and the real number b the ***imaginary part of*** z. These numbers are denoted by $\text{Re}(z)$ and $\text{Im}(z)$, respectively. Thus,

$$\text{Re}(4 - 3i) = 4 \quad \text{and} \quad \text{Im}(4 - 3i) = -3$$

When complex numbers are represented geometrically in an xy-coordinate system, the x-axis is called the ***real axis***, the y-axis the ***imaginary axis***, and the plane is called the ***complex plane*** (Figure B.1.3).

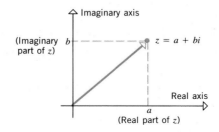

Figure B.1.3

Just as two vectors in R^2 are defined to be equal if they have the same components, so we define two complex numbers to be equal if their real parts are equal and their imaginary parts are equal:

Definition B.1.2. Two complex numbers, $a + bi$ and $c + di$, are defined to be *equal*, written

$$a + bi = c + di$$

if $a = c$ and $b = d$.

If $b = 0$, then the complex number $a + bi$ reduces to $a + 0i$, which we write simply as a. Thus, for any real number a,

$$a = a + 0i$$

so that the real numbers can be regarded as complex numbers with an imaginary part of zero. Geometrically, the real numbers correspond to points on the real axis. If $a = 0$, then $a + bi$ reduces to $0 + bi$, which we usually write as bi. These complex numbers, which correspond to points on the imaginary axis, are called *pure imaginary numbers*.

Just as vectors in R^2 are added by adding corresponding components, so complex numbers are added by adding their real parts and adding their imaginary parts:

$$(a + bi) + (c + di) = (a + c) + (b + d)i \tag{1}$$

The operations of subtraction and multiplication by a *real* number are also similar to the corresponding vector operations in R^2:

$$(a + bi) - (c + di) = (a - c) + (b - d)i \tag{2}$$

$$k(a + bi) = (ka) + (kb)i, \quad (k \text{ real}) \tag{3}$$

Because the operations of addition, subtraction, and multiplication of a complex number by a real number parallel the corresponding operations for vectors in R^2, the familiar geometric interpretations of these operations hold for complex numbers (Figure B.1.4).

It follows from (3) that $(-1)z + z = 0$ (verify), so we denote $(-1)z$ as $-z$ and call it the *negative of z*.

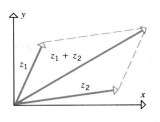

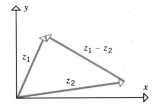

The sum of two complex numbers.

The difference of two complex numbers.

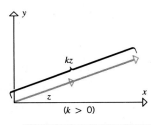

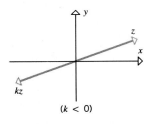

$(k > 0)$

$(k < 0)$

Figure B.1.4

The product of a complex number z and a real number k.

Example 3 If $z_1 = 4 - 5i$ and $z_2 = -1 + 6i$, find $z_1 + z_2$, $z_1 - z_2$, $3z_1$, and $-z_2$.

Solution.

$$z_1 + z_2 = (4 - 5i) + (-1 + 6i) = (4 - 1) + (-5 + 6)i = 3 + i$$
$$z_1 - z_2 = (4 - 5i) - (-1 + 6i) = (4 + 1) + (-5 - 6)i = 5 - 11i$$
$$3z_1 = 3(4 - 5i) = 12 - 15i$$
$$-z_2 = (-1)z_2 = (-1)(-1 + 6i) = 1 - 6i \quad \blacktriangleleft$$

So far, there has been a parallel between complex numbers and vectors in R^2. However, we now define multiplication of complex numbers, an operation with no vector analog in R^2. To motivate the definition, we expand the product

$$(a + bi)(c + di)$$

following the usual rules of algebra, but treating i^2 as -1. This yields

$$(a + bi)(c + di) = ac + bdi^2 + adi + bci$$
$$= (ac - bd) + (ad + bc)i$$

which suggests the following *definition*:

$$(a + bi)(c + di) = (ac - bd) + (ad + bc)i \tag{4}$$

Example 4

$$(3 + {}^{\cdot}2i)(4 + 5i) = (3 \cdot 4 - 2 \cdot 5) + (3 \cdot 5 + 2 \cdot 4)i$$
$$= 2 + 23i$$
$$(4 - i)(2 - 3i) = [4 \cdot 2 - (-1)(-3)] + [(4)(-3) + (-1)(2)]i$$
$$= 5 - 14i$$
$$i^2 = (0 + i)(0 + i) = (0 \cdot 0 - 1 \cdot 1) + (0 \cdot 1 + 1 \cdot 0)i = -1 \quad \blacktriangleleft$$

We leave it as an exercise to verify the following rules of complex arithmetic:

$$z_1 + z_2 = z_2 + z_1$$
$$z_1 z_2 = z_2 z_1$$
$$z_1 + (z_2 + z_3) = (z_1 + z_2) + z_3$$
$$z_1(z_2 z_3) = (z_1 z_2)z_3$$
$$z_1(z_2 + z_3) = z_1 z_2 + z_1 z_3$$
$$0 + z = z$$
$$z + (-z) = 0$$
$$1 \cdot z = z$$

These rules make it possible to multiply complex numbers without using Formula (4) directly. Following the procedure used to motivate that formula, we can simply multiply each term of $a + bi$ by each term of $c + di$, set $i^2 = -1$, and simplify.

Example 5

$$(3 + 2i)(4 + i) = 12 + 3i + 8i + 2i^2 = 12 + 11i - 2 = 10 + 11i$$
$$(5 - \tfrac{1}{2}i)(2 + 3i) = 10 + 15i - i - \tfrac{3}{2}i^2 = 10 + 14i + \tfrac{3}{2} = \tfrac{23}{2} + 14i$$
$$i(1 + i)(1 - 2i) = i(1 - 2i + i - 2i^2) = i(3 - i) = 3i - i^2 = 1 + 3i \quad \blacktriangleleft$$

REMARK.   Unlike the real numbers, there is no size ordering for the complex numbers. Thus, the order symbols $<$, $\leq$, $>$, and $\geq$ are not used with complex numbers.

▶ Exercise Set B.1

1. In each part plot the point and sketch the vector that corresponds to the given complex number.

(a) $2 + 3i$ (b) -4 (c) $-3 - 2i$ (d) $-5i$

2. Express each complex number in Exercise 1 as an ordered pair of real numbers.

3. In each part use the given information to find the real numbers x and y.
   (a) $x - iy = -2 + 3i$ (b) $(x + y) + (x - y)i = 3 + i$

4. Given that $z_1 = 1 - 2i$ and $z_2 = 4 + 5i$, find
   (a) $z_1 + z_2$ (b) $z_1 - z_2$ (c) $4z_1$ (d) $-z_2$ (e) $3z_1 + 4z_2$ (f) $\frac{1}{2}z_1 - \frac{3}{2}z_2$

5. In each part solve for z.
   (a) $z + (1 - i) = 3 + 2i$ (b) $-5z = 5 + 10i$ (c) $(i - z) + (2z - 3i) = -2 + 7i$

6. In each part sketch the vectors z_1, z_2, $z_1 + z_2$, and $z_1 - z_2$.
   (a) $z_1 = 3 + i$, $z_2 = 1 + 4i$ (b) $z_1 = -2 + 2i$, $z_2 = 4 + 5i$

7. In each part sketch the vectors z and kz.
   (a) $z = 1 + i$, $k = 2$ (b) $z = -3 - 4i$, $k = -2$ (c) $z = 4 + 6i$, $k = \frac{1}{2}$

8. In each part find real numbers k_1 and k_2 that satisfy the equation.
   (a) $k_1 i + k_2(1 + i) = 3 - 2i$ (b) $k_1(2 + 3i) + k_2(1 - 4i) = 7 + 5i$

9. In each part find $z_1 z_2$, z_1^2, and z_2^2.
   (a) $z_1 = 3i$, $z_2 = 1 - i$ (b) $z_1 = 4 + 6i$, $z_2 = 2 - 3i$ (c) $z_1 = \frac{1}{3}(2 + 4i)$, $z_2 = \frac{1}{2}(1 - 5i)$

10. Given that $z_1 = 2 - 5i$ and $z_2 = -1 - i$, find
    (a) $z_1 - z_1 z_2$ (b) $(z_1 + 3z_2)^2$ (c) $[z_1 + (1 + z_2)]^2$ (d) $iz_2 - z_1^2$

In Exercises 11–18 perform the calculations and express the result in the form $a + bi$.

11. $(1 + 2i)(4 - 6i)^2$

12. $(2 - i)(3 + i)(4 - 2i)$

13. $(1 - 3i)^3$

14. $i(1 + 7i) - 3i(4 + 2i)$

15. $[(2 + i)(\frac{1}{2} + \frac{3}{4}i)]^2$

16. $(\sqrt{2} + i) - i\sqrt{2}(1 + \sqrt{2}i)$

17. $(1 + i + i^2 + i^3)^{100}$

18. $(3 - 2i)^2 - (3 + 2i)^2$

19. Show that
    (a) $\text{Im}(iz) = \text{Re}(z)$ (b) $\text{Re}(iz) = -\text{Im}(z)$

20. In each part solve the equation by the quadratic formula and check your results by substituting the solutions into the given equation.
    (a) $z^2 + 2z + 2 = 0$ (b) $z^2 - z + 1 = 0$

21. (a) Show that if n is a positive integer, then the only possible values for i^n are 1, -1, i, and $-i$.
    (b) Find i^{2509}. [**Hint.** The value of i^n can be determined from the remainder when n is divided by 4.]

22. Prove: If $z_1 z_2 = 0$, then $z_1 = 0$ or $z_2 = 0$.

23. Use the result of Exercise 22 to prove: If $zz_1 = zz_2$ and $z \neq 0$, then $z_1 = z_2$.

24. Prove that for all complex numbers z_1, z_2, and z_3
   (a) $z_1 + z_2 = z_2 + z_1$ (b) $z_1 + (z_2 + z_3) = (z_1 + z_2) + z_3$

25. Prove that for all complex numbers z_1, z_2, and z_3
   (a) $z_1 z_2 = z_2 z_1$ (b) $z_1(z_2 z_3) = (z_1 z_2)z_3$

26. Prove that $z_1(z_2 + z_3) = z_1 z_2 + z_1 z_3$ for all complex numbers z_1, z_2, and z_3.

B.2 MODULUS; COMPLEX CONJUGATE; DIVISION

> *Our main objective in this section is to define the division of complex numbers.*

COMPLEX CONJUGATES

We will begin with some preliminary ideas.

If $z = a + bi$ is any complex number, then the **conjugate of z**, denoted by $\bar{z}$ (read "z bar"), is defined by

$$\bar{z} = a - bi$$

In words, $\bar{z}$ is obtained by reversing the sign of the imaginary part of z. Geometrically, $\bar{z}$ is the reflection of z about the real axis (Figure B.2.1).

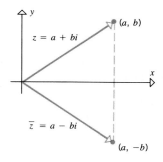

Figure B.2.1 | The conjugate of a complex number.

Example 1

$$z = 3 + 2i \qquad \bar{z} = 3 - 2i$$
$$z = -4 - 2i \qquad \bar{z} = -4 + 2i$$
$$z = i \qquad \bar{z} = -i$$
$$z = 4 \qquad z = 4 \qquad \blacktriangleleft$$

REMARK. The last line in Example 1 illustrates the fact that a real number is the same as its conjugate. More precisely, it can be shown (Exercise 22) that $z = \bar{z}$ if and only if z is a real number.

If a complex number z is viewed as a vector in R^2, then the norm or length of the vector is called the modulus (or *absolute value*) of z. More precisely:

MODULUS

Definition B.2.1.   The *modulus* of a complex number $z = a + bi$, denoted by $|z|$, is defined by

$$|z| = \sqrt{a^2 + b^2} \tag{1}$$

If $b = 0$, then $z = a$ is a real number, and

$$|z| = \sqrt{a^2 + 0^2} = \sqrt{a^2} = |a|$$

so the modulus of a real number is simply its absolute value. Thus, the modulus of z is also called the **absolute value** of z.

Example 2   Find $|z|$ if $z = 3 - 4i$.

Solution.   From (1) with $a = 3$ and $b = -4$, it follows that $|z| = \sqrt{(3)^2 + (-4)^2} = \sqrt{25} = 5$.   ◀

The following theorem establishes a basic relationship between $\bar{z}$ and $|z|$.

Theorem B.2.2.   *For any complex number z,*

$$z\bar{z} = |z|^2$$

Proof. If $z = a + bi$, then

$$z\bar{z} = (a + bi)(a - bi) = a^2 - abi + bai - b^2i^2$$
$$= a^2 + b^2 = |z|^2 \qquad \blacksquare$$

We now turn to the division of complex numbers. Our objective is to define division as the inverse of multiplication. Thus, if $z_2 \neq 0$, then our definition of $z = z_1/z_2$ should be such that

$$z_1 = z_2 z \tag{2}$$

Our procedure will be to prove that (2) has a unique solution for z if $z_2 \neq 0$, and then define z_1/z_2 to be this value of z. As with real numbers, division by zero is not allowed.

Theorem B.2.3. *If $z_2 \neq 0$, then Equation (2) has a unique solution, which is*

$$z = \frac{1}{|z_2|^2} z_1 \bar{z}_2 \tag{3}$$

Proof. Let $z = x + iy$, $z_1 = x_1 + iy_1$, and $z_2 = x_2 + iy_2$. Then (2) can be written as

$$x_1 + iy_1 = (x_2 + iy_2)(x + iy)$$

or

$$x_1 + iy_1 = (x_2 x - y_2 y) + i(y_2 x + x_2 y)$$

or, on equating real and imaginary parts,

$$\begin{aligned} x_2 x - y_2 y &= x_1 \\ y_2 x + x_2 y &= y_1 \end{aligned} \tag{4}$$

Since $z_2 = x_2 + iy_2 \neq 0$, it follows that x_2 and y_2 are not both zero, so

$$\begin{vmatrix} x_2 & -y_2 \\ y_2 & x_2 \end{vmatrix} = x_2^2 + y_2^2 \neq 0$$

Thus, by Cramer's Rule (Theorem A.1.1), system (4) has the unique solution

$$x = \frac{\begin{vmatrix} x_1 & -y_2 \\ y_1 & x_2 \end{vmatrix}}{\begin{vmatrix} x_2 & -y_2 \\ y_2 & x_2 \end{vmatrix}} = \frac{x_1 x_2 + y_1 y_2}{x_2^2 + y_2^2} = \frac{x_1 x_2 + y_1 y_2}{|z_2|^2}$$

$$y = \frac{\begin{vmatrix} x_2 & x_1 \\ y_2 & y_1 \end{vmatrix}}{\begin{vmatrix} x_2 & -y_2 \\ y_2 & x_2 \end{vmatrix}} = \frac{y_1 x_2 - x_1 y_2}{x_2^2 + y_2^2} = \frac{y_1 x_2 - x_1 y_2}{|z_2|^2}$$

Thus,

$$z = x + iy = \frac{1}{|z_2|^2} \left[(x_1 x_2 + y_1 y_2) + i(y_1 x_2 - x_1 y_2) \right]$$

$$= \frac{1}{|z_2|^2} (x_1 + iy_1)(x_2 - iy_2)$$

$$= \frac{1}{|z_2|^2} z_1 \bar{z}_2 \qquad \blacksquare$$

Thus, for $z_2 \neq 0$ we define

$$\frac{z_1}{z_2} = \frac{1}{|z_2|^2} z_1 \bar{z}_2 \tag{5}$$

REMARK. To remember this formula, multiply numerator and denominator of z_1/z_2 by $\bar{z}_2$:

$$\frac{z_1}{z_2} = \frac{z_1 \bar{z}_2}{z_2 \bar{z}_2} = \frac{z_1 \bar{z}_2}{|z_2|^2} = \frac{1}{|z_2|^2} z_1 \bar{z}_2$$

Example 3 Express

$$\frac{3 + 4i}{1 - 2i}$$

in the form $a + bi$.

Solution. From (5) with $z_1 = 3 + 4i$ and $z_2 = 1 - 2i$,

$$\frac{3 + 4i}{1 - 2i} = \frac{1}{|1 - 2i|^2} (3 + 4i)(\overline{1 - 2i})$$

$$= \frac{1}{5}(3 + 4i)(1 + 2i)$$

$$= \frac{1}{5}(-5 + 10i)$$

$$= -1 + 2i$$

Alternative Solution. As in the remark above, multiply numerator and denominator by the conjugate of the denominator:

$$\frac{3 + 4i}{1 - 2i} = \frac{3 + 4i}{1 - 2i} \cdot \frac{1 + 2i}{1 + 2i} = \frac{-5 + 10i}{5} = -1 + 2i \qquad \blacktriangleleft$$

**PROPERTIES
OF THE
COMPLEX
CONJUGATE**

We conclude this section by listing some properties of the complex conjugate that will be useful in the next section.

Theorem B.2.4. *For any complex numbers* z, z_3, *and* z_2

(a) $\overline{z_1 + z_2} = \bar{z}_1 + \bar{z}_2$

(b) $\overline{z_1 - z_2} = \bar{z}_1 - \bar{z}_2$

(c) $\overline{z_1 z_2} = \bar{z}_1 \bar{z}_2$

(d) $\overline{(z_1/z_2)} = \bar{z}_1/\bar{z}_2$

(e) $\bar{\bar{z}} = z$

We prove (a) and leave the rest as exercises.

Proof (a). Let $z_1 = a_1 + b_1 i$ and $z_2 = a_2 + b_2 i$; then

$$\overline{z_1 + z_2} = \overline{(a_1 + a_2) + (b_1 + b_2)i}$$
$$= (a_1 + a_2) - (b_1 + b_2)i$$
$$= (a_1 - b_1 i) + (a_2 - b_2 i)$$
$$= \bar{z}_1 + \bar{z}_2 \quad \blacksquare$$

REMARK. It is possible to extend part (a) of Theorem B.2.4 to n terms and part (c) to n factors. More precisely,

$$\overline{z_1 + z_2 + \cdots + z_n} = \bar{z}_1 + \bar{z}_2 + \cdots + \bar{z}_n$$
$$\overline{z_1 z_2 \cdots z_n} = \bar{z}_1 \bar{z}_2 \cdots \bar{z}_n$$

▶ Exercise Set B.2

1. In each part find $\bar{z}$.
   (a) $z = 2 + 7i$ (b) $z = -3 - 5i$ (c) $z = 5i$ (d) $z = -i$ (e) $z = -9$ (f) $z = 0$

2. In each part find $|z|$.
   (a) $z = i$ (b) $z = -7i$ (c) $z = -3 - 4i$ (d) $z = 1 + i$ (e) $z = -8$ (f) $z = 0$

3. Verify that $z\bar{z} = |z|^2$ for
   (a) $z = 2 - 4i$ (b) $z = -3 + 5i$ (c) $z = \sqrt{2} - \sqrt{2}i$

4. Given that $z_1 = 1 - 5i$ and $z_2 = 3 + 4i$, find
   (a) z_1/z_2 (b) $\bar{z}_1/z_2$ (c) $z_1/\bar{z}_2$ (d) $\overline{(z_1/z_2)}$ (e) $z_1/|z_2|$ (f) $|z_1/z_2|$

5. In each part find $1/z$.

   (a) $z = i$ (b) $z = 1 - 5i$ (c) $z = \dfrac{-i}{7}$

6. Given that $z_1 = 1 + i$ and $z_2 = 1 - 2i$, find

(a) $z_1 - \left(\dfrac{z_1}{z_2}\right)$ (b) $\dfrac{z_1 - 1}{z_2}$ (c) $z_1^2 - \left(\dfrac{iz_1}{z_2}\right)$ (d) $\dfrac{z_1}{iz_2}$

In Exercises 7–14 perform the calculations and express the result in the form $a + bi$.

7. $\dfrac{i}{1 + i}$ **8.** $\dfrac{2}{(1 - i)(3 + i)}$ **9.** $\dfrac{1}{(3 + 4i)^2}$

10. $\dfrac{2 + i}{i(-3 + 4i)}$ **11.** $\dfrac{\sqrt{3} + i}{(1 - i)(\sqrt{3} - i)}$ **12.** $\dfrac{1}{i(3 - 2i)(1 + i)}$

13. $\dfrac{i}{(1 - i)(1 - 2i)(1 + 2i)}$ **14.** $\dfrac{1 - 2i}{3 + 4i} - \dfrac{2 + i}{5i}$

15. In each part solve for z.
(a) $iz = 2 - i$ (b) $(4 - 3i)\bar{z} = i$

16. Use Theorem B.2.4 to prove the following identities:

(a) $\overline{\bar{z} + 5i} = z - 5i$ (b) $\overline{iz} = -i\bar{z}$ (c) $\dfrac{\overline{i + \bar{z}}}{i - z} = -1$

17. In each part sketch the set of points in the complex plane that satisfy the equation.
(a) $|z| = 2$ (b) $|z - (1 + i)| = 1$ (c) $|z - i| = |z + i|$ (d) $\text{Im}(\bar{z} + i) = 3$

18. In each part sketch the set of points in the complex plane that satisfy the given condition(s).
(a) $|z + i| \le 1$ (b) $1 < |z| < 2$ (c) $|2z - 4i| < 1$ (d) $|z| \le |z + i|$

19. Given that $z = x + iy$, find
(a) $\text{Re}(\overline{iz})$ (b) $\text{Im}(\overline{iz})$ (c) $\text{Re}(i\bar{z})$ (d) $\text{Im}(i\bar{z})$

20. (a) Show that if n is a positive integer, then the only possible values for $(1/i)^n$ are 1, -1, i, and $-i$.
(b) Find $(1/i)^{2509}$. [**Hint.** See Exercise 21(b) of Section B.1.]

21. Prove:

(a) $\dfrac{1}{2}(z + \bar{z}) = \text{Re}(z)$ (b) $\dfrac{1}{2i}(z - \bar{z}) = \text{Im}(z)$

22. Prove: $z = \bar{z}$ if and only if z is a real number.

23. Given that $z_1 = x_1 + iy_1$ and $z_2 = x_2 + iy_2$, find

(a) $\text{Re}\left(\dfrac{z_1}{z_2}\right)$ (b) $\text{Im}\left(\dfrac{z_1}{z_2}\right)$

24. Prove: If $(\bar{z})^2 = z^2$, then z is either real or pure imaginary.

25. Prove that $|z| = |\bar{z}|$.

26. Prove:
   (a) $\overline{z_1 - z_2} = \bar{z}_1 - \bar{z}_2$ (b) $\overline{z_1 z_2} = \bar{z}_1 \bar{z}_2$ (c) $\overline{(z_1/z_2)} = \bar{z}_1/\bar{z}_2$ (d) $\bar{\bar{z}} = z$

27. (a) Prove that $\overline{z^2} = (\bar{z})^2$.
   (b) Prove that if n is a positive integer, then $\overline{z^n} = (\bar{z})^n$.
   (c) Is the result in (b) true if n is a negative integer? Explain.

28. Let $p(x) = a_0 + a_1 x + a_2 x^2 + \cdots + a_n x^n$ be a polynomial for which the coefficients $a_0, a_1, a_2, \ldots, a_n$ are real. Prove that if z is a solution of the equation $p(x) = 0$, then so is $\bar{z}$.

29. Prove: For any complex number z, $|\mathrm{Re}(z)| \leq |z|$ and $|\mathrm{Im}(z)| \leq |z|$.

30. Prove that

$$\frac{|\mathrm{Re}(z)| + |\mathrm{Im}(z)|}{\sqrt{2}} \leq |z|$$

[**_Hint._** Let $z = x + iy$ and use the fact that $(|x| - |y|)^2 \geq 0$.]

■ **B.3** POLAR FORM; DEMOIVRE'S THEOREM

> *In this section we shall discuss a way to represent complex numbers using trigonometric properties. Our work will lead to a fundamental formula for powers of complex numbers and a method for finding nth roots of complex numbers.*

POLAR FORM OF A COMPLEX NUMBER

If $z = x + iy$ is a nonzero complex number, $r = |z|$, and θ measures the angle from the positive real axis to the vector z, then, as suggested by Figure B.3.1,

$$x = r \cos \theta$$
$$y = r \sin \theta \tag{1}$$

so that $z = x + iy$ can be written as

$$z = r \cos \theta + ir \sin \theta$$

or

$$z = r(\cos \theta + i \sin \theta) \tag{2}$$

This is called a **_polar form of z_**.

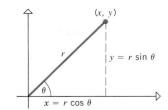

Figure B.3.1

ARGUMENT OF
A COMPLEX
NUMBER

The angle θ is called an **argument of** z and is denoted by

$$\theta = \arg z$$

The argument of z is not uniquely determined because we can add or subtract any multiple of 2π from θ to produce another value of the argument. However, there is only one value of the argument in radians that satisfies

$$-\pi < \theta \le \pi$$

This is called the **principal argument of** z and is denoted by

$$\theta = \text{Arg } z$$

Example 1 Express the following complex numbers in polar form using their principal arguments:

(a) $z = 1 + \sqrt{3}i$ (b) $z = -1 - i$

Solution (a). The value of r is

$$r = |z| = \sqrt{1^2 + (\sqrt{3})^2} = \sqrt{4} = 2$$

and since $x = 1$ and $y = \sqrt{3}$, it follows from (1) that

$$1 = 2 \cos \theta$$
$$\sqrt{3} = 2 \sin \theta$$

so $\cos \theta = 1/2$ and $\sin \theta = \sqrt{3}/2$. The only value of θ that satisfies these relations and meets the requirement $-\pi < \theta \le \pi$ is $\theta = \pi/3$ ($= 60°$) (see Figure B.3.2a). Thus, a polar form of z is

$$z = 2\left(\cos \frac{\pi}{3} + i \sin \frac{\pi}{3}\right)$$

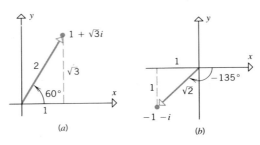

Figure B.3.2

Solution (b). The value of r is

$$r = |z| = \sqrt{(-1)^2 + (-1)^2} = \sqrt{2}$$

and since $x = -1$, $y = -1$, it follows from (1) that

$$-1 = \sqrt{2} \cos \theta$$
$$-1 = \sqrt{2} \sin \theta$$

so $\cos \theta = -1/\sqrt{2}$ and $\sin \theta = -1/\sqrt{2}$. The only value of θ that satisfies these relations and meets the requirement $-\pi < \theta \leq \pi$ is $\theta = -3\pi/4$ $(= -135°)$ (Figure B.3.2b). Thus, a polar form of z is

$$z = \sqrt{2}\left(\cos \frac{-3\pi}{4} + i \sin \frac{-3\pi}{4}\right) \qquad \blacktriangleleft$$

GEOMETRIC INTERPRETATION OF MULTIPLICATION AND DIVISION

We now show how polar forms can be used to give geometric interpretations of multiplication and division of complex numbers. Let

$$z_1 = r_1(\cos \theta_1 + i \sin \theta_1) \qquad \text{and} \qquad z_2 = r_2(\cos \theta_2 + i \sin \theta_2)$$

Multiplying, we obtain

$$z_1 z_2 = r_1 r_2[(\cos \theta_1 \cos \theta_2 - \sin \theta_1 \sin \theta_2) + i(\sin \theta_1 \cos \theta_2 + \cos \theta_1 \sin \theta_2)]$$

Recalling the trigonometric identities

$$\cos(\theta_1 + \theta_2) = \cos \theta_1 \cos \theta_2 - \sin \theta_1 \sin \theta_2$$
$$\sin(\theta_1 + \theta_2) = \sin \theta_1 \cos \theta_2 + \cos \theta_1 \sin \theta_2$$

We obtain

$$z_1 z_2 = r_1 r_2[\cos(\theta_1 + \theta_2) + i \sin(\theta_1 + \theta_2)] \tag{3}$$

which is a polar form of the complex number with modulus $r_1 r_2$ and argument $\theta_1 + \theta_2$. Thus, we have shown that

$$|z_1 z_2| = |z_1| |z_2| \tag{4}$$

and

$$\arg(z_1 z_2) = \arg z_1 + \arg z_2$$

(Why?)

In words, *the product of two complex numbers is obtained by multiplying their moduli and adding their arguments* (Figure B.3.3).

We leave it as an exercise to show that if $z_2 \neq 0$, then

$$\frac{z_1}{z_2} = \frac{r_1}{r_2} \left[\cos(\theta_1 - \theta_2) + i \sin(\theta_1 - \theta_2) \right] \tag{5}$$

from which it follows that

$$\left| \frac{z_1}{z_2} \right| = \frac{|z_1|}{|z_2|} \qquad \text{if } z_2 \neq 0$$

and

$$\arg\left(\frac{z_1}{z_2} \right) = \arg z_1 - \arg z_2$$

In words, *the quotient of two complex numbers is obtained by dividing their moduli and subtracting their arguments (in the appropriate order).*

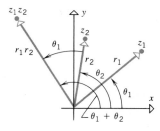

Figure B.3.3   The product of two complex numbers.

Example 2   Let

$$z_1 = 1 + \sqrt{3}i \qquad \text{and} \qquad z_2 = \sqrt{3} + i$$

Polar forms of these complex numbers are

$$z_1 = 2\left(\cos\frac{\pi}{3} + i \sin\frac{\pi}{3} \right)$$

$$z_2 = 2\left(\cos\frac{\pi}{6} + i \sin\frac{\pi}{6} \right)$$

(verify) so that from (3)

$$z_1 z_2 = 4\left[\cos\left(\frac{\pi}{3} + \frac{\pi}{6}\right) + i \sin\left(\frac{\pi}{3} + \frac{\pi}{6}\right)\right]$$

$$= 4\left[\cos\frac{\pi}{2} + i \sin\frac{\pi}{2}\right]$$

$$= 4[0 + i] = 4i$$

and from (5)

$$\frac{z_1}{z_2} = 1 \cdot \left[\cos\left(\frac{\pi}{3} - \frac{\pi}{6}\right) + i \sin\left(\frac{\pi}{3} - \frac{\pi}{6}\right)\right]$$

$$= \cos\frac{\pi}{6} + i \sin\frac{\pi}{6}$$

$$= \frac{\sqrt{3}}{2} + \frac{1}{2}i$$

As a check, we calculate $z_1 z_2$ and z_1/z_2 directly without using polar forms for z_1 and z_2:

$$z_1 z_2 = (1 + \sqrt{3}i)(\sqrt{3} + i) = (\sqrt{3} - \sqrt{3}) + (3 + 1)i = 4i$$

$$\frac{z_1}{z_2} = \frac{1 + \sqrt{3}i}{\sqrt{3} + i} \cdot \frac{\sqrt{3} - i}{\sqrt{3} - i} = \frac{(\sqrt{3} + \sqrt{3}) + (-i + 3i)}{4} = \frac{\sqrt{3}}{2} + \frac{1}{2}i$$

which agrees with our previous results.   ◀

The complex number i has a modulus of 1 and an argument of $\pi/2$ ($= 90°$), so the product iz has the same modulus as z, but its argument is $90°$ greater than that of z. In short, *multiplying z by i rotates z counterclockwise by $90°$* (Figure B.3.4).

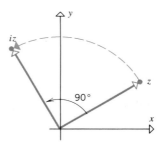

Figure B.3.4 | Multiplying by i rotates z counterclockwise by $90°$.

DEMOIVRE'S FORMULA

If n is a positive integer and $z = r(\cos\theta + i\sin\theta)$, then from Formula (3),

$$z^n = \underbrace{z \cdot z \cdot z \cdots z}_{n\text{-factors}} = r^n[\cos(\underbrace{\theta + \theta + \cdots + \theta}_{n\text{-terms}}) + i\sin(\underbrace{\theta + \theta + \cdots + \theta}_{n\text{-terms}})]$$

or

$$z^n = r^n(\cos n\theta + i \sin n\theta) \tag{6}$$

In the special case where $r = 1$, we have $z = \cos \theta + i \sin \theta$, so (6) becomes

$$(\cos \theta + i \sin \theta)^n = \cos n\theta + i \sin n\theta \tag{7}$$

which is called **DeMoivre's* Formula**. Although we derived (7) assuming n to be a positive integer, it will be shown in the exercises that this formula is valid for all integers n.

**FINDING
nTH ROOTS**

We now show how DeMoivre's Formula can be used to obtain roots of complex numbers. If n is a positive integer, and z is any complex number, then we define an **nth root of z** to be any complex number w that satisfies the equation

$$w^n = z \tag{8}$$

We denote an nth root of z by $z^{1/n}$. If $z \neq 0$, then we can derive formulas for the nth roots of z as follows. Let

$$w = \rho(\cos \alpha + i \sin \alpha) \qquad \text{and} \qquad z = r(\cos \theta + i \sin \theta)$$

If we assume that w satisfies (8), then it follows from (7) that

$$\rho^n(\cos n\alpha + i \sin n\alpha) = r(\cos \theta + i \sin \theta) \tag{9}$$

Comparing the moduli of the two sides, we see that $\rho^n = r$ or

$$\rho = \sqrt[n]{r}$$

where $\sqrt[n]{r}$ denotes the real positive nth root of r. Moreover, in order to have $\cos n\alpha = \cos \theta$ and $\sin n\alpha = \sin \theta$ in (9), the angles $n\alpha$ and θ must either be equal or differ by a multiple of 2π. That is,

$$n\alpha = \theta + 2k\pi, \qquad k = 0, \pm1, \pm2, \ldots$$

or

$$\alpha = \frac{\theta}{n} + \frac{2k\pi}{n}, \qquad k = 0, \pm1, \pm2, \ldots$$

Thus, the values of $w = \rho(\cos \alpha + i \sin \alpha)$ that satisfy (8) are given by

$$w = \sqrt[n]{r}\left[\cos\left(\frac{\theta}{n} + \frac{2k\pi}{n}\right) + i \sin\left(\frac{\theta}{n} + \frac{2k\pi}{n}\right)\right], \qquad k = 0, \pm1, \pm2, \ldots$$

* *Abraham DeMoivre* (1667–1754) was a French mathematician who made important contributions to probability, statistics, and trigonometry. He developed the concept of statistically independent events, wrote a major influential treatise on probability, and helped transform trigonometry from a branch of geometry into a branch of analysis through his use of complex numbers. In spite of his important work, he barely managed to eke out a living as a tutor and a consultant on gambling and insurance.

Although there are infinitely many values of k, it can be shown (Exercise 16) that $k = 0, 1, 2, \ldots, n - 1$ produce distinct values of w satisfying (8), but all other choices of k yield duplicates of these. Therefore, there are exactly n different nth roots of $z = r(\cos \theta + i \sin \theta)$, and these are given by

$$z^{1/n} = \sqrt[n]{r}\left[\cos\left(\frac{\theta}{n} + \frac{2k\pi}{n}\right) + i \sin\left(\frac{\theta}{n} + \frac{2k\pi}{n}\right)\right], \qquad k = 0, 1, 2, \ldots, n - 1 \qquad (10)$$

Example 3   Find all cube roots of -8.

Solution.   Since -8 lies on the negative real axis, we can use $\theta = \pi$ as an argument. Moreover, $r = |z| = |-8| = 8$, so a polar form of -8 is

$$-8 = 8(\cos \pi + i \sin \pi)$$

From (10) with $n = 3$ it follows that

$$(-8)^{1/3} = \sqrt[3]{8}\left[\cos\left(\frac{\pi}{3} + \frac{2k\pi}{3}\right) + i \sin\left(\frac{\pi}{3} + \frac{2k\pi}{3}\right)\right], \qquad k = 0, 1, 2$$

Thus, the cube roots of -8 are

$$2\left(\cos\frac{\pi}{3} + i \sin\frac{\pi}{3}\right) = 2\left(\frac{1}{2} + \frac{\sqrt{3}}{2}i\right) = 1 + \sqrt{3}i$$

$$2(\cos \pi + i \sin \pi) = 2(-1) = -2$$

$$2\cos\left(\frac{5\pi}{3} + i \sin\frac{5\pi}{3}\right) = 2\left(\frac{1}{2} - \frac{\sqrt{3}}{2}i\right) = 1 - \sqrt{3}i \qquad \blacktriangleleft$$

As shown in Figure B.3.5, the three cube roots of -8 obtained in Example 3 are equally spaced $\pi/3$ radians ($= 120°$) apart around the circle of radius 2 centered at the origin. This is not accidental. In general, it follows from Formula (10) that the nth roots of z lie on the circle of radius $\sqrt[n]{r}$ ($= \sqrt[n]{|z|}$), and are equally spaced $2\pi/n$ radians apart. (Can you see why?) Thus, once one nth root of z is found, the remaining $n - 1$ roots can be generated by rotating this root successively through increments of $2\pi/n$ radians.

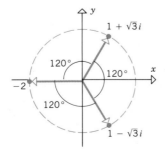

Figure B.3.5   The cube roots of -8.

Example 4 Find all fourth roots of 1.

Solution. We could apply Formula (10). Instead, we observe that $w = 1$ is one fourth root of 1, so that the remaining three roots can be generated by rotating this root through increments of $2\pi/4 = \pi/2$ radians $(= 90°)$. From Figure B.3.6 we see that the fourth roots of 1 are

$$1, \quad i, \quad -1, \quad -i \quad \blacktriangleleft$$

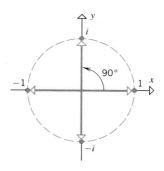

Figure B.3.6   The fourth roots of 1.

COMPLEX EXPONENTS

We conclude this section with some comments on notation.

In more detailed studies of complex numbers, complex exponents are defined, and it is proved that

$$\cos\theta + i\sin\theta = e^{i\theta} \tag{11}$$

where e is an irrational real number given approximately by $e \approx 2.71828 \ldots$. (Exercise 18.)

It follows from (11) that the polar form

$$z = r(\cos\theta + i\sin\theta)$$

can be written more briefly as

$$z = re^{i\theta} \tag{12}$$

Example 5 In Example 1 it was shown that

$$1 + \sqrt{3}i = 2\left(\cos\frac{\pi}{3} + i\sin\frac{\pi}{3}\right)$$

From (12) this can also be written as

$$1 + \sqrt{3}i = 2e^{i\pi/3} \quad \blacktriangleleft$$

It can be proved that complex exponents follow the same laws as real exponents, so that if

$$z_1 = r_1 e^{i\theta_1} \quad \text{and} \quad z_2 = r_2 e^{i\theta_2}$$

then

$$z_1 z_2 = r_1 r_2 e^{i\theta_1 + i\theta_2} = r_1 r_2 e^{i(\theta_1 + \theta_2)}$$

$$\frac{z_1}{z_2} = \frac{r_1}{r_2} e^{i\theta_1 - i\theta_2} = \frac{r_1}{r_2} e^{i(\theta_1 - \theta_2)}$$

But these formulas are just Formulas (3) and (5) in a different notation.

We conclude this section with a useful formula for $\bar{z}$ in polar notation. If

$$z = re^{i\theta} = r(\cos \theta + i \sin \theta)$$

then

$$\bar{z} = r(\cos \theta - i \sin \theta) \tag{13}$$

Recalling the trigonometric identities

$$\sin(-\theta) = -\sin \theta \quad \text{and} \quad \cos(-\theta) = \cos \theta$$

we can rewrite (13) as

$$\bar{z} = r[\cos(-\theta) + i \sin(-\theta)] = re^{i(-\theta)}$$

or equivalently

$$\bar{z} = re^{-i\theta} \tag{14}$$

In the special case where $r = 1$, the polar form of z is $z = e^{i\theta}$, and (14) yields the formula

$$\overline{e^{i\theta}} = e^{-i\theta} \tag{15}$$

▶ **Exercises B.3**

---

1. In each part find the principal argument of z.
  (a) $z = 1$ (b) $z = i$ (c) $z = -i$ (d) $z = 1 + i$ (e) $z = -1 + \sqrt{3}i$ (f) $z = 1 - i$

2. In each part find the value of $\theta = \arg(1 - \sqrt{3}i)$ that satisfies the given condition.

  (a) $0 \le \theta < 2\pi$ (b) $-\pi < \theta \le \pi$ (c) $-\frac{\pi}{6} \le \theta < \frac{11\pi}{6}$

3. In each part express the complex number in polar form using its principal argument.
  (a) $2i$ (b) -4 (c) $5 + 5i$ (d) $-6 + 6\sqrt{3}i$ (e) $-3 - 3i$ (f) $2\sqrt{3} - 2i$

4. Given that $z_1 = 2(\cos \pi/4 + i \sin \pi/4)$ and $z_2 = 3(\cos \pi/6 + i \sin \pi/6)$, find a polar form of

(a) $z_1 z_2$ (b) $\dfrac{z_1}{z_2}$ (c) $\dfrac{z_2}{z_1}$ (d) $\dfrac{z_1^5}{z_2^2}$

5. Express $z_1 = i$, $z_2 = 1 - \sqrt{3}i$, and $z_3 = \sqrt{3} + i$ in polar form, and use your results to find $z_1 z_2 / z_3$. Check your results by performing the calculations without using polar forms.

6. Use Formula (6) to find

(a) $(1 + i)^{12}$ (b) $\left(\dfrac{1}{\sqrt{2}} - \dfrac{1}{\sqrt{2}} i \right)^{-6}$ (c) $(\sqrt{3} + i)^7$ (d) $(1 - i\sqrt{3})^{-10}$

7. In each part find all the roots and sketch them as vectors in the complex plane.

(a) $(-i)^{1/2}$ (b) $(1 + \sqrt{3}i)^{1/2}$ (c) $(-27)^{1/3}$ (d) $(i)^{1/3}$ (e) $(-1)^{1/4}$ (f) $(-8 + 8\sqrt{3}i)^{1/4}$

8. Use the method of Example 4 to find all cube roots of 1.

9. Use the method of Example 4 to find all sixth roots of 1.

10. Find all square roots of $1 + i$ and express your results in polar form.

11. In each part find all solutions of the equation.

(a) $z^4 - 16 = 0$ (b) $z^{4/3} = -4$

12. Find four solutions of the equation $z^4 + 8 = 0$ and use your results to factor $z^4 + 8$ into two quadratic factors with real coefficients.

13. It was shown in the text that multiplying z by i rotates z counterclockwise by 90°. What is the geometric effect of dividing z by i?

14. In each part use (6) to calculate the given power.

(a) $(1 + i)^8$ (b) $(-2\sqrt{3} + 2i)^{-9}$

15. In each part find $\text{Re}(z)$ and $\text{Im}(z)$.

(a) $z = 3e^{i\pi}$ (b) $z = 3e^{-i\pi}$ (c) $\bar{z} = \sqrt{2}e^{\pi i/2}$ (d) $\bar{z} = -3e^{-2\pi i}$

16. (a) Show that the values of $z^{1/n}$ in Formula (10) are all different.
(b) Show that integer values of k other than $k = 0, 1, 2, \ldots, n - 1$ produce values of $z^{1/n}$ that are duplicates of those in Formula (10).

17. Show that Formula (7) is valid if $n = 0$ or n is a negative integer.

18. To motivate Formula (11), recall the Maclaurin series for e^x:

$$e^x = 1 + x + \frac{x^2}{2!} + \cdots + \frac{x^n}{n!} + \cdots$$

(a) By substituting $x = i\theta$ in this series and simplifying, obtain the formula

$$e^{i\theta} = \left(1 - \frac{\theta^2}{2!} + \frac{\theta^4}{4!} - \frac{\theta^6}{6!} + \cdots \right) + i \left(\theta - \frac{\theta^3}{3!} + \frac{\theta^5}{5!} - \frac{\theta^7}{7!} + \cdots \right)$$

(b) Use the result in (a) to obtain (11).

19. Derive Formula (5).

C

Appendix

11
Infinite Series

Brook Taylor   (1685-1731)

■ 11.1 SEQUENCES

This chapter is concerned with the study of "infinite series," which, loosely speaking, are sums with infinitely many terms. The material in this chapter has far-reaching applications in engineering and science and is the cornerstone for many branches of mathematics. In this initial section we shall develop some preliminary results that are important in their own right.

□ DEFINITION OF A
SEQUENCE

In everyday language, we use the term "sequence" to suggest a succession of objects or events given in a specified order. *Informally* speaking, the term "sequence" in mathematics is used to describe an unending succession of numbers. Some possibilities are

$$1, 2, 3, 4, \ldots$$
$$2, 4, 6, 8, \ldots$$
$$1, \tfrac{1}{2}, \tfrac{1}{3}, \tfrac{1}{4}, \ldots$$
$$1, -1, 1, -1, \ldots$$

In each case, the three dots are used to suggest that the sequence continues indefinitely, following the obvious pattern. The numbers in a sequence are called the *terms* of the sequence. The terms may be described according to the positions they occupy. Thus, a sequence has a *first term*, a *second term*, a *third term*, and so forth. Because a sequence continues indefinitely, there is no last term.

The most common way to specify a sequence is to give a formula for the terms. To illustrate the idea, we have listed the terms in the sequence

$$2, 4, 6, 8, \ldots$$

together with their term numbers:

TERM NUMBER	1	2	3	4	...
TERM	2	4	6	8	...

There is a clear relationship here; each term is twice its term number. Thus, for each positive integer n the nth term in the sequence $2, 4, 6, 8, \ldots$ is given by the formula $2n$. This is denoted by writing

$$2, 4, 6, 8, \ldots, 2n, \ldots$$

or more compactly in **bracket notation** as

$$\{2n\}_{n=1}^{+\infty}$$

From the bracket notation, the terms in the sequence can be generated by successively substituting the integer values $n = 1, 2, 3, \ldots$ into the formula $2n$.

Example 1 List the first five terms of the sequence $\{2^n\}_{n=1}^{+\infty}$

Solution. Substituting $n = 1, 2, 3, 4, 5$ into the formula 2^n yields

$$2^1, 2^2, 2^3, 2^4, 2^5, \ldots$$

or, equivalently,

$$2, 4, 8, 16, 32, \ldots \quad \blacktriangleleft$$

Example 2 Express the following sequences in bracket notation.

(a) $\dfrac{1}{2}, \dfrac{2}{3}, \dfrac{3}{4}, \dfrac{4}{5}, \ldots$ (b) $\dfrac{1}{2}, \dfrac{1}{4}, \dfrac{1}{8}, \dfrac{1}{16}, \ldots$

(c) $1, -1, 1, -1, \ldots$ (d) $\dfrac{1}{2}, -\dfrac{2}{3}, \dfrac{3}{4}, -\dfrac{4}{5}, \ldots$

(e) $1, 3, 5, 7, \ldots$

Solution.

(a) Begin by comparing terms and term numbers:

TERM NUMBER	1	2	3	4	...
TERM	$\dfrac{1}{2}$	$\dfrac{2}{3}$	$\dfrac{3}{4}$	$\dfrac{4}{5}$	...

In each term, the numerator is the same as the term number, and the denominator is one greater than the term number. Thus, the nth term is $n/(n + 1)$ and the sequence can be written as

$$\left\{\frac{n}{n + 1}\right\}_{n=1}^{+\infty}$$

(b) Observe that the sequence can be rewritten as

$$\frac{1}{2}, \frac{1}{2^2}, \frac{1}{2^3}, \frac{1}{2^4}, \ldots$$

and construct a table comparing terms and term numbers:

TERM NUMBER	1	2	3	4	...
TERM	$\dfrac{1}{2}$	$\dfrac{1}{2^2}$	$\dfrac{1}{2^3}$	$\dfrac{1}{2^4}$	...

From the table we see that the nth term is $1/2^n$, so the sequence can be written as

$$\left\{\frac{1}{2^n}\right\}_{n=1}^{+\infty}$$

(c)   Observe first that $(-1)^r$ is either 1 or -1 according to whether r is an even integer or an odd integer. In the sequence 1, -1, 1, -1, ... the odd-numbered terms are 1's and the even-numbered terms are -1's. Thus, a formula for the nth term can be obtained by raising -1 to a power that will be even when n is odd and odd when n is even. This is accomplished by the formula $(-1)^{n+1}$, so the sequence can be written as

$$\{(-1)^{n+1}\}_{n=1}^{+\infty}$$

(d)   Combining the results in parts (a) and (c), we can write this sequence as

$$\left\{(-1)^{n+1}\,\frac{n}{n+1}\right\}_{n=1}^{+\infty}$$

(e)   Begin by comparing terms and term numbers:

TERM NUMBER	1	2	3	4	...
TERM	1	3	5	7	...

From the table we see that each term is one less than twice the term number. Thus, the nth term is $2n - 1$, and the sequence can be written as

$$\{2n - 1\}_{n=1}^{+\infty} \qquad \blacktriangleleft$$

Frequently we shall want to write down a sequence without specifying the numerical values of the terms. We do this by writing

$$a_1, a_2, \ldots, a_n, \ldots$$

or in bracket notation

$$\{a_n\}_{n=1}^{+\infty}$$

or sometimes simply

$$\{a_n\}$$

(There is nothing special about the letters a and n; any other letters may be used.)

In the beginning of this section we stated informally that a sequence is an unending succession of numbers. However, this is not a satisfactory mathematical definition because the word "succession" is itself an undefined term. The time has come to formally define the term "sequence." When we write a sequence such as

$$2, 4, 6, 8, \ldots, 2n, \ldots$$

in bracket notation

$$\{2n\}_{n=1}^{+\infty} \tag{1}$$

we are specifying a rule that tells us how to associate a numerical value, namely $2n$, with each positive integer n. Stated another way, $\{2n\}_{n=1}^{+\infty}$ may be regarded as a formula for a function whose independent variable n ranges over the positive integers. Indeed, we could rewrite (1) in functional notation as

$$f(n) = 2n, \quad n = 1, 2, 3, \ldots$$

From this point of view, the notation

$$2, 4, 6, 8, \ldots, 2n, \ldots$$

represents a listing of the function values

$$f(1), f(2), f(3), \ldots, f(n), \ldots$$

This suggests the following definition.

11.1.1   DEFINITION.   A *sequence* (or *infinite sequence*) is a function whose domain is the set of positive integers.

☐ **GRAPHS OF SEQUENCES**

Because sequences are functions, we may inquire about the graph of a sequence. For example, the graph of the sequence

$$\left\{ \frac{1}{n} \right\}_{n=1}^{+\infty}$$

is the graph of the equation

$$y = \frac{1}{n}, \quad n = 1, 2, 3, \ldots$$

Because the right side of this equation is defined only for positive integer values of n, the graph consists of a succession of isolated points (Figure 11.1.1*a*). This is in marked distinction to the graph of

$$y = \frac{1}{x}, \quad x \geq 1$$

which is a continuous curve (Figure 11.1.1*b*).

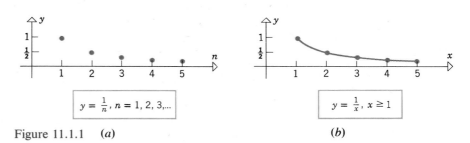

Figure 11.1.1   (*a*)   (*b*)

In Figure 11.1.2 we have sketched the graphs of four sequences:

$$\{n + 1\}_{n=1}^{+\infty}, \quad \{(-1)^{n+1}\}_{n=1}^{+\infty}, \quad \left\{\frac{n}{n+1}\right\}_{n=1}^{+\infty}, \quad \left\{1 + \left(-\frac{1}{2}\right)^n\right\}_{n=1}^{+\infty}$$

Each of these sequences behaves differently as n gets larger and larger. In the sequence $\{n + 1\}$, the terms grow without bound; in the sequence $\{(-1)^{n+1}\}$ the terms oscillate between 1 and -1; in the sequence $\left\{\dfrac{n}{n+1}\right\}$ the terms increase toward a "limit" of 1; and finally, in the sequence $\{1 + (-\frac{1}{2})^n\}$ the terms also tend toward a "limit" of 1, but do so in an oscillatory fashion.

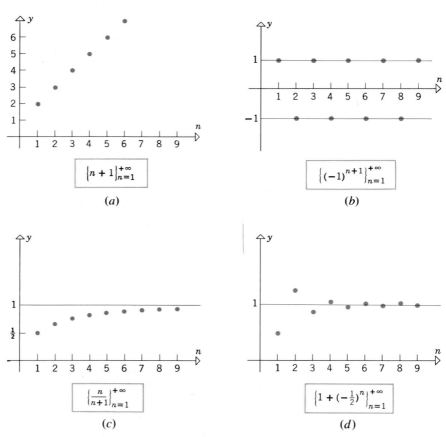

Figure 11.1.2

☐ **LIMIT OF A SEQUENCE**   Let us try to make the term "limit" more precise. To say that a sequence $\{a_n\}_{n=1}^{+\infty}$ approaches a limit L as n gets large is intended to mean that eventually the terms in the sequence become arbitrarily close to the number L. Thus, if we choose *any* positive number ϵ, the terms in the sequence will eventually be within ϵ units of L. Geometrically, this means that if we sketch the lines $y = L + \epsilon$ and $y = L - \epsilon$, the terms in the sequence will eventually be trapped within the band between these lines, and thus be within ϵ units of L (Figure 11.1.3).

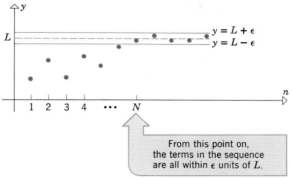

Figure 11.1.3

The following definition expresses this idea precisely.

11.1.2 DEFINITION. A sequence $\{a_n\}_{n=1}^{+\infty}$ is said to have the *limit L* if given any $\epsilon > 0$ there is a positive integer N such that $|a_n - L| < \epsilon$ when $n \geq N$.

If a sequence $\{a_n\}_{n=1}^{+\infty}$ has a limit L, we say that the sequence *converges* to L and write

$$\lim_{n \to +\infty} a_n = L$$

A sequence that does not have a finite limit is said to *diverge*.

Example 3 Figure 11.1.2 suggests that the sequences $\{n + 1\}_{n=1}^{+\infty}$ and $\{(-1)^{n+1}\}_{n=1}^{+\infty}$ diverge, while the sequences

$$\left\{\frac{n}{n + 1}\right\}_{n=1}^{+\infty} \quad \text{and} \quad \left\{1 + \left(-\frac{1}{2}\right)^n\right\}_{n=1}^{+\infty}$$

converge to 1; that is,

$$\lim_{n \to +\infty} \frac{n}{n + 1} = 1 \quad \text{and} \quad \lim_{n \to +\infty} \left(1 + \left(-\frac{1}{2}\right)^n\right) = 1 \quad \blacktriangleleft$$

Many familiar properties of limits apply to limits of sequences.

11.1.3 THEOREM. *Suppose that the sequences $\{a_n\}$ and $\{b_n\}$ converge to limits L_1 and L_2, respectively, and c is a constant. Then*

(a) $\quad \lim\limits_{n \to +\infty} c = c$

(b) $\quad \lim\limits_{n \to +\infty} ca_n = c \lim\limits_{n \to +\infty} a_n = cL_1$

(c) $\quad \lim\limits_{n \to +\infty} (a_n + b_n) = \lim\limits_{n \to +\infty} a_n + \lim\limits_{n \to +\infty} b_n = L_1 + L_2$

(d) $\quad \lim\limits_{n \to +\infty} (a_n - b_n) = \lim\limits_{n \to +\infty} a_n - \lim\limits_{n \to +\infty} b_n = L_1 - L_2$

(e) $\quad \lim\limits_{n \to +\infty} (a_n b_n) = \lim\limits_{n \to +\infty} a_n \cdot \lim\limits_{n \to +\infty} b_n = L_1 L_2$

(f) $\quad \lim\limits_{n \to +\infty} \left(\dfrac{a_n}{b_n}\right) = \dfrac{\lim\limits_{n \to +\infty} a_n}{\lim\limits_{n \to +\infty} b_n} = \dfrac{L_1}{L_2} \quad (\text{if } L_2 \neq 0)$

(We omit the proof.)

Example 4 In each part, determine whether the given sequence converges or diverges. If it converges, find the limit.

(a) $\quad \left\{ \dfrac{n}{2n + 1} \right\}_{n=1}^{+\infty}$
(b) $\quad \left\{ (-1)^{n+1} \dfrac{n}{2n + 1} \right\}_{n=1}^{+\infty}$

(c) $\quad \left\{ (-1)^{n+1} \dfrac{1}{n} \right\}_{n=1}^{+\infty}$
(d) $\quad \{8 - 2n\}_{n=1}^{+\infty}$
(e) $\quad \left\{ \dfrac{n}{e^n} \right\}_{n=1}^{+\infty}$

Solution.

(a) Dividing numerator and denominator by n yields

$$\lim_{n \to +\infty} \frac{n}{2n + 1} = \lim_{n \to +\infty} \frac{1}{\left(2 + \dfrac{1}{n}\right)} = \frac{\lim\limits_{n \to +\infty} 1}{\lim\limits_{n \to +\infty} \left(2 + \dfrac{1}{n}\right)} = \frac{\lim\limits_{n \to +\infty} 1}{\lim\limits_{n \to +\infty} 2 + \lim\limits_{n \to +\infty} \dfrac{1}{n}}$$

$$= \frac{1}{2 + 0} = \frac{1}{2}$$

Thus, $\left\{ \dfrac{n}{2n + 1} \right\}_{n=1}^{+\infty}$ converges to $\dfrac{1}{2}$.

(b) From part (a),

$$\lim_{n \to +\infty} \frac{n}{2n + 1} = \frac{1}{2}$$

Thus, since $(-1)^{n+1}$ oscillates between $+1$ and -1, the product $(-1)^{n+1} \dfrac{n}{2n + 1}$ oscillates between positive and negative values, with the odd-numbered terms approaching $\frac{1}{2}$ and the even-numbered terms approaching $-\frac{1}{2}$. Therefore, the sequence $\left\{(-1)^{n+1}\dfrac{n}{2n+1}\right\}_{n=1}^{+\infty}$ approaches no limit—it diverges.

(c)   Since $\lim\limits_{n\to+\infty} 1/n = 0$, the product $(-1)^{n+1}(1/n)$ oscillates between positive and negative values, with the odd-numbered terms approaching 0 through positive values and the even-numbered terms approaching 0 through negative values. Thus,

$$\lim_{n\to+\infty}(-1)^{n+1}\frac{1}{n} = 0$$

so the sequence converges to 0.

(d) $\lim\limits_{n\to+\infty}(8 - 2n) = -\infty$, so the sequence $\{8 - 2n\}_{n=1}^{+\infty}$ diverges.

(e)   We want to find $\lim\limits_{n\to+\infty} n/e^n$, which is an indeterminate form of type ∞/∞.

Unfortunately, we cannot apply L'Hôpital's rule directly since e^n and n are not differentiable functions (n assumes only integer values). However, we can apply L'Hôpital's rule to the related problem $\lim\limits_{x\to+\infty} x/e^x$ to obtain

$$\lim_{x\to+\infty}\frac{x}{e^x} = \lim_{x\to+\infty}\frac{1}{e^x} = 0$$

We conclude from this that $\lim\limits_{n\to+\infty} n/e^n = 0$ since the values of n/e^n and x/e^x are the same when x is a positive integer.   ◀

Example 5   Show that $\lim\limits_{n\to+\infty} \sqrt[n]{n} = 1$.

Solution.   With the aid of L'Hôpital's rule, $\lim\limits_{n\to+\infty}\dfrac{1}{n}\ln n = 0$. Thus,

$$\lim_{n\to+\infty} \sqrt[n]{n} = \lim_{n\to+\infty} n^{1/n} = \lim_{n\to+\infty} e^{(1/n)\ln n} = e^0 = 1 \quad ◀$$

Exercise Set 11.1

In Exercises 1–18, show the first five terms of the sequence, determine whether the sequence converges, and if so find the limit. (When writing out the terms of the sequence, you need not find numerical values; leave the terms in the first form you obtain.)

1. $\left\{\dfrac{n}{n+2}\right\}_{n=1}^{+\infty}.$

2. $\left\{\dfrac{n^2}{2n+1}\right\}_{n=1}^{+\infty}.$

3. $\{2\}_{n=1}^{+\infty}.$

4. $\left\{\ln\left(\dfrac{1}{n}\right)\right\}_{n=1}^{+\infty}.$

5. $\left\{\dfrac{\ln n}{n}\right\}_{n=1}^{+\infty}.$

6. $\left\{n\sin\dfrac{\pi}{n}\right\}_{n=1}^{+\infty}.$

7. $\{1 + (-1)^n\}_{n=1}^{+\infty}.$

8. $\left\{\dfrac{(-1)^{n+1}}{n^2}\right\}_{n=1}^{+\infty}.$

9. $\left\{(-1)^n\dfrac{2n^3}{n^3+1}\right\}_{n=1}^{+\infty}.$

10. $\left\{\dfrac{n}{2^n}\right\}_{n=1}^{+\infty}.$

11. $\left\{\dfrac{(n+1)(n+2)}{2n^2}\right\}_{n=1}^{+\infty}.$

12. $\left\{\dfrac{\pi^n}{4^n}\right\}_{n=1}^{+\infty}.$

13. $\{\cos (3/n)\}_{n=1}^{+\infty}.$

14. $\left\{\cos \dfrac{\pi n}{2}\right\}_{n=1}^{+\infty}.$

15. $\{n^2 e^{-n}\}_{n=1}^{+\infty}.$

16. $\{\sqrt{n^2 + 3n} - n\}_{n=1}^{+\infty}.$

17. $\left\{\left(\dfrac{n+3}{n+1}\right)^n\right\}_{n=1}^{+\infty}.$

18. $\left\{\left(1 - \dfrac{2}{n}\right)^n\right\}_{n=1}^{+\infty}.$

In Exercises 19–26, express the sequence in the notation $\{a_n\}_{n=1}^{+\infty}$, determine whether the sequence converges, and if so find its limit.

---

19. $\dfrac{1}{2}, \dfrac{3}{4}, \dfrac{5}{6}, \dfrac{7}{8}, \ldots\ .$

20. $0, \dfrac{1}{2^2}, \dfrac{2}{3^2}, \dfrac{3}{4^2}, \ldots\ .$

21. $\dfrac{1}{3}, \dfrac{1}{9}, \dfrac{1}{27}, \dfrac{1}{81}, \ldots\ .$

22. $-1, 2, -3, 4, -5, \ldots\ .$

23. $\left(1 - \dfrac{1}{2}\right), \left(\dfrac{1}{2} - \dfrac{1}{3}\right), \left(\dfrac{1}{3} - \dfrac{1}{4}\right), \left(\dfrac{1}{4} - \dfrac{1}{5}\right), \ldots\ .$

24. $3, \dfrac{3}{2}, \dfrac{3}{2^2}, \dfrac{3}{2^3}, \ldots\ .$

25. $(\sqrt{2} - \sqrt{3}), (\sqrt{3} - \sqrt{4}), (\sqrt{4} - \sqrt{5}), \ldots\ .$

26. $\dfrac{1}{3^5}, -\dfrac{1}{3^6}, \dfrac{1}{3^7}, -\dfrac{1}{3^8}, \ldots\ .$

27. Let $\{a_n\}$ be the sequence for which $a_1 = \sqrt{6}$ and $a_{n+1} = \sqrt{6 + a_n}$ for $n \geq 1$.
(a) Find the first three terms of the sequence.
(b) It can be shown that the sequence $\{a_n\}$ converges. Assuming this to be so, find its limit L. [*Hint:* $\lim\limits_{n \to +\infty} a_n = \lim\limits_{n \to +\infty} a_{n+1} = L$.]

28. Let a_1 and k be any positive real numbers and let $\{a_n\}$ be the sequence for which $a_{n+1} = \frac{1}{2}(a_n + k/a_n)$ for $n \geq 1$. Assuming that this sequence converges, find its limit using the hint in Exercise 27(b).

29. The *Fibonacci sequence* is defined by $a_{n+2} = a_n + a_{n+1}$ for $n \geq 1$, where $a_1 = a_2 = 1$.
(a) Find the first eight terms of the sequence.
(b) Find $\lim\limits_{n \to +\infty} (a_{n+1}/a_n)$ assuming that this limit exists.
[*Hint:* $\lim\limits_{n \to +\infty} (a_{n+1}/a_n) = \lim\limits_{n \to +\infty} (a_{n+2}/a_{n+1})$.]

30. The *n*th term a_n of the sequence $1, 2, 1, 4, 1, 6, \ldots$ is best written in the form

$$a_n = \begin{cases} 1, & \text{if } n \text{ is odd} \\ n, & \text{if } n \text{ is even} \end{cases}$$

since a single formula applicable to all terms would

be too complicated to be useful. By considering even and odd terms separately, find a similar expression for the *n*th term of the sequence

(a) $1, \dfrac{1}{2^2}, 3, \dfrac{1}{2^4}, 5, \dfrac{1}{2^6}, \ldots$

(b) $1, \dfrac{1}{3}, \dfrac{1}{3}, \dfrac{1}{5}, \dfrac{1}{5}, \dfrac{1}{7}, \dfrac{1}{7}, \dfrac{1}{9}, \dfrac{1}{9}, \ldots\ .$

31. Consider the sequence $\{a_n\}_{n=1}^{+\infty}$ where

$$a_n = \dfrac{1}{n^2} + \dfrac{2}{n^2} + \cdots + \dfrac{n}{n^2}$$

(a) Write out the first four terms of the sequence.
(b) Find the limit of the sequence. [*Hint:* Sum up the terms in the formula for a_n.]

32. Follow the directions in Exercise 31 with

$$a_n = \dfrac{1^2}{n^3} + \dfrac{2^2}{n^3} + \cdots + \dfrac{n^2}{n^3}$$

33. If we accept the fact that the sequence $\{1/n\}_{n=1}^{+\infty}$ converges to the limit $L = 0$, then according to Definition 11.1.2, for every $\epsilon > 0$, there exists an integer N such that $|a_n - L| = |(1/n) - 0| < \epsilon$ when $n \geq N$. In each part, find the smallest possible value of N for the given value of ϵ.
(a) $\epsilon = 0.5$ (b) $\epsilon = 0.1$ (c) $\epsilon = 0.001$.

34. If we accept the fact that the sequence

$$\left\{\dfrac{n}{n+1}\right\}_{n=1}^{+\infty}$$

converges to the limit $L = 1$, then according to Definition 11.1.2, for every $\epsilon > 0$ there exists an integer N such that

$$|a_n - L| = \left|\dfrac{n}{n+1} - 1\right| < \epsilon$$

when $n \geq N$. In each part, find the smallest value of N for the given value of ϵ.
(a) $\epsilon = 0.25$ (b) $\epsilon = 0.1$ (c) $\epsilon = 0.001$.

35. Prove:
(a) The sequence $\{1/n\}_{n=1}^{+\infty}$ converges to 0.
(b) The sequence $\left\{\dfrac{n}{n+1}\right\}_{n=1}^{+\infty}$ converges to 1.

36. Consider the sequence $\{a_n\}_{n=1}^{+\infty}$ whose *n*th term is

$$a_n = \sum_{k=0}^{n-1} \dfrac{1}{1 + k/n} \cdot \dfrac{1}{n}$$

Show that $\lim\limits_{n \to +\infty} a_n = \ln 2$. [*Hint:* Interpret $\lim\limits_{n \to +\infty} a_n$ as a definite integral.]

37. (a) Show that a polygon with n equal sides inscribed in a circle of radius r has perimeter $p_n = 2rn \sin(\pi/n)$.

  (b) By finding the limit of the sequence $\{p_n\}_{n=1}^{+\infty}$, derive the formula for the circumference of the circle.

38. Find $\lim\limits_{n \to +\infty} r^n$, where r is a real number. [*Hint:* Consider the cases $|r| < 1$, $|r| > 1$, $r = 1$, and $r = -1$ separately.]

39. Find the limit of the sequence $\{(2^n + 3^n)^{1/n}\}_{n=1}^{+\infty}$.

■ **11.2** MONOTONE SEQUENCES

Sometimes the critical information about a sequence is whether it converges or not, with the limit being of lesser interest. In this section we shall discuss results that are used to study convergence of sequences.

□ **TERMINOLOGY**

We begin with some terminology.

11.2.1 DEFINITION. A sequence $\{a_n\}$ is called

increasing if	$a_1 < a_2 < a_3 < \cdots < a_n < \cdots$
nondecreasing if	$a_1 \le a_2 \le a_3 \le \cdots \le a_n \le \cdots$
decreasing if	$a_1 > a_2 > a_3 > \cdots > a_n > \cdots$
nonincreasing if	$a_1 \ge a_2 \ge a_3 \ge \cdots \ge a_n \ge \cdots$

A sequence that is either nondecreasing or nonincreasing is called **monotone**, and a sequence that is increasing or decreasing is called **strictly monotone**. Observe that a strictly monotone sequence is monotone, but not conversely. (Why?)

Example 1

$$\frac{1}{2}, \frac{2}{3}, \frac{3}{4}, \ldots, \frac{n}{n+1}, \ldots \qquad \text{is increasing}$$

$$1, \frac{1}{2}, \frac{1}{3}, \ldots, \frac{1}{n}, \ldots \qquad \text{is decreasing}$$

$$1, 1, 2, 2, 3, 3, \ldots \qquad \text{is nondecreasing}$$

$$1, 1, \frac{1}{2}, \frac{1}{2}, \frac{1}{3}, \frac{1}{3}, \ldots \qquad \text{is nonincreasing}$$

All four of these sequences are monotone, but the sequence

$$1, -\frac{1}{2}, \frac{1}{3}, -\frac{1}{4}, \ldots, (-1)^{n+1}\frac{1}{n}, \ldots$$

is not. The first and second sequences are strictly monotone. ◄

☐ **TESTING FOR**
MONOTONICITY

In order for a sequence to be increasing, *all* pairs of successive terms, a_n and a_{n+1}, must satisfy $a_n < a_{n+1}$, or equivalently, $a_n - a_{n+1} < 0$. More generally, monotone sequences can be classified as follows:

DIFFERENCE BETWEEN SUCCESSIVE TERMS	CLASSIFICATION
$a_n - a_{n+1} < 0$	Increasing
$a_n - a_{n+1} > 0$	Decreasing
$a_n - a_{n+1} \leq 0$	Nondecreasing
$a_n - a_{n+1} \geq 0$	Nonincreasing

Frequently, one can *guess* whether a sequence is increasing, decreasing, nondecreasing, or nonincreasing after writing out some of the initial terms. However, to be certain that the guess is correct, a precise mathematical proof is needed. The following example illustrates a method for doing this.

Example 2 Show that

$$\frac{1}{2}, \frac{2}{3}, \frac{3}{4}, \ldots, \frac{n}{n+1}, \ldots$$

is an increasing sequence.

Solution. It is intuitively clear that the sequence is increasing. To prove that this is so, let

$$a_n = \frac{n}{n+1}$$

We can obtain a_{n+1} by replacing n by $n + 1$ in this formula. This yields

$$a_{n+1} = \frac{n+1}{(n+1)+1} = \frac{n+1}{n+2}$$

Thus, for $n \geq 1$

$$a_n - a_{n+1} = \frac{n}{n+1} - \frac{n+1}{n+2} = \frac{n^2 + 2n - n^2 - 2n - 1}{(n+1)(n+2)}$$

$$= -\frac{1}{(n+1)(n+2)} < 0$$

This proves that the sequence is increasing. ◄

If a_n and a_{n+1} are any successive terms in an increasing sequence, then $a_n < a_{n+1}$. If the terms in the sequence are all positive, then we can divide both sides of this inequality by a_n to obtain $1 < a_{n+1}/a_n$ or equivalently $a_{n+1}/a_n > 1$. More generally, monotone sequences with *positive* terms can be classified as follows:

RATIO OF SUCCESSIVE TERMS	CLASSIFICATION
$a_{n+1}/a_n > 1$	Increasing
$a_{n+1}/a_n < 1$	Decreasing
$a_{n+1}/a_n \geq 1$	Nondecreasing
$a_{n+1}/a_n \leq 1$	Nonincreasing

Example 3 Show that the sequence in Example 2 is increasing by examining the ratio of successive terms.

Solution. As shown in the solution of Example 2,

$$a_n = \frac{n}{n+1} \quad \text{and} \quad a_{n+1} = \frac{n+1}{n+2}$$

Thus,

$$\frac{a_{n+1}}{a_n} = \frac{(n+1)/(n+2)}{n/(n+1)} = \frac{n+1}{n+2} \cdot \frac{n+1}{n} = \frac{n^2+2n+1}{n^2+2n} \tag{1}$$

Since the numerator in (1) exceeds the denominator, the ratio exceeds 1, that is, $a_{n+1}/a_n > 1$ for $n \geq 1$. This proves that the sequence is increasing. ◄

In our subsequent work we will encounter sequences involving **factorials**. The reader will recall that if n is a positive integer, then $n!$ (n factorial) is the product of the first n positive integers, that is, $n! = 1 \cdot 2 \cdot 3 \cdots n$. Furthermore, it is agreed that $0! = 1$.

Example 4 Show that the sequence

$$\frac{e}{2!}, \frac{e^2}{3!}, \frac{e^3}{4!}, \ldots, \frac{e^n}{(n+1)!}, \ldots$$

is decreasing.

Solution. We shall examine the ratio of successive terms. Since

$$a_n = \frac{e^n}{(n+1)!}$$

it follows on replacing n by $n+1$ that

$$a_{n+1} = \frac{e^{n+1}}{[(n+1)+1]!} = \frac{e^{n+1}}{(n+2)!}$$

Thus,

$$\frac{a_{n+1}}{a_n} = \frac{e^{n+1}/(n+2)!}{e^n/(n+1)!} = \frac{e^{n+1}}{e^n} \cdot \frac{(n+1)!}{(n+2)!} = \frac{e}{n+2}$$

For $n \geq 1$, it follows that $n + 2 \geq 3 > e \; (\approx 2.718\ldots)$, so

$$\frac{a_{n+1}}{a_n} = \frac{e}{n + 2} < 1$$

for $n \geq 1$. This proves that the sequence is decreasing. ◄

The following example illustrates still a third technique for determining whether a sequence is increasing or decreasing.

Example 5 In Examples 2 and 3 we proved that the sequence

$$\frac{1}{2}, \frac{2}{3}, \frac{3}{4}, \ldots, \frac{n}{n + 1}, \ldots$$

is increasing by considering the difference and ratio of successive terms. Alternatively, we can proceed as follows. Let

$$f(x) = \frac{x}{x + 1}$$

so the nth term in the given sequence is $a_n = f(n)$. The function f is increasing for $x \geq 1$ since

$$f'(x) = \frac{(x + 1)(1) - x(1)}{(x + 1)^2} = \frac{1}{(x + 1)^2} > 0$$

Thus,

$$a_n = f(n) < f(n + 1) = a_{n+1}$$

which proves that the given sequence is increasing. ◄

In general, if $f(n) = a_n$ is the nth term of a sequence, and if f is differentiable for $x \geq 1$, then we have the following results:

DERIVATIVE OF f FOR $x \geq 1$	CLASSIFICATION OF THE SEQUENCE WITH nTH TERM $a_n = f(n)$
$f'(x) > 0$	Increasing
$f'(x) < 0$	Decreasing
$f'(x) \geq 0$	Nondecreasing
$f'(x) \leq 0$	Nonincreasing

We omit the proof.

☐ **CONVERGENCE OF MONOTONE SEQUENCES**

The following two theorems, whose optional proofs are discussed at the end of this section, show that a monotone sequence either converges or it becomes infinite—divergence by oscillation cannot occur.

11.2.2 THEOREM. *If $a_1 \le a_2 \le a_3 \le \cdots \le a_n \le \cdots$ is a nondecreasing sequence, then there are two possibilities:*

(a) *There is a constant M such that $a_n \le M$ for all n, in which case the sequence converges to a limit L satisfying $L \le M$.*

(b) *No such constant exists, in which case $\lim\limits_{n \to +\infty} a_n = +\infty$.*

11.2.3 THEOREM. *If $a_1 \ge a_2 \ge a_3 \ge \cdots \ge a_n \ge \cdots$ is a nonincreasing sequence, then there are two possibilities:*

(a) *There is a constant M such that $a_n \ge M$ for all n, in which case the sequence converges to a limit L satisfying $L \ge M$.*

(b) *No such constant exists, in which case $\lim\limits_{n \to +\infty} a_n = -\infty$.*

It should be noted that these results do not give a method for obtaining limits; they tell us only whether a limit exists. To prove these theorems we need a preliminary result that takes us to the very foundations of the real number system. In this text we have not been concerned with a logical development of the real numbers; our approach has been to accept the familiar properties of real numbers and to work with them. Indeed, we have not even attempted to define the term "real number." However, by the late nineteenth century, the study of limits and functions in calculus necessitated a precise axiomatic formulation of the real numbers in much the same way that Euclidean geometry is developed from axioms. While we will not attempt to pursue this development, we shall have need for the following axiom about real numbers.

11.2.4 AXIOM (*The Completeness Axiom*). *If S is a nonempty set of real numbers, and if there is some real number that is greater than or equal to every number in S, then there is a smallest real number that is greater than or equal to every number in S.*

For example, let S be the set of numbers in the interval $(1, 3)$. It is true that there exists a number u greater than or equal to every number in S; some examples are $u = 10$, $u = 100$, and $u = 3.2$. The smallest number u that is greater than or equal to every number in S is $u = 3$.

There is an alternative phrasing of the completeness axiom that is useful to know. Let us call a number u an ***upper bound*** for a set S if u is greater than or equal to every number in S; and if S has a smallest upper bound, call it the ***least upper bound*** of S. Using this terminology the Completeness Axiom states:

11.2.5 AXIOM (*The Completeness Axiom, Alternative Form*). *If a nonempty set S of real numbers has an upper bound, then S has a least upper bound.*

Example 6   As shown in Examples 2 and 3, the sequence $\left\{\dfrac{n}{n+1}\right\}_{n=1}^{+\infty}$ is increasing (hence nondecreasing). Since

$$a_n = \frac{n}{n+1} < 1, \quad n = 1, 2, \ldots$$

the terms in the sequence have $M = 1$ as an upper bound. By Theorem 11.2.2 the sequence must converge to a limit $L \leq M$. This is indeed the case since

$$\lim_{n \to +\infty} \frac{n}{n+1} = \lim_{n \to +\infty} \frac{1}{1 + \dfrac{1}{n}} = 1 \quad \blacktriangleleft$$

Because the limit of a sequence $\{a_n\}$ describes the behavior of the terms as n gets *large,* one can alter or even delete a *finite* number of terms in a sequence without affecting either the convergence or the value of the limit. That is, the original and the modified sequence will both converge or both diverge, and in the case of convergence both will have the same limit. We omit the proof.

Example 7   Show that the sequence

$$\left\{\frac{5^n}{n!}\right\}_{n=1}^{+\infty}$$

converges.

Solution.   It is tedious to determine convergence directly from the limit

$$\lim_{n \to +\infty} \frac{5^n}{n!}$$

Thus, we shall proceed indirectly. If we let

$$a_n = \frac{5^n}{n!}$$

then

$$a_{n+1} = \frac{5^{n+1}}{(n+1)!}$$

so that

$$\frac{a_{n+1}}{a_n} = \frac{5^{n+1}/(n+1)!}{5^n/n!} = \frac{5^{n+1}}{5^n} \cdot \frac{n!}{(n+1)!} = \frac{5}{n+1}$$

For $n = 1$, 2, and 3, the value of the ratio a_{n+1}/a_n is greater than 1, so $a_{n+1} > a_n$. Thus,

$$a_1 < a_2 < a_3 < a_4$$

For $n = 4$ the value of a_{n+1}/a_n is 1, so

$$a_4 = a_5$$

For $n \geq 5$ the value of a_{n+1}/a_n is less than 1, so

$$a_5 > a_6 > a_7 > a_8 > \cdots$$

Thus, if we discard the first four terms of the given sequence (which will not affect convergence), the resulting sequence will be decreasing. Moreover, each term in the sequence is positive, so by Theorem 11.2.3 with $M = 0$ the sequence converges to some limit that is ≥ 0. ◀

■ OPTIONAL

Proof of Theorem 11.2.2.

(a) Assume there exists a number M such that $a_n \leq M$ for $n = 1, 2, \ldots$. Then M is an upper bound for the set of terms in the sequence. By the Completeness Axiom there is a least upper bound for the terms, call it L. Now let ϵ be any positive number. Since L is the least upper bound for the terms, $L - \epsilon$ is not an upper bound for the terms, which means that there is at least one term a_N such that

$$a_N > L - \epsilon$$

Moreover, since $\{a_n\}$ is a nondecreasing sequence, we must have

$$a_n \geq a_N > L - \epsilon \tag{2}$$

when $n \geq N$. But a_n cannot exceed L since L is an upper bound for the terms. This observation together with (2) tells us that $L \geq a_n > L - \epsilon$ for $n \geq N$, so all terms from the Nth on are within ϵ units of L. This is exactly the requirement to have

$$\lim_{n \to +\infty} a_n = L$$

Finally, $L \leq M$ since M is an upper bound for the terms and L is the least upper bound. This proves part (a).

(b) If there is no number M such that $a_n \leq M$ for $n = 1, 2, \ldots$, then no matter how large we choose M, there is a term a_N such that

$$a_N > M$$

and, since the sequence is nondecreasing,

$$a_n \geq a_N > M$$

when $n \geq N$. Thus, the terms in the sequence become arbitrarily large as n increases. That is,

$$\lim_{n \to +\infty} a_n = +\infty \qquad ∎$$

The proof of Theorem 11.2.3 will be omitted since it is similar to the proof of 11.2.2.

▶ Exercise Set 11.2

In Exercises 1–6, determine whether the given sequence $\{a_n\}$ is monotone by examining $a_n - a_{n+1}$. If so, classify it as increasing, decreasing, nonincreasing, or nondecreasing.

1. $\left\{\dfrac{1}{n}\right\}_{n=1}^{+\infty}$.

2. $\left\{1 - \dfrac{1}{n}\right\}_{n=1}^{+\infty}$.

3. $\left\{\dfrac{n}{2n+1}\right\}_{n=1}^{+\infty}$.

4. $\left\{\dfrac{n}{4n-1}\right\}_{n=1}^{+\infty}$.

5. $\{n - 2^n\}_{n=1}^{+\infty}$.

6. $\{n - n^2\}_{n=1}^{+\infty}$.

In Exercises 7–18, determine whether the given sequence $\{a_n\}$ is monotone by examining a_{n+1}/a_n. If so, classify it as increasing, decreasing, nonincreasing, or nondecreasing.

7. $\left\{\dfrac{n}{2n+1}\right\}_{n=1}^{+\infty}$.

8. $\left\{\dfrac{n}{2^n}\right\}_{n=1}^{+\infty}$.

9. $\{ne^{-n}\}_{n=1}^{+\infty}$.

10. $\left\{\dfrac{n^2}{3^n}\right\}_{n=1}^{+\infty}$.

11. $\left\{\dfrac{2^n}{n!}\right\}_{n=1}^{+\infty}$.

12. $\left\{\dfrac{e^n}{n!}\right\}_{n=1}^{+\infty}$.

13. $\left\{\dfrac{n!}{3^n}\right\}_{n=1}^{+\infty}$.

14. $\left\{\dfrac{n^2}{n!}\right\}_{n=1}^{+\infty}$.

15. $\left\{\dfrac{10^n}{(2n)!}\right\}_{n=1}^{+\infty}$.

16. $\left\{\dfrac{2^n}{1+2^n}\right\}_{n=1}^{+\infty}$.

17. $\left\{\dfrac{n^n}{n!}\right\}_{n=1}^{+\infty}$.

18. $\left\{\dfrac{10^n}{2^{(n^2)}}\right\}_{n=1}^{+\infty}$.

In Exercises 19–24, use differentiation to show that the sequence is strictly monotone and classify it as increasing or decreasing.

19. $\left\{\dfrac{n}{2n+1}\right\}_{n=1}^{+\infty}$.

20. $\left\{3 - \dfrac{1}{n}\right\}_{n=1}^{+\infty}$.

21. $\left\{\dfrac{1}{n + \ln n}\right\}_{n=1}^{+\infty}$.

22. $\{ne^{-2n}\}_{n=1}^{+\infty}$.

23. $\left\{\dfrac{\ln(n+2)}{n+2}\right\}_{n=1}^{+\infty}$.

24. $\{\tan^{-1} n\}_{n=1}^{+\infty}$.

In Exercises 25–30, show that the sequence is monotone and apply Theorem 11.2.2 or 11.2.3 to determine whether it converges.

25. $\left\{\dfrac{n}{5^n}\right\}_{n=1}^{+\infty}$.

26. $\left\{\dfrac{2^n}{(n+1)!}\right\}_{n=1}^{+\infty}$.

27. $\left\{n - \dfrac{1}{n}\right\}_{n=1}^{+\infty}$.

28. $\left\{\cos\dfrac{\pi}{2n}\right\}_{n=1}^{+\infty}$.

29. $\left\{2 + \dfrac{1}{n}\right\}_{n=1}^{+\infty}$.

30. $\left\{\dfrac{4n-1}{5n+2}\right\}_{n=1}^{+\infty}$.

31. Find the limit (if it exists) of the sequence.

(a) $1, -1, 1, \dfrac{1}{2}, \dfrac{1}{3}, \dfrac{1}{4}, \dfrac{1}{5}, \ldots$

(b) $-\dfrac{1}{2}, 0, 0, 0, 1, 2, 3, 4, \ldots$.

32. (a) Is $\{100^n/n!\}_{n=1}^{+\infty}$ a monotone sequence?

(b) Is it a convergent sequence? Justify your answer.

33. Show that $\left\{\dfrac{3^n}{1+3^{2n}}\right\}_{n=1}^{+\infty}$ is a decreasing sequence.

34. Show that $\left\{\dfrac{1 \cdot 3 \cdot 5 \cdots (2n-1)}{n!}\right\}_{n=1}^{+\infty}$ is an increasing sequence.

35. Let $\{a_n\}$ be the sequence for which $a_1 = \sqrt{2}$ and $a_{n+1} = \sqrt{2 + a_n}$ for $n \geq 1$.

(a) Find the first three terms of the sequence.

(b) Show that $a_n < 2$ for $n \geq 1$.

(c) Show that $a_{n+1}^2 - a_n^2 = (2 - a_n)(1 + a_n)$ for $n \geq 1$.

(d) Use the results in parts (b) and (c) to show that $\{a_n\}$ is an increasing sequence. [*Hint:* If x and y are positive real numbers such that $x^2 - y^2 > 0$, then it follows by factoring that $x - y > 0$.]

(e) Show that $\{a_n\}$ converges and find its limit L.

36. Let $\{a_n\}$ be the sequence for which $a_1 = \sqrt{3}$ and $a_{n+1} = \sqrt{3a_n}$ for $n \geq 1$.

(a) Find the first three terms of the sequence.

(b) Show that $a_n < 3$ for $n \geq 1$.

(c) Show that $a_{n+1}^2 - a_n^2 = a_n(3 - a_n)$ for $n \geq 1$.

(d) Use the hint in Exercise 35(d) to show that $\{a_n\}$ is an increasing sequence.

(e) Show that $\{a_n\}$ converges and find its limit L.

37. (a) Show that if $\{a_n\}_{n=1}^{+\infty}$ is a nonincreasing sequence, then $\{-a_n\}_{n=1}^{+\infty}$ is a nondecreasing sequence.

(b) Use part (a) and Theorem 11.2.2 to help prove Theorem 11.2.3.

38. (a) Deduce the inequalities

$$\int_1^n \ln x \, dx < \ln n! < \int_1^{n+1} \ln x \, dx$$

from Figure 11.2.1 for $n \geq 2$ by comparing appropriate areas.

(b) Use the result in part (a) to show that

$$\frac{n^n}{e^{n-1}} < n! < \frac{(n+1)^{n+1}}{e^n}, \quad n > 1$$

39. Use the Squeezing Theorem (Theorem 2.8.2) and the result in Exercise 38(b) to show that

$$\lim_{n \to +\infty} \frac{\sqrt[n]{n!}}{n} = \frac{1}{e}$$

40. Use the left inequality in Exercise 38(b) to show that

$$\lim_{n \to +\infty} \sqrt[n]{n!} = +\infty.$$

41. (a) Show that $\left\{\dfrac{n^n}{n! e^n}\right\}_{n=1}^{+\infty}$ is a decreasing sequence.

[*Hint:* From Exercise 99(d) of Section 7.4 we have $(1 + 1/x)^x < e$ for $x > 0$.]

(b) Does the sequence converge? Explain.

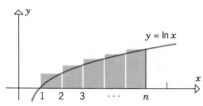
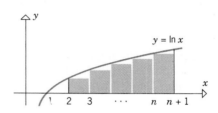

Figure 11.2.1

11.3 INFINITE SERIES

The purpose of this section is to discuss sums that contain infinitely many terms. The most familiar examples of such sums occur in the decimal representation of real numbers. For example, when we write $\frac{1}{3}$ in the decimal form $\frac{1}{3} = 0.3333 \ldots$ we mean

$$\frac{1}{3} = 0.3 + 0.03 + 0.003 + 0.0003 + \cdots$$

which suggests that the decimal representation of $\frac{1}{3}$ can be viewed as a sum of infinitely many real numbers.

☐ SUMS OF INFINITE SERIES

Our first objective is to define what is meant by the "sum" of infinitely many real numbers. We begin with some terminology.

11.3.1 DEFINITION. An *infinite series* is an expression of the form

$$u_1 + u_2 + u_3 + \cdots + u_k + \cdots$$

or in sigma notation

$$\sum_{k=1}^{\infty} u_k$$

The numbers $u_1, u_2, u_3, \ldots$ are called the *terms* of the series.

Since it is physically impossible to add infinitely many numbers together, sums of infinite series are defined and computed by an indirect limiting process. To motivate the basic idea, consider the infinite decimal

$$0.3333\ldots \tag{1}$$

which can be viewed as the infinite series

$$0.3 + 0.03 + 0.003 + 0.0003 + \cdots$$

or equivalently

$$\frac{3}{10} + \frac{3}{10^2} + \frac{3}{10^3} + \frac{3}{10^4} + \cdots \tag{2}$$

Since (1) is the decimal expansion of $\frac{1}{3}$, any reasonable definition for the sum of an infinite series should yield $\frac{1}{3}$ for the sum of (2). To obtain such a definition, consider the following sequence of (finite) sums:

$$s_1 = \frac{3}{10} = 0.3$$

$$s_2 = \frac{3}{10} + \frac{3}{10^2} = 0.33$$

$$s_3 = \frac{3}{10} + \frac{3}{10^2} + \frac{3}{10^3} = 0.333$$

$$s_4 = \frac{3}{10} + \frac{3}{10^2} + \frac{3}{10^3} + \frac{3}{10^4} = 0.3333$$

$$\vdots$$

The sequence of numbers $s_1, s_2, s_3, s_4, \ldots$ can be viewed as a succession of approximations to the "sum" of the infinite series, which we want to be $\frac{1}{3}$. As we progress through the sequence, more and more terms of the infinite series are used, and the approximations get better and better, suggesting that the desired sum of $\frac{1}{3}$ might be the *limit* of this sequence of approximations. To see that this is so, we must calculate the limit of the general term in the sequence of approximations, namely

$$s_n = \frac{3}{10} + \frac{3}{10^2} + \cdots + \frac{3}{10^n} \tag{3}$$

The problem of calculating

$$\lim_{n \to +\infty} s_n = \lim_{n \to +\infty} \left\{ \frac{3}{10} + \frac{3}{10^2} + \cdots + \frac{3}{10^n} \right\}$$

is complicated by the fact that both the last term and the number of terms in the sum change with n. It is best to rewrite such limits in a closed form in which the number of terms does not vary, if possible. (See the remark following

Example 3 in Section 5.4.) To do this, we multiply both sides of (3) by $\frac{1}{10}$ to obtain

$$\frac{1}{10} s_n = \frac{3}{10^2} + \frac{3}{10^3} + \cdots + \frac{3}{10^n} + \frac{3}{10^{n+1}} \tag{4}$$

and then subtract (4) from (3) to obtain

$$s_n - \frac{1}{10} s_n = \frac{3}{10} - \frac{3}{10^{n+1}}$$

$$\frac{9}{10} s_n = \frac{3}{10} \left(1 - \frac{1}{10^n} \right)$$

$$s_n = \frac{1}{3} \left(1 - \frac{1}{10^n} \right)$$

Since $1/10^n \to 0$ as $n \to +\infty$, it follows that

$$\lim_{n \to +\infty} s_n = \lim_{n \to +\infty} \frac{1}{3} \left(1 - \frac{1}{10^n} \right) = \frac{1}{3}$$

which we denote by writing

$$\frac{1}{3} = \frac{3}{10} + \frac{3}{10^2} + \frac{3}{10^3} + \cdots + \frac{3}{10^n} + \cdots$$

Motivated by the foregoing example, we are now ready to define the general concept of the "sum" of an infinite series

$$\sum_{k=1}^{\infty} u_k$$

We begin with some terminology: Let s_n denote the sum of the first n terms of the series. Thus,

$$s_1 = u_1$$
$$s_2 = u_1 + u_2$$
$$s_3 = u_1 + u_2 + u_3$$
$$\vdots$$
$$s_n = u_1 + u_2 + u_3 + \cdots + u_n = \sum_{k=1}^{n} u_k$$

The number s_n is called the **nth partial sum** of the series and the sequence $\{s_n\}_{n=1}^{+\infty}$ is called the **sequence of partial sums**.

WARNING.   In everyday English the words "sequence" and "series" are often used interchangeably. However, this is not so in mathematics—mathematically, a sequence is a *succession* and a series is a *sum*. It is essential that you keep this distinction in mind.

As n increases, the partial sum $s_n = u_1 + u_2 + \cdots + u_n$ includes more and more terms of the series. Thus, if s_n tends toward a limit as $n \to +\infty$, it is reasonable to view this limit as the sum of *all* the terms in the series. This suggests the following definition.

11.3.2   DEFINITION.   Let $\{s_n\}$ be the sequence of partial sums of the series $\sum_{k=1}^{\infty} u_k$. If the sequence $\{s_n\}$ converges to a limit S, then the series is said to **converge** and S is called the **sum** of the series. We denote this by writing

$$S = \sum_{k=1}^{\infty} u_k$$

If the sequence of partial sums diverges, then the series is said to **diverge.** A divergent series has no sum.

Example 1   Determine whether the series

$$1 - 1 + 1 - 1 + 1 - 1 + \cdots$$

converges or diverges. If it converges, find the sum.

Solution.   It is tempting to conclude that the sum of the series is zero by arguing that the positive and negative terms cancel one another. However, this is *not correct;* the problem is that algebraic operations that hold for finite sums do not carry over to infinite series in all cases. Later, we shall discuss conditions under which familiar algebraic operations can be applied to infinite series, but for this example we turn directly to Definition 11.3.2. The partial sums are

$$s_1 = 1$$
$$s_2 = 1 - 1 = 0$$
$$s_3 = 1 - 1 + 1 = 1$$
$$s_4 = 1 - 1 + 1 - 1 = 0$$

and so forth. Thus, the sequence of partial sums is

$$1, 0, 1, 0, 1, 0, \ldots$$

Since this is a divergent sequence, the given series diverges and consequently has no sum.   ◀

☐ **GEOMETRIC SERIES**

The series encountered thus far are examples of **geometric series.** A geometric series is one of the form

$$a + ar + ar^2 + ar^3 + \cdots + ar^{k-1} + \cdots \quad (a \neq 0)$$

where each term is obtained by multiplying the previous one by a constant r. The multiplier r is called the **ratio** for the series. Some examples of geometric series are

$$1 + 2 + 4 + 8 + \cdots + 2^{k-1} + \cdots \qquad \boxed{a = 1, r = 2}$$

$$\frac{3}{10} + \frac{3}{10^2} + \frac{3}{10^3} + \cdots + \frac{3}{10^{k-1}} + \cdots \qquad \boxed{a = \tfrac{3}{10}, r = \tfrac{1}{10}}$$

$$\frac{1}{2} - \frac{1}{4} + \frac{1}{8} - \frac{1}{16} + \cdots + (-1)^{k+1}\frac{1}{2^k} + \cdots \qquad \boxed{a = \tfrac{1}{2}, r = -\tfrac{1}{2}}$$

$$1 + 1 + 1 + \cdots + 1 + \cdots \qquad \boxed{a = 1, r = 1}$$

$$1 - 1 + 1 - 1 + \cdots + (-1)^{k+1} + \cdots \qquad \boxed{a = 1, r = -1}$$

The following theorem is the fundamental result on convergence of geometric series.

11.3.3 THEOREM. *A geometric series*

$$a + ar + ar^2 + \cdots + ar^{k-1} + \cdots \qquad (a \neq 0)$$

converges if $|r| < 1$ and diverges if $|r| \geq 1$. If the series converges, then the sum is

$$\frac{a}{1 - r} = a + ar + ar^2 + \cdots + ar^{k-1} + \cdots$$

Proof. Let us treat the case $|r| = 1$ first. If $r = 1$, then the series is

$$a + a + a + \cdots + a + \cdots$$

so the nth partial sum is $s_n = na$ and $\lim_{n \to +\infty} s_n = \lim_{n \to +\infty} na = \pm\infty$ (the sign depending on whether a is positive or negative). This proves divergence. If $r = -1$, the series is

$$a - a + a - a + \cdots$$

so the sequence of partial sums is

$$a, 0, a, 0, a, 0, \ldots$$

which diverges.

Now let us consider the case where $|r| \neq 1$. The nth partial sum of the series is

$$s_n = a + ar + ar^2 + \cdots + ar^{n-1} \tag{5}$$

Multiplying both sides of (5) by r yields

$$rs_n = ar + ar^2 + \cdots + ar^{n-1} + ar^n \tag{6}$$

and subtracting (6) from (5) gives

$$s_n - rs_n = a - ar^n$$

or

$$(1 - r)s_n = a - ar^n \tag{7}$$

Since $r \neq 1$ in the case we are considering, this can be rewritten as

$$s_n = \frac{a - ar^n}{1 - r} = \frac{a}{1 - r} - \frac{ar^n}{1 - r} \tag{8}$$

If $|r| < 1$, then $\lim\limits_{n \to +\infty} r^n = 0$, so $\{s_n\}$ converges. From (8)

$$\lim_{n \to +\infty} s_n = \frac{a}{1 - r}$$

If $|r| > 1$, then either $r > 1$ or $r < -1$. In the case $r > 1$, $\lim\limits_{n \to +\infty} r^n = +\infty$, and in the case $r < -1$, r^n oscillates between positive and negative values that grow in magnitude, so $\{s_n\}$ diverges in both cases. ∎

Example 2 The series

$$5 + \frac{5}{4} + \frac{5}{4^2} + \cdots + \frac{5}{4^{k-1}} + \cdots$$

is a geometric series with $a = 5$ and $r = \frac{1}{4}$. Since $|r| = \frac{1}{4} < 1$, the series converges and the sum is

$$\frac{a}{1 - r} = \frac{5}{1 - \frac{1}{4}} = \frac{20}{3} \qquad \blacktriangleleft$$

Example 3 Find the rational number represented by the repeating decimal

$$0.784784784 \ldots$$

Solution. We can write

$$0.784784784 \ldots = 0.784 + 0.000784 + 0.000000784 + \cdots$$

so the given decimal is the sum of a geometric series with $a = 0.784$ and $r = 0.001$. Thus,

$$0.784784784 \ldots = \frac{a}{1 - r} = \frac{0.784}{1 - 0.001} = \frac{0.784}{0.999} = \frac{784}{999} \qquad \blacktriangleleft$$

Example 4 Determine whether the series

$$\sum_{k=1}^{\infty} \frac{1}{k(k + 1)} = \frac{1}{1 \cdot 2} + \frac{1}{2 \cdot 3} + \frac{1}{3 \cdot 4} + \frac{1}{4 \cdot 5} + \cdots$$

converges or diverges. If it converges, find the sum.

Solution. The nth partial sum of the series is

$$s_n = \sum_{k=1}^{n} \frac{1}{k(k+1)} = \frac{1}{1 \cdot 2} + \frac{1}{2 \cdot 3} + \frac{1}{3 \cdot 4} + \cdots + \frac{1}{n(n+1)}$$

To calculate $\lim_{n \to +\infty} s_n$ we shall rewrite s_n in closed form. This may be accomplished by using the method of partial fractions to obtain (verify)

$$\frac{1}{k(k+1)} = \frac{1}{k} - \frac{1}{k+1}$$

from which it follows that

$$s_n = \sum_{k=1}^{n} \left(\frac{1}{k} - \frac{1}{k+1} \right)$$

$$= \left(1 - \frac{1}{2} \right) + \left(\frac{1}{2} - \frac{1}{3} \right) + \left(\frac{1}{3} - \frac{1}{4} \right) + \cdots + \left(\frac{1}{n} - \frac{1}{n+1} \right)$$

$$= 1 + \left(-\frac{1}{2} + \frac{1}{2} \right) + \left(-\frac{1}{3} + \frac{1}{3} \right) + \cdots + \left(-\frac{1}{n} + \frac{1}{n} \right) - \frac{1}{n+1} \qquad (9)$$

The sum in (9) is an example of a **telescoping sum,** which means that each term cancels part of the next term, thereby collapsing the sum (like a folding telescope) into only two terms. After the cancellation, (9) can be written as

$$s_n = 1 - \frac{1}{n+1}$$

so

$$\lim_{n \to +\infty} s_n = \lim_{n \to +\infty} \left(1 - \frac{1}{n+1} \right) = 1$$

and therefore

$$1 = \sum_{k=1}^{\infty} \frac{1}{k(k+1)} \qquad \blacktriangleleft$$

☐ **HARMONIC SERIES**

One of the most famous and important of all diverging series is the **harmonic series,**

$$\sum_{k=1}^{\infty} \frac{1}{k} = 1 + \frac{1}{2} + \frac{1}{3} + \frac{1}{4} + \frac{1}{5} + \cdots$$

which arises in connection with the overtones produced by a vibrating musical string. It is not immediately evident that this series diverges. However, the divergence will become apparent when we examine the partial sums in detail.

Because the terms in the series are all positive, the partial sums

$$s_1 = 1, \quad s_2 = 1 + \frac{1}{2}, \quad s_3 = 1 + \frac{1}{2} + \frac{1}{3}, \quad s_4 = 1 + \frac{1}{2} + \frac{1}{3} + \frac{1}{4}, \ldots$$

form an increasing sequence

$$s_1 < s_2 < s_3 < \cdots < s_n < \cdots$$

Thus, by Theorem 11.2.2 we can prove divergence by demonstrating that there is no constant M that is greater than or equal to *every* partial sum. To this end, we shall consider some selected partial sums, namely $s_2, s_4, s_8, s_{16}, s_{32}, \ldots$. Note that the subscripts are successive powers of 2, so that these are the partial sums of the form s_{2^n}. These partial sums satisfy the inequalities

$$s_2 = 1 + \frac{1}{2} > \frac{1}{2} + \frac{1}{2} = \frac{2}{2}$$

$$s_4 = s_2 + \frac{1}{3} + \frac{1}{4} > s_2 + \left(\frac{1}{4} + \frac{1}{4}\right) = s_2 + \frac{1}{2} > \frac{3}{2}$$

$$s_8 = s_4 + \frac{1}{5} + \frac{1}{6} + \frac{1}{7} + \frac{1}{8} > s_4 + \left(\frac{1}{8} + \frac{1}{8} + \frac{1}{8} + \frac{1}{8}\right) = s_4 + \frac{1}{2} > \frac{4}{2}$$

$$s_{16} = s_8 + \frac{1}{9} + \frac{1}{10} + \frac{1}{11} + \frac{1}{12} + \frac{1}{13} + \frac{1}{14} + \frac{1}{15} + \frac{1}{16}$$

$$> s_8 + \left(\frac{1}{16} + \frac{1}{16} + \frac{1}{16} + \frac{1}{16} + \frac{1}{16} + \frac{1}{16} + \frac{1}{16} + \frac{1}{16}\right) = s_8 + \frac{1}{2} > \frac{5}{2}$$

$$\vdots$$

$$s_{2^n} > \frac{n+1}{2}$$

Now if M is any constant, we can certainly find a positive integer n such that $(n + 1)/2 > M$. But for this n

$$s_{2^n} > \frac{n+1}{2} > M$$

so that no constant M is greater than or equal to *every* partial sum of the harmonic series. This proves divergence.

▶ Exercise Set 11.3

1. In each part, find the first four partial sums; find a closed form for the nth partial sum; and determine whether the series converges (if so, give the sum).

   (a) $\displaystyle\sum_{k=1}^{\infty} \frac{2}{5^{k-1}}$ (b) $\displaystyle\sum_{k=1}^{\infty} \frac{1}{(k+1)(k+2)}$

   (c) $\displaystyle\sum_{k=1}^{\infty} \frac{2^{k-1}}{4}$.

In Exercises 2–16, determine whether the series converges or diverges. If it converges, find the sum.

2. $\displaystyle\sum_{k=1}^{\infty} \frac{1}{5^k}$.

3. $\displaystyle\sum_{k=1}^{\infty} \left(-\frac{3}{4}\right)^{k-1}$.

4. $\displaystyle\sum_{k=1}^{\infty} \left(\frac{2}{3}\right)^{k+2}$.

5. $\displaystyle\sum_{k=1}^{\infty} (-1)^{k-1} \frac{7}{6^{k-1}}$.

6. $\displaystyle\sum_{k=1}^{\infty} 4^{k-1}.$

7. $\displaystyle\sum_{k=1}^{\infty} \left(-\frac{3}{2}\right)^{k+1}.$

8. $\displaystyle\sum_{k=1}^{\infty} \left(\frac{1}{k+3} - \frac{1}{k+4}\right).$

9. $\displaystyle\sum_{k=1}^{\infty} \frac{1}{(k+2)(k+3)}.$

10. $\displaystyle\sum_{k=1}^{\infty} \left(\frac{1}{2^k} - \frac{1}{2^{k+1}}\right).$

11. $\displaystyle\sum_{k=1}^{\infty} \frac{1}{9k^2 + 3k - 2}.$

12. $\displaystyle\sum_{k=2}^{\infty} \frac{1}{k^2 - 1}.$

13. $\displaystyle\sum_{k=1}^{\infty} \frac{4^{k+2}}{7^{k-1}}.$

14. $\displaystyle\sum_{k=1}^{\infty} (e/\pi)^{k-1}.$

15. $\displaystyle\sum_{k=1}^{\infty} (-1/2)^k.$

16. $\displaystyle\sum_{k=3}^{\infty} \frac{5}{k-2}.$

In Exercises 17–22, express the repeating decimal as a fraction.

17. $0.4444\ldots.$

18. $0.9999\ldots.$

19. $5.373737\ldots.$

20. $0.159159159\ldots.$

21. $0.782178217821\ldots.$

22. $0.451141414\ldots.$

23. Find a closed form for the nth partial sum of the series

$$\ln\frac{1}{2} + \ln\frac{2}{3} + \ln\frac{3}{4} + \cdots + \ln\frac{n}{n+1} + \cdots$$

and determine whether the series converges.

24. A ball is dropped from a height of 10 m. Each time it strikes the ground it bounces vertically to a height that is $\frac{3}{4}$ of the previous height. Find the total distance the ball will travel if it is allowed to bounce indefinitely.

25. Show: $\displaystyle\sum_{k=2}^{\infty} \ln(1 - 1/k^2) = -\ln 2.$

26. Show: $\displaystyle\sum_{k=1}^{\infty} \frac{\sqrt{k+1} - \sqrt{k}}{\sqrt{k^2 + k}} = 1.$

27. Show: $\displaystyle\sum_{k=1}^{\infty} \left(\frac{1}{k} - \frac{1}{k+2}\right) = \frac{3}{2}.$

28. (a) Find A and B such that

$$\frac{6^k}{(3^{k+1} - 2^{k+1})(3^k - 2^k)} = \frac{2^k A}{3^k - 2^k} + \frac{2^k B}{3^{k+1} - 2^{k+1}}$$

(b) Use the result in part (a) to help show that

$$\sum_{k=1}^{\infty} \frac{6^k}{(3^{k+1} - 2^{k+1})(3^k - 2^k)} = 2$$

[This problem appeared in the Forty-Fifth Annual William Lowell Putnam Mathematical Competition.]

29. Show: $\dfrac{1}{1\cdot 3} + \dfrac{1}{3\cdot 5} + \dfrac{1}{5\cdot 7} + \cdots = \dfrac{1}{2}.$

30. Show: $\dfrac{1}{1\cdot 3} + \dfrac{1}{2\cdot 4} + \dfrac{1}{3\cdot 5} + \cdots = \dfrac{3}{4}.$

31. Use geometric series to show that

(a) $\displaystyle\sum_{k=0}^{\infty} (-1)^k x^k = \frac{1}{1+x}$ if $-1 < x < 1$

(b) $\displaystyle\sum_{k=0}^{\infty} (x-3)^k = \frac{1}{4-x}$ if $2 < x < 4$

(c) $\displaystyle\sum_{k=0}^{\infty} (-1)^k x^{2k} = \frac{1}{1+x^2}$ if $-1 < x < 1.$

In Exercises 32–35, find all values of x for which the series converges, and for these values find its sum.

32. $x - x^3 + x^5 - x^7 + x^9 - \cdots.$

33. $\dfrac{1}{x^2} + \dfrac{2}{x^3} + \dfrac{4}{x^4} + \dfrac{8}{x^5} + \dfrac{16}{x^6} + \cdots.$

34. $e^{-x} + e^{-2x} + e^{-3x} + e^{-4x} + e^{-5x} + \cdots.$

35. $\sin x - \frac{1}{2}\sin^2 x + \frac{1}{4}\sin^3 x - \frac{1}{8}\sin^4 x + \cdots.$

36. Prove the following decimal equality assuming that $a_n \neq 9$:

$$0.a_1 a_2 \ldots a_n 9999\ldots = 0.a_1 a_2 \ldots (a_n + 1)0000\ldots$$

37. Let a_1 be any real number and define

$$a_{n+1} = \tfrac{1}{2}(a_n + 1) \text{ for } n = 1, 2, 3, \ldots$$

Show that the sequence $\{a_n\}_{n=1}^{+\infty}$ converges and find its limit. [*Hint:* Express a_n in terms of a_1.]

38. Lines L_1 and L_2 form an angle θ, $0 < \theta < \pi/2$, at their point of intersection P (Figure 11.3.1). A point P_0 is chosen that is on L_1 and a units from P. Starting from P_0 a zig-zag path is constructed by successively going back and forth between L_1 and L_2 along a perpendicular from one line to the other. Find the following sums in terms of θ:

(a) $P_0 P_1 + P_1 P_2 + P_2 P_3 + \cdots$
(b) $P_0 P_1 + P_2 P_3 + P_4 P_5 + \cdots$
(c) $P_1 P_2 + P_3 P_4 + P_5 P_6 + \cdots.$

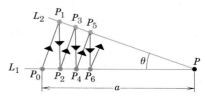

Figure 11.3.1

39. As shown in Figure 11.3.2, suppose that an angle θ is bisected using a straightedge and compass to produce ray R_1, then the angle between R_1 and the initial side is bisected to produce ray R_2. Thereafter, rays $R_3, R_4, R_5, \ldots$ are constructed in succession by bisecting the angle between the previous two rays. Show that the sequence of angles that these rays make with the initial side has a limit of $\theta/3$. [This problem is based on *Trisection of an Angle in an Infinite Number of Steps* by Eric Kincannon, which appeared in *The College Mathematics Journal*, Vol. 21, No. 5, November 1990.]

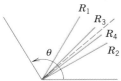

Figure 11.3.2

40. In his *Treatise on the Configurations of Qualities and Motions* (written in the 1350s), the French Bishop of Lisieux, Nicole Oresme, used a geometric method to show that

$$\sum_{k=1}^{\infty} \frac{k}{2^k} = \frac{1}{2} + \frac{2}{4} + \frac{3}{8} + \frac{4}{16} + \cdots = 2$$

In Figure 11.3.3a each term in the series is represented by the area of a rectangle. In Figure 11.3.3b the configuration in part (a) has been divided into rectangles with areas $A_1, A_2, A_3, \ldots$. Show that $A_1 + A_2 + A_3 + \cdots = 2$. [For convenience, the horizontal and vertical scales in Figure 11.3.3 are different.]

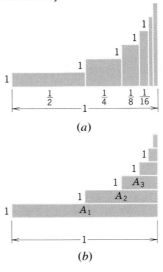

(a)

(b)   Figure 11.3.3

41. The great Swiss mathematician, Leonhard Euler, (biography on p. 71) made occasional errors in his pioneering work on infinite series. For example, Euler deduced that

$$\tfrac{1}{2} = 1 - 1 + 1 - 1 + \cdots$$

and

$$-1 = 1 + 2 + 4 + 8 + \cdots$$

by substituting $x = -1$ and $x = 2$ in the formula

$$\frac{1}{1 - x} = 1 + x + x^2 + x^3 + \cdots$$

What was the error in his reasoning?

11.4 CONVERGENCE; THE INTEGRAL TEST

In the previous section we found sums of series and investigated convergence by first writing the nth partial sum s_n in closed form and then examining its limit. However, it is relatively rare that the nth partial sum of a series can be written in closed form; for most series, convergence or divergence is determined by using convergence tests, some of which we shall introduce in this section. Once it is established that a series converges, the sum of the series can generally be approximated to any degree of accuracy by a partial sum with sufficiently many terms.

☐ **THE DIVERGENCE TEST**

Our first theorem states that the terms of an infinite series must tend toward zero if the series is to converge.

11.4.1 THEOREM. *If the series Σu_k converges, then $\lim\limits_{k\to+\infty} u_k = 0$.*

Proof. The term u_k can be written as

$$u_k = s_k - s_{k-1} \tag{1}$$

where s_k is the sum of the first k terms and s_{k-1} is the sum of the first $k - 1$ terms. If S denotes the sum of the series, then $\lim\limits_{k\to+\infty} s_k = S$, and since $(k - 1) \to +\infty$ as $k \to +\infty$, we also have $\lim\limits_{k\to+\infty} s_{k-1} = S$. Thus, from (1)

$$\lim_{k\to+\infty} u_k = \lim_{k\to+\infty} (s_k - s_{k-1}) = S - S = 0 \qquad \blacksquare$$

The following result is just an alternative phrasing of the above theorem and needs no additional proof.

11.4.2 THEOREM (*The Divergence Test*). *If $\lim\limits_{k\to+\infty} u_k \neq 0$, then the series Σu_k diverges.*

Example 1 The series

$$\sum_{k=1}^{\infty} \frac{k}{k+1} = \frac{1}{2} + \frac{2}{3} + \frac{3}{4} + \cdots + \frac{k}{k+1} + \cdots$$

diverges since

$$\lim_{k\to+\infty} \frac{k}{k+1} = \lim_{k\to+\infty} \frac{1}{1 + 1/k} = 1 \neq 0 \qquad \blacktriangleleft$$

WARNING. The converse of Theorem 11.4.1 is false. To prove that a series converges it does not suffice to show that $\lim\limits_{k\to+\infty} u_k = 0$, since this property may hold for divergent as well as convergent series. For example, the kth term of the divergent harmonic series $1 + 1/2 + 1/3 + \cdots + 1/k + \cdots$ approaches zero as $k \to +\infty$, and the kth term of the convergent geometric series $1/2 + 1/2^2 + \cdots + 1/2^k + \cdots$ tends to zero as $k \to +\infty$.

☐ **ALGEBRAIC PROPERTIES OF INFINITE SERIES**

For brevity, the proof of the following result is left for the exercises.

11.4.3   THEOREM.

(a)   *If Σu_k and Σv_k are convergent series, then $\Sigma(u_k + v_k)$ and $\Sigma(u_k - v_k)$ are convergent series and the sums of these series are related by*

$$\sum_{k=1}^{\infty}(u_k + v_k) = \sum_{k=1}^{\infty} u_k + \sum_{k=1}^{\infty} v_k$$

$$\sum_{k=1}^{\infty}(u_k - v_k) = \sum_{k=1}^{\infty} u_k - \sum_{k=1}^{\infty} v_k$$

(b)   *If c is a nonzero constant, then the series Σu_k and $\Sigma c u_k$ both converge or both diverge. In the case of convergence, the sums are related by*

$$\sum_{k=1}^{\infty} c u_k = c \sum_{k=1}^{\infty} u_k$$

(c)   *Convergence or divergence is unaffected by deleting a finite number of terms from the beginning of a series; that is, for any positive integer K, the series*

$$\sum_{k=1}^{\infty} u_k = u_1 + u_2 + u_3 + \cdots$$

and

$$\sum_{k=K}^{\infty} u_k = u_K + u_{K+1} + u_{K+2} + \cdots$$

both converge or both diverge.

REMARK.   Do not read too much into part (c) of this theorem. Although the convergence is not affected when a finite number of terms is deleted from the beginning of a convergent series, the *sum* of the series is changed by the removal of these terms.

Example 2   Find the sum of the series

$$\sum_{k=1}^{\infty}\left(\frac{3}{4^k} - \frac{2}{5^{k-1}}\right)$$

Solution.   The series

$$\sum_{k=1}^{\infty}\frac{3}{4^k} = \frac{3}{4} + \frac{3}{4^2} + \frac{3}{4^3} + \cdots$$

is a convergent geometric series ($a = \frac{3}{4}, r = \frac{1}{4}$), and the series

$$\sum_{k=1}^{\infty}\frac{2}{5^{k-1}} = 2 + \frac{2}{5} + \frac{2}{5^2} + \frac{2}{5^3} + \cdots$$

is also a convergent geometric series ($a = 2, r = \frac{1}{5}$). Thus, from Theorems 11.4.3(a) and 11.3.3 the given series converges and

$$\sum_{k=1}^{\infty} \left(\frac{3}{4^k} - \frac{2}{5^{k-1}} \right) = \sum_{k=1}^{\infty} \frac{3}{4^k} - \sum_{k=1}^{\infty} \frac{2}{5^{k-1}} = \frac{\frac{3}{4}}{1 - \frac{1}{4}} - \frac{2}{1 - \frac{1}{5}}$$

$$= 1 - \frac{5}{2} = -\frac{3}{2} \qquad \blacktriangleleft$$

Example 3 The series

$$\sum_{k=1}^{\infty} \frac{5}{k} = 5 + \frac{5}{2} + \frac{5}{3} + \cdots + \frac{5}{k} + \cdots$$

diverges by part (b) of Theorem 11.4.3, since

$$\sum_{k=1}^{\infty} \frac{5}{k} = \sum_{k=1}^{\infty} 5 \left(\frac{1}{k} \right)$$

which shows that each term is a constant times the corresponding term of the divergent harmonic series. $\blacktriangleleft$

Example 4 The series

$$\sum_{k=10}^{\infty} \frac{1}{k} = \frac{1}{10} + \frac{1}{11} + \frac{1}{12} + \cdots$$

diverges by part (c) of Theorem 11.4.3, since this series results by deleting the first nine terms from the divergent harmonic series. $\blacktriangleleft$

☐ **CONVERGENCE TESTS**

If an infinite series $u_1 + u_2 + u_3 + \cdots + u_k + \cdots$ has *nonnegative terms*, then the partial sums $s_1 = u_1$, $s_2 = u_1 + u_2$, $s_3 = u_1 + u_2 + u_3, \ldots$ form a nondecreasing sequence, that is,

$$s_1 \le s_2 \le s_3 \le \cdots \le s_n \le \cdots$$

If there is a finite constant M such that $s_n \le M$ for all n, then according to Theorem 11.2.2, the sequence of partial sums will converge to a limit S, satisfying $S \le M$. If no such constant exists, then $\lim_{n \to +\infty} s_n = +\infty$. This yields the following theorem.

11.4.4 THEOREM. *If Σu_k is a series with nonnegative terms, and if there is a constant M such that*

$$s_n = u_1 + u_2 + \cdots + u_n \le M$$

for every n, then the series converges and the sum S satisfies $S \le M$. If no such M exists, then the series diverges.

□ **THE INTEGRAL TEST**   If we have a series with positive terms, say

$$\sum_{k=1}^{\infty} \frac{1}{k^2}$$

and if we form the improper integral

$$\int_{1}^{+\infty} \frac{1}{x^2}\, dx$$

whose integrand is obtained by replacing the summation index k by x, then there is a relationship between convergence of the series and convergence of the improper integral.

11.4.5   THEOREM (*The Integral Test*).   *Let Σu_k be a series with positive terms, and let $f(x)$ be the function that results when k is replaced by x in the formula for u_k. If f is decreasing and continuous for $x \geq 1$, then*

$$\sum_{k=1}^{\infty} u_k \quad and \quad \int_{1}^{+\infty} f(x)\, dx$$

both converge or both diverge.

We shall defer the proof to the end of the section and proceed with some examples.

Example 5   Determine whether

$$\sum_{k=1}^{\infty} \frac{1}{k^2}$$

converges or diverges.

Solution.   If we replace k by x in the formula for u_k, we obtain the function

$$f(x) = \frac{1}{x^2}$$

which satisfies the hypotheses of the integral test. (Verify.) Since

$$\int_{1}^{+\infty} \frac{1}{x^2}\, dx = \lim_{l \to +\infty} \int_{1}^{l} \frac{dx}{x^2} = \lim_{l \to +\infty} \left[-\frac{1}{x} \right]_{1}^{l} = \lim_{l \to +\infty} \left[1 - \frac{1}{l} \right] = 1$$

the integral converges and consequently the series converges.   ◀

REMARK.   In the above example, do *not* erroneously conclude that $\displaystyle\sum_{k=1}^{\infty} \frac{1}{k^2} = 1$ from the fact that $\displaystyle\int_{1}^{+\infty} \frac{1}{x^2}\, dx = 1$. (In advanced texts it is proved that the sum of this series is $\pi^2/6$; and, in fact, the sum of the first two terms alone exceeds 1.)

Example 6 The integral test provides another way to demonstrate divergence of the harmonic series $\sum\limits_{k=1}^{\infty} \dfrac{1}{k}$. If we replace k by x in the formula for u_k, we obtain the function $f(x) = 1/x$, which satisfies the hypotheses of the integral test. (Verify.) Since

$$\int_1^{+\infty} \frac{1}{x}\,dx = \lim_{l \to +\infty} \int_1^l \frac{1}{x}\,dx = \lim_{l \to +\infty} [\ln l - \ln 1] = +\infty$$

the integral diverges and consequently so does the series. ◀

Example 7 Determine whether the series

$$\frac{1}{e} + \frac{2}{e^4} + \frac{3}{e^9} + \cdots + \frac{k}{e^{k^2}} + \cdots$$

converges or diverges.

Solution. If we replace k by x in the formula for u_k, we obtain the function

$$f(x) = \frac{x}{e^{x^2}} = xe^{-x^2}$$

For $x \geq 1$, this function has positive values and is continuous. Moreover, for $x \geq 1$ the derivative

$$f'(x) = e^{-x^2} - 2x^2 e^{-x^2} = e^{-x^2}(1 - 2x^2)$$

is negative, so that f is decreasing for $x \geq 1$. Thus, the hypotheses of the integral test are met. But

$$\int_1^{+\infty} xe^{-x^2}\,dx = \lim_{l \to +\infty} \int_1^l xe^{-x^2}\,dx = \lim_{l \to +\infty} \left[-\frac{1}{2} e^{-x^2} \right]_1^l$$

$$= \left(-\frac{1}{2} \right) \lim_{l \to +\infty} \left[e^{-l^2} - e^{-1} \right] = \frac{1}{2e}$$

Thus, the improper integral and the series converge. ◀

□ *p*-SERIES

The harmonic series and the series in Example 5 are special cases of a class of series called ***p*-series** or ***hyperharmonic series***. A *p*-series is an infinite series of the form

$$\sum_{k=1}^{\infty} \frac{1}{k^p} = 1 + \frac{1}{2^p} + \frac{1}{3^p} + \cdots + \frac{1}{k^p} + \cdots$$

where $p > 0$. Examples of *p*-series are

$$\sum_{k=1}^{\infty} \frac{1}{k} = 1 + \frac{1}{2} + \frac{1}{3} + \cdots + \frac{1}{k} + \cdots \qquad \boxed{p = 1}$$

$$\sum_{k=1}^{\infty} \frac{1}{k^2} = 1 + \frac{1}{2^2} + \frac{1}{3^2} + \cdots + \frac{1}{k^2} + \cdots \qquad \boxed{p = 2}$$

$$\sum_{k=1}^{\infty} \frac{1}{\sqrt{k}} = 1 + \frac{1}{\sqrt{2}} + \frac{1}{\sqrt{3}} + \cdots + \frac{1}{\sqrt{k}} + \cdots \qquad \boxed{p = \frac{1}{2}}$$

The following theorem tells when a p-series converges.

11.4.6 THEOREM (*Convergence of p-Series*).

$$\sum_{k=1}^{\infty} \frac{1}{k^p} = 1 + \frac{1}{2^p} + \frac{1}{3^p} + \cdots + \frac{1}{k^p} + \cdots$$

converges if $p > 1$ and diverges if $0 < p \leq 1$.

Proof. To establish this result when $p \neq 1$, we shall use the integral test.

$$\int_1^{+\infty} \frac{1}{x^p}\, dx = \lim_{l \to +\infty} \int_1^l x^{-p}\, dx$$

$$= \lim_{l \to +\infty} \frac{x^{1-p}}{1-p} \bigg]_1^l$$

$$= \lim_{l \to +\infty} \left[\frac{l^{1-p}}{1-p} - \frac{1}{1-p} \right]$$

If $p > 1$, then $1 - p < 0$, so $l^{1-p} \to 0$ as $l \to +\infty$. Thus, the integral converges [its value is $-1/(1-p)$] and consequently the series also converges. For $0 < p < 1$, it follows that $1 - p > 0$ and $l^{1-p} \to +\infty$ as $l \to +\infty$, so the integral and the series diverge. The case $p = 1$ is the harmonic series, which was previously shown to diverge. ∎

Example 8

$$1 + \frac{1}{\sqrt[3]{2}} + \frac{1}{\sqrt[3]{3}} + \cdots + \frac{1}{\sqrt[3]{k}} + \cdots$$

diverges since it is a p-series with $p = \frac{1}{3} < 1$. ◀

□ **PROOF OF THE INTEGRAL TEST**

We conclude this section by proving Theorem 11.4.5 (the integral test).

Proof. Let $f(x)$ satisfy the hypotheses of the theorem. Since

$$f(1) = u_1,\ f(2) = u_2, \ldots,\ f(n) = u_n, \ldots$$

the rectangles in Figures 11.4.1a and 11.4.1b have areas $u_1, u_2, \ldots, u_n$ as

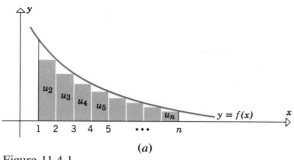
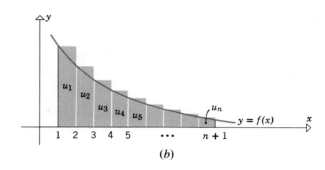

(a)   *(b)*

Figure 11.4.1

indicated. In Figure 11.4.1*a*, the total area of the rectangles is less than the area under the curve from $x = 1$ to $x = n$, so

$$u_2 + u_3 + \cdots + u_n < \int_1^n f(x)\, dx$$

Therefore,

$$u_1 + u_2 + u_3 + \cdots + u_n < u_1 + \int_1^n f(x)\, dx \tag{2}$$

In Figure 11.4.1*b*, the total area of the rectangles is greater than the area under the curve from $x = 1$ to $x = n + 1$, so that

$$\int_1^{n+1} f(x)\, dx < u_1 + u_2 + \cdots + u_n \tag{3}$$

If we let $s_n = u_1 + u_2 + \cdots + u_n$ be the nth partial sum of the series, then (2) and (3) yield

$$\int_1^{n+1} f(x)\, dx < s_n < u_1 + \int_1^n f(x)\, dx \tag{4}$$

If the integral $\int_1^{+\infty} f(x)\, dx$ converges to a finite value L, then from the right-hand inequality in (4)

$$s_n < u_1 + \int_1^n f(x)\, dx < u_1 + \int_1^{+\infty} f(x)\, dx = u_1 + L$$

Thus, each partial sum is less than the finite constant $u_1 + L$, and the series converges by Theorem 11.4.4. On the other hand, if the integral $\int_1^{+\infty} f(x)\, dx$ diverges, then

$$\lim_{n \to +\infty} \int_1^{n+1} f(x)\, dx = +\infty$$

so that from the left-hand inequality in (4), $\displaystyle\lim_{n \to +\infty} s_n = +\infty$. This means that the series also diverges. ▐

REMARK.   If the summation index in a series Σu_k does not begin with $k = 1$, a variation of the integral test may still apply. It can be shown that

$$\sum_{k=K}^{\infty} u_k \quad \text{and} \quad \int_{K}^{+\infty} f(x) \, dx$$

both converge or both diverge provided the hypotheses of Theorem 11.4.5 hold for $x \geq K$.

▶ Exercise Set 11.4 $\boxed{C}$ 37, 38, 39, 40

In Exercises 1–4, use Theorem 11.4.3 to find the sum of the series.

1. $\displaystyle\sum_{k=1}^{\infty} \left[\frac{1}{2^k} + \frac{1}{4^k} \right].$ 2. $\displaystyle\sum_{k=1}^{\infty} \left[\frac{1}{5^k} - \frac{1}{k(k + 1)} \right].$

3. $\displaystyle\sum_{k=2}^{\infty} \left[\frac{1}{k^2 - 1} - \frac{7}{10^{k-1}} \right].$

4. $\displaystyle\sum_{k=1}^{\infty} \left[\frac{7}{3^k} + \frac{6}{(k + 3)(k + 4)} \right].$

5.   In each part, determine whether the given p-series converges or diverges.

   (a) $\displaystyle\sum_{k=1}^{\infty} \frac{1}{k^3}$ (b) $\displaystyle\sum_{k=1}^{\infty} \frac{1}{\sqrt{k}}$

   (c) $\displaystyle\sum_{k=1}^{\infty} k^{-1}$ (d) $\displaystyle\sum_{k=1}^{\infty} k^{-2/3}$

   (e) $\displaystyle\sum_{k=1}^{\infty} k^{-4/3}$ (f) $\displaystyle\sum_{k=1}^{\infty} \frac{1}{\sqrt[4]{k}}$

   (g) $\displaystyle\sum_{k=1}^{\infty} \frac{1}{\sqrt[3]{k^5}}$ (h) $\displaystyle\sum_{k=1}^{\infty} \frac{1}{k^{\pi}}.$

In Exercises 6–8, use the divergence test to show that the series diverges.

6.   (a) $\displaystyle\sum_{k=1}^{\infty} \frac{k + 1}{k + 2}$ (b) $\displaystyle\sum_{k=1}^{\infty} \ln k.$

7.   (a) $\displaystyle\sum_{k=1}^{\infty} \frac{k^2 + k + 3}{2k^2 + 1}$ (b) $\displaystyle\sum_{k=1}^{\infty} \left(1 + \frac{1}{k} \right)^k.$

8.   (a) $\displaystyle\sum_{k=1}^{\infty} \cos k\pi$ (b) $\displaystyle\sum_{k=1}^{\infty} \frac{e^k}{k}.$

In Exercises 9–30, determine whether the series converges or diverges.

9. $\displaystyle\sum_{k=1}^{\infty} \frac{1}{k + 6}.$ 10. $\displaystyle\sum_{k=1}^{\infty} \frac{3}{5k}.$

11. $\displaystyle\sum_{k=1}^{\infty} \frac{1}{5k + 2}.$ 12. $\displaystyle\sum_{k=1}^{\infty} \frac{k}{1 + k^2}.$

13. $\displaystyle\sum_{k=1}^{\infty} \frac{1}{1 + 9k^2}.$ 14. $\displaystyle\sum_{k=1}^{\infty} \frac{1}{(4 + 2k)^{3/2}}.$

15. $\displaystyle\sum_{k=1}^{\infty} \frac{1}{\sqrt{k + 5}}.$ 16. $\displaystyle\sum_{k=1}^{\infty} \frac{1}{\sqrt[k]{e}}.$

17. $\displaystyle\sum_{k=1}^{\infty} \frac{1}{\sqrt[3]{2k - 1}}.$ 18. $\displaystyle\sum_{k=3}^{\infty} \frac{\ln k}{k}.$

19. $\displaystyle\sum_{k=1}^{\infty} \frac{k}{\ln (k + 1)}.$ 20. $\displaystyle\sum_{k=1}^{\infty} ke^{-k^2}.$

21. $\displaystyle\sum_{k=1}^{\infty} \frac{1}{(k + 1)[\ln (k + 1)]^2}.$

22. $\displaystyle\sum_{k=1}^{\infty} \frac{k^2 + 1}{k^2 + 3}.$ 23. $\displaystyle\sum_{k=1}^{\infty} \left(1 + \frac{1}{k} \right)^{-k}.$

24. $\displaystyle\sum_{k=1}^{\infty} \frac{1}{\sqrt{k^2 + 1}}.$ 25. $\displaystyle\sum_{k=1}^{\infty} \frac{\tan^{-1} k}{1 + k^2}.$

26. $\displaystyle\sum_{k=1}^{\infty} \text{sech}^2 k.$ 27. $\displaystyle\sum_{k=5}^{\infty} 7k^{-p} \quad (p > 1).$

28. $\displaystyle\sum_{k=1}^{\infty} 7(k + 5)^{-p} \quad (p \leq 1).$

29. $\displaystyle\sum_{k=1}^{\infty} k^2 \sin^2 \left(\frac{1}{k} \right).$ 30. $\displaystyle\sum_{k=1}^{\infty} k^2 e^{-k^3}.$

31.   Prove: $\displaystyle\sum_{k=2}^{\infty} \frac{1}{k(\ln k)^p}$ converges if $p > 1$ and diverges if $p \leq 1$.

32.   Prove: $\displaystyle\sum_{k=3}^{\infty} \frac{1}{k(\ln k)[\ln (\ln k)]^p}$ converges if $p > 1$ and diverges if $p \leq 1$.

33.   Prove: If Σu_k converges and Σv_k diverges, then $\Sigma(u_k + v_k)$ diverges and $\Sigma(u_k - v_k)$ diverges. [*Hint:* Assume that $\Sigma(u_k + v_k)$ converges and use Theorem

11.4.3 to obtain a contradiction. Similarly, for $\Sigma(u_k - v_k)$.]

34. Find examples to show that $\Sigma(u_k + v_k)$ and $\Sigma(u_k - v_k)$ may converge or may diverge if Σu_k and Σv_k both diverge.

35. With the help of Exercise 33, determine whether the given series in parts (a)–(d) converge or diverge.

(a) $\displaystyle\sum_{k=1}^{\infty} \left[\left(\frac{2}{3}\right)^{k-1} + \frac{1}{k} \right]$

(b) $\displaystyle\sum_{k=1}^{\infty} \left[\frac{k^2}{1 + k^2} + \frac{1}{k(k + 1)} \right]$

(c) $\displaystyle\sum_{k=1}^{\infty} \left[\frac{1}{3k + 2} + \frac{1}{k^{3/2}} \right]$

(d) $\displaystyle\sum_{k=2}^{\infty} \left[\frac{1}{k(\ln k)^2} - \frac{1}{k^2} \right]$.

36. If the sum S of a convergent series $\sum_{k=1}^{\infty} u_k$ of positive terms is approximated by its nth partial sum s_n, then the error in the approximation is $S - s_n$. Let $f(x)$ be the function that results when k is replaced by x in the formula for u_k. Show that if f is decreasing for $x \geq n$, then

$$\int_{n+1}^{+\infty} f(x)\, dx < S - s_n < \int_{n}^{+\infty} f(x)\, dx$$

The result in Exercise 36 can be written as

$$s_n + \int_{n+1}^{+\infty} f(x)\, dx < S < s_n + \int_{n}^{+\infty} f(x)\, dx$$

which provides an upper and lower bound on the sum S of the series. Use this result in part (a) of Exercises 37 and 38.

---

37. (a) Find the partial sum s_{10} of the series $\sum_{k=1}^{\infty} 1/k^3$, and use it to obtain an upper and lower bound on the sum S of the series. Express your results to four decimal places.

(b) Use the right-hand inequality in Exercise 36 to find a value of n to ensure that the error in approximating S by s_n is less than 10^{-3}.

38. (a) Find the partial sum s_6 of the series $\sum_{k=1}^{\infty} 1/k^4$, and use it to obtain an upper and lower bound on the sum S of the series. Express your results to five decimal places. [*Note:* Compare your bounds with the exact value of S, which is $\pi^4/90$.]

(b) Use the right-hand inequality in Exercise 36 to find a value of n to ensure that the error in approximating S by s_n is less than 10^{-5}.

39. Let s_n be the nth partial sum of the divergent series $\sum_{k=1}^{\infty} 1/k$.

(a) Use inequality (4) to show that for $n \geq 2$

$$\ln(n + 1) < s_n < 1 + \ln n$$

Then find integer upper and lower bounds for $s_{1,000,000}$.

(b) Find a value of n to ensure that $s_n > 100$.

40. Let s_n be the nth partial sum of the divergent series $\sum_{k=1}^{\infty} 1/\sqrt{k}$.

(a) Use inequality (4) to show that for $n \geq 2$

$$2\sqrt{n + 1} - 2 < s_n < 2\sqrt{n} - 1$$

Then find integer upper and lower bounds for $s_{10,000}$.

(b) Find a value of n to ensure that $s_n > 100$.

■ 11.5 ADDITIONAL CONVERGENCE TESTS

In this section we shall develop some additional convergence tests for series with positive terms.

☐ **THE COMPARISON TEST**   Our first result is not only a practical test for convergence, but is also a theoretical tool that will be used to develop other convergence tests.

11.5.1   THEOREM (*The Comparison Test*).   *Let* Σa_k *and* Σb_k *be series with nonnegative terms and suppose*

$$a_1 \leq b_1, \ a_2 \leq b_2, \ a_3 \leq b_3, \ \ldots, \ a_k \leq b_k, \ \ldots$$

(a)   *If the "bigger series"* Σb_k *converges, then the "smaller series"* Σa_k *also converges.*

(b)   *On the other hand, if the "smaller series"* Σa_k *diverges, then the "bigger series"* Σb_k *also diverges.*

Proof (a).   Suppose that the series Σb_k converges and its sum is B. Then for all n

$$b_1 + b_2 + \cdots + b_n \leq \sum_{k=1}^{\infty} b_k = B.$$

From our hypothesis it follows that

$$a_1 + a_2 + \cdots + a_n \leq b_1 + b_2 + \cdots + b_n$$

so that

$$a_1 + a_2 + \cdots + a_n \leq B$$

Thus, each partial sum of the series Σa_k is less than or equal to B, so that Σa_k converges by Theorem 11.4.4.

Proof (b).   This part is really just an alternative phrasing of part (a). If Σa_k diverges, then Σb_k must diverge since convergence of Σb_k would imply convergence of Σa_k, contrary to the hypothesis.   ∎

☐ THE RATIO TEST

Since the comparison test requires a little ingenuity to use, we shall wait until the next section before applying it. For now, we shall use the comparison test to develop some other tests that are easier to apply.

11.5.2   THEOREM (*The Ratio Test*).   *Let* Σu_k *be a series with positive terms and suppose*

$$\lim_{k \to +\infty} \frac{u_{k+1}}{u_k} = \rho$$

(a)   *If* $\rho < 1$, *the series converges.*

(b)   *If* $\rho > 1$ *or* $\rho = +\infty$, *the series diverges.*

(c)   *If* $\rho = 1$, *the series may converge or diverge, so that another test must be tried.*

Proof (a). Assume that $\rho < 1$, and let $r = \frac{1}{2}(1 + \rho)$. Thus, $\rho < r < 1$, since r is the midpoint between 1 and ρ. It follows that the number

$$\epsilon = r - \rho \tag{1}$$

is positive. Since

$$\rho = \lim_{k \to +\infty} \frac{u_{k+1}}{u_k}$$

it follows that for k sufficiently large, say $k \geq K$, the ratios u_{k+1}/u_k are within ϵ units of ρ. Thus, we will have

$$\frac{u_{k+1}}{u_k} < \rho + \epsilon \quad \text{when} \quad k \geq K$$

or on substituting (1)

$$\frac{u_{k+1}}{u_k} < r \quad \text{when} \quad k \geq K$$

that is,

$$u_{k+1} < r u_k \quad \text{when} \quad k \geq K$$

This yields the inequalities

$$u_{K+1} < r u_K$$
$$u_{K+2} < r u_{K+1} < r^2 u_K$$
$$u_{K+3} < r u_{K+2} < r^3 u_K$$
$$u_{K+4} < r u_{K+3} < r^4 u_K \tag{2}$$
$$\vdots$$

But $|r| < 1$ (why?), so

$$r u_K + r^2 u_K + r^3 u_K + \cdots$$

is a convergent geometric series. From the inequalities in (2) and the comparison test it follows that

$$u_{K+1} + u_{K+2} + u_{K+3} + \cdots$$

must also be a convergent series. Thus, $u_1 + u_2 + u_3 + \cdots + u_k + \cdots$ converges by Theorem 11.4.3(c).

Proof (b). Assume that $\rho > 1$. Thus,

$$\epsilon = \rho - 1 \tag{3}$$

is a positive number. Since

$$\rho = \lim_{k \to +\infty} \frac{u_{k+1}}{u_k}$$

it follows that for k sufficiently large, say $k \geq K$, the ratio u_{k+1}/u_k is within ϵ units of ρ. Thus,

$$\frac{u_{k+1}}{u_k} > \rho - \epsilon \quad \text{when} \quad k \geq K$$

or on substituting (3)

$$\frac{u_{k+1}}{u_k} > 1 \quad \text{when} \quad k \geq K$$

that is,

$$u_{k+1} > u_k \quad \text{when} \quad k \geq K$$

This yields the inequalities

$$u_{K+1} > u_K$$
$$u_{K+2} > u_{K+1} > u_K$$
$$u_{K+3} > u_{K+2} > u_K$$
$$u_{K+4} > u_{K+3} > u_K \tag{4}$$
$$\vdots$$

Since $u_K > 0$, it follows from the inequalities in (4) that $\lim\limits_{k \to +\infty} u_k \neq 0$, so $u_1 + u_2 + \cdots + u_k + \cdots$ diverges by Theorem 11.4.2. The proof in the case where $\rho = +\infty$ is omitted.

Proof (c). The series

$$\sum_{k=1}^{\infty} \frac{1}{k} \quad \text{and} \quad \sum_{k=1}^{\infty} \frac{1}{k^2}$$

both have $\rho = 1$ (verify). Since the first is the divergent harmonic series and the second is a convergent p-series, the ratio test does not distinguish between convergence and divergence when $\rho = 1$. ▊

Example 1   The series

$$\sum_{k=1}^{\infty} \frac{1}{k!}$$

converges by the ratio test since

$$\rho = \lim_{k \to +\infty} \frac{u_{k+1}}{u_k} = \lim_{k \to +\infty} \frac{1/(k+1)!}{1/k!} = \lim_{k \to +\infty} \frac{k!}{(k+1)!} = \lim_{k \to +\infty} \frac{1}{k+1} = 0$$

so $\rho < 1$. ◄

Example 2 The series

$$\sum_{k=1}^{\infty} \frac{k}{2^k}$$

converges by the ratio test since

$$\rho = \lim_{k \to +\infty} \frac{u_{k+1}}{u_k} = \lim_{k \to +\infty} \frac{k+1}{2^{k+1}} \cdot \frac{2^k}{k} = \frac{1}{2} \lim_{k \to +\infty} \frac{k+1}{k} = \frac{1}{2}$$

so $\rho < 1$. ◄

Example 3 The series

$$\sum_{k=1}^{\infty} \frac{k^k}{k!}$$

diverges by the ratio test since

$$\rho = \lim_{k \to +\infty} \frac{u_{k+1}}{u_k} = \lim_{k \to +\infty} \frac{(k+1)^{k+1}}{(k+1)!} \cdot \frac{k!}{k^k}$$

$$= \lim_{k \to +\infty} \frac{(k+1)^k}{k^k}$$

$$= \lim_{k \to +\infty} \left(1 + \frac{1}{k}\right)^k = e \qquad \boxed{\text{See Section 7.2, Formula (5).}}$$

Since $\rho = e > 1$, the series diverges. ◄

Example 4 Determine whether the series

$$1 + \frac{1}{3} + \frac{1}{5} + \frac{1}{7} + \cdots + \frac{1}{2k-1} + \cdots$$

converges or diverges.

Solution. The ratio test is of no help since

$$\rho = \lim_{k \to +\infty} \frac{u_{k+1}}{u_k} = \lim_{k \to +\infty} \frac{1}{2(k+1)-1} \cdot \frac{2k-1}{1} = \lim_{k \to +\infty} \frac{2k-1}{2k+1} = 1$$

However, the integral test proves that the series diverges since

$$\int_1^{+\infty} \frac{dx}{2x-1} = \lim_{l \to +\infty} \int_1^l \frac{dx}{2x-1} = \lim_{l \to +\infty} \frac{1}{2} \ln(2x-1) \Big]_1^l = +\infty \qquad ◄$$

Example 5   The series

$$\frac{2!}{4} + \frac{4!}{4^2} + \frac{6!}{4^3} + \cdots + \frac{(2k)!}{4^k} + \cdots$$

diverges since

$$\rho = \lim_{k \to +\infty} \frac{u_{k+1}}{u_k} = \lim_{k \to +\infty} \frac{[2(k+1)]!}{4^{k+1}} \cdot \frac{4^k}{(2k)!} = \lim_{k \to +\infty} \left(\frac{(2k+2)!}{(2k)!} \cdot \frac{1}{4} \right)$$

$$= \frac{1}{4} \lim_{k \to +\infty} (2k+2)(2k+1) = +\infty \quad \blacktriangleleft$$

☐ **THE ROOT TEST**

Sometimes the following result is easier to apply than the ratio test.

11.5.3   THEOREM (*The Root Test*).   *Let Σu_k be a series with positive terms and suppose*

$$\rho = \lim_{k \to +\infty} \sqrt[k]{u_k} = \lim_{k \to +\infty} (u_k)^{1/k}$$

(a)   *If $\rho < 1$, the series converges.*
(b)   *If $\rho > 1$, or $\rho = +\infty$, the series diverges.*
(c)   *If $\rho = 1$, the series may converge or diverge, so that another test must be tried.*

Since the proof of the root test is similar to the proof of the ratio test, we shall omit it.

Example 6   The series

$$\sum_{k=2}^{\infty} \left(\frac{4k - 5}{2k + 1} \right)^k$$

diverges by the root test since

$$\rho = \lim_{k \to +\infty} (u_k)^{1/k} = \lim_{k \to +\infty} \frac{4k - 5}{2k + 1} = 2 > 1 \quad \blacktriangleleft$$

Example 7   The series

$$\sum_{k=1}^{\infty} \frac{1}{(\ln (k + 1))^k}$$

converges by the root test, since

$$\lim_{k \to +\infty} (u_k)^{1/k} = \lim_{k \to +\infty} \frac{1}{\ln (k + 1)} = 0 < 1 \quad \blacktriangleleft$$

◻ **COMMENTS ON NOTATION**

We conclude this section with a remark about notation. Until now we have written most of our infinite series in the form

$$\sum_{k=1}^{\infty} u_k \tag{5}$$

with the summation index beginning at 1. If the summation index begins at some other integer, it is always possible to rewrite the series in form (5). Thus, for example, the series

$$\sum_{k=0}^{\infty} \frac{2^k}{k!} = 1 + 2 + \frac{2^2}{2!} + \frac{2^3}{3!} + \cdots \tag{6}$$

can be written as

$$\sum_{k=1}^{\infty} \frac{2^{k-1}}{(k-1)!} = 1 + 2 + \frac{2^2}{2!} + \frac{2^3}{3!} + \cdots \tag{7}$$

However, for purposes of applying the convergence tests, it is not necessary that the series have form (5). For example, we can apply the ratio test to (6) without converting to the more complicated form (7). Doing so yields

$$\rho = \lim_{k \to +\infty} \frac{u_{k+1}}{u_k} = \lim_{k \to +\infty} \frac{2^{k+1}}{(k+1)!} \cdot \frac{k!}{2^k} = \lim_{k \to +\infty} \frac{2}{k+1} = 0$$

which shows that the series converges since $\rho < 1$.

▶ Exercise Set 11.5   Ⓒ *41, 42, 43, 44*

In Exercises 1–6, apply the ratio test. According to the test, does the series converge, does the series diverge, or are the results inconclusive?

1. $\sum_{k=1}^{\infty} \frac{3^k}{k!}$.

2. $\sum_{k=1}^{\infty} \frac{4^k}{k^2}$.

3. $\sum_{k=2}^{\infty} \frac{1}{5k}$.

4. $\sum_{k=1}^{\infty} k\left(\frac{1}{2}\right)^k$.

5. $\sum_{k=1}^{\infty} \frac{k!}{k^3}$.

6. $\sum_{k=1}^{\infty} \frac{k}{k^2 + 1}$.

In Exercises 7–10, apply the root test. According to the test, does the series converge, does the series diverge, or are the results inconclusive?

7. $\sum_{k=1}^{\infty} \left(\frac{3k+2}{2k-1}\right)^k$.

8. $\sum_{k=1}^{\infty} \left(\frac{k}{100}\right)^k$.

9. $\sum_{k=1}^{\infty} \frac{k}{5^k}$.

10. $\sum_{k=1}^{\infty} (1 + e^{-k})^k$.

In Exercises 11–32, use any appropriate test to determine whether the series converges.

11. $\sum_{k=1}^{\infty} \frac{2^k}{k^3}$.

12. $\sum_{k=1}^{\infty} \frac{1}{k^2}$.

13. $\sum_{k=0}^{\infty} \frac{7^k}{k!}$.

14. $\sum_{k=1}^{\infty} \frac{1}{2k+1}$.

15. $\sum_{k=1}^{\infty} \frac{k^2}{5^k}$.

16. $\sum_{k=1}^{\infty} \frac{k! \, 10^k}{3^k}$.

17. $\sum_{k=1}^{\infty} k^{50} e^{-k}$.

18. $\sum_{k=1}^{\infty} \frac{k^2}{k^3 + 1}$.

19. $\sum_{k=1}^{\infty} k\left(\frac{2}{3}\right)^k$.

20. $\sum_{k=1}^{\infty} k^k$.

21. $\sum_{k=2}^{\infty} \frac{1}{k \ln k}$.

22. $\sum_{k=1}^{\infty} \frac{2^k}{k^3 + 1}$.

23. $\sum_{k=1}^{\infty} \left(\frac{4}{7k-1}\right)^k$.

24. $\sum_{k=1}^{\infty} \frac{(k!)^2 2^k}{(2k+2)!}$.

25. $\displaystyle\sum_{k=0}^{\infty} \frac{(k!)^2}{(2k)!}$.

26. $\displaystyle\sum_{k=1}^{\infty} \frac{1}{k^2 + 25}$.

27. $\displaystyle\sum_{k=1}^{\infty} \frac{1}{1 + \sqrt{k}}$.

28. $\displaystyle\sum_{k=1}^{\infty} \frac{k!}{k^k}$.

29. $\displaystyle\sum_{k=1}^{\infty} \frac{\ln k}{e^k}$.

30. $\displaystyle\sum_{k=1}^{\infty} \frac{k!}{e^{k^2}}$.

31. $\displaystyle\sum_{k=0}^{\infty} \frac{(k + 4)!}{4!k!4^k}$.

32. $\displaystyle\sum_{k=1}^{\infty} \left(\frac{k}{k + 1}\right)^{k^2}$.

> In Exercises 33–35, show that the series converges.

33. $1 + \dfrac{1 \cdot 2}{1 \cdot 3} + \dfrac{1 \cdot 2 \cdot 3}{1 \cdot 3 \cdot 5} + \dfrac{1 \cdot 2 \cdot 3 \cdot 4}{1 \cdot 3 \cdot 5 \cdot 7} + \cdots$.

34. $1 + \dfrac{1 \cdot 3}{3!} + \dfrac{1 \cdot 3 \cdot 5}{5!} + \dfrac{1 \cdot 3 \cdot 5 \cdot 7}{7!} + \cdots$.

35. $\dfrac{2!}{1} + \dfrac{3!}{1 \cdot 4} + \dfrac{4!}{1 \cdot 4 \cdot 7} + \dfrac{5!}{1 \cdot 4 \cdot 7 \cdot 10} + \cdots$.

36. For which positive values of α does $\sum_{k=1}^{\infty} \alpha^k/k^\alpha$ converge?

37. (a) Show: $\lim\limits_{k \to +\infty} (\ln k)^{1/k} = 1$. [*Hint:* Let
$y = (\ln x)^{1/x}$ and find $\lim\limits_{x \to +\infty} \ln y$.]

    (b) Use the result in part (a) and the root test to show that $\sum_{k=1}^{\infty} (\ln k)/3^k$ converges.

    (c) Show that the series converges using the ratio test.

38. Prove: $\lim\limits_{k \to +\infty} k!/k^k = 0$. [*Hint:* Use Theorem 11.4.1.]

39. Prove: $\lim\limits_{k \to +\infty} \dfrac{a^k}{k!} = 0$ for every real number a. [See the hint in Exercise 38.]

40. If the sum S of a convergent series $\sum_{k=1}^{\infty} u_k$ of positive terms is approximated by the nth partial sum s_n, then the error in the approximation is defined as

$$S - s_n = \sum_{k=n+1}^{\infty} u_k = u_{n+1} + u_{n+2} + u_{n+3} + \cdots$$

Let $r_k = u_{k+1}/u_k$.

(a) Prove: If r_k is *decreasing* for all $k \geq n + 1$ and if $r_{n+1} < 1$, then

$$S - s_n < \frac{u_{n+1}}{1 - r_{n+1}}$$

[*Hint:* Show that the sum of the series $\sum_{k=n+1}^{\infty} u_k$ is less than the sum of the convergent geometric series

$$u_{n+1} + r_{n+1}u_{n+1} + r_{n+1}^2 u_{n+1} + \cdots$$

(b) Prove: If r_k is *increasing* for all $k \geq n + 1$ and if $\lim\limits_{k \to +\infty} r_k = \rho < 1$, then

$$S - s_n < \frac{u_{n+1}}{1 - \rho}$$

[*Hint:* Show that the sum of the series $\sum_{k=n+1}^{\infty} u_k$ is less than the sum of the convergent geometric series

$$u_{n+1} + \rho u_{n+1} + \rho^2 u_{n+1} + \cdots$$

> In Exercises 41–44, use the results in Exercise 40. For each exercise do the following:
> (a) Compute the stated partial sum and find an upper bound on the error in approximating S by the partial sum. Express your results to five decimal places.
> (b) Find a value of n to ensure that s_n will approximate S with an error that is less than 10^{-5}.

41. $S = \displaystyle\sum_{k=1}^{\infty} \frac{1}{k!}; \ s_5$.

42. $S = \displaystyle\sum_{k=1}^{\infty} \frac{k}{3^k}; \ s_8$.

43. $S = \displaystyle\sum_{k=1}^{\infty} \frac{1}{k2^k}; \ s_7$.

44. $S = \displaystyle\sum_{k=1}^{\infty} \frac{1}{\sqrt{k}3^k}; \ s_4$.

▪ 11.6 APPLYING THE COMPARISON TEST

> *In this section we shall discuss procedures for applying the comparison test and we shall state an alternative version of this test which is easier to work with. Before starting, we remind the reader that the comparison test applies only to series with positive terms.*

□ SOME USEFUL
INFORMAL PRINCIPLES

There are two basic steps required to apply the comparison test to a series Σu_k of positive terms:

- Guess at whether the series Σu_k converges or diverges.
- Find a series that proves the guess to be correct. Thus, if the guess is divergence, we must find a divergent series whose terms are "smaller" than the corresponding terms of Σu_k, and if the guess is convergence, we must find a convergent series whose terms are "bigger" than the corresponding terms of Σu_k.

To help with the guessing process in the first step we have formulated some principles that sometimes *suggest* whether a series is likely to converge or diverge. We have called these "informal principles" because they are not intended as formal theorems. In fact, we shall not guarantee that they *always* work. However, they work often enough to be useful as a starting point for the comparison test.

11.6.1 INFORMAL PRINCIPLE. *Constant terms in the denominator of u_k can usually be deleted without affecting the convergence or divergence of the series.*

Example 1 Use the above principle to help guess whether the following series converge or diverge.

(a) $\displaystyle\sum_{k=1}^{\infty} \frac{1}{2^k + 1}$ (b) $\displaystyle\sum_{k=5}^{\infty} \frac{1}{\sqrt{k} - 2}$ (c) $\displaystyle\sum_{k=1}^{\infty} \frac{1}{(k + \frac{1}{2})^3}$

Solution.

(a) Deleting the constant 1 suggests that

$$\sum_{k=1}^{\infty} \frac{1}{2^k + 1} \quad \text{behaves like} \quad \sum_{k=1}^{\infty} \frac{1}{2^k}$$

The modified series is a convergent geometric series, so the given series is likely to converge.

(b) Deleting the -2 suggests that

$$\sum_{k=5}^{\infty} \frac{1}{\sqrt{k} - 2} \quad \text{behaves like} \quad \sum_{k=5}^{\infty} \frac{1}{\sqrt{k}}$$

The modified series is a portion of a divergent p-series ($p = \frac{1}{2}$), so the given series is likely to diverge.

(c) Deleting the $\frac{1}{2}$ suggests that

$$\sum_{k=1}^{\infty} \frac{1}{(k + \frac{1}{2})^3} \quad \text{behaves like} \quad \sum_{k=1}^{\infty} \frac{1}{k^3}$$

The modified series is a convergent p-series ($p = 3$), so the given series is likely to converge. ◄

> **11.6.2** INFORMAL PRINCIPLE. *If a polynomial in k appears as a factor in the numerator or denominator of u_k, all but the highest power of k in the polynomial may usually be deleted without affecting the convergence or divergence of the series.*

Example 2 Use the above principle to help guess whether the following series converge or diverge.

(a) $\displaystyle\sum_{k=1}^{\infty} \frac{1}{\sqrt{k^3 + 2k}}$ (b) $\displaystyle\sum_{k=1}^{\infty} \frac{6k^4 - 2k^3 + 1}{k^5 + k^2 - 2k}$

Solution (a). Deleting the term $2k$ suggests that

$$\sum_{k=1}^{\infty} \frac{1}{\sqrt{k^3 + 2k}} \text{ behaves like } \sum_{k=1}^{\infty} \frac{1}{\sqrt{k^3}} = \sum_{k=1}^{\infty} \frac{1}{k^{3/2}}$$

Since the modified series is a convergent p-series $(p = \frac{3}{2})$, the given series is likely to converge.

Solution (b). Deleting all but the highest powers of k in the numerator and also in the denominator suggests that

$$\sum_{k=1}^{\infty} \frac{6k^4 - 2k^3 + 1}{k^5 + k^2 - 2k} \text{ behaves like } \sum_{k=1}^{\infty} \frac{6k^4}{k^5} = \sum_{k=1}^{\infty} 6\left(\frac{1}{k}\right)$$

Since each term in the modified series is a constant times the corresponding term in the divergent harmonic series, the given series is likely to diverge. ◀

Once it is decided whether a series is likely to converge or diverge, the second step in applying the comparison test is to produce a series with which the given series can be compared to substantiate the guess. Let us consider the case of convergence first. To prove Σa_k converges by the comparison test, we must find a *convergent* series Σb_k such that

$$a_k \leq b_k$$

for all k. Frequently, b_k is derived from the formula for a_k by either increasing the numerator of a_k, or decreasing the denominator of a_k, or both.

Example 3 Use the comparison test to determine whether

$$\sum_{k=1}^{\infty} \frac{1}{2k^2 + k}$$

converges or diverges.

Solution. Using Principle 11.6.2, the given series behaves like the series

$$\sum_{k=1}^{\infty} \frac{1}{2k^2} = \frac{1}{2} \sum_{k=1}^{\infty} \frac{1}{k^2}$$

which is a constant times a convergent p-series. Thus, the given series is likely to converge. To prove the convergence, observe that when we discard the k from the denominator of $1/(2k^2 + k)$, the denominator decreases and the ratio increases, so that

$$\frac{1}{2k^2 + k} < \frac{1}{2k^2}$$

for $k = 1, 2, \ldots$. Since

$$\sum_{k=1}^{\infty} \frac{1}{2k^2} = \frac{1}{2} \sum_{k=1}^{\infty} \frac{1}{k^2}$$

converges, so does $\displaystyle\sum_{k=1}^{\infty} \frac{1}{2k^2 + k}$ by the comparison test.　◄

Example 4　Use the comparison test to determine whether

$$\sum_{k=1}^{\infty} \frac{1}{2k^2 - k}$$

converges or diverges.

Solution.　Using Principle 11.6.2, the series behaves like the convergent series

$$\sum_{k=1}^{\infty} \frac{1}{2k^2} = \frac{1}{2} \sum_{k=1}^{\infty} \frac{1}{k^2}$$

Thus, the given series is likely to converge. However, if we discard the k from the denominator of $1/(2k^2 - k)$, the denominator increases and the ratio decreases, so that

$$\frac{1}{2k^2 - k} > \frac{1}{2k^2}$$

Unfortunately, this inequality is in the wrong direction to prove convergence of the given series. A different approach is needed; we must do something to decrease the denominator, not increase it. We accomplish this by replacing k by k^2 to obtain

$$\frac{1}{2k^2 - k} \leq \frac{1}{2k^2 - k^2} = \frac{1}{k^2}$$

Since $\displaystyle\sum_{k=1}^{\infty} \frac{1}{k^2}$ is a convergent p-series, the given series converges by the comparison test.　◄

To prove that a series Σa_k diverges by the comparison test, we must produce a divergent series Σb_k of positive terms such that $a_k \geq b_k$ for all k.

Example 5   Use the comparison test to determine whether

$$\sum_{k=1}^{\infty} \frac{1}{k - \frac{1}{4}}$$

converges or diverges.

Solution.   Using Principle 11.6.1, the series behaves like the divergent harmonic series

$$\sum_{k=1}^{\infty} \frac{1}{k}$$

Thus, the given series is likely to diverge. Since

$$\frac{1}{k - \frac{1}{4}} > \frac{1}{k} \quad \text{for } k = 1, 2, \ldots$$

and since $\sum_{k=1}^{\infty} \frac{1}{k}$ diverges, the given series diverges by the comparison test. ◄

Example 6   Use the comparison test to determine whether

$$\sum_{k=1}^{\infty} \frac{1}{\sqrt{k} + 5}$$

converges or diverges.

Solution.   Using Principle 11.6.1, the series behaves like the divergent *p*-series

$$\sum_{k=1}^{\infty} \frac{1}{\sqrt{k}}$$

Thus, the given series is likely to diverge. For $k \geq 25$ we have

$$\frac{1}{\sqrt{k} + 5} \geq \frac{1}{\sqrt{k} + \sqrt{k}} = \frac{1}{2\sqrt{k}}$$

and since

$$\sum_{k=25}^{\infty} \frac{1}{2\sqrt{k}}$$

diverges (why?), the series

$$\sum_{k=25}^{\infty} \frac{1}{\sqrt{k} + 5}$$

diverges by the comparison test; consequently, the given series diverges by Theorem 11.4.3(*c*). ◄

☐ **THE LIMIT COMPARISON TEST**

As the last four examples show, some tricky work with inequalities may be required to apply the comparison test. Fortunately, there is an alternative version of this test, which avoids this complication. We shall state the result, but defer its proof to the end of the section so that we can proceed immediately to some examples.

> **11.6.3** THEOREM (*The Limit Comparison Test*). *Let Σa_k and Σb_k be series with positive terms and suppose*
>
> $$\rho = \lim_{k \to +\infty} \frac{a_k}{b_k}$$
>
> *If ρ is finite and $\rho \neq 0$, then the series both converge or both diverge.*

To illustrate how this test works we shall reconsider the problems in Examples 4 and 5.

Example 7   Use the limit comparison test to determine whether the following series converge or diverge:

$$\text{(a)} \quad \sum_{k=1}^{\infty} \frac{1}{2k^2 - k} \qquad \text{(b)} \quad \sum_{k=1}^{\infty} \frac{1}{k - \frac{1}{4}}$$

Solution (a).   As in Example 4, we guess that the given series behaves like the convergent series

$$\sum_{k=1}^{\infty} \frac{1}{2k^2} \tag{1}$$

Thus, from Theorem 11.6.3 with

$$a_k = \frac{1}{2k^2 - k} \quad \text{and} \quad b_k = \frac{1}{2k^2}$$

we obtain

$$\rho = \lim_{k \to +\infty} \frac{a_k}{b_k} = \lim_{k \to +\infty} \frac{2k^2}{2k^2 - k} = \lim_{k \to +\infty} \frac{2}{2 - 1/k} = 1$$

Since (1) converges, so does the given series since ρ is finite and positive.

Solution (b).   As in Example 5, we guess that the given series behaves like the divergent series

$$\sum_{k=1}^{\infty} \frac{1}{k} \tag{2}$$

Thus, from Theorem 11.6.3 with

$$a_k = \frac{1}{k - \frac{1}{4}} \quad \text{and} \quad b_k = \frac{1}{k}$$

we obtain

$$\rho = \lim_{k \to +\infty} \frac{a_k}{b_k} = \lim_{k \to +\infty} \frac{k}{k - \frac{1}{4}} = \lim_{k \to +\infty} \frac{1}{1 - \frac{1}{4k}} = 1$$

Since (2) diverges, so does the given series since ρ is finite and nonzero.   ◀

Example 8   Use the limit comparison test to determine whether

$$\sum_{k=1}^{\infty} \frac{3k^3 - 2k^2 + 4}{k^5 - k^3 + 2}$$

converges or diverges.

Solution.   From Principle 11.6.2, the series behaves like

$$\sum_{k=1}^{\infty} \frac{3k^3}{k^5} = \sum_{k=1}^{\infty} \frac{3}{k^2} \tag{3}$$

which converges since it is a constant times a convergent p-series. Thus, the given series is likely to converge. To substantiate this, we apply the limit comparison test to series (3) and the given series. We obtain

$$\rho = \lim_{k \to +\infty} \frac{\dfrac{3k^3 - 2k^2 + 4}{k^5 - k^3 + 2}}{\dfrac{3}{k^2}} = \lim_{k \to +\infty} \frac{3k^5 - 2k^4 + 4k^2}{3k^5 - 3k^3 + 6} = 1$$

Since $\rho \neq 0$, the given series converges because series (3) converges.   ◀

Unlike the comparison test, the limit comparison test does not require any tricky manipulation of inequalities. However, the test only applies when $0 < \rho < +\infty$. We note, though, that if $\rho = 0$ or $\rho = +\infty$, conclusions about convergence or divergence may be drawn in certain cases (see Exercise 44).

REMARK.   As a practical matter you should generally try the limit comparison test before the comparison test because it is easier to apply.

■ OPTIONAL

We conclude this section with a proof of the limit comparison test.

Proof of Theorem 11.6.3.   We must show that Σb_k converges when Σa_k converges and conversely. The idea of the proof is to apply the comparison test to Σa_k and suitable multiples of Σb_k. To this end, let

$$\epsilon = \frac{\rho}{2} \tag{4}$$

so that $\epsilon > 0$ because $\rho > 0$ by hypothesis. From the assumption that

$$\rho = \lim_{k \to +\infty} \frac{a_k}{b_k}$$

it follows that for k sufficiently large, say $k \geq K$, the ratio a_k/b_k will be within ϵ units of ρ. Thus,

$$\rho - \epsilon < \frac{a_k}{b_k} < \rho + \epsilon \quad \text{when} \quad k \geq K$$

or on substituting (4)

$$\frac{1}{2}\rho < \frac{a_k}{b_k} < \frac{3}{2}\rho \quad \text{when} \quad k \geq K$$

or

$$\frac{1}{2}\rho b_k < a_k < \frac{3}{2}\rho b_k \quad \text{when} \quad k \geq K \tag{5}$$

If $\sum\limits_{k=1}^{\infty} a_k$ converges, then $\sum\limits_{k=K}^{\infty} a_k$ converges by Theorem 11.4.3(c). It follows from the comparison test and the left-hand inequality in (5) that $\sum\limits_{k=K}^{\infty} \frac{1}{2}\rho b_k$ converges. Thus, $\sum\limits_{k=K}^{\infty} b_k$ converges [Theorem 11.4.3(b)] and $\sum\limits_{k=1}^{\infty} b_k$ converges [Theorem 11.4.3(c)].

Conversely, if $\sum\limits_{k=1}^{\infty} b_k$ converges, then $\sum\limits_{k=K}^{\infty} \frac{3}{2}\rho b_k$ converges [Theorem 11.4.3(b), (c)], so $\sum\limits_{k=K}^{\infty} a_k$ converges by the comparison test and the right-hand inequality in (5). Thus, $\sum\limits_{k=1}^{\infty} a_k$ converges. ∎

▶ Exercise Set 11.6

In Exercises 1–6, prove that the series converges by the comparison test.

In Exercises 7–12, prove that the series diverges by the comparison test.

1. $\sum\limits_{k=1}^{\infty} \dfrac{1}{3^k + 5}$.

2. $\sum\limits_{k=1}^{\infty} \dfrac{2}{k^4 + k}$.

7. $\sum\limits_{k=1}^{\infty} \dfrac{3}{k - \frac{1}{4}}$.

8. $\sum\limits_{k=1}^{\infty} \dfrac{1}{\sqrt{k} + 8}$.

3. $\sum\limits_{k=1}^{\infty} \dfrac{1}{5k^2 - k}$.

4. $\sum\limits_{k=1}^{\infty} \dfrac{k}{8k^3 + 2k^2 - 1}$.

9. $\sum\limits_{k=1}^{\infty} \dfrac{9}{\sqrt{k} + 1}$.

10. $\sum\limits_{k=2}^{\infty} \dfrac{k + 1}{k^2 - k}$.

5. $\sum\limits_{k=1}^{\infty} \dfrac{2^k - 1}{3^k + 2k}$.

6. $\sum\limits_{k=1}^{\infty} \dfrac{5 \sin^2 k}{k!}$.

11. $\sum\limits_{k=1}^{\infty} \dfrac{k^{4/3}}{8k^2 + 5k + 1}$.

12. $\sum\limits_{k=1}^{\infty} \dfrac{k^{-1/2}}{2 + \sin^2 k}$.

In Exercises 13–18, use the limit comparison test to determine whether the series converges or diverges.

13. $\sum_{k=1}^{\infty} \dfrac{4k^2 - 2k + 6}{8k^7 + k - 8}$.

14. $\sum_{k=1}^{\infty} \dfrac{1}{9k + 6}$.

15. $\sum_{k=1}^{\infty} \dfrac{5}{3^k + 1}$.

16. $\sum_{k=1}^{\infty} \dfrac{k(k + 3)}{(k + 1)(k + 2)(k + 5)}$.

17. $\sum_{k=1}^{\infty} \dfrac{1}{\sqrt[3]{8k^2 - 3k}}$.

18. $\sum_{k=1}^{\infty} \dfrac{1}{(2k + 3)^{17}}$.

In Exercises 19–34, use any method to determine whether the series converges or diverges. In some cases, you may have to use tests from earlier sections.

19. $\sum_{k=1}^{\infty} \dfrac{1}{k^3 + 2k + 1}$.

20. $\sum_{k=1}^{\infty} \dfrac{1}{(3 + k)^{2/5}}$.

21. $\sum_{k=1}^{\infty} \dfrac{1}{9k - 2}$.

22. $\sum_{k=1}^{\infty} \dfrac{\ln k}{k}$.

23. $\sum_{k=1}^{\infty} \dfrac{\sqrt{k}}{k^3 + 1}$.

24. $\sum_{k=1}^{\infty} \dfrac{4}{2 + 3^k k}$.

25. $\sum_{k=1}^{\infty} \dfrac{1}{\sqrt{k(k + 1)}}$.

26. $\sum_{k=1}^{\infty} \dfrac{2 + (-1)^k}{5^k}$.

27. $\sum_{k=1}^{\infty} \dfrac{2 + \sqrt{k}}{(k + 1)^3 - 1}$.

28. $\sum_{k=1}^{\infty} \dfrac{4 + |\cos k|}{k^3}$.

29. $\sum_{k=1}^{\infty} \dfrac{1}{4 + 2^{-k}}$.

30. $\sum_{k=1}^{\infty} \dfrac{\sqrt{k} \ln k}{k^3 + 1}$.

31. $\sum_{k=1}^{\infty} \dfrac{\tan^{-1} k}{k^2}$.

32. $\sum_{k=1}^{\infty} \dfrac{5^k + k}{k! + 3}$.

33. $\sum_{k=1}^{\infty} \dfrac{\ln k}{k\sqrt{k}}$.

34. $\sum_{k=1}^{\infty} \dfrac{\cos(1/k)}{k^2}$.

35. Use the limit comparison test to show that the series

$\sum_{k=1}^{\infty} (1 - \cos(1/k))$ converges.

$\left[\textit{Hint:} \text{ Compare with the series } \sum_{k=1}^{\infty} 1/k^2. \right]$

36. Use the limit comparison test to show that the series

$\sum_{k=1}^{\infty} \sin(\pi/k)$ diverges.

$\left[\textit{Hint:} \text{ Compare with the series } \sum_{k=1}^{\infty} \pi/k. \right]$

37. Use the comparison test to determine whether $\sum_{k=1}^{\infty} \dfrac{\ln k}{k^2}$ converges or diverges. [*Hint:* $\ln x < \sqrt{x}$ by Exercise 66 of Section 7.3.]

38. Determine whether $\sum_{k=2}^{\infty} \dfrac{1}{(\ln k)^2}$ converges or diverges. [*Hint:* See the hint in Exercise 37.]

39. Let a, b, and p be positive constants. For which values of p does the series $\sum_{k=1}^{\infty} \dfrac{1}{(a + bk)^p}$ converge?

40. (a) Show that $k^k \geq k!$ and use this result to prove that the series $\sum_{k=1}^{\infty} k^{-k}$ converges by the comparison test.
   (b) Prove convergence using the root test.

41. Use the limit comparison test to investigate convergence of $\sum_{k=1}^{\infty} \dfrac{(k + 1)^2}{(k + 2)!}$.

42. Use the limit comparison test to investigate convergence of the series $1 + \frac{1}{3} + \frac{1}{5} + \frac{1}{7} + \cdots$

43. Prove that $\sum_{k=1}^{\infty} 1/k!$ converges by comparison with a suitable geometric series.

44. Let Σa_k and Σb_k be series with positive terms. Prove:
   (a) If $\lim\limits_{k \to +\infty} (a_k/b_k) = 0$ and Σb_k converges, then Σa_k converges.
   (b) If $\lim\limits_{k \to +\infty} (a_k/b_k) = +\infty$ and Σb_k diverges, then Σa_k diverges.

■ 11.7 ALTERNATING SERIES; CONDITIONAL CONVERGENCE

So far our emphasis has been on series with positive terms. In this section we shall discuss series containing both positive and negative terms.

□ ALTERNATING SERIES

Of special importance are series whose terms are alternately positive and negative. These are called *alternating series*. Some examples are

$$1 - 1 + 1 - 1 + \cdots + (-1)^{k+1} + \cdots$$

$$1 - \frac{1}{2} + \frac{1}{3} - \frac{1}{4} + \cdots + (-1)^{k+1}\frac{1}{k} + \cdots$$

$$-1 + 2 - 3 + 4 - \cdots + (-1)^k k + \cdots$$

$$-1 + \frac{1}{2!} - \frac{1}{3!} + \frac{1}{4!} - \cdots + (-1)^k\frac{1}{k!} + \cdots$$

In general, an alternating series has one of two possible forms:

$$\sum_{k=1}^{\infty} (-1)^{k+1} a_k = a_1 - a_2 + a_3 - a_4 + \cdots \tag{1}$$

or

$$\sum_{k=1}^{\infty} (-1)^{k} a_k = -a_1 + a_2 - a_3 + a_4 - \cdots \tag{2}$$

where the a_k's are assumed to be positive in both cases.

The following theorem is the key result on convergence of alternating series.

11.7.1 THEOREM (*Alternating Series Test*). *An alternating series of either form (1) or form (2) converges if the following two conditions are satisfied:*

   (a) $a_1 > a_2 > a_3 > \cdots > a_k > \cdots$

   (b) $\lim_{k \to +\infty} a_k = 0$

Proof. Our proof of this theorem will use the following result about sequences which is intuitively obvious, but which we will not actually prove:

If the even-numbered terms of a sequence converge to a limit L, and if the odd-numbered terms of the sequence converge to the same limit L, then the entire sequence converges to the limit L.

To prove convergence of an alternating series, we shall show that its even partial sums and odd partial sums converge to the same limit S; it will then follow from the foregoing result that the entire sequence of partial sums converges to S, which will imply convergence of the series.

We shall consider the case

$$a_1 - a_2 + a_3 - a_4 + \cdots + (-1)^{k+1} a_k + \cdots$$

The other case is left as an exercise. Consider the partial sums

$$s_2 = a_1 - a_2$$
$$s_4 = (a_1 - a_2) + (a_3 - a_4)$$
$$s_6 = (a_1 - a_2) + (a_3 - a_4) + (a_5 - a_6)$$
$$s_8 = (a_1 - a_2) + (a_3 - a_4) + (a_5 - a_6) + (a_7 - a_8)$$
$$\vdots$$

By condition (a), each of the differences appearing in the parentheses is positive, so

$$s_2 < s_4 < s_6 < s_8 < \cdots$$

Moreover, the terms in this sequence are all less than or equal to a_1 since we can write

$$s_2 = a_1 - a_2$$
$$s_4 = a_1 - (a_2 - a_3) - a_4$$
$$s_6 = a_1 - (a_2 - a_3) - (a_4 - a_5) - a_6$$
$$s_8 = a_1 - (a_2 - a_3) - (a_4 - a_5) - (a_6 - a_7) - a_8$$
$$\vdots$$

Thus, the sequence

$$s_2, s_4, s_6, s_8, \ldots, s_{2n}, \ldots$$

is increasing and has an upper bound (namely, a_1), so it converges to some limit S by Theorem 11.2.2; that is,

$$\lim_{n \to +\infty} s_{2n} = S \tag{3}$$

We shall show next that the sequence

$$s_1, s_3, s_5, \ldots, s_{2n-1}, \ldots$$

also has limit S. Since the $(2n)$-th term in the given alternating series is $-a_{2n}$, it follows that $s_{2n} - s_{2n-1} = -a_{2n}$, which can be written as

$$s_{2n-1} = s_{2n} + a_{2n} \tag{4}$$

But $2n \to +\infty$ as $n \to +\infty$, so $\lim_{n \to +\infty} a_{2n} = 0$ by condition (b). Thus, from (3) and (4)

$$\lim_{n \to +\infty} s_{2n-1} = \lim_{n \to +\infty} s_{2n} + \lim_{n \to +\infty} a_{2n} = S + 0 = S \qquad (5)$$

From (3) and (5), the sequence $s_1, s_2, s_3, \ldots, s_n, \ldots$ converges to S, so the series converges. ∎

Example 1 The series

$$1 - \frac{1}{2} + \frac{1}{3} - \frac{1}{4} + \cdots + (-1)^{k+1} \frac{1}{k} + \cdots$$

is called the **alternating harmonic series**. Since

$$a_k = \frac{1}{k} > \frac{1}{k+1} = a_{k+1}$$

and

$$\lim_{k \to +\infty} a_k = \lim_{k \to +\infty} \frac{1}{k} = 0$$

this series converges by the alternating series test. ◄

Example 2 Determine whether the alternating series

$$\sum_{k=1}^{\infty} (-1)^{k+1} \frac{k+3}{k(k+1)}$$

converges or diverges.

Solution. Condition (*b*) of the alternating series test is satisfied since

$$\lim_{k \to +\infty} a_k = \lim_{k \to +\infty} \frac{k+3}{k(k+1)} = \lim_{k \to +\infty} \frac{\dfrac{1}{k} + \dfrac{3}{k^2}}{1 + \dfrac{1}{k}} = 0$$

To see if condition (*a*) is met, we must determine whether the sequence

$$\{a_k\}_{k=1}^{+\infty} = \left\{ \frac{k+3}{k(k+1)} \right\}_{k=1}^{+\infty}$$

is decreasing. Since

$$\frac{a_{k+1}}{a_k} = \frac{k+4}{(k+1)(k+2)} \cdot \frac{k(k+1)}{k+3} = \frac{k^2 + 4k}{k^2 + 5k + 6}$$

$$= \frac{k^2 + 4k}{(k^2 + 4k) + (k+6)} < 1$$

we have $a_k > a_{k+1}$, so the series converges by the alternating series test. ◄

REMARK. If an alternating series violates condition (*b*) of the alternating series test, then the series must diverge by the divergence test (11.4.2). However, if condition (*b*) is satisfied, but condition (*a*) is not, the series can either converge or diverge.*

Figure 11.7.1 provides some insight into the way in which an alternating series

$$a_1 - a_2 + a_3 - a_4 + \cdots + (-1)^{k+1} a_k + \cdots$$

converges to its sum S when the conditions of the alternating series test are satisfied.

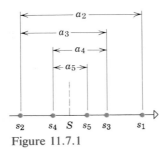

Figure 11.7.1

In the figure we have plotted the successive partial sums on a horizontal. Because

$$a_1 > a_2 > a_3 > a_4 > \cdots$$

and

$$\lim_{k \to +\infty} a_k = 0$$

the successive partial sums oscillate in smaller and smaller steps, closing in on the sum S. It is of interest to note that the even-numbered partial sums are less than or equal to S and the odd-numbered partial sums are greater than S. Thus, the sum S falls between any two successive partial sums; that is, for any positive integer n

$$s_n < S < s_{n+1} \quad \text{or} \quad s_{n+1} < S < s_n$$

depending on whether n is even or odd. In either case,

$$|S - s_n| < |s_{n+1} - s_n| \tag{6}$$

---

*The interested reader will find some nice examples in an article by R. Lariviere, "On a Convergence Test for Alternating Series," *Mathematics Magazine*, Vol. 29, 1956, p. 88.

But $s_{n+1} - s_n = \pm a_{n+1}$ (the sign depending on whether n is even or odd), so (6) yields

$$|S - s_n| < a_{n+1} \tag{7}$$

Since $|S - s_n|$ represents the magnitude of error that results when we approximate the sum of the entire series by the sum of the first n terms, (7) tells us that this error is less than the magnitude of the $(n + 1)$-st term in the series. The reader can check that (7) also holds for alternating series of the form

$$-a_1 + a_2 - a_3 + a_4 - \cdots + (-1)^k a_k + \cdots$$

In summary, we have the following result.

11.7.2 THEOREM. *If an alternating series satisfies the conditions of the alternating series test, and if the sum S of the series is approximated by the nth partial sum s_n, thereby resulting in an error of $S - s_n$, then*

$$|S - s_n| < a_{n+1}$$

Moreover, the sign of the error is the same as that of the coefficient of a_{n+1} in the series.

Example 3 As shown in Example 1, the alternating harmonic series

$$1 - \frac{1}{2} + \frac{1}{3} - \frac{1}{4} + \cdots + (-1)^{k+1} \frac{1}{k} + \cdots$$

satisfies the conditions of the alternating series test; hence, the series has a sum S, which we know must lie between any two successive partial sums. In particular, it must lie between

$$s_7 = 1 - \frac{1}{2} + \frac{1}{3} - \frac{1}{4} + \frac{1}{5} - \frac{1}{6} + \frac{1}{7} = \frac{319}{420}$$

and

$$s_8 = 1 - \frac{1}{2} + \frac{1}{3} - \frac{1}{4} + \frac{1}{5} - \frac{1}{6} + \frac{1}{7} - \frac{1}{8} = \frac{533}{840}$$

so

$$\frac{533}{840} < S < \frac{319}{420} \tag{8}$$

Later in this chapter we shall show that the sum S of the alternating harmonic series is $\ln 2$. If we accept this to be so for now, it follows from (8) that

$$\frac{533}{840} < \ln 2 < \frac{319}{420}$$

or with the help of a calculator (verify)

$$0.6345 < \ln 2 < 0.7596$$

The value of $\ln 2$, rounded to four decimal places, is 0.6931, which is consistent with these inequalities. It follows from Theorem 11.7.2 that

$$|\ln 2 - s_7| = \left| \ln 2 - \frac{319}{420} \right| < a_8 = \frac{1}{8}$$

and

$$|\ln 2 - s_8| = \left| \ln 2 - \frac{533}{840} \right| < a_9 = \frac{1}{9}$$

We leave it for the reader to verify that these error estimates are correct by using a calculator. ◄

ABSOLUTE AND CONDITIONAL CONVERGENCE

The series

$$1 - \frac{1}{2} - \frac{1}{2^2} + \frac{1}{2^3} + \frac{1}{2^4} - \frac{1}{2^5} - \frac{1}{2^6} + \cdots$$

does not fit in any of the categories studied so far—it has mixed signs, but is not alternating. We shall now develop some convergence tests that can be applied to such series.

11.7.3   DEFINITION.   A series

$$\sum_{k=1}^{\infty} u_k = u_1 + u_2 + \cdots + u_k + \cdots$$

is said to *converge absolutely* if the series of absolute values

$$\sum_{k=1}^{\infty} |u_k| = |u_1| + |u_2| + \cdots + |u_k| + \cdots$$

converges.

Example 4   The series

$$1 - \frac{1}{2} - \frac{1}{2^2} + \frac{1}{2^3} + \frac{1}{2^4} - \frac{1}{2^5} - \frac{1}{2^6} + \cdots$$

converges absolutely since the series of absolute values

$$1 + \frac{1}{2} + \frac{1}{2^2} + \frac{1}{2^3} + \frac{1}{2^4} + \frac{1}{2^5} + \frac{1}{2^6} + \cdots$$

is a convergent geometric series. On the other hand, the alternating harmonic series

$$1 - \frac{1}{2} + \frac{1}{3} - \frac{1}{4} + \frac{1}{5} - \cdots$$

does not converge absolutely since the series of absolute values

$$1 + \frac{1}{2} + \frac{1}{3} + \frac{1}{4} + \frac{1}{5} + \cdots$$

diverges.

Absolute convergence is of importance because of the following theorem.

11.7.4 THEOREM.   *If the series*

$$\sum_{k=1}^{\infty} |u_k| = |u_1| + |u_2| + \cdots + |u_k| + \cdots$$

converges, then so does the series

$$\sum_{k=1}^{\infty} u_k = u_1 + u_2 + \cdots + u_k + \cdots$$

In other words, if a series converges absolutely, then it converges.

Proof.   Our proof is based on a trick. We shall show that the series

$$\sum_{k=1}^{\infty} (u_k + |u_k|) \tag{9}$$

converges. Since $\Sigma |u_k|$ is assumed to converge, it will then follow from Theorem 11.4.3(a) that Σu_k converges, since

$$\sum_{k=1}^{\infty} u_k = \sum_{k=1}^{\infty} [(u_k + |u_k|) - |u_k|]$$

For all k, the value of $u_k + |u_k|$ is either 0 or $2|u_k|$, depending on whether u_k is negative or not. Thus, for all values of k

$$0 \le u_k + |u_k| \le 2|u_k| \tag{10}$$

But $\Sigma 2|u_k|$ is a convergent series since it is a constant times the convergent series $\Sigma |u_k|$. Thus, from (10), series (9) converges by the comparison test. ∎

Example 5   In Example 4 we showed that

$$1 - \frac{1}{2} - \frac{1}{2^2} + \frac{1}{2^3} + \frac{1}{2^4} - \frac{1}{2^5} - \frac{1}{2^6} + \cdots$$

converges absolutely. It follows from Theorem 11.7.4 that the series converges.   ◄

Example 6   Show that the series

$$\sum_{k=1}^{\infty} \frac{\cos k}{k^2}$$

converges.

Solution.   Since $|\cos k| \leq 1$ for all k,

$$\left| \frac{\cos k}{k^2} \right| \leq \frac{1}{k^2}$$

Thus,

$$\sum_{k=1}^{\infty} \left| \frac{\cos k}{k^2} \right|$$

converges by the comparison test, and consequently

$$\sum_{k=1}^{\infty} \frac{\cos k}{k^2}$$

converges.   ◄

If $\Sigma|u_k|$ *diverges*, no conclusion can be drawn about the convergence or divergence of Σu_k. For example, consider the two series

$$1 - \frac{1}{2} + \frac{1}{3} - \frac{1}{4} + \cdots + (-1)^{k+1}\frac{1}{k} + \cdots \tag{11}$$

$$-1 - \frac{1}{2} - \frac{1}{3} - \frac{1}{4} - \cdots - \frac{1}{k} - \cdots \tag{12}$$

Series (11), the alternating harmonic series, converges, whereas series (12), being a constant times the harmonic series, diverges. Yet in each case the series of absolute values is

$$1 + \frac{1}{2} + \frac{1}{3} + \cdots + \frac{1}{k} + \cdots$$

which diverges. A series such as (11), which is convergent, but not absolutely convergent, is called *conditionally convergent*.

The following version of the ratio test is useful for investigating absolute convergence.

11.7.5 THEOREM (*Ratio Test for Absolute Convergence*). *Let Σu_k be a series with nonzero terms and suppose*

$$\lim_{k \to +\infty} \frac{|u_{k+1}|}{|u_k|} = \rho$$

(a) *If $\rho < 1$, the series Σu_k converges absolutely and therefore converges.*
(b) *If $\rho > 1$ or if $\rho = +\infty$, then the series Σu_k diverges.*
(c) *If $\rho = 1$, no conclusion about convergence or absolute convergence can be drawn from this test.*

The proof is discussed in the exercises.

Example 7 The series

$$\sum_{k=1}^{\infty} (-1)^k \frac{2^k}{k!}$$

converges absolutely since

$$\rho = \lim_{k \to +\infty} \frac{|u_{k+1}|}{|u_k|} = \lim_{k \to +\infty} \frac{2^{k+1}}{(k+1)!} \cdot \frac{k!}{2^k} = \lim_{k \to +\infty} \frac{2}{k+1} = 0 < 1 \quad \blacktriangleleft$$

Example 8 We proved earlier (Theorem 11.3.3) that a geometric series

$$a + ar + ar^2 + \cdots + ar^{k-1} + \cdots$$

converges if $|r| < 1$ and diverges if $|r| \geq 1$. However, a stronger statement can be made—the series converges *absolutely* if $|r| < 1$. This follows from Theorem 11.7.5 since

$$\rho = \lim_{k \to +\infty} \frac{|u_{k+1}|}{|u_k|} = \lim_{k \to +\infty} \frac{|ar^k|}{|ar^{k-1}|} = \lim_{k \to +\infty} |r| = |r|$$

so that $\rho < 1$ if $|r| < 1$. $\quad \blacktriangleleft$

The following review is included as a ready reference to convergence tests.

Review of Convergence Tests

NAME	STATEMENT	COMMENTS
Divergence Test (11.4.2)	If $\lim\limits_{k \to +\infty} u_k \neq 0$, then Σu_k diverges.	If $\lim\limits_{k \to +\infty} u_k = 0$, Σu_k may or may not converge.
Integral Test (11.4.5)	Let Σu_k be a series with positive terms, and let $f(x)$ be the function that results when k is replaced by x in the formula for u_k. If f is decreasing and continuous for $x \geq 1$, then $$\sum_{k=1}^{\infty} u_k \quad \text{and} \quad \int_{1}^{+\infty} f(x)\, dx$$ both converge or both diverge.	Use this test when $f(x)$ is easy to integrate. This test only applies to series that have positive terms.
Comparison Test (11.5.1)	Let Σa_k and Σb_k be series with nonnegative terms such that $$a_1 \leq b_1,\ a_2 \leq b_2,\ \ldots,\ a_k \leq b_k,\ \ldots$$ If Σb_k converges, then Σa_k converges, and if Σa_k diverges, then Σb_k diverges.	Use this test as a last resort; other tests are often easier to apply. This test only applies to series with nonnegative terms.
Ratio Test (11.5.2)	Let Σu_k be a series with positive terms and suppose $$\lim_{k \to +\infty} \frac{u_{k+1}}{u_k} = \rho$$ (a) Series converges if $\rho < 1$. (b) Series diverges if $\rho > 1$ or $\rho = +\infty$. (c) No conclusion if $\rho = 1$.	Try this test when u_k involves factorials or kth powers.
Root Test (11.5.3)	Let Σu_k be a series with positive terms such that $$\rho = \lim_{k \to +\infty} \sqrt[k]{u_k}$$ (a) Series converges if $\rho < 1$. (b) Series diverges if $\rho > 1$ or $\rho = +\infty$. (c) No conclusion if $\rho = 1$.	Try this test when u_k involves kth powers.
Limit Comparison Test (11.6.3)	Let Σa_k and Σb_k be series with positive terms such that $$\rho = \lim_{k \to +\infty} \frac{a_k}{b_k}$$ If $0 < \rho < +\infty$, then both series converge or both diverge.	This is easier to apply than the comparison test, but still requires some skill in choosing the series Σb_k for comparison.
Alternating Series Test (11.7.1)	The series $$a_1 - a_2 + a_3 - a_4 + \cdots$$ $$-a_1 + a_2 - a_3 + a_4 - \cdots$$ converge if (a) $a_1 > a_2 > a_3 > \cdots$ (b) $\lim\limits_{k \to +\infty} a_k = 0$	This test applies only to alternating series. It is assumed that $a_k > 0$ for all k.

Review of Convergence Tests (Continued)

NAME	STATEMENT	COMMENTS
Ratio Test for Absolute Convergence (11.7.5)	Let Σu_k be a series with nonzero terms such that $$\rho = \lim_{k \to +\infty} \frac{|u_{k+1}|}{|u_k|}$$ (a) Series converges absolutely if $\rho < 1$. (b) Series diverges if $\rho > 1$ or $\rho = +\infty$. (c) No conclusion if $\rho = 1$.	The series need not have positive terms and need not be alternating to use this test.

▶ Exercise Set 11.7 $\boxed{C}$ 38–46, 56

In Exercises 1–6, use the alternating series test to determine whether the series converges or diverges.

1. $\displaystyle\sum_{k=1}^{\infty} \frac{(-1)^{k+1}}{2k+1}$.

2. $\displaystyle\sum_{k=1}^{\infty} (-1)^{k+1} \frac{k}{3^k}$.

3. $\displaystyle\sum_{k=1}^{\infty} (-1)^{k+1} \frac{k+1}{3k+1}$.

4. $\displaystyle\sum_{k=1}^{\infty} (-1)^{k+1} \frac{k+4}{k^2+k}$.

5. $\displaystyle\sum_{k=1}^{\infty} (-1)^{k+1} e^{-k}$.

6. $\displaystyle\sum_{k=3}^{\infty} (-1)^k \frac{\ln k}{k}$.

In Exercises 7–12, use the ratio test for absolute convergence to determine whether the series converges absolutely or diverges.

7. $\displaystyle\sum_{k=1}^{\infty} \left(-\frac{3}{5}\right)^k$.

8. $\displaystyle\sum_{k=1}^{\infty} (-1)^{k+1} \frac{2^k}{k!}$.

9. $\displaystyle\sum_{k=1}^{\infty} (-1)^{k+1} \frac{3^k}{k^2}$.

10. $\displaystyle\sum_{k=1}^{\infty} (-1)^k \left(\frac{k}{5^k}\right)$.

11. $\displaystyle\sum_{k=1}^{\infty} (-1)^k \left(\frac{k^3}{e^k}\right)$.

12. $\displaystyle\sum_{k=1}^{\infty} (-1)^{k+1} \frac{k^k}{k!}$.

In Exercises 13–30, classify the series as absolutely convergent, conditionally convergent, or divergent.

13. $\displaystyle\sum_{k=1}^{\infty} \frac{(-1)^{k+1}}{3k}$.

14. $\displaystyle\sum_{k=1}^{\infty} \frac{(-1)^{k+1}}{k^{4/3}}$.

15. $\displaystyle\sum_{k=1}^{\infty} \frac{(-4)^k}{k^2}$.

16. $\displaystyle\sum_{k=1}^{\infty} \frac{(-1)^{k+1}}{k!}$.

17. $\displaystyle\sum_{k=1}^{\infty} \frac{\cos k\pi}{k}$.

18. $\displaystyle\sum_{k=3}^{\infty} \frac{(-1)^k \ln k}{k}$.

19. $\displaystyle\sum_{k=1}^{\infty} (-1)^{k+1} \left(\frac{k+2}{3k-1}\right)^k$.

20. $\displaystyle\sum_{k=1}^{\infty} \frac{(-1)^{k+1}}{k^2+1}$.

21. $\displaystyle\sum_{k=1}^{\infty} (-1)^{k+1} \frac{k+2}{k(k+3)}$.

22. $\displaystyle\sum_{k=1}^{\infty} \frac{(-1)^{k+1}k^2}{k^3+1}$.

23. $\displaystyle\sum_{k=1}^{\infty} \sin\frac{k\pi}{2}$.

24. $\displaystyle\sum_{k=1}^{\infty} \frac{\sin k}{k^3}$.

25. $\displaystyle\sum_{k=2}^{\infty} \frac{(-1)^k}{k \ln k}$.

26. $\displaystyle\sum_{k=1}^{\infty} \frac{(-1)^k}{\sqrt{k(k+1)}}$.

27. $\displaystyle\sum_{k=2}^{\infty} \left(-\frac{1}{\ln k}\right)^k$.

28. $\displaystyle\sum_{k=1}^{\infty} \frac{(-1)^{k+1}}{\sqrt{k+1}+\sqrt{k}}$.

29. $\displaystyle\sum_{k=2}^{\infty} \frac{(-1)^k(k^2+1)}{k^3+2}$.

30. $\displaystyle\sum_{k=1}^{\infty} \frac{k \cos k\pi}{k^2+1}$.

In Exercises 31–34, the series satisfies the conditions of the alternating series test. For the stated value of n, use Theorem 11.7.2 to find an upper bound on the magnitude of the error that results if the sum of the series is approximated by the nth partial sum.

31. $\displaystyle\sum_{k=1}^{\infty} \frac{(-1)^{k+1}}{k}$; $n = 7$.

32. $\displaystyle\sum_{k=1}^{\infty} \frac{(-1)^{k+1}}{k!}$; $n = 5$.

33. $\displaystyle\sum_{k=1}^{\infty} \frac{(-1)^{k+1}}{\sqrt{k}}$; $n = 99$.

34. $\displaystyle\sum_{k=1}^{\infty} \frac{(-1)^{k+1}}{(k+1)\ln(k+1)}$; $n = 3$.

In Exercises 35–38, the series satisfies the conditions of the alternating series test. Use Theorem 11.7.2 to find a value of n for which the nth partial sum is ensured to approximate the sum of the series to the stated accuracy.

35. $\displaystyle\sum_{k=1}^{\infty}\frac{(-1)^{k+1}}{k}$; $|\text{error}| < 0.0001$.

36. $\displaystyle\sum_{k=1}^{\infty}\frac{(-1)^{k+1}}{k!}$; $|\text{error}| < 0.00001$.

37. $\displaystyle\sum_{k=1}^{\infty}\frac{(-1)^{k+1}}{\sqrt{k}}$; $|\text{error}| < 0.005$.

38. $\displaystyle\sum_{k=1}^{\infty}\frac{(-1)^{k+1}}{(k+1)\ln(k+1)}$; $|\text{error}| < 0.1$.

In Exercises 39 and 40, use Theorem 11.7.2 to find an upper bound on the magnitude of the error that results if s_{10} is used to approximate the sum of the given *geometric* series. Compute s_{10} rounded to four decimal places and compare this value with the exact sum of the series.

39. $\dfrac{3}{4} - \dfrac{3}{8} + \dfrac{3}{16} - \dfrac{3}{32} + \cdots$.

40. $1 - \dfrac{2}{3} + \dfrac{4}{9} - \dfrac{8}{27} + \cdots$.

In Exercises 41–44, the series satisfies the conditions of the alternating series test. Use Theorem 11.7.2 to find a value of n for which s_n is ensured to approximate the sum of the series with an error that is less than 10^{-4} in magnitude. Compute s_n rounded to five decimal places and compare this value with the sum of the series.

41. $\sin 1 = 1 - \dfrac{1}{3!} + \dfrac{1}{5!} - \dfrac{1}{7!} + \cdots$.

42. $\cos 1 = 1 - \dfrac{1}{2!} + \dfrac{1}{4!} - \dfrac{1}{6!} + \cdots$.

43. $\ln\frac{3}{2} = \dfrac{1}{1\cdot 2} - \dfrac{1}{2\cdot 2^2} + \dfrac{1}{3\cdot 2^3} - \dfrac{1}{4\cdot 2^4} + \cdots$.

44. $\dfrac{\pi}{16} = \dfrac{1}{1^5 + 4\cdot 1} - \dfrac{1}{3^5 + 4\cdot 3}$
$\qquad\qquad + \dfrac{1}{5^5 + 4\cdot 5} - \dfrac{1}{7^5 + 4\cdot 7} + \cdots$.

45. For the series $\dfrac{\pi^2}{12} = 1 - \dfrac{1}{2^2} + \dfrac{1}{3^2} - \dfrac{1}{4^2} + \cdots$

(a) use Theorem 11.7.2 to find a value of n for which s_n is ensured to approximate the sum of the series with an error that is less than 5×10^{-3} in magnitude

(b) compute s_{10} and show that the magnitude of the error is less than 5×10^{-3}, thus showing that the value of n obtained from Theorem 11.7.2 is a conservative estimate.

46. For the series $\dfrac{\pi}{4} = 1 - \frac{1}{3} + \frac{1}{5} - \frac{1}{7} + \cdots$

(a) use Theorem 11.7.2 to find a value of n for which s_n is ensured to approximate the sum of the series with an error that is less than 10^{-2} in magnitude

(b) compute s_{26} and show that the magnitude of the error is less than 10^{-2}, thus showing that the value of n obtained from Theorem 11.7.2 is a conservative estimate.

47. Prove: If Σa_k converges absolutely, then Σa_k^2 converges.

48. Show that the converse of the result in Exercise 47 is false by finding a series for which Σa_k^2 converges, but $\Sigma|a_k|$ diverges.

49. Prove Theorem 11.7.1 for series of the form
$$-a_1 + a_2 - a_3 + a_4 - \cdots + (-1)^k a_k + \cdots$$

50. Prove Theorem 11.7.5. [*Hint:* Theorem 11.7.4 will help in part (*a*). For part (*b*), it may help to review the proof of Theorem 11.5.2.]

51. The sum of an absolutely convergent series is independent of the order in which the terms are added, but the terms of a conditionally convergent series can be rearranged to converge to any given value, or even diverge. For example, let S be the sum of the conditionally convergent alternating harmonic series,
$$S = 1 - \frac{1}{2} + \frac{1}{3} - \frac{1}{4} + \frac{1}{5} - \frac{1}{6} + \cdots$$

Rearrange the terms in this series to get
$$\left(1 - \frac{1}{2} - \frac{1}{4}\right) + \left(\frac{1}{3} - \frac{1}{6} - \frac{1}{8}\right)$$
$$+ \left(\frac{1}{5} - \frac{1}{10} - \frac{1}{12}\right) + \cdots$$

Show that this rearrangement results in a series that converges to $S/2$. [*Hint:* Add the first two terms within each pair of parentheses.]

52. Based on the discussion in Exercise 51, rearrange the terms in the convergent series

$$1 - \frac{1}{\sqrt{2}} + \frac{1}{\sqrt{3}} - \frac{1}{\sqrt{4}} + \frac{1}{\sqrt{5}} - \frac{1}{\sqrt{6}} + \cdots$$

as

$$\left(1 + \frac{1}{\sqrt{3}} - \frac{1}{\sqrt{2}}\right) + \left(\frac{1}{\sqrt{5}} + \frac{1}{\sqrt{7}} - \frac{1}{\sqrt{4}}\right) + \cdots$$

$$= \sum_{k=1}^{\infty} \left(\frac{1}{\sqrt{4k-3}} + \frac{1}{\sqrt{4k-1}} - \frac{1}{\sqrt{2k}}\right)$$

Show that this rearrangement results in a series that diverges to $+\infty$. [*Hint:* Note that $1/\sqrt{4k-3} > 1/\sqrt{4k}$ and $1/\sqrt{4k-1} > 1/\sqrt{4k}$. Show that the kth term in the rearranged series is greater than $(1 - 1/\sqrt{2})/\sqrt{k}$.]

In Exercises 53–55, use parts (*a*) and (*b*) of Theorem 11.4.3 and the fact that the sum of an absolutely convergent series is independent of the order in which the terms are added.

53. Given: $\dfrac{\pi^2}{6} = 1 + \dfrac{1}{2^2} + \dfrac{1}{3^2} + \dfrac{1}{4^2} + \cdots$.

Show: $\dfrac{\pi^2}{8} = 1 + \dfrac{1}{3^2} + \dfrac{1}{5^2} + \dfrac{1}{7^2} + \cdots$.

54. Given: $\dfrac{\pi^4}{90} = 1 + \dfrac{1}{2^4} + \dfrac{1}{3^4} + \dfrac{1}{4^4} + \cdots$.

Show: $\dfrac{\pi^4}{96} = 1 + \dfrac{1}{3^4} + \dfrac{1}{5^4} + \dfrac{1}{7^4} + \cdots$.

55. Given: $\dfrac{\pi^2}{6} = 1 + \dfrac{1}{2^2} + \dfrac{1}{3^2} + \dfrac{1}{4^2} + \cdots$.

Show: $\dfrac{\pi^2}{12} = 1 - \dfrac{1}{2^2} + \dfrac{1}{3^2} - \dfrac{1}{4^2} + \cdots$.

56. A small bug moves back and forth along a straight line as follows: it walks D units, stops and reverses direction, walks $D/2$ units, stops and reverses direction, walks $D/3$ units, stops and reverses direction, walks $D/4$ units, stops and reverses direction, and so forth, until it stops for the 1000th time. Given that $D = 180$ cm, find upper and lower bounds for

(a) the final distance between the bug and its starting point; [*Hint:* Use Theorem 11.7.2 and the fact that $1 - \frac{1}{2} + \frac{1}{3} - \frac{1}{4} + \cdots = \ln 2$.]

(b) the total distance traveled by the bug. [*Hint:* Use inequality (4) in Section 11.4.]

▪ 11.8 POWER SERIES

In previous sections we studied series with constant terms. In this section we shall consider series whose terms involve variables. Such series are of fundamental importance in many branches of mathematics and the physical sciences.

☐ **POWER SERIES IN** *x*

If $c_0, c_1, c_2, \ldots$ are constants and x is a variable, then a series of the form

$$\sum_{k=0}^{\infty} c_k x^k = c_0 + c_1 x + c_2 x^2 + \cdots + c_k x^k + \cdots$$

is called a ***power series in x.*** Some examples are

$$\sum_{k=0}^{\infty} x^k = 1 + x + x^2 + x^3 + \cdots$$

$$\sum_{k=0}^{\infty} \frac{x^k}{k!} = 1 + x + \frac{x^2}{2!} + \frac{x^3}{3!} + \cdots$$

$$\sum_{k=0}^{\infty} (-1)^k \frac{x^{k+1}}{k+1} = x - \frac{x^2}{2} + \frac{x^3}{3} - \frac{x^4}{4} + \cdots$$

$$\sum_{k=0}^{\infty} (-1)^k \frac{x^{2k}}{(2k)!} = 1 - \frac{x^2}{2!} + \frac{x^4}{4!} - \frac{x^6}{6!} + \cdots$$

$$\sum_{k=0}^{\infty} (-1)^k \frac{x^{2k+1}}{(2k+1)!} = x - \frac{x^3}{3!} + \frac{x^5}{5!} - \frac{x^7}{7!} + \cdots$$

If a numerical value is substituted for x in a power series $\Sigma c_k x^k$, then we obtain a series of constants that may either converge or diverge. This leads to the following basic problem.

A FUNDAMENTAL PROBLEM. *For what values of x does a given power series, $\Sigma c_k x^k$, converge?*

The following theorem is the fundamental result on convergence of power series. The proof can be found in most advanced calculus texts.

11.8.1 THEOREM. *For any power series in x, exactly one of the following is true:*

(a) *The series converges only for $x = 0$.*

(b) *The series converges absolutely for all real values of x.*

(c) *The series converges absolutely for all x in some finite open interval $(-R, R)$, and diverges if $x < -R$ or $x > R$ (Figure 11.8.1). At the points $x = R$ and $x = -R$ the series may converge absolutely, converge conditionally, or diverge, depending on the particular series.*

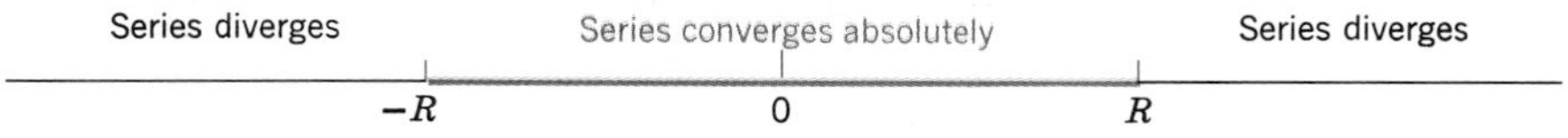

Series diverges   Series converges absolutely   Series diverges

$-R$ 0 R

Figure 11.8.1

□ RADIUS AND INTERVAL OF CONVERGENCE

In case (c), where the power series converges absolutely for $|x| < R$ and diverges for $|x| > R$, we call R the *radius of convergence*. In case (a), where the series converges only for $x = 0$, we define the radius of convergence to be $R = 0$; and in case (b), where the series converges absolutely for all x, we define the radius of convergence to be $R = +\infty$. The set of all values of x for which a power series converges is called the *interval of convergence*.

Example 1 Find the interval of convergence and radius of convergence of the power series

$$\sum_{k=0}^{\infty} x^k = 1 + x + x^2 + \cdots + x^k + \cdots$$

Solution. For every x, the given series is a geometric series with ratio $r = x$. Thus, by Example 8 of Section 11.7, the series converges absolutely if $-1 < x < 1$ and diverges if $|x| \geq 1$. Therefore, the interval of convergence is $(-1, 1)$ and the radius of convergence is $R = 1$. ◄

Example 2 Find the interval of convergence and radius of convergence of

$$\sum_{k=0}^{\infty} \frac{x^k}{k!}$$

Solution. We shall apply the ratio test for absolute convergence (11.7.5). For every real number x,

$$\rho = \lim_{k \to +\infty} \left| \frac{u_{k+1}}{u_k} \right| = \lim_{k \to +\infty} \left| \frac{x^{k+1}}{(k+1)!} \cdot \frac{k!}{x^k} \right| = \lim_{k \to +\infty} \left| \frac{x}{k+1} \right| = 0$$

Since $\rho < 1$ for all x, the series converges absolutely for all x. Thus, the interval of convergence is $(-\infty, +\infty)$ and the radius of convergence is $R = +\infty$. ◄

REMARK. There is a useful consequence of Example 2. Since

$$\sum_{k=0}^{\infty} \frac{x^k}{k!}$$

converges for all x, Theorem 11.4.1 implies that for all values of x

$$\lim_{k \to +\infty} \frac{x^k}{k!} = 0 \tag{1}$$

We shall need this result later.

Example 3 Find the interval of convergence and radius of convergence of

$$\sum_{k=0}^{\infty} k! x^k$$

Solution. If $x = 0$, the series has only one nonzero term and therefore converges. If $x \neq 0$, the ratio test yields

$$\rho = \lim_{k \to +\infty} \left| \frac{u_{k+1}}{u_k} \right| = \lim_{k \to +\infty} \left| \frac{(k+1)! x^{k+1}}{k! x^k} \right| = \lim_{k \to +\infty} |(k+1)x| = +\infty$$

Therefore, the series converges if $x = 0$, but diverges for all other x. Consequently, the interval of convergence is the single point $x = 0$ and the radius of convergence is $R = 0$. ◄

Example 4 Find the interval of convergence and radius of convergence of

$$\sum_{k=0}^{\infty} \frac{(-1)^k x^k}{3^k (k+1)}$$

Solution. Since $|(-1)^k| = |(-1)^{k+1}| = 1$, we obtain

$$\rho = \lim_{k \to +\infty} \left| \frac{u_{k+1}}{u_k} \right| = \lim_{k \to +\infty} \left| \frac{x^{k+1}}{3^{k+1}(k+2)} \cdot \frac{3^k(k+1)}{x^k} \right|$$

$$= \lim_{k \to +\infty} \left[\frac{|x|}{3} \cdot \left(\frac{k+1}{k+2} \right) \right]$$

$$= \frac{|x|}{3} \lim_{k \to +\infty} \left(\frac{1 + 1/k}{1 + 2/k} \right) = \frac{|x|}{3}$$

The ratio test for absolute convergence implies that the series converges absolutely if $|x| < 3$ and diverges if $|x| > 3$. The ratio test fails to provide any information when $|x| = 3$, so the cases $x = -3$ and $x = 3$ need separate analyses. Substituting $x = -3$ in the given series yields

$$\sum_{k=0}^{\infty} \frac{(-1)^k(-3)^k}{3^k(k+1)} = \sum_{k=0}^{\infty} \frac{(-1)^k(-1)^k 3^k}{3^k(k+1)} = \sum_{k=0}^{\infty} \frac{1}{k+1}$$

which is the divergent harmonic series $1 + \frac{1}{2} + \frac{1}{3} + \frac{1}{4} + \cdots$. Substituting $x = 3$ in the given series yields

$$\sum_{k=0}^{\infty} \frac{(-1)^k 3^k}{3^k(k+1)} = \sum_{k=0}^{\infty} \frac{(-1)^k}{k+1} = 1 - \frac{1}{2} + \frac{1}{3} - \frac{1}{4} + \cdots$$

which is the conditionally convergent alternating harmonic series. Thus, the interval of convergence for the given series is $(-3, 3]$ and the radius of convergence is $R = 3$. ◀

Example 5 Find the interval and radius of convergence of the series

$$\sum_{k=0}^{\infty} (-1)^k \frac{x^{2k}}{(2k)!}$$

Solution. Since $|(-1)^k| = |(-1)^{k+1}| = 1$, we have

$$\rho = \lim_{k \to +\infty} \left| \frac{u_{k+1}}{u_k} \right| = \lim_{k \to +\infty} \left| \frac{x^{2(k+1)}}{[2(k+1)]!} \cdot \frac{(2k)!}{x^{2k}} \right| = \lim_{k \to +\infty} \left| \frac{x^{2k+2}}{(2k+2)!} \cdot \frac{(2k)!}{x^{2k}} \right|$$

$$= \lim_{k \to +\infty} \left| \frac{x^2}{(2k+2)(2k+1)} \right| = x^2 \lim_{k \to +\infty} \frac{1}{(2k+2)(2k+1)} = x^2 \cdot 0 = 0$$

Thus, $\rho < 1$ for all x, which means that the interval of convergence is $(-\infty, +\infty)$ and the radius of convergence is $R = +\infty$. ◀

☐ **POWER SERIES IN** $x - a$

In addition to power series in x, we shall be interested in series of the form

$$\sum_{k=0}^{\infty} c_k(x-a)^k = c_0 + c_1(x-a) + c_2(x-a)^2 + \cdots + c_k(x-a)^k + \cdots$$

where $c_0, c_1, c_2, \ldots$ and a are constants. Such a series is called a ***power series in*** $x - a$. Some examples are

$$\sum_{k=0}^{\infty} \frac{(x-1)^k}{k+1} = 1 + \frac{(x-1)}{2} + \frac{(x-1)^2}{3} + \frac{(x-1)^3}{4} + \cdots$$

$$\sum_{k=0}^{\infty} \frac{(-1)^k(x+3)^k}{k!} = 1 - (x+3) + \frac{(x+3)^2}{2!} - \frac{(x+3)^3}{3!} + \cdots$$

The first is a power series in $x - 1$ ($a = 1$) and the second a power series in $x + 3$ ($a = -3$).

The convergence properties of a power series $\Sigma c_k(x - a)^k$ may be obtained from Theorem 11.8.1 by substituting $X = x - a$ to obtain

$$\sum_{k=0}^{\infty} a_k X^k$$

which is a power series in X. There are three possibilities for this series: it converges only when $X = 0$, or equivalently only when $x = a$; it converges absolutely for all values of X, or equivalently for all values of x; or finally, it converges absolutely for all X satisfying

$$-R < X < R \tag{2}$$

and diverges when $X < -R$ or $X > R$. But (2) can be written as

$$-R < x - a < R$$

or

$$a - R < x < a + R$$

Thus, we are led to the following result.

11.8.2 THEOREM. *For a power series $\Sigma c_k(x - a)^k$, exactly one of the following is true:*

(a) *The series converges only for $x = a$.*

(b) *The series converges absolutely for all real values of x.*

(c) *The series converges absolutely for all x in some finite open interval $(a - R, a + R)$ and diverges if $x < a - R$ or $x > a + R$ (Figure 11.8.2). At the points $x = a - R$ and $x = a + R$, the series may converge absolutely, converge conditionally, or diverge, depending on the particular series.*

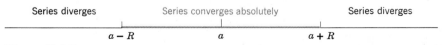

Figure 11.8.2

In cases (a), (b), and (c) of Theorem 11.8.2 the series is said to have **radius of convergence** 0, $+\infty$, and R, respectively. The set of all values of x for which the series converges is called the **interval of convergence**.

Example 6   Find the interval of convergence and radius of convergence of the series

$$\sum_{k=1}^{\infty} \frac{(x-5)^k}{k^2}$$

Solution.   We apply the ratio test for absolute convergence.

$$\rho = \lim_{k \to +\infty} \left| \frac{u_{k+1}}{u_k} \right| = \lim_{k \to +\infty} \left| \frac{(x-5)^{k+1}}{(k+1)^2} \cdot \frac{k^2}{(x-5)^k} \right|$$

$$= \lim_{k \to +\infty} \left[|x-5| \left(\frac{k}{k+1} \right)^2 \right]$$

$$= |x-5| \lim_{k \to +\infty} \left(\frac{1}{1+1/k} \right)^2 = |x-5|$$

Thus, the series converges absolutely if $|x-5| < 1$, or $-1 < x - 5 < 1$, or $4 < x < 6$. The series diverges if $x < 4$ or $x > 6$.

To determine the convergence behavior at the endpoints $x = 4$ and $x = 6$, we substitute these values in the given series. If $x = 6$, the series becomes

$$\sum_{k=1}^{\infty} \frac{1^k}{k^2} = \sum_{k=1}^{\infty} \frac{1}{k^2} = 1 + \frac{1}{2^2} + \frac{1}{3^2} + \frac{1}{4^2} + \cdots$$

which is a convergent p-series ($p = 2$). If $x = 4$, the series becomes

$$\sum_{k=1}^{\infty} \frac{(-1)^k}{k^2} = -1 + \frac{1}{2^2} - \frac{1}{3^2} + \frac{1}{4^2} - \cdots$$

Since this series converges absolutely, the interval of convergence for the given series is $[4, 6]$. The radius of convergence is $R = 1$ (Figure 11.8.3).   ◀

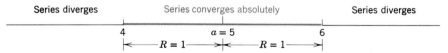

Figure 11.8.3

REMARK.   The ratio test should never be used for testing for convergence at the endpoints of the interval of convergence, since the ratio ρ is always 1 at those points (why?).

▶ **Exercise Set 11.8**

In Exercises 1–24, find the radius of convergence and the interval of convergence.

1. $\sum_{k=0}^{\infty} \frac{x^k}{k+1}$.

2. $\sum_{k=0}^{\infty} 3^k x^k$.

3. $\sum_{k=0}^{\infty} \frac{(-1)^k x^k}{k!}$.

4. $\sum_{k=0}^{\infty} \frac{k!}{2^k} x^k$.

5. $\sum_{k=1}^{\infty} \frac{5^k}{k^2} x^k$.

6. $\sum_{k=2}^{\infty} \frac{x^k}{\ln k}$.

7. $\displaystyle\sum_{k=1}^{\infty} \frac{x^k}{k(k+1)}$.

8. $\displaystyle\sum_{k=0}^{\infty} \frac{(-2)^k x^{k+1}}{k+1}$.

9. $\displaystyle\sum_{k=1}^{\infty} (-1)^{k-1} \frac{x^k}{\sqrt{k}}$.

10. $\displaystyle\sum_{k=0}^{\infty} \frac{(-1)^k x^{2k}}{(2k)!}$.

11. $\displaystyle\sum_{k=0}^{\infty} (-1)^k \frac{x^{2k+1}}{(2k+1)!}$.

12. $\displaystyle\sum_{k=1}^{\infty} (-1)^k \frac{x^{3k}}{k^{3/2}}$.

13. $\displaystyle\sum_{k=0}^{\infty} \frac{3^k}{k!} x^k$.

14. $\displaystyle\sum_{k=2}^{\infty} (-1)^{k+1} \frac{x^k}{k(\ln k)^2}$.

15. $\displaystyle\sum_{k=0}^{\infty} \frac{x^k}{1+k^2}$.

16. $\displaystyle\sum_{k=0}^{\infty} \frac{(x-3)^k}{2^k}$.

17. $\displaystyle\sum_{k=1}^{\infty} (-1)^{k+1} \frac{(x+1)^k}{k}$.

18. $\displaystyle\sum_{k=0}^{\infty} (-1)^k \frac{(x-4)^k}{(k+1)^2}$.

19. $\displaystyle\sum_{k=0}^{\infty} \left(\frac{3}{4}\right)^k (x+5)^k$.

20. $\displaystyle\sum_{k=1}^{\infty} \frac{(2k+1)!}{k^3} (x-2)^k$.

21. $\displaystyle\sum_{k=1}^{\infty} (-1)^k \frac{(x+1)^{2k+1}}{k^2+4}$.

22. $\displaystyle\sum_{k=1}^{\infty} \frac{(\ln k)(x-3)^k}{k}$.

23. $\displaystyle\sum_{k=0}^{\infty} \frac{\pi^k(x-1)^{2k}}{(2k+1)!}$.

24. $\displaystyle\sum_{k=0}^{\infty} \frac{(2x-3)^k}{4^{2k}}$.

In Exercises 25–27, show the first four terms of the series, and find the radius of convergence.

25. $\displaystyle\sum_{k=1}^{\infty} \frac{1 \cdot 2 \cdot 3 \cdots k}{1 \cdot 4 \cdot 7 \cdots (3k-2)} x^k$.

26. $\displaystyle\sum_{k=1}^{\infty} (-1)^k \frac{1 \cdot 2 \cdot 3 \cdots k}{1 \cdot 3 \cdot 5 \cdots (2k-1)} x^{2k+1}$.

27. $\displaystyle\sum_{k=1}^{\infty} \frac{1 \cdot 3 \cdot 5 \cdots (2k-1)}{(2k-2)!} x^k$.

28. Use the root test to find the interval of convergence of $\displaystyle\sum_{k=2}^{\infty} \frac{x^k}{(\ln k)^k}$.

29. Find the interval of convergence of $\displaystyle\sum_{k=0}^{\infty} \frac{(x-a)^k}{b^k}$, where $b > 0$.

30. Find the radius of convergence of the power series $\displaystyle\sum_{k=0}^{\infty} \frac{(pk)!}{(k!)^p} x^k$, where p is a positive integer.

31. Find the radius of convergence of the power series $\displaystyle\sum_{k=0}^{\infty} \frac{(k+p)!}{k!(k+q)!} x^k$, where p and q are positive integers.

32. Prove: If $\lim_{k \to +\infty} |c_k|^{1/k} = L$, where $L \neq 0$, then $1/L$ is the radius of convergence of the power series $\sum_{k=0}^{\infty} c_k x^k$.

33. Prove: If the power series $\sum_{k=0}^{\infty} c_k x^k$ has radius of convergence R, then the series $\sum_{k=0}^{\infty} c_k x^{2k}$ has radius of convergence $\sqrt{R}$.

34. Prove: If the interval of convergence of the series $\sum_{k=0}^{\infty} c_k(x-a)^k$ is $(a-R, a+R]$, then the series converges conditionally at $a+R$.

11.9 TAYLOR AND MACLAURIN SERIES

One of the early applications of calculus was the computation of approximate numerical values for functions such as $\sin x$, $\ln x$, and e^x. One common method for obtaining such values is to approximate the function by a polynomial, then use the polynomial to compute the desired numerical values. In this section we shall discuss procedures for approximating functions by polynomials, and in the next section we shall investigate the errors in these approximations. In this section we shall also see how polynomial approximations lead naturally to the important problem of finding a power series that converges to a specified function.

☐ **APPROXIMATING FUNCTIONS BY POLYNOMIALS**

The problem of primary interest in this section can be phrased informally as follows:

> **PROBLEM.**   *Given a function f and a point a on the x-axis, find a polynomial of specified degree that best approximates the function f in the "vicinity" of the point a.*

As stated, the problem is somewhat vague in that we have not specified any requirements on f such as continuity, differentiability, and so forth, and it is not at all evident what we mean by the "best approximation in the vicinity of a point." However, the problem is suggestive enough to get us started, and we shall resolve the ambiguities as we progress.

Suppose that we are interested in approximating a function f in the vicinity of the point $a = 0$ by a polynomial

$$p(x) = c_0 + c_1 x + \cdots + c_n x^n \tag{1}$$

Because $p(x)$ has $n + 1$ coefficients, it seems reasonable that we should be able to impose $n + 1$ conditions on this polynomial to achieve a good approximation to $f(x)$. Because the point $a = 0$ is the center of interest, our strategy will be to choose the coefficients of $p(x)$ so that the value of p and its first n derivatives are the same as the value of f and its first n derivatives at $a = 0$. By forcing this high degree of "match" at $a = 0$, it is reasonable to hope that $f(x)$ and $p(x)$ will remain close over some interval (possibly quite small) centered at $a = 0$. Thus, we shall assume that f can be differentiated n times at 0, and we shall try to find the coefficients in (1) such that

$$f(0) = p(0), \quad f'(0) = p'(0), \quad f''(0) = p''(0), \quad \ldots, \quad f^{(n)}(0) = p^{(n)}(0) \tag{2}$$

The first n derivatives of (1) are

$$p(x) = c_0 + c_1 x + c_2 x^2 + c_3 x^3 + \cdots + c_n x^n$$
$$p'(x) = c_1 + 2c_2 x + 3c_3 x^2 + \cdots + nc_n x^{n-1}$$
$$p''(x) = 2c_2 + 3 \cdot 2c_3 x + \cdots + n(n-1)c_n x^{n-2}$$
$$p'''(x) = 3 \cdot 2c_3 + \cdots + n(n-1)(n-2)c_n x^{n-3}$$
$$\vdots$$
$$p^{(n)}(x) = n(n-1)(n-2) \cdots (1)c_n$$

from which it follows that

$$p(0) = c_0$$
$$p'(0) = c_1$$
$$p''(0) = 2c_2 = 2!c_2$$
$$p'''(0) = 3 \cdot 2c_3 = 3!c_3$$
$$\vdots$$
$$p^{(n)}(0) = n(n-1)(n-2) \cdots (1)c_n = n!c_n$$

Thus, to satisfy the conditions in (2) we must have

$$f(0) = c_0$$

$$f'(0) = c_1$$

$$f''(0) = 2!c_2$$

$$f'''(0) = 3!c_3$$

$$\vdots$$

$$f^{(n)}(0) = n!c_n$$

which yields the following values for the coefficients of $p(x)$:

$$c_0 = f(0), \quad c_1 = f'(0), \quad c_2 = \frac{f''(0)}{2!}, \quad c_3 = \frac{f'''(0)}{3!}, \quad \ldots, c_n = \frac{f^{(n)}(0)}{n!}$$

☐ MACLAURIN
POLYNOMIALS

The polynomial that results by using these coefficients in (1) is called the *nth Maclaurin* polynomial for f*. In summary, we have the following definition.

11.9.1 DEFINITION.   If f can be differentiated n times at 0, then we define the ***nth Maclaurin polynomial for f*** to be

$$p_n(x) = f(0) + f'(0)x + \frac{f''(0)}{2!}x^2 + \frac{f'''(0)}{3!}x^3 + \cdots + \frac{f^{(n)}(0)}{n!}x^n \qquad (3)$$

This polynomial has the property that its value and the values of its first n derivatives match the value of $f(x)$ and its first n derivatives when $x = 0$.

Example 1   Find the Maclaurin polynomials p_0, p_1, p_2, p_3, and p_n for e^x.

Solution.   Let $f(x) = e^x$. Thus,

$$f'(x) = f''(x) = f'''(x) = \cdots = f^{(n)}(x) = e^x$$

---

* COLIN MACLAURIN(1698–1746). Scottish mathematician. Maclaurin's father, a minister, died when the boy was only six months old, and his mother when he was nine years old. He was then raised by an uncle who was also a minister.

   Maclaurin entered Glasgow University as a divinity student, but transferred to mathematics after one year. He received his Master's degree at age 17 and, in spite of his youth, began teaching at Marischal College in Aberdeen, Scotland.

   Maclaurin met Isaac Newton during a visit to London in 1719 and from that time on he became Newton's disciple. During that era, some of Newton's analytic methods were bitterly attacked by major mathematicians and much of Maclaurin's important mathematical work resulted from his efforts to defend Newton's ideas geometrically. Maclaurin's work, *A Treatise of Fluxions* (1742), was the first systematic formulation of Newton's methods. The treatise was so carefully done that it was a standard of mathematical rigor in calculus until the work of Cauchy in 1821.

   Maclaurin was an outstanding experimentalist. He devised numerous ingenious mechanical devices, made important astronomical observations, performed actuarial computations for insurance societies, and helped to improve maps of the islands around Scotland.

and
$$f(0) = f'(0) = f''(0) = f'''(0) = \cdots = f^{(n)}(0) = e^0 = 1$$

Therefore,

$$p_0(x) = f(0) = 1$$

$$p_1(x) = f(0) + f'(0)x = 1 + x$$

$$p_2(x) = f(0) + f'(0)x + \frac{f''(0)}{2!}x^2 = 1 + x + \frac{x^2}{2!} = 1 + x + \frac{1}{2}x^2$$

$$p_3(x) = f(0) + f'(0)x + \frac{f''(0)}{2!}x^2 + \frac{f'''(0)}{3!}x^3$$

$$= 1 + x + \frac{x^2}{2!} + \frac{x^3}{3!} = 1 + x + \frac{1}{2}x^2 + \frac{1}{6}x^3$$

$$p_n(x) = f(0) + f'(0)x + \frac{f''(0)}{2!}x^2 + \cdots + \frac{f^{(n)}(0)}{n!}x^n$$

$$= 1 + x + \frac{x^2}{2!} + \cdots + \frac{x^n}{n!} \quad \blacktriangleleft$$

Figure 11.9.1 shows the graphs of e^x and its first four Maclaurin polynomials. Note that the graphs of e^x and $p_3(x)$ are virtually indistinguishable over the interval from $-\frac{1}{2}$ to $\frac{1}{2}$. (In the next section we shall investigate in detail the accuracy of Maclaurin polynomial approximations.)

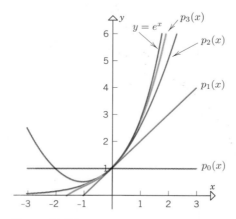

Figure 11.9.1

Example 2   Find the nth Maclaurin polynomial for $\ln(x + 1)$.

Solution.   Let $f(x) = \ln(x + 1)$ and arrange the computations as follows:

$$f(x) = \ln(x + 1) \qquad\qquad f(0) = \ln 1 = 0$$

$$f'(x) \; = \frac{1}{x+1} \qquad\qquad f'(0) \; = 1$$

$$f''(x) \; = -\frac{1}{(x+1)^2} \qquad\qquad f''(0) \; = -1$$

$$f'''(x) \; = \frac{2}{(x+1)^3} \qquad\qquad f'''(0) \; = 2$$

$$f^{(4)}(x) \; = -\frac{3 \cdot 2}{(x+1)^4} \qquad\qquad f^{(4)}(0) \; = -3!$$

$$f^{(5)}(x) \; = \frac{4 \cdot 3 \cdot 2}{(x+1)^5} \qquad\qquad f^{(5)}(0) \; = 4!$$

$$\vdots \qquad\qquad\qquad\qquad \vdots$$

$$f^{(n)}(x) = (-1)^{n+1} \frac{(n-1)!}{(x+1)^n} \qquad f^{(n)}(0) = (-1)^{n+1}(n-1)!$$

Substituting these values in (3) yields

$$p_n(x) = x - \frac{x^2}{2} + \frac{x^3}{3} - \cdots + (-1)^{n+1} \frac{x^n}{n}$$

The graphs of $f(x) = \ln(x+1)$, $p_1(x)$, $p_2(x)$, $p_3(x)$, and $p_4(x)$ are shown in Figure 11.9.2. Observe that the approximations tend to deteriorate rapidly as x moves away from the origin, but that the higher-degree Maclaurin polynomials remain good approximations over larger intervals. ◀

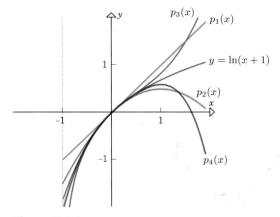

Figure 11.9.2

Example 3   In the Maclaurin polynomials for $\sin x$, only the odd powers of x appear explicitly. To see this, let $f(x) = \sin x$; thus,

$$f(x) \; = \sin x \qquad f(0) \; = 0$$
$$f'(x) \; = \cos x \qquad f'(0) = 1$$
$$f''(x) \; = -\sin x \qquad f''(0) = 0$$
$$f'''(x) \; = -\cos x \qquad f'''(0) = -1$$

Since $f^{(4)}(x) = \sin x = f(x)$, the pattern 0, 1, 0, −1 will repeat over and over as we evaluate successive derivatives at 0. Therefore, the successive Maclaurin polynomials for $\sin x$ are

$$p_1(x) = 0 + x = x$$

$$p_2(x) = 0 + x + 0 = x$$

$$p_3(x) = 0 + x + 0 - \frac{x^3}{3!} = x - \frac{x^3}{3!}$$

$$p_4(x) = 0 + x + 0 - \frac{x^3}{3!} + 0 = x - \frac{x^3}{3!}$$

$$p_5(x) = 0 + x + 0 - \frac{x^3}{3!} + 0 + \frac{x^5}{5!} = x - \frac{x^3}{3!} + \frac{x^5}{5!}$$

$$p_6(x) = 0 + x + 0 - \frac{x^3}{3!} + 0 + \frac{x^5}{5!} + 0 = x - \frac{x^3}{3!} + \frac{x^5}{5!}$$

$$p_7(x) = 0 + x + 0 - \frac{x^3}{3!} + 0 + \frac{x^5}{5!} + 0 - \frac{x^7}{7!} = x - \frac{x^3}{3!} + \frac{x^5}{5!} - \frac{x^7}{7!}$$

$$\vdots$$

In general, the Maclaurin polynomials for $\sin x$ are

$$p_{2n+1}(x) = p_{2n+2}(x) = x - \frac{x^3}{3!} + \frac{x^5}{5!} - \frac{x^7}{7!} + \cdots + (-1)^n \frac{x^{2n+1}}{(2n + 1)!}$$

$$(n = 0, 1, 2, \ldots)$$

The graphs of $f(x) = \sin x$, $p_1(x)$, $p_3(x)$, $p_5(x)$, and $p_7(x)$ are shown in Figure 11.9.3. Observe that $p_7(x)$ is such a good approximation to $\sin x$ over the interval $[-\pi, \pi]$ that its graph is nearly indistinguishable from that of $\sin x$ over this interval. ◄

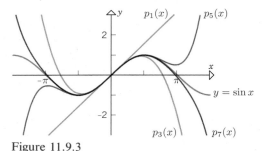

Figure 11.9.3

Example 4   In the Maclaurin polynomials for $\cos x$, only the even powers of x appear explicitly. Using computations similar to those in Example 3, the reader should be able to show that the successive Maclaurin polynomials for $\cos x$ are

$$p_0(x) = p_1(x) = 1$$

$$p_2(x) = p_3(x) = 1 - \frac{x^2}{2!}$$

$$p_4(x) = p_5(x) = 1 - \frac{x^2}{2!} + \frac{x^4}{4!}$$

$$p_6(x) = p_7(x) = 1 - \frac{x^2}{2!} + \frac{x^4}{4!} - \frac{x^6}{6!}$$

$$p_8(x) = p_9(x) = 1 - \frac{x^2}{2!} + \frac{x^4}{4!} - \frac{x^6}{6!} + \frac{x^8}{8!}$$

In general, the Maclaurin polynomials for $\cos x$ are

$$p_{2n}(x) = p_{2n+1}(x) = 1 - \frac{x^2}{2!} + \frac{x^4}{4!} - \cdots + (-1)^n \frac{x^{2n}}{(2n)!}$$

$$(n = 0, 1, 2, \ldots)$$

(See Figure 11.9.4.) ◀

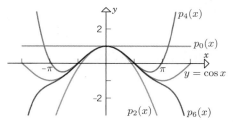

Figure 11.9.4

□ **TAYLOR POLYNOMIALS**   Because the Maclaurin polynomials for a function f are obtained by matching their values and the values of their derivatives with those of f at the point $a = 0$, the accuracy of the Maclaurin polynomials as approximations to f is best in the vicinity of the origin. Generally, the farther x is from the origin, the less accurate the Maclaurin polynomial is as an approximation to f. If we are interested in a polynomial approximation to f that has its best accuracy in the vicinity of a general point a (possibly different from 0), then we should choose the approximating polynomial p so that its value and the values of its derivatives match those of f at the point $x = a$. The computations are simplest if $p(x)$ is expressed in powers of $x - a$, that is,

$$p(x) = c_0 + c_1(x - a) + c_2(x - a)^2 + \cdots + c_n(x - a)^n \tag{4}$$

We leave it as an exercise for the reader to calculate the first n derivatives of $p(x)$ and show that

$$p(a) = c_0, \quad p'(a) = c_1, \quad p''(a) = 2!c_2, \quad p'''(a) = 3!c_3, \quad \ldots, \quad p^{(n)}(a) = n!c_n$$

Thus, if we want the values of $p(x)$ and its first n derivatives to match the values of $f(x)$ and its first n derivatives at $x = a$, we must have

$$c_0 = f(a), \quad c_1 = f'(a), \quad c_2 = \frac{f''(a)}{2!}, \quad c_3 = \frac{f'''(a)}{3!}, \quad \ldots, \quad c_n = \frac{f^{(n)}(a)}{n!}$$

Substituting these values in (4) we obtain a polynomial called the *nth Taylor* polynomial about $x = a$ for f.*

11.9.2   DEFINITION.   If f can be differentiated n times at a, then we define the ***nth Taylor polynomial for f about $x = a$*** to be

$$p_n(x) = f(a) + f'(a)(x - a) + \frac{f''(a)}{2!}(x - a)^2$$

$$+ \frac{f'''(a)}{3!}(x - a)^3 + \cdots + \frac{f^{(n)}(a)}{n!}(x - a)^n \quad (5)$$

REMARK.   Observe that with $a = 0$, the nth Taylor polynomial for f is the nth Maclaurin polynomial for f; that is, the Maclaurin polynomials are special cases of the Taylor polynomials.

Example 5   Find the Taylor polynomials $p_1(x)$, $p_2(x)$, and $p_3(x)$ for $\sin x$ about $x = \pi/3$.

Solution.   Let $f(x) = \sin x$. Thus,

$$f(x) \;\; = \sin x \qquad f(\pi/3) \;\; = \sin \frac{\pi}{3} = \frac{\sqrt{3}}{2}$$

$$f'(x) = \cos x \qquad f'(\pi/3) = \cos \frac{\pi}{3} = \frac{1}{2}$$

$$f''(x) = -\sin x \qquad f''(\pi/3) = -\sin \frac{\pi}{3} = -\frac{\sqrt{3}}{2}$$

---

*BROOK TAYLOR (1685–1731). English mathematician. Taylor was born of well-to-do parents. Musicians and artists were entertained frequently in the Taylor home, which undoubtedly had a lasting influence on young Brook. In later years, Taylor published a definitive work on the mathematical theory of perspective and obtained major mathematical results about the vibrations of strings. There also exists an unpublished work, *On Musick*, that was intended to be part of a joint paper with Isaac Newton.

Taylor's life was scarred with unhappiness, illness, and tragedy. Because his first wife was not rich enough to suit his father, the two men argued bitterly and parted ways. Subsequently, his wife died in childbirth. Then, after he remarried, his second wife also died in childbirth, though his daughter survived.

Taylor's most productive period was from 1714 to 1719, during which time he wrote on a wide range of subjects—magnetism, capillary action, thermometers, perspective, and calculus. In his final years, Taylor devoted his writing efforts to religion and philosophy. According to Taylor, the results that bear his name were motivated by coffeehouse conversations about works of Newton on planetary motion and works of Halley ("Halley's comet") on roots of polynomials.

Taylor's writing style was so terse and hard to understand that he never received credit for many of his innovations.

$$f'''(x) = -\cos x \qquad f'''(\pi/3) = -\cos\frac{\pi}{3} = -\frac{1}{2}$$

Substituting in (5) with $a = \pi/3$ yields

$$p_1(x) = f(\pi/3) + f'(\pi/3)\left(x - \frac{\pi}{3}\right) = \frac{\sqrt{3}}{2} + \frac{1}{2}\left(x - \frac{\pi}{3}\right)$$

$$p_2(x) = f(\pi/3) + f'(\pi/3)\left(x - \frac{\pi}{3}\right) + \frac{f''(\pi/3)}{2!}\left(x - \frac{\pi}{3}\right)^2$$

$$= \frac{\sqrt{3}}{2} + \frac{1}{2}\left(x - \frac{\pi}{3}\right) - \frac{\sqrt{3}}{2\cdot 2!}\left(x - \frac{\pi}{3}\right)^2$$

$$p_3(x) = f(\pi/3) + f'(\pi/3)\left(x - \frac{\pi}{3}\right) + \frac{f''(\pi/3)}{2!}\left(x - \frac{\pi}{3}\right)^2 + \frac{f'''(\pi/3)}{3!}\left(x - \frac{\pi}{3}\right)^3$$

$$= \frac{\sqrt{3}}{2} + \frac{1}{2}\left(x - \frac{\pi}{3}\right) - \frac{\sqrt{3}}{2\cdot 2!}\left(x - \frac{\pi}{3}\right)^2 - \frac{1}{2\cdot 3!}\left(x - \frac{\pi}{3}\right)^3 \quad \blacktriangleleft$$

☐ **SIGMA NOTATION FOR TAYLOR AND MACLAURIN POLYNOMIALS**

Frequently, it is convenient to express the defining formula for the Taylor polynomial in sigma notation. To do this, we use the notation $f^{(k)}(a)$ to denote the kth derivative of f at $x = a$, and we make the added convention that $f^{(0)}(a)$ denotes $f(a)$. This enables us to write

$$\sum_{k=0}^{n} \frac{f^{(k)}(a)}{k!}(x - a)^k = f(a) + f'(a)(x - a)$$

$$+ \frac{f''(a)}{2!}(x - a)^2 + \cdots + \frac{f^{(n)}(a)}{n!}(x - a)^n$$

In particular, the nth Maclaurin polynomial for $f(x)$ may be written as

$$\sum_{k=0}^{n} \frac{f^{(k)}(0)}{k!}x^k = f(0) + f'(0)x + \frac{f''(0)}{2!}x^2 + \cdots + \frac{f^{(n)}(0)}{n!}x^n \qquad (6)$$

☐ **TAYLOR AND MACLAURIN SERIES**

If we assume that the function f has derivatives of all orders at $x = 0$, then the nth Maclaurin polynomial of f can be regarded as a partial sum of the infinite series

$$\sum_{k=0}^{\infty} \frac{f^{(k)}(0)}{k!}x^k = f(0) + f'(0)x + \frac{f''(0)}{2!}x^2 + \cdots + \frac{f^{(k)}(0)}{k!}x^k + \cdots$$

(A similar conclusion can be drawn for the Taylor polynomials about $x = a$.) As n increases, the Maclaurin polynomials that form the partial sums have more and more derivatives whose values match those of f at the origin, so for each fixed x in the vicinity of the origin, it is reasonable to expect that the partial sums will become better and better approximations of $f(x)$ and that they might actually converge to $f(x)$. Should this happen, we would have

$$f(x) = \lim_{n\to+\infty} \sum_{k=0}^{n} \frac{f^{(k)}(0)}{k!}x^k = \sum_{k=0}^{\infty} \frac{f^{(k)}(0)}{k!}x^k$$

which expresses $f(x)$ as the sum of a power series in the vicinity of the origin. A similar conclusion can be drawn for the Taylor polynomials at any point $x = a$. We make the following definitions.

11.9.3 DEFINITION.   If f has derivatives of all orders at a, then we define the *Taylor series for f about x = a* to be

$$\sum_{k=0}^{\infty} \frac{f^{(k)}(a)}{k!}(x - a)^k = f(a) + f'(a)(x - a)$$

$$+ \frac{f''(a)}{2!}(x - a)^2 + \cdots + \frac{f^{(k)}(a)}{k!}(x - a)^k + \cdots \qquad (7)$$

11.9.4 DEFINITION.   If f has derivatives of all orders at 0, then we define the *Maclaurin series for f* to be

$$\sum_{k=0}^{\infty} \frac{f^{(k)}(0)}{k!}x^k = f(0) + f'(0)x + \frac{f''(0)}{2!}x^2 + \cdots + \frac{f^{(k)}(0)}{k!}x^k + \cdots \qquad (8)$$

Observe that the Maclaurin series for f is just the Taylor series for f about $a = 0$.

Example 6 In Example 1, we found the nth Maclaurin polynomial for the function e^x to be

$$\sum_{k=0}^{n} \frac{x^k}{k!} = 1 + x + \frac{x^2}{2!} + \cdots + \frac{x^n}{n!}$$

Thus, the Maclaurin series for e^x is

$$\sum_{k=0}^{\infty} \frac{x^k}{k!} = 1 + x + \frac{x^2}{2!} + \frac{x^3}{3!} + \cdots + \frac{x^n}{n!} + \cdots$$

Similarly, from Example 3 the Maclaurin series for $\sin x$ is

$$\sum_{k=0}^{\infty} (-1)^k \frac{x^{2k+1}}{(2k + 1)!} = x - \frac{x^3}{3!} + \frac{x^5}{5!} - \frac{x^7}{7!} + \cdots$$

and from Example 4 the Maclaurin series for $\cos x$ is

$$\sum_{k=0}^{\infty} (-1)^k \frac{x^{2k}}{(2k)!} = 1 - \frac{x^2}{2!} + \frac{x^4}{4!} - \frac{x^6}{6!} + \cdots \qquad \blacktriangleleft$$

Example 7   Find the Taylor series about $x = 1$ for $1/x$.

Solution. Let $f(x) = 1/x$ so that

$$f(x) = \frac{1}{x} \qquad\qquad f(1) = 1$$

$$f'(x) = -\frac{1}{x^2} \qquad\qquad f'(1) = -1$$

$$f''(x) = \frac{2}{x^3} \qquad\qquad f''(1) = 2!$$

$$f'''(x) = -\frac{3\cdot 2}{x^4} \qquad\qquad f'''(1) = -3!$$

$$f^{(4)}(x) = \frac{4\cdot 3\cdot 2}{x^5} \qquad\qquad f^{(4)}(1) = 4!$$

$$\vdots \qquad\qquad\qquad \vdots$$

$$f^{(k)}(x) = (-1)^k\frac{k!}{x^{k+1}} \qquad f^{(k)}(1) = (-1)^k k!$$

$$\vdots \qquad\qquad\qquad \vdots$$

Thus, substituting in (7) with $a = 1$ yields

$$\sum_{k=0}^{\infty} \frac{(-1)^k k!}{k!}(x-1)^k = \sum_{k=0}^{\infty} (-1)^k(x-1)^k$$

$$= 1 - (x-1) + (x-1)^2 - (x-1)^3 + \cdots \qquad \blacktriangleleft$$

We conclude this section by emphasizing again that we have no guarantee that the Maclaurin and Taylor series of a function f converge, or if they do converge, that the sum is $f(x)$. Convergence questions will be addressed in the next section.

▶ Exercise Set 11.9

In Exercises 1–12, find the fourth Maclaurin polynomial ($n = 4$) for the given function.

1. e^{-2x}.

2. $\dfrac{1}{1+x}$.

3. $\sin 2x$.

4. $e^x \cos x$.

5. $\tan x$.

6. $x^3 - x^2 + 2x + 1$.

7. xe^x.

8. $\tan^{-1} x$.

9. $\sec x$.

10. $\sqrt{1+x}$.

11. $\ln(3 + 2x)$.

12. $\sinh x$.

In Exercises 13–22, find the third Taylor polynomial ($n = 3$) about $x = a$ for the given function.

13. e^x; $a = 1$.

14. $\ln x$; $a = 1$.

15. $\sqrt{x}$; $a = 4$.

16. $x^4 + x - 3$; $a = -2$.

17. $\cos x$; $a = \dfrac{\pi}{4}$.

18. $\tan x$; $a = \dfrac{\pi}{3}$.

19. $\sin \pi x$; $a = -\dfrac{1}{3}$.

20. $\csc x$; $a = \dfrac{\pi}{2}$.

21. $\tan^{-1} x$; $a = 1$.

22. $\cosh x$; $a = \ln 2$.

In Exercises 23–31, find the Maclaurin series for the given function. Express your answer in sigma notation.

23. e^{-x}.

24. e^{ax}.

25. $\dfrac{1}{1 + x}$.

26. xe^x.

27. $\ln(1 + x)$.

28. $\sin \pi x$.

29. $\cos\left(\dfrac{x}{2}\right)$.

30. $\sinh x$.

31. $\cosh x$.

In Exercises 32–39, find the Taylor series about $x = a$ for the given function. Express your answer in sigma notation.

32. $\dfrac{1}{x}$; $a = 3$.

33. $\dfrac{1}{x}$; $a = -1$.

34. e^x; $a = 2$.

35. $\ln x$; $a = 1$.

36. $\cos x$; $a = \pi/2$.

37. $\sin \pi x$; $a = 1/2$.

38. $\dfrac{1}{x + 2}$; $a = 3$.

39. $\sinh x$; $a = \ln 4$.

40. Prove: The value of $p_n(x) = \displaystyle\sum_{k=0}^{n} \dfrac{f^{(k)}(a)}{k!}(x - a)^k$ and its first n derivatives match the value of $f(x)$ and its first n derivatives at $x = a$.

▪ 11.10 TAYLOR FORMULA WITH REMAINDER; CONVERGENCE OF TAYLOR SERIES

In this section we shall analyze the error that results when a function f is approximated by a Taylor or Maclaurin polynomial. This work will enable us to investigate convergence properties of Taylor and Maclaurin series.

☐ **TAYLOR'S THEOREM**

If we approximate a function f by its nth Taylor polynomial p_n, then the error at a point x is represented by the difference $f(x) - p_n(x)$. This difference is commonly called the ***nth remainder*** and is denoted by

$$R_n(x) = f(x) - p_n(x)$$

The following theorem gives an important explicit formula for this remainder.

11.10.1   THEOREM (*Taylor's Theorem*). *Suppose that a function f can be differentiated n + 1 times at each point in an interval containing the point a, and let*

$$p_n(x) = f(a) + f'(a)(x - a) + \frac{f''(a)}{2!}(x - a)^2 + \cdots + \frac{f^{(n)}(a)}{n!}(x - a)^n$$

be the nth Taylor polynomial about x = a for f. Then for each x in the interval, there is at least one point c between a and x such that

$$R_n(x) = f(x) - p_n(x) = \frac{f^{(n+1)}(c)}{(n + 1)!}(x - a)^{n+1} \tag{1}$$

In the foregoing theorem, the statement that c is "between" a and x means that c is in the interval (a, x) if $a < x$, or in (x, a) if $x < a$, or $c = a$ if $x = a$.

Rewriting (1) as

$$f(x) = p_n(x) + \frac{f^{(n+1)}(c)}{(n+1)!}(x - a)^{n+1}$$

then substituting the expression for $p_n(x)$ stated in Theorem 11.10.1 yields the following result, called **Taylor's formula with remainder**.

$$f(x) = f(a) + f'(a)(x - a) + \frac{f''(a)}{2!}(x - a)^2 + \cdots$$
$$+ \frac{f^{(n)}(a)}{n!}(x - a)^n + \frac{f^{(n+1)}(c)}{(n+1)!}(x - a)^{n+1} \qquad (2)$$

□ **LAGRANGE'S FORM OF THE REMAINDER**

This formula expresses $f(x)$ as the sum of its nth Taylor polynomial about $x = a$ plus a remainder or error term,

$$R_n(x) = \frac{f^{(n+1)}(c)}{(n+1)!}(x - a)^{n+1} \qquad (3)$$

in which c is some unspecified number between a and x. The value of c depends on a, x, and n. Historically, (3) was not discovered by Taylor, but rather by Joseph Louis Lagrange.* For this reason, (3) is frequently called **Lagrange's form of the remainder**. Other formulas for the remainder can be found in ad-

---

*JOSEPH LOUIS LAGRANGE (1736–1813). French–Italian mathematician and astronomer. Lagrange, the son of a public official, was born in Turin, Italy. (Baptismal records list his name as Giuseppe Lodovico Lagrangia.) Although his father wanted him to be a lawyer, Lagrange was attracted to mathematics and astronomy after reading a memoir by the astronomer Halley. At age 16 he began to study mathematics on his own and by age 19 was appointed to a professorship at the Royal Artillery School in Turin. The following year Lagrange sent Euler solutions to some famous problems using new methods that eventually blossomed into a branch of mathematics called calculus of variations. These methods and Lagrange's applications of them to problems in celestial mechanics were so monumental that by age 25 he was regarded by many of his contemporaries as the greatest living mathematician.

In 1776, on the recommendation of Euler, he was chosen to succeed Euler as the director of the Berlin Academy. During his stay in Berlin, Lagrange distinguished himself not only in celestial mechanics, but also in algebraic equations and the theory of numbers. After twenty years in Berlin, he moved to Paris at the invitation of Louis XVI. He was given apartments in the Louvre and treated with great honor, even during the revolution.

Napoleon was a great admirer of Lagrange and showered him with honors—count, senator, and Legion of Honor. The years Lagrange spent in Paris were devoted primarily to didactic treatises summarizing his mathematical conceptions. One of Lagrange's most famous works is a memoir, *Mécanique Analytique*, in which he reduced the theory of mechanics to a few general formulas from which all other necessary equations could be derived.

It is an interesting historical fact that Lagrange's father speculated unsuccessfully in several financial ventures, so his family was forced to live quite modestly. Lagrange himself stated that if his family had money, he would not have made mathematics his vocation.

In spite of his fame, Lagrange was always a shy and modest man. On his death, he was buried with honor in the Pantheon.

vanced calculus texts. The proof of Taylor's Theorem will be deferred until the end of the section.

For Maclaurin polynomials ($a = 0$), (2) and (3) become

$$f(x) = f(0) + f'(0)x + \frac{f''(0)}{2!}x^2 + \cdots + \frac{f^{(n)}(0)}{n!}x^n + \frac{f^{(n+1)}(c)}{(n + 1)!}x^{n+1} \quad (4)$$

$$R_n(x) = \frac{f^{(n+1)}(c)}{(n + 1)!}x^{n+1} \quad \text{(where } c \text{ is between 0 and } x) \quad (5)$$

Example 1   From Example 1 of the previous section, the nth Maclaurin polynomial for $f(x) = e^x$ is

$$1 + x + \frac{x^2}{2!} + \cdots + \frac{x^n}{n!}$$

Since $f^{(n+1)}(x) = e^x$, it follows that

$$f^{(n+1)}(c) = e^c$$

Thus, from Taylor's formula with remainder (4)

$$e^x = 1 + x + \frac{x^2}{2!} + \cdots + \frac{x^n}{n!} + \frac{e^c}{(n + 1)!}x^{n+1} \quad (6)$$

where c is between 0 and x. Note that (6) is valid for all real values of x since the hypotheses of Taylor's Theorem are satisfied on the interval $(-\infty, +\infty)$ (verify).   ◀

Example 2   In Example 5 of the previous section, we found the Taylor polynomial of degree 3 about $x = \pi/3$ for $f(x) = \sin x$ to be

$$\frac{\sqrt{3}}{2} + \frac{1}{2}\left(x - \frac{\pi}{3}\right) - \frac{\sqrt{3}}{4}\left(x - \frac{\pi}{3}\right)^2 - \frac{1}{12}\left(x - \frac{\pi}{3}\right)^3$$

Since $f^{(4)}(x) = \sin x$ (verify), it follows that $f^{(4)}(c) = \sin c$. Thus, from Taylor's formula with remainder (2) in the case $a = \pi/3$ and $n = 3$ we obtain

$$\sin x = \frac{\sqrt{3}}{2} + \frac{1}{2}\left(x - \frac{\pi}{3}\right) - \frac{\sqrt{3}}{4}\left(x - \frac{\pi}{3}\right)^2 - \frac{1}{12}\left(x - \frac{\pi}{3}\right)^3 + \frac{\sin c}{4!}\left(x - \frac{\pi}{3}\right)^4$$

where c is some number between $\pi/3$ and x.   ◀

□ CONVERGENCE OF
TAYLOR SERIES

In the previous section we raised the problem of finding those values of x, if any, for which the Taylor series for f about $x = a$ converges to $f(x)$, that is,

$$f(x) = f(a) + f'(a)(x - a) + \frac{f''(a)}{2!}(x - a)^2$$

$$+ \cdots + \frac{f^{(k)}(a)}{k!}(x - a)^k + \cdots \quad (7)$$

To solve this problem, let us write (7) in sigma notation:

$$f(x) = \sum_{k=0}^{\infty} \frac{f^{(k)}(a)}{k!}(x - a)^k$$

This equality is equivalent to

$$f(x) = \lim_{n \to +\infty} \sum_{k=0}^{n} \frac{f^{(k)}(a)}{k!}(x - a)^k$$

or

$$\lim_{n \to +\infty} \left[f(x) - \sum_{k=0}^{n} \frac{f^{(k)}(a)}{k!}(x - a)^k \right] = 0 \quad (8)$$

But the bracketed expression in (8) is $R_n(x)$, since it is the difference between $f(x)$ and its nth Taylor polynomial about $x = a$. Thus, (8) can be written as

$$\lim_{n \to +\infty} R_n(x) = 0$$

which leads us to the following result.

11.10.2 THEOREM. *The equality*

$$f(x) = \sum_{k=0}^{\infty} \frac{f^{(k)}(a)}{k!}(x - a)^k$$

holds if and only if $\lim_{n \to +\infty} R_n(x) = 0$.

To paraphrase this theorem, the *Taylor series for f converges to f(x) at precisely those points where the remainder approaches zero.*

Example 3   Show that the Maclaurin series for e^x converges to e^x for all x.

Solution.   We want to show that

$$e^x = 1 + x + \frac{x^2}{2!} + \frac{x^3}{3!} + \cdots + \frac{x^n}{n!} + \cdots \quad (9)$$

holds for all x. From (6),

$$e^x = 1 + x + \frac{x^2}{2!} + \cdots + \frac{x^n}{n!} + R_n(x)$$

where

$$R_n(x) = \frac{e^c}{(n + 1)!} x^{n+1}$$

and c is between 0 and x. Thus, we must show that for all x

$$\lim_{n \to +\infty} R_n(x) = \lim_{n \to +\infty} \frac{e^c}{(n + 1)!} x^{n+1} = 0 \qquad (10)$$

(Keep in mind that c is a function of x, not a constant.) Our proof of this result will hinge on the limit

$$\lim_{n \to +\infty} \frac{x^{n+1}}{(n + 1)!} = 0 \qquad (11)$$

which follows from Formula (1) of Section 11.8. We shall consider three cases: $x > 0$, $x < 0$, and $x = 0$. If $x > 0$, then

$$0 < c < x$$

from which it follows that

$$0 < e^c < e^x$$

and consequently

$$0 < \frac{e^c}{(n + 1)!} x^{n+1} < \frac{e^x}{(n + 1)!} x^{n+1} \qquad (12)$$

From (11) we obtain

$$\lim_{n \to +\infty} \frac{e^x}{(n + 1)!} x^{n+1} = e^x \lim_{n \to +\infty} \frac{x^{n+1}}{(n + 1)!} = e^x \cdot 0 = 0$$

Thus, from (12) and the Squeezing Theorem (2.8.2), result (10) follows.

In the case $x < 0$, we have $c < 0$ since c is between 0 and x. Thus, $0 < e^c < 1$ and consequently

$$0 < e^c \left| \frac{x^{n+1}}{(n + 1)!} \right| < \left| \frac{x^{n+1}}{(n + 1)!} \right|$$

or

$$0 < \left| \frac{e^c}{(n + 1)!} x^{n+1} \right| < \left| \frac{x^{n+1}}{(n + 1)!} \right|$$

or

$$0 < |R_n(x)| < \left| \frac{x^{n+1}}{(n+1)!} \right|$$

From (11) and the Squeezing Theorem it follows that $\lim\limits_{n \to +\infty} |R_n(x)| = 0$, and so $\lim\limits_{n \to +\infty} R_n(x) = 0$.

Convergence in the case $x = 0$ is obvious since (9) reduces to

$$e^0 = 1 + 0 + 0 + 0 + \cdots \quad \blacktriangleleft$$

Example 4   Show that the Maclaurin series for $\sin x$ converges to $\sin x$ for all x.

Solution.   We must show that

$$\sin x = x - \frac{x^3}{3!} + \frac{x^5}{5!} - \frac{x^7}{7!} + \cdots$$

holds for all x. Let $f(x) = \sin x$, so that for all x either

$$f^{(n+1)}(x) = \pm \cos x \quad \text{or} \quad f^{(n+1)}(x) = \pm \sin x$$

In all of these cases $|f^{(n+1)}(x)| \leq 1$, so that for all possible values of c

$$|f^{(n+1)}(c)| \leq 1$$

Hence,

$$0 \leq |R_n(x)| = \left| \frac{f^{(n+1)}(c)}{(n+1)!} x^{n+1} \right| \leq \frac{|x|^{n+1}}{(n+1)!} \tag{13}$$

From (11) with $|x|$ replacing x,

$$\lim_{n \to +\infty} \frac{|x|^{n+1}}{(n+1)!} = 0$$

so from (13) and the Squeezing Theorem

$$\lim_{n \to +\infty} |R_n(x)| = 0$$

and consequently

$$\lim_{n \to +\infty} R_n(x) = 0$$

for all x. $\blacktriangleleft$

Example 5   Find the Taylor series for $\sin x$ about $x = \pi/2$ and show that the series converges to $\sin x$ for all x.

Solution. Let $f(x) = \sin x$. Thus,

$$f(x) = \sin x \qquad f\left(\frac{\pi}{2}\right) = \sin\frac{\pi}{2} = 1$$

$$f'(x) = \cos x \qquad f'\left(\frac{\pi}{2}\right) = \cos\frac{\pi}{2} = 0$$

$$f''(x) = -\sin x \qquad f''\left(\frac{\pi}{2}\right) = -\sin\frac{\pi}{2} = -1$$

$$f'''(x) = -\cos x \qquad f'''\left(\frac{\pi}{2}\right) = -\cos\frac{\pi}{2} = 0$$

Since $f^{(4)}(x) = \sin x$, the pattern $1, 0, -1, 0$ will repeat over and over as we evaluate successive derivatives at $\pi/2$. Thus, the Taylor series representation about $x = \pi/2$ for $\sin x$ is

$$\sin x = 1 - \frac{1}{2!}\left(x - \frac{\pi}{2}\right)^2 + \frac{1}{4!}\left(x - \frac{\pi}{2}\right)^4 - \frac{1}{6!}\left(x - \frac{\pi}{2}\right)^6 + \cdots \qquad (14)$$

which we are trying to show is valid for all x. As shown in Example 4, $|f^{(n+1)}(c)| \leq 1$, so

$$0 \leq |R_n(x)| = \left|\frac{f^{(n+1)}(c)}{(n+1)!}\left(x - \frac{\pi}{2}\right)^{n+1}\right| \leq \frac{\left|x - \frac{\pi}{2}\right|^{n+1}}{(n+1)!}$$

From (11) with $|x - \pi/2|$ replacing x it follows that

$$\lim_{n \to +\infty} \frac{\left|x - \frac{\pi}{2}\right|^{n+1}}{(n+1)!} = 0$$

so that by the same argument given in Example 4, $\lim_{n \to +\infty} R_n(x) = 0$ for all x. This shows that (14) is valid for all x. ◄

REMARK. It can be shown that the Taylor series for e^x, $\sin x$, and $\cos x$ about any point $x = a$ converges to these functions for all x.

□ **CONSTRUCTING MACLAURIN SERIES BY SUBSTITUTION**

Sometimes Maclaurin series can be obtained by substituting in other Maclaurin series.

Example 6 Using the Maclaurin series

$$e^x = 1 + x + \frac{x^2}{2!} + \frac{x^3}{3!} + \frac{x^4}{4!} + \cdots \qquad -\infty < x < +\infty$$

we can derive the Maclaurin series for e^{-x} by substituting $-x$ for x to obtain

$$e^{-x} = 1 + (-x) + \frac{(-x)^2}{2!} + \frac{(-x)^3}{3!} + \frac{(-x)^4}{4!} + \cdots \qquad -\infty < -x < +\infty$$

or

$$e^{-x} = 1 - x + \frac{x^2}{2!} - \frac{x^3}{3!} + \frac{x^4}{4!} - \cdots \qquad -\infty < x < +\infty$$

From the Maclaurin series for e^x and e^{-x} we can obtain the Maclaurin series for $\cosh x$ by writing

$$\cosh x = \frac{1}{2}(e^x + e^{-x}) = \frac{1}{2}\left(\left[1 + x + \frac{x^2}{2!} + \frac{x^3}{3!} + \frac{x^4}{4!} + \cdots\right]\right.$$
$$\left.+ \left[1 - x + \frac{x^2}{2!} - \frac{x^3}{3!} + \frac{x^4}{4!} + \cdots\right]\right)$$

or

$$\cosh x = 1 + \frac{x^2}{2!} + \frac{x^4}{4!} + \cdots \qquad -\infty < x < +\infty \qquad \blacktriangleleft$$

REMARK.   We could have derived the foregoing Maclaurin series directly. However, sometimes indirect methods are useful when it is messy to calculate the higher derivatives required for a Maclaurin series. There is, however, a loose thread in the logic of Example 6. We have produced a power series in x that converges to $\cosh x$ for all x. But isn't it conceivable that we have produced a power series in x *different* from the Maclaurin series? In Section 11.12 we shall show that if a power series in $x - a$ converges to $f(x)$ on some interval containing a, then the series must be the Taylor series for f about $x = a$. Thus, we are assured the series obtained for $\cosh x$ is, in fact, the Maclaurin series.

Example 7   Using the Maclaurin series

$$\frac{1}{1 - x} = 1 + x + x^2 + x^3 + \cdots \qquad -1 < x < 1$$

we can derive the Maclaurin series for $1/(1 - 2x^2)$ by substituting $2x^2$ for x to obtain

$$\frac{1}{1 - 2x^2} = 1 + (2x^2) + (2x^2)^2 + (2x^2)^3 + \cdots \qquad -1 < 2x^2 < 1$$

or

$$\frac{1}{1 - 2x^2} = 1 + 2x^2 + 4x^4 + 8x^6 + \cdots = \sum_{k=0}^{\infty} 2^k x^{2k} \qquad -1 < 2x^2 < 1$$

Since $2x^2 \geq 0$ for all x, the convergence condition $-1 < 2x^2 < 1$ can be written in the equivalent form $0 \leq 2x^2 < 1$ or $0 \leq x^2 < 1/2$ or $-1/\sqrt{2} < x < 1/\sqrt{2}$.

$\blacktriangleleft$

□ **BINOMIAL SERIES**

If m is a real number, then the Maclaurin series for $(1 + x)^m$ is called the **binomial series;** it is given by (verify)

$$1 + mx + \frac{m(m - 1)}{2!} x^2 + \frac{m(m - 1)(m - 2)}{3!} x^3 + \cdots$$

REMARK.   If m is a nonnegative integer, then $f(x) = (1 + x)^m$ is a polynomial of degree m, so

$$f^{(m+1)}(0) = f^{(m+2)}(0) = f^{(m+3)}(0) = \cdots = 0$$

and the binomial series reduces to the familiar binomial expansion

$$(1 + x)^m = 1 + mx + \frac{m(m-1)}{2!}x^2 + \frac{m(m-1)(m-2)}{3!}x^3 + \cdots + x^m$$

which is valid for $-\infty < x < +\infty$.

It is proved in advanced calculus that if m is not a nonnegative integer, then the binomial series converges to $(1 + x)^m$ if $|x| < 1$. Thus, for such values of x

$$(1 + x)^m = 1 + mx + \frac{m(m-1)}{2!}x^2 + \cdots$$
$$+ \frac{m(m-1)(m-2)\cdots(m-k+1)}{k!}x^k + \cdots \qquad (15)$$

or in sigma notation

$$(1 + x)^m = 1 + \sum_{k=1}^{\infty} \frac{m(m-1)\cdots(m-k+1)}{k!}x^k \quad \text{if } |x| < 1 \qquad (16)$$

Example 8   Express $1/\sqrt{1 + x}$ as a binomial series.

Solution.   Substituting $m = -\frac{1}{2}$ in (15) yields

$$\frac{1}{\sqrt{1 + x}} = 1 - \frac{1}{2}x + \frac{(-\frac{1}{2})(-\frac{1}{2}-1)}{2!}x^2 + \frac{(-\frac{1}{2})(-\frac{1}{2}-1)(-\frac{1}{2}-2)}{3!}x^3$$
$$+ \cdots + \frac{(-\frac{1}{2})(-\frac{3}{2})(-\frac{5}{2})\cdots(-\frac{1}{2}-k+1)}{k!}x^k + \cdots$$

$$= 1 - \frac{1}{2}x + \frac{1\cdot 3}{2^2\cdot 2!}x^2 - \frac{1\cdot 3\cdot 5}{2^3\cdot 3!}x^3 + \cdots$$
$$+ (-1)^k\frac{1\cdot 3\cdot 5\cdots(2k-1)}{2^k k!}x^k + \cdots \qquad \blacktriangleleft$$

For reference, we have listed in Table 11.10.1 the Maclaurin series for a number of important functions, and we have indicated the interval over which the series converges to the function. It should be noted that the intervals of convergence stated for $\ln(1 + x)$ and $\tan^{-1}x$ are somewhat difficult to obtain directly. However, these intervals can be obtained by indirect methods that we shall study in the last section of this chapter.

Table 11.10.1

MACLAURIN SERIES	INTERVAL OF VALIDITY
$\dfrac{1}{1-x} = \displaystyle\sum_{k=0}^{\infty} x^k = 1 + x + x^2 + x^3 + \cdots$	$-1 < x < 1$
$e^x = \displaystyle\sum_{k=0}^{\infty} \dfrac{x^k}{k!} = 1 + x + \dfrac{x^2}{2!} + \dfrac{x^3}{3!} + \dfrac{x^4}{4!} + \cdots$	$-\infty < x < +\infty$
$\sin x = \displaystyle\sum_{k=0}^{\infty} (-1)^k \dfrac{x^{2k+1}}{(2k+1)!} = x - \dfrac{x^3}{3!} + \dfrac{x^5}{5!} - \dfrac{x^7}{7!} + \cdots$	$-\infty < x < +\infty$
$\cos x = \displaystyle\sum_{k=0}^{\infty} (-1)^k \dfrac{x^{2k}}{(2k)!} = 1 - \dfrac{x^2}{2!} + \dfrac{x^4}{4!} - \dfrac{x^6}{6!} + \cdots$	$-\infty < x < +\infty$
$\ln(1+x) = \displaystyle\sum_{k=0}^{\infty} (-1)^k \dfrac{x^{k+1}}{k+1} = x - \dfrac{x^2}{2} + \dfrac{x^3}{3} - \dfrac{x^4}{4} + \cdots$	$-1 < x \le 1$
$\tan^{-1} x = \displaystyle\sum_{k=0}^{\infty} (-1)^k \dfrac{x^{2k+1}}{2k+1} = x - \dfrac{x^3}{3} + \dfrac{x^5}{5} - \dfrac{x^7}{7} + \cdots$	$-1 \le x \le 1$
$\sinh x = \displaystyle\sum_{k=0}^{\infty} \dfrac{x^{2k+1}}{(2k+1)!} = x + \dfrac{x^3}{3!} + \dfrac{x^5}{5!} + \dfrac{x^7}{7!} + \cdots$	$-\infty < x < +\infty$
$\cosh x = \displaystyle\sum_{k=0}^{\infty} \dfrac{x^{2k}}{(2k)!} = 1 + \dfrac{x^2}{2!} + \dfrac{x^4}{4!} + \dfrac{x^6}{6!} + \cdots$	$-\infty < x < +\infty$
$(1+x)^m = 1 + \displaystyle\sum_{k=1}^{\infty} \dfrac{m(m-1)\cdots(m-k+1)}{k!} x^k$	$-1 < x < 1$

■ OPTIONAL

We conclude this section with a proof of Taylor's Theorem.

Proof of Theorem 11.10.1. By hypothesis, f can be differentiated $n + 1$ times at each point in an interval containing the point a. Choose any point b in this interval. To be specific, we shall assume that $b > a$. (The cases $b < a$ and $b = a$ are left to the reader.) Let $p_n(x)$ be the nth Taylor polynomial for $f(x)$ about $x = a$ and define

$$h(x) = f(x) - p_n(x) \tag{17}$$

$$g(x) = (x - a)^{n+1} \tag{18}$$

Because $f(x)$ and $p_n(x)$ have the same value and the same first n derivatives at $x = a$, it follows that

$$h(a) = h'(a) = h''(a) = \cdots = h^{(n)}(a) = 0 \tag{19}$$

Also, we leave it for the reader to show that

$$g(a) = g'(a) = g''(a) = \cdots = g^{(n)}(a) = 0 \tag{20}$$

and that $g(x)$ and its first n derivatives are nonzero when $x \neq a$.

It is straightforward to check that h and g satisfy the hypotheses of the Extended Mean-Value Theorem (10.2.2) on the interval $[a, b]$, so that there is a point c_1 with $a < c_1 < b$ such that

$$\frac{h(b) - h(a)}{g(b) - g(a)} = \frac{h'(c_1)}{g'(c_1)} \tag{21}$$

or, from (19) and (20),

$$\frac{h(b)}{g(b)} = \frac{h'(c_1)}{g'(c_1)} \tag{22}$$

If we now apply the Extended Mean-Value Theorem to h' and g' over the interval $[a, c_1]$, we may deduce that there is a point c_2 with $a < c_2 < c_1 < b$ such that

$$\frac{h'(c_1) - h'(a)}{g'(c_1) - g'(a)} = \frac{h''(c_2)}{g''(c_2)}$$

or, from (19) and (20),

$$\frac{h'(c_1)}{g'(c_1)} = \frac{h''(c_2)}{g''(c_2)}$$

which, when combined with (22), yields

$$\frac{h(b)}{g(b)} = \frac{h''(c_2)}{g''(c_2)}$$

It should now be clear that if we continue in this way, applying the Extended Mean-Value Theorem to the successive derivatives of h and g, we shall eventually obtain a relationship of the form

$$\frac{h(b)}{g(b)} = \frac{h^{(n+1)}(c_{n+1})}{g^{(n+1)}(c_{n+1})} \tag{23}$$

where $a < c_{n+1} < b$. However, $p_n(x)$ is a polynomial of degree n, so that its $(n + 1)$-st derivative is zero. Thus, from (17)

$$h^{(n+1)}(c_{n+1}) = f^{(n+1)}(c_{n+1}) \tag{24}$$

Also, from (18), the $(n + 1)$-st derivative of $g(x)$ is the constant $(n + 1)!$, so that

$$g^{(n+1)}(c_{n+1}) = (n + 1)! \tag{25}$$

Substituting (24) and (25) into (23) yields

$$\frac{h(b)}{g(b)} = \frac{f^{(n+1)}(c_{n+1})}{(n+1)!}$$

Letting $c = c_{n+1}$, and using (17) and (18), it follows that

$$f(b) - p_n(b) = \frac{f^{(n+1)}(c)}{(n+1)!}(b-a)^{n+1}$$

But this is precisely Formula (1) in Taylor's Theorem (Theorem 11.10.1), with the exception that the variable here is b rather than x. Thus, to finish we need only replace b by x. ■

▶ **Exercise Set 11.10**

For the functions in Exercises 1–14, find Lagrange's form of the remainder for the given values of a and n.

1. e^{2x}; $a = 0$; $n = 5$.   2. $\cos x$; $a = 0$; $n = 8$.

3. $\dfrac{1}{x+1}$; $a = 0$; $n = 4$.   4. $\tan x$; $a = 0$; $n = 2$.

5. xe^x; $a = 0$; $n = 3$.

6. $\ln(1+x)$; $a = 0$; $n = 5$.

7. $\tan^{-1}x$; $a = 0$; $n = 2$.   8. $\sinh x$; $a = 0$; $n = 6$.

9. $\sqrt{x}$; $a = 4$; $n = 3$.   10. $\dfrac{1}{x}$; $a = 1$; $n = 5$.

11. $\sin x$; $a = \dfrac{\pi}{6}$; $n = 4$.   12. $\cos \pi x$; $a = \dfrac{1}{2}$; $n = 2$.

13. $\dfrac{1}{(1+x)^2}$; $a = -2$; $n = 5$.

14. $\csc x$; $a = \dfrac{\pi}{2}$; $n = 1$.

In Exercises 15–18, find Lagrange's form of the remainder $R_n(x)$ when $a = 0$.

15. $f(x) = \dfrac{1}{1-x}$.   16. $f(x) = e^{-x}$.

17. $f(x) = e^{2x}$.   18. $f(x) = \ln(1+x)$.

19. Prove: The Maclaurin series for $\cos x$ converges to $\cos x$ for all x.

20. Prove: The Taylor series for $\sin x$ about $x = \pi/4$ converges to $\sin x$ for all x.

21. Prove: The Taylor series for e^x about $x = 1$ converges to e^x for all x.

22. (a) Prove: The Maclaurin series for $\ln(1+x)$ converges to $\ln(1+x)$ if $0 \le x \le 1$. [*Remark:* The

Maclaurin series actually converges to $\ln(1+x)$ on the interval $(-1, 1]$, but the Lagrange form of the remainder is not strong enough to establish this fact.]

(b) Use $x = 1$ in the Maclaurin series for $\ln(1+x)$, Table 11.10.1, to show that

$$\ln 2 = 1 - \frac{1}{2} + \frac{1}{3} - \frac{1}{4} + \cdots$$

23. Prove: The Taylor series for e^x about any point $x = a$ converges to e^x for all x.

24. Prove: The Taylor series for $\sin x$ about any point $x = a$ converges to $\sin x$ for all x.

25. Prove: The Taylor series for $\cos x$ about any point $x = a$ converges to $\cos x$ for all x.

26. Derive (15). (Do not try to prove convergence.)

In Exercises 27–50, use the Maclaurin series in Table 11.10.1 to obtain the Maclaurin series for the given function. In each case give the first four terms of the series and specify the interval on which the series converges to the function.

27. e^{-2x}.   28. $x^2 e^x$.

29. xe^{-x}.   30. e^{x^2}.

31. $\sin 2x$.   32. $\cos 2x$.

33. $x^2 \cos x$.   34. $\sin(x^2)$.

35. $\sin^2 x$. [*Hint:* $\sin^2 x = \frac{1}{2}(1 - \cos 2x)$.]

36. $\cos^2 x$. [*Hint:* $\cos^2 x = \frac{1}{2}(1 + \cos 2x)$.]

37. $\ln(1 - x^2)$.   38. $\ln(1 + 2x)$.

39. $\dfrac{1}{1 - 4x^2}$.   40. $\dfrac{x}{1 - x}$.

41. $\dfrac{x^2}{1 + 3x}$.

42. $\dfrac{x}{1 + x^2}$.

43. $x \sinh 2x$.

44. $\cosh (x^2)$.

45. $\sqrt{1 + 3x}$.

46. $\sqrt{1 + x^2}$.

47. $\dfrac{1}{(1 - 2x)^2}$.

48. $\dfrac{x}{(1 + 2x)^3}$.

49. $\dfrac{x}{\sqrt{1 - x^2}}$.

50. $x(1 - x^2)^{3/2}$.

51. Use the Maclaurin series for $1/(1 - x)$ to express $1/x$ in powers of $x - 1$. Find the interval of convergence. $\left[Hint: \dfrac{1}{x} = \dfrac{1}{1 + (x - 1)}. \right]$

52. Show that the series

$$1 - \frac{x}{2!} + \frac{x^2}{4!} - \frac{x^3}{6!} + \cdots$$

converges to the function

$$f(x) = \begin{cases} \cos \sqrt{x}, & x \geq 0 \\ \cosh \sqrt{-x}, & x < 0 \end{cases}$$

[*Hint:* Use the Maclaurin series for $\cos x$ and $\cosh x$ to obtain series for $\cos \sqrt{x}$ where $x \geq 0$, and $\cosh \sqrt{-x}$ where $x \leq 0$.]

53. If m is any real number, and k is a nonnegative integer, then we define the **binomial coefficients** $\dbinom{m}{k}$ by the formulas $\dbinom{m}{0} = 1$ and

$$\binom{m}{k} = \frac{m(m - 1)(m - 2) \cdots (m - k + 1)}{k!}$$

for $k \geq 1$.

Express Formula (16) in terms of binomial coefficients.

In Exercises 54–57, use the known Maclaurin series for e^x, $\sin x$, and $\cos x$ to help find the sum of the given series.

54. $2 + \dfrac{4}{2!} + \dfrac{8}{3!} + \dfrac{16}{4!} + \cdots$.

55. $\pi - \dfrac{\pi^3}{3!} + \dfrac{\pi^5}{5!} - \dfrac{\pi^7}{7!} + \cdots$.

56. $1 - \dfrac{e^2}{2!} + \dfrac{e^4}{4!} - \dfrac{e^6}{6!} + \cdots$.

57. $1 - \ln 3 + \dfrac{(\ln 3)^2}{2!} - \dfrac{(\ln 3)^3}{3!} + \cdots$.

58. (a) Use the Maclaurin series for $\dfrac{1}{1 - x}$ to help find the Maclaurin series for the function

$$f(x) = \frac{x}{1 - x^2}$$

(b) Use the result in part (a) to help find $f^{(5)}(0)$ and $f^{(6)}(0)$.

59. (a) Use the Maclaurin series for $\cos x$ to find the Maclaurin series for $f(x) = x^2 \cos 2x$.

(b) Use the result in part (a) to help find $f^{(5)}(0)$.

60. The purpose of this exercise is to show that the Taylor series of a function f may possibly converge to a value different from $f(x)$ for certain x. Let

$$f(x) = \begin{cases} e^{-1/x^2}, & x \neq 0 \\ 0, & x = 0 \end{cases}$$

(a) Use the definition of a derivative to show that $f'(0) = 0$.

(b) With some difficulty it can be shown that $f^{(n)}(0) = 0$ for $n \geq 2$. Accepting this fact, show that the Maclaurin series of f converges for all x, but converges to $f(x)$ only at the point $x = 0$.

■ 11.11 COMPUTATIONS USING TAYLOR SERIES

In this section we shall show how Taylor and Maclaurin series can be used to obtain approximate values for trigonometric functions and logarithms.

□ **AN UPPER BOUND ON**
THE REMAINDER

In the previous section we showed that for all x

$$e^x = 1 + x + \frac{x^2}{2!} + \frac{x^3}{3!} + \frac{x^4}{4!} + \cdots + \frac{x^k}{k!} + \cdots$$

In particular, if we let $x = 1$, we obtain the following expression for e as the sum of an infinite series

$$e = 1 + 1 + \frac{1}{2!} + \frac{1}{3!} + \frac{1}{4!} + \cdots + \frac{1}{k!} + \cdots$$

Thus, we can approximate e to any degree of accuracy using an appropriate partial sum

$$e \approx 1 + 1 + \frac{1}{2!} + \frac{1}{3!} + \cdots + \frac{1}{n!} \tag{1}$$

The value of n in this formula is at our disposal—the larger we choose n the more accurate the approximation. This is illustrated in Table 11.11.1, where we have used (1) to approximate e for various values of n. To nine decimal-place accuracy, the value of e is 2.718281828. This level of accuracy is obtained with $n = 13$.

Table 11.11.1

n	$1 + 1 + \dfrac{1}{2!} + \dfrac{1}{3!} + \cdots + \dfrac{1}{n!}$
0	1.000000000
1	2.000000000
3	2.500000000
4	2.666666667
5	2.708333333
10	2.718281526
11	2.718281801

In practical applications where approximate values of e are needed, one would be interested in approximating e to a specified degree of accuracy, in which case it would be important to determine how large n should be chosen to attain the desired accuracy. For example, how large should n be taken in (1) to ensure an error that is less than 0.00005? We shall now show how Lagrange's remainder formula can be used to answer such questions.

According to Taylor's Theorem, the absolute value of the error that results when $f(x)$ is approximated by its nth Taylor polynomial $p_n(x)$ about $x = a$ is

$$|R_n(x)| = |f(x) - p_n(x)| = \left| \frac{f^{(n+1)}(c)}{(n+1)!} (x - a)^{n+1} \right| \tag{2}$$

In this formula, c is an unknown number between a and x, so that the value of $f^{(n+1)}(c)$ usually cannot be determined. However, it is frequently possible to determine an upper bound on the size of $|f^{(n+1)}(c)|$; that is, one can often find a constant M such that $|f^{(n+1)}(c)| \leq M$. For such an M, it follows from (2) that

$$|R_n(x)| = |f(x) - p_n(x)| \leq \frac{M}{(n+1)!}|x - a|^{n+1} \tag{3}$$

which gives an upper bound on the magnitude of the error $R_n(x)$. In the examples to follow we shall give applications of this formula, but first it will be helpful to introduce some terminology relating to the accuracy of approximations.

An approximation is said to be **accurate to n decimal places** if the magnitude of the error is less than 0.5×10^{-n}. For example,

DESCRIPTION	MAGNITUDE OF THE ERROR IS LESS THAN	
1 decimal-place accuracy	0.05	$= 0.5 \times 10^{-1}$
2 decimal-place accuracy	0.005	$= 0.5 \times 10^{-2}$
3 decimal-place accuracy	0.0005	$= 0.5 \times 10^{-3}$
4 decimal-place accuracy	0.00005	$= 0.5 \times 10^{-4}$

□ APPROXIMATING e

Example 1   Use (1) to approximate e to four decimal-place accuracy.

Solution.   It follows from Taylor's formula with remainder that

$$e^x = 1 + x + \frac{x^2}{2!} + \cdots + \frac{x^n}{n!} + \frac{e^c}{(n+1)!}x^{n+1}$$

where c is between 0 and x (see Example 1, Section 11.10). Thus, in the case $x = 1$ we obtain

$$e = 1 + 1 + \frac{1}{2!} + \cdots + \frac{1}{n!} + \frac{e^c}{(n+1)!}$$

where c is between 0 and 1. This tells us that the magnitude of the error in approximation (1) is

$$|R_n| = \left|\frac{e^c}{(n+1)!}\right| = \frac{e^c}{(n+1)!} \tag{4}$$

where $0 < c < 1$. Since $c < 1$, it follows that

$$e^c < e^1 = e$$

so that from (4)

$$|R_n| < \frac{e}{(n+1)!} \tag{5}$$

This inequality provides an upper bound on the magnitude of the error R_n. Unfortunately, this inequality is not very useful since the right side involves the quantity e, which we are trying to estimate. However, if we use the fact that $e < 3$, then we can replace (5) with the following less precise but more useful result:

$$|R_n| < \frac{3}{(n + 1)!} \tag{5a}$$

It follows that if we choose n so that

$$|R_n| < \frac{3}{(n + 1)!} < 0.5 \times 10^{-4} = 0.00005 \tag{6}$$

then approximation (1) will be accurate to four decimal places. An appropriate value for n may be found by trial and error. For example, using a calculator one can evaluate $3/(n + 1)!$ for $n = 0, 1, 2, \ldots$ until a value of n satisfying (6) is obtained. We leave it for the reader to show that $n = 8$ is the first positive integer satisfying (6). Thus, to four decimal-place accuracy

$$e \approx 1 + 1 + \frac{1}{2!} + \frac{1}{3!} + \frac{1}{4!} + \frac{1}{5!} + \frac{1}{6!} + \frac{1}{7!} + \frac{1}{8!} \approx 2.7183 \tag{7}$$

◀

REMARK.   It should be noted that there are two types of errors that result when computing with series. The first, called **truncation error,** is the error that results when the entire series is approximated by a partial sum. The second kind of error, called **roundoff error,** results when decimal approximations are used. For example, (7) involves a truncation error of at most 0.5×10^{-4} (four decimal-place accuracy). However, to obtain the numerical value in (7) we used a calculator to evaluate the left side, resulting in roundoff error due to the limitations of the calculator. The problem of controlling roundoff error is surprisingly difficult and is studied in courses in a branch of mathematics called **numerical analysis.** As a rule of thumb, to achieve n decimal-place accuracy in a final result, one should use more than $n + 1$ decimal places in each intermediate computation, then round off to n decimal places at the end. However, even this procedure may occasionally not produce n decimal-place accuracy. For purposes of this text, we recommend that you perform all intermediate calculations with the maximum number of decimal places that your calculator will allow, then round off the final result.

□ APPROXIMATING
TRIGONOMETRIC
FUNCTIONS

Example 2   Use the Maclaurin series for $\sin x$ to approximate $\sin 3°$ to five decimal-place accuracy.

Solution.   In the Maclaurin series

$$\sin x = x - \frac{x^3}{3!} + \frac{x^5}{5!} - \frac{x^7}{7!} + \cdots \tag{8}$$

the angle x is assumed to be in radians (because the differentiation formulas for the trigonometric functions were derived with this assumption). Since $3° = \pi/60$ radians, it follows from (8) that

$$\sin 3° = \sin \frac{\pi}{60} = \left(\frac{\pi}{60}\right) - \frac{\left(\frac{\pi}{60}\right)^3}{3!} + \frac{\left(\frac{\pi}{60}\right)^5}{5!} - \frac{\left(\frac{\pi}{60}\right)^7}{7!} + \cdots \qquad (9)$$

We must now decide how many terms in this series must be kept in order to obtain five decimal-place accuracy. We shall consider two possible approaches, one using Lagrange's remainder formula, and the other exploiting the fact that (9) satisfies the conditions of the alternating series test.

If we let $f(x) = \sin x$, then the magnitude of the error that results when $\sin x$ is approximated by its nth Maclaurin polynomial is

$$|R_n| = \left| \frac{f^{(n+1)}(c)}{(n+1)!} x^{n+1} \right|$$

where c is between 0 and x. Since $f^{(n+1)}(c)$ is either $\pm\sin c$ or $\pm\cos c$, it follows that $|f^{(n+1)}(c)| \le 1$, so

$$|R_n| \le \frac{|x|^{n+1}}{(n+1)!}$$

In particular, if $x = \pi/60$, then

$$|R_n| \le \frac{\left(\frac{\pi}{60}\right)^{n+1}}{(n+1)!}$$

Thus, for five decimal-place accuracy, we must choose n so that

$$\frac{\left(\frac{\pi}{60}\right)^{n+1}}{(n+1)!} < 0.5 \times 10^{-5} = 0.000005$$

By trial and error with the help of a calculator or computer, the reader can check that $n = 3$ is the smallest n that works. Thus, in (9) we need only keep terms up to the third power for five decimal-place accuracy, that is,

$$\sin 3° \approx \left(\frac{\pi}{60}\right) - \frac{\left(\frac{\pi}{60}\right)^3}{3!} \approx 0.05234 \qquad (10)$$

An alternative approach to determining n uses the fact that (9) satisfies the conditions of the alternating series test, Theorem 11.7.1. (Verify.) Thus, by Theorem 11.7.2, if we use only those terms up to and including

$$\pm \frac{\left(\dfrac{\pi}{60}\right)^m}{m!} \qquad \boxed{m \text{ is an odd positive integer.}}$$

then the magnitude of the error will be at most

$$\frac{\left(\dfrac{\pi}{60}\right)^{m+2}}{(m + 2)!}$$

Thus, for five decimal-place accuracy, we look for the first positive odd integer m such that

$$\frac{\left(\dfrac{\pi}{60}\right)^{m+2}}{(m + 2)!} < 0.5 \times 10^{-5} = 0.000005$$

By trial and error, $m = 3$ is the first such integer. Thus, to five decimal-place accuracy

$$\sin 3° \approx \left(\frac{\pi}{60}\right) - \frac{\left(\dfrac{\pi}{60}\right)^3}{3!} \approx 0.05234 \tag{11}$$

which is the same as our earlier result.   ◄

   To approximate the value of a function f at a point x_0 using a Taylor series, two factors enter into the selection of the point $x = a$ for the series. First, the point $x = a$ must be selected so that f and its derivatives can be evaluated at a, since those values are needed to find Taylor series. Second, it is desirable to choose a as close as possible to x_0 since the "rate of convergence" of a Taylor series is usually most rapid close to a; that is, fewer terms are required in a partial sum to achieve a given level of accuracy.

Example 3   Approximate $\sin 92°$ to five decimal-place accuracy.

Solution.   We will use the Taylor series for $\sin x$ with $a = \pi/2 \ (= 90°)$, since $\sin x$ and its derivatives can be readily evaluated at this point and this value of a is reasonably close to the point $x = 92° = \frac{23}{45}\pi$ where we want to make the approximation. From Example 5 in Section 11.10, the Taylor series for $\sin x$ about $x = \pi/2$ is

$$\sin x = 1 - \frac{1}{2!}\left(x - \frac{\pi}{2}\right)^2 + \frac{1}{4!}\left(x - \frac{\pi}{2}\right)^4 - \frac{1}{6!}\left(x - \frac{\pi}{2}\right)^6 + \cdots \tag{12}$$

Substituting $x = \frac{23}{45}\pi$ in (12) yields

$$\sin 92° = \sin \frac{23}{45}\pi = 1 - \frac{1}{2!}\left(\frac{\pi}{90}\right)^2 + \frac{1}{4!}\left(\frac{\pi}{90}\right)^4 - \frac{1}{6!}\left(\frac{\pi}{90}\right)^6 + \cdots \tag{13}$$

If we let $f(x) = \sin x$, then the magnitude of the error in approximating $\sin x$ by its nth Taylor polynomial about $a = \pi/2$ is

$$|R_n| = \left| \frac{f^{(n+1)}(c)}{(n+1)!} \left(x - \frac{\pi}{2} \right)^{n+1} \right|$$

where c is between $\pi/2$ and x. As was the case in the previous example, $|f^{(n+1)}(c)| \leq 1$, so

$$|R_n| \leq \frac{\left| x - \dfrac{\pi}{2} \right|^{n+1}}{(n+1)!}$$

In particular, if $x = \frac{23}{45}\pi$, then

$$|R_n| \leq \frac{\left(\dfrac{\pi}{90} \right)^{n+1}}{(n+1)!}$$

For five decimal-place accuracy, we must choose n so that

$$\frac{\left(\dfrac{\pi}{90} \right)^{n+1}}{(n+1)!} < 0.5 \times 10^{-5} = 0.000005$$

By trial and error, the smallest such integer is $n = 3$. Thus, in (13) we need only keep terms up to the third power for five decimal-place accuracy, that is,

$$\sin 92° \approx 1 + 0 \cdot \left(\frac{\pi}{90} \right) - \frac{1}{2!} \left(\frac{\pi}{90} \right)^2 + 0 \cdot \left(\frac{\pi}{90} \right)^3$$

$$= 1 - \frac{1}{2!} \left(\frac{\pi}{90} \right)^2 \approx 0.99939$$

As in Example 2, we could also have obtained this result by exploiting the fact that (13) is an alternating series.   ◀

☐ **APPROXIMATING LOGARITHMS**

The Maclaurin series

$$\ln(1 + x) = x - \frac{x^2}{2} + \frac{x^3}{3} - \frac{x^4}{4} + \cdots \quad -1 < x \leq 1 \tag{14}$$

is the starting point for the approximation of natural logarithms. Unfortunately, the usefulness of this series is limited because of its slow convergence and the restriction $-1 < x \leq 1$. However, if we replace x by $-x$ in this series, we obtain

$$\ln(1 - x) = -x - \frac{x^2}{2} - \frac{x^3}{3} - \frac{x^4}{4} - \cdots \quad -1 \leq x < 1 \tag{15}$$

and on subtracting (15) from (14) we obtain

$$\ln\left(\frac{1+x}{1-x}\right) = 2\left(x + \frac{x^3}{3} + \frac{x^5}{5} + \frac{x^7}{7} + \cdots\right) \quad -1 < x < 1 \tag{16}$$

Series (16), first obtained by James Gregory* in 1668, can be used to compute the natural logarithm of any positive number y by letting

$$y = \frac{1+x}{1-x}$$

or equivalently

$$x = \frac{y-1}{y+1} \tag{17}$$

and noting that $-1 < x < 1$. For example, to compute $\ln 2$ we let $y = 2$ in (17), which yields $x = 1/3$. Substituting this value in (16) gives

$$\ln 2 = 2\left[(1/3) + \frac{(1/3)^3}{3} + \frac{(1/3)^5}{5} + \frac{(1/3)^7}{7} + \cdots\right]$$

Adding the four terms shown and then rounding to four decimal places at the end yields

$$\ln 2 \approx 0.6931$$

It is of interest to note that in the case $x = 1$, series (14) yields

$$\ln 2 = 1 - \frac{1}{2} + \frac{1}{3} - \frac{1}{4} + \frac{1}{5} - \cdots$$

This result is noteworthy because it is the first time we have been able to obtain the sum of the alternating harmonic series. However, this series converges too slowly to be of any computational value.

□ APPROXIMATING π

If we let $x = 1$ in the Maclaurin series

$$\tan^{-1} x = x - \frac{x^3}{3} + \frac{x^5}{5} - \frac{x^7}{7} + \cdots \quad -1 \leq x \leq 1 \tag{18}$$

---

*JAMES GREGORY (1638–1675). Scottish mathematician and astronomer. Gregory, the son of a minister, was famous in his time as the inventor of the Gregorian reflecting telescope, so named in his honor. Although he is not generally ranked with the great mathematicians, much of his work relating to calculus was studied by Leibniz and Newton and undoubtedly influenced some of their discoveries. There is a manuscript, discovered posthumously, which shows that Gregory had anticipated Taylor series well before Taylor.

we obtain

$$\frac{\pi}{4} = \tan^{-1} 1 = 1 - \frac{1}{3} + \frac{1}{5} - \frac{1}{7} + \cdots$$

or

$$\pi = 4\left[1 - \frac{1}{3} + \frac{1}{5} - \frac{1}{7} + \cdots \right]$$

This famous series, obtained by Leibniz in 1674, converges too slowly to be of computational importance. A more practical procedure for approximating π uses the identity

$$\frac{\pi}{4} = \tan^{-1} \frac{1}{2} + \tan^{-1} \frac{1}{3} \tag{19}$$

By using this identity and series (18) to approximate $\tan^{-1}\frac{1}{2}$ and $\tan^{-1}\frac{1}{3}$, the value of π can be effectively approximated to any degree of accuracy.

▶ Exercise Set 11.11   [C] *1–14, 19, 21*

1. Use inequality (5a) to find a value of n to ensure that (1) will approximate e to
   (a) five decimal-place accuracy
   (b) ten decimal-place accuracy.

In Exercises 2–8, apply Lagrange's form of the remainder.

2. Use $x = -1$ in the Maclaurin series for e^x to approximate $1/e$ to three decimal-place accuracy.

3. Use $x = \frac{1}{2}$ in the Maclaurin series for e^x to approximate $\sqrt{e}$ to four decimal-place accuracy.

4. Use the Maclaurin series for $\sin x$ to approximate $\sin 4°$ to five decimal-place accuracy.

5. Use the Maclaurin series for $\cos x$ to approximate $\cos(\pi/20)$ to four decimal-place accuracy.

6. Use an appropriate Taylor series for $\sin x$ to approximate $\sin 85°$ to four decimal-place accuracy.

7. Use an appropriate Taylor series for $\cos x$ to approximate $\cos 58°$ to four decimal-place accuracy.

8. Use an appropriate Taylor series for $\sin x$ to approximate $\sin 35°$ to four decimal-place accuracy.

9. Use the first two terms of series (16) to approximate $\ln 1.25$. Round your answer to three decimal places.

10. Use the first six terms of series (16) to approximate $\ln 3$. Round your answer to four decimal places.

11. Use the Maclaurin series for $\tan^{-1} x$ to approximate $\tan^{-1} 0.1$ to three decimal-place accuracy. [*Hint:* Use the fact that (18) is an alternating series.]

12. Use the Maclaurin series for $\sinh x$ to approximate $\sinh 0.5$ to three decimal-place accuracy.

13. Use the Maclaurin series for $\cosh x$ to approximate $\cosh 0.1$ to four decimal-place accuracy.

14. Use an appropriate Taylor series for $\sqrt[3]{x}$ to approximate $\sqrt[3]{28}$ to three decimal-place accuracy.

15. Find an interval of values for x containing $x = 0$ over which $\sin x$ can be approximated by $x - x^3/3!$ with three decimal-place accuracy ensured.

16. Find an interval of values for x over which e^x can be approximated by $1 + x + x^2/2!$ with three decimal-place accuracy ensured.

17. Find an upper bound on the magnitude of the error in the approximation $\cos x \approx 1 - x^2/2! + x^4/4!$ if $-0.2 \leq x \leq 0.2$.

18. Find an upper bound on the magnitude of the error in the approximation $\ln(1 + x) \approx x$ if $|x| < 0.01$.

19. In each part find the number of terms of the stated series that are required to ensure that the sum of the terms approximates $\ln 2$ to six decimal-place accuracy.

(a) The alternating harmonic series
(b) Series (16).

20. Prove identity (19).

21. Approximate $\tan^{-1}\frac{1}{2}$ and $\tan^{-1}\frac{1}{3}$ to three decimal-place accuracy, and then use identity (19) to approximate π.

■ **11.12 DIFFERENTIATION AND INTEGRATION OF POWER SERIES**

In this section we shall show that if a power series in $x - a$ converges in some interval and has a sum of $f(x)$ in that interval, then the power series must be the Taylor series about $x = a$ for the function f. This result is important for many reasons, but it is important for computational purposes because it will enable us to find Taylor series without using the defining formula for such series. This is often necessary in cases where successive derivatives of f are prohibitively complicated to compute. In this section we shall also consider conditions under which a Taylor series can be differentiated or integrated term by term.

☐ **POWER SERIES REPRESENTATIONS OF FUNCTIONS**

If a function f is expressed as the sum of a power series for all x in some interval, then we shall say that the power series *represents* f on the interval or that the series is a *power series representation of f* on the interval. For example, we know from our study of geometric series that

$$\frac{1}{1-x} = 1 + x + x^2 + x^3 + \cdots \quad -1 < x < 1$$

Thus, the series $1 + x + x^2 + x^3 + \cdots$ represents the function

$$f(x) = \frac{1}{1-x}$$

on the interval $(-1, 1)$.

The following theorems show that if a function f is represented by a power series on an interval, then differentiating the series term by term produces a power series representation of f' on the interval, and integrating the series term by term produces a power series representation of the integral of f on the interval.

11.12.1   THEOREM (***Differentiation of Power Series***).   *If a function f is repre-sented by a power series, say*

$$f(x) = \sum_{k=0}^{\infty} c_k(x - a)^k$$

where the series has a nonzero radius of convergence R, then:

(*a*)   *The series of differentiated terms*

$$\sum_{k=0}^{\infty} \frac{d}{dx}[c_k(x - a)^k] = \sum_{k=1}^{\infty} kc_k(x - a)^{k-1}$$

has radius of convergence R.

(*b*)   *The function f is differentiable on the interval* $(a - R, a + R)$*, and for every x in this interval*

$$f'(x) = \sum_{k=0}^{\infty} \frac{d}{dx}[c_k(x - a)^k]$$

To paraphrase this theorem informally, *a power series representation of a function can be differentiated term by term on any open interval within the interval of convergence.*

Example 1   To illustrate this theorem, we shall use the Maclaurin series

$$\sin x = x - \frac{x^3}{3!} + \frac{x^5}{5!} - \frac{x^7}{7!} + \cdots \qquad -\infty < x < +\infty$$

$$\cos x = 1 - \frac{x^2}{2!} + \frac{x^4}{4!} - \frac{x^6}{6!} + \cdots \qquad -\infty < x < +\infty$$

$$e^x = 1 + x + \frac{x^2}{2!} + \frac{x^3}{3!} + \frac{x^4}{4!} + \cdots \quad -\infty < x < +\infty$$

to obtain the familiar derivative formulas

$$\frac{d}{dx}[\sin x] = \cos x \quad \text{and} \quad \frac{d}{dx}[e^x] = e^x$$

Differentiating the Maclaurin series for $\sin x$ and e^x term by term yields

$$\frac{d}{dx}[\sin x] = \frac{d}{dx}\left[x - \frac{x^3}{3!} + \frac{x^5}{5!} - \frac{x^7}{7!} + \cdots\right] = 1 - 3\frac{x^2}{3!} + 5\frac{x^4}{5!} - 7\frac{x^6}{7!} + \cdots$$

$$= 1 - \frac{x^2}{2!} + \frac{x^4}{4!} - \frac{x^6}{6!} + \cdots$$

$$= \cos x$$

$$\frac{d}{dx}[e^x] = \frac{d}{dx}\left[1 + x + \frac{x^2}{2!} + \frac{x^3}{3!} + \cdots\right] = 1 + 2\frac{x}{2!} + 3\frac{x^2}{3!} + 4\frac{x^3}{4!} + \cdots$$

$$= 1 + x + \frac{x^2}{2!} + \frac{x^3}{3!} + \cdots$$

$$= e^x \quad \blacktriangleleft$$

11.12.2   THEOREM (*Integration of Power Series*).   *If a function f is represented by a power series, say*

$$f(x) = \sum_{k=0}^{\infty} c_k(x - a)^k$$

where the series has a nonzero radius of convergence R, then:

(a)   The series of integrated terms

$$\sum_{k=0}^{\infty}\left[\int c_k(x - a)^k \, dx\right] = \sum_{k=0}^{\infty}\frac{c_k}{k + 1}(x - a)^{k+1}$$

has radius of convergence R.

(b)   The function f is continuous on the interval $(a - R, a + R)$ and for all x in this interval

$$\int f(x) \, dx = \sum_{k=0}^{\infty}\left[\int c_k(x - a)^k \, dx\right] + C$$

(c)   For all α and β in the interval $(a - R, a + R)$, the series

$$\sum_{k=0}^{\infty}\left[\int_{\alpha}^{\beta} c_k(x - a)^k \, dx\right]$$

converges absolutely and

$$\int_{\alpha}^{\beta} f(x) \, dx = \sum_{k=0}^{\infty}\left[\int_{\alpha}^{\beta} c_k(x - a)^k \, dx\right]$$

To paraphrase this theorem informally, *a power series representation of a function can be integrated term by term on any interval within the interval of convergence.*

REMARK.   Note that in part (*b*) a separate constant of integration is not introduced for each term in the series; rather, a single constant C is added to the entire series.

Example 2 To illustrate part (b) of the foregoing theorem, we shall use the Maclaurin series for $\sin x$ and $\cos x$ to obtain the familiar integration formula

$$\int \cos x \, dx = \sin x + C$$

Integrating the Maclaurin series for $\cos x$ term by term yields

$$\int \cos x \, dx = \int \left[1 - \frac{x^2}{2!} + \frac{x^4}{4!} - \frac{x^6}{6!} + \cdots \right] dx$$

$$= \left[x - \frac{x^3}{3(2!)} + \frac{x^5}{5(4!)} - \frac{x^7}{7(6!)} + \cdots \right] + C$$

$$= \left[x - \frac{x^3}{3!} + \frac{x^5}{5!} - \frac{x^7}{7!} + \cdots \right] + C$$

$$= \sin x + C \quad ◀$$

Example 3 The integral

$$\int_0^1 e^{-x^2} \, dx$$

cannot be evaluated directly because there is no elementary antiderivative of e^{-x^2}. However, it is possible to approximate the integral by some numerical technique such as Simpson's rule. Still, another possibility is to represent e^{-x^2} by its Maclaurin series and then integrate term by term in accordance with part (c) of Theorem 11.12.2. This produces a series that converges to the integral.

The simplest way to obtain the Maclaurin series for e^{-x^2} is to replace x by $-x^2$ in the Maclaurin series

$$e^x = 1 + x + \frac{x^2}{2!} + \frac{x^3}{3!} + \frac{x^4}{4!} + \cdots$$

to obtain

$$e^{-x^2} = 1 - x^2 + \frac{x^4}{2!} - \frac{x^6}{3!} + \frac{x^8}{4!} - \cdots$$

Therefore,

$$\int_0^1 e^{-x^2} \, dx = \int_0^1 \left[1 - x^2 + \frac{x^4}{2!} - \frac{x^6}{3!} + \frac{x^8}{4!} - \cdots \right] dx$$

$$= \left[x - \frac{x^3}{3} + \frac{x^5}{5(2!)} - \frac{x^7}{7(3!)} + \frac{x^9}{9(4!)} - \cdots \right]_0^1$$

$$= 1 - \frac{1}{3} + \frac{1}{5 \cdot 2!} - \frac{1}{7 \cdot 3!} + \frac{1}{9 \cdot 4!} - \cdots \qquad (1)$$

Thus, we have found a series that converges to the value of the integral $\int_0^1 e^{-x^2}\, dx$. Using the first three terms in this series we obtain the approximation

$$\int_0^1 e^{-x^2}\, dx \approx 1 - \frac{1}{3} + \frac{1}{10} = \frac{23}{30} \approx 0.767$$

Since series (1) satisfies the conditions of the alternating series test, it follows from Theorem 11.7.2 that the magnitude of the error in this approximation is at most $1/(7 \cdot 3!) = 1/42 \approx 0.0238$. Greater accuracy can be obtained by using more terms in the series. ◄

☐ **POWER SERIES REPRESENTATIONS MUST BE TAYLOR SERIES**

The following theorem shows that if a function f is represented by a power series in $x - a$ on an interval, then that series must be the Taylor series for f; thus, Taylor series are the only power series that can represent functions on an interval.

11.12.3 THEOREM. *If*

$$f(x) = c_0 + c_1(x - a) + c_2(x - a)^2 + \cdots + c_n(x - a)^n + \cdots$$

for all x in some open interval containing a, then the series is the Taylor series for f about a.

Proof. By repeated application of Theorem 11.12.1(b) we obtain

$$f(x) = c_0 + c_1(x - a) + c_2(x - a)^2 + c_3(x - a)^3 + c_4(x - a)^4 + \cdots$$
$$f'(x) = c_1 + 2c_2(x - a) + 3c_3(x - a)^2 + 4c_4(x - a)^3 + \cdots$$
$$f''(x) = 2!c_2 + (3 \cdot 2)c_3(x - a) + (4 \cdot 3)c_4(x - a)^2 + \cdots$$
$$f'''(x) = 3!c_3 + (4 \cdot 3 \cdot 2)c_4(x - a) + \cdots$$
$$\vdots$$

On substituting $x = a$, all the powers of $x - a$ drop out leaving

$$f(a) = c_0 \qquad\qquad c_0 = f(a)$$
$$f'(a) = c_1 \qquad\qquad c_1 = f'(a)$$
$$f''(a) = 2!c_2 \quad \text{or} \quad c_2 = \frac{f''(a)}{2!}$$
$$f'''(a) = 3!c_3 \qquad\qquad c_3 = \frac{f'''(a)}{3!}$$
$$\vdots$$

which shows that the coefficients $c_0, c_1, c_2, c_3, \ldots$ are precisely the coefficients in the Taylor series about a for $f(x)$. ∎

REMARK.   The foregoing theorem tells us that no matter how we arrive at a power series in $x - a$ converging to $f(x)$, be it by substitution, by integration, by differentiation, or by algebraic manipulation, the resulting series will be the Taylor series about a for $f(x)$.

Example 4   Find the Maclaurin series for $\tan^{-1} x$.

Solution.   We could calculate this Maclaurin series directly. However, we can also exploit Theorem 11.12.2 by first observing that

$$\int \frac{1}{1 + x^2}\, dx = \tan^{-1} x + C$$

and then integrating the Maclaurin series for $1/(1 + x^2)$ term by term. Since

$$\frac{1}{1 - x} = 1 + x + x^2 + x^3 + x^4 + \cdots \qquad -1 < x < 1$$

it follows on replacing x by $-x^2$ that

$$\frac{1}{1 + x^2} = 1 - x^2 + x^4 - x^6 + x^8 - \cdots \qquad -1 < x < 1 \qquad (2)$$

Thus,

$$\tan^{-1} x + C = \int \frac{1}{1 + x^2}\, dx = \int [1 - x^2 + x^4 - x^6 + x^8 - \cdots]\, dx$$

or

$$\tan^{-1} x = \left[x - \frac{x^3}{3} + \frac{x^5}{5} - \frac{x^7}{7} + \frac{x^9}{9} - \cdots \right] - C$$

The constant of integration may be evaluated by substituting $x = 0$ and using the condition $\tan^{-1} 0 = 0$. This gives $C = 0$, so that

$$\tan^{-1} x = x - \frac{x^3}{3} + \frac{x^5}{5} - \frac{x^7}{7} + \frac{x^9}{9} - \cdots \qquad (3)$$

We are guaranteed by Theorem 11.12.3 that the series we have produced is the Maclaurin series for $\tan^{-1} x$ and that it actually converges to $\tan^{-1} x$ for $-1 < x < 1$.   ◀

REMARK.   Theorems 11.12.1 and 11.12.2 say nothing about the behavior of the differentiated and integrated series at the endpoints $a - R$ and $a + R$. Indeed, by differentiating termwise, convergence may be lost at one or both endpoints; and by integrating termwise, convergence may be gained at one or both endpoints. As an illustration, we derived series (3) for $\tan^{-1} x$ by integrating series (2) for $1/(1 + x^2)$. We omit the proof that the series for $\tan^{-1} x$

converges to $\tan^{-1} x$ on the interval $[-1, 1]$, while the series for $1/(1 + x^2)$ converges to $1/(1 + x^2)$ only on $(-1, 1)$. Thus, convergence was gained at both endpoints by integrating.

☐ **MISCELLANEOUS TECHNIQUES FOR OBTAINING TAYLOR SERIES**

We conclude this section with some techniques for obtaining Taylor series that would be messy to obtain directly.

Example 5 Find the first three terms that occur in the Maclaurin series for $e^{-x^2} \tan^{-1} x$.

Solution. Using the series for e^{-x^2} and $\tan^{-1} x$ obtained in Examples 3 and 4 gives

$$e^{-x^2} \tan^{-1} x = \left(1 - x^2 + \frac{x^4}{2} - \cdots \right)\left(x - \frac{x^3}{3} + \frac{x^5}{5} - \cdots \right)$$

We now multiply out, following a familiar format from elementary algebra:

$$
\begin{array}{r}
1 - x^2 + \dfrac{x^4}{2} - \cdots \\[2mm]
\times \quad x - \dfrac{x^3}{3} + \dfrac{x^5}{5} - \cdots \\[1mm]
\hline
x - x^3 + \dfrac{x^5}{2} - \cdots \\[2mm]
- \dfrac{x^3}{3} + \dfrac{x^5}{3} - \dfrac{x^7}{6} + \cdots \\[2mm]
\dfrac{x^5}{5} - \dfrac{x^7}{5} + \cdots \\[1mm]
\hline
x - \tfrac{4}{3} x^3 + \tfrac{31}{30} x^5 - \cdots
\end{array}
$$

Thus,

$$e^{-x^2} \tan^{-1} x = x - \frac{4}{3} x^3 + \frac{31}{30} x^5 - \cdots \qquad -1 < x < 1$$

If desired, more terms in the series can be obtained by including more terms in the factors. ◄

Example 6 Find the first three nonzero terms in the Maclaurin series for $\tan x$.

Solution. Instead of computing the series directly, we write

$$\tan x = \frac{\sin x}{\cos x} = \frac{x - \dfrac{x^3}{3!} + \dfrac{x^5}{5!} - \cdots}{1 - \dfrac{x^2}{2!} + \dfrac{x^4}{4!} - \cdots} \qquad -\frac{\pi}{2} < x < \frac{\pi}{2}$$

and follow the familiar "long division" process to obtain

$$
1 - \frac{x^2}{2} + \frac{x^4}{24} - \cdots \overline{\Big)\; x - \frac{x^3}{6} + \frac{x^5}{120} - \cdots}
$$

with quotient $x + \dfrac{x^3}{3} + \dfrac{2x^5}{15} + \cdots$

$$
x - \frac{x^3}{2} + \frac{x^5}{24} - \cdots
$$

$$
\frac{x^3}{3} - \frac{x^5}{30} + \cdots
$$

$$
\frac{x^3}{3} - \frac{x^5}{6} + \cdots
$$

$$
\frac{2x^5}{15} + \cdots
$$

Thus,

$$
\tan x = x + \frac{x^3}{3} + \frac{2x^5}{15} + \cdots \qquad -\frac{\pi}{2} < x < \frac{\pi}{2} \qquad \blacktriangleleft
$$

▶ Exercise Set 11.12   C 14–21

In Exercises 1–4, obtain the stated results by differentiating or integrating Maclaurin series term by term.

1. (a) $\dfrac{d}{dx}[e^x] = e^x$ (b) $\displaystyle\int e^x\, dx = e^x + C.$

2. (a) $\dfrac{d}{dx}[\cos x] = -\sin x$

(b) $\displaystyle\int \sin x\, dx = -\cos x + C.$

3. (a) $\dfrac{d}{dx}[\sinh x] = \cosh x$

(b) $\displaystyle\int \sinh x\, dx = \cosh x + C.$

4. (a) $\dfrac{d}{dx}[\ln(1+x)] = \dfrac{1}{1+x}$

(b) $\displaystyle\int \dfrac{1}{1+x}\, dx = \ln(1+x) + C.$

5. Derive the Maclaurin series for $1/(1+x)^2$ by differentiating an appropriate Maclaurin series term by term.

6. By differentiating an appropriate series, show that

$$\sum_{k=1}^{\infty} kx^k = \frac{x}{(1-x)^2} \qquad \text{for } -1 < x < 1$$

$$\left[\textit{Hint: } \text{Consider } x\,\frac{d}{dx}\left[\frac{1}{1-x}\right]. \right]$$

7. By integrating an appropriate series, show that

$$\sum_{k=1}^{\infty} \frac{x^k}{k} = \ln\left(\frac{1}{1-x}\right) \qquad \text{for } -1 < x < 1$$

$$\left[\textit{Hint: } \ln\left(\frac{1}{1-x}\right) = -\ln(1-x). \right]$$

8. Use the result of Exercise 6 to find the sum of the series

$$\frac{1}{3} + \frac{2}{3^2} + \frac{3}{3^3} + \frac{4}{3^4} + \cdots$$

9. Use the result of Exercise 7 to find the sum of the series

$$\frac{1}{4} + \frac{1}{2(4^2)} + \frac{1}{3(4^3)} + \frac{1}{4(4^4)} + \cdots$$

10. Find the sum

$$\sum_{k=0}^{\infty} \frac{k+1}{k!} = 1 + 2 + \frac{3}{2!} + \frac{4}{3!} + \frac{5}{4!} + \cdots$$

[*Hint:* Differentiate the Maclaurin series for xe^x.]

11. Find the sum of the series

$$2 + 6x + 12x^2 + 20x^3 + \cdots$$

[*Hint:* Find the second derivative of the Maclaurin series for $1/(1-x)$.]

12. Find the sum $\sum_{k=1}^{\infty} \dfrac{k^2}{4^k}$. [*Hint:* Differentiate the Maclaurin series for $1/(1 - x)$, multiply by x, differentiate, and multiply by x again.]

13. Let $f(x) = \sum_{k=0}^{\infty} (-1)^k \dfrac{x^{k+1}}{k + 1}$

$$= x - \frac{x^2}{2} + \frac{x^3}{3} - \frac{x^4}{4} + \cdots$$

(a) Use the ratio test to show that the series converges for all x in the interval $(-1, 1)$.

(b) Use part (a) of Theorem 11.12.1 to find a power series for $f'(x)$. What is its interval of convergence?

(c) From the series obtained in part (b), deduce that $f'(x) = 1/(1 + x)$ and hence that

$$f(x) = \ln (1 + x) \text{ for } -1 < x < 1$$

[*Remark:* The Lagrange form of the remainder can be used to show that the Maclaurin series for $\ln (1 + x)$ converges to $\ln (1 + x)$ for $x = 1$ as well.]

In Exercises 14–21, use series to approximate the value of the integral to three decimal-place accuracy.

14. $\displaystyle\int_0^1 \sin x^2 \, dx$.   **15.** $\displaystyle\int_0^1 \cos \sqrt{x} \, dx$.

16. $\displaystyle\int_0^{0.1} \dfrac{\sin x}{x} \, dx$.   **17.** $\displaystyle\int_0^{1/2} \dfrac{dx}{1 + x^4}$.

18. $\displaystyle\int_0^{1/2} \tan^{-1} 2x^2 \, dx$.   **19.** $\displaystyle\int_0^{0.1} e^{-x^3} \, dx$.

20. $\displaystyle\int_0^{0.2} \sqrt[3]{1 + x^4} \, dx$.   **21.** $\displaystyle\int_0^{1/2} \dfrac{dx}{\sqrt[4]{x^2 + 1}}$.

In Exercises 22–29, use any method to find the first four nonzero terms in the Maclaurin series of the given function.

22. $x^4 e^x$.   **23.** $e^{-x^2} \cos x$.

24. $\dfrac{x^2}{1 + x^4}$.   **25.** $\dfrac{\sin x}{e^x}$.

26. $\tanh x$.   **27.** $x \ln (1 - x^2)$.

28. $\dfrac{\ln (1 + x)}{1 - x}$.   **29.** $x^2 e^{4x} \sqrt{1 + x}$.

30. Obtain the familiar result, $\lim_{x \to 0} (\sin x)/x = 1$, by finding a power series for $(\sin x)/x$ and taking the limit term by term.

31. Use the method of Exercise 30 to find the limits.

(a) $\displaystyle\lim_{x \to 0} \dfrac{1 - \cos x}{\sin x}$

(b) $\displaystyle\lim_{x \to 0} \dfrac{\ln \sqrt{1 + x} - \sin 2x}{x}$.

32. (a) Use the relationship

$$\int \frac{1}{\sqrt{1 - x^2}} \, dx = \sin^{-1} x + C$$

to find the first four nonzero terms in the Maclaurin series for $\sin^{-1} x$.

(b) Express the series in sigma notation.

(c) What is the radius of convergence?

33. (a) Use the relationship

$$\int \frac{1}{\sqrt{1 + x^2}} \, dx = \sinh^{-1} x + C$$

to find the first four nonzero terms in the Maclaurin series for $\sinh^{-1} x$.

(b) Express the series in sigma notation.

(c) What is the radius of convergence?

34. Prove: If the power series $\sum_{k=0}^{\infty} a_k x^k$ and $\sum_{k=0}^{\infty} b_k x^k$ have the same sum on an interval $(-r, r)$, then $a_k = b_k$ for all values of k.

▶ SUPPLEMENTARY EXERCISES   Ⓒ 26, 50, 51, 52

In Exercises 1–6, find $L = \lim_{n \to +\infty} a_n$ if it exists.

1. $a_n = (-1)^n / e^n$.   **2.** $a_n = e^{1/n}$.

3. $a_n = \dfrac{1}{\sqrt{n}} - \dfrac{1}{\sqrt{n + 1}}$.   **4.** $a_n = \sin (\pi n)$.

5. $a_n = \sin \left(\dfrac{(2n - 1)\pi}{2} \right)$.   **6.** $a_n = \dfrac{n + 1}{n(n + 2)}$.

7. Which of the sequences $\{a_n\}_{n=1}^{+\infty}$ in Exercises 1–6 are (a) decreasing, (b) nondecreasing, and (c) alternating?

8. Suppose $f(x)$ satisfies

$$f'(x) > 0 \quad \text{and} \quad f(x) \leq 1 - e^{-x}$$

for all $x \geq 1$. What can you conclude about the convergence of $\{a_n\}$ if $a_n = f(n)$, $n = 1, 2, \ldots$?

9. Use your knowledge of geometric series and p-series to determine all values of q for which the following series converge.

(a) $\displaystyle\sum_{k=0}^{\infty} \pi^k/q^{2k}$ (b) $\displaystyle\sum_{k=1}^{\infty} (1/k^q)^3$

(c) $\displaystyle\sum_{k=2}^{\infty} 1/(\ln q^k)$ (d) $\displaystyle\sum_{k=2}^{\infty} 1/(\ln q)^k$.

10. (a) Use a suitable test to find all values of q for which $\displaystyle\sum_{k=2}^{\infty} 1/[k (\ln k)^q]$ converges.

(b) Why can't you use the integral test for the series $\displaystyle\sum_{k=1}^{\infty} (2 + \cos k)/k^2$? Test for convergence using a test that does apply.

11. Express $1.3636\ldots$ as (a) an infinite series in sigma notation, and (b) a ratio of integers.

12. In parts (a)–(d), use the comparison test to determine whether the series converges.

(a) $\displaystyle\sum_{k=1}^{\infty} \frac{2k - 1}{3k^2 - k}$ (b) $\displaystyle\sum_{k=1}^{\infty} \frac{2k + 1}{3k^2 + k}$

(c) $\displaystyle\sum_{k=1}^{\infty} \frac{2k - 1}{3k^3 - k^2}$ (d) $\displaystyle\sum_{k=1}^{\infty} \frac{2k + 1}{3k^3 + k^2}$.

13. Find the sum of the series (if it converges).

(a) $\displaystyle\sum_{k=1}^{\infty} \frac{2^k + 3^k}{6^{k+1}}$ (b) $\displaystyle\sum_{k=2}^{\infty} \ln\left(1 + \frac{1}{k}\right)$

(c) $\displaystyle\sum_{k=1}^{\infty} [k^{-1/2} - (k + 1)^{-1/2}]$.

In Exercises 14–21, determine whether the series converges or diverges. You may use the following limits without proof:

$$\lim_{k \to +\infty} (1 + 1/k)^k = e, \quad \lim_{k \to +\infty} \sqrt[k]{k} = 1, \quad \lim_{k \to +\infty} \sqrt[k]{a} = 1$$

14. $\displaystyle\sum_{k=0}^{\infty} e^{-k}$. 15. $\displaystyle\sum_{k=1}^{\infty} ke^{-k^2}$.

16. $\displaystyle\sum_{k=1}^{\infty} \frac{k}{k^2 + 2k + 7}$. 17. $\displaystyle\sum_{k=1}^{\infty} \frac{\sqrt{k}}{k^2 + 7}$.

18. $\displaystyle\sum_{k=1}^{\infty} \left(\frac{k}{k + 1}\right)^k$. 19. $\displaystyle\sum_{k=0}^{\infty} \frac{3^k k!}{(2k)!}$.

20. $\displaystyle\sum_{k=0}^{\infty} \frac{k^6 3^k}{(k + 1)!}$. 21. $\displaystyle\sum_{k=1}^{\infty} \left(\frac{5k}{2k + 1}\right)^{3k}$.

In Exercises 22–25, determine whether the given series is absolutely convergent, conditionally convergent, or divergent.

22. $\displaystyle\sum_{k=1}^{\infty} (-1)^k/e^{1/k}$. 23. $\displaystyle\sum_{k=0}^{\infty} (-2)^k/(3^k + 1)$.

24. $\displaystyle\sum_{k=0}^{\infty} (-1)^k/(2k + 1)$. 25. $\displaystyle\sum_{k=0}^{\infty} (-1)^k 3^k/2^{k+1}$.

26. Find a value of n to ensure that the nth partial sum approximates the sum of the series to the stated accuracy.

(a) $\displaystyle\sum_{k=1}^{\infty} \frac{(-1)^k}{k^2 + 1}$; $|\text{error}| < 0.0001$

(b) $\displaystyle\sum_{k=1}^{\infty} \frac{(-1)^k}{5^k + 1}$; $|\text{error}| < 0.00005$.

In Exercises 27–32, determine the radius of convergence and the interval of convergence of the given power series.

27. $\displaystyle\sum_{k=1}^{\infty} \frac{(x - 1)^k}{k\sqrt{k}}$. 28. $\displaystyle\sum_{k=1}^{\infty} \frac{(2x)^k}{3k}$.

29. $\displaystyle\sum_{k=1}^{\infty} \frac{(1 - x)^{2k}}{4^k k}$. 30. $\displaystyle\sum_{k=1}^{\infty} \frac{k^2(x + 2)^k}{(k + 1)!}$.

31. $\displaystyle\sum_{k=1}^{\infty} \frac{k!(x - 1)^k}{5^k}$. 32. $\displaystyle\sum_{k=1}^{\infty} \frac{(2k)! x^k}{(2k + 1)!}$.

In Exercises 33–35, find
(a) the nth Taylor polynomial for f about $x = a$ (for the stated values of n and a);
(b) Lagrange's form of $R_n(x)$ (for the stated values of n and a);
(c) an upper bound on the absolute value of the error if $f(x)$ is approximated over the given interval by the Taylor polynomial obtained in part (a).

33. $f(x) = \ln (x - 1)$; $a = 2$; $n = 3$; $[\frac{3}{2}, 2]$.

34. $f(x) = e^{x/2}$; $a = 0$; $n = 4$; $[-1, 0]$.

35. $f(x) = \sqrt{x}$; $a = 1$; $n = 2$; $[\frac{4}{9}, 1]$.

36. (a) Use the identity $a - x = a(1 - x/a)$ to find the Maclaurin series for $1/(a - x)$ from the geometric series. What is its radius of convergence?

(b) Find the Maclaurin series and radius of convergence of $1/(3 + x)$.

(c) Find the Maclaurin series and radius of convergence of $2x/(4 + x^2)$.

(d) Use partial fractions to find the Maclaurin series and radius of convergence of

$$\frac{1}{(1 - x)(2 - x)}$$

37. Use the known Maclaurin series for $\ln(1 + x)$ to find the Maclaurin series and radius of convergence of $\ln(a + x)$ for $a > 0$.

38. Use the identity $x = a + (x - a)$ and the known Maclaurin series for e^x, $\sin x$, $\cos x$, and $1/(1 - x)$ to find the Taylor series about $x = a$ for (a) e^x, (b) $\sin x$, and (c) $1/x$.

39. Use the series of Example 8 in Section 11.10 to find the Maclaurin series and radius of convergence of $1/\sqrt{9 + x}$.

In Exercises 40–45, use any method to find the first three nonzero terms of the Maclaurin series.

40. $e^{\tan x}$.

41. $\sec x$.

42. $(\sin x)/(e^x - x)$.

43. $\sqrt{\cos x}$.

44. $e^x \ln(1 - x)$.

45. $\ln(1 + \sin x)$.

46. Find a power series for $\dfrac{1 - \cos 3x}{x^2}$ and use it to evaluate $\lim\limits_{x \to 0} \dfrac{1 - \cos 3x}{x^2}$.

47. Find a power series for $\dfrac{\ln(1 - 2x)}{x}$ and use it to evaluate $\lim\limits_{x \to 0} \dfrac{\ln(1 - 2x)}{x}$.

48. How many decimal places of accuracy can be guaranteed if we approximate $\cos x$ by $1 - x^2/2$ for $-0.1 < x < 0.1$?

49. For what values of x can $\sin x$ be replaced by $x - x^3/6 + x^5/120$ with an ensured accuracy of 6×10^{-4}?

In Exercises 50–52, approximate the indicated quantity to three decimal-place accuracy.

50. $\cos(10°)$.

51. $\displaystyle\int_0^1 \frac{(1 - e^{-t/2})}{t}\, dt$.

52. $\displaystyle\int_0^1 \frac{\sin x}{\sqrt{x}}\, dx$.

53. Show that $y = \displaystyle\sum_{n=0}^{\infty} k^n x^n/n!$ satisfies $y' - ky = 0$ for any fixed k.

■ **7.7** FIRST-ORDER DIFFERENTIAL EQUATIONS AND APPLICATIONS

> *In this section we shall begin to study equations that involve an unknown function and its derivatives. These are called **differential equations**. Such equations describe many of the fundamental principles in science and engineering, and their study constitutes a major field of mathematics. Our work in this section is limited to the most basic types of differential equations. In the final chapter of this text we shall go a little further into this topic.*

☐ TERMINOLOGY

Some examples of differential equations are

$$\frac{dy}{dx} = 3y \tag{1}$$

$$\frac{d^2y}{dx^2} - 6\frac{dy}{dx} + 8y = 0 \tag{2}$$

$$y' - y = e^{2x} \tag{3}$$

$$\frac{d^3y}{dt^3} - t\frac{dy}{dt} + (t^2 - 1)y = e^t \tag{4}$$

In the first three equations, $y = y(x)$ is an unknown function of x and in the last equation $y = y(t)$ is an unknown function of t. The **order** of a differential equation is the order of the highest derivative that appears in the equation. Thus, (1) and (3) are first-order equations, (2) is second order, and (4) is third order.

A function $y = y(x)$ is a **solution** of a differential equation if the equation is satisfied when $y(x)$ and its derivatives are substituted. For example,

$$y = e^{2x} \tag{5}$$

is a solution of the equation

$$\frac{dy}{dx} - y = e^{2x} \tag{6}$$

since

$$\frac{dy}{dx} - y = 2e^{2x} - e^{2x} = e^{2x}$$

Similarly,

$$y = e^x + e^{2x} \tag{7}$$

is also a solution of (6). More generally,

$$y = Ce^x + e^{2x} \tag{8}$$

is a solution of (6) for any constant C (verify). This solution is of special importance since it can be proved that *all* solutions of (6) can be obtained by substituting values for the arbitrary constant C. For example, $C = 0$ yields solution (5) and $C = 1$ yields solution (7). A solution of a differential equation from which all other solutions can be derived by substituting values for arbitrary constants is called the **general solution** of the equation. Usually, the general solution of an nth-order equation contains n arbitrary constants. Thus, (8), which is the general solution of the first-order equation (6), contains one arbitrary constant.

Often, solutions of differential equations are expressed as implicitly defined functions. For example,

$$\ln y = xy + C \tag{9}$$

defines a solution of

$$\frac{dy}{dx} = \frac{y^2}{1 - xy} \tag{10}$$

for any value of the constant C, since implicit differentiation of (9) yields

$$\frac{1}{y}\frac{dy}{dx} = x\frac{dy}{dx} + y$$

or

$$\frac{dy}{dx} - xy\frac{dy}{dx} = y^2$$

from which (10) follows.

☐ FIRST-ORDER
SEPARABLE EQUATIONS

A first-order differential equation is called **separable** if it is expressible in the form

$$\frac{dy}{dx} = \frac{g(x)}{h(y)}$$

To solve such an equation we rewrite it in the differential form

$$h(y)\,dy = g(x)\,dx \tag{11}$$

and integrate both sides, thereby obtaining the general solution

$$\int h(y)\,dy = \int g(x)\,dx + C$$

where C is an arbitrary constant.

In (11), the x and y variables are "separated" from each other, hence the term *separable* differential equation.

Example 1   Solve the equation $\dfrac{dy}{dx} = \dfrac{x}{y^2}$.

Solution.   Changing to differential form and integrating yields

$$y^2 \, dy = x \, dx$$

$$\int y^2 \, dy = \int x \, dx$$

$$\frac{y^3}{3} = \frac{x^2}{2} + C$$

This expresses the solution implicitly. If desired, we can solve explicitly for y to obtain

$$y = (\tfrac{3}{2}x^2 + 3C)^{1/3} \qquad \text{or} \qquad y = (\tfrac{3}{2}x^2 + K)^{1/3}$$

where $K \, (= 3C)$ is an arbitrary constant.   ◀

Example 2   Solve the equation $x(y - 1) \dfrac{dy}{dx} = y$.

Solution.   Separating the variables and integrating yields

$$x(y - 1) \, dy = y \, dx$$

$$\frac{y - 1}{y} \, dy = \frac{dx}{x}$$

$$\int \frac{y - 1}{y} \, dy = \int \frac{dx}{x}$$

$$\int \left(1 - \frac{1}{y}\right) dy = \int \frac{dx}{x}$$

$$y - \ln|y| = \ln|x| + C$$

or, using properties of logarithms,

$$y = \ln|xy| + C$$

This gives a solution implicitly as a function of x; in this case there is no simple formula for y explicitly as a function of x.   ◀

☐ **INITIAL-VALUE PROBLEMS**

When a physical problem leads to a differential equation, there are usually conditions in the problem that determine specific values for the arbitrary constants in the general solution of the equation. For a first-order equation, a condition that specifies the value of the unknown function $y(x)$ at some point $x = x_0$ is called an ***initial condition***. A first-order differential equation together with one initial condition constitutes a ***first-order initial-value problem***.

Example 3   Solve the initial-value problem

$$\frac{dy}{dx} = -4xy^2, \quad y(0) = 1$$

Solution. We first solve the differential equation:

$$\frac{dy}{y^2} = -4x \, dx$$

$$\int \frac{dy}{y^2} = \int -4x \, dx$$

$$-\frac{1}{y} = -2x^2 + C_1$$

or on multiplying by -1, taking reciprocals, and writing C in place of $-C_1$,

$$y = \frac{1}{2x^2 + C} \tag{12}$$

The initial condition, $y(0) = 1$, requires that $y = 1$ when $x = 0$. Substituting these values in (12) yields $C = 1$. Thus, the solution of the initial-value problem is

$$y = \frac{1}{2x^2 + 1} \qquad \blacktriangleleft$$

☐ **FIRST-ORDER LINEAR EQUATIONS**

Not every first-order differential equation is separable. For example, it is impossible to separate the variables in the equation

$$\frac{dy}{dx} + x^2y = e^x$$

However, this equation can be solved by a different method that we shall now consider.

A first-order differential equation is called **linear** if it is expressible in the form

$$\frac{dy}{dx} + p(x)y = q(x) \tag{13}$$

where the functions $p(x)$ and $q(x)$ may or may not be constant. Some examples are

$$\frac{dy}{dx} + x^2y = e^x \qquad \boxed{p(x) = x^2, \, q(x) = e^x}$$

$$y' - 3e^xy = 0 \qquad \boxed{p(x) = -3e^x, \, q(x) = 0}$$

$$\frac{dy}{dx} + 5y = 2 \qquad \boxed{p(x) = 5, \, q(x) = 2}$$

$$\frac{dy}{dx} + (\sin x)y + x^3 = 0 \qquad \boxed{p(x) = \sin x, \, q(x) = -x^3}$$

One procedure for solving (13) is based on the observation that if we define $\rho = \rho(x)$ by

$$\rho = e^{\int p(x)\,dx}$$

then

$$\frac{d\rho}{dx} = e^{\int p(x)\,dx} \cdot \frac{d}{dx}\int p(x)\,dx = \rho p(x)$$

Thus,

$$\frac{d}{dx}(\rho y) = \rho\frac{dy}{dx} + \frac{d\rho}{dx}y = \rho\frac{dy}{dx} + \rho p(x)y \tag{14}$$

If (13) is multiplied through by ρ, it becomes

$$\rho\frac{dy}{dx} + \rho p(x)y = \rho q(x)$$

or from (14),

$$\frac{d}{dx}(\rho y) = \rho q(x)$$

This equation can be solved by integrating both sides to obtain

$$\rho y = \int \rho q(x)\,dx + C$$

or

$$y = \frac{1}{\rho}\left[\int \rho q(x)\,dx + C\right]$$

To summarize, (13) can be solved in three steps:

> **Step 1.** Calculate
>
> $$\rho = e^{\int p(x)\,dx}$$
>
> This is called the ***integrating factor***. Since any ρ will suffice, we can take the constant of integration to be zero in this step.
>
> **Step 2.** Multiply both sides of (13) by ρ and express the result as
>
> $$\frac{d}{dx}(\rho y) = \rho q(x)$$
>
> **Step 3.** Integrate both sides of the equation obtained in Step 2 and then solve for y. Be sure to include a constant of integration in this step.

Example 4 Solve the equation

$$\frac{dy}{dx} - 4xy = x \tag{15}$$

Solution. Since $p(x) = -4x$, the integrating factor is

$$\rho = e^{\int(-4x)\,dx} = e^{-2x^2}$$

If we multiply (15) by ρ, we obtain

$$\frac{d}{dx}(e^{-2x^2}y) = xe^{-2x^2}$$

Integrating both sides of this equation yields

$$e^{-2x^2}y = \int xe^{-2x^2}\,dx = -\frac{1}{4}e^{-2x^2} + C$$

and then multiplying both sides by e^{2x^2} yields

$$y = -\frac{1}{4} + Ce^{2x^2} \quad \blacktriangleleft$$

Example 5 Solve the equation

$$x\frac{dy}{dx} - y = x \quad (x > 0)$$

Solution. To put the equation in form (13), we divide through by x to obtain

$$\frac{dy}{dx} - \frac{1}{x}y = 1 \tag{16}$$

Since $p(x) = -1/x$, the integrating factor is

$$\rho = e^{\int-(1/x)\,dx} = e^{-\ln|x|} = \frac{1}{|x|} = \frac{1}{x}$$

(The absolute value was dropped because of the assumption that $x > 0$.) If we multiply (16) by ρ, we obtain

$$\frac{d}{dx}\left(\frac{1}{x}y\right) = \frac{1}{x}$$

Integrating both sides of this equation yields

$$\frac{1}{x}y = \int\frac{1}{x}\,dx = \ln x + C$$

or

$$y = x\ln x + Cx \quad \blacktriangleleft$$

☐ **APPLICATIONS**

We conclude this section with some applications of first-order differential equations.

Example 6 (*Geometry*)   Find a curve in the xy-plane that passes through $(0, 3)$ and whose tangent line at a point (x, y) has slope $2x/y^2$.

Solution.   Since the slope of the tangent line is dy/dx, we have

$$\frac{dy}{dx} = \frac{2x}{y^2} \tag{17}$$

and, since the curve passes through $(0, 3)$, we have the initial condition

$$y(0) = 3 \tag{18}$$

Equation (17) is separable and can be written as

$$y^2\, dy = 2x\, dx$$

so

$$\int y^2\, dy = \int 2x\, dx \qquad \text{or} \qquad \tfrac{1}{3} y^3 = x^2 + C$$

From the initial condition, (18), it follows that $C = 9$, and so the curve has the equation

$$\tfrac{1}{3} y^3 = x^2 + 9 \qquad \text{or} \qquad y = (3x^2 + 27)^{1/3} \qquad \blacktriangleleft$$

Example 7 (*Mixing Problems*)   At time $t = 0$, a tank contains 4 lb of salt dissolved in 100 gal of water. Suppose that brine containing 2 lb of salt per gallon of water is allowed to enter the tank at a rate of 5 gal/min and that the mixed solution is drained from the tank at the same rate. Find the amount of salt in the tank after 10 min.

Solution.   Let $y(t)$ be the amount of salt (in pounds) at time t. We are interested in finding $y(10)$, the amount of salt at time $t = 10$. We will begin by finding an expression for dy/dt, the rate of change of the amount of salt in the tank at time t. Clearly,

$$\frac{dy}{dt} = \text{rate in} - \text{rate out} \tag{19}$$

where *rate in* is the rate at which salt enters the tank and *rate out* is the rate at which salt leaves the tank. But

$$\text{rate in} = (2 \text{ lb/gal}) \cdot (5 \text{ gal/min}) = 10 \text{ lb/min}$$

At time t, the mixture contains $y(t)$ lb of salt in 100 gal of water; thus, the concentration of salt at time t is $y(t)/100$ lb/gal and

$$\text{rate out} = \left(\frac{y(t)}{100} \text{ lb/gal} \right) \cdot (5 \text{ gal/min}) = \frac{y(t)}{20} \text{ lb/min}$$

Therefore, (19) can be written as

$$\frac{dy}{dt} = 10 - \frac{y}{20}$$

or

$$\frac{dy}{dt} + \frac{y}{20} = 10 \qquad (20)$$

which is a first-order linear differential equation. We also have the initial condition

$$y(0) = 4 \qquad (21)$$

since the tank contains 4 lb of salt at time $t = 0$.

Multiplying both sides of (20) by the integrating factor

$$\rho = e^{\int(1/20)\,dt} = e^{t/20}$$

yields

$$\frac{d}{dt}(e^{t/20}y) = 10e^{t/20}$$

so

$$e^{t/20}y = \int 10e^{t/20}\,dt = 200e^{t/20} + C$$

or

$$y(t) = 200 + Ce^{-t/20}$$

From the initial condition, (21), it follows that

$$4 = 200 + C \quad \text{or} \quad C = -196$$

so

$$y(t) = 200 - 196e^{-t/20}$$

Thus, after 10 min ($t = 10$), the amount of salt in the tank is

$$y(10) = 200 - 196e^{-0.5} \approx 81.1 \text{ lb}$$

□ **EXPONENTIAL GROWTH**   Many quantities increase or decrease with time in proportion to the amount of the quantity present. Some examples are human population, bacteria in a culture, drug concentration in the bloodstream, radioactivity, and the values of certain kinds of investments. We shall show how differential equations can be used to study the growth and decay of such quantities.

7.7.1 DEFINITION.   A quantity is said to have an *exponential growth (decay) model* if at each instant of time its rate of increase (decrease) is proportional to the amount of the quantity present.

Consider a quantity with an exponential growth or decay model and denote by $y(t)$ the amount of the quantity present at time t. We assume that $y(t) > 0$ for all t. Since the rate of change of $y(t)$ is proportional to the amount present, it follows that $y(t)$ satisfies

$$\frac{dy}{dt} = ky \tag{22}$$

where k is a constant of proportionality.

The constant k in (22) is called the ***growth constant*** if $k > 0$ and the ***decay constant*** if $k < 0$. If $k > 0$, then $dy/dt > 0$, so that y is increasing with time (a growth model). If $k < 0$, then $dy/dt < 0$, so that y is decreasing with time (a decay model).

REMARK.   The constant k in (22) is sometimes called the ***growth rate*** (a negative growth rate meaning that y is decreasing). Strictly speaking, this is not correct since the growth rate of y is not k, but rather $dy/dt\ (= ky)$. Although it is rarely done, it would be more accurate to call k the ***relative growth rate***, since it follows from (22) that

$$k = \frac{dy/dt}{y}$$

However, it is so common to call k the growth rate that we shall use this standard terminology in this section. The growth rate k is usually expressed as a percentage; thus, a growth rate of 3% means $k = 0.03$ and a growth rate of -500% means $k = -5$.

If we are given a quantity y with an exponential growth or decay model, and if we know the amount y_0 present at some initial time $t = 0$, then we can find the amount present at any time t by solving the initial-value problem

$$\frac{dy}{dt} = ky, \quad y(0) = y_0 \tag{23}$$

This differential equation is both separable and linear and thus can be solved by either of the methods we have studied in this section. We shall treat it as a linear equation. Rewriting the differential equation as

$$\frac{dy}{dt} - ky = 0$$

and multiplying through by the integrating factor

$$\rho = e^{\int -k\,dt} = e^{-kt}$$

yields

$$\frac{d}{dt}(e^{-kt}y) = 0$$

After integrating,

$$e^{-kt}y = C \quad \text{or} \quad y = Ce^{kt}$$

From the initial condition, $y(0) = y_0$, it follows that $C = y_0$; thus, the solution of (23) is

$$y(t) = y_0 e^{kt} \tag{24}$$

Exponential models have proved useful in studies of population growth. Although populations (e.g., people, bacteria, and flowers) grow in discrete steps, we can apply the results of this section if we are willing to approximate the population graph by a continuous curve $y = y(t)$ (Figure 7.7.1).

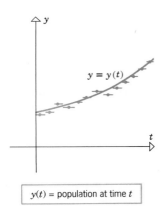

$y(t) =$ population at time t

Figure 7.7.1

Example 8 (*Population Growth*)   According to United Nations data, the world population at the beginning of 1990 was approximately 5.3 billion and growing at a rate of about 2% per year. Assuming an exponential growth model, estimate the world population at the beginning of the year 2015.

Solution.   Let

$t = $ time elapsed from the beginning of 1990 (in years)
$y = $ world population (in billions)

Since the beginning of 1990 corresponds to $t = 0$, it follows from the given data that

$$y_0 = y(0) = 5.3 \text{ (billion)}$$

Since the growth rate is 2% ($k = 0.02$), it follows from (24) that the world population at time t will be

$$y(t) = y_0 e^{kt} = 5.3 e^{0.02t} \tag{25}$$

Since the beginning of the year 2015 corresponds to an elapsed time of $t = 25$ years ($2015 - 1990 = 25$), it follows from (25) that the world population by the year 2015 will be

$$y(25) = 5.3 e^{0.02(25)} = 5.3 e^{0.5} \approx 5.3(1.6487) = 8.7381 \text{ (billion)}$$

which is a population of approximately 8.7 billion.   ◀

☐ **DOUBLING AND HALVING TIME**   If a quantity has an exponential growth model, then the time required for it to double in size is called the ***doubling time***. Similarly, if a quantity has an exponential decay model, then the time required for it to reduce in value by half is called the ***halving time***. As it turns out, doubling and halving times depend only on the growth rate and not on the amount present initially. To see why, suppose y has an exponential growth model so that

$$y = y_0 e^{kt} \quad (k > 0)$$

At any fixed time t_1 let

$$y_1 = y_0 e^{kt_1} \tag{26}$$

be the value of y, and let T denote the amount of time required for y to double in size. Thus, at time $t_1 + T$ the value of y will be $2y_1$, so

$$2y_1 = y_0 e^{k(t_1 + T)} = y_0 e^{kt_1} e^{kT}$$

or, from (26),

$$2y_1 = y_1 e^{kT}$$

Thus,

$$2 = e^{kT} \quad \text{and} \quad \ln 2 = kT$$

Therefore, the doubling time T is

> **Doubling Time**
> $$T = \frac{1}{k} \ln 2 \tag{27}$$

which does not depend on y_0 or t_1. We leave it as an exercise to show that the halving time for a quantity with an exponential decay model ($k < 0$) is

> **Halving Time**
> $$T = -\frac{1}{k} \ln 2 \tag{28}$$

Example 9 It follows from (27) that at the current 2% annual growth rate, the doubling time for the world population is

$$T = \frac{1}{0.02} \ln 2 \approx \frac{1}{0.02} (0.6931) = 34.655$$

or approximately 35 years. Thus, with a continued 2% annual growth rate the population of 5.3 billion in 1990 will double to 10.6 billion by the year 2025 and will double again to 21.2 billion by 2060. ◀

Radioactive elements continually undergo a process of disintegration called *radioactive decay*. It is a physical fact that at each instant of time the rate of decay is proportional to the amount of the element present. Consequently, the amount of any radioactive element has an exponential decay model. For radioactive elements, halving time is called *half-life*.

Example 10 (*Radioactive Decay*) The radioactive element carbon-14 has a half-life of 5750 years. If 100 grams of this element are present initially, how much will be left after 1000 years?

Solution.   From (28) the decay constant is

$$k = -\frac{1}{T} \ln 2 \approx -\frac{1}{5750} (0.6931) \approx -0.00012$$

Thus, if we take $t = 0$ to be the present time, then $y_0 = y(0) = 100$, so that (24) implies that the amount of carbon-14 after 1000 years will be

$$y(1000) = 100e^{-0.00012(1000)} = 100e^{-0.12} \approx 100(0.88692) = 88.692$$

Thus, about 88.69 grams of carbon-14 will remain.   ◀

▶ Exercise Set 7.7   Ⓒ 27, 29–32, 34–41, 46

In Exercises 1–6, solve the given separable differential equation. Where convenient, express the solution explicitly as a function of x.

1. $\dfrac{dy}{dx} = \dfrac{y}{x}$.

2. $\dfrac{dy}{dx} = \dfrac{x^3}{(1 + x^4)y}$.

3. $\sqrt{1 + x^2}\, y' + x(1 + y) = 0$.

4. $3 \tan y - \dfrac{dy}{dx} \sec x = 0$.

5. $e^{-y} \sin x - y' \cos^2 x = 0$.

6. $\dfrac{dy}{dx} = 1 - y + x^2 - yx^2$.

In Exercises 7–12, solve the given first-order linear differential equation. Where convenient, express the solution explicitly as a function of x.

7. $\dfrac{dy}{dx} + 3y = e^{-2x}$.

8. $\dfrac{dy}{dx} - \dfrac{5}{x} y = x$ $(x > 0)$.

9. $y' + y = \cos(e^x)$.   **10.** $2\dfrac{dy}{dx} + 4y = 1$.

11. $x^2 y' + 3xy + 2x^5 = 0$ $(x > 0)$.

12. $\dfrac{dy}{dx} + y - \dfrac{1}{1 + e^x} = 0$.

In Exercises 13–18, solve the initial-value problems.

13. $\dfrac{dy}{dx} - xy = x$, $y(0) = 3$.

14. $2y\dfrac{dy}{dx} = 3x^2(x^3 + 1)^{-1/2}$, $y(2) = 1$.

15. $\dfrac{dy}{dt} + y = 2$, $y(0) = 1$.

16. $y' - xe^y = 2e^y$, $y(0) = 0$.

17. $y^2 t \dfrac{dy}{dt} - t + 1 = 0$, $y(1) = 3$ $(t > 0)$.

18. $y' \cosh x + y \sinh x = \cosh^2 x$, $y(0) = \frac{1}{4}$.

In Exercises 19–22, find an equation of the curve in the xy-plane that passes through the given point and whose tangent at (x, y) has the given slope.

19. $(1, -1)$; slope $= \dfrac{y^2}{3\sqrt{x}}$.

20. $(1, 1)$; slope $= \dfrac{3x^2}{2y}$.   **21.** $(2, 0)$; slope $= xe^y$.

22. $(1, -1)$; slope $= 2y + 3$, $y > -\frac{3}{2}$.

The following discussion is needed for Exercises 23 and 24. Suppose that a tank containing a liquid is vented to the air at the top and has an outlet at the bottom through which the liquid can drain. It follows from *Torricelli's law* in physics that if the outlet is opened at time $t = 0$, then at each instant the depth of the liquid $h(t)$ and the area $A(h)$ of the liquid's surface are related by

$$A(h) \frac{dh}{dt} = -k\sqrt{h}$$

where k is a positive constant that depends on such factors as the viscosity of the liquid and the cross-sectional area of the outlet. Use this result in Exercises 23 and 24, assuming that h is in feet, $A(h)$ is in square feet, and t is in seconds. A calculator will be useful.

23. Suppose that the cylindrical tank in Figure 7.7.2 is filled to a depth of 4 ft at time $t = 0$ and that the constant in Torricelli's law is $k = 0.025$.
   (a)   Find $h(t)$.
   (b)   How many minutes will it take for the tank to drain completely?

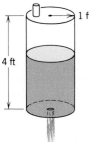

Figure 7.7.2

24. Follow the directions of Exercise 23 for the cylindrical tank in Figure 7.7.3, assuming that the tank is filled to a depth of 4 ft at time $t = 0$ and that the constant in Torricelli's law is $k = 0.025$.

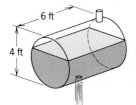

Figure 7.7.3

25. A particle moving along the x-axis encounters a resisting force that results in an acceleration of $a = dv/dt = -0.04v^2$. Given that $x = 0$ cm and $v = 50$ cm/sec at time $t = 0$, find the velocity v and position x as a function of t for $t \geq 0$.

26. A particle moving along the x-axis encounters a resisting force that results in an acceleration of $a = dv/dt = -0.02\sqrt{v}$. Given that $x = 0$ cm and $v = 9$ cm/sec at time $t = 0$, find the velocity v and position x as a function of t for $t \geq 0$.

27. A rocket, fired upward from rest at time $t = 0$, has an initial mass of m_0 (including its fuel). Assuming that the fuel is consumed at a constant rate k, the mass m of the rocket, while fuel is being burned, will be given by $m = m_0 - kt$. It can be shown that if air resistance is neglected and the fuel gases are expelled at a constant speed c relative to the rocket,

then the velocity v of the rocket will satisfy the equation

$$m\frac{dv}{dt} = ck - mg$$

where g is the acceleration due to gravity.
   (a)   Find $v(t)$ keeping in mind that the mass m is a function of t.
   (b)   Suppose that the fuel accounts for 80% of the initial mass of the rocket and that all of the fuel is consumed in 100 sec. Find the velocity of the rocket in meters/second at the instant the fuel is exhausted. (Use $g = 9.8$ m/sec^2 and $c = 2500$ m/sec.)

28. An object of mass m is dropped from rest at time $t = 0$. If we assume that the only forces acting on the object are the constant force due to gravity and a retarding force of air resistance, which is proportional to the velocity $v(t)$ of the object, then the velocity will satisfy the equation

$$m\frac{dv}{dt} = mg - kv$$

where k is a positive constant and g is the acceleration due to gravity.
   (a)   Find $v(t)$.
   (b)   Use the result in part (a) to find $\lim\limits_{t \to +\infty} v(t)$.
   (c)   Find the distance $x(t)$ that the object has fallen at time t given that $x = 0$ when $t = 0$.

29. An object of constant mass is projected upward from the surface of the earth. Neglecting air resistance, the gravitational force exerted by the earth on the object results in an acceleration of

$$a = dv/dt = -gR^2/x^2$$

where x is the distance of the object from the center of the earth, R is the radius of the earth, and g is the acceleration due to gravity at the surface of the earth. From the chain rule $dv/dt = v\,dv/dx$, so v and x must satisfy the equation

$$v\,dv/dx = -gR^2/x^2$$

   (a)   Find v^2 in terms of x given that $v = v_0$ when $x = R$.
   (b)   Use the result in part (a) to show that the velocity cannot reach zero if $v_0 \geq \sqrt{2gR}$.
   (c)   The minimum value of v_0 that is required for the object not to fall back to the earth is called the *escape velocity*. Assuming that $g = 32$ ft/sec^2 and $R = 3960$ mi, show that the escape velocity is approximately 6.9 mi/sec.

30. A bullet of mass m, fired straight up with an initial velocity of v_0, is slowed by the force of gravity and a drag force of air resistance kv^2, where g is the constant acceleration due to gravity and k is a positive constant. As the bullet moves upward, its velocity v satisfies the equation

$$m\frac{dv}{dt} = -(kv^2 + mg)$$

(a) Show that if x is the position of the bullet at time t (Figure 7.7.4), then

$$mv\frac{dv}{dx} = -(kv^2 + mg)$$

(b) Express x in terms of v given that $x = 0$ when $v = v_0$.

(c) Assuming that $v_0 = 988$ m/sec, $g = 9.8$ m/sec^2, $m = 3.56 \times 10^{-3}$ kg, and $k = 7.3 \times 10^{-6}$, use the result in part (b) to find out how high the bullet rises. [*Hint:* Find the velocity of the bullet at its highest point.]

Figure 7.7.4

31. At time $t = 0$, a tank contains 25 lb of salt dissolved in 50 gal of water. Then brine containing 4 lb of salt per gallon of water is allowed to enter the tank at a rate of 2 gal/min and the mixed solution is drained from the tank at the same rate.
(a) How much salt is in the tank at an arbitrary time t?
(b) How much salt is in the tank after 25 min?

32. A tank initially contains 200 gal of pure water. Then at time $t = 0$ brine containing 5 lb of salt per gallon of water is allowed to enter the tank at a rate of 10 gal/min and the mixed solution is drained from the tank at the same rate.
(a) How much salt is in the tank at an arbitrary time t?
(b) How much salt is in the tank after 30 min?

33. A tank with a 1000-gal capacity initially contains 500 gal of brine containing 50 lb of salt. At time $t = 0$, pure water is added at a rate of 20 gal/min and the mixed solution is drained off at a rate of 10 gal/min. How much salt is in the tank when it reaches the point of overflowing?

34. The number of bacteria in a certain culture grows exponentially at a rate of 1% per hour. Assuming that 10,000 bacteria are present initially, find
(a) the number of bacteria present at any time t
(b) the number of bacteria present after 5 hr
(c) the time required for the number of bacteria to reach 45,000.

35. Polonium-210 is a radioactive element with a half-life of 140 days. Assume that a sample weighs 10 mg initially.
(a) Find a formula for the amount that will remain after t days.
(b) How much will remain after 10 weeks?

36. In a certain chemical reaction a substance decomposes at a rate proportional to the amount present. Tests show that under appropriate conditions 15,000 grams will reduce to 5000 grams in 10 hours.
(a) Find a formula for the amount that will remain from a 15,000-gram sample after t hours.
(b) How long will it take for 50% of an initial sample of y_0 grams to decompose?

37. One hundred fruit flies are placed in a breeding container that can support a population of at most 5000 flies. If the population grows exponentially at a rate of 2% per day, how long will it take for the container to reach capacity?

38. In 1960 the American scientist W. F. Libby won the Nobel prize for his discovery of carbon dating, a method for determining the age of certain fossils. Carbon dating is based on the fact that nitrogen is converted to radioactive carbon-14 by cosmic radiation in the upper atmosphere. This radioactive carbon is absorbed by plant and animal tissue through the life processes while the plant or animal lives. However, when the plant or animal dies the absorption process stops and the amount of carbon-14 decreases through radioactive decay. Suppose that tests on a fossil show that 70% of its carbon-14 has decayed. Estimate the age of the fossil, assuming a half-life of 5750 years for carbon-14.

39. Forty percent of a radioactive substance decays in 5 years. Find the half-life of the substance.

40. Assume that if the temperature is constant, then the atmospheric pressure p varies with the altitude h (above sea level) in such a way that $dp/dh = kp$, where k is a constant.
   (a) Find a formula for p in terms of k, h, and the atmospheric pressure p_0 at sea level.
   (b) Given that p measures 15 lb/in^2 at sea level and 12 lb/in^2 at 5000 ft above sea level, find the pressure at 10,000 ft (assuming temperature is constant).

41. The town of Grayrock had a population of 10,000 in 1980 and 12,000 in 1990.
   (a) Assuming an exponential growth model, estimate the population in 2000.
   (b) What is the doubling time for the town's population?

42. Prove: If a quantity A has an exponential growth or decay model and A has values A_1 and A_2 at times t_1 and t_2, respectively, then the growth rate k is

$$k = \frac{1}{t_1 - t_2} \ln\left(\frac{A_1}{A_2}\right)$$

43. Newton's law of cooling states that the rate at which an object cools is proportional to the difference in temperature between the object and the surrounding medium. Show that if C is the constant temperature of a surrounding medium, then the temperature $T(t)$ at time t of a cooling object is given by

$$T(t) = (T_0 - C)e^{kt} + C$$

where T_0 is the temperature of the object at $t = 0$ and k is a negative constant.

44. A liquid with an initial temperature of 200° is enclosed in a metal container that is held at a constant temperature of 80°. If the liquid cools to 120° in 30 min, what will the temperature be after 1 hr? (Use the result of Exercise 43.)

45. Suppose P dollars is invested at an annual interest rate of $r \times 100\%$. If the accumulated interest is cred-

ited to the account at the end of the year, then the interest is said to be *compounded annually;* if it is credited at the end of each six-month period, then it is said to be *compounded semiannually;* and if it is credited at the end of each three-month period, then it is said to be *compounded quarterly.* The more frequently the interest is compounded, the better it is for the investor since more of the interest is itself earning interest.
   (a) Show that if interest is compounded n times a year at equally spaced intervals, then the value A of the investment after t years is

$$A = P\left(1 + \frac{r}{n}\right)^{nt}$$

   (b) One can imagine interest to be compounded each day, each hour, each minute, and so forth. Carried to the limit one can conceive of interest compounded at each instant of time; this is called *continuous compounding.* Thus, from part (a), the value A of P dollars after t years when invested at an annual rate of $r \times 100\%$, compounded continuously, is

$$A = \lim_{n \to +\infty} P\left(1 + \frac{r}{n}\right)^{nt}$$

   Use the fact that $\lim_{x \to 0}(1 + x)^{1/x} = e$ to prove that $A = Pe^{rt}$.
   (c) Use the result in part (b) to show that money invested at continuous compound interest increases at a rate proportional to the amount present.

46. (a) If $1000 is invested at 8% per year compounded continuously (Exercise 45), what will the investment be worth after 5 years?
   (b) If it is desired that an investment at 8% per year compounded continuously should have a value of $10,000 after 10 years, how much should be invested now?
   (c) How long does it take for an investment at 8% per year compounded continuously to double in value?

47. Derive Formula (28) for halving time.

48. Let a quantity have an exponential growth model with growth rate k. How long does it take for the quantity to triple in size?

Answers to Odd-Numbered Exercises

▶ Exercise Set 14.1 **(Page 843)**

1. (a) $\sqrt{14}$; $(1, \frac{1}{2}, \frac{3}{2})$ (b) $\sqrt{11}$; $(\frac{9}{2}, \frac{3}{2}, \frac{9}{2})$
    (c) $\sqrt{30}$; $(\frac{1}{2}, -\frac{1}{2}, 4)$ (d) $\sqrt{42}$; $(\frac{3}{2}, 1, -\frac{5}{2})$

3. $(-6, 2, 1)$, $(-6, 2, -2)$, $(-6, 1, -2)$, $(4, 2, 1)$,
    $(4, 1, 1)$, $(4, 1, -2)$, $(4, 2, -2)$, $(-6, 1, 1)$

5. (b) $(2, 1, 6)$ (c) 49

7. distance to x-axis is $\sqrt{y_0{}^2 + z_0{}^2}$
    distance to y-axis is $\sqrt{x_0{}^2 + z_0{}^2}$

9. $(x + 2)^2 + (y - 4)^2 + (z + 1)^2 = 36$

11. $x^2 + (y - 1)^2 + z^2 = 9$

13. (a) $(x - 2)^2 + (y + 1)^2 + (z + 3)^2 = 9$
    (b) $(x - 2)^2 + (y + 1)^2 + (z + 3)^2 = 1$
    (c) $(x - 2)^2 + (y + 1)^2 + (z + 3)^2 = 4$

15. $(x + \frac{1}{2})^2 + (y - 2)^2 + (z - 2)^2 = \frac{5}{4}$

17. $(x - 3)^2 + (y + 2)^2 + (z - 4)^2 = 41$

19. $x^2 + y^2 + z^2 = 30 \pm 2\sqrt{29}$

21. sphere, center $(-5, -2, -1)$, radius 7

23. sphere, center $(\frac{1}{2}, \frac{3}{4}, -\frac{5}{4})$, radius $\frac{3}{4}\sqrt{6}$

25. no graph

27. largest, $3 + \sqrt{6}$; smallest, $3 - \sqrt{6}$

29. all points outside the circular cylinder
    $(y + 3)^2 + (z - 2)^2 = 16$

31. $(2 - \sqrt{3})R$

33. (a)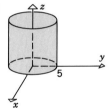
(b)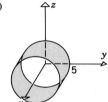
(c)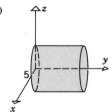

35. (a)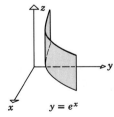
$y = e^x$
(b)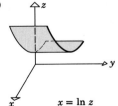
$x = \ln z$
(c)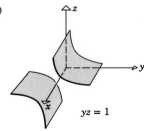
$yz = 1$

37. (a)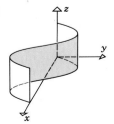
(b)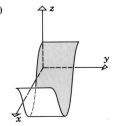

39. (a)
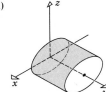

(b)
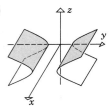

▶ Exercise Set 14.2 **(Page 853)**

1.
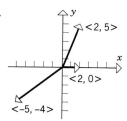

3.
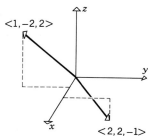

5.
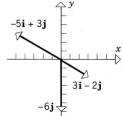

7.
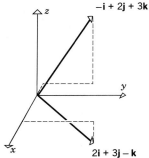

9. (a) $\langle -1, 3 \rangle$ (b) $\langle -7, 2 \rangle$
(c) $\langle 2, 1 \rangle$ (d) $\langle -8, 7 \rangle$

11. (a) $\langle -3, 6, 1 \rangle$ (b) $\langle 1, -3, -5 \rangle$

13. $(4, -4)$ **15.** $(4, -4)$

17. $(8, -1, -3)$

19. (a) $-5i - 2j$ (b) $8i + 10j$
(c) $-2i + 4j$ (d) $40i + 36j$
(e) $-20i + 22j$ (f) $-8i - 8j$

21. (a) $-i + 4j - 2k$ (b) $18i + 12j - 6k$
(c) $-i - 5j - 2k$ (d) $40i - 4j - 4k$
(e) $-2i - 16j - 18k$ (f) $-i + 13j - 2k$

23. (a) $\sqrt{2}$ (b) 2   (c) 3

25. (a) $\sqrt{14}$ (b) 3

27. (a) $\sqrt{41}$ (b) 7
(c) $3\sqrt{29} + 8$ (d) $3\sqrt{10}$
(e) $\langle \frac{3}{5}, \frac{4}{5} \rangle$ (f) 1

29. (a) $2\sqrt{3}$ (b) $\sqrt{14} + \sqrt{2}$
(c) $2\sqrt{14} + 2\sqrt{2}$ (d) $2\sqrt{37}$
(e) $\dfrac{1}{\sqrt{6}}i + \dfrac{1}{\sqrt{6}}j - \dfrac{2}{\sqrt{6}}k$ (f) 1

31. $\langle -\frac{2}{3}, 1 \rangle$ **33.** $\mathbf{u} = \frac{5}{7}i + \frac{2}{7}j + \frac{1}{7}k,$
$\mathbf{v} = \frac{8}{7}i - \frac{1}{7}j - \frac{1}{7}k$

43. $-\frac{3}{5}i + \frac{4}{5}j$ **45.** $-\dfrac{1}{\sqrt{14}}(3i - 2j + k)$

47. $\dfrac{1}{3\sqrt{2}}(4i + j - k)$

49. $\langle -\frac{3}{2}, 2 \rangle$ **51.** $6i - 8j - 2k$

53. circle; radius 1, center at (x_0, y_0)

55. (a) sphere; radius 2, center at $(0, 0, 0)$
(b) sphere; radius 3, center at (x_0, y_0, z_0)
(c) all points on and inside the sphere of radius 1, centered at (x_0, y_0, z_0)

57. (a) $\left\langle \dfrac{1}{\sqrt{2}}, -\dfrac{1}{\sqrt{2}} \right\rangle$, $\left\langle -\dfrac{1}{\sqrt{2}}, \dfrac{1}{\sqrt{2}} \right\rangle$
(b) $\left\langle \dfrac{1}{\sqrt{2}}, \dfrac{1}{\sqrt{2}} \right\rangle$, $\left\langle -\dfrac{1}{\sqrt{2}}, -\dfrac{1}{\sqrt{2}} \right\rangle$

59. $(\frac{13}{4}, \frac{21}{4})$

61. (a) $\langle 1/2, \sqrt{3}/2 \rangle$ (b) $\langle -2\sqrt{2}, 2\sqrt{2} \rangle$

63. $(1, 1), (-1, -1)$ **77.** (a) $(\frac{3}{2}, \frac{1}{2})$

▶ Exercise Set 14.3 (Page 864)

1. (a) -10 (b) -3 (c) 0   (d) -20

3. (a) obtuse   (b) acute
   (c) obtuse   (d) orthogonal

5. (a) $\frac{14}{13}\mathbf{i} + \frac{21}{13}\mathbf{j}$ (b) $\langle 2, 6 \rangle$
   (c) $-\frac{11}{13}\mathbf{i} + \mathbf{j} + \frac{55}{13}\mathbf{k}$ (d) $\langle -\frac{32}{89}, -\frac{12}{89}, \frac{73}{89} \rangle$

9. (a) 6   (b) 36   (c) $24\sqrt{5}$ (d) $24\sqrt{5}$

11. $\dfrac{1}{5\sqrt{2}}, \dfrac{4}{\sqrt{65}}, \dfrac{9}{\sqrt{130}}$

13. The right angle is at vertex B.

15. (a) the line through the origin and perpendicular to $\mathbf{r}_0$
    (b) the line through (x_0, y_0) and perpendicular to $\mathbf{r}_0$
    (c) circle; center at $(\frac{1}{2}x_0, \frac{1}{2}y_0)$, radius $\frac{1}{2}\|\mathbf{r}_0\|$

17. (a) $-\frac{3}{4}$ (b) $\frac{1}{7}$
    (c) $\dfrac{48 + 25\sqrt{3}}{11}$ (d) $\frac{4}{3}$

21. (a) $\frac{2}{5}$ (b) $2/\sqrt{5}$ (c) $2/\sqrt{5}$

23. (a) $\frac{4}{3}$ (b) $\frac{1}{3}\sqrt{137}$

25. -12 ft $\cdot$ lb   31. $71°$

▶ Exercise Set 14.4 (Page 873)

1. $\langle 7, 10, 9 \rangle$ 3. $\langle -4, -6, -3 \rangle$

5. (a) $\langle -20, -67, -9 \rangle$ (b) $\langle -78, 52, -26 \rangle$
   (c) $\langle 24, 0, -16 \rangle$ (d) $\langle -12, -22, -8 \rangle$
   (e) $\langle 0, -56, -392 \rangle$ (f) $\langle 0, 56, 392 \rangle$

9. $\pm \dfrac{1}{\sqrt{5}}(2\mathbf{j} + \mathbf{k})$ 11. $2\mathbf{v} \times \mathbf{u}$

13. (a) $\frac{1}{2}\sqrt{374}$ (b) $9\sqrt{13}$ 15. (a) $\frac{1}{2}\sqrt{26}$ (b) $\frac{1}{3}\sqrt{26}$

17. (a) $2\sqrt{\dfrac{141}{29}}$ (b) $\frac{1}{3}\sqrt{137}$

19. ambiguous, needs parentheses

21. 80   23. 1

25. (a) 16   (b) 45

27. (a) 9   (b) $\sqrt{122}$ (c) $\sin^{-1}(\frac{9}{14})$

37. (a) $\frac{2}{3}$ (b) $\frac{1}{2}$

▶ Exercise Set 14.5 (Page 880)

1. $x = 3 + 2t, y = -2 + 3t$

3. $x = 4, y = 1 + 2t$

5. $x = 5 - 3t, y = -2 + 6t, z = 1 + t$

7. $x = -t, y = 6t, z = t$

9. same as Exercise 1 with $0 \leq t \leq 1$

11. same as Exercise 3 with $0 \leq t \leq 1$

13. same as Exercise 5 with $0 \leq t \leq 1$

15. same as Exercise 7 with $0 \leq t \leq 1$

17. $x = -5 + 2t, y = 2 - 3t$

19. $x = 3 + 4t, y = -4 + 3t$

21. $x = -1 + 3t, y = 2 - 4t, z = 4 + t$

23. $x = -2 + 2t, y = -t, z = 5 + 2t$

25. $x = 3 + t, y = 7, z = 0$ 27. $(-1, 1), (3, 9)$

29. (a) $(-2, 10, 0)$ (b) $(-2, 0, -5)$
    (c) does not intersect yz-plane

31. $(0, 4, -2), (4, 0, 6)$

33. $x = x_1 + at, y = y_1 + bt, z = z_1 + ct$

35. $(1, -1, 2)$ 39. (a) no   (b) yes

41. $(1, \frac{14}{3}, -\frac{5}{3})$

45. (a) $\langle x, y \rangle = \langle 2, -1 \rangle + t\langle -7, 4 \rangle$
    (b) $\langle x, y \rangle = \langle 0, 3 \rangle + t\langle 4, 0 \rangle$

47. the line segment joining the points $(1, 0)$ and $(-3, 6)$

49. $2\sqrt{5}$ 51. $\sqrt{35/6}$

53. (b) $84°$ (c) $x = 7 + t, y = -1, z = -2 + t$

55. $x = t, y = 2 + t, z = 1 - t$

57. (a) $\sqrt{17}$ cm   (b) $\frac{1}{2}\sqrt{14}$ cm

▶ Exercise Set 14.6 (Page 888)

1. $x + 4y + 2z = 28$ 3. $z = 0$

5. (a) $2y - z = 1$
   (b) $x + 9y - 5z = 16$

7. (a) yes   (b) no   9. (a) yes   (b) no

11. (a) $35°$ (b) $79°$

13. (a) $z = 0$ (b) $y = 0$ (c) $x = 0$

15. $4x - 2y + 7z = 0$ 17. $4x - 13y + 21z = -14$

19. $x = 5 - 2t, y = 5t, z = -2 + 11t$

21. $x + y - 3z = 6$ 23. $7x + y + 9z = 25$

25. $2x + 4y + 8z = 29$

29. (a) $x = -\frac{11}{7} - 23t, y = -\frac{12}{7} + t, z = -7t$
    (b) $x = -5t, y = -3t, z = 0$

31. $\frac{5}{3}$ 33. $\frac{137}{21}$

35. $\dfrac{5}{3\sqrt{6}}$ 37. $\dfrac{25}{\sqrt{126}}$ 39. $\dfrac{95}{\sqrt{1817}}$

43. $(x - 2)^2 + (y - 1)^2 + (z + 3)^2 = \frac{121}{14}$

► Exercise Set 14.7 **(Page 900)**

1. (a) $4x^2 + y^2 = 4$; ellipse   (b) $y^2 + z^2 = 3$; circle
   (c) $4x^2 + z^2 = 3$; ellipse

3. (a) $9x^2 - z^2 = 16$; hyperbola   (b) $y^2 + z^2 = 20$; circle
   (c) $9x^2 - y^2 = 20$; hyperbola

5. (a) $z = 4y^2$; parabola
   (b) $z = 9x^2 + 16$; parabola
   (c) $9x^2 + 4y^2 = 4$; ellipse

7.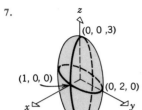

Ellipsoid

9.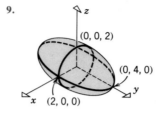

Ellipsoid

11.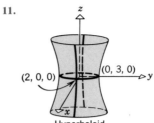

Hyperboloid
of one sheet

13.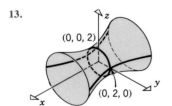

Hyperboloid
of one sheet

15.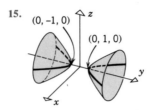

Hyperboloid
of two sheets

17.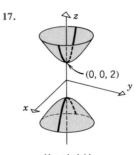

Hyperboloid
of two sheets

19.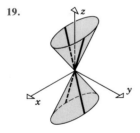

Elliptic cone

21.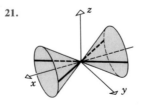

Circular cone

23.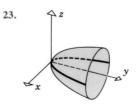

Circular paraboloid

25.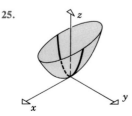

Elliptic paraboloid

27.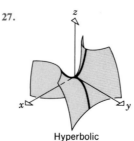

Hyperbolic
paraboloid

(a)

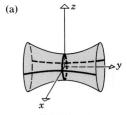

Hyperboloid
of one sheet

(b)

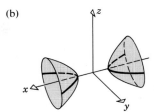

Hyperboloid
of two sheets

(c)

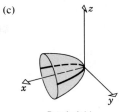

Paraboloid

(d)

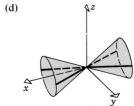

Cone

(e)

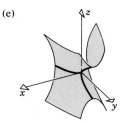

Hyperbolic
paraboloid

(f)

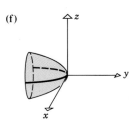

Paraboloid

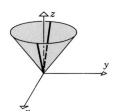

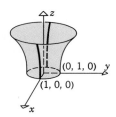

(0, 1, 0)
(1, 0, 0)

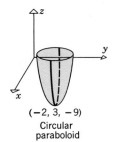

(−2, 3, −9)
Circular
paraboloid

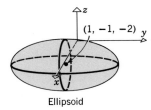

(1, −1, −2)

Ellipsoid

(0, −1, 5)

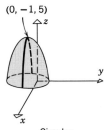

Circular
paraboloid

53. (a) focus: $\left(k, 0, \dfrac{k^2}{9} + 1\right)$, vertex: $\left(k, 0, \dfrac{k^2}{9}\right)$

(b) foci: $(\sqrt{5k}, 0, k)$, $(-\sqrt{5k}, 0, k)$;
endpoints of major axis: $(3\sqrt{k}, 0, k)$,
$(-3\sqrt{k}, 0, k)$;
endpoints of minor axis: $(0, 2\sqrt{k}, k)$,
$(0, -2\sqrt{k}, k)$

45. $x^2 + y^2 = 2$; circle

47. $x^2 + y^2 - 2x = 0$; circle

49. $y = \frac{1}{2}x^2 - \frac{1}{2}$; parabola

51. $x^2 + y^2 + 4x = 5$, $x \geq 0$; circular arc

55. $z = \frac{1}{4}(x^2 + y^2)$; paraboloid

▶ Exercise Set 14.8 **(Page 906)**

1. (a) $(8, \pi/6, -4)$ (b) $(5\sqrt{2}, 3\pi/4, 6)$
   (c) $(2, \pi/2, 0)$ (d) $(8, 5\pi/3, 6)$
   (e) $(2, 7\pi/4, 1)$ (f) $(0, 0, 1)$

3. (a) $(2\sqrt{2}, \pi/3, 3\pi/4)$ (b) $(2, 7\pi/4, \pi/4)$
   (c) $(6, \pi/2, \pi/3)$ (d) $(10, 5\pi/6, \pi/2)$
   (e) $(8\sqrt{2}, \pi/4, \pi/6)$ (f) $(2\sqrt{2}, 5\pi/3, 3\pi/4)$

5. (a) $(2\sqrt{3}, \pi/6, \pi/6)$ (b) $(\sqrt{2}, \pi/4, 3\pi/4)$
   (c) $(2, 3\pi/4, \pi/2)$ (d) $(4\sqrt{3}, 1, 2\pi/3)$
   (e) $(4\sqrt{2}, 5\pi/6, \pi/4)$ (f) $(2\sqrt{2}, 0, 3\pi/4)$

7.
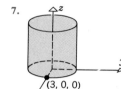
$x^2 + y^2 = 9$
(3, 0, 0)

9.
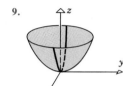
$z = x^2 + y^2$

11.
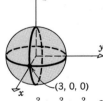
$x^2 + (y-2)^2 = 4$
(0, 4, 0)

13.
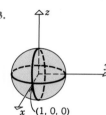
$x^2 + y^2 + z^2 = 1$
(1, 0, 0)

15.
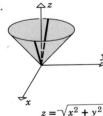
(3, 0, 0)
$x^2 + y^2 + z^2 = 9$

17.
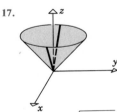
$z = \sqrt{x^2 + y^2}$

19.
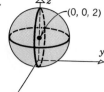
(0, 0, 2)
$x^2 + y^2 + (z - 2)^2 = 4$

21.
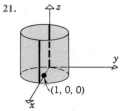
(1, 0, 0)
$(x - 1)^2 + y^2 = 1$

23. (a) $z = 3$ (b) $\rho = 3 \sec \phi$
25. (a) $z = 3r^2$ (b) $\rho = \frac{1}{3} \csc \phi \cot \phi$
27. (a) $r = 2$ (b) $\rho = 2 \csc \phi$
29. (a) $r^2 + z^2 = 9$ (b) $\rho = 3$
31. (a) $r(2 \cos \theta + 3 \sin \theta) + 4z = 1$
    (b) $\rho(2 \sin \phi \cos \theta + 3 \sin \phi \sin \theta + 4 \cos \phi) = 1$
33. (a) $r^2 \cos^2 \theta = 16 - z^2$
    (b) $\rho^2(1 - \sin^2 \phi \sin^2 \theta) = 16$
35. all points on or above the paraboloid $z = x^2 + y^2$ that are also on or below the plane $z = 4$
37. all points on or between the concentric spheres $\rho = 1$ and $\rho = 3$
39. spherical: $(4000, \pi/6, \pi/6)$
    rectangular: $(1000\sqrt{3}, 1000, 2000\sqrt{3})$

41.
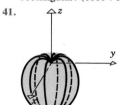

▶ Chapter 14 Supplementary Exercises **(Page 907)**

1. (a) $3\mathbf{i} - 4\mathbf{j}$, 5   (b) $-\mathbf{i} + 4\mathbf{j}$, $\sqrt{17}$
3. $-3\mathbf{i} + 4\mathbf{j}$ 5. $\frac{3}{5}(4\mathbf{i} + 3\mathbf{j})$
7. $-6\mathbf{i} + 6\sqrt{3}\mathbf{j}$ 9. $\mathbf{i} + \frac{1}{2}\mathbf{j}$
11. $\mathbf{F}_3 = -(\mathbf{F}_1 + \mathbf{F}_2) = \mathbf{i} + 5\mathbf{j}$
13. (a) $\sqrt{6}$ (b) -3
    (c) $\langle 5, -5, -5 \rangle$ (d) $\langle -5, 5, 5 \rangle$
    (e) $\frac{5}{2}\sqrt{3}$ (f) $\langle -1, 8, -9 \rangle$
15. (a) $\frac{2}{3}$ (b) $\frac{2}{5}$
    (c) $\cos^{-1}(-\frac{2}{15})$ (d) $\frac{2}{5}, -\frac{4}{5}, 0$
17. Each side reduces to $2\mathbf{i} - 2\mathbf{j} + \mathbf{k}$.

19. $\langle -3/\sqrt{2}, 0, 3/\sqrt{2} \rangle$
21. (a) $-\frac{1}{2}\mathbf{v}$ (b) $\frac{1}{2}(3\mathbf{i} + 5\mathbf{j} - 4\mathbf{k})$
23. $62°$ 25. (a) $\|\mathbf{u}\| = \|\mathbf{v}\|$
27. $\pm(5\mathbf{i} + 7\mathbf{j} - \mathbf{k})/(5\sqrt{3})$
29. $x - y - z + 4 = 0$ 31. $x - y - z = -1$
33. (a) $k = -2, l = 12$ (b) yes, at $(7, 0, 0)$
    (c) $(\frac{136}{13}, -\frac{15}{13}, -\frac{60}{13})$
35. (a) $x = 2t + 1, y = 3t - 1, z = -3t + 2$
    (b) $x = 1, y = 5t - 3, z = -7t + 4$

37. (a) $x = t, y = -t + 2, z = t - 1$
    (b) $60°$

39. (a) the region inside the elliptic paraboloid $z = 4x^2 + 9y^2$
    (b) the point $(0, 0, 0)$

41. elliptic cone   43. ellipsoid

45. hyperboloid of two sheets   47. 13 ft · lb

49. (a) $(1, 1, 1)$ (b) $(\sqrt{3}, \pi/4, \tan^{-1}\sqrt{2})$

51. (a) $z = x^2 - y^2$ (b) $xz = 1$

▶ Exercise Set 15.1 **(Page 917)**

1. $(-\infty, +\infty); -\mathbf{i} - 3\pi\mathbf{j}$

3. $[2, +\infty); -\mathbf{i} - \ln 3\mathbf{j} + \mathbf{k}$

5. $\mathbf{r} = 3 \cos t\mathbf{i} + (t + \sin t)\mathbf{j}$

7. $\mathbf{r} = 2t\mathbf{i} + 2 \sin 3t\mathbf{j} + 5 \cos 3t\mathbf{k}$

9. $x = 3t^2, y = -2, z = 0$

11. $x = 2t - 1, y = -3\sqrt{t}, z = \sin 3t$

13. line in the xy-plane through $(2, 0)$, parallel to $-3\mathbf{i} - 4\mathbf{j}$

15. line through $(0, -3, 1)$, parallel to $2\mathbf{i} + 3\mathbf{k}$

17. ellipse in the plane $z = -1$, center at $(0, 0, -1)$, major axis of length 6 parallel to x-axis, minor axis of length 4 parallel to y-axis

19. $-\frac{3}{2}$

21. $(\frac{3}{2}, 0, \frac{3}{2})$

23.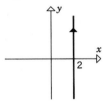

25.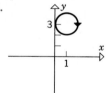

27.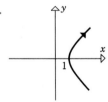

29.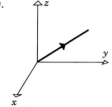

31.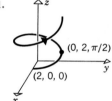

33.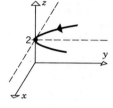

37. lies on the sphere $x^2 + y^2 + z^2 = 4$ and the plane $z = \sqrt{3}x$; center $(0, 0, 0)$, radius 2

39. $3/(2\pi)$ 41. conical helix

43. $x = t, y = \frac{1}{4}t^2 - 1, z = \frac{1}{4}t^2 + 1$; parabola

45. $x = 3 \cos t, y = 3 \sin t, z = 9 \cos^2 t$

47. $x = 1 + \cos t, y = \frac{1}{2} \sin t, z = 2 + 2 \cos t$

▶ Exercise Set 15.2 **(Page 926)**

1. $9\mathbf{i} + 6\mathbf{j}$

3. $\mathbf{j}$

5. $2\mathbf{i} - 3\mathbf{j} + 4\mathbf{k}$

7. $\frac{1}{2}\pi\mathbf{i} + \mathbf{k}$

11. $5\mathbf{i} + (1 - 2t)\mathbf{j}$

13. $-\dfrac{1}{t^2}\mathbf{i} + \sec^2 t\mathbf{j} + 2e^{2t}\mathbf{k}$

15.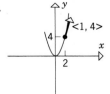

17.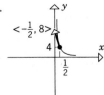

19.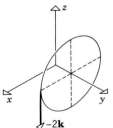

21.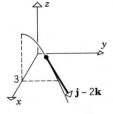

23. $x = 1 + 2t, y = 2 - t, z = 0$

25. $x = 1 - \sqrt{3}\pi t, y = \sqrt{3} + \pi t, z = 1 + 3t$

27. $\mathbf{r} = -\mathbf{i} + 2\mathbf{j} + t(2\mathbf{i} + \frac{3}{4}\mathbf{j})$

29. $\mathbf{r} = 4\mathbf{i} + \mathbf{j} + t(-4\mathbf{i} + \mathbf{j} + 4\mathbf{k})$

31. $2y + z = \pi/2$

33. (a) $(-2, 4, 6)$, $(1, 1, -3)$　　(b) $76°$, $71°$

35. $68°$　　　**39.** smooth　　**41.** not smooth

43. (a) $7t^6$

(b) $12t \sec t \tan t + 12 \sec t - \dfrac{\sin t}{t} - (\cos t) \ln t$

45. $4\mathbf{i} + 8(4u + 1)\mathbf{j}$　　　**47.** $2ue^{u^2}\mathbf{i} - 8ue^{-u^2}\mathbf{j}$

▶ **Exercise Set 15.3 (Page 935)**

1. $3t\mathbf{i} + 2t^2\mathbf{j} + \mathbf{C}$　　　**3.** $\langle 0, -\frac{2}{3} \rangle$

5. $\frac{52}{3}\mathbf{i} + 4\mathbf{j}$

7. $\langle (t - 1)e^t, t(\ln t - 1) \rangle + \mathbf{C}$

9. $\frac{1}{3}t^3\mathbf{i} - t^2\mathbf{j} + \ln|t|\mathbf{k} + \mathbf{C}$

11. $\frac{1}{2}(e^2 - 1)\mathbf{i} + (1 - e^{-1})\mathbf{j} + \frac{1}{2}\mathbf{k}$

13. (a) $\mathbf{F} = 3t\mathbf{i} - 2\mathbf{j} - 3t^2\mathbf{k}$　　(b) 30

15. $(1 + \sin t)\mathbf{i} - (\cos t)\mathbf{j}$

17. $(t^4 + 2)\mathbf{i} - (t^2 + 4)\mathbf{j}$

19. $2(t - 1)\mathbf{i} + \frac{1}{2}\ln\frac{1}{2}(t^2 + 1)\mathbf{j} + \frac{1}{2}(t^2 - 1)\mathbf{k}$

21. $\sqrt{14}$　　　　　　　**23.** 28

25. $e - e^{-1}$　　　　　　**27.** $t_0\sqrt{a^2 + c^2}$

29. $x = \frac{3}{5}s - 2$, $y = \frac{4}{5}s + 3$

31. $x = 3 + \cos s$, $y = 2 + \sin s$; $0 \le s \le 2\pi$

33. $x = \frac{1}{3}[(3s + 1)^{2/3} - 1]^{3/2}$
$y = \frac{1}{2}[(3s + 1)^{2/3} - 1]$; $s \ge 0$

35. $x = \left(\dfrac{s}{\sqrt{2}} + 1\right) \cos\left[\ln\left(\dfrac{s}{\sqrt{2}} + 1\right)\right]$

$y = \left(\dfrac{s}{\sqrt{2}} + 1\right) \sin\left[\ln\left(\dfrac{s}{\sqrt{2}} + 1\right)\right]$

where $0 \le s \le \sqrt{2}(e^{\pi/2} - 1)$

37. $x = (\sqrt{2s + 1} - 1) \cos(\sqrt{2s + 1} - 1)$,
$y = (\sqrt{2s + 1} - 1) \sin(\sqrt{2s + 1} - 1)$,
$z = \frac{2}{3}\sqrt{2}[\sqrt{2s + 1} - 1]^{3/2}$; $s \ge 0$

41. (b) $x = a(\cos\sqrt{2s/a} + \sqrt{2s/a} \sin\sqrt{2s/a})$
$y = a(\sin\sqrt{2s/a} - \sqrt{2s/a} \cos\sqrt{2s/a})$, $s \ge 0$

43. (a) $\frac{9}{2}$　　(b) $9 - 2\sqrt{6}$

45. (a) $\sqrt{3}(1 - e^{-2})$　　(b) $4\sqrt{5}$

▶ **Exercise Set 15.4 (Page 941)**

1.
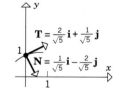

3.

5.
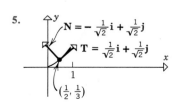

7.

9. $\mathbf{T} = -\dfrac{4}{\sqrt{17}}\mathbf{i} + \dfrac{1}{\sqrt{17}}\mathbf{k}$　　**11.** $\mathbf{T} = \mathbf{i}$

$\mathbf{N} = -\mathbf{j}$　　　　　　　　　　　$\mathbf{N} = \mathbf{j}$

13. $\mathbf{T} = -\dfrac{3}{\sqrt{10}}\mathbf{i} + \dfrac{1}{\sqrt{10}}\mathbf{k}$

$\mathbf{N} = -\mathbf{j}$

15. $\mathbf{T} = \dfrac{1}{\sqrt{5}}\mathbf{j} + \dfrac{2}{\sqrt{5}}\mathbf{k}$

$\mathbf{N} = -\dfrac{2}{\sqrt{5}}\mathbf{j} + \dfrac{1}{\sqrt{5}}\mathbf{k}$

19. See Exercise 1.　　　　**21.** See Exercise 3.

23. See Exercise 9.　　　　**25.** See Exercise 11.

27. $(\frac{4}{5}\cos t)\mathbf{i} - (\frac{4}{5}\sin t)\mathbf{j} - \frac{3}{5}\mathbf{k}$

▶ Exercise Set 15.5 **(Page 950)**

1. $\frac{96}{125}$

3. $\frac{6}{5\sqrt{10}}$

5. $\frac{3}{2\sqrt{2}}$

7. $\frac{4}{17}$

9. 1

11. $\frac{2}{5}$

13. $\frac{2}{5\sqrt{5}}$

17. 1

19. $\frac{1}{\sqrt{2}}$

21. $\frac{4}{5\sqrt{5}}$

23. $\cos x$; maximum for $x = 0$

25. See Exercise 1.

27. See Exercise 3.

29. See Exercise 5.

31. $\kappa(0) = a/b^2$, $\kappa(\pi/2) = b/a^2$

33. 1

35. $\frac{3}{2\sqrt{2}a}$

37. $\kappa(t) = \frac{1}{4a}\csc\frac{t}{2}$

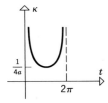

39.

$\rho = 1$

41.

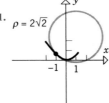
$\rho = 2\sqrt{2}$

43.

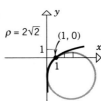
$\rho = 2\sqrt{2}$ (1, 0)

45.

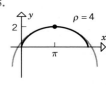
$\rho = 4$

47. $\rho(0) = \frac{1}{2}$

$\rho\left(\frac{\pi}{2}\right) = 4$

49. $2|p|$

51. $(3, 0)$, $(-3, 0)$

53. maximum 2, minimum $1/\sqrt{2}$

55. $\kappa = \frac{1}{\sqrt{1 + a^2}\,r}$

57. $\sim 2.86°/\text{cm}$

61. $\frac{1}{2r}$

63. (a) $2t + \frac{1}{t}$ (b) $2t + \frac{1}{t}$

(c) $8 + \ln 3$

65. (a) $\mathbf{T} = \frac{3}{5}\mathbf{i} + \frac{4}{5}\mathbf{j}$

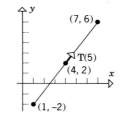
(7, 6)

T(5)

(4, 2)

(1, −2)

73. $\frac{2}{(t^2 + 2)^2}$

75. $-\frac{\sqrt{2}}{(e^t + e^{-t})^2}$

▶ Exercise Set 15.6 **(Page 966)**

1.

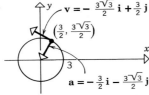
$\mathbf{v} = -\frac{3\sqrt{3}}{2}\mathbf{i} + \frac{3}{2}\mathbf{j}$

$\left(\frac{3}{2}, \frac{3\sqrt{3}}{2}\right)$

$\mathbf{a} = -\frac{3}{2}\mathbf{i} - \frac{3\sqrt{3}}{2}\mathbf{j}$

$\mathbf{v}(t) = -3\sin t\,\mathbf{i} + 3\cos t\,\mathbf{j}$
$\mathbf{a}(t) = -3\cos t\,\mathbf{i} - 3\sin t\,\mathbf{j}$
$\|\mathbf{v}(t)\| = 3$

3.

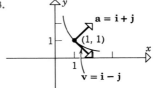
$\mathbf{a} = \mathbf{i} + \mathbf{j}$

(1, 1)

$\mathbf{v} = \mathbf{i} - \mathbf{j}$

$\mathbf{v}(t) = e^t\mathbf{i} - e^{-t}\mathbf{j}$
$\mathbf{a}(t) = e^t\mathbf{i} + e^{-t}\mathbf{j}$
$\|\mathbf{v}(t)\| = \sqrt{e^{2t} + e^{-2t}}$

5.

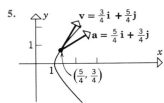

$$\mathbf{v} = \tfrac{3}{4}\mathbf{i} + \tfrac{5}{4}\mathbf{j}$$
$$\mathbf{a} = \tfrac{5}{4}\mathbf{i} + \tfrac{3}{4}\mathbf{j}$$

$$\left(\tfrac{5}{4}, \tfrac{3}{4}\right)$$

$$\mathbf{v}(t) = \sinh t\,\mathbf{i} + \cosh t\,\mathbf{j}$$
$$\mathbf{a}(t) = \cosh t\,\mathbf{i} + \sinh t\,\mathbf{j}$$
$$\|\mathbf{v}(t)\| = \sqrt{\sinh^2 t + \cosh^2 t}$$

7. $\mathbf{v} = \mathbf{i} + \mathbf{j} + \mathbf{k}$
$\|\mathbf{v}\| = \sqrt{3}$
$\mathbf{a} = \mathbf{j} + 2\mathbf{k}$

9. $\mathbf{v} = -\sqrt{2}\mathbf{i} + \sqrt{2}\mathbf{j} + \mathbf{k}$
$\|\mathbf{v}\| = \sqrt{5}$
$\mathbf{a} = -\sqrt{2}\mathbf{i} - \sqrt{2}\mathbf{j}$

11. $\mathbf{v} = e^{\pi/2}\mathbf{i} - e^{\pi/2}\mathbf{j} + \mathbf{k}$
$\|\mathbf{v}\| = \sqrt{2e^{\pi} + 1}$
$\mathbf{a} = -2e^{\pi/2}\mathbf{j}$

13. $3\sqrt{2};\ \mathbf{r} = 24\mathbf{i} + 8\mathbf{j}$

15. maximum 6, minimum 3

17. $15°$

19. $\mathbf{r} = -\tfrac{19}{16}\mathbf{i} + \tfrac{3}{2}\mathbf{j} + \tfrac{3}{16}\mathbf{k}$

21. The particles move counterclockwise around the circle $x^2 + y^2 = 4$; $\|\mathbf{r}_1'\| = 6$, $\|\mathbf{r}_2'\| = 4t$.

23. $\mathbf{r}(t) = -16t^2\mathbf{j}$
$\mathbf{v}(t) = -32t\mathbf{j}$

25. $\mathbf{r}(t) = (t + \cos t - 1)\mathbf{i} + (\sin t - t + 1)\mathbf{j}$
$\mathbf{v}(t) = (1 - \sin t)\mathbf{i} + (\cos t - 1)\mathbf{j}$

27. $\mathbf{r}(t) = \tfrac{1}{2}t^2\mathbf{i} + \mathbf{j} + \tfrac{1}{6}t^3\mathbf{k}$
$\mathbf{v}(t) = t\mathbf{i} + \tfrac{1}{2}t^2\mathbf{k}$

29. $\mathbf{r}(t) = (t - \sin t - 1)\mathbf{i} + (1 - \cos t)\mathbf{j} + e^t\mathbf{k}$
$\mathbf{v}(t) = (1 - \cos t)\mathbf{i} + \sin t\,\mathbf{j} + e^t\mathbf{k}$

33. $8\mathbf{i} + \tfrac{26}{3}\mathbf{j},\ \tfrac{1}{3}(13\sqrt{13} - 5\sqrt{5})$

35. $\mathbf{0},\ 12\pi$

37. $2\mathbf{i} - \tfrac{2}{3}\mathbf{j} + \sqrt{2}\ln 3\mathbf{k},\ \tfrac{8}{3}$

39. $a_T = 0,\ a_N = 2$

41. $a_T = 0,\ a_N = \sqrt{2}$

43. $a_T = 2\sqrt{5},\ a_N = 2\sqrt{5}$

45. $a_T = \dfrac{22}{\sqrt{14}},\ a_N = \sqrt{\dfrac{38}{7}}$

47. $a_T = 0,\ a_N = 3$

49. $a_T = -3,\ a_N = 2;\ \mathbf{T} = -\mathbf{j},\ \mathbf{N} = \mathbf{i}$

51. $a_T = \tfrac{4}{3},\ a_N = \tfrac{1}{3}\sqrt{29};$
$\mathbf{T} = \tfrac{2}{3}\mathbf{i} + \tfrac{2}{3}\mathbf{j} + \tfrac{1}{3}\mathbf{k},$
$\mathbf{N} = \dfrac{1}{3\sqrt{29}}(\mathbf{i} - 8\mathbf{j} + 14\mathbf{k})$

53. $\tfrac{3}{2}$

55. $-\pi/\sqrt{2}$

57. $\tfrac{1}{8}$

59. $\sqrt{29}/27$

61. 9×10^{10} km/sec^2

63. $\dfrac{18}{(1 + 4x^2)^{3/2}}$

65. (a) $x = 160t$
$y = 160\sqrt{3}t - 16t^2,\ t \geq 0$
(b) 1200 ft
(c) $1600\sqrt{3}$ ft
(d) 320 ft/sec

67. $40\sqrt{3}$ ft

69. 800 ft/sec

71. $15°,\ 75°$

73. (b) $45°;\ v_0^2/g$

75. (a) 2.62 sec   (b) 181.5 ft

▶ Chapter 15 Supplementary Exercises (**Page 968**)

1. (a) $\tfrac{1}{2}(t + 4)^{-1/2}\mathbf{i} + 2\mathbf{j},\ -\tfrac{1}{4}(t + 4)^{-3/2}\mathbf{i}$
(b) The graph is the part of the parabola $y = 2(x^2 - 4)$ starting at $(0, -8)$ and directed up and toward the right. At $(1, -6)[t = -3],\ \mathbf{r}' = \tfrac{1}{2}\mathbf{i} + 2\mathbf{j},\ \mathbf{r}'' = -\tfrac{1}{4}\mathbf{i};$ at $(2, 0)[t = 0],$
$\mathbf{r}' = \tfrac{1}{4}\mathbf{i} + 2\mathbf{j},\ \mathbf{r}'' = -\tfrac{1}{32}\mathbf{i}.$

3. (a) $\langle 6t^2, 3t^2 \rangle,\ \langle 12t, 6t \rangle$
(b) The graph is the straight line $x - 2y + 3 = 0$, directed to the right and up; at $(-1, 1)[t = 0],\ \mathbf{r}' = \mathbf{r}'' = \langle 0, 0 \rangle$; at $(-\tfrac{5}{4}, \tfrac{7}{8})[t = -\tfrac{1}{2}],$
$\mathbf{r}' = \langle \tfrac{3}{2}, \tfrac{3}{4} \rangle,\ \mathbf{r}'' = \langle -6, -3 \rangle.$

5. (a) $t(k\mathbf{i} + m\mathbf{j}) + \mathbf{C}$ (b) $\langle 4, 4 \rangle$
(c) 2   (d) $\sqrt{t^2 + 3}\mathbf{i} + \ln(\sin t)\mathbf{j} + \mathbf{C}$

7. (a) $1 + 1/t^2$
(b) $x = \sqrt{s^2 + 4},\ y = 2\ln[\tfrac{1}{2}(s + \sqrt{s^2 + 4})]$

9. the parabola $y = x^2 + 1$ in the plane $z = 1$, traced so that x increases with t

11. $\mathbf{v} = a\mathbf{i},\ \|\mathbf{v}\| = a,\ \mathbf{a} = -a(\mathbf{j} + \mathbf{k}),\ \mathbf{T} = \mathbf{i},$
$\mathbf{N} = -(\mathbf{j} + \mathbf{k})/\sqrt{2},\ \kappa = \sqrt{2}/a$

13. $4\pi\sqrt{37}$ **15.** $x = 1 - t,\ y = 1 + 2t,\ z = 1$

17. (a) $\langle 9, 9, -9 \rangle$ (b) $\langle 3 + 4t^3, -4t, 6t^2 \rangle$

19. (a) $(2\mathbf{i} - \mathbf{j})/\sqrt{5}$ (b) $6/5^{3/2}$

21. (a) $\mathbf{j}$ (b) 2

23. $\tfrac{2}{25}$ **25.** 1

27. $(x - 1)^2 + (y - \tfrac{1}{2})^2 = \tfrac{1}{4}$; at $(1, 0),\ dy/dx = 0,\ d^2y/dx^2 = 2$

29. $(1/u)\langle e^t, 4e^{2t} \rangle = \langle 1, 4u \rangle$

31. $a_T = 1,\ a_N = t$

33. (a) $\langle -1, 1 \rangle,\ \langle 1, 1 \rangle,\ \sqrt{2}$
(b) $1/\sqrt{2}$ (c) $0,\ \sqrt{2}$
(d) trajectory: the branch of the hyperbola $xy = 1$ in the first quadrant, traced so that y increases with t
(e) $C(2, 2)$

35. $(2t - \sin t)\mathbf{i} + (e^{2t} - 1)\mathbf{j}$

37. (a) $a_T = 0,\quad a_N = 32$
(b) $a_T = -64/\sqrt{5},\quad a_N = 32/\sqrt{5}$

39. $a_T = 0,\ a_N = 8\pi^2$ m/sec^2

41. (a) $\sqrt{5}e^t$ (b) $2\sqrt{5}$

▶ Exercise Set 16.1 **(Page 982)**

1. (a) 5   (b) 3   (c) 1   (d) -2
   (e) $9a^3 + 1$ (f) $a^3b^2 - a^2b^3 + 1$

3. (a) $x^2 - y^2 + 3$ (b) $3x^3y^4 + 3$

5. $x^3e^{x^3(3y+1)}$

7. (a) $t^2 + 3t^{10}$ (b) 0   (c) 3076

9. (a) 19   (b) -9 (c) 3
   (d) $a^6 + 3$ (e) $-t^8 + 3$
   (f) $(a + b)(a - b)^2b^3 + 3$

11. $(y + 1)e^{x^2(y+1)z^2}$

13. (a) t^{14} (b) 0   (c) 16,384

15.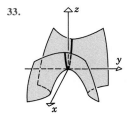

17.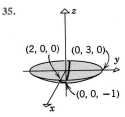

19.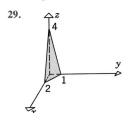

21. all points above or on the line $y = -2$

23. all points above the line $y = 2x$

25. all points not on the plane $x + y + z = 0$

27. all points within the cylinder $x^2 + y^2 = 1$

29.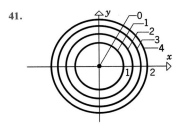

31.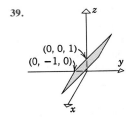

33.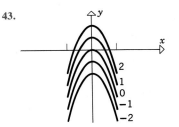

35.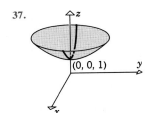

37.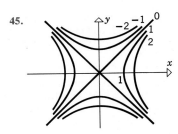

39.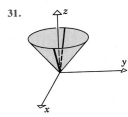

41.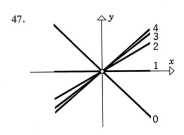

43.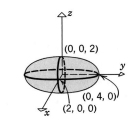

45.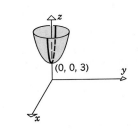

47.

49.

51.

53. concentric spheres, center at $(2, 0, 0)$

55. concentric cylinders, common axis the y-axis

57. (a) $x^2 - 2x^3 + 3xy = 0$
(b) $x^2 - 2x^3 + 3xy = 0$
(c) $x^2 - 2x^3 + 3xy = -18$

59. (a) $x^2 + y^2 - z = 5$
(b) $x^2 + y^2 - z = -2$
(c) $x^2 + y^2 - z = 0$

61.

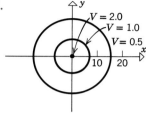

63. (a) A (b) B

65. (a) decrease   (b) increase
(c) increase   (d) decrease

67. (a) open   (b) neither
(c) closed   (d) closed

69. (a) bounded   (b) unbounded
(c) unbounded   (d) unbounded

▶ Exercise Set 16.2 **(Page 993)**

1.

 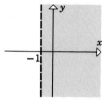

3.

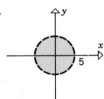

5.

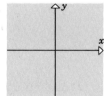

7.

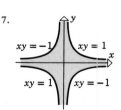

9. all of 3-space

11. all points not on the cylinder $x^2 + z^2 = 1$

13. 35   **15.** -8

17. 0   **19.** does not exist

21. 1   **23.** 0

25. does not exist   **27.** 0

29. 0   **31.** $\frac{8}{3}$

33. 0   **35.** does not exist

41. $-\pi/2$

43. no

45. choose $\delta = \sqrt{\epsilon}$ **47.** choose $\delta = \sqrt{\epsilon}$

▶ Exercise Set 16.3 **(Page 1002)**

1. $\dfrac{\partial z}{\partial x} = 9x^2y^2, \dfrac{\partial z}{\partial y} = 6x^3y$

3. $\dfrac{\partial z}{\partial x} = 8xy^3e^{x^2y^3}, \dfrac{\partial z}{\partial y} = 12x^2y^2e^{x^2y^3}$

5. $\dfrac{\partial z}{\partial x} = \dfrac{x^3}{y^{3/5} + x} + 3x^2 \ln(1 + xy^{-3/5})$,
$\dfrac{\partial z}{\partial y} = -\dfrac{3x^4}{5(y^{8/5} + xy)}$

7. $f_x(x, y) = \frac{3}{2}x^2y(5x^2 - 7)(3x^5y - 7x^3y)^{-1/2}$,
$f_y(x, y) = \frac{1}{2}x^3(3x^2 - 7)(3x^5y - 7x^3y)^{-1/2}$

9. $f_x(x, y) = \dfrac{y^{-1/2}}{y^2 + x^2}$,
$f_y(x, y) = -\dfrac{xy^{-3/2}}{y^2 + x^2} - \dfrac{3}{2}y^{-5/2} \tan^{-1}\left(\dfrac{x}{y}\right)$

11. $f_x(x, y) = -\frac{4}{3}y^2 \sec^2 x(y^2 \tan x)^{-7/3}$,
$f_y(x, y) = -\frac{8}{3}y \tan x(y^2 \tan x)^{-7/3}$

13. (a) -6 (b) -21

15. (a) $1/\sqrt{17}$ (b) $8/\sqrt{17}$

17. $\dfrac{\partial z}{\partial x} = -\dfrac{x}{z}, \dfrac{\partial z}{\partial y} = -\dfrac{y}{z}$

19. $\dfrac{\partial z}{\partial x} = -\dfrac{2x + yz^2 \cos(xyz)}{xyz \cos(xyz) + \sin(xyz)}$
$\dfrac{\partial z}{\partial y} = -\dfrac{xz^2 \cos(xyz)}{xyz \cos(xyz) + \sin(xyz)}$

21. $f_{xx} = 8$,
$f_{yy} = -96xy^2 + 140y^3$,
$f_{xy} = f_{yx} = -32y^3$

23. $f_{xx} = e^x \cos y,$
$\quad f_{yy} = -e^x \cos y,$
$\quad f_{xy} = f_{yx} = -e^x \sin y$

25. $f_{xx} = -\dfrac{16}{(4x - 5y)^2},$
$\quad f_{yy} = -\dfrac{25}{(4x - 5y)^2},$
$\quad f_{xy} = f_{yx} = \dfrac{20}{(4x - 5y)^2}$

27. (a) $30xy^4 - 4$　　(b) $60x^2y^3$
　　(c) $60x^3y^2$

29. (a) -30　　(b) -125　　(c) 150

31. (a) $\dfrac{\partial^3 f}{\partial x^3}$　　(b) $\dfrac{\partial^3 f}{\partial y^2 \partial x}$
　　(c) $\dfrac{\partial^4 f}{\partial x^2 \partial y^2}$　　(d) $\dfrac{\partial^4 f}{\partial y^3 \partial x}$

33. $\dfrac{\partial w}{\partial x} = 2xy^4z^3 + y,\ \dfrac{\partial w}{\partial y} = 4x^2y^3z^3 + x,$
$\quad \dfrac{\partial w}{\partial z} = 3x^2y^4z^2 + 2z$

35. $\dfrac{\partial w}{\partial x} = \dfrac{2x}{y^2 + z^2},\ \dfrac{\partial w}{\partial y} = -\dfrac{2y(x^2 + z^2)}{(y^2 + z^2)^2},$
$\quad \dfrac{\partial w}{\partial z} = \dfrac{2z(y^2 - x^2)}{(y^2 + z^2)^2}$

37. $\dfrac{\partial w}{\partial x} = \dfrac{x}{\sqrt{x^2 + y^2 + z^2}},\ \dfrac{\partial w}{\partial y} = \dfrac{y}{\sqrt{x^2 + y^2 + z^2}},$
$\quad \dfrac{\partial w}{\partial z} = \dfrac{z}{\sqrt{x^2 + y^2 + z^2}}$

39. $f_x = -\dfrac{y^2z^3}{x^2y^4z^6 + 1},\ f_y = -\dfrac{2xyz^3}{x^2y^4z^6 + 1},$
$\quad f_z = -\dfrac{3xy^2z^2}{x^2y^4z^6 + 1}$

41. $f_x = 4xyz \cosh\sqrt{z} \sinh(x^2yz) \cosh(x^2yz),$
$\quad f_y = 2x^2z \cosh\sqrt{z} \sinh(x^2yz) \cosh(x^2yz),$
$\quad f_z = 2x^2y \cosh\sqrt{z} \sinh(x^2yz) \cosh(x^2yz)$
$\qquad\qquad\qquad\qquad + \dfrac{\sinh\sqrt{z}\ \sinh^2(x^2yz)}{2\sqrt{z}}$

43. (a) -80　　(b) 40　　(c) -60

45. (a) $2/\sqrt{7}$　　(b) $4/\sqrt{7}$　　(c) $1/\sqrt{7}$

47. $\dfrac{\partial w}{\partial x} = -\dfrac{x}{w},\ \dfrac{\partial w}{\partial y} = -\dfrac{y}{w},\ \dfrac{\partial w}{\partial z} = -\dfrac{z}{w}$

49. $\dfrac{\partial w}{\partial x} = -\dfrac{yzw \cos(xyz)}{2w + \sin(xyz)}$
$\quad \dfrac{\partial w}{\partial y} = -\dfrac{xzw \cos(xyz)}{2w + \sin(xyz)}$
$\quad \dfrac{\partial w}{\partial z} = -\dfrac{xyw \cos(xyz)}{2w + \sin(xyz)}$

51. (a) $15x^2y^4z^7 + 2y$　　(b) $35x^3y^4z^6 + 3y^2$
　　(c) $21x^2y^5z^6$　　(d) $42x^3y^5z^5$
　　(e) $140x^3y^3z^6 + 6y$　　(f) $30xy^4z^7$
　　(g) $105x^2y^4z^6$　　(h) $210xy^4z^6$

55. 6　　　　　　　　57. (a) 8　　(b) -2

59. (a) $\dfrac{\partial V}{\partial r} = 2\pi r h$　　(b) $\dfrac{\partial V}{\partial h} = \pi r^2$
　　(c) 48π　　(d) 64π

61. (a) $\frac{1}{5}$　　(b) $-\frac{25}{8}$

63. $\dfrac{\partial z}{\partial x} = -1 - \dfrac{\cos(x - y)}{\cos(x + z)},$
$\quad \dfrac{\partial z}{\partial y} = \dfrac{\cos(x - y)}{\cos(x + z)},$
$\quad \dfrac{\partial^2 z}{\partial x \partial y} = -\dfrac{\cos^2(x + z)\sin(x - y) + \cos^2(x - y)\sin(x + z) + \cos(x - y)\sin(x + z)\cos(x + z)}{\cos^3(x + z)}$

65. (a) 4 degrees per centimeter
　　(b) 8 degrees per centimeter

69. (a) Both are positive; $\partial T/\partial x$.
　　(b) All are negative.

► Exercise Set 16.4 (Page 1016)

1. 0.872　　　　　　　3. $\frac{7}{2}$

5. $\epsilon_1 = 0,\ \epsilon_2 = \Delta x$

7. $\epsilon_1 = y\Delta x + \Delta x \Delta y,\ \epsilon_2 = 2x\Delta x$

13. $f_{xy} = f_{yx} = 12x^2 + 6x$

15. $f_{xy} = f_{yx} = -6xy^2 \sin(x^2 + y^3)$

17. (a) 4　　(b) 5

19. $42t^{13}$　　　　　21. $\dfrac{3 \sin(1/t)}{t^2}$

23. $-\frac{10}{3} t^{7/3} e^{1 - t^{10/3}}$

25. $\dfrac{\partial z}{\partial u} = 24u^2v^2 - 16uv^3 - 2v + 3,$
$\quad \dfrac{\partial z}{\partial v} = 16u^3v - 24u^2v^2 - 2u - 3$

27. $\dfrac{\partial z}{\partial u} = -\dfrac{2 \sin u}{3 \sin v},\ \dfrac{\partial z}{\partial v} = -\dfrac{2 \cos u \cos v}{3 \sin^2 v}$

29. $\dfrac{\partial z}{\partial u} = e^u,\ \dfrac{\partial z}{\partial v} = 0$

31. $\dfrac{\partial z}{\partial u} = \dfrac{2e^{2u}}{1 + e^{4u}}, \dfrac{\partial z}{\partial v} = 0$

33. $\dfrac{\partial T}{\partial r} = 3r^2 \sin\theta \cos^2\theta - 4r^3 \sin^3\theta \cos\theta,$

$\dfrac{\partial T}{\partial \theta} = -2r^3 \sin^2\theta \cos\theta + r^4 \sin^4\theta$
$+ r^3 \cos^3\theta - 3r^4 \cos^2\theta \sin^2\theta$

35. $\dfrac{\partial t}{\partial x} = \dfrac{x^2 + y^2}{4x^2 y^3}, \dfrac{\partial t}{\partial y} = \dfrac{y^2 - 3x^2}{4xy^4}$ **37.** $-\pi$

39. $\dfrac{\partial z}{\partial r}\bigg|_{r=2,\,\theta=\pi/6} = \sqrt{3}e^{\sqrt{3}},$

$\dfrac{\partial z}{\partial \theta}\bigg|_{r=2,\,\theta=\pi/6} = (2 - 4\sqrt{3})e^{\sqrt{3}}$

41. $\dfrac{x^2 - y^2}{2xy - y^2}$

43. $\dfrac{2\sqrt{xy} - y}{x - 6\sqrt{xy}}$

45. -39 mi/hr

47. $-\frac{7}{36}\sqrt{3}$ rad/sec

49. $\frac{4}{3} - 16 \ln 3$ °C/sec

55. (b) $z = C_1 \ln r + C_2$

65. (a) $n = 2$ (b) $n = 1$
(c) $n = 3$ (d) $n = -4$

67. (a) $\dfrac{\partial w}{\partial x} = \dfrac{\partial f}{\partial x} + \dfrac{\partial f}{\partial y}\dfrac{\partial y}{\partial x}$

(b) $\dfrac{\partial w}{\partial z} = \dfrac{\partial f}{\partial y}\dfrac{\partial y}{\partial z}$

▶ **Exercise Set 16.5** (Page 1026)

1. $48x - 14y - z = 64;$
$x = 1 + 48t, y = -2 - 14t, z = 12 - t$

3. $x - y - z = 0;$
$x = 1 + t, y = -t, z = 1 - t$

5. $3y - z = -1;$
$x = \pi/6, y = 3t, z = 1 - t$

7. $3x - 4z = -25;$
$x = -3 + \frac{3}{4}t, y = 0, z = 4 - t$

9. $7\,dx - 2\,dy$

11. $\dfrac{y}{1 + x^2 y^2}\,dx + \dfrac{x}{1 + x^2 y^2}\,dy$

13. 0.10

15. 0.03

17. (a) all points on the x-axis and y-axis
(b) $(0, -2, -4)$

19. $(\frac{1}{2}, -2, -\frac{3}{4})$ **23.** 0.088 cm

25. 8% **27.** $r\%$

29. 2% **31.** 0.004 rad   **33.** 0.3%

35. (a) $(r + s)\%$ (b) $(r + s)\%$
(c) $(2r + 3s)\%$ (d) $(3r + \frac{1}{2}s)\%$

37. (a) $(-2, 1, 5), (0, 3, 9)$
(b) $4/(3\sqrt{14})$ at $(-2, 1, 5)$, $4/\sqrt{222}$ at $(0, 3, 9)$

▶ **Exercise Set 16.6** (Page 1035)

1. $4\mathbf{i} - 8\mathbf{j}$ **3.** $\dfrac{x}{x^2 + y^2}\mathbf{i} + \dfrac{y}{x^2 + y^2}\mathbf{j}$

5. $-36\mathbf{i} - 12\mathbf{j}$ **7.** $4\mathbf{i} + 4\mathbf{j}$

9. $6\sqrt{2}$ **11.** $-3/\sqrt{10}$

13. 0 **15.** $-8\sqrt{2}$

17. $\dfrac{1}{2\sqrt{2}}$ **19.** $\dfrac{1}{2} + \dfrac{\sqrt{3}}{8}$ **21.** $2\sqrt{2}$

23.

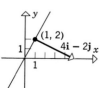

25.

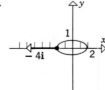

27. $\dfrac{3}{\sqrt{13}}\mathbf{i} - \dfrac{2}{\sqrt{13}}\mathbf{j}, 4\sqrt{13}$

29. $\frac{4}{5}\mathbf{i} - \frac{3}{5}\mathbf{j}, 1$

31. $-\dfrac{1}{\sqrt{10}}\mathbf{i} - \dfrac{3}{\sqrt{10}}\mathbf{j}, -2\sqrt{10}$

33. $\dfrac{3}{\sqrt{10}}\mathbf{i} - \dfrac{1}{\sqrt{10}}\mathbf{j}, -\sqrt{5}$

35. $1/\sqrt{5}$ **37.** $-\frac{3}{2}e$

39. $-\dfrac{4}{\sqrt{17}}\mathbf{i} + \dfrac{1}{\sqrt{17}}\mathbf{j}$

41. (a) 5 (b) 10 (c) $-5\sqrt{5}$

43. $8/\sqrt{29}$

45. all points on the ellipse $9x^2 + y^2 = 9$

47. $36/\sqrt{17}$ **49.** (a) $2e^{-\pi/2}\mathbf{i}$

▶ Exercise Set 16.7 **(Page 1043)**

1. $165t^{32}$

3. $-2t \cos(t^2)$

5. 3264

7. -320

9. $-\frac{314}{741}$

11. $72/\sqrt{14}$

13. $-\frac{8}{63}$

15. $\frac{1}{\sqrt{2}}\mathbf{i} - \frac{1}{\sqrt{2}}\mathbf{j}, 3\sqrt{2}$

17. $-\frac{1}{\sqrt{2}}\mathbf{i} + \frac{1}{\sqrt{2}}\mathbf{j}, \frac{\sqrt{2}}{2}$

19. $\frac{1}{\sqrt{266}}\mathbf{i} - \frac{11}{\sqrt{266}}\mathbf{j} + \frac{12}{\sqrt{266}}\mathbf{k}, -\sqrt{266}$

21. $3/\sqrt{11}$

23. $-\frac{10}{3}\mathbf{i} + \frac{5}{3}\mathbf{j} + \frac{10}{3}\mathbf{k}$

25. $3x - 2y + 6z = -49$;
$x = -3 + 3t, y = 2 - 2t, z = -6 + 6t$

27. $x - y - 15z = -17$;
$x = 3 + t, y = 5 - t, z = 1 - 15t$

31. $(1, \frac{2}{3}, \frac{2}{3}), (-1, -\frac{2}{3}, -\frac{2}{3})$

33. $8\,dx - 3\,dy + 4\,dz$

35. $\frac{yz}{1 + (xyz)^2}\,dx + \frac{xz}{1 + (xyz)^2}\,dy + \frac{xy}{1 + (xyz)^2}\,dz$

37. 0.96

39. $2.35\ \text{cm}^3$

41. $39\ \text{ft}^2$

43. 14%

▶ Exercise Set 16.8 **(Page 1050)**

1. $\frac{\partial f}{\partial v} = 8vw^3x^4y^5, \frac{\partial f}{\partial w} = 12v^2w^2x^4y^5,$
$\frac{\partial f}{\partial x} = 16v^2w^3x^3y^5, \frac{\partial f}{\partial y} = 20v^2w^3x^4y^4$

3. $\frac{\partial f}{\partial v_1} = \frac{2v_1}{v_3^2 + v_4^2}, \frac{\partial f}{\partial v_2} = -\frac{2v_2}{v_3^2 + v_4^2},$
$\frac{\partial f}{\partial v_3} = -\frac{2v_3(v_1^2 - v_2^2)}{(v_3^2 + v_4^2)^2}, \frac{\partial f}{\partial v_4} = -\frac{2v_4(v_1^2 - v_2^2)}{(v_3^2 + v_4^2)^2}$

5. $f_v(1, -2, 4, 8) = 128, \quad f_w(1, -2, 4, 8) = -512,$
$f_x(1, -2, 4, 8) = 32, \quad f_y(1, -2, 4, 8) = \frac{64}{3}$

7. $210t^{29}$

9. $\frac{\partial z}{\partial r} = \frac{2r \cos^2 \theta}{r^2 \cos^2 \theta + 1}, \frac{\partial z}{\partial \theta} = -\frac{2r^2 \sin \theta \cos \theta}{r^2 \cos^2 \theta + 1}$

11. $\frac{\partial w}{\partial \rho} = 2\rho(4 \sin^2 \phi + \cos^2 \phi),$
$\frac{\partial w}{\partial \phi} = 6\rho^2 \sin \phi \cos \phi,$
$\frac{\partial w}{\partial \theta} = 0$

13. $\frac{-4 \cos 2y \sin 2y + 2y + 1}{2\sqrt{\cos^2 2y + y^2 + y}}$

15. $\sqrt{3} + \pi/6$ cm/sec, increasing

17. $\frac{\partial z}{\partial x} = -\frac{2x + 4z}{4x + 2z - 3y}, \frac{\partial z}{\partial y} = \frac{3z}{4x + 2z - 3y}$

27. $\frac{\partial z}{\partial x} = \frac{ye^x}{15 \cos 3z + 3}, \frac{\partial z}{\partial y} = \frac{e^x}{15 \cos 3z + 3}$

33. (a) $\frac{dw}{dt} = \frac{\partial w}{\partial x_1}\frac{dx_1}{dt} + \frac{\partial w}{\partial x_2}\frac{dx_2}{dt} + \frac{\partial w}{\partial x_3}\frac{dx_3}{dt} + \frac{\partial w}{\partial x_4}\frac{dx_4}{dt}$

(b) $\frac{\partial w}{\partial v_1} = \frac{\partial w}{\partial x_1}\frac{\partial x_1}{\partial v_1} + \frac{\partial w}{\partial x_2}\frac{\partial x_2}{\partial v_1} + \frac{\partial w}{\partial x_3}\frac{\partial x_3}{\partial v_1} + \frac{\partial w}{\partial x_4}\frac{\partial x_4}{\partial v_1},$
$\frac{\partial w}{\partial v_2} = \frac{\partial w}{\partial x_1}\frac{\partial x_1}{\partial v_2} + \frac{\partial w}{\partial x_2}\frac{\partial x_2}{\partial v_2} + \frac{\partial w}{\partial x_3}\frac{\partial x_3}{\partial v_2} + \frac{\partial w}{\partial x_4}\frac{\partial x_4}{\partial v_2},$
$\frac{\partial w}{\partial v_3} = \frac{\partial w}{\partial x_1}\frac{\partial x_1}{\partial v_3} + \frac{\partial w}{\partial x_2}\frac{\partial x_2}{\partial v_3} + \frac{\partial w}{\partial x_3}\frac{\partial x_3}{\partial v_3} + \frac{\partial w}{\partial x_4}\frac{\partial x_4}{\partial v_3}$

▶ Exercise Set 16.9 **(Page 1063)**

1. $(0, 0)$ relative min

3. $(1, -2)$ saddle

5. $(2, -1)$ relative min

7. $(2, 1), (-2, 1)$ saddle;
$(0, 0)$ relative min

9. $(-1, -1), (1, 1)$ relative min

11. $(0, 0)$ saddle

13. none

15. $(0, 0), (4, 0), (0, -2)$ saddle;
$(\frac{4}{3}, -\frac{2}{3})$ relative min

17. $(-1, 0)$ relative max

19. $(\pi/2, \pi/2)$ relative max

21. (b) $(0, 0)$ relative min

23. critical point $(1, 0)$

25. absolute maximum 0 at $(0, 0)$;
absolute minimum -12 at $(0, 4)$

27. absolute maximum 3 at $(0, 1), (2, 1)$;
absolute minimum -1 at $(1, 0), (1, 2)$

29. absolute maximum $\frac{33}{4}$ at $(-\frac{1}{2}, \pm\frac{1}{2}\sqrt{15})$;
absolute minimum $-\frac{1}{4}$ at $(\frac{1}{2}, 0)$

33. $9, 9, 9$ **35.** $(\sqrt{5}, 0, 0), (-\sqrt{5}, 0, 0)$

37. $\frac{1}{27}$ **39.** length and width 2 ft, height 4 ft

41. $\frac{3}{2}\sqrt{6}$

43. length and width $\sqrt[3]{2V}$, height $\frac{1}{2}\sqrt[3]{2V}$

47. $y = 0.5x + 0.8$

49. for $0 \le x \le 1$ let $f(x, y) = \begin{cases} y, & 0 < y < 1 \\ \frac{1}{2}, & y = 0 \text{ or } y = 1 \end{cases}$;
for $-\infty < x < +\infty$ and $y > 0$ let $f(x, y) = y$

▶ Exercise Set 16.10 **(Page 1072)**

1. max $\sqrt{2}$ at $(\sqrt{2}, 1)$ and $(-\sqrt{2}, -1)$,
   min $-\sqrt{2}$ at $(-\sqrt{2}, 1)$ and $(\sqrt{2}, -1)$

3. max $\sqrt{2}$ at $(1/\sqrt{2}, 0)$,
   min $-\sqrt{2}$ at $(-1/\sqrt{2}, 0)$

5. max 6 at $(\frac{4}{3}, \frac{2}{3}, -\frac{4}{3})$,
   min -6 at $(-\frac{4}{3}, -\frac{2}{3}, \frac{4}{3})$

7. max $\dfrac{1}{3\sqrt{3}}$ at $\left(\dfrac{1}{\sqrt{3}}, \dfrac{1}{\sqrt{3}}, \dfrac{1}{\sqrt{3}}\right)$,
   $\left(\dfrac{1}{\sqrt{3}}, -\dfrac{1}{\sqrt{3}}, -\dfrac{1}{\sqrt{3}}\right)$, $\left(-\dfrac{1}{\sqrt{3}}, \dfrac{1}{\sqrt{3}}, -\dfrac{1}{\sqrt{3}}\right)$,
   $\left(-\dfrac{1}{\sqrt{3}}, -\dfrac{1}{\sqrt{3}}, \dfrac{1}{\sqrt{3}}\right)$;

min $-\dfrac{1}{3\sqrt{3}}$ at $\left(\dfrac{1}{\sqrt{3}}, \dfrac{1}{\sqrt{3}}, -\dfrac{1}{\sqrt{3}}\right)$,
$\left(\dfrac{1}{\sqrt{3}}, -\dfrac{1}{\sqrt{3}}, \dfrac{1}{\sqrt{3}}\right)$, $\left(-\dfrac{1}{\sqrt{3}}, \dfrac{1}{\sqrt{3}}, \dfrac{1}{\sqrt{3}}\right)$,
$\left(-\dfrac{1}{\sqrt{3}}, -\dfrac{1}{\sqrt{3}}, -\dfrac{1}{\sqrt{3}}\right)$

9. $(\frac{2}{5}, \frac{19}{5})$ 11. $(1, -1, 1)$ 13. $\frac{1}{2}$

15. $(3, 6)$ closest, $(-3, -6)$ farthest

17. 9, 9, 9

19. $(\sqrt{5}, 0, 0)$, $(-\sqrt{5}, 0, 0)$

21. length and width 2 ft, height 4 ft

▶ Chapter 16 Supplementary Exercises **(Page 1073)**

1. (a) 1   (b) yx (c) $e^{r+s} \ln (rs)$

3. (a) the upper nappe of the elliptical cone $z^2 = x^2 + 4y^2$
   (b) the plane with x-, y-, z-intercepts $a, b, 1$

5. $\dfrac{1}{x \sin yz} - \dfrac{3y \ln (xy) \cos yz}{\sin^2 yz}$

7. $\pi/2; 0; 1, -\pi^2/4$ 9. 2

11. $(2x - 2y + 4r)/(x^2 + y^2 + 2z) = 2/(r + s)$

17. (a) increasing 12 N/m²/min
    (b) increasing 240 N/m²/min

19. (a) $0 = f(0, 0)$ (b) yes

21. (a) $-(6x - 5y + y \sec^2 xy)/(-5x + x \sec^2 xy)$
    (b) $-\dfrac{\ln y + \cos (x - y)}{\dfrac{x}{y} - \cos (x - y)}$

23. $\dfrac{dV}{dt} = R\left[\dfrac{dE/dt}{r + R} - \dfrac{dr}{dt}\dfrac{E}{(r + R)^2}\right]$

25. (a) $\langle 6, 45 \rangle$ (b) $-174/\sqrt{17}$

27. (a) $\langle \frac{1}{3}, \frac{1}{3}, \frac{1}{6} \rangle$ (b) $\sqrt{3}/9$

29. (a) $\langle 1, 3, 0 \rangle$ (b) $43/15$

31. (a) $\pm(\mathbf{i} + \mathbf{j})/\sqrt{2}$ (b) $\pm(2\mathbf{i} + \mathbf{j})/\sqrt{5}$

33. $-7/\sqrt{5}$

35. (a) $\mathbf{N} = \langle 8, -6, -5 \rangle$
    (b) $8(x - 4) - 6(y + 3) - 5(z - 10) = 0$

37. $(1, 1, 1)$, $(-1, -1, 1)$, $(0, 0, 2)$

39. $(-\frac{1}{3}, -\frac{1}{2}, 2)$

41. $dV = -\frac{1}{3}(0.2)$ m³, $\Delta V = -\frac{1}{3}(0.218)$ m³

43. relative min at $(15, -8)$

45. relative min at $(3, 9)$; saddle point at $(0, 0)$

47. $\dfrac{2a}{\sqrt{3}} \times \dfrac{2b}{\sqrt{3}} \times \dfrac{2c}{\sqrt{3}}$

49. $I_1:I_2:I_3 = R_1^{-1}:R_2^{-1}:R_3^{-1}$

▶ Exercise Set 17.1 **(Page 1084)**

1. 7   3. 2   5. 2

7. $\frac{2}{15}(31 - 9\sqrt{3})$ 9. 3   11. $1 - \ln 2$

13. $\frac{1}{2}(1 - \ln 2)$ 15. 0

17. $\frac{1}{3}$ 19. 1

21.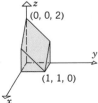

23.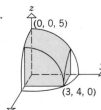

25.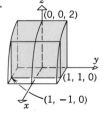

27. 172   29. 8

▶ Exercise Set 17.2 **(Page 1094)**

1. $\frac{1}{40}$ 3. 9   5. $\pi/2$

7. 1   9. $\frac{2}{3}a^3$ 11. $\frac{1}{12}$

13. 32   15. $-2/\pi$

17. 576   19. $\frac{1}{2}(\sqrt{17} - 1)$

21. 0   23. $\pi/4 - \frac{1}{2}\ln 2$

25. $\frac{50}{3}$ 27. $-\frac{1}{2}$

29. $\frac{25}{2}$ 31. $\sqrt{2} - 1$

33. 32   35. $\frac{125}{12}$

37. $\frac{56}{15}$ 39. 170

41. $\frac{27}{2}\pi$ 43. $\frac{20}{3}$

45. $\pi/2$ 47. $\frac{2000}{3}$

49. $\displaystyle\int_0^{\sqrt{2}}\int_{y^2}^2 f(x, y)\,dx\,dy$

51. $\displaystyle\int_1^{e^2}\int_{\ln x}^2 f(x, y)\,dy\,dx$

53. $\displaystyle\int_{-1}^1\int_{-2\sqrt{1-y^2}}^{2\sqrt{1-y^2}} f(x, y)\,dx\,dy$

55. $\displaystyle\int_0^{\pi/2}\int_0^{\sin x} f(x, y)\,dy\,dx$

57. $\frac{1}{8}(1 - e^{-16})$ 59. $\frac{1}{3}(e^8 - 1)$

61. $\frac{1}{8}\pi$ 63. $\frac{1}{4}(1 - \cos 8)$

65. (a) 0   (b) $-\frac{603}{40}$ 67. π 69. 0

▶ Exercise Set 17.3 **(Page 1103)**

1. $\frac{1}{6}$ 3. 0   5. 0

7. $3\pi/2$ 9. $\pi/16$ 11. $4\pi/3 + 2\sqrt{3}$

13. $\frac{4}{3}(27 - 16\sqrt{2})\pi$ 15. $\frac{32}{9}$

17. $5\pi/32$ 19. $(1 - e^{-1})\pi$

21. $\dfrac{\pi}{8}\ln 5$ 23. $\pi/8$

25. $\frac{16}{9}$ 27. $\dfrac{\pi}{2}\left(1 - \dfrac{1}{\sqrt{1 + a^2}}\right)$

29. $\frac{1}{4}\pi(\sqrt{5} - 1)$ 31. $\frac{1}{9}(3\pi - 4)a^2 c$

33. $\dfrac{4\pi}{3} + 2\sqrt{3} - 2$ 35. (b) $\pi/4$ (c) $\sqrt{\pi}/2$

▶ Exercise Set 17.4 **(Page 1109)**

1. 6π 3. $\sqrt{5}/6$ 9. 8π 11. $2(\pi - 2)a^2$

5. $\dfrac{\pi}{6}(5\sqrt{5} - 1)$ 7. $\dfrac{\pi}{18}(10\sqrt{10} - 1)$ 13. 128

▶ Exercise Set 17.5 **(Page 1118)**

1. 8   3. 7   5. $\frac{81}{5}$

7. $\frac{128}{15}$ 9. $\dfrac{\pi}{2}(\pi - 3)$ 11. $\frac{1}{6}$

13. 4   15. $\frac{256}{15}$ 17. 9π

19. 2π 21. $\dfrac{\pi}{6}(8\sqrt{2} - 7)a^3$

23. (a)   (b)

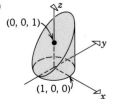

$(0, 0, 1)$, $(1, 0, 0)$

$(0, 9, 9)$, $(3, 9, 0)$

25. (a)

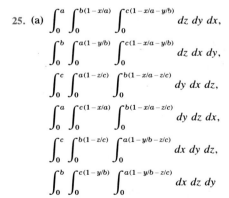

$\displaystyle\int_0^a\int_0^{b(1-x/a)}\int_0^{c(1-x/a-y/b)} dz\,dy\,dx,$

$\displaystyle\int_0^b\int_0^{a(1-y/b)}\int_0^{c(1-x/a-y/b)} dz\,dx\,dy,$

$\displaystyle\int_0^c\int_0^{a(1-z/c)}\int_0^{b(1-x/a-z/c)} dy\,dx\,dz,$

$\displaystyle\int_0^a\int_0^{c(1-x/a)}\int_0^{b(1-x/a-z/c)} dy\,dz\,dx,$

$\displaystyle\int_0^c\int_0^{b(1-z/c)}\int_0^{a(1-y/b-z/c)} dx\,dy\,dz,$

$\displaystyle\int_0^b\int_0^{c(1-y/b)}\int_0^{a(1-y/b-z/c)} dx\,dz\,dy$

27. $\frac{1}{6}a^3$

29. (a) 0   (b) $\frac{1}{2}(e^2 - 1)$

31. (a) 10   (b) 0   33. 0

▶ Exercise Set 17.6 **(Page 1131)**

1. 10 feet to the right of m_1
3. $\frac{13}{20}$, $(\frac{190}{273}, \frac{6}{13})$
21. $(\frac{1}{2}, 0, \frac{3}{5})$
23. $\left(\frac{3a}{8}, \frac{3a}{8}, \frac{3a}{8}\right)$

5. $\frac{a^4}{8}$, $\left(\frac{8a}{15}, \frac{8a}{15}\right)$
7. $(\frac{2}{3}, \frac{1}{3})$
25. $\frac{2}{3}\pi k a^3$
29. $\left(\frac{128}{105\pi}, \frac{128}{105\pi}\right)$

9. $(-\frac{1}{2}, \frac{2}{5})$
11. $\left(0, \frac{4(b^3 - a^3)}{3\pi(b^2 - a^2)}\right)$
33. $2\pi^2 kab$
35. $\left(\frac{a}{3}, \frac{b}{3}\right)$

13. $(0, \frac{3}{5})$
15. $\frac{a^4}{2}$, $\left(\frac{a}{3}, \frac{a}{2}, \frac{a}{2}\right)$
37. $\frac{1}{12}M(a^2 + b^2)$
39. $\frac{1}{4}MR^2$

17. $\frac{1}{6}$, $(0, \frac{16}{35}, \frac{1}{2})$
19. $(\frac{1}{4}, \frac{1}{4}, \frac{1}{4})$
41. $\frac{3}{2}MR^2$
43. $\frac{2}{3}Ma^2$

▶ Exercise Set 17.7 **(Page 1143)**

1. $\pi/4$
3. $\pi/16$
33. $\left(0, 0, \frac{2a}{5}\right)$
35. $(0, 0, \frac{11}{30})$

5. $81\pi/2$
7. $\frac{8}{3}(10\sqrt{5} - 19)\pi$

9. $\frac{32}{9}(3\pi - 4)$
11. $\frac{2}{3}(\sqrt{3} - 1)\pi$
37. $\frac{63}{8}\pi$
41. $M(\frac{1}{4}R^2 + \frac{1}{3}h^2)$

13. $9\sqrt{2}\pi$
15. $\frac{4}{3}\pi a^3$
43. $\frac{2}{5}MR^2$

17. $\frac{11}{6}\pi$
19. $\frac{1}{4}(2 - \sqrt{2})\pi$
45. (a) $2\pi k\delta(\sqrt{R^2 + a^2} - \sqrt{R^2 + (a + h)^2} + h)$

21. $\pi k a^4$
23. $\left(0, 0, \frac{7}{16\sqrt{2} - 14}\right)$
   (b) Other components are zero because of symmetry.

25. $\left(\frac{3a}{8}, \frac{3a}{8}, \frac{3a}{8}\right)$
27. $\left(0, \frac{195}{152}, 0\right)$
47. $2\pi k\delta h\left(1 - \frac{h}{\sqrt{R^2 + h^2}}\right)$

29. $\frac{4}{3}(1 - e^{-1})\pi$
31. 81π

▶ Chapter 17 Supplementary Exercises **(Page 1146)**

1. $-1/(\sqrt{2}\pi)$
3. $-\frac{9}{14}$
27. (a) $\int_0^{1/2} \int_0^{1-2y} \int_0^{1-z-2y} z \, dx \, dz \, dy$

5. $\int_0^1 \int_{2y}^2 e^x e^y \, dx \, dy$
7. $\frac{3}{2}$
   (b) $\int_0^4 \int_0^{4-y} \int_0^{\sqrt{y}} 3 \, dx \, dz \, dy$

9. (a) 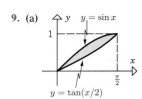 (b)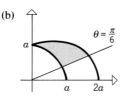
29. (a) $\int_0^{2\pi} \int_0^{\pi/3} \int_0^a (\rho^4 \sin^3 \phi) \, d\rho \, d\phi \, d\theta$

   (b) $\int_0^{2\pi} \int_0^{\sqrt{3a/2}} \int_{r/\sqrt{3}}^{\sqrt{a^2 - r^2}} r^3 \, dz \, dr \, d\theta$

   (c) $4\int_0^{\sqrt{3a/2}} \int_0^{\sqrt{(3a^2/4) - x^2}} \int_{\sqrt{(x^2+y^2)/3}}^{\sqrt{a^2 - x^2 - y^2}} (x^2 + y^2) \, dz \, dy \, dx$

11. $(1 - \cos 64)/3$
13. 6
15. $\pi/4$

17. $\int_0^{2a} \int_0^{\sqrt{2ay - y^2}} \frac{2xy}{x^2 + y^2} \, dx \, dy = a^2$
31. $8\pi a^3/3$
33. $\pi a^3/9$

35. $\bar{x} = 0, \bar{y} = 5a/6$
37. $\bar{x} = 0, \bar{y} = b/4$

19. $\pi/4$
21. $(37^{3/2} - 1)\pi/54$
39. $kabc^2/24$
41. $\bar{x} = 0, \bar{y} = \frac{12}{7}, \bar{z} = \frac{8}{7}$

23. $15\pi\sqrt{2}$
25. $512\pi/3$
43. $\bar{x} = \bar{y} = 0, \bar{z} = h/4$

▶ Exercise Set 18.1 **(Page 1158)**

1. (a) $\frac{4}{3}$ (b) 0 (c) $\frac{4}{3}$
3. $-\frac{8}{3}$
11. 0
13. $1 - e^{-1}$
15. $1 - e^3$

5. $1 - \pi$
7. 3
9. $-1 - \pi/4$
17. $\frac{23}{6}$
19. $\frac{3}{5}$

21. (a) $\frac{5}{4} - \pi/8 + \frac{1}{2}\tan^{-1}2$ **23.** $-\frac{37}{2}$ **25.** -12

(b) $\frac{1}{3}\tan^{-1}2 - \frac{2}{3}\tan^{-1}\frac{1}{2}$ (c) $\frac{3}{4}$ **27.** 0

▶ Exercise Set 18.2 **(Page 1168)**

1. $\frac{1}{2}x^2 + \frac{1}{2}y^2 + K$ **3.** not conservative   **11.** $9e^2$ **13.** 32

5. $x\cos y + y\sin x + K$ **15.** $-\frac{1}{2}$ **17.** $\pi/4$

7. 13   **9.** -6 **19.** (a) $-2a$ (b) $-2a$ (c) $-2a$ (d) 0

▶ Exercise Set 18.3 **(Page 1174)**

1. 0   **3.** 0   **5.** 0   **23.** (c) $A = \frac{1}{2}[(x_1y_2 - x_2y_1) + (x_2y_3 - x_3y_2)$

7. 8π **9.** -4 **11.** -1 $+ \cdots + (x_ny_1 - x_1y_n)]$

13. 0   **15.** πab **17.** $\frac{1}{2}ab$ (d) 8

19. $\frac{250}{3}$ **21.** (b) $\left(0, \dfrac{4a}{3\pi}\right)$

▶ Exercise Set 18.4 **(Page 1185)**

1. $\dfrac{\pi\delta_0}{6}(5\sqrt{5} - 1)$ **3.** $\frac{4}{3}\pi\delta_0$

5. $15\pi/\sqrt{2}$ **7.** $\pi/4$

9. $3/\sqrt{2}$ **11.** 9

13. $4\pi/3$ **15.** $\frac{8}{3}\pi a^4$

17. $\frac{1}{4}[37\sqrt{37} - 1]$

19. (a) $\dfrac{\sqrt{29}}{16}\displaystyle\int_0^6\int_0^{(12-2x)/3} xy(12 - 2x - 3y)\,dy\,dx$

(b) $\dfrac{\sqrt{29}}{4}\displaystyle\int_0^3\int_0^{(12-4z)/3} yz(12 - 3y - 4z)\,dy\,dz$

(c) $\dfrac{2\sqrt{29}}{9}\displaystyle\int_0^3\int_0^{6-2z} xz(6 - x - 2z)\,dx\,dz$

21. (a) $\displaystyle\int_0^4\int_1^2 y^3z\sqrt{1 + 4y^2}\,dy\,dz$

(b) $\dfrac{1}{2}\displaystyle\int_0^4\int_1^4 xz\sqrt{1 + 4x}\,dx\,dz$

▶ Exercise Set 18.5 **(Page 1194)**

1. $\dfrac{2}{\sqrt{29}}\mathbf{i} + \dfrac{3}{\sqrt{29}}\mathbf{j} + \dfrac{4}{\sqrt{29}}\mathbf{k}$

5. 2π **7.** $\pi/2$ **9.** 54π

11. $\frac{14}{3}\pi$ **13.** 0   **15.** $4\pi a^3$

3. (a) $-\dfrac{2}{\sqrt{21}}\mathbf{i} - \dfrac{4}{\sqrt{21}}\mathbf{j} + \dfrac{1}{\sqrt{21}}\mathbf{k}$

(b) $\dfrac{3}{5\sqrt{2}}\mathbf{i} - \dfrac{4}{5\sqrt{2}}\mathbf{j} + \dfrac{1}{\sqrt{2}}\mathbf{k}$ (c) $\frac{3}{5}\mathbf{i} - \frac{4}{5}\mathbf{k}$

19. (a) $\displaystyle\iint_R \mathbf{F}\cdot\left(\mathbf{i} - \frac{\partial x}{\partial y}\mathbf{j} - \frac{\partial x}{\partial z}\mathbf{k}\right)dA$ $\begin{bmatrix}\text{Forward} \\ R = \text{projection on } yz\text{-plane}\end{bmatrix}$

$\displaystyle\iint_R \mathbf{F}\cdot\left(-\mathbf{i} + \frac{\partial x}{\partial y}\mathbf{j} + \frac{\partial x}{\partial z}\mathbf{k}\right)dA$ $\begin{bmatrix}\text{Backward} \\ R = \text{projection on } yz\text{-plane}\end{bmatrix}$

(b) $\displaystyle\iint_R \mathbf{F}\cdot\left(-\frac{\partial y}{\partial x}\mathbf{i} + \mathbf{j} - \frac{\partial y}{\partial z}\mathbf{k}\right)dA$ $\begin{bmatrix}\text{Right} \\ R = \text{projection on } xz\text{-plane}\end{bmatrix}$

$\displaystyle\iint_R \mathbf{F}\cdot\left(\frac{\partial y}{\partial x}\mathbf{i} - \mathbf{j} + \frac{\partial y}{\partial z}\mathbf{k}\right)dA$ $\begin{bmatrix}\text{Left} \\ R = \text{projection on } xz\text{-plane}\end{bmatrix}$

21. 2π

▶ Exercise Set 18.6 **(Page 1202)**

1. $\text{div } \mathbf{F} = z^3 + 8y^3x^2 + 10zy$

3. $\text{div } \mathbf{F} = ye^{xy} + \sin y + 2\sin z \cos z$

5. $\text{div } \mathbf{F} = \dfrac{1}{x} + xz\, e^{xyz} + \dfrac{x}{x^2 + z^2}$

7. 8

11. 180π

15. $\pi/2$

19. $4\pi a^3$

9. $3\pi a^2$

13. $\frac{1}{24}$

17. $\frac{4608}{35}$

23. $\pi a^2 h$

▶ Exercise Set 18.7 **(Page 1207)**

1. $\text{curl } \mathbf{F} = \mathbf{0}$

3. $\text{curl } \mathbf{F} = -xe^{xy}\mathbf{k}$

5. $\text{curl } \mathbf{F} = -xye^{xyz}\mathbf{i} + \dfrac{z}{x^2 + z^2}\mathbf{j} + yz\, e^{xyz}\mathbf{k}$

7. 16π

11. πa^2

15. 0

9. 0

13. $\frac{3}{2}$

17. 2π

▶ Exercise Set 18.8 **(Page 1215)**

1. 2π

5. $4\pi a^3$

9. no sources or sinks ($\text{div } \mathbf{F} = 0$)

3. 54π

7. 180π

11. sources at all points except the origin; no sinks

13. (a) $\frac{3}{2}$ (b) -1 (c) $\mathbf{n} = -\dfrac{1}{\sqrt{2}}\mathbf{j} - \dfrac{1}{\sqrt{2}}\mathbf{k}$

▶ Chapter 18 Supplementary Exercises **(Page 1215)**

1. 4

5. 2

9. $\dfrac{y^2}{x} + 2y^2 + x^3 + C$

3. 6

7. not conservative

11. $-7\pi^2/16$ 13. $\frac{3}{2}$

17. 32

21. $\frac{12}{5}$

25. $(0, 0, \frac{149}{65})$

15. 0

19. $-45\pi/4$

23. 2

27. $\pi/2$

▶ Exercise Set 19.1 **(Page 1223)**

3. $y = c_1 e^x + c_2 e^{-4x}$ 5. $y = c_1 e^x + c_2 x e^x$

7. $y = c_1 \cos \sqrt{5}x + c_2 \sin \sqrt{5}x$

9. $y = c_1 + c_2 e^x$

11. $y = c_1 e^{-2t} + c_2 t e^{-2t}$

13. $y = e^{2x}(c_1 \cos 3x + c_2 \sin 3x)$

15. $y = c_1 e^{-x/4} + c_2 e^{x/2}$

17. $y = 2e^x - e^{-3x}$ 19. $y = (2 - 5x)e^{3x}$

21. $y = -e^{-2x}(3\cos x + 6\sin x)$

23. (a) $y'' - 3y' - 10y = 0$

(b) $y'' - 8y' + 16y = 0$

(c) $y'' + 2y' + 17y = 0$

25. (a) $k < 0$ or $k > 4$

(b) $0, 4$ (c) $0 < k < 4$

27. (a) $y = \dfrac{1}{x}\,[c_1 \cos(\ln x) + c_2 \sin(\ln x)]$

(b) $y = c_1 x^{1+\sqrt{3}} + c_2 x^{1-\sqrt{3}}$

▶ Exercise Set 19.2 **(Page 1231)**

1. $y = c_1 e^{-x} + c_2 e^{-5x} + \frac{1}{16}e^{3x}$

3. $y = c_1 e^{4x} + c_2 e^{5x} - 3xe^{5x}$

5. $y = (c_1 + c_2 x)e^{-x} + \frac{1}{2}x^2 e^{-x}$

7. $y = c_1 e^{3x} + c_2 e^{-4x} - \frac{13}{216} - \frac{1}{18}x - \frac{1}{3}x^2$

9. $y = c_1 + c_2 e^{6x} + \frac{5}{36}x - \frac{1}{12}x^2$

11. $y = c_1 + c_2 x - \frac{1}{2}x^2 + \frac{1}{20}x^5$

13. $y = c_1 e^{-x} + c_2 e^{2x} - 3\cos x - \sin x$

15. $y = c_1 e^{-2x} + c_2 e^{2x} - \frac{3}{8}\cos 2x - \frac{1}{4}\sin 2x$

17. $y = c_1 \cos x + c_2 \sin x - \frac{1}{2}x \cos x$

19. $y = c_1 e^x + c_2 e^{2x} + \frac{3}{4} + \frac{1}{2}x$

21. $y = e^{-2x}(c_1 \cos \sqrt{5}x + c_2 \sin \sqrt{5}x) - \frac{94}{729} + \frac{19}{81}x + \frac{1}{9}x^2$

23. $y = c_1 \cos 2x + c_2 \sin 2x - \frac{1}{8}x \cos 2x$

25. (b) $-\frac{3}{16} - \frac{1}{4}x + \frac{1}{5}xe^x$

27. $y = c_1 e^{-x} + c_2 e^x - 1 + \frac{1}{2}xe^x$

29. $y = c_1 \cos 2x + c_2 \sin 2x + \frac{1}{4} + \frac{1}{4}x + \frac{1}{3}\sin x$

31. $y = (c_1 + c_2 x)e^x + \frac{1}{4}x^2 e^x + \frac{1}{8}e^{-x}$

33. $y = c_1 \cos x + c_2 \sin x + 6 - 2\cos 2x$

35. (a) $y = c_1 \cos \mu x + c_2 \sin \mu x + \dfrac{a}{\mu^2 - b^2}\sin bx$

(b) $y = c_1 \cos \mu x + c_2 \sin \mu x + \displaystyle\sum_{k=1}^{n} \dfrac{a_k}{\mu^2 - k^2\pi^2}\sin k\pi x$

37. $x_0 = 1$; $y = 2x^2 - 4e^{x-1} + c$, c arbitrary

▶ Exercise Set 19.3 **(Page 1235)**

1. $y = c_1 \cos x + c_2 \sin x + x^2 - 2$

3. $y = c_1 e^x + c_2 e^{-2x} + \frac{2}{3}xe^x$

5. $y = c_1 \cos 2x + c_2 \sin 2x - \frac{1}{4}x\cos 2x$

7. $y = c_1 \cos x + c_2 \sin x - (\cos x)\ln|\sec x + \tan x|$

9. $y = c_1 e^x + c_2 xe^x + xe^x \ln|x|$

11. $y = c_1 \cos x + c_2 \sin x + \frac{3}{2} + \frac{1}{2}\cos 2x$

13. $y = c_1 \cos x + c_2 \sin x - x\cos x + (\sin x)\ln|\sin x|$

15. $y = c_1 \cos x + c_2 \sin x + x\cos x - (\sin x)\ln|\cos x|$

17. $y = c_1 e^{-x} + c_2 xe^{-x} - e^{-x}\ln|x|$

19. $y = c_1 e^{-2x} + c_2 xe^{-2x} + (x-2)e^{-x}$

21. $y = c_1 \cos x + c_2 \sin x - 1 + (\sin x)\ln|\sec x + \tan x|$

23. $y = c_1 e^x + c_2 xe^x - e^x \ln|x|$

25. $y = c_1 e^x + c_2 e^{-x} + \frac{1}{5}e^x(2\sin x - \cos x)$

27. $y = c_1 e^{-x} + c_2 xe^{-x} + \frac{1}{2}x^2 e^{-x}\ln|x| - \frac{3}{4}x^2 e^{-x}$

▶ Exercise Set 19.4 **(Page 1244)**

1. (a) $y'' + 4y = 0$, $y(0) = 1$, $y'(0) = 0$
   (b) $y = \cos 2t$

3. (a) $y'' + 196y = 0$, $y(0) = -10$, $y'(0) = 0$
   (b) $y = -10\cos 14t$

5. (a) $y = 2\cos 8t$ (b) 2
   (c) $\pi/4$ (d) $4/\pi$

7. (a) $y = -\frac{1}{4}\cos 8\sqrt{6}t$ (b) $\frac{1}{4}$
   (c) $\dfrac{\pi}{4\sqrt{6}}$ (d) $\dfrac{4\sqrt{6}}{\pi}$

9. (a) $y'' + 4y' + 8y = 0$, $y(0) = -3$, $y'(0) = 0$
   (b) $y = -3e^{-2t}(\cos 2t + \sin 2t)$

11. (a) $y = \frac{5}{2}\sqrt{6}e^{-t/5}\cos\left(\dfrac{\sqrt{2}}{5}t - \tan^{-1}\dfrac{1}{\sqrt{2}}\right)$
    (b) $\dfrac{10\pi}{\sqrt{2}}$ (c) $\dfrac{\sqrt{2}}{10\pi}$

13. (a) $y = -\frac{1}{4}\sin 8t$ (b) $\frac{1}{4}$
    (c) $\pi/4$ (d) $4/\pi$

15. $y = \frac{1}{3}e^{-t}(\cos 7t - 2\sin 7t)$

17. (a) $\frac{1}{32}\pi^2$ (b) $\frac{9}{4}$

19. $T = 2\pi\sqrt{\dfrac{\delta h}{\rho g}}$

APPENDIX EXERCISES

▶ Exercise Set A.1 **(Page A7)**

1. $x = 1$, $y = 2$

3. $x_1 = -\frac{7}{32}$, $x_2 = \frac{5}{16}$

5. $x = 2$, $y = -1$, $z = 3$

7. $x_1 = \frac{26}{21}$, $x_2 = \frac{25}{21}$, $x_3 = \frac{5}{7}$

9. $x' = x\cos\theta + y\sin\theta$
   $y' = -x\sin\theta + y\cos\theta$

► Exercise Set B.1 (page B6)

1. (a–d)

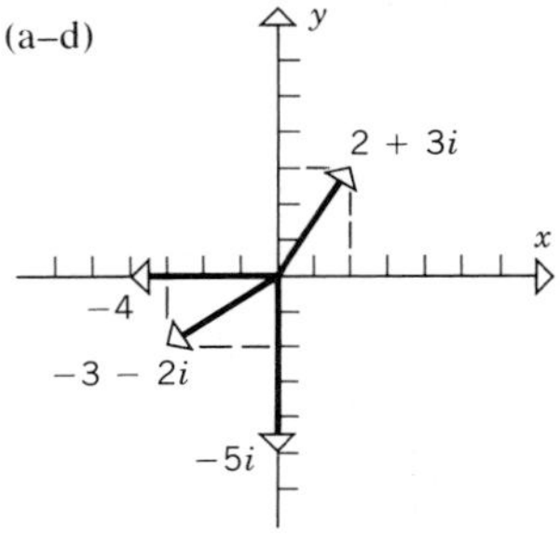

y
2 + 3i
x
−4
−3 − 2i
−5i

2. (a) (2, 3) (b) (−4, 0) (c) (−3, −2) (d) (0, −5)
3. (a) $x = -2, y = -3$ (b) $x = 2, y = 1$
4. (a) $5 + 3i$ (b) $-3 - 7i$ (c) $4 - 8i$ (d) $-4 - 5i$ (e) $19 + 14i$ (f) $-\frac{11}{2} - \frac{17}{2}i$
5. (a) $2 + 3i$ (b) $-1 - 2i$ (c) $-2 + 9i$
6. (a)

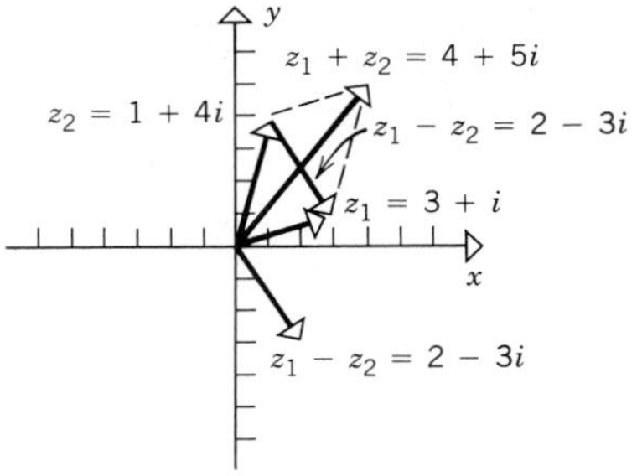

y
z1 + z2 = 4 + 5i
z2 = 1 + 4i
z1 − z2 = 2 − 3i
z1 = 3 + i
x
z1 − z2 = 2 − 3i

(b)

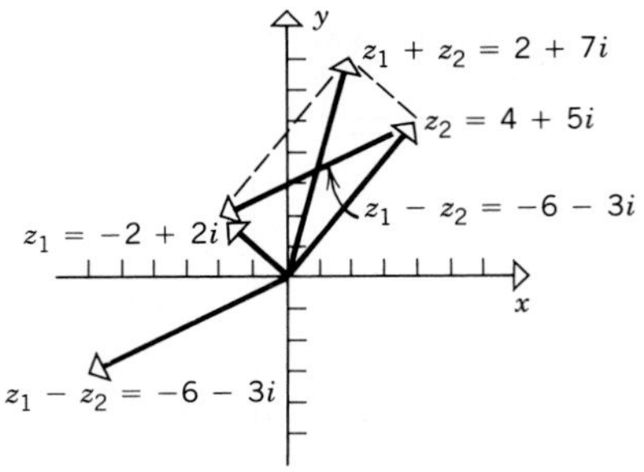

y
z1 + z2 = 2 + 7i
z2 = 4 + 5i
z1 − z2 = −6 − 3i
z1 = −2 + 2i
x
z1 − z2 = −6 − 3i

7. (a)

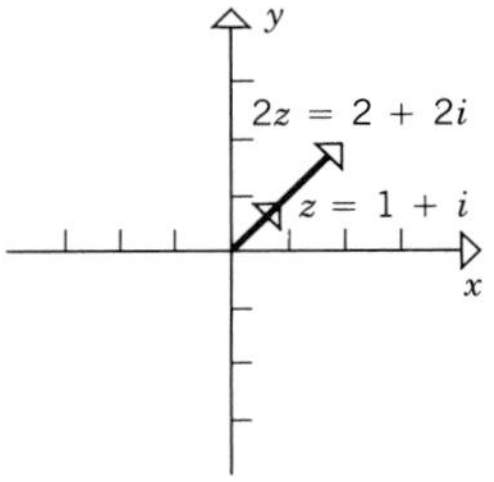

y
2z = 2 + 2i
z = 1 + i
x

(b)

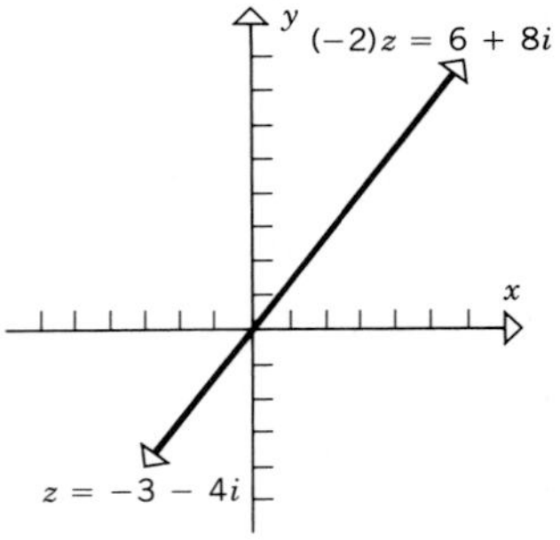

y
(−2)z = 6 + 8i
x
z = −3 − 4i

(c)

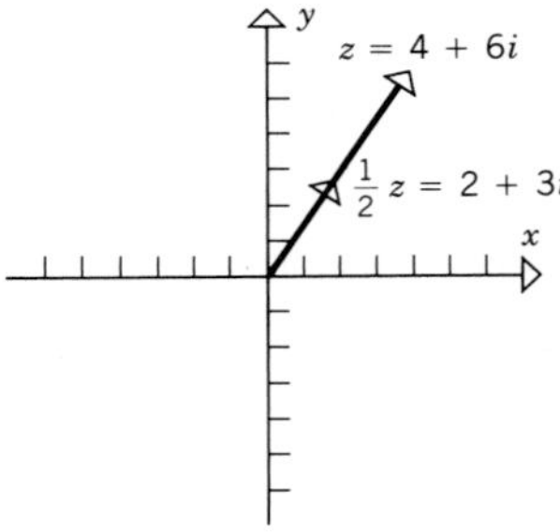

y
z = 4 + 6i
1/2 z = 2 + 3i
x

8. (a) $k_1 = -5, k_2 = 3$
   (b) $k_1 = 3, k_2 = 1$
9. (a) $z_1z_2 = 3 + 3i, z_1^2 = -9, z_2^2 = -2i$
   (b) $z_1z_2 = 26, z_1^2 = -20 + 48i, z_2^2 = -5 - 12i$
   (c) $z_1z_2 = \frac{11}{3} - i, z_1^2 = \frac{4}{9}(-3 + 4i), z_2^2 = -6 - \frac{5}{2}i$

10. (a) $9 - 8i$ (b) $-63 + 16i$ (c) $-32 - 24i$ (d) $22 + 19i$
11. $76 - 88i$
12. $26 - 18i$
13. $-26 + 18i$
14. $-1 - 11i$
15. $-\frac{63}{16} + i$
16. $(2 + \sqrt{2}) + i(1 - \sqrt{2})$
17. 0
18. $-24i$

► Exercise Set B.2 (page B12)

1. (a) $2 - 7i$ (b) $-3 + 5i$ (c) $-5i$ (d) i (e) -9 (f) 0
2. (a) 1 (b) 7 (c) 5 (d) $\sqrt{2}$ (e) 8 (f) 0

4. (a) $-\frac{17}{25} - \frac{19}{25}i$ (b) $\frac{23}{25} + \frac{11}{25}i$ (c) $\frac{23}{25} - \frac{11}{25}i$ (d) $-\frac{17}{25} + \frac{19}{25}i$ (e) $\frac{1}{5} - i$ (f) $\frac{\sqrt{26}}{5}$

5. (a) $-i$ (b) $\frac{1}{26} + \frac{5}{26}i$ (c) $7i$ 6. (a) $\frac{6}{5} + \frac{2}{5}i$ (b) $-\frac{2}{5} + \frac{1}{5}i$ (c) $\frac{3}{5} + \frac{11}{5}i$ (d) $\frac{3}{5} + \frac{1}{5}i$ 7. $\frac{1}{2} + \frac{1}{2}i$

8. $\frac{2}{5} + \frac{1}{5}i$ 9. $-\frac{7}{625} - \frac{24}{625}i$ 10. $-\frac{11}{25} + \frac{2}{25}i$ 11. $\dfrac{1 - \sqrt{3}}{4} + \dfrac{1 + \sqrt{3}}{4}i$ 12. $-\frac{1}{26} - \frac{5}{26}i$ 13. $-\frac{1}{10} + \frac{1}{10}i$

14. $-\frac{2}{5}$ 15. (a) $-1 - 2i$ (b) $-\frac{3}{25} - \frac{4}{25}i$

17. (a)

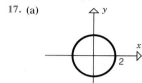

(b)

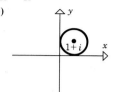

(c)

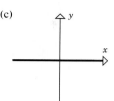

(d)

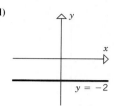

18. (a)

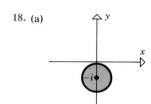

(b)

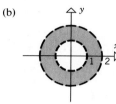

(c)

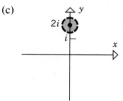

(d)

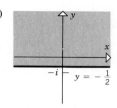

19. (a) $-y$ (b) $-x$ (c) y (d) x 20. (b) $-i$ 23. (a) $\dfrac{x_1 x_2 + y_1 y_2}{x_2^2 + y_2^2}$ (b) $\dfrac{x_2 y_1 - x_1 y_2}{x_2^2 + y_2^2}$

27. (c) Yes, if $z \neq 0$.

▶ Exercise Set B.3 **(page B22)**

1. (a) 0 (b) $\pi/2$ (c) $-\pi/2$ (d) $\pi/4$ (e) $2\pi/3$ (f) $-\pi/4$ 2. (a) $5\pi/3$ (b) $-\pi/3$ (c) $5\pi/3$

3. (a) $2\left[\cos\left(\dfrac{\pi}{2}\right) + i \sin\left(\dfrac{\pi}{2}\right)\right]$ (b) $4[\cos \pi + i \sin \pi]$

   (c) $5\sqrt{2}\left[\cos\left(\dfrac{\pi}{4}\right) + i \sin\left(\dfrac{\pi}{4}\right)\right]$ (d) $12\left[\cos\left(\dfrac{2\pi}{3}\right) + i \sin\left(\dfrac{2\pi}{3}\right)\right]$

   (e) $3\sqrt{2}\left[\cos\left(-\dfrac{3\pi}{4}\right) + i \sin\left(-\dfrac{3\pi}{4}\right)\right]$ (f) $4\left[\cos\left(-\dfrac{\pi}{6}\right) + i \sin\left(-\dfrac{\pi}{6}\right)\right]$

4. (a) $6\left[\cos\left(\dfrac{5\pi}{12}\right) + i \sin\left(\dfrac{5\pi}{12}\right)\right]$ (b) $\dfrac{2}{3}\left[\cos\left(\dfrac{\pi}{12}\right) + i \sin\left(\dfrac{\pi}{12}\right)\right]$

   (c) $\dfrac{3}{2}\left[\cos\left(-\dfrac{\pi}{12}\right) + i \sin\left(-\dfrac{\pi}{12}\right)\right]$ (d) $\dfrac{32}{9}\left[\cos\left(\dfrac{11\pi}{12}\right) + i \sin\left(\dfrac{11\pi}{12}\right)\right]$

5. 1   6. (a) -64 (b) $-i$ (c) $-64\sqrt{3} - 64i$ (d) $-\dfrac{1 + \sqrt{3}i}{2048}$

7. (a)

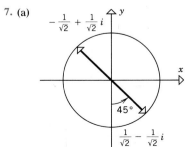

(b)

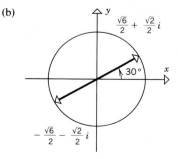

(c)

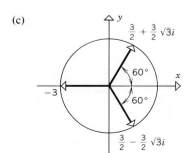

(d)

(e)
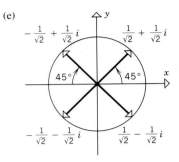

(f)
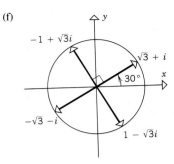

8.
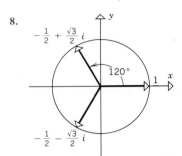

9.
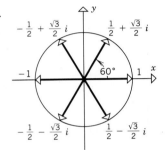

10. $\sqrt[4]{2}\left[\cos\left(\dfrac{\pi}{8}\right) + i\sin\left(\dfrac{\pi}{8}\right)\right]$, $\sqrt[4]{2}\left[\cos\left(\dfrac{9\pi}{8}\right) + i\sin\left(\dfrac{9\pi}{8}\right)\right]$ 11. (a) $\pm 2, \pm 2i$ (b) $\pm(2+2i), \pm(2-2i)$

12. The roots are $\pm(2^{1/4} + 2^{1/4}i)$, $\pm(2^{1/4} - 2^{1/4}i)$ and the factorization is
$z^4 + 8 = (z^2 - 2^{5/4}z + 2^{3/2})\cdot(z^2 + 2^{5/4}z + 2^{3/2})$.

13. Rotates z clockwise by 90°.   14. (a) 16 (b) $\dfrac{i}{4^9}$

15. (a) $\mathrm{Re}(z) = -3$, $\mathrm{Im}(z) = 0$ (b) $\mathrm{Re}(z) = -3$, $\mathrm{Im}(z) = 0$
(c) $\mathrm{Re}(z) = 0$, $\mathrm{Im}(z) = -\sqrt{2}$ (d) $\mathrm{Re}(z) = -3$, $\mathrm{Im}(z) = 0$

▶ Exercise Set 11.1 **(Page 649)**

1. $\frac{1}{3}, \frac{2}{4}, \frac{3}{5}, \frac{4}{6}, \frac{5}{7}$; converges to 1

3. 2, 2, 2, 2, 2; converges to 2

5. $\dfrac{\ln 1}{1}, \dfrac{\ln 2}{2}, \dfrac{\ln 3}{3}, \dfrac{\ln 4}{4}, \dfrac{\ln 5}{5}$; converges to 0

7. 0, 2, 0, 2, 0; diverges

9. $-1, \frac{16}{9}, -\frac{54}{28}, \frac{128}{65}, -\frac{250}{126}$; diverges

11. $\frac{6}{2}, \frac{12}{8}, \frac{20}{18}, \frac{30}{32}, \frac{42}{50}$; converges to $\frac{1}{2}$

13. $\cos 3, \cos\frac{3}{2}, \cos 1, \cos\frac{3}{4}, \cos\frac{3}{5}$; converges to 1

15. $e^{-1}, 4e^{-2}, 9e^{-3}, 16e^{-4}, 25e^{-5}$; converges to 0

17. $2, (\frac{5}{3})^2, (\frac{6}{4})^3, (\frac{7}{5})^4, (\frac{8}{6})^5$; converges to e^2

19. $\left\{\dfrac{2n-1}{2n}\right\}_{n=1}^{+\infty}$; converges to 1

21. $\left\{\dfrac{1}{3^n}\right\}_{n=1}^{+\infty}$; converges to 0

23. $\left\{\dfrac{1}{n} - \dfrac{1}{n+1}\right\}_{n=1}^{+\infty}$; converges to 0

25. $\{\sqrt{n+1} - \sqrt{n+2}\}_{n=1}^{+\infty}$; converges to 0

27. (a) $\sqrt{6}, \sqrt{6+\sqrt{6}}, \sqrt{6+\sqrt{6+\sqrt{6}}}$ (b) 3

29. (a) 1, 1, 2, 3, 5, 8, 13, 21 (b) $(1+\sqrt{5})/2$

31. (a) $1, \frac{3}{4}, \frac{2}{3}, \frac{5}{8}$ (b) $\frac{1}{2}$

33. (a) 3 (b) 11 (c) 1001 **39.** 3

▶ Exercise Set 11.2 **(Page 658)**

1. decreasing
3. increasing
5. decreasing
7. increasing
9. decreasing
11. nonincreasing
13. not monotone
15. decreasing
17. increasing
19. increasing
21. decreasing
23. decreasing

25. converges
27. diverges
29. converges
31. (a) 0   (b) $+\infty$ (does not exist)
35. (a) $\sqrt{2}, \sqrt{2 + \sqrt{2}}, \sqrt{2 + \sqrt{2 + \sqrt{2}}}$ (e) 2
41. (b) converges (decreasing and bounded below by 0)

▶ Exercise Set 11.3 **(Page 666)**

1. (a) converges to $\frac{5}{2}$ (b) converges to $\frac{1}{2}$
   (c) diverges
3. $\frac{4}{7}$
5. 6
7. diverges
9. $\frac{1}{3}$
11. $\frac{1}{6}$
13. $\frac{448}{3}$
15. $-\frac{1}{3}$
17. $\frac{4}{9}$
19. $\frac{532}{99}$

21. $\frac{869}{1111}$
23. diverges
33. $\dfrac{1}{x^2 - 2x}, |x| > 2$
35. $\dfrac{2 \sin x}{2 + \sin x}, -\infty < x < +\infty$ 37. 1
41. The series converges to $1/(1 - x)$ only for $-1 < x < 1$.

▶ Exercise Set 11.4 **(Page 676)**

1. $\frac{4}{3}$
3. $-\frac{1}{36}$

9. diverges
11. diverges
13. converges
15. diverges
17. diverges
19. diverges
21. converges
23. diverges
25. converges
27. converges
29. diverges

5. (a) converges   (b) diverges   (c) diverges
   (d) diverges   (e) converges   (f) diverges
   (g) converges   (h) converges
35. (a) diverges   (b) diverges
    (c) diverges   (d) converges
37. (a) 1.1975; $1.2016 < S < 1.2026$
    (b) 23
39. (a) $13 < s_{1,000,000} < 15$
    (b) 2.69×10^{43}

▶ Exercise Set 11.5 **(Page 683)**

1. converges
3. inconclusive
5. diverges
7. diverges
9. converges
11. diverges
13. converges
15. converges
17. converges
19. converges

21. diverges
23. converges
25. converges
27. diverges
29. converges
31. converges
41. (a) 1.71667; error < 0.00163 (b) 8
43. (a) 0.69226; error < 0.00098 (b) 13

▶ Exercise Set 11.6 **(Page 691)**

13. converges
15. converges
17. diverges
19. converges
21. diverges
23. converges
25. diverges
27. converges

29. diverges
31. converges
33. converges
37. converges
39. $p > 1$
41. converges

▶ Exercise Set 11.7 **(Page 703)**

1. converges
3. diverges
5. converges
7. absolutely
9. diverges
11. absolutely
13. conditionally
15. divergent
17. conditionally
19. absolutely
21. conditionally
23. divergent
25. conditionally
27. absolutely

29. conditionally   31. 0.125
33. 0.1   35. 9,999   37. 39,999
39. $|\text{error}| < 0.00074$; $s_{10} \approx 0.4995$; exact sum $= 0.5$
41. $n = 4$, $s_4 \approx 0.84147$; $\sin(1) \approx 0.841470985$
43. $n = 9$, $s_9 \approx 0.40553$; $\ln \frac{3}{2} \approx 0.405465108$
45. (a) 14
     (b) 0.817962176; $|\text{error}| \approx 0.004504858$

▶ Exercise Set 11.8 **(Page 710)**

1. $1, [-1, 1)$
3. $+\infty, (-\infty, +\infty)$
5. $\frac{1}{5}, [-\frac{1}{5}, \frac{1}{5}]$
7. $1, [-1, 1]$
9. $1, (-1, 1]$
11. $+\infty, (-\infty, +\infty)$
13. $+\infty, (-\infty, +\infty)$
15. $1, [-1, 1]$
17. $1, (-2, 0]$
19. $\frac{4}{3}, (-\frac{19}{3}, -\frac{11}{3})$

21. $1, [-2, 0]$
23. $+\infty, (-\infty, +\infty)$
25. $x + \frac{1}{2}x^2 + \frac{3}{14}x^3 + \frac{3}{35}x^4 + \cdots$; $R = 3$
27. $x + \frac{3}{2}x^2 + \frac{5}{8}x^3 + \frac{7}{48}x^4 + \cdots$; $R = +\infty$
29. $(a - b, a + b)$
31. $+\infty$

▶ Exercise Set 11.9 **(Page 721)**

1. $1 - 2x + 2x^2 - \frac{4}{3}x^3 + \frac{2}{3}x^4$

3. $2x - \frac{4}{3}x^3$

5. $x + \frac{1}{3}x^3$

7. $x + x^2 + \frac{x^3}{2!} + \frac{x^4}{3!}$

9. $1 + \frac{1}{2}x^2 + \frac{5}{24}x^4$

17. $\frac{\sqrt{2}}{2} - \frac{\sqrt{2}}{2}\left(x - \frac{\pi}{4}\right) - \frac{\sqrt{2}}{4}\left(x - \frac{\pi}{4}\right)^2 + \frac{\sqrt{2}}{12}\left(x - \frac{\pi}{4}\right)^3$

19. $-\frac{\sqrt{3}}{2} + \frac{\pi}{2}\left(x + \frac{1}{3}\right) + \frac{\sqrt{3}\pi^2}{4}\left(x + \frac{1}{3}\right)^2 - \frac{\pi^3}{12}\left(x + \frac{1}{3}\right)^3$

21. $\frac{\pi}{4} + \frac{1}{2}(x - 1) - \frac{1}{4}(x - 1)^2 + \frac{1}{12}(x - 1)^3$

23. $\sum_{k=0}^{\infty} (-1)^k \frac{x^k}{k!}$

25. $\sum_{k=0}^{\infty} (-1)^k x^k$

27. $\sum_{k=1}^{\infty} (-1)^{k+1} \frac{x^k}{k}$

29. $\sum_{k=0}^{\infty} (-1)^k \frac{x^{2k}}{4^k(2k)!}$

11. $\ln 3 + \frac{2}{3}x - \frac{2}{9}x^2 + \frac{8}{81}x^3 - \frac{4}{81}x^4$

13. $e + e(x - 1) + \frac{e}{2!}(x - 1)^2 + \frac{e}{3!}(x - 1)^3$

15. $2 + \frac{1}{4}(x - 4) - \frac{1}{64}(x - 4)^2 + \frac{1}{512}(x - 4)^3$

31. $\sum_{k=0}^{\infty} \frac{x^{2k}}{(2k)!}$

33. $\sum_{k=0}^{\infty} (-1)(x + 1)^k$

35. $\sum_{k=1}^{\infty} (-1)^{k+1} \frac{(x - 1)^k}{k}$

37. $\sum_{k=0}^{\infty} (-1)^k \frac{\pi^{2k}}{(2k)!}\left(x - \frac{1}{2}\right)^{2k}$

39. $\sum_{k=0}^{\infty} \left(\frac{16 + (-1)^{k+1}}{8}\right)\frac{(x - \ln 4)^k}{k!}$

▶ Exercise Set 11.10 **(Page 733)**

1. $\frac{2^6 e^{2c}}{6!}x^6$

3. $-\frac{x^5}{(c + 1)^6}$

5. $\frac{(4 + c)e^c}{4!}x^4$

7. $-\frac{(1 - 3c^2)}{3(1 + c^2)^3}x^3$

9. $-\frac{5}{128c^{7/2}}(x - 4)^4$

11. $\frac{\cos c}{5!}\left(x - \frac{\pi}{6}\right)^5$

13. $\frac{7}{(1 + c)^8}(x + 2)^6$

15. $\frac{x^{n+1}}{(1 - c)^{n+2}}$

17. $\frac{2^{n+1}e^{2c}}{(n + 1)!}x^{n+1}$

27. $1 - 2x + 2x^2 - \frac{4}{3}x^3 + \cdots$; $(-\infty, +\infty)$

29. $x - x^2 + \dfrac{1}{2!}x^3 - \dfrac{1}{3!}x^4 + \cdots; (-\infty, +\infty)$

31. $2x - \dfrac{2^3}{3!}x^3 + \dfrac{2^5}{5!}x^5 - \dfrac{2^7}{7!}x^7 + \cdots; (-\infty, +\infty)$

33. $x^2 - \dfrac{1}{2!}x^4 + \dfrac{1}{4!}x^6 - \dfrac{1}{6!}x^8 + \cdots; (-\infty, +\infty)$

35. $x^2 - \dfrac{2^3}{4!}x^4 + \dfrac{2^5}{6!}x^6 - \dfrac{2^7}{8!}x^8 + \cdots; (-\infty, +\infty)$

37. $-x^2 - \frac{1}{2}x^4 - \frac{1}{3}x^6 - \frac{1}{4}x^8 - \cdots; (-1, 1)$

39. $1 + 4x^2 + 16x^4 + 64x^6 + \cdots; (-\frac{1}{2}, \frac{1}{2})$

41. $x^2 - 3x^3 + 9x^4 - 27x^5 + \cdots; (-\frac{1}{3}, \frac{1}{3})$

43. $2x^2 + \dfrac{2^3}{3!}x^4 + \dfrac{2^5}{5!}x^6 + \dfrac{2^7}{7!}x^8 + \cdots; (-\infty, +\infty)$

45. $1 + \frac{3}{2}x - \frac{9}{8}x^2 + \frac{27}{16}x^3 - \cdots; (-\frac{1}{3}, \frac{1}{3})$

47. $1 + 4x + 12x^2 + 32x^3 + \cdots; (-\frac{1}{2}, \frac{1}{2})$

49. $x + \frac{1}{2}x^3 + \frac{3}{8}x^5 + \frac{5}{16}x^7 + \cdots; (-1, 1)$

51. $\displaystyle\sum_{k=0}^{\infty} (-1)^k (x-1)^k, (0, 2)$ 53. $\displaystyle\sum_{k=0}^{\infty} \binom{m}{k} x^k$

55. $\sin \pi = 0$ 57. $e^{-\ln 3} = \frac{1}{3}$

59. (a) $x^2 - 2x^4 + \frac{2}{3}x^6 - \frac{4}{45}x^8 + \cdots$ (b) 0

▶ Exercise Set 11.11 **(Page 742)**

1. (a) 9   (b) 13

3. 1.6487

5. 0.9877

7. 0.5299

9. 0.223

11. 0.100

13. 1.0050

15. $|x| < 0.569$

17. 9×10^{-8}

19. (a) 1,999,999   (b) 8   21. 3.140

▶ Exercise Set 11.12 **(Page 750)**

5. $\displaystyle\sum_{k=1}^{\infty} (-1)^{k+1} k x^{k-1}$

9. $\ln \frac{4}{3}$ 11. $\dfrac{2}{(1-x)^3}$

13. (b) $\displaystyle\sum_{k=0}^{\infty} (-1)^k x^k; (-1, 1)$

15. 0.764   17. 0.494

19. 0.100   21. 0.491

23. $1 - \frac{3}{2}x^2 + \frac{25}{24}x^4 - \frac{331}{720}x^6 + \cdots$

25. $x - x^2 + \frac{1}{3}x^3 - \frac{1}{30}x^5 + \cdots$

27. $-x^3 - \frac{1}{2}x^5 - \frac{1}{3}x^7 - \frac{1}{4}x^9 - \cdots$

29. $x^2 + \frac{9}{2}x^3 + \frac{79}{8}x^4 + \frac{683}{48}x^5 + \cdots$

31. (a) 0   (b) $-\frac{3}{2}$

33. (a) $x - \frac{1}{6}x^3 + \frac{3}{40}x^5 - \frac{5}{112}x^7$

(b) $x + \displaystyle\sum_{k=1}^{\infty} (-1)^k \dfrac{1 \cdot 3 \cdot 5 \cdots (2k-1)}{2^k k!(2k+1)} x^{2k+1}$ (c) 1

▶ Chapter 11 Supplementary Exercises **(Page 751)**

1. $L = 0$ 3. $L = 0$

5. does not exist

7. (a) 2, 3, 6   (b) 4   (c) 1, 5

9. (a) $|q| > \sqrt{\pi}$ (b) $q > \frac{1}{3}$
   (c) none $(p = 1)$ (d) $q > e$ or $0 < q < 1/e$

11. (a) $1 + 36 \displaystyle\sum_{k=1}^{\infty} (0.01)^k$ (b) $\frac{15}{11}$

13. (a) $\frac{1}{4}$ (b) diverges   (c) 1

15. converges   17. converges   19. converges

21. diverges (general term does not approach zero)

23. converges absolutely

25. diverges   27. $R = 1; 0 \le x \le 2$

29. $R = 2; -1 < x < 3$ 31. $R = 0; x = 1$

33. (a) $(x-2) - (x-2)^2/2 + (x-2)^3/3$

(b) $\dfrac{-1}{4(c-1)^4}(x-2)^4$, c between 2 and x (c) $\dfrac{(\frac{1}{2})^4}{4(\frac{1}{2})^4} = \frac{1}{4}$

35. (a) $1 + (x-1)/2 - (x-1)^2/8$

(b) $\dfrac{(x-1)^3}{16c^{5/2}}$, c between 1 and x

(c) $\dfrac{(5/9)^3}{16(2/3)^5} < 0.0814$

37. $\ln a + \displaystyle\sum_{k=0}^{\infty} (-1)^k \dfrac{(x/a)^{k+1}}{k+1}; R = a$

39. $\dfrac{1}{3}\left\{1 + \displaystyle\sum_{k=1}^{\infty} (-1)^k \dfrac{1 \cdot 3 \cdots (2k-1)}{2 \cdot 4 \cdots (2k)} (x/9)^k\right\};$
$R = 9$

41. $1 + x^2/2 + 5x^4/24 + \cdots$

43. $1 - x^2/4 - x^4/96 + \cdots$

45. $x - \frac{1}{2}x^2 + \frac{1}{6}x^3 + \cdots$

47. $-2 - 2x - \frac{8}{3}x^3 - \cdots; -2$

49. $|x| < 1.17$ 51. 0.444

▶ Exercise Set 7.7 **(Page 514)**

1. $y = Cx$

3. $y = Ce^{-\sqrt{1+x^2}} - 1$

5. $y = \ln(\sec x + C)$

7. $y = e^{-2x} + Ce^{-3x}$

9. $y = e^{-x}\sin(e^x) + Ce^{-x}$

11. $y = -\frac{2}{7}x^4 + Cx^{-3}$

13. $y = -1 + 4e^{x^2/2}$

15. $y = 2 - e^{-t}$

17. $y = \sqrt[3]{3t - 3\ln t + 24}$

19. $y = -3/(2\sqrt{x} + 1)$

21. $y = -\ln(3 - x^2/2)$

23. (a) $h(t) = (2 - 0.003979t)^2$
    (b) about 8.4 min

25. $v = 50/(2t + 1)$, $x = 25\ln(2t + 1)$

27. (a) $v(t) = c\ln\dfrac{m_0}{m_0 - kt} - gt$
    (b) 3044 m/sec

29. (a) $v^2 = 2gR^2/x + v_0^2 - 2gR$

31. (a) $y = 200 - 175e^{-t/25}$ 33. 25 lb
    (b) 136 lb

35. (a) $y = 10e^{-0.005t}$ 37. 196 days
    (b) 7 mg

39. 6.8 years   41. (a) 14,400
                               (b) 38 years

Index

INTEGRALS CONTAINING $2au - u^2$

50. $\displaystyle\int \sqrt{2au - u^2}\, du = \frac{u-a}{2}\sqrt{2au-u^2} + \frac{a^2}{2}\sin^{-1}\left(\frac{u-a}{a}\right) + C$

51. $\displaystyle\int u\sqrt{2au - u^2}\, du = \frac{2u^2 - au - 3a^2}{6}\sqrt{2au-u^2} + \frac{a^3}{2}\sin^{-1}\left(\frac{u-a}{a}\right) + C$

52. $\displaystyle\int \frac{\sqrt{2au-u^2}\, du}{u} = \sqrt{2au-u^2} + a\sin^{-1}\left(\frac{u-a}{a}\right) + C$

53. $\displaystyle\int \frac{\sqrt{2au-u^2}\, du}{u^2} = -\frac{2\sqrt{2au-u^2}}{u} - \sin^{-1}\left(\frac{u-a}{a}\right) + C$

54. $\displaystyle\int \frac{du}{\sqrt{2au-u^2}} = \sin^{-1}\left(\frac{u-a}{a}\right) + C$

55. $\displaystyle\int \frac{u\, du}{\sqrt{2au-u^2}} = -\sqrt{2au-u^2} + a\sin^{-1}\left(\frac{u-a}{a}\right) + C$

56. $\displaystyle\int \frac{u^2\, du}{\sqrt{2au-u^2}} = -\frac{(u+3a)}{2}\sqrt{2au-u^2} + \frac{3a^2}{2}\sin^{-1}\left(\frac{u-a}{a}\right) + C$

57. $\displaystyle\int \frac{du}{u\sqrt{2au-u^2}} = -\frac{\sqrt{2au-u^2}}{au} + C$

58. $\displaystyle\int \frac{du}{(2au-u^2)^{3/2}} = \frac{u-a}{a^2\sqrt{2au-u^2}} + C$

59. $\displaystyle\int \frac{u\, du}{(2au-u^2)^{3/2}} = \frac{u}{a\sqrt{2au-u^2}} + C$

INTEGRALS CONTAINING TRIGONOMETRIC FUNCTIONS

60. $\displaystyle\int \sin u\, du = -\cos u + C$

61. $\displaystyle\int \cos u\, du = \sin u + C$

62. $\displaystyle\int \tan u\, du = \ln|\sec u| + C$

63. $\displaystyle\int \cot u\, du = \ln|\sin u| + C$

64. $\displaystyle\int \sec u\, du = \ln|\sec u + \tan u| + C$
$= \ln|\tan(\tfrac{1}{4}\pi + \tfrac{1}{2}u)| + C$

65. $\displaystyle\int \csc u\, du = \ln|\csc u - \cot u| + C$
$= \ln|\tan\tfrac{1}{2}u| + C$

66. $\displaystyle\int \sec^2 u\, du = \tan u + C$

67. $\displaystyle\int \csc^2 u\, du = -\cot u + C$

68. $\displaystyle\int \sec u \tan u\, du = \sec u + C$

69. $\displaystyle\int \csc u \cot u\, du = -\csc u + C$

70. $\displaystyle\int \sin^2 u\, du = \tfrac{1}{2}u - \tfrac{1}{4}\sin 2u + C$

71. $\displaystyle\int \cos^2 u\, du = \tfrac{1}{2}u + \tfrac{1}{4}\sin 2u + C$

72. $\displaystyle\int \tan^2 u\, du = \tan u - u + C$

73. $\displaystyle\int \cot^2 u\, du = -\cot u - u + C$

74. $\displaystyle\int \sin^n u\, du = -\frac{1}{n}\sin^{n-1} u \cos u + \frac{n-1}{n}\int \sin^{n-2} u\, du$

75. $\displaystyle\int \cos^n u\, du = \frac{1}{n}\cos^{n-1} u \sin u + \frac{n-1}{n}\int \cos^{n-2} u\, du$

76. $\displaystyle\int \tan^n u\, du = \frac{1}{n-1}\tan^{n-1} u - \int \tan^{n-2} u\, du$

77. $\displaystyle\int \cot^n u\, du = -\frac{1}{n-1}\cot^{n-1} u - \int \cot^{n-2} u\, du$

78. $\displaystyle\int \sec^n u\, du = \frac{1}{n-1}\sec^{n-2} u \tan u + \frac{n-2}{n-1}\int \sec^{n-2} u\, du$

79. $\displaystyle\int \csc^n u\, du = -\frac{1}{n-1}\csc^{n-2} u \cot u + \frac{n-2}{n-1}\int \csc^{n-2} u\, du$

80. $\displaystyle\int \sin mu \sin nu\, du = -\frac{\sin(m+n)u}{2(m+n)} + \frac{\sin(m-n)u}{2(m-n)} + C$

81. $\displaystyle\int \cos mu \cos nu\, du = \frac{\sin(m+n)u}{2(m+n)} + \frac{\sin(m-n)u}{2(m-n)} + C$

82. $\displaystyle\int \sin mu \cos nu\, du = -\frac{\cos(m+n)u}{2(m+n)} - \frac{\cos(m-n)u}{2(m-n)} + C$

83. $\displaystyle\int u \sin u\, du = \sin u - u \cos u + C$

84. $\displaystyle\int u \cos u\, du = \cos u + u \sin u + C$

85. $\displaystyle\int u^2 \sin u\, du = 2u \sin u + (2 - u^2)\cos u + C$

86. $\displaystyle\int u^2 \cos u\, du = 2u \cos u + (u^2 - 2)\sin u + C$

87. $\displaystyle\int u^n \sin u\, du = -u^n \cos u + n\int u^{n-1}\cos u\, du$

88. $\displaystyle\int u^n \cos u\, du = u^n \sin u - n\int u^{n-1}\sin u\, du$

89. $\displaystyle\int \sin^m u \cos^n u\, du = -\frac{\sin^{m-1} u \cos^{n+1} u}{m+n} + \frac{m-1}{m+n}\int \sin^{m-2} u \cos^n u\, du$
$= \frac{\sin^{m+1} u \cos^{n-1} u}{m+n} + \frac{n-1}{m+n}\int \sin^m u \cos^{n-2} u\, du$